Anti-Aging Medicine

Hidekazu Yamada • Yuji Naito
Editors

Anti-Aging Medicine

Basics and Clinical Practice

Springer

Editors
Hidekazu Yamada
Anti-Aging Center
Kindai University
Higashiosaka, Osaka, Japan

Yuji Naito
Department of Human Immunology
and Nutrition Science
Kyoto Prefectural University of Medicine
Kamigyoku, Kyoto, Japan

ISBN 978-981-97-8516-2 ISBN 978-981-97-8514-8 (eBook)
https://doi.org/10.1007/978-981-97-8514-8

The original submitted manuscript has been translated into English. The translation was done using artificial intelligence. A subsequent revision was performed by the author(s) to further refine the work and to ensure that the translation is appropriate concerning content and scientific correctness. It may, however, read stylistically different from a conventional translation.

Translation from the Japanese language edition: "Anti-Aging Igaku no Kiso to Rinsho" by Hidekazu Yamada and Yuji Naito, © MEDICAL VIEW CO., LTD. under exclusive license to Springer Nature Singapore Pte Ltd. 2023. Published by MEDICAL VIEW CO., LTD.. All Rights Reserved.

This Springer imprint is published by the registered company Springer Nature Singapore Pte Ltd.
The registered company address is: 152 Beach Road, #21-01/04 Gateway East, Singapore 189721, Singapore

Contents

The original version of the book has been revised. A correction to this book can be found at https://doi.org/10.1007/978-981-97-8514-8_164

Part I

Understanding and Prospects of Anti-aging Medicine

1 What Is Anti-Aging Medicine?

Hidekazu Yamada

Anti-aging medicine is a form of preventive medicine and a part of longevity science. It views aging as a disease and aims for healthy longevity through interdisciplinary practice based on longevity research. Furthermore, in recent years, treatments that reverse aging by reprogramming using Yamanaka factors have begun to be performed. Therefore, the field of anti-aging medicine has become a field aiming not only for interventions in aging-related diseases and extending the maximum lifespan but also for true rejuvenation using regenerative medicine. About 20% of aging is related to genes, and about 80% is said to be environmental factors. In the past 100 years or so, the maximum lifespan of humans has not changed much, around 120 years, but the average lifespan has rapidly extended. Japan has a long average lifespan, but there is a gap in the healthy lifespan. Anti-aging medicine aims to reduce this gap and extend the healthy lifespan. In other words, the primary goal is to aim for a sudden death in old age (PPK).

1.1 From the Perspective of Lifespan [1]

It is known that many species of organisms face death at a certain chronological age. Lifespan is thought to be the disappearance of bioresilience (recovery power). With therapeutic intervention and an appropriate lifestyle, it is possible to prevent aging-related diseases until around the age of 60. Even the healthiest people eventually become extremely weak and sick, and then die. According to the lifespan limit theory, following Gompertz's law on mortality (after the age of 35–40, the mortality rate increases exponentially with age. It is said to double every 8 years after the age of 35), 120–150 years is the limit of maximum lifespan. The basic idea is to control the decline in biological resilience (the ability to bounce back and recover), continue to respond to frailty, and extend the healthy lifespan. On the other hand, it is becoming possible to "rejuvenate" using reprogramming with Yamanaka's four factors, and to reverse the decline in bioresilience, so we need to think about how to perceive the lifespan of the next generation (Fig. 1.1). In longevity science, we will challenge the maximum lifespan.

H. Yamada (✉)
Anti-Aging Center, Kindai University, Osaka, Japan
e-mail: yamada@mitameonline.com

H. Yamada, Y. Naito (eds.), *Anti-Aging Medicine*, https://doi.org/10.1007/978-981-97-8514-8_1

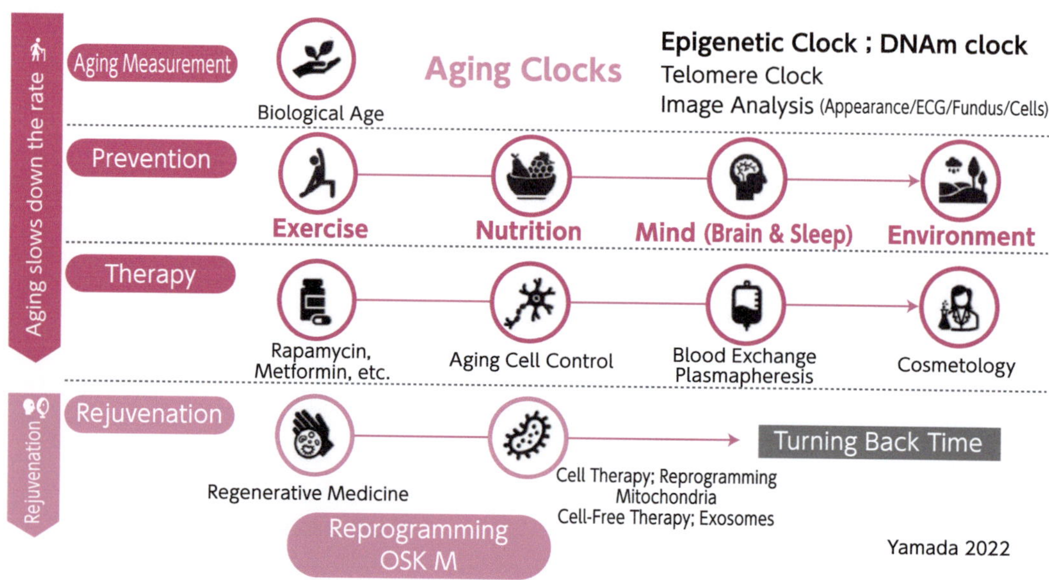

Fig. 1.1 Strategy of anti-aging medicine

1.2 Aging Is a Disease [2]

The International Statistical Classification of Diseases and Related Health Problems (ICD) 11 has been adopted from 2022. This classification is a critical standard for public health, such as policy formulation, budget determination, and insurance policy decisions. In ICD-11, a new extension code, XT9T; "age-related" was added. XT9T means "caused by a pathological process that persistently brings about the loss of adaptation and progress of organisms in old age." The next related one is MG2A. ICD-10 had a code that could be called "senile debility," but in ICD-11, instead, the code MG2A, which originally meant "old age," was adopted. MG2A originally means "old This was defined as "age", and after some discussion [3], the definition was changed from "old age" to "decline in inherent abilities with aging," and "old age" was classified as a "general symptom" rather than a specific disease. From the standpoint of longevity medicine, the MG2A code should have changed the definition of an individual's age from a chronological one to a biological one, but in the end, it did not go as far as to make aging a disease. We should create a society where the elderly do not have to endure health problems that young people can receive treatment for, due to their age, and we should promote anti-ageism. Recent aging research [4] has shown that aging is coded in genes preserved across species, with epigenetic environmental factors involved. The main features of aging [5] such as, genome instability, telomere shortening, epigenetic changes, loss of protein homeostasis, inactivation of macroautophagy, deregulation of nutrient sensing, mitochondrial dysfunction, cellular senescence, stem cell exhaustion, changes in intercellular communication, chronic inflammation, and dysbiosis, suggest that important pathways in cell biology (IGF-1/AKT/FOXO3A, TP53/P21/P16, mTOR/S6K) can be said to be the cause of age-related diseases such as cancer and diabetes. Supplements like quercetin can remove aging cells (senolytics), and through the activity of SIR-1SIR-6, DNA repair, autophagy, mitochondrial biosynthesis, and improvement in response to stress and calorie restriction are possible. Mitochondrial transplantation therapy, plasma exchange therapy, and thymus regeneration can also slow aging. Furthermore, with already medically approved drugs like metformin, rapamycin, and statins, we are at a point where we can consider repurposing disease indications to extend a healthy lifespan. In this way, interventions in molecules common to aging and disease pathways are expected to extend a healthy lifespan. Also, these anti-aging drugs may also be useful for the immune decline in COVID-19. In particular, the information theory of aging, which assumes loss due to temporal transcription networks and epigenetic information, may allow for the observation of short-term changes by examining DNA methylation, for example, in the field of preventive medicine that intervenes in exercise, nutrition, mind (brain, sleep), and environment. In the future, clinical trials with appropriate endpoints targeting aging will be necessary. For this, a numerical evaluation index of "human aging progression," which can be called an aging clock, is necessary. For clinical trials, it is important to establish a measurement method (biomarker) for the aging rate that is accurate and shows a significant difference in a short period of time. The Japan Anti-Aging Medical Association aims to be preventive, interdisciplinary, and practical from this standpoint. From the standpoint of longevity medicine, we should look at comprehensive aging changes and provide better care to society. Interventions in exercise, nutrition, mind (brain, sleep), and environment from a young age will slow down the aging speed and aim for a society where we can live a healthy long life with well-being. Creating an era where aging control is possible is the form of longevity medicine and anti-aging medicine (Hidekazu Yamada).

References

1. Ferrucci L, Gonzalez-Freire M, et al. Measuring biological aging in humans: a quest. Aging Cell. 2020;19:e13080.
2. Khaltourina D, Matveyev Y, et al. Aging fits the disease criteria of the international classification of diseases. Mech Ageing Dev. 2020;189:111230.
3. Rabheru K, Byles J, et al. How "old age" was withdrawn as a diagnosis from ICD-11. Lancet Healthy Longev. 2022;3:e457–9.
4. Campisi J, Kapahi P, et al. From discoveries in ageing research to therapeutics for healthy ageing. Nature. 2019;571:183–92.
5. López-Otín C, Blasco MA, et al. Hallmarks of aging: an expanding universe. Cell. 2023;186:243–78.

2 The Philosophy and Spirituality of Anti-Aging Medicine

Shinzo Kato

It seems that anti-aging medicine has been primarily aimed at preventing physical aging. This is probably because the underlying assumption was that medicine deals with physical problems, and that aging is something bad and should be eliminated. In this section, I would like to return to the basic question of how we should think about health and consider the future direction of anti-aging medicine.

2.1 Not Just Physical, But Toward Four-Dimensional Medicine

When thinking about human health, it is becoming accepted in fields such as palliative care and holistic care to consider not only the physical aspect, but also the psychological, social, and spiritual dimensions, or to add the intellectual dimension and consider it in five dimensions [1]. Therefore, I believe that anti-aging medicine, which deals with aging, will also be required to include care in the psychological, social, and spiritual aspects. In this context, the reconstruction of communities, city planning, social relationships [2], the review of relationships between people, the importance of life education, and the impact of artistic activities on health may also become future themes as society ages as a whole. The fifth M in the five M's proposed by Mary Tinetti in the field of geriatric medicine, "Matters most to me," is thought to advocate the importance of life education [3] (see Table 2.1).

S. Kato (✉)
MOA Takanawa Clinic, Tokyo, Japan
e-mail: Katos@myad.jp

Table 2.1 To live healthily

Reduce causes of illness	Increase vitality
Improve test data Improve diet Encourage exercise Reduce alcohol consumption Quit smoking Reduce stress Deal with anxiety	Purpose in life Goals, dreams, hopes Comfort and tranquility Contribute to society Family and friends Trust and solidarity Have confidence

2.2 Does a "Healthy Mind Dwell in a Healthy Body"?

The phrase "a healthy mind dwells in a healthy body" is widely circulated. However, have you ever imagined how this phrase would be received by people with disabilities or illnesses? If interpreted literally, it would mean that people with disabilities or illnesses cannot have healthy minds, which would be a cruel word for them. On the other hand, there are many examples of Olympic athletes and soldiers not necessarily having a healthy mind. It is still fresh in our memory that the coach of the Nihon University American football team and a boxing coach

H. Yamada, Y. Naito (eds.), *Anti-Aging Medicine*, https://doi.org/10.1007/978-981-97-8514-8_2

caused incidents. On the other hand, there are not a few, or rather many, kind-hearted people among the sick and disabled, which many people may have felt.

This phrase was originally borrowed from a line in the tenth song of the Roman Satires by the ancient Roman poet Juvenal [4]. Originally, humans have many desires such as money, power, status, eloquence, talent, honor, prosperity, longevity, and beauty, and often pray to God for these things, but all of them result in vain. Therefore, the words to pray for are "to have a sound body and a sound mind" in the context of "sound body in sound mind." The word "mind" is being spoken (Table 2.1). However, as militarism advanced during the Second World War, for the sake of enriching the country and strengthening the military, it was proclaimed that "a healthy mind dwells in a healthy body," and it was used as a phrase to encourage physical training.At the end of the tenth song, it is said, "If you still want to ask something of the gods, and if you want to offer the pure white innards of a pig and small sausages to a small shrine, pray for a healthy mind in a healthy body. Pray for a strong mind that cuts off the fear of death. A mind that considers the end of life as a blessing of nature. A mind that can endure any suffering. A mind that knows no anger, is desireless and tranquil. Pray for a mind that believes that the hardships and efforts of Hercules are more desirable than the lustful, drunken, and luxurious life of King Sardanapalus. What I recommend to you is something you can give to yourself with your own power." In other words, we should pray to have a strong and grateful mind, and that is something you can do with your own power. Perhaps anti-aging medicine could also move in this direction.

2.3 Is Aging a Bad Thing?

In modern society, the negative aspects of aging, such as senility and frailty, are often emphasized. However, shouldn't we be aware that there are both good and bad aspects to aging? For example, words like "old teacher," "elder," "senior," "councilor," "ripe," "experienced," "long-established," and idioms like "mature and ripe," "anxious old woman," clearly show that aging was not always seen as a bad thing. As we age, we gain experience, our thinking matures, and we become able to see things fairly, or from a bird's-eye view. Shouldn't we pay more attention to the positive aspects of aging, make use of them, or try to enhance them, as part of geriatric medicine? In the medical model, we investigate the cause of the disease, treat it by removing it, and think of disease and health as separate. It is disease-centered, and the goal is to remove the disease and achieve absolute health. On the other hand, Antonovsky's health generation theory (salutogenesis) is patient-centered and aims to solve problems by nurturing the healthy part [5]. Health, disease, and death are a continuum, and we think about health at the center. Even if there is a disease or disability, if the overall order as a human being is maintained, that person is relatively healthy and good. We try to find and utilize the remaining resources. With this kind of thinking, looking at aging from a bird's-eye view may also be a challenge for future anti-aging medicine. In that sense, the author hopes that the goal should be well-aging or good-aging, rather than anti-aging.

References

1. Hutchinson TA (Translated by Tsuneto A). Whole person care practice edition—tune, open, and devote your heart in the medical AI era. Tokyo: Miwa Bookstore; 2020.
2. Holt-Lunstad J, Smith TB, et al. Social relationships and mortality risk: a meta-analytic review. PLoS Med. 2010;27:e1000316.
3. The geriatric 5Ms—the five simple words every geriatrician needs to know (the new mantra). British Geriatrics Society, 2017. https://www.bgs.org.uk/blog/the-geriatric-5ms-%E2%80%93-the-five-simple-words-every-geriatrician-needs-to-know.
4. Persius J (Translated by Kokuhara Y). Roman satire collection (Iwanami Bunko). Tokyo: Iwanami Shoten; 2012.
5. Antonovsky A (Supervised Translation by Yamasaki K, Yoshii K). Solving the mystery of health—the mechanism of stress coping and health maintenance. Tokyo: Yushindo Kobunsha; 2001.

3 Global Trends in Anti-Aging Medicine

Yoshikazu Yonei

In recent years, the world has been experiencing a rapidly aging population, and in the medical field, various problems such as soaring medical costs and specialization-related issues have become a concern. As Japan is one of the world's leading longevity countries, we believe that anti-aging medicine, which aims to "promote health, improve quality of life, and achieve healthy longevity," should be promoted as a national policy. While it is important to import excellent medical technology from abroad, we must not forget to disseminate Japanese technology abroad. This section discusses the history and current status of anti-aging medicine from its inception to the present.

3.1 History

As of 1948, the World Health Organization (WHO) charter defined health as "Health is a state of complete physical, mental and social well-being and not merely the absence of disease or infirmity." Health has two elements: positive health and negative health. It is natural to be free from disease and disability, which corresponds to negative health. Positive health is proclaimed to be a state of well-being both mentally, physically, and socially. Half a century passed before anti-aging medicine, a next-generation preventive medicine that aims for positive health, appeared in Japan. In 1992, the American Academy of Anti-Aging Medicine (A4M) was established, and the first decade (1990s), which can be called the founding period, passed. I bought the copyright from A4M representative R. Goldman and published a publication for publicizing anti-aging medicine [1, 2]. A4M has made a significant contribution to the spread of anti-aging medicine in Japan.

In the 2000s (development period), anti-aging medicine rapidly developed and became recognized in countries around the world. It can be said to be the dynamic period of anti-aging medicine. Research groups and medical associations for anti-aging medicine were established in various countries. In countries facing rapid aging while still developing, anti-aging medicine has become an important national policy.

Under such circumstances, the first Aesthetic and Anti-Aging Medicine World Congress (AMWC) was held in Paris, France, in 2003, and has been held regularly since then. AMWC is sponsored by the international publisher (Informa), regularly publishes the journal "Prime," and I have been the information disseminator from Japan [3, 4]. The first AMWC will be held in 2022. The event was held in Japan. Participants from Asia and North and South America attended widely, and it had a significant presence and influence. Such international conferences are meaningful in that they aim to share information at a global level

Y. Yonei (✉)
Anti-Aging Medical Research Center, Glycative Stress Research Center, Graduate School of Life and Medical Sciences, Kyoto, Japan
e-mail: yyonei@mail.doshisha.ac.jp

H. Yamada, Y. Naito (eds.), *Anti-Aging Medicine*, https://doi.org/10.1007/978-981-97-8514-8_3

and contribute to patient treatment. On the other hand, the Japan Anti-Aging Medical Association, established as a research group in 2001, maintains cooperation with existing domestic medical associations, and leading experts from various specialized fields participate as directors and councilors, focusing on research and education. It is a conference where doctors and medical practitioners can participate with confidence. The activities of the Japan Anti-Aging Medical Association, which boasts nearly 10,000 members, are attracting attention as a very unique presence worldwide, and it has become an important place to disseminate Japan's medical technology to the world. The journal "Anti-Aging Medicine (English version)," which has been disseminating valuable research results of the association members domestically and internationally, has been published in two forms since 2014: the association's official journal "NPJAging" and "Glycative StressResearch" (https://www.toukastress.jp/webj), which focuses on social implementation. I hope that the characteristics of these academic journals will be fully utilized to disseminate the research results of the association members to the world.

3.2 World Education System

The education system for certified specialists and instructors at the Japan Anti-Aging Medical Association is well known. Here, I will introduce two education systems overseas. The European Course in Anti-Aging Medicine (ECAAM) is held in conjunction with the AMWC, and study groups are held around the world. The Dr. Hertoghe Medical School (https://hertoghemedicalschool.eu/) offers a program focused on anti-aging endocrinology [5].

3.3 Conclusion

In order to continue leading the world, it is important to view global trends from a high perspective and with a wide field of view, and to continue to disseminate our activities as medical data one by one to the world (Fig. 3.1). The 2000s were the dawn of the Japan Anti-Aging Medical Association, and the 2010s could be called the exploratory period. I am looking forward to the 2020s becoming a soaring period.

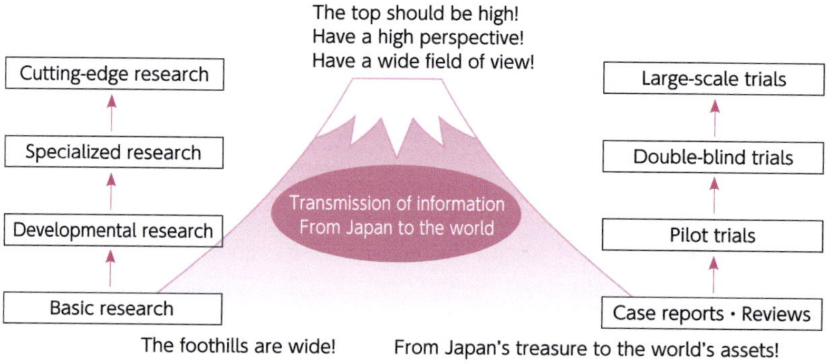

Fig. 3.1 Let's build a beautiful evidence mountain loved by the world

References

1. Klatz R (Translated by Fujino K, Supervised by Yonei Yand others). How to rejuvenate in 10 weeks. Tokyo: Koseido Publishing; 2001.
2. Goldman R, Kratz R, et al. (Supervised by Yonei Y, Translated by Fujino K). Brain fitness: people whose brains age, and those who don't. Tokyo: Koseido Publishing; 2001.
3. Yonei Y, Yagi M, et al. Glycative stress and sleep quality. Prime. 2018;8(6):19–23.
4. Yonei Y, Yagi M, et al. The vicious cycles induced by glycative stress. Prime. 2020;10(6):34–6.
5. Dalle C, Donatini B. Unlocking the lifespan: healthy solutions for intelligent aging. Independently Published; 2020.

4 Anti-Aging Medicine in a Longevity Society

Masahiro Akishita

As we enter a longevity society, often referred to as the era of living to 100, the scope of anti-aging is broad, and the approach can significantly differ between youth and old age. In youth, since there is no observable decline in organ and tissue function, the focus is on inhibiting the aging of each organ, i.e., degenerative changes, and age-related changes in tissue components as much as possible. On the other hand, in old age, where age-related changes in organs and tissues have already progressed to some extent, it becomes more important to suppress the progression of aging while preventing pathological aging and frailty (frailty) that come with functional decline, and to maintain healthy longevity. Taking into account the current situation of the longevity society, this section will provide an overview of the key points of anti-aging in old age.

4.1 Physiological Aging and Pathological Aging

The process of aging can be conceptually divided into physiological aging (physiological aging) and pathological aging (pathological aging). Taking the change in bone mass (bone density) with age as an example (Fig. 4.1), bone mass generally increases until maturity in both men and women, peaks in the 20–30 s, and then decreases with age. When bone mass falls below 70% of the average for young adults, the risk of fracture increases, leading to the domain of pathological aging diagnosed as osteoporosis. In other words, osteoporosis can be considered a state where the decrease in bone mass with age (physiological aging) is accelerated by factors such as menopause, genetic factors, and environmental factors (pathological aging).Another

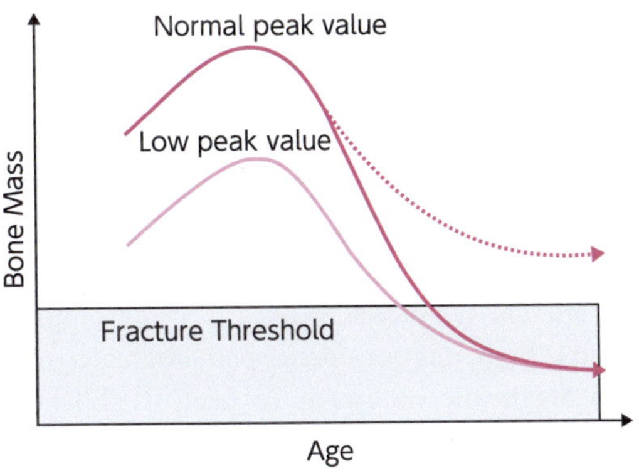

Fig. 4.1 Physiological aging and pathological aging from the perspective of bone mass reduction. If the peak value is normal and the rate of bone mass reduction is small, it will not decrease to the fracture threshold (physiological aging). On the other hand, even if the peak value is normal and the rate of bone mass reduction is large, or if the rate of bone mass reduction is normal but the peak value is low, it will decrease to the fracture threshold (pathological aging)

M. Akishita (✉)
Department of Geriatric Medicine, Graduate School of Medicine, The University of Tokyo, Tokyo, Japan

Tokyo Metropolitan Institute for Geriatrics and Gerontology, Tokyo, Japan
e-mail: akishita@m.u-tokyo.ac.jp

H. Yamada, Y. Naito (eds.), *Anti-Aging Medicine*, https://doi.org/10.1007/978-981-97-8514-8_4

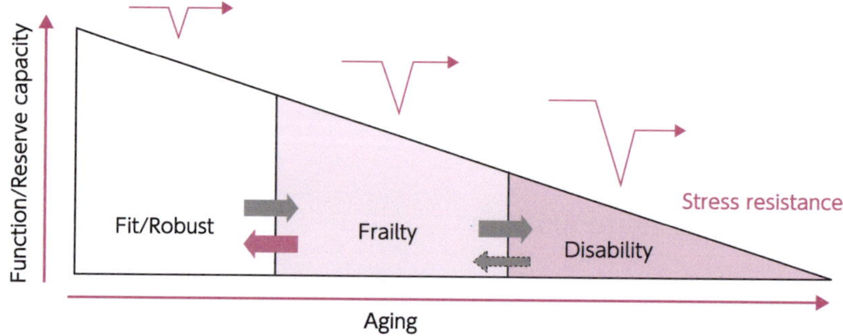

Fig. 4.2 Scientific definition of frailty: organ frailty. A state of reduced stress resistance due to the decline in physiological reserve capacity with aging. At the organ level, it is a state of reduced reserve capacity that cannot adequately respond to stress, against the backdrop of aging of constituent cells. (Example) Low cardiac output during exercise and high blood sugar after meals

example of pathological aging is when an organ does not fully develop due to genetic factors or environmental factors during the growth period. In terms of bone mass, this corresponds to cases where the maximum bone mass falls below the average due to insufficient female hormone action during adolescence or lack of nutrition and exercise (Fig. 4.1). In this case, even at a normal rate of bone loss due to aging, bone mass can easily decrease to the fracture threshold, resulting in a pathological condition called osteoporosis. The same concept applies to skeletal muscle mass and higher brain functions. Thus, physiological aging and pathological aging are different, but the boundary is ambiguous, and it is a fact that there are states that can be considered both. Therefore, clinically, if there are no significant clinical symptoms or the risk of developing them is low, it is considered physiological aging, and if there are pathological clinical symptoms or the risk of developing them is high, it is considered pathological aging, and diagnostic criteria have been established. For example, the diagnostic criterion for hypertension is a blood pressure of 140/90 mmHg or higher, but blood pressure increases with age, and many cases of mild hypertension are asymptomatic. However, because it has been medically proven that the risk of developing stroke and heart disease increases if hypertension continues for many years, even mild hypertension is considered a high-risk state and is treated as pathological aging.

4.2 Scientific Concept of Frailty

As mentioned above, there is a boundary area between physiological aging and pathological aging. This state fits well with the scientific concept of frailty (Fig. 4.2). At the individual level, frailty is defined as a state in which vulnerability to stress increases due to a decrease in physiological reserve capacity in old age, leading to outcomes such as functional impairment, need for care, and death (refer to "IV. I. 20. Evaluation of Frailty"). At the organ level, it is a state where the reserve capacity decreases due to the aging of the constituent cells and organ networks, and a sufficient response to load cannot be made, and it can be considered as a pre-stage organ dysfunction. The background of frailty is organ/tissue frailty, and there is no doubt that the development of effective intervention methods at this stage is a future task for frailty prevention. The same applies to old age.

4.3 Healthy Life Expectancy

As longevity and aging in society accelerate, the importance of extending healthy life expectancy and achieving healthy longevity rather than extending life expectancy have been emphasized. Therefore, clinical research on aging has shifted from outcomes such as mortality and life expectancy to outcomes such as health status and frailty/need for care. Even in basic research on aging using experimental animals, it is not enough to simply evaluate life expectancy. Healthy life expectancy (healthy life "Expectancy" generally refers to the "period of time one can live without being limited in daily life due to health problems." The WHO reported the importance of this concept in 2000 and has been calculating data from member countries. According to the 2017 data (1), the healthy life expectancy of Japanese people is 71.4 years for men (second in the world) and 74.6 years for women (first in the world). Although the definition and calculation method are slightly different from the WHO, the Ministry of Health, Labour and Welfare announced that the healthy life expectancy based on the 2016 statistics was 72.1 years for men and 74.8 years for women, with an average life expectancy of 80.9 years for men and 87.1 years for women. Therefore, the difference between healthy life expectancy and average life expectancy (the period of needing care) is 8.8 years for men and 12.3 years for women, and internationally, the longer the average life expectancy, the longer this period tends to be. Therefore, the goal of aging research should be to extend not only life expectancy but also healthy life expectancy. Also, while women live longer, there is a problem that the period of needing care is long, and research on aging and geriatric diseases that are conscious of gender differences, such as the difference in the cause of diseases requiring care between men and women, is also necessary.

4.4 New Definition of the Elderly

In developed countries, people over 65 have generally been called elderly, but as societies like Japan have become more long-lived, there have been more objections to uniformly calling people over 65 elderly. Therefore, the Definition Review Working Group on the Elderly of the Japan Gerontological Society and the Japan Geriatrics Society proposed to revise the definition of the elderly to be 75 and over, and published a report (2). Looking at the results of surveys on disease incidence rates and physical and mental aging changes, the appearance of changes in physical and psychological functions due to aging in current Japanese elderly people is delayed by 5–10 years compared to 10–20 years ago, and this "rejuvenation" phenomenon is a major basis. Especially, in the age group of 65–74, the majority of people are able to maintain their physical and mental health and engage in active social activities. Also, according to the results of various awareness surveys, there is a strong negative opinion in society in general about considering people over 65 as elderly, and in a survey by the Cabinet Office, there are many opinions that consider people over 70 or 75 as elderly. Looking at recent epidemiological studies and national survey results, further rejuvenation can be seen in physical and mental functions and disease incidence rates. However, as stated in the above report, there are many issues with a uniform definition based on chronological age, and considering the large individual differences in aging degree with aging, it will be more important to consider biological age and make individual judgments.

References

1. GBD. DALYs and HALE collaborators: global, regional, and national disability-adjusted life-years (DALYs) for 359 diseases and injuries and healthy life expectancy (HALE) for 195 countries and territories, 1990–2017: systematic analysis for the global burden of disease study 2017. Lancet. 2017;2018(392):1859–922.
2. Japan Gerontological Society & Japan Geriatrics Society. Report of the "working group on definitions concerning the elderly". Tokyo: Japan Gerontological Society & Japan Geriatrics Society; 2017. http://geront.jp/news/pdf/topic_170420_01_01.pdf

5 Anti-Aging Medicine and Well-being

Shigeo Horie

- Individual qualities and environmental factors are related to well-being.
- Enhancing well-being can reduce disease risk and contribute to anti-aging.
- Globally, measures to enhance well-being are attracting attention.

The development of industrial society has brought about an abundance of consumer goods and extended life expectancy, but it is hard to say that it has necessarily improved the quality of individual life, satisfaction, or happiness. Furthermore, the COVID-19 crisis has significantly changed our understanding of health and life's purpose, and the importance of achieving well-being, which means healthy and sustainable happiness, not only for individuals but also for society, is being widely shared worldwide. This is also being reflected in policies and corporate health management. It is known that high well-being reduces the risk of age-related diseases and slows down the decline in physical function [1]. Especially when one is aware of a purpose in life, it can be said that they are in a state of high well-being, being both physically and mentally healthy and having healthy lifestyle habits. Being able to feel a purpose in life comes from being able to acknowledge the significance of one's past life, having life goals, and having a sense of living from the present to the future toward those goals. Meta-analysis shows that having a clear purpose in life reduces the risk of death (hazard ratio = 0.83; 95% CI = 0.75–0.9 [1] has been reported [2]. Well-being is related to individual qualities and social environmental factors. Mental capital is a measure of the ability to achieve sustainable happiness, researched primarily by American business scholar Professor Fred Luthans, and is composed of four elements: hope, efficacy, resilience, and optimism, collectively referred to as HERO. Hope, or aspiration, requires (1) exciting future goals, (2) the power and resources (pathways) to achieve these goals, and (3) a supportive environment. The ability of hope to enhance survival even in desperate situations is detailed in the book "Night and Fog" by Viktor Frankl, a Jew who miraculously survived a concentration camp. Hope can be enhanced by setting specific goals such as "eating vegetables every day" or "exercising three times a week." Moreover, the more pathways there are to the goal you are trying to achieve, the better. For example, for the goal of "exercising three times a week," there could be multiple pathways such as jogging or going to the gym, and the more resources/pathways there are, the more effective they are in strengthening hope. Efficacy is the

S. Horie (✉)
Department of Urology, Advanced Informatics of Genetic Diseases, Digital Therapeutics, Juntendo University Graduate School of Medicine, Tokyo, Japan
e-mail: shorie@juntendo.ac.jp; secretary-uroprof@juntendo.ac.jp

H. Yamada, Y. Naito (eds.), *Anti-Aging Medicine*, https://doi.org/10.1007/978-981-97-8514-8_5

belief that one can bring about positive change. We learn from our experiences in social relationships and acquire social skills. The belief that one is useful in society is important for well-being. The key to enhancing efficacy is successful experiences. Not only the individual's successful experiences based on practice and proficiency are important, but also the successful experiences of others can enhance efficacy. Observing the success of respected seniors or role model colleagues can give one the confidence that "I can do it too." Furthermore, social persuasion and positive feedback are also important. Affirmative words from others, such as "you can do it" or "you have achieved excellent results," are effective in enhancing efficacy. Resilience, in a broad sense, means the ability to maintain social adaptability even in adversity or stress, or "flexibility." With resilience, one can recover quickly even from social damage. To strengthen resilience, it is helpful to observe the environment around oneself and anticipate potential obstacles in advance. Furthermore, rather than avoiding risks such as failure, facing them as opportunities for challenge and development also strengthens resilience. Optimism, or the ability to be optimistic about the future, is known to be related to the risk of age-related diseases and the prognosis of chronic diseases such as cancer. Having a clear purpose in life and accompanying optimism reduces the risk of cognitive decline [3]. In a study that prospectively observed over 70,000 elderly women for 8 years, those with higher levels of optimism had a 38% lower risk of death from heart disease and a 39% lower risk of death from stroke [4]. To enhance optimism, it is important to be generous about the past and grateful for the present environment. Generosity toward the past involves accepting oneself, even if the outcome is not what one wanted, without blaming oneself too much (self-compassion). Gratitude for the current environment involves being able to appreciate the bonds with the people around oneself, and remembering to be grateful every day leads to optimism for the future and enhances well-being. The purpose of life, resilience, optimism, and the actions and results of oneself can be controlled by oneself (internal). It is known that having a sense of control (locus of control) and social bonds can lead to better physical function and health status, and less frequent visits to medical institutions. As a social environment, social bonds are important for well-being. A meta-analysis summarizing over 70 clinical studies has shown that a thinning of social bonds increases the risk of death. In Ogimi Village, Okinawa, one of the world's longest-lived regions known as the blue zone, it is introduced that residents are deeply connected through a cooperative work called Moai. Social bonds are not just naturally occurring in the community. About 30 years ago, Spiegel and his colleagues randomly assigned metastatic breast cancer patients, with the control group receiving only standard treatment, and the intervention group receiving a 90-minute group session once a week for a year in addition to standard treatment. The sessions were led by a psychiatrist or psychotherapist who is a breast cancer survivor, and the patients discussed their feelings about having cancer, the impact on their lives, and the side effects of chemotherapy and radiation therapy. The sessions were designed to deepen the interaction between members so that they would not feel lonely, and to think about how to convey what they experienced in the tragedy of having cancer to other patients and their families. The average survival time of the control group was 18 months, whereas it was significantly extended to 36 months in the intervention group that conducted group sessions (Fig. 5.1). It is worth noting that when patients feel less lonely feel connected by bonds, and are in a state of high well-being, they have the power to greatly change the prognosis of metastatic cancer.

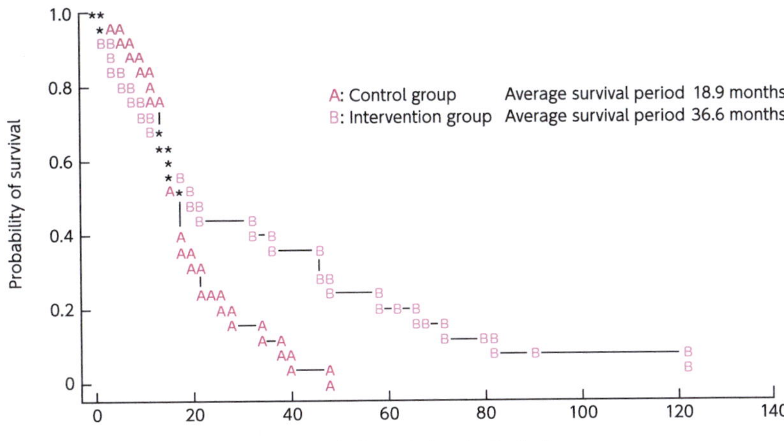

Fig. 5.1 Social bonds extend the prognosis of cancer! (**a**) Control group average survival period 18.9 months. (**b**) Intervention group average survival period 36.6 months. (Quoted from [5])

References

1. Kubzansky LD, Huffman JC, et al. Positive psychological well-being and cardiovascular disease: JACC health promotion series. J Am Coll Cardiol. 2018;72:1382–96.
2. Cohen R, Bavishi C, et al. Purpose in life and its relationship to all-cause mortality and cardiovascular events: a meta-analysis. Psychosom Med. 2016;78:122–33.
3. Boyle PA, Buchman AS, et al. Effect of a purpose in life on risk of incident Alzheimer disease and mild cognitive impairment in community-dwelling older persons. Arch Gen Psychiatry. 2010;67:304–10.
4. Kim ES, Hagan KA, et al. Optimism and cause-specific mortality: a prospective cohort study. Am J Epidemiol. 2017;185:21–9.
5. Spiegel D, Bloom JR, et al. Effect of psychosocial treatment on survival of patients with metastatic breast cancer. Lancet. 1989;2:888–91.

Epidemiological Research in Anti-Aging Medicine

6

Teruhide Koyama

6.1 Regional Resident Observation Study on Longevity [1, 2]

There are reports of investigations into what is named "Blue Zones," the world's longevity regions. The Blue Zones include the world's four major longevity regions, including Okinawa in Japan (Sardinia in Italy, Loma Linda in the United States, and the Nicoya Peninsula in Costa Rica). Adventurer Dan Buettner investigated these regional residents and identified the following nine rules as common factors for longevity [1].

Rule 1: Continue moderate exercise.
Rule 2: Control calorie intake by eating until you are 80% full.
Rule 3: Eat plant-based foods.
Rule 4: Drink red wine in moderation.
Rule 5: Have a clear sense of purpose.
Rule 6: Slow down in life.
Rule 7: Have faith.
Rule 8: Prioritize family.
Rule 9: Connect with people.

These rules, which were clarified through careful interviews with residents of regions with a high demographic of long-lived individuals, have been shown to be valid in many subsequent epidemiological studies.

6.2 Changes in Epidemiological Research on Aging from the Past to the Present

In the Blue Zones, common principles were discovered by examining the factors of longevity in the culture and lifestyle of the residents. On the other hand, past epidemiological studies have focused on health outcomes according to the clinical expertise of the participating researchers. Furthermore, when presented in papers, many epidemiological indicators were developed primarily to evaluate a single health status. For example, studies on cause-specific mortality rates depend on a single cause of death. Such an approach may be inappropriate for studies targeting elderly people who may suffer from multiple diseases or functional disabilities. Depression can lead to a decrease in activity and a decrease in muscle strength, which can lead to an increase in joint pain and osteoarthritis. Also, a history of myocardial infarction may weaken the association with diseases where blood pressure is a risk factor, such as stroke, due to subsequent blood pressure control and changes in behavior (lifestyle habits such as diet and exercise). Considering that multiple diseases coexist in old age, research focusing on a single group of diagnoses may lose

T. Koyama (✉)
Department of Epidemiology for Community Health and Medicine, Kyoto Prefectural University of Medicine, Kyoto, Japan
e-mail: tkoyama@koto.kpu-m.ac.jp

H. Yamada, Y. Naito (eds.), *Anti-Aging Medicine*, https://doi.org/10.1007/978-981-97-8514-8_6

sight of the essence. Overcoming this problem cannot be achieved by treating multiple different health outcomes separately, and there was a demand for the development of an integrated approach focusing on functional disabilities and new methods to deal with multiple health outcomes. Therefore, in epidemiological research, prospective cohort studies that require many years to track the occurrence of endpoints such as death and disease have evolved. For example, prospective cohort studies for specific diseases, such as the Framingham study, have also been applied to aging research in the process of long-term follow-up of study participants. As a result, the role of epidemiology in aging research has rapidly developed and expanded, helping to characterize the changes and burdens of diseases and health conditions that affect the elderly. The interest of epidemiologists involved in aging research has shifted from a narrow focus on disease to research on longevity and the quality of healthy aging (successful aging).

6.3 Epidemiological Research as Social Medicine

People are born with a foundation of genetics, upon which lifestyle habits such as diet, sleep, and exercise accumulate. Sometimes, health is judged by the presence or absence of disease, but when considering a person's life, factors such as where they live, income, education, and community also affect health. This is expressed as socio-economic status (SES). Furthermore, social environmental factors such as climate change and war, which we cannot resist, also affect health. As science advances and new cultures such as cars and smartphones are born, life becomes convenient, but traffic accidents occur and new diseases such as game disorders are born. The research field that includes not only genetics and individual lifestyle habits but also the impact on health due to social activities and the times is called the social determinants of health (Fig. 6.1). It is widely recognized that people's health and diseases are influenced by social, economic, political, and environmental conditions, and today's epidemiology is established as social medicine that deals with social factors as an integrated activity of a wide range of research using medicine, biology and genetics, social sciences, demography, geography, economics, etc.

6.4 To Connect Epidemiological Research to the Practice of Anti-Aging Medicine

To solve the research problems of aging changes, it is expected that evidence will accumulate through large-scale longitudinal surveys at the regional, national, or international level based on cohort studies that require a time axis. These studies steadily provide abundant data, and the contribution to aging research increases as the number of years of follow-up surveys increases. In recent years, the Japan Multi-Institutional

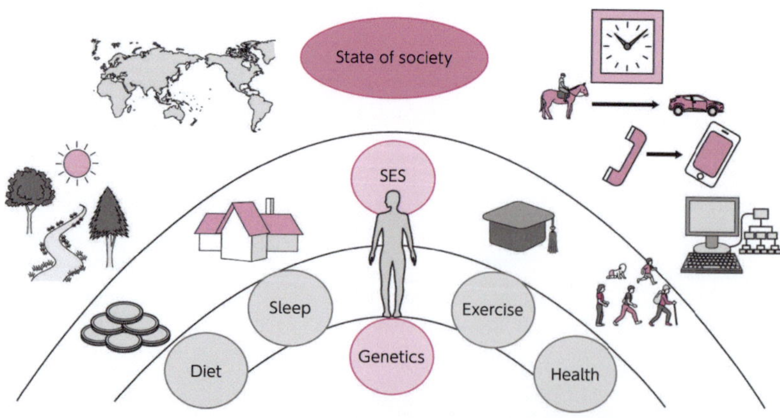

Fig. 6.1 Social determinants of health. The field of a study conducting scientific inquiries into the social determinants of health is called "Social Medicine (Epidemiology)"

Collaborative Cohort Study (J-MICC study), the Tohoku Medical Megabank Organization, the Iwate Tohoku Medical Megabank Organization, and the Next Generation Multipurpose Cohort Study [JPHC (Japan Public Health Center-based prospective)—NEXT Study] are proceeding as large-scale cohorts exceeding 100,000 people in Japan, and the use of data and biological samples, including genome information, is being promoted. It is desirable for anti-aging researchers to actively use cohort study data and biological samples. The current goal of epidemiological aging research is to establish scientific evidence to minimize the disease burden associated with aging, including social and environmental factors. This should compress the incidence and disability to the shortest period of human life, improving the quality of life for current and future generations of the elderly. In the future, interdisciplinary approaches that consider the determinants of health throughout life, including the treatment of aging, are expected.

References

1. Buettner D. Blue zones learning about health and longevity rules from centenarians around the world. Tokyo: Discover Twenty One; 2010.
2. Fried LP. Epidemiology of aging. Epidemiol Rev. 2000;22:95–106.

7 Policy Promotion of Anti-Aging Medicine

Toshiharu Furukawa

In Japan, as the average lifespan increases and the birthrate decreases, the proportion of people over 65 years old reached 28.4% in fiscal 2019, and further aging is expected in the future. By fiscal 2036, it is estimated that one in three people will be elderly. On the other hand, the total amount of Japan's accumulated deficit bonds exceeds twice the gross domestic product, a level that stands out globally. The most important and difficult problem is how to control the social security expenses that continue to increase every year due to aging. According to future estimates of social security expenses, pensions will hardly increase, and most of the increase will be in medical and nursing care costs [1]. Even as longevity progresses, it is necessary to extend the healthy lifespan as much as possible to control the growth of medical and nursing care costs, and to realize a surplus in household budgets up to the early 70s by supporting the employment of the elderly, and to raise the pensionable age. The significant role of anti-aging medicine in the future super-aged society is believed to be in extending the healthy lifespan of the population and enabling the elderly to work in good health. In Japan, due to the rapid aging of the population and changes in lifestyle habits, the proportion of lifestyle-related diseases such as cancer, ischemic heart disease, cerebrovascular disease, and diabetes in all diseases is increasing, and the proportion of medical expenses related to these lifestyle-related diseases in national medical expenses is about 30%. Therefore, since fiscal 2000, the government has selected goals for issues related to the improvement of lifestyle-related diseases and their causes, and has been promoting a new national health promotion movement in which citizens take the initiative. Among them, the central issue is the extension of healthy life expectancy. Note 1: "National Health Promotion Movement in the 21st Century (Healthy Japan 21)" from fiscal 2000, "Second National Health Promotion Movement in the 21st Century (Healthy Japan 21 (Second))" from fiscal 2013, etc. Note 2: In the "Healthy Life Extension Plan" of the first year of Reiwa (2019), efforts are being made to extend healthy life expectancy. The concept diagram of the difference between the average lifespan and healthy lifespan (unhealthy period) when the average lifespan is extended is shown in Fig. 7.1. Compared to the present, A is when the unhealthy period becomes longer, B is when the unhealthy period remains the same, and C is when the unhealthy period becomes shorter. In the pattern of B, it is thought that medical and nursing care costs will not change much from the present, but the total amount of social security expenses will increase because the pension receipt period will be extended. In order to suppress social security expenses, it is necessary to realize the pattern of C. So far, there is no high-level scientific evidence that the promotion of prevention and

T. Furukawa (✉)
Law School and Medical School, Keio University, Tokyo, Japan
e-mail: toshiharu_furukawa1@sangiin.go.jp

H. Yamada, Y. Naito (eds.), *Anti-Aging Medicine*, https://doi.org/10.1007/978-981-97-8514-8_7

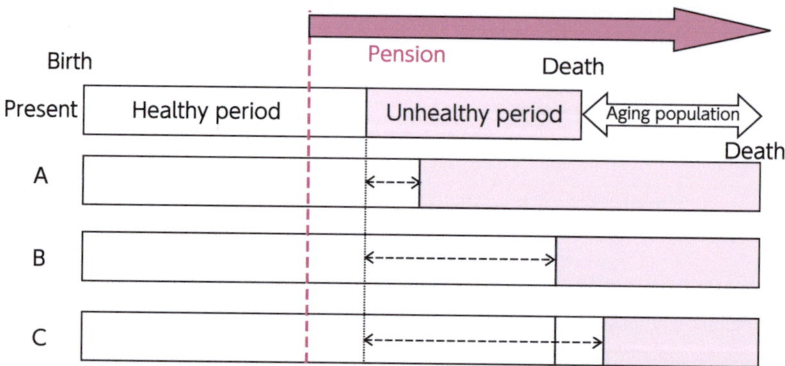

Fig. 7.1 Longevity and medical/nursing care costs. There is no clear evidence that the prevention of lifestyle diseases reduces medical and nursing care costs. Medical and nursing care costs are determined by the length of the unhealthy period and the density of medical and nursing care

health creation reduces medical costs. In 1997, it was reported that promoting a nonsmoking policy would reduce per capita medical costs, but as a result of more people living longer, the total medical costs would increase [2]. In a 2016 study estimating the historical transition of the difference between average lifespan and healthy lifespan in advanced countries (North America, Europe, and high-income Asian countries), people born in 1990, 2006, and 2016 have both average lifespan and healthy lifespan extending over the years, but the difference (unhealthy period) is expanding for both men and women over the years [3]. The promotion of prevention and health creation may also increase the number of people living longer, potentially leading to an increase in medical costs. Also, the implementation cost of disease prevention policies needs to be considered. A systematic review of 599 papers conducting a cost-effectiveness analysis of preventive medicine found that only about 20% of preventive medicine has a cost-suppressing effect, that is, both health promotion effect and medical cost suppression effect [4], and it is important to carefully analyze the intervention cost and benefits of preventive medicine. There are often discussions without evidence that "prevention and health creation lead to suppression of medical costs," but rather, it is important to make everyone work long and healthy through prevention and health creation, create social added value, and cover the increasing medical costs from that perspective. Looking at the movement of average lifespan and healthy lifespan in Japan from 2001 to 2019 (Fig. 7.2), both men and women's average lifespan and healthy lifespan are extended over the years, but from 2001 to 2010, the difference between average lifespan and healthy lifespan (unhealthy period) was expanding for both men and women, but in 2010–2019, both men and women continued to see a decrease. Compared to 2001, the unhealthy period in 2019 increased by 0.06 years for men and decreased by 0.21 years for women. However, what's important is that the healthy lifespan has extended by 3.28 years for men and 2.73 years for women. During this extended healthy lifespan, we should promote the employment of the elderly so they can contribute to social security expenses. In reality, medical and nursing care costs change not only with the length of the unhealthy period but also with the density of medical and nursing care during that period. Looking at studies in our country on this point, although somewhat outdated, it has been reported that medical costs decrease as the age of the elderly increases, but nursing care costs increase as age increases. Therefore, as the average lifespan extends, the density of elderly medical care decreases, but conversely, the density of nursing care is expected to increase [5, 6]. From this point as well, it is necessary to strengthen efforts to prevent the need for nursing care. Reflecting the changes in Japan's population structure, changes occur in medical needs, and the impact of the population structure on medical costs is estimated to

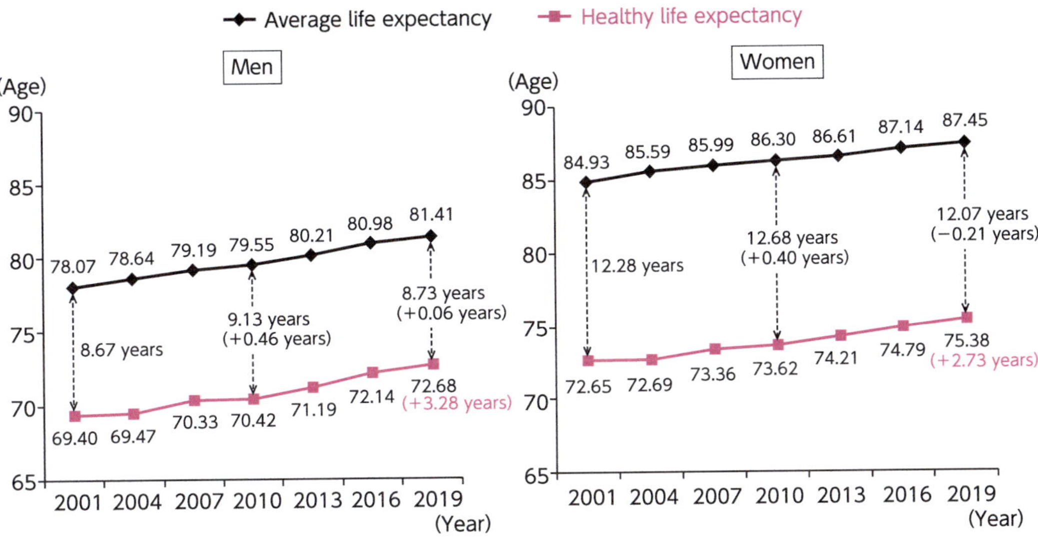

Fig. 7.2 Transition of average life expectancy and healthy life expectancy in Japan. (Average life expectancy: Heisei 13, 16, 19, 25, 28, and Reiwa first year are based on the Ministry of Health, Labor and Welfare's "Simple Life Table," and Heisei 22nd year is based on the "Complete Life Table")

increase until around 2030, with the increase due to aging exceeding the decrease due to population decline. However, after 2030, although aging will continue, the decrease due to population decline will exceed, and it is estimated to decrease (Fig. 7.3). In contrast, the cost of nursing care is estimated to continue to increase beyond 2040, with the increase due to aging exceeding the decrease due to population decline. The main causes of becoming in need of high-level care include cerebrovascular diseases (stroke), joint diseases, dementia, fractures and falls, and weakness due to old age (Fig. 7.4). Among these, the incidence rate per population of cerebrovascular diseases is continuing to decrease due to lifestyle disease measures. On the other hand, joint diseases, fractures and falls, and senility are mainly due to musculoskeletal disorders, especially walking is important. In the elderly aged 65 and over, it has been reported that the faster the walking speed, the higher the 10-year survival rate [7]. Dementia is the disease with the greatest health burden in Japan by 2040 [8], and there are not a few developing treatments, but it has been reported that the longer the usual walking time, the less cognitive decline, so walking is important [9]. Measures to maintain musculoskeletal function are considered necessary along with lifestyle disease measures. Anti-aging medicine so far has mainly targeted lifestyle diseases such as cerebrovascular disorders, hypertension, hyperlipidemia, ischemic heart disease, and diabetes. These metabolic diseases can be intervened with diet therapy and exercise therapy, and their preventive effects can be expected. On the other hand, in the future, we would like to expect that scientific evidence will accumulate for cancer prevention and dementia prevention. Cancer is the top cause of death and a disease that half of the people suffer from, and dementia, including its reserve army, is said to be nearly 30% of the elderly, if effective intervention is possible for both diseases, the impact on future medical and nursing care systems and social security finances will be extremely large. Furthermore, in recent years, with the development of information and communication technology (ICT), the advancement of health and medical data utilization, the spread of smartphones and wearable devices, the standardization of data such as health checkups and screenings, and the spread of personal health record (PHR) services by private companies, the movement to utilize ICT in the field of health promotion is spreading. We hope that anti-aging medicine will contribute to a further extension of healthy life expectancy by providing a theoretical basis for the utilization of data at these pre-disease stages.

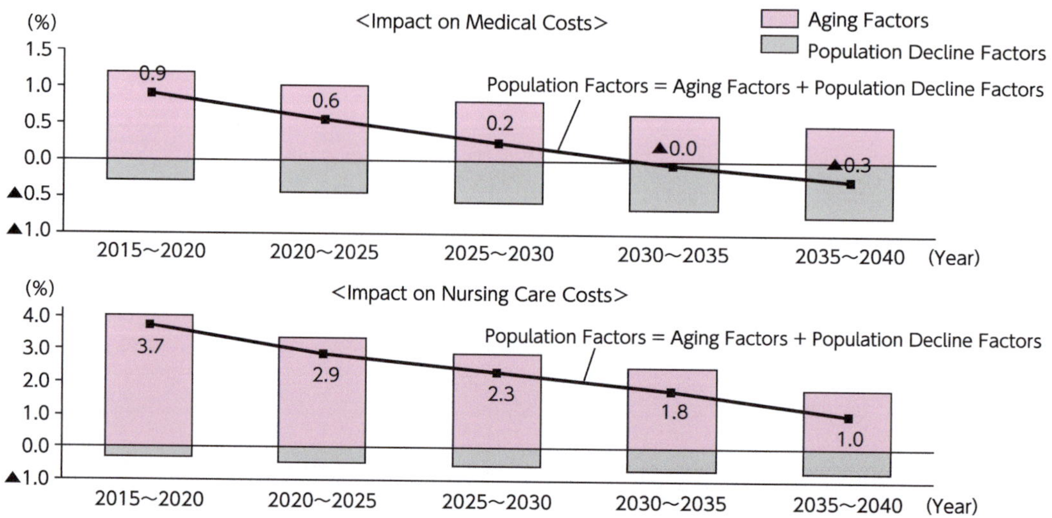

Fig. 7.3 The impact of population structure changes on medical and nursing care costs. (Calculation Method) Based on the actual per capita medical and nursing care expenses by age group and the future population by age group, the growth rate of medical and nursing care expenses based on the future population by age group, assuming that the per capita medical and nursing care expenses by age group are fixed, is considered as the growth rate due to "population factors." Then, the decrease rate of the total population is considered the "population decline factor," and the amount obtained by subtracting the "population decline factor" from the "population factor" is considered as the "aging factor." (Data Used) Ministry of Health, Labour and Welfare "Basic Materials on Health Insurance," "Survey on the Actual Situation of Nursing Care Benefits", National Institute of Social Security and Population Problems "Future Estimated Population of Japan"

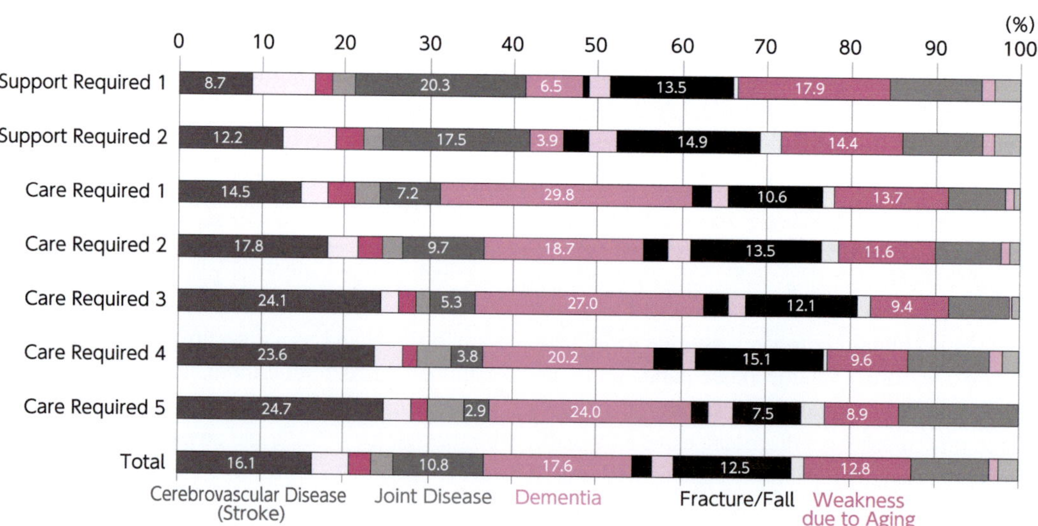

Fig. 7.4 Percentage of main causes of need for care by degree of need for care (2019). Made from the Ministry of Health, Labour and Welfare: 2019 National Livelihood Basic Survey)

References

1. Ministry of Health, Labour and Welfare: Report of the National Council for Social Security System Reform (August 6, 2013). 2013, p 23–24.
2. Barendregt JJ, Bonneux L, et al. The health care cost of smoking. N Engl J Med. 1997;337:1052–7.
3. GBD. DALYs and HALE collaborators: global, regional, and national disability-adjusted life-years (DALYs) for 333 diseases and injuries and healthy life expectancy (HALE) for 195 countries and territories, 1990-2016: a systematic analysis for the global burden of disease study 2016. Lancet. 2016;2017(390):1260–344.
4. Cohen JT, Neumann PJ, et al. Does preventive care save money? Health economics and the presidential candidates. N Engl J Med. 2008;358:661–3.
5. Tetsuo F, Atsushi G. Medical expenses of elderly deceased. Medi Econ Res. 1994;1:107–18.
6. Hashimoto H, Horiguchi H, et al. Micro data analysis of medical and long-term care utilization among the elderly in Japan. Int J Environ Res Public Health. 2010;7:3022–37.
7. Studenski S, Perera S, et al. Gait speed and survival in older adults. JAMA. 2011;305:50–8.
8. GBD. DALYs and HALE collaborators: global, regional, and national disability-adjusted life-years (DALYs) for 359 diseases and injuries and healthy life expectancy (HALE) for 195 countries and territories, 1990-2017: a systematic analysis for the global burden of disease study 2017. Lancet. 2017;2018(392):1859–922.
9. Weuve J, Kang JH, et al. Physical activity, including walking, and cognitive function in older women. JAMA. 2004;292:1454–61.

8 Anti-Aging Medicine and Medical Staff

Yukiko Minamiyama and Tastsuto Ashizawa

The definition of medical staff refers to "medical professionals who have national qualifications related to medical care, nursing care, welfare, etc." other than physicians and dentists. Specifically, nurses, radiological technologists, pharmacists, physical therapists, occupational therapists, clinical laboratory technicians, clinical engineering technicians, social workers, mental health welfare workers, etc. are considered medical staff, also known as co-medicals, covering 34 professions (Fig. 8.1). This chapter discusses how these medical staff can be involved in anti-aging medicine, considering both the current and future prospects.

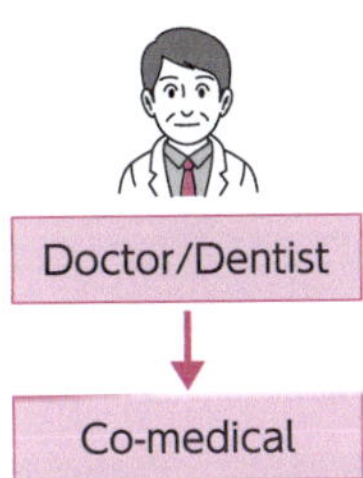

Nurse	Certified Psychologist	Care Support Specialist	Medical Physicist
Pharmacist	Clinical Psychologist	Care Welfare Worker	Dental Assistant
Clinical Laboratory Technician	Registered Dietitian	Nursery Teacher	Mental Health Welfare Worker
Radiological Technologist	Nutritionist	Dental Technician	Social Worker
Clinical Engineer	Dental Hygienist	Acupuncturist	Prosthetist
Physical Therapist	Public Health Nurse	Moxibustion Practitioner	Medical Office Assistant
Occupational Therapist	Midwife	Massage Therapist	Medical Insurance Clerk
Speech-Language-Hearing Therapist	Sanitary Inspector	Judo Orthopedist	
Orthoptist	Emergency Medical Technician	Medical Information Manager	

Fig. 8.1 Medical professions

Y. Minamiyama (✉)
Food Hygiene and Environmental Health, Division of Applied Life Science, Graduate School of Life and Environmental Sciences, Kyoto Prefectural University, Kyoto, Japan
e-mail: yukiko-m@kpu.ac.jp

T. Ashizawa
Tokyo Medical University, Tokyo, Japan

H. Yamada, Y. Naito (eds.), *Anti-Aging Medicine*, https://doi.org/10.1007/978-981-97-8514-8_8

8.1 Anti-Aging Checkup

The anti-aging checkup aims to slow down the unavoidable aging clock as much as possible by evaluating the degree of aging from bone age, nerve age, vascular age, muscle age, hormone age, and measuring factors such as oxidative stress as aging risk factors, and understanding the physical and mental health status (Fig. 8.2). In other words, biological age and actual age are not the same, and in unhealthy people, biological age is often rated higher than actual age. On the other hand, healthy people are often seen as younger than their actual age and are rated lower. Anti-aging medicine focuses on treating and preventing diseases associated with aging at the molecular and cellular level and slowing down the process of biological aging. The goal is to slow down the progression of pathological conditions associated with aging by utilizing the latest scientific research, advances in biotechnology, and clinical preventive medicine. Therefore, doctors and co-medicals at anti-aging clinics need to help slow down the progression of diseases and functional impairments associated with aging through preventive screening for early detection of diseases, rapid intervention, and optimal nutritional guidance. Furthermore, anti-aging medicine emphasizes not just maintaining the status quo of disabilities, but enhancing health, optimizing performance, and improving the quality of life of patients.

8.2 The Essence of Anti-aging Medicine

8.2.1 Nutrition and Supplements

Hormonal balance and metabolic balance are both extremely important for health, happiness, and optimal performance. To evaluate these in an individual, it is necessary to measure whether the body is functioning correctly and assess the intake and absorption of appropriate nutrients. In this process, the evaluation is carried out not only by doctors but also by various paramedical staff such as clinical laboratory technicians, physical therapists, and dietitians. For example, clinical laboratory technicians measure things like the concentration of amino acids, vitamins, and minerals in the blood, physical therapists measure exercise function, and dietitians determine nutritional deficiencies using the latest nutritional tests. By adjusting the diet and supplements to suit the individual, they can make a significant contribution to improving health and vitality. Therefore, in anti-aging medicine, it is important to design treatment programs based on professional examinations by paramedical staff, individual risk assessments, and unique health histories. This customized treatment helps patients achieve and maintain optimal levels of sexual, physical, and mental performance.

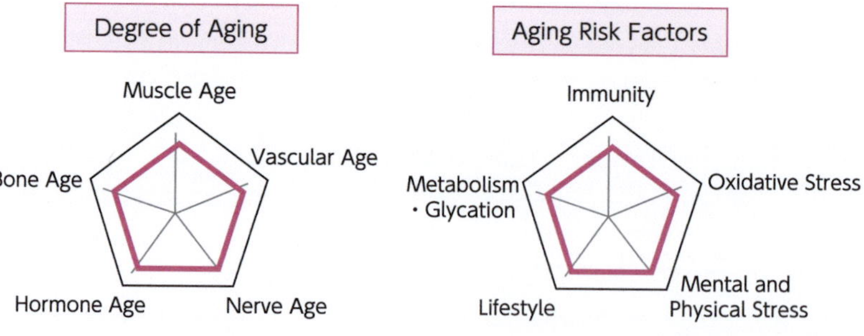

Fig. 8.2 Anti-aging checkup

8.2.2 Preventive Medicine

In anti-aging medicine, preventing the onset of future diseases is as important as treating existing diseases. Therefore, proactive screening tests such as the latest specimen tests and imaging diagnoses, and rapid interventions, are basic methods for detecting serious diseases that threaten the patient's life. The implementation of these tests is indispensable to the presence of paramedical staff, including radiological technicians and clinical laboratory technicians. Also, by optimizing health, improving nutrition, and positively adjusting lifestyle, it is possible to minimize the risk of major diseases such as heart disease, hypertension, diabetes, cancer, cerebrovascular disorders, and dementia.

8.2.3 Integrative Medicine

The stress of modern society is changing in various ways, but it is needless to say that it is definitely increasing. Anti-aging medicine focuses on optimizing health from an integrative and holistic approach to treating the body and mind, and therefore often combines various diagnostic and treatment methods. In this case, the active involvement of acupuncturists, certified psychologists, clinical psychologists, and fitness trainers, along with doctors, is necessary.

8.2.4 Functional Medicine

The aim is to view the patient's body as an integrated system, discover the fundamental causes of chronic conditions through a system-oriented approach, and provide treatment. The conditions in question include organ fatigue, aging, allergies and food sensitivities, autoimmune diseases, chronic fatigue syndrome, depression and anxiety neurosis, eating disorders, infertility, and menopausal disorders. In understanding and treating these conditions, the best treatment plan for the patient is determined by considering both external factors (environment) and internal factors (lifestyle).

In recent years, it has been found that abnormalities in time genes and proteins can disrupt the body's internal clock, leading to aging and immune abnormalities. In addition to the tests and evaluations in previous anti-aging checks, a better consultation is needed through multidisciplinary collaboration, including not only doctors and dentists but also medical staff, to measure new evaluation indicators and provide treatment or guidance for chronic functional failures (Fig. 8.3).

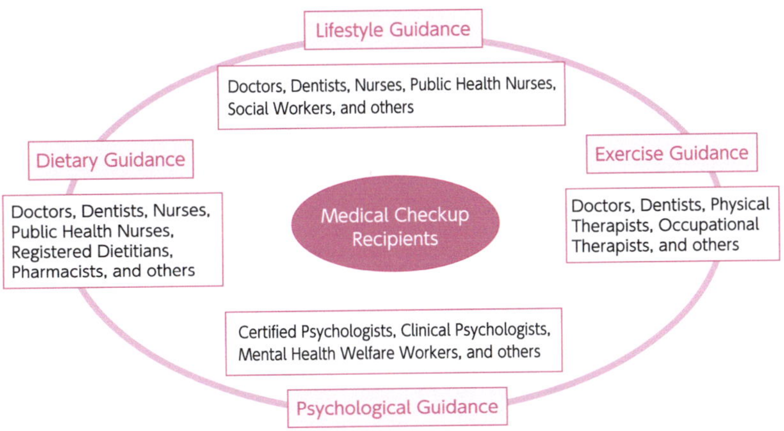

Fig. 8.3 Mechanism of consultation through multidisciplinary collaboration

9 Occupational Medicine and Anti-Aging Medicine

Seichi Horie

9.1 The Purpose of Occupational Health

The International Labour Organization (ILO) and the World Health Organization (WHO) define the purpose of occupational health as (1) preventing the occurrence and exacerbation of harmful health effects caused by work (prevention of occupational diseases) and (2) ensuring that even those with diseases or disabilities can work (securing employability). In addition, the WHO recommends activities to reduce workplace environmental risks, social disparities, high-risk behaviors, and health service utilization barriers that affect the health of workers, and refers to cardiovascular diseases, chronic respiratory diseases, musculoskeletal disorders, etc., which are widely prevalent among the general public and whose incidence rates increase or worsen due to working conditions and the state of the working environment, as work-related diseases [1].

S. Horie (✉)
Department of Health Policy and Management, Institute of Industrial Ecological Sciences, University of Occupational and Environmental Health, Fukuoka, Japan
e-mail: horie@med.uoeh-u.ac.jp; J-shkkan@mbox.med.uoeh-u.ac.jp

9.2 Occupational Safety and Health Act and the Duties of Occupational Physicians

The Occupational Safety and Health Act stipulates measures to prevent occupational accidents that businesses (corporations) should implement, with the aim of "ensuring the safety and health of workers in the workplace and promoting the formation of a comfortable working environment." The Act requires businesses to conduct health examinations of workers and measurements of the working environment, and based on the results of each, to take measures aimed at promoting the health of workers and making the workplace more comfortable (Fig. 9.1). Workers are required to undergo health examinations, which encourage health behaviors. In recent years, risk factors for cardiovascular diseases have been added to the items of regular health examinations, and the rate of findings has been increasing. In addition, businesses that employ 50 or more people at all times are required to appoint an occupational physician, who has a role in expressing opinions on the measures that the business should take based on the health examination results of each worker. Occupational physicians have the authority to make recommendations to businesses from an independent standpoint, and it is expected that they will utilize this for the purpose of occupational health.

H. Yamada, Y. Naito (eds.), *Anti-Aging Medicine*, https://doi.org/10.1007/978-981-97-8514-8_9

Fig. 9.1 Occupational health activities conducted by employers based on the Occupational Safety and Health Act

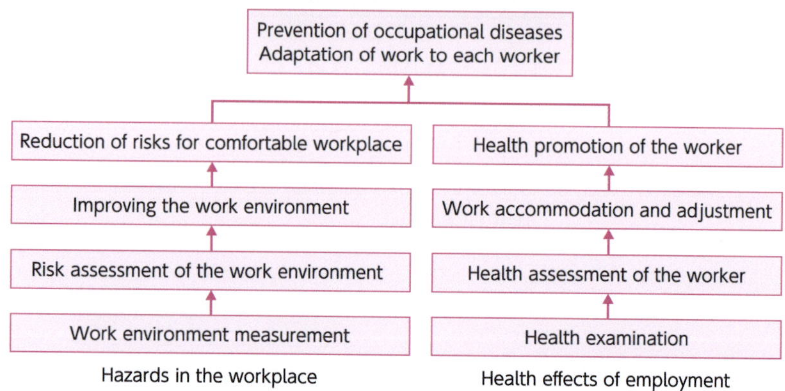

9.3 Total Health Promotion Plan (THP)

The Occupational Safety and Health Act requires businesses to make efforts to take measures to promote the health of workers. The Ministry of Health, Labour and Welfare has introduced the THP (Total Healthpromotion Plan) is being promoted [2]. This guideline encourages businesses to carry out continuous and planned health maintenance and promotion activities, specifically, (1) exercise guidance (fun and effective exercise guidance considering the living conditions and wishes of workers), (2) mental health care (guidance on stress awareness and relaxation), (3) nutritional guidance (guidance on improving dietary habits and eating behaviors), (4) oral health guidance (guidance on dental and oral health), and (5) health guidance (guidance on sleep, smoking, drinking, etc. in workplace life). It also encourages awareness and habituation of exercise from a young age, and encourages the use of resources inside and outside the workplace to prevent lifestyle-related diseases through collaboration with health insurers.

9.4 Employment of the Elderly

In recent years, the proportion of people aged 65 and over in the total number of workers has been increasing year by year, reaching 13.6% in 2020 (9.1% in 2010). Also, the employment rate of the elderly has been increasing year by year, reaching 25.1% in 2020 (19.4% in the same year). On the other hand, the number of work-related accidents resulting in four or more days off per thousand workers (thousand rate) was 2.42 for men and 1.50 for women aged 25–29 in 2021, compared to 4.23 and 4.49 for those aged 65–69, and 4.35 and 4.52 for those aged 70–74, with the elderly having about twice the incidence rate for men and three times for women compared to younger people, with falls being particularly common. The physical functions necessary for employment are reported to be affected by aging as follows. The values for ages 55–59, when the highest level at young age is set at 100, are reported to be dark adaptation 36, balance function 48, motor coordination ability 59, visual acuity 63, leg extension strength 63, and instantaneous reaction 71. Muscle strength begins to decline from the lower leg strength in a body with raised arms, and grip strength and fingertip muscle strength are relatively easy to maintain. Reaction time is maintained for simple movements, but tends to slow down for choices depending on conditions or complex movements. The adjusting power and contrast vision (vision at low contrast) of vision decrease. Hearing decreases in the ability to hear inserted sounds in conversation and sounds in the high-frequency band. Flexibility peaks in the late teens and drops sharply until around 40 for men and around 30 for women, but the decline slows down after that. Fatigue recovery takes longer with age. Short-term memory tends to decline, but long-term memory is relatively easy to maintain. Knowledge and skills acquired through learning and training are maintained the longer they are used. And individual differences expand with age. Also, the elderly tend to underestimate the decline in their physical functions. In occupational health, it is expected that understanding these characteristics of the elderly will promote improvements in the workplace environment and work content so that the elderly can work safely and securely.

9.5 Age-Friendly Guidelines

The Ministry of Health, Labour and Welfare has compiled age-friendly guidelines to ensure that the elderly can work safely and healthily [3]. In these guidelines, employers are encouraged to (1) establish a safety and health management system, (2) improve the workplace environment [4] (Table 9.1), (3) understand physical functions [5] (Table 9.2), (4) respond according to health and physical functions, and (5) promote safety and health education. Workers are also encouraged to maintain their physical functions, such as strengthening their lower body muscles. Many good examples have been reported that promote the active participation of the elderly [4]. It is expected that by improving these workplace environments and maintaining physical functions, future labor accidents involving the elderly,

Table 9.1 Examples of improving the workplace environment for the elderly

Enlarging text	Securing illumination
Reducing and eliminating steps	Installing handrails and anti-slip measures
Wearing shoes with slip resistance	Thorough implementation of 4S (sorting, setting in order, sweeping, and standardizing) in passages and workplaces
Installing high work platforms, etc.	Limiting the weight and size of transported items, materials, and tools
Utilizing carts	Creating work standards for the elderly

Central Labour Accident Prevention Association Education Promotion Department: Chapter 2 Safety and Health Measures for Promoting the Active Participation of Older Workers. Safety and Health Measures for Promoting the Active Participation of Older Workers—Collection of Advanced Company Initiatives, Central Labour Accident Prevention Association, Tokyo, 2017, p17–30. Created by the author

Table 9.2 Self-check sheet for fall risk assessment for the elderly [5]

	Evaluation results (1 or 2 indicates a high risk of falling)				
Evaluation Index	1	2	3	4	5
Two-step test	~1.24	1.25–1.38	1.39–1.46	1.47–1.65	1.66 ~
Seated stepping test (times)	~24	25–28	29–43	44–47	48 ~
Functional reach (cm)	~19	20–29	30–35	36–39	40 ~
Closed-eye one-leg stand (seconds)	~7.0	7.1–17.0	17.1–55.0	55.1–90.0	90.1 ~
Open-eye one-leg stand (seconds)	~15.0	15.1–30.0	30.1–84.0	84.1–120.0	120.1 ~

Two-step Test (Walking Ability and Muscle Strength): the maximum value of two-step length divided by height
Seated Stepping Test (Agility): The number of times you can open and close your legs in 20 s while sitting on a chair
Functional Reach (Dynamic Balance): The distance you can reach horizontally by tilting your upper body while standing
Closed-eye One-leg Stand (Static Balance): The number of seconds you can stand on one leg with your eyes closed
Open-eye One-leg Stand (Static Balance): The number of seconds you can stand on one leg with your eyes open

which are expected to increase further in our country, will be prevented, and health promotion in the workplace will be promoted.

References

1. Horie M. International trends in industrial health policy and academia. Health Indicators. 2015;62:1–8.
2. Ministry of Health, Labor and Welfare: Guidelines for promoting the health maintenance of workers at workplaces. Showa 63 September 1st Health Maintenance Promotion Guidelines Public Notice No. 1, Reiwa 4 March 31st Revised Public Notice No. 10, 2022.
3. Ministry of Health, Labor and Welfare Labor Standards Bureau Safety and Health Department: Guidelines for ensuring the safety and health of elderly workers. Reiwa 2 March 16th Basic Safety Issue 0316 No. 1, 2020.
4. Central Labor Accident Prevention Association Education Promotion Department. Chapter 2 Safety and health measures for promoting the active participation of elderly workers, Chapter 3 Examples of advanced companies. Safety and Health Measures for Promoting the Active Participation of Elderly Workers—Collection of Examples of Advanced Companies, Central Labor Accident Prevention Association, Tokyo, 2017, p 17–134.
5. Central Labor Accident Prevention Association Education Zero Disaster Promotion Department: Explanation of the checklist for ensuring the safety and health of elderly workers, VI_ Situation of labor accidents of elderly workers, VII_ Status of physical and mental functions with aging. Age Action 100 (Revised Edition) ~Towards workplace improvement for ensuring the safety and health of elderly workers leading to the realization of a lifelong active society~, Central Labor Accident Prevention Association, Tokyo, 2021, p 21–122.

10 Anti-Aging Medicine in the Post/with-Corona Era

Yoshihiro Kitamura

The infectious disease (COVID-19) of the new coronavirus (SARS-CoV-2) begins when SARS-CoV-2 binds to the virus receptor in the upper respiratory tract. The virus receptor of SARS-CoV-2 is angiotensin-converting enzyme 2 (ACE2). ACE2 is widely expressed on the endothelial cell membrane of blood vessels throughout the body, as well as on the airway epithelium and type II epithelium of the alveoli. Therefore, COVID-19 is a complex disease state that should be described as a systemic multi-organ inflammatory disease accompanied by vasculitis/coagulation disorders/cytokine storms due to vascular endothelial injury, rather than a mere respiratory disease. In fact, in most severe cases of COVID-19, the focus has been on calming uncontrollable inflammatory responses. However, exposure to SARS-CoV-2 does not necessarily lead to the onset and severity of the disease. This is because an appropriate natural immune response suppresses the onset or leads to recovery even if the disease does occur.

Y. Kitamura (✉)
Nippon Medical School, Tokyo, Japan
e-mail: y-kitamura@nms.ac.jp

10.1 Old Age Is a Risk Factor for Severe COVID-19

The main risk factor for severe COVID-19 is "old age." The mechanism by which old age (aging) becomes a risk for severity is not fully understood, but the hypothesis of immune decline in the elderly, namely "immunosenescence," has been proposed. Among them, aging of natural immunity and chronic inflammation are important [1, 2]. The initial natural immune response is weak in the elderly. Specifically, the decrease in plasmacytoid dendritic cells (pDC), the main players in natural immunity, the delay in type 1/3 interferon responses, the decrease in antigen presentation function, and the decrease in tissue resident macrophages are the main conditions [1]. In the elderly, low-level chronic inflammation (inflammaging, inflammatory aging) is normalized throughout the body, especially in vascular endothelium [2]. Inflammaging is a continuous increase in inflammatory cytokines [IL (interleukin)-1β, IL-18, etc.] due to various causes (Fig. 10.1). Serum levels of several inflammatory biomarkers, such as C-reactive protein (CRP), IL-6, IL-18, and TNF-α (tumor necrosis factor-α), moderately increase. First, the increase in aging cells due to inappropriate nutrition and the resulting senescence-associated secretory phenotype (SASP) are cited as causes. Inappropriate nutrition specifically refers to a diet with a high dietary inflammatory index (DII) and malnutrition.

H. Yamada, Y. Naito (eds.), *Anti-Aging Medicine*, https://doi.org/10.1007/978-981-97-8514-8_10

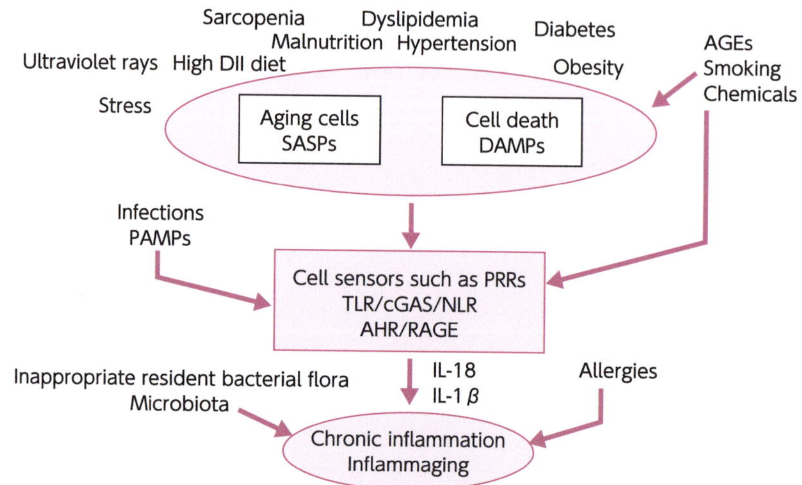

Fig. 10.1 Conceptual diagram of inflammaging. Infections by microbes, lifestyle habits, diseases, physical stimuli, and chemicals cause low-level chronic inflammation through cell damage. Inappropriate diet, lack of exercise, and unfavorable microbiota may also potentially cause inflammaging. *AHR* aryl hydrocarbon receptors, *cGAS* cyclic GMP-AMP synthase, *NLR* NOD-like receptor, *PAMPs* pathogen-associated molecular patterns, *PRRs* pattern recognition receptors, *RAGE* receptor for advanced glycation end products, *TLR* Toll-like receptor

Excessive intake of carbohydrates can also be considered an inappropriate diet as it leads to an increase in advanced glycation end products (AGEs). In an evaluation of severe COVID-19 risk factors specific to the elderly, care facility admission/daily life activities (activities of daily Decreased Activities of Daily Living (ADL)) are being highlighted as factors for severe disease progression, so sarcopenia and low activity may be involved in chronic inflammation, although the mechanism is unknown. Cell death throughout the body due to hypertension, diabetes, obesity, hyperlipidemia, etc., especially vascular endothelial cell injury, recognizes various molecules released from dead cells as damage-associated molecular patterns (DAMPs) in crisis management, leading to an increase in the release of inflammatory cytokines [3].

10.2 Building a Body Strong Against Infectious Diseases Is Anti-Inflammatory Aging

In the future, the COVID-19 pandemic will end due to the attenuation of the virus. Instead, the infectivity (transmissibility) of the coronavirus is predicted to increase. Therefore, in the era of living with the coronavirus, it is expected that we will commonly get infected with the coronavirus, and occasionally severe cases will be seen in the elderly with the above-mentioned risk factors. In other words, anti-aging in the era of living with the coronavirus can be rephrased as the need to build a body that can easily recover from a cold rather than not getting infected with the coronavirus. To do this, it is important to remove the causes of inflammatory aging (anti-inflammatory aging). However, unfortunately, it is difficult to completely remove the causes shown in Fig. 10.1, and life cannot be sustained without them. Therefore, dietary improvement is recommended first. Excessive intake of saturated fats, meat, and refined sugar should be improved. The Mediterranean diet is recommended because it can improve inflammatory processes and oxidative stress, maintain good gut microbiota, and potentially reduce circulating levels of inflammatory biomarkers. Correcting a low vitamin D status may reduce chronic inflammation, and anti-inflammatory aging with vitamin D is probably effective. It is certain that probiotics are effective for immune anti-aging. There are scattered reports of immune regulation and activation by so-called lactic acid bacteria, including reports of Lactococcus species related to the activation of pDCs and reports of improved immune responses by *Lacticaseibacillus* species [4].

10.3 Moderate Exercise Is Also Recommended

In the analysis of elderly infected people in the sixth wave, being admitted to a long-term care facility was a risk factor for severe disease. Furthermore, a decrease in ADL was a more important risk factor than facility admission. In the analysis of infected people in Niigata Prefecture in the seventh wave, "obesity" was a risk factor for severe disease in young people, but in contrast, "thinness (frailty)" was a risk factor in the elderly. Whether the state of frailty or decreased ADL is a risk factor, or whether the lifestyle habits that led to that state are also risk factors, is unclear, but maintaining and increasing muscle strength and mass from a young age and moderately exercising can be considered anti-inflammatory aging [5].

References

1. Merad M, Blish CA, et al. The immunology and immunopathology of COVID-19. Science. 2022;375:1122–7.
2. Chen Y, Klein SL, et al. Aging in COVID-19: vulnerability, immunity and intervention. Ageing Res Rev. 2021;65:101205.
3. Shenoy S. Gut microbiome, vitamin D, ACE2 interactions are critical factors in immune-senescence and inflammaging: key for vaccine response and severity of COVID-19 infection. Inflamm Res. 2022;71:13–26.
4. Omaki T. Examination of risk factors for severe omicron strain cases in COVIREGI-JP/REBIND registered patients. 83rd Advisory Board on Measures against New Coronavirus Infections (May 11, 2022), Document 3–8, Ministry of Health, Labour and Welfare, 2022. https://www.mhlw.go.jp/content/10900000/000937663.pdf
5. Takahashi M, Ayamegawa Y, et al. New coronavirus infection in the 7th wave: about risk factors for severe disease (preliminary report). 95th Advisory Board on Measures against New Coronavirus Infections (August 18, 2022), Document 3–9, Ministry of Health, Labour and Welfare, 2022. https://www.mhlw.go.jp/content/10900000/000977550.pdf

11 The Future of Antiaging Medicine

Kazuo Tsubota

The national medical expenses in Japan for the first year of Reiwa (fiscal year 2019) amounted to 44 trillion 3895 billion yen, an increase of 994.6 billion yen, or 2.3%, compared to the previous fiscal year's 43 trillion 3949 billion yen. The national medical expenses per capita were 351,800 yen, an increase of 8600 yen, or 2.5%, compared to the previous fiscal year's 343,200 yen. Despite efforts such as the active introduction of generic drugs and the implementation of the late-stage elderly medical care system, the current situation is that medical expenses cannot be suppressed. Therefore, the urgent introduction of preventive medicine has become an issue, and the eradication of metabolic syndrome is considered the first step. There is a need to shift from the traditional disease treatment-centered medical policy, which starts treatment after becoming ill, to a medical policy that can intervene before becoming ill. Antiaging medicine, which focuses on aging, can be considered preventive medicine and is becoming a major trend in medical science that aligns with the current medical policy in our country. Currently, the calorie restriction (CR) hypothesis and the oxidative stress hypothesis are the aging theories for evidence-based antiaging medicine, and various interventions are being carried out based on these [1–4]. What will this trend look like in the future?

K. Tsubota (✉)
Department of Ophthalmology, Keio University School of Medicine, Tokyo, Japan

Tsubota Laboratory, Inc., Tokyo, Japan
e-mail: tsubota@tsubota-lab.com

11.1 The Science of Aging Advances Further

The reason why the CR hypothesis and the oxidative stress hypothesis occupy a large position in antiaging medicine is that actual interventions are possible based on these theories [5, 6]. There is no doubt that various factors such as epigenetics, genetic damage, increase in senescent cells, telomere shortening, and protein cross linking are involved in the mechanism of aging, but intervention in these is difficult at the current stage. What about the future? It is quite conceivable that the mechanism of aging will be further elucidated and many new molecular targets will be found. In addition to sirtuins, nicotinamide mononucleotide (NMN), and nicotinamide riboside (NR), which are at the center of the CR hypothesis, there are heavyweight gene candidates such as forkhead box (FOXO), mammalian target of rapamycin (mTOR), nuclear respiratory factor (NRF), and peroxisome proliferator-activated receptor (PPAR)-γ coactivator 1 (PGC1), [7–13] and new intervention possibilities will emerge. Even now, Dr. de Cabo and others from the National Institute on Aging (NIA), who are very cautious about announcing research results, have listed what is currently recognized as anti-aging research under strict conditions in a review in Cell magazine titled

H. Yamada, Y. Naito (eds.), *Anti-Aging Medicine*, https://doi.org/10.1007/978-981-97-8514-8_11

"The Search for Antiaging Interventions: From Elixirs to Fasting Regimens" [14]. The conditions are (1) extending lifespan and healthy lifespan, (2) being confirmed by at least three types of animals, and (3) being confirmed by at least three research institutes worldwide. The seven items that meet these conditions are CR, fasting, exercise, resveratrol, rapamycin, spermidine, and metformin. Most of them are related to the CR hypothesis, and it is clear that the central thesis of anti-aging medical research comes from the CR hypothesis.

11.2 Development of the Hormesis Hypothesis

What is currently attracting attention as a molecular mechanism that is beneficial to antiaging medicine, such as CR and exercise, is the so-called "hormesis hypothesis." Hormesis refers to the improvement of the response to the next stress when a non-lethal stress is applied to the organism. It's a theory that if you're in a slightly tough situation, you won't die, but if a tougher situation comes along, you'll have resistance. It is well known that reactive oxygen species are produced by exercise. Why is exercise good for the body even though reactive oxygen species promote aging? It is thought to be because the group of enzymes that remove reactive oxygen species are expressed by the reactive oxygen species produced by exercise. Also, the function of mitochondria increases with exercise, and the same effect as CR can be obtained [15]. Temporary stress produces a good overall result. Conversely, it can be easily inferred from this theory that if you continue to exercise all day, there is a possibility that it will work negatively. Looking at retinal disorders, bright light causes light damage. However, if you apply a certain amount of light, stress response genes such as nuclear respiratory factor 2 (NRF2) work, activating the system for removing reactive oxygen species such as catalase and suppressing light damage [16]. If these molecular mechanisms further develop, not only taking vitamins that simply reduce reactive oxygen species, but also improving the body's defense response will lead to new intervention methods for antiaging.

11.3 Calorie Restriction (CR) Mimetics

Various studies are being conducted with the idea of wanting to achieve the same effects as CR without doing CR. One method is to try to achieve the same or better effects as CR without strict CR. Professor Valter Longo of the University of California advocates the concept of a fasting mimicking diet (FMD), which includes basic research and clinical applications [17–19]. In humans, it involves reducing calorie intake by about 50% for only 5 days a month, particularly limiting the intake of proteins and carbohydrates during this period. This allows one to enjoy the positive effects of CR while suppressing the negatives such as lowered body temperature, muscle loss, and decreased happiness. It has also been suggested that it could even rejuvenate stem cells, attracting attention [20]. Another method is called CR mimetics, which is induced by supplements or drugs [21]. Among these, resveratrol and related substances are currently attracting attention. Since it is believed that sirtuins are activated by CR, which is a major pathway for the expression of longevity genes, the idea is to ingest resveratrol, which can activate sirtuins, to achieve the same effect. Substances with over 1000 times the effect of resveratrol have already been developed, and this field is expected to develop significantly in the future [22, 23]. Furthermore, since sirtuins are nicotinamide adeninedinucleotide (NAD)-dependent enzymes, methods to increase NAD concentration have begun to be proposed. One is to provide precursors, with NMN and NR being promising. Another is the use of inhibitors of the poly ADP-ribose polymerase (PARP) enzyme, which competitively uses NAD. NR is already being sold as a supplement in the United States and is highly anticipated. Also, the application of ketone bodies as CR mimetics is expected. Among ketone bodies, hydroxybutyrate (HBA) [24, 25] is expected to control oxidative stress and inflammation through FOXO and inflammasome. We are also developing therapeutic drugs applying antiaging medicine, positioning HBA eye drops as local CR mimetics.

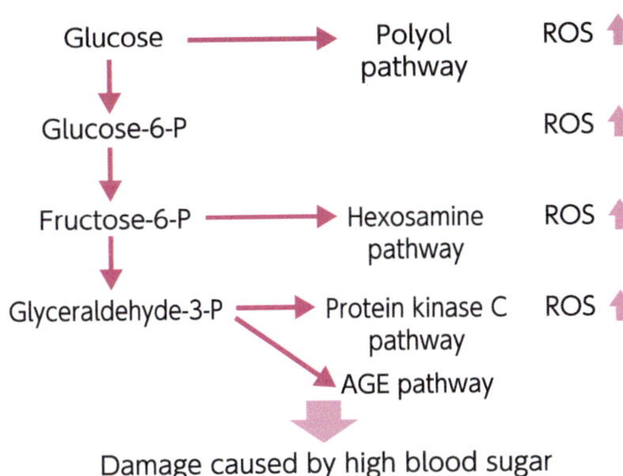

Fig. 11.1 The role of oxidative stress in diabetes. Reactive oxygen species (ROS) are involved in all four pathways

11.4 Advances in Oxidative Stress Control

Even if we take diabetes as an example, almost all the damage caused by high blood sugar is thought to be due to oxidative stress. Active oxygen is involved in all four seemingly unrelated famous pathways [26] (Fig. 11.1). Therefore, vitamins A, C, and E and various supplements that suppress oxidative stress should be effective in preventing complications of diabetes, but no definitive treatment method by oxidative stress control has been developed yet. This is thought to be because no substance capable of sufficient oxidative stress control has been developed yet. Also, since oxidative stress is also used as a signal within cells [27], it is expected that methods to selectively eliminate harmful hydroxyl radicals, such as using hydrogen water, without indiscriminately eliminating hydrogen peroxide, will be further developed.

11.5 Antiaging Medicine as Disease Treatment

Whether it's CR mimetics or oxidative stress control, they were traditionally developed for the somewhat vague goal of anti-aging, but a recent major shift in thinking is to use antiaging medicine for disease treatment (Fig. 11.2). It's a groundbreaking idea to use antiaging medicine in disease treatment medicine beyond preventive medicine.

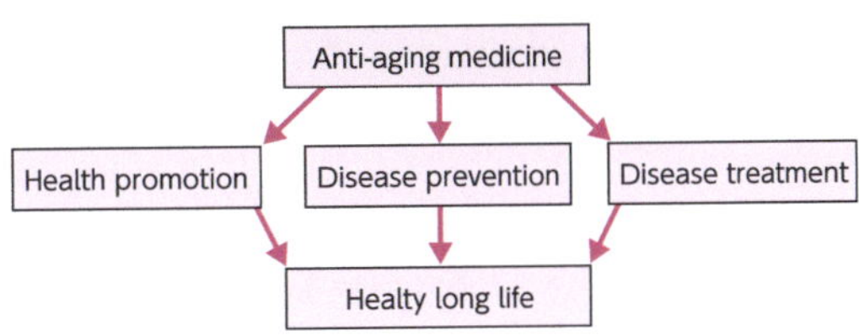

Fig. 11.2 Anti-aging medicine as a treatment for diseases. It will be used not only for promoting health and longevity and preventing diseases

In fact, resveratrol was originally developed from aging research on yeast and nematodes that activate sirtuins, but in 2008, it was approved as an orphan drug for mitochondrial encephalomyopathy by the U.S. Food and Drug Administration (FDA). NMN, NR, and active oxygen are involved in age-related hearing loss and dry eye, and it is being explored whether treatment can be achieved through oxidative stress control [28–30].

11.6 Future Prospects

The future of antiaging medicine is bright [31]. It will not only establish an important position as preventive medicine but will also be incorporated into disease treatment. However, there are still hurdles. One is that while longevity genes such as sirtuins have the ability to suppress the aging of somatic cells, some of the seven sirtuins found in mammals have been found to also suppress the aging of cancer cells and protect cancer cells from anticancer drugs, which can have a negative effect on the whole body. Even with antioxidant supplements, excessive intake can potentially eliminate necessary oxidative stress, and it has become important to determine the necessary amount for each individual. Nevertheless, the future of antiaging medicine is bright. As the mechanism of aging is elucidated, relatively simple and inexpensive methods of antiaging, such as increasing the concentration of NAD, increasing the concentration of ketone bodies, and introducing FMD, have come into view [18, 32]. Even now, we see a few months of life extension every year, and if we can push this further and extend the lifespan of the Japanese, or even all of humanity, the social impact will be enormous. This is because it is certain that a significant part of the current increase in Gross Domestic Product (GDP) is due to life extension [33], and if we can withdraw from the cost of treating diseases after they occur through preventive medicine, we can expect economic effects. It is expected to become an even more important medicine not only for individuals but also for society. Recently, we reported that not only antiaging but also reverse aging (rejuvenation) is possible in mammals by manipulating epigenetics [34]. This could usher in a new era.

References

1. Kirkwood TB. Understanding the odd science of aging. Cell. 2005;120:437–47.
2. Kenyon C. The plasticity of aging: insights from long-lived mutants. Cell. 2005;120:449–60.
3. Kennedy BK, Smith ED, et al. The enigmatic role of Sir2 in aging. Cell. 2005;123:548–50.
4. Hadley EC, Lakatta EG, et al. The future of aging therapies. Cell. 2005;120:557–67.
5. Roth GS, Lane MA, et al. Biomarkers of caloric restriction may predict longevity in humans. Science. 2002;297:811.
6. Khaw KT, Wareham N, et al. Combined impact of health behaviours and mortality in men and women: the EPIC-Norfolk prospective population study. PLoS Med. 2008;5:e12.
7. Wang MC, Bohmann D, et al. JNK extends life span and limits growth by antagonizing cellular and organism-wide responses to insulin signaling. Cell. 2005;121:115–25.
8. Essers MA, de Vries-Smits LM, et al. Functional interaction between beta-catenin and FOXO in oxidative stress signaling. Science. 2005;308:1181–4.
9. Hwangbo DS, Gershman B, et al. Drosophila dFOXO controls lifespan and regulates insulin signalling in brain and fat body. Nature. 2004;429:562–6.
10. Brunet A, Sweeney LB, et al. Stress-dependent regulation of FOXO transcription factors by the SIRT1 deacetylase. Science. 2004;303:2011–5.
11. Liu Y, Kern JT, et al. A genomic screen for activators of the antioxidant response element. Proc Natl Acad Sci USA. 2007;104:5205–10.
12. Jäger S, Handschin C, et al. AMP-activated protein kinase (AMPK) action in skeletal muscle via direct phosphorylation of PGC-1alpha. Proc Natl Acad Sci USA. 2007;104:12017–22.
13. Yamauchi T, Noi Y, et al. Targeted disruption of AdipoR1 and AdipoR2 causes abrogation of adiponectin binding and metabolic actions. Nat Med. 2007;13:332–9.
14. de Cabo R, Carmona-Gutierrez D, et al. The search for antiaging interventions: from elixirs to fasting regimens. Cell. 2014;157:1515–26.
15. Schulz TJ, Zarse K, et al. Glucose restriction extends Caenorhabditis elegans life span by inducing mitochondrial respiration and increasing oxidative stress. Cell Metab. 2007;6:280–93.
16. Tanito M, Masutani H, et al. Sulforaphane induces thioredoxin through the antioxidant-responsive element and attenuates retinal light damage in mice. Invest Ophthalmol Vis Sci. 2005;46:979–87.
17. Minhas PS, Latif-Hernandez A, et al. Restoring metabolism of myeloid cells reverses cognitive decline in ageing. Nature. 2021;590:122–8.
18. Longo VD, Anderson RM. Nutrition, longevity and disease: from molecular mechanisms to interventions. Cell. 2022;185:1455–70.
19. Vernieri C, Fucà G, et al. Fasting-mimicking diet is safe and reshapes metabolism and antitumor immunity in patients with cancer. Cancer Discov. 2022;12:90–107.
20. Cheng CW, Villani V, et al. Fasting-mimicking diet promotes Ngn3-driven β-cell regeneration to reverse diabetes. Cell. 2017;168:775–88.e12.
21. Curtis R, Geesaman BJ, et al. Ageing and metabolism: drug discovery opportunities. Nat Rev Drug Discov. 2005;4:569–80.
22. Milne JC, Lambert PD, et al. Small molecule activators of SIRT1 as therapeutics for the treatment of type 2 diabetes. Nature. 2007;450:712–6.
23. Baur JA, Pearson KJ, et al. Resveratrol improves health and survival of mice on a high-calorie diet. Nature. 2006;444:337–42.
24. Izuta Y, Imada T, et al. Ketone body 3-hydroxybutyrate mimics calorie restriction via the Nrf2 activator, fumarate, in the retina. Aging Cell. 2018;17:e12699.
25. Nakamura S, Hisamura R, et al. Fasting mitigates immediate hypersensitivity: a pivotal role of endogenous D-beta-hydroxybutyrate. Nutr Metab (Lond). 2014;11:40.
26. Brownlee M. Biochemistry and molecular cell biology of diabetic complications. Nature. 2001;414:813–20.
27. Veal EA, Day AM, et al. Hydrogen peroxide sensing and signaling. Mol Cell. 2007;26:1–14.
28. Imai S, Guarente L. NAD+ and sirtuins in aging and disease. Trends Cell Biol. 2014;24:464–71.
29. Brown KD, Maqsood S, et al. Activation of SIRT3 by the NAD+ precursor nicotinamide riboside protects from noise-induced hearing loss. Cell Metab. 2014;20:1059–68.
30. Brenner C. Boosting NAD to spare hearing. Cell Metab. 2014;20:926–7.
31. Fontana L, Partridge L, et al. Extending healthy lifespan—from yeast to humans. Science. 2010;328:321–6.
32. Shimazu T, Hirschey MD, et al. Suppression of oxidative stress by β-hydroxybutyrate, an endogenous histone deacetylase inhibitor. Science. 2013;339:211–4.
33. Scott AJ, Ellison M, et al. The economic value of targeting aging. Nat Aging. 2021;1:616–23.
34. Yang JH, Hayano M, et al. Loss of epigenetic information as a cause of mammalian aging. Cell. 2023;186:305–26.

Part II

Basic of Anti-aging Medicine

Biological Age and Aging Clock: Aging Clock

12

Hidekazu Yamada

12.1 Aging Clock [1]

The term "Aging Clock" refers to a concept used to measure or predict biological age. It is based on the idea that there are biomarkers that serve as indicators of aging. By using these, it is possible to estimate an individual's functional age, regardless of their chronological age. These biomarkers could include a variety of physiological, biochemical, and molecular biological indicators, such as the length of telomeres, patterns of DNA methylation, and levels of specific proteins or enzymes in the body.

There are several different methods and approaches to measuring the aging clock, and the specific markers used may vary depending on the approach. Aging clocks can be based on a single marker or a combination of multiple markers (Table 12.1). There are also aging clocks designed to be used as general indicators of biological age, as well as more specialized clocks that focus on specific aspects of aging, such as cardiovascular health or cognitive function. Advances in omics research have revealed that aging biology is extremely complex, with tens of thousands of molecular features changing with age. Some aging clocks utilize omics data, expanding the scope of aging biomarker research by several orders of magnitude.

Aging clocks are used in research on aging and in the development of interventions to promote healthy aging. Of course, individuals can also use them to track and improve their health and well-being over the long term (some companies offer this as a service). However, it should be noted that they are not always accurate and may not necessarily reflect an individual's actual biological age. They operate on the assumption that a certain part (organ, tissue, cell) will die or become diseased when it reaches its lifespan.

H. Yamada (✉)
Anti-Aging Center, Kindai University, Osaka, Japan
e-mail: yamada@mitameonline.com

H. Yamada, Y. Naito (eds.), *Anti-Aging Medicine*, https://doi.org/10.1007/978-981-97-8514-8_12

Table 12.1 Various biological aging measurements

		Example	
Conventional biomarkers	Changes in individual values	Blood lipid proteins, blood cell count, sugar, albumin, liver enzymes, CRP, inflammation markers	Simplified chronokines would be best if found
Physical and physiological	Fluctuations in system biology internal networks	Blood pressure, grip strength, walking speed, frailty index, cognitive function, psychological/subjective age evaluation, cardiopulmonary parameters, voice analysis	Easy to recognize, generally easy to understand stress measurement
Microorganisms	Changes in types and functions of microorganisms	Intestinal flora, skin flora, oral flora	Gage (intestinal flora clock)
Imaging and its quantification	Measurement of visual changes	Body shape (BMI and liver fat), appearance (photos), skin, MRI (thymus atrophy)	Easy intuitive understanding
Genes	Characteristics of tissue aging, gene polymorphism, GWAS, risk score	APOE-3, FOXO3, IL-6, IGF-1, SIRT6	Polygenic age score
Others	Multi-omics resilience	Epigenetic (methylation) clock, glycation clock, transcriptomics clock, proteomics clock, metabolomics clock, telomere clock, dynamic biological state indicator (DOSI) clock, psychological clock	Big data processing such as machine learning

Biological age can be derived from biometric information. Studies are being conducted from single items to multiple items. Research on multiomics clocks at the organ and cellular level is particularly advanced. Necessary for the development of treatment methods in antiaging medicine. In the future, to conduct clinical trials and receive medical approval, a ruler that can estimate involvement in the mechanism of aging as a biomarker, understand changes in a short period, and can be used as an endpoint is necessary

12.2 Biological Age [2]

Biological age, unlike chronological age, which is determined by the number of times the earth has orbited the sun, refers to the functional age of an organism. Biological age is determined by various factors such as genetics, lifestyle, and environmental exposure, and is thought to be influenced by factors such as physical activity, nutrition, mental state (brain and sleep: stress), and exposure to the environment. In longevity science, biological age indicators are important (Table 12.2). Health status and factors of the exposome [3] (the total amount of external environmental factors over a lifetime; education, income, human relationships, chemicals, bacterial flora, infections, radiation, ultraviolet rays, etc.) can make biological age younger than chronological age, or accelerate aging. There are various methods and biomarkers for measuring biological age, including cardiovascular health, muscle mass and strength, cognitive function, telomere length, and the degree of DNA methylation. When the focus is on disease incidence or mortality associated with aging, it is referred to as an "aging clock". Also, the term "biological clock" is sometimes used to refer to internal systems that regulate specific physiological processes in the body, such as sleep-wake cycles, hormone production, and women's menstrual cycles. At present, the term "body clock" is used more broadly to refer to any internal system or process controlled by internal timing mechanisms. The aging clock can be considered a type of biological clock that measures the functional age of an organism. (In the epigenetic clock [4], the circadian rhythm gene region may be included in the adopted CpG, which is thought to be related to lifespan.) Using reprogramming at the level of mice or cells, it is possible to "rejuvenate" in the aging program. For the development of treatments, classification and scaling equivalent to the TNM classification in cancer are necessary. Biological aging is very complex and is thought to be caused by the interaction of multiple dysregulated cells and biochemical processes. Over the past decade, the emergence of omics approaches has made it possible to begin addressing the full molecular com-

Table 12.2 Research methods for predicting biological age (currently, machine learning is mainstream)

Data collection	• Gather data from multiple sources • Data from traditional biomarkers, genomics, proteomics, metabolomics, etc. • Preferably, a prospective cohort study across multiple time points is ideal to minimize the impact of the exposome (following the same individuals)
Data preprocessing	• Cleaning, transforming, and normalizing data for collected data analysis
Marker identification	• Identify markers or indicators related to aging from the data • These may include DNA methylation patterns, protein expression levels, or metabolic pathways
Model development	• Develop a statistical model using the identified markers to predict an individual's biological age • The model should be based on machine learning algorithms or other statistical methods (capable of analyzing images, such as neural networks) • Confirm the model's validity by testing and validating it with an independent dataset
Using the model	• Predict an individual's biological age and identify potential interventions

Each organ, part, and cell is said to have an age (brain age, lung age, skin age, intestinal age, hormone age, etc.). By examining the changes in these indicators, we can identify factors related to the adjustment of aging speed. We seek indicators that correlate highly with biological age as a standard. The accuracy is improved by using evaluations of indicators at multiple points in time. For the evaluation of the accuracy of the indicators, it is necessary to consider measurement errors, individual differences, and gender differences

plexity of this field in aging biology. The new generation of tools for measuring biological aging has made it possible to quantitatively assess the characteristics of aging at the molecular resolution. They use data from the epigenome, transcriptomics, proteomics, and metabolomics. "Aging clocks" using machine learning and neural networks are being created. It is becoming a portfolio combining several aging clocks from a simple aging clock. Furthermore, age measurement and acceleration are being done with next-generation AI including deep neural networks. All current aging clocks are correlation statistical models, so it is natural that they are not causal. Furthermore, if the multi-layer omics clock [5] can be used for predicting aging and applied to interventions, it would be clinically useful, but many are cross-sectional studies (cross-sectional) = measurements of people of the same chronological age, longitudinal. Study (longitudinal) = Since it is not a measurement of the same person, it is impossible to exclude environmental factors (exposome) that should be considered, so research with prospective cohort data is important (meaning the importance of the Dunedin Study [2]. In summary, in anti-aging medicine, the concept of aging clocks is used to estimate biological age using indicators that can better understand aging mechanisms (for example, epigenetic clocks), and it is considered whether this can be used for intervention. In the future, under the premise of treating aging as a disease, staging is always necessary. Also, to respond to (1) symptoms with aging, (2) acceleration of aging, (3) deceleration, stop, reversal of aging due to therapeutic intervention, it would be good to use the concept of aging clocks to consider what is the endpoint of treatment. For example, if the criteria for frailty are set as a biological clock, it can be said to be effective if improvement is shown by intervention.

References

1. Tarkhov AE, Denisov KA, et al. Aging clocks, entropy, and the limits of age-reversal. bioRxiv. 2022; https://doi.org/10.1101/2022.02.06.479300.
2. Belsky DW, Caspi A, et al. DunedinPACE, a DNA methylation biomarker of the pace of aging. elife. 2022;11:e73420.
3. Vermeulen R, Schymanski EL, et al. The exposome and health: where chemistry meets biology. Science. 2020;367:392–6.
4. Seale K, Horvath S, et al. Making sense of the ageing methylome. Nat Rev Genet. 2022;23:585–605.
5. Sturm G, Monzel AS, et al. A multi-omics longitudinal aging dataset in primary human fibroblasts with mitochondrial perturbations. Sci Data. 2022;9:751.

13 Biological Age and Aging Clock: DNA Methylation and Biological Age

Hidekazu Yamada

Epigenetic changes are one of the characteristics of aging. Aging is likely to be largely due to the loss of epigenomic information. Therefore, the epigenetic clock (EC) is measuring aging by targeting DNA methylation [1]. It aims to measure biological age and can strongly predict morbidity and mortality rates associated with aging, along with various health risks. It represents biological age beyond chronological age and can be called a life-course clock. Currently, there are ECs that only determine based on the degree of DNA methylation in a predetermined CpG region (such as the Horvath clock) and ECs as aging clocks that include multifactorial factors including clinical blood and EC (GrimAge V2, DunedinPACE). In 2011, it was shown that age prediction in chronological order is possible by examining DNA methylation, and in 2013, the concept of an epigenetic aging clock was announced by Hannum and others and has continued to the present (Fig. 13.1). While other biomarkers often measure organ

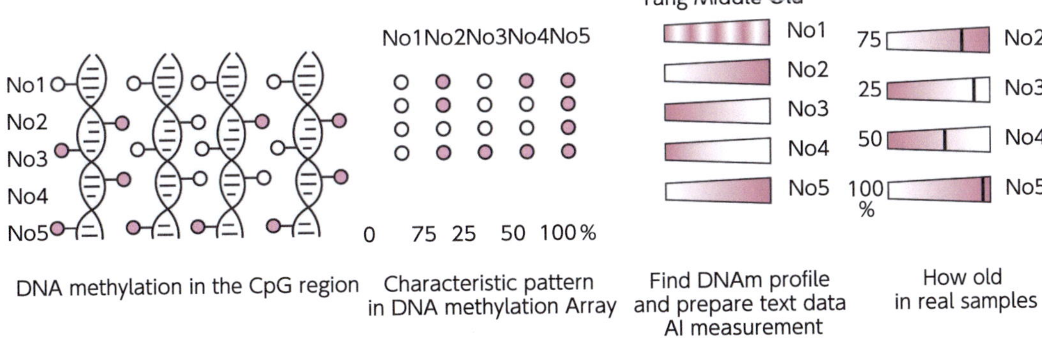

Fig. 13.1 Overview of the measurement method. EC is created by examining the cytosine methylation (CpG methylation) of CpG dinucleotides in regions rich in DNA CG. Initially, it started from the method of selecting the most changing CpG sites over time from millions of (bisulfite sequencing) sites, but now it is measured with several hundred thousand (methylation arrays). From tissue or blood cells, average the methylation of each cell, and based on this, input the degree of CpG methylation necessary for each clock algorithm to output biological age and acceleration

H. Yamada (✉)
Anti-Aging Center, Kindai University, Osaka, Japan
e-mail: yamada@mitameonline.com

H. Yamada, Y. Naito (eds.), *Anti-Aging Medicine*, https://doi.org/10.1007/978-981-97-8514-8_13

dysfunction, EC is an aging biomarker that can also be applied to fetal tissues. As of 2022, the so-called third-generation EC is being used. Because it uses machine learning, the process is black-boxed, and it has a characteristic of weak connection with traditional biological mechanisms (such as enzymatic reactions). Currently, DunedinPACE, which was developed by adding EC to blood data in prospective cohorts, is being used clinically. Furthermore, each individual's biological age and its deformation rate (acceleration) are shown. It is widely used in human aging intervention trials and cohort studies. The problem is accuracy, and discussions about sensitivity and specificity are being held. The accuracy of the error between the same samples (intraclass correlation coefficients; ICC) is also under consideration. It is said that it has not yet reached the level of disease treatment judgment.

13.1 Characteristics of DNA Methylation Clock

It can predict biological age across various tissue types. In other words, it is measuring aging signals shared between cell types. Age acceleration of EC has been found to be associated with COVID-19 and HIV infection, surgery, obesity, menopause, diet, heart disease, anxious socioeconomic status, drug treatment, lifestyle, environmental exposure, and stress (Table 13.1). A clock that can be applied to multiple species at the same time, such as the DNA methylation clock (Horvath, etc.), has been announced [16]. There are universal clocks for all mammals and clocks that can be commonly used in mice, rats, and humans, which can also be used for drug development. The current EC shows the average methylation of thousands of cells, and it varies by cell even at the same age. Each cell has its own age. In the future, measurement with a methylation clock of a single cell will progress.

Table 13.1 Epigenetic clock

First generation DNA mClock	[2–5]
Second generation DNA mClock	[6–11]
Second generation human DNA mClock	[12, 13], Telomere length clocks
Third generation mClock	DunedinPACE, [14, 15]
Next-generation: Higher accuracy and closer to mechanisms	• DunedinPACE (PCA) (Levine); 0.99–0.998 and ICCs of 0.97–0.99 longitudinal study • Genomic and ribosomal DNAme PRC2 clock (2022); a watch with Polycomb 2 • IntinClock (Varden 2023); tissue independent; naïve CD8 T cell; young • TIME seq (Sinclair)
Others	For cell culture: Minteer (2022) Enhanced EC: AltumAge (2022) Universal mammalian clock: Lui (2021) Multi-species clock: Horvath, Goya, Katcher (2022); Applying human and rat clocks to the study of young rat plasma (dual clock)

The third generation is currently mainstream, reaching clinical use level in terms of accuracy, sensitivity, and intra-class correlation coefficient ICC. DNA methylation is said to be relatively linear from 20 to 80 years old. In EC, it is said to be usable from the fetal stage. Multispecies clocks are particularly useful for testing whether mouse or rat data can be adapted to humans

13.2 Mechanism [17, 18]

The EC appears to be looking at the epigenome maintenance system. In many cancers and multiple cohorts, cancer tissues are accelerated in the Horvath clock. From research on premature aging, it is known that cytosines that are methylated are often close to the binding sites of the Polycomb Repressive Complex 2, and that DNA methyltransferases and TET enzymes play important roles. EC is younger in stem cells, and differentiated cells lose their identity as they age. Metabolic stress and HIV, COVID-19 infection stress are factors that promote aging. On the other hand, EC is not related to aging cells or telomeres (Table 13.2). Cellular aging is cell type-specific, so it's unclear if it's all cellular aging, but at least in the system of replicative aging and cancer gene-induced aging, it's said to have no relation with EC in vitro. Naked mole rats are said to be resistant to aging, but from the perspective of EC, they are similar to human models and start aging after calendar age slows down.

Table 13.2 Characteristics of aging from the cellular level perspective and the epigenetic clock

Relevant	Not relevant
(3) Epigenetic changes (5) Nutrient sensing—via mTOR; SNP in MLST8—rapamycin—calorie restriction (6) Mitochondrial function(8) Stem cell exhaustion(9) Intercellular communication	(1) Genomic instability (2) Telomere attrition (7) Cellular senescence

Based on the 2013 paper by Carlos López-Otín titled "The Hallmarks of Aging" published in Cell, which outlines 9 hallmarks of aging, the number was later expanded to 12. Among these hallmarks, Epigenetic Alterations, Genomic Instability, Stem Cell Exhaustion, and Chronic Inflammation are considered to be primarily involved in DNA methylation

13.3 Germline reset [19]

The embryo reset phenomenon is attracting attention in the treatment of aging. Embryonic stem cells (ESC) and induced pluripotent stem cells (iPSC) are completely young in EC. The passage of these cells does not affect epigenetic aging. EC can be used a few days after conception. Cellular initialization is one of the promising methods in aging treatment, but to restore the information of the epigenome, EC is essential for the judgment of rejuvenation. From the perspective of longevity science, it is necessary to create a theory of aging formation to include aging in the regulatory process of therapeutic development of the Food and Drug Administration (FDA), and consideration in the epigenome including DNA methylation will be useful.

References

1. Yamada H. Aging and epigenetics: new approaches—epigenetic clock. Prog Med. 2021;277:617–8.
2. Horvath S. DNA methylation age of human tissues and cell types. Genome Biology. 2013;14(10):R115.
3. Hannum G, Guinney J, Zhao L, Zhang L, Hughes G, Shover A, et al. Genome-wide methylation profiles reveal quantitative views of human aging rates. Molecular Cell. 2013;49(2):359–67.
4. Horvath S, Raj K. DNA methylation-based biomarkers and the epigenetic clock theory of aging. Nature Reviews Genetics. 2018;19(6):371–84.
5. Galkin F, Zhang B, Dmitriev SE, Gladyshev VN. DNAmAgeClocks: a bioconductor package for reference-based and reference-free epigenetic age estimation. Bioinformatics. 2021;37(12);1759–65.
6. Wang T, Tsui B, Kreisberg JF, Robertson NA, Gross AM, Yu MK, et al. Epigenetic aging signatures in mice livers are slowed by dwarfism, calorie restriction and rapamycin treatment. Genome Biology. 2017;18(1):57.
7. Mozhui K, Pandey AK. Conservation of epigenetic aging signals in mice and humans. GeroScience. 2021;43(5):2301–14.

8. Meer MV, Podolskiy DI, Tyshkovskiy A, Gladyshev VN. Aging as a sequence of damage, signaling, and remodeling events. Cell Metabolism. 2018;27(6):1176–87.
9. Petkovich DA, Podolskiy DI, Lobanov AV, Lee SG, Miller RA, Gladyshev VN. Using DNA methylation profiling to evaluate biological age and longevity interventions. Cell Metabolism. 2017;25(4):954–60. e6.
10. Stubbs TM, Bonder MJ, Stark AK, Krueger F, BI Ageing Clock Team, von Meyenn F, et al. Multi-tissue DNA methylation age predictor in mouse. Genome Biology. 2017;18(1):68.
11. Thompson MJ, Chwialkowska K, Rubbi L, Lusis AJ, Davis RC, Srivastava A, et al. A multi-tissue full lifespan epigenetic clock for mice. Aging. 2018;10(10):2832–54.
12. Levine ME, Lu AT, Quach A, Chen BH, Assimes TL, Bandinelli S, et al. An epigenetic biomarker of aging for lifespan and healthspan. Aging. 2018;10(4):573–91.
13. Lu AT, Quach A, Wilson JG, Reiner AP, Aviv A, Raj K, et al. DNA methylation GrimAge strongly predicts lifespan and healthspan. Aging. 2019;11(2):303–27.
14. Belsky DW, Caspi A, Arseneault L, Baccarelli A, Corcoran DL, Gao X, et al. Quantification of the pace of biological aging in humans through a blood test: the DunedinPoAm DNA methylation algorithm. eLife. 2020;9:e54870.
15. Belsky DW, Caspi A, Corcoran DL, Sugden K, Poulton R, Arseneault L, et al. DunedinPACE, a DNA methylation biomarker of the pace of aging. eLife. 2022;11:e73420.
16. Horvath S, Singh K, et al. Reversing age: dual species measurement of epigenetic age with a single clock. https://doi.org/10.1101/2020.05.07.082917.
17. Horvath S, Raj K. DNA methylation-based biomarkers and the epigenetic clock theory of ageing. Nat Rev Genet. 2018;19:371–84.
18. Horvath S, Haghani A, et al. DNA methylation clocks tick in naked mole rats but queens age more slowly than nonbreeders. Nat Aging. 2022;2:46–59.
19. Kerepesi C, Zhang B, et al. Epigenetic clocks reveal a rejuvenation event during embryogenesis followed by aging. Sci Adv. 2021;7:eabg6082.

Biological Age and Aging Clock: Inflammation and Biological Age

14

Kazuhiko Uchiyama and Tomohisa Takagi

14.1 Biological Age

Among the "aging indicators" that objectively evaluate the progression of aging, the most straightforward is chronological age. However, when based on chronological age, the degree of aging progression and the manifestation of symptoms vary from person to person. That is, human aging does not necessarily occur at the same speed, and in recent years, the concept of "biological age," which estimates the degree of aging progression from the degree of physical function decline, has been attracting attention. Biological age reflects the degree of various organ function declines with aging, separate from chronological age, and is considered the age estimated based on various physical function declines. In the future, it is clear that the proportion of elderly people will increase in many countries, including Japan, and how to slow down aging and prolong the period of health and productivity has become an important research theme.

14.2 Indicators of Biological Age

According to the American Federation for Aging Research, aging indicators must meet the following criteria [1]: (1) They must be better predictors than chronological age for predicting the speed of aging. (2) They must reflect the process of aging without being affected by disease. (3) They must have no harmful effects on the human body and can be measured repeatedly. (4) They must be applicable not only to humans but also to animals like mice. No aging index that meets all these criteria has been found so far, but many studies so far have used the so-called "epigenetic clock," or DNA methylation age (DNAmAge), as an indicator of biological age. This is because a strong correlation has been reported between DNA methylation levels and biological age [2]. Especially Horvath's aging the indicator called "clock" is the most commonly used, and it has been clarified that diseases including Huntington's disease, Down syndrome, and HIV-1 infection promote aging in biological age. Also, telomeres are parts that repeat the DNA base sequence "TTAGGG" located at the end of chromosomes, and each time normal somatic cells divide, a part of this telomere gradually shortens. When the telomeres shorten to a certain extent, the ability to replicate the chain of DNA is lost, and somatic cell division stops. Therefore, it was expected that the length of telomeres could be an aging indicator, but it does not meet the above criteria, and it has not been concluded whether the length of telomeres can be used as an aging indicator. In recent years, the concept of "inflammation aging clock (iAge)" that measures inflammatory load and predicts aging of multiple comorbidities,

K. Uchiyama (✉) · T. Takagi
Molecular Gastroenterology and Hepatology, Kyoto Prefectural University of Medicine, Kyoto, Japan
e-mail: k-uchi@koto.kpu-m.ac.jp

H. Yamada, Y. Naito (eds.), *Anti-Aging Medicine*, https://doi.org/10.1007/978-981-97-8514-8_14

frailty, immune and cardiovascular systems have been reported. It is stated that CXCL9, a chemokine, controls the expression of multiple genes involved in inflammation, and is involved in cellular aging, vascular aging, and heart remodeling, as proven in humans and mice. This report suggests that inflammation associated with aging should be considered as an indicator of biological age in the future.

14.3 Inflammation and Cellular Aging [3–5]

The process of inflammation includes mechanisms of acute inflammation induced by infection or tissue damage, and mechanisms of chronic inflammation persistently induced by sustained stress to the tissue. Chronic inflammation is believed to promote tissue remodeling such as fibrosis and functional impairment, and is deeply involved not only in autoimmune diseases such as bronchial asthma, atopic dermatitis, and chronic rheumatoid arthritis, which are caused by immune and inflammatory control abnormalities, but also in various age-related diseases such as arteriosclerosis, heart failure, diabetes, obesity, dementia, cancer, or even aging itself. With aging, the immune system shows a decrease in acquired immune response reactions, but in fact, it has been reported that elderly people have high levels of cytokines in their blood, even if they are healthy, and that inflammatory signals are constantly activated. This state of chronic inflammation observed with aging is called inflammaging (inflammation + aging). In addition, chronic inflammation is not only caused by aging, but also by overeating-induced obesity. It has been experimentally proven that calorie restriction suppresses chronic inflammation and improves biological age. Recently, attention has been focused on "cellular aging" as the essence of individual aging. Cellular aging is a state in which cells acquire resistance to proliferative stimuli and stably stop the cell cycle, a phenomenon discovered by Hayflick about 50 years ago. Aged cells express various inflammatory cytokines, extracellular matrix-degrading enzymes, and other secretory proteins with inflammatory and cancer-promoting effects, and these phenomena are called senescence-associated secretory phenotype (SASP), and the proteins that cause these phenomena are collectively called SASP factors. Excessive SASP caused by the accumulation of aged cells is believed to cause various diseases that can be caused by age-related chronic inflammation, such as cancer, cardiovascular disease, diabetes, cataracts, Alzheimer's disease, osteoporosis, and osteoarthritis. In fact, in studies using mice, it has been reported that aging cells that cause chronic inflammation accumulate in the body with age, and that chronic inflammation due to aging cells is important in the pathology of kidney aging and non-alcoholic steatohepatitis, and that removing aging cells reduces inflammation markers and improves these pathologies. It has also been reported that mice with removed aging cells live longer, while those with accumulated cells live shorter lives. However, the induction of inflammation by the SASP mechanism of aged cells is an essential physiological program to remove unnecessary cells and maintain tissue homeostasis, and its activation is not necessarily directly related to individual aging, and it is not clear to what extent chronic inflammation induced by SASP contributes to individual aging.

14.4 Summary

Chronic infectious diseases, lack of exercise, visceral obesity, diet, psychological stress, and lack of sleep are considered causes of chronic inflammation. However, it has been reported that in the elderly, the persistence of inflammation, accompanied by a local and systemic increase in inflammatory mediators such as tumor necrosis factor α (TNFα) and interleukin-1β (IL-1β), is observed. Various changes within an individual due to aging make it easier for chronic inflammation to occur, and the impact of chronic inflammation on an individual's biological age is gradually becoming clear. Future research is expected to not only make chronic inflammation a marker of biological age, but also to control biological age by controlling chronic inflammation (Fig. 14.1).

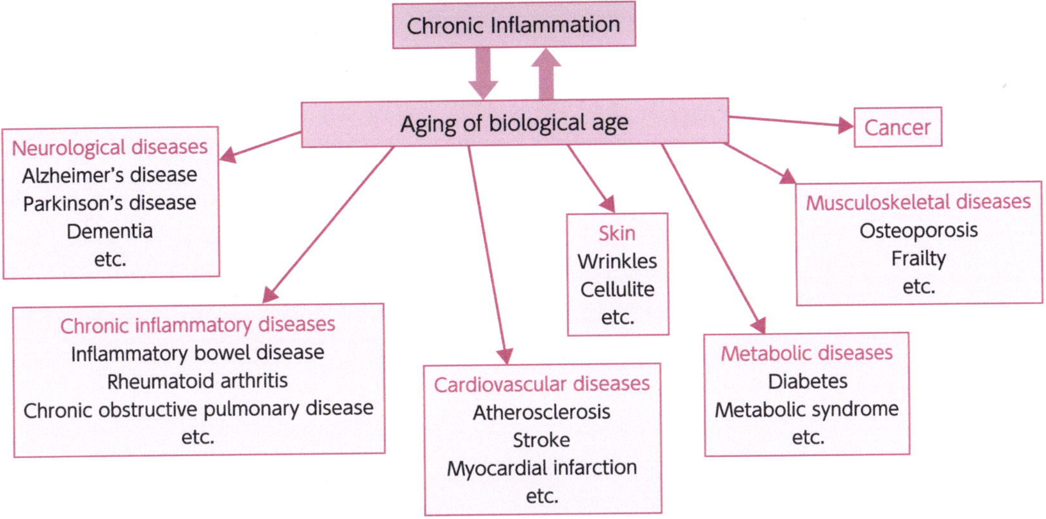

Fig. 14.1 Diseases related to chronic inflammation and biological age

References

1. Johnson TE. Recent results: biomarkers of aging. Exp Gerontol. 2006;41:1243–6.
2. Horvath S. DNA methylation age of human tissues and cell types. Genome Biol. 2013;14:R115.
3. Sayed N, Huang Y, et al. An inflammatory aging clock (iAge) based on deep learning tracks multimorbidity, immunosenescence, frailty and cardiovascular aging. Nat Aging. 2021;1:598–615.
4. Franceschi C, Bonafè M, et al. Inflamm-aging. An evolutionary perspective on immunosenescence. Ann N Y Acad Sci. 2000;908:244–54.
5. Chambers ES, Akbar AN. Can blocking inflammation enhance immunity during aging? J Allergy Clin Immunol. 2020;145:1323–31.

Genes and Antiaging Medicine: Genomes and Longevity

15

Tomohiro Katsuya

15.1 Average Lifespan and Longevity Records of Japanese People

The average lifespan of Japanese people announced by the Ministry of Health, Labour and Welfare in 2021 is 81.47 years for men and 87.57 years for women, making Japan the third longest-lived country for men and the first for women. The population of centenarians, those over 100 years old, is also on the rise, from just 153 people in 1963 to 90,256 nationwide as of September 2022. The "genetic program theory," which posits that lifespan is determined by the genome, has long been known as one of the twin pillars of gerontology, based on twin studies and the existence of long-lived families. The four daughters of Gin Kanie, known as "Kin-san and Gin-san," famous centenarian twins featured in TV commercials, are also long-lived, ranging in age from 95 to 108 (including those still alive), which is considered a good example of the existence of longevity genes. In the Guinness Book of Records, the longest lifespan is held by Jeanne-Louise Calment of France, who lived for 122 years and 164 days. The second longest is Rikiko Tanaka of Japan, who lived for 119 years and 107 days, and the longest for a man is Jiroemon Kimura of Japan, who lived for 116 years and 54 days.

15.2 Aging-Related Diseases Proposed by Professor Martin G

The existence of premature aging syndromes, where aging progresses early due to genetic factors, provides clues about the relationship between the genome and lifespan. Professor Martin G of the University of Seattle proposed "pathological and cytological criteria for aging" in "Mendelian Inheritance in Man" in 1975, with Down syndrome in first place, followed by famous premature aging syndromes such as Werner syndrome and Cockayne syndrome in second place, and Hutchinson-Gilford progeria syndrome (HGPS) in third place. It is widely known that Down syndrome is caused by trisomy of chromosome 21, but the causative gene for Werner syndrome, which is common in Japanese people, is related to DNA repair and is a helicase. It has been clarified that WRN, which has a motif, and HGPS, which is caused by a single base mutation in the gene of lamin A, one of the nuclear membrane backing proteins, resulting in a loss of 50 amino acids, are revealing the roles of aging-related genes one after another. The authors have reported a case where the lifespan was extended by 30 years due to the reduction of

T. Katsuya (✉)
Katsuya Clinic, Hyogo, Japan
e-mail: tkatsuya@iris.eonet.ne.jp

H. Yamada, Y. Naito (eds.), *Anti-Aging Medicine*, https://doi.org/10.1007/978-981-97-8514-8_15

progerin deposition and the stop of 50 amino acid loss at 35 by slightly shifting the position of the single base mutation of LMNA [1]. In the genetic analysis of muscular dystrophy known as progeria, it has been shown that the severity of the disease and lifespan vary depending on the length of the repeat sequence of three bases (CTG), and it is suggested that the extension of CTG repeats changes the chromatin structure around it and hinders post-transcriptional modifications in muscle cells. ATM, which has been identified as the causative gene for ataxia-telangiectasia, one of the progerias, is now known to be involved in the phosphorylation of p53, BRCA1, etc., and is known as a representative of tumor suppressor genes for stopping the cell cycle during DNA damage and inducing apoptosis.

15.3 From Longevity Model Animals to Sirtuins

The relationship between lifespan and genes was first studied in flies, and models that extend lifespan by about twice have been developed. Subsequently, it was shown in nematodes (Caenorhabditis elegans) that mutations in age-1 result in a 40–60% extension of lifespan, and currently nearly 900 genes related to lifespan have been reported, which can be seen on the GenAge database (https://genomics.senescence.info/genes/search.php?organism=Caenorhabditis+elegans&show=4). Age-1 is a gene that codes for PICKCS (phosphatidylinositol 3-kinase catalytic subunit), and PI3K-deficient adults show quite high resistance to stress, although their heat resistance is somewhat inferior. The improvement in survival rate is thought to be due to the removal of residual PI3KCS and its products and the modification of the kinase signal transduction cascade associated with it [2]. In 1995, an epigenetic "silencing" factor was reported as a longevity gene, and 5 years later, Sir2 was reported as a gene controlling the replicative lifespan of yeast, showing that it mediates histone deacetylation in a NAD+ (nicotinamide adenine dinucleotide)-dependent manner and that lifespan extension is recognized under calorie restriction. This discovery led to the development of research on the Sirtuin family, known as Sirt 1–7, which not only deacetylates, but also works as a deacylase that removes various acyl groups, and is now known as a representative of longevity genes that protect against age-related diseases and extend healthy lifespan.

15.4 Biological Methylation and Lifespan Clock

In the latest analysis targeting mammals, it has been shown that the density of CpG islands at the promoter sites of genes targeted for DNA methylation correlates with lifespan, and the maximum lifespan of vertebrates is R2 A lifespan clock that predicts with 0.76 accuracy has been developed [3]. According to this clock, the longest-lived mammal is the bowhead whale at 268 years, while the African elephant is 65 years, and modern humans are calculated to be 38 years, demonstrating that we are achieving longevity against the laws of nature with the development of civilization. In studies searching for sites of DNA methylation related to aging throughout the genome, it has been reported that 147 sites are CpG islands related to lifespan, equivalent to a lifespan clock [4]. On the other hand, it has also been reported that protein methylation and demethylation are involved in the lifespan control of nematodes, and differences in methyl group transfer regulate the phosphorylation of the transcription factor DAF-16, affecting lifespan [5].

15.5 Search for Longevity Genes by Genome-Wide Studies (GWAS)

The heritability of adult death age is estimated to be about 25%, and the heritability is said to increase with age. Linkage analysis of long-lived families supports a longevity gene locus on chromosome 3, but the reproducible lifespan-related genes in candidate gene approaches are APOE and FOXO3A. APOE is considered the strongest genetic risk for Alzheimer's disease, and having one ε4 allele of APOE (ε2/ε4, ε3/ε4) makes it 3.2 times more likely to develop the disease, and having two (ε4/ε4) makes it 11.6 times more likely. Therefore, the risk assessment for Alzheimer's disease is always done by APOE genotype, and naturally, it has a significant impact on lifespan. FOXO (forkhead box O) is a transcription factor related to aging speed and lifespan, and rs2892292, a single nucleotide polymorphism (SNPs) of FOXO3A, has been reported multiple times in relation to longevity. In the Japanese elderly longitudinal study (Septuagenarians, Octogenarians, Nonagenarians Investigation with Centenarians Study; SONIC study) conducted by Osaka University and others, an increase in the G allele of rs2892292 has been observed in the 90-year-old group. Genome-wide studies (genome-wide association study; GWAS) and whole-genome sequence data show that risk alleles defined in GWAS of multifactorial diseases are found in long-lived individuals, suggesting that these individuals may be tolerating them due to protective genetic factors. Such protective factors may "buffer" the effects of specific risk alleles, and rare alleles may also contribute significantly to healthy aging and longevity [6]. Recently, attempts have been made to identify factors related to lifespan from the analysis of a large number of samples registered in biobanks, and a study examining the relationship between PRS (polygenic risk score), a health indicator and treatable biomarker, and lifespan in 676,000 people across races showed that the strongest factors shortening the lifespan of people in the world today are hypertension and obesity [7]. Furthermore, a study evaluating the genetic risk of 80 diseases in 735,748 people found that the number of years subtracted from lifespan to healthy lifespan is DALYs (disability-adjusted life it is evaluated as years, and it has been shown that the rs3798220 of the lipoprotein A gene (LPA) has the strongest influence at the individual level, and there is a difference of 1.18 DALYs depending on the presence or absence of risk allele possession. In addition, there is a difference of 3.63 DALYs between the top 10% and the bottom 90% of the multigene score related to chronic pain [8]. In addition to these reports, significant associations have been reported in GWAS using centenarians, such as LMNA and WRN reported in progeria, and CDKN2A/B known for their association with coronary artery disease.

References

1. Fukuchi K, Katsuya T, et al. LMNA mutation in a 45 year old Japanese subject with Hutchinson-Gilford progeria syndrome. J Med Genet. 2004;41:e67.
2. Campisi J, Kapahi P, et al. From discoveries in aging research to therapeutics for healthy aging. Nature. 2019;571:183–92.
3. Mayne B, Berry B, et al. A genomic predictor of lifespan in vertebrates. Sci Rep. 2019;9:17866.
4. McCartney DL, Min JL, et al. Genome-wide association studies identify 137 genetic loci for DNA methylation biomarkers of aging. Genome Biol. 2021;22:194.
5. Fukumizu A. Lifespan and aging through biological methylation. Basic Aging Res. 2017;41:19–21.
6. Brooks Wilson AR. Genetics of healthy aging and longevity. Hum Genet. 2013;132:1323–38.
7. Sakaue S, Kanai M, et al. Trans-biobank analysis with 676,000 individuals elucidates the association of polygenic risk scores of complex traits with human lifespan. Nat Med. 2020;26:542–8.
8. Jukarainen S, Kiiskinen T, et al. Genetic risk factors have a substantial impact on healthy life years. Nat Med. 2022;28:1893–901.

16 Genes and Antiaging Medicine: Genome Instability and Antiaging

Tomohiro Katsuya

16.1 What Genome Instability Means

Genome instability has been continuously highlighted as one of the main causes of aging. Our body cells are constantly exposed to various external factors that cause DNA damage, such as reactive oxygen species, ultraviolet rays, radiation, and chemicals. However, through complex genome repair mechanisms, damage is removed, restored to the correct base sequence, and homeostasis is maintained. However, when this mechanism does not work well, and the replication of the genome is not correctly done during cell division, or abnormalities occur in epigenetic modifications, it is known that mutations accumulate in various organs and tissues of animals such as humans, mice, and flies, which also serves as corroborative evidence for the aging theory of "error accumulation." However, it is not clear whether the frequency of these random changes is sufficient to influence phenotypes generally associated with aging. On the other hand, it is clear that cancer is an aging-related disease caused by the accumulation of mutations and epimutations (changes in epigenetic modifications without changes in the DNA sequence itself), and it needs to be considered separately from general aging. The accumulation of genetic mutations with aging disrupts cellular homeostasis, brings about genome instability, and all body cell mutations, chromosomal aneuploidy, and copy mutations are factors that exacerbate DNA damage.

16.2 Genome Instability Revealed from Research on Premature Aging

With aging, defects appear in the DNA repair mechanism, and it has become clear from research on premature aging that this affects the expression of essential genes and transcription pathways. In Werner's syndrome, which is common in Japanese people, before the age of 30, clear signs of aging such as gray hair, type 2 diabetes, osteoporosis, and cataracts are shown, and the probability of developing multiple interstitial cancers increases. It is known that the causative gene of this syndrome is WRN, but in the repair mechanism of double-strand breaks (DSB), one of the nonhomologous end, it is thought that the mechanism by which the Ku protein, which controls nonhomologous end joining (NHEJ), and the WRN protein combine to reduce DNA repair ability and increase genome instability [1]. The WRN protein has a RecQ-type DNA helicase motif, and Bloom syndrome and Rothmund-Thomson syndrome are also caused by mutations in the same motif that functions in DNA recombination, repli-

T. Katsuya (✉)
Katsuya Clinic, Hyogo, Japan
e-mail: tkatsuya@iris.eonet.ne.jp

H. Yamada, Y. Naito (eds.), *Anti-Aging Medicine*, https://doi.org/10.1007/978-981-97-8514-8_16

cation, repair, and telomere maintenance, positioning them as a series of disease groups with similar mechanisms. Even in the same premature aging, in Cockayne syndrome, typical age-related pathologies such as neurodegeneration, arteriosclerosis, and osteoporosis occur before the age of 10, which is caused by a defect in transcription-coupled repair. Defects in genome nucleotide excision repair (NER) significantly increase the incidence of skin cancer in patients with xeroderma pigmentosum by increasing their sensitivity to sunlight, and it is known that children with the disease avoid daylight and go to school at night. Defects in DSB repair are also observed in ataxia telangiectasia (AT), and defects in DNA cross-link repair are said to cause Fanconi anemia. The cause of Hutchinson-Gilford progeria syndrome is due to abnormalities in LMNA, a backing protein of the nuclear membrane, and nuclear membrane dysfunction is thought to be associated with nuclear genome instability. Mutations in aged mitochondrial DNA and deficiencies in nuclear lamina also cause genome instability. A review [2] published in Nature in 2021 suggests that changes in phenotype associated with general aging converge on DNA damage both causally and mechanistically, i.e., DNA instability is the main cause of aging.

16.3 DNA Damage Plays a Central Role in Aging Changes

Genome instability in a broad sense means the inherent property of DNA to undergo chemical modifications, called DNA damage, and change its structure and functional characteristics. DNA damage includes subtle chemical modifications such as spontaneous deamination and hydrolysis, breaks, kinks, gaps, adducts, DNA-protein cross-links, as well as abnormal DNA structures such as R-loops, and intermediates that change from DNA such as stalled transcription, replication, recombination complexes. The integrity of DNA is maintained only by constant repair, and highly sophisticated DNA repair and damage response (DNA damage response: DDR). The intricate network system counteracts the damage to genetic information that comes with aging and various external exposures (such as ultraviolet rays and radiation). Congenital defects in these maintenance systems can cause cancer and premature aging, and it is clear that there is a close relationship between genome integrity, cancer, and aging [3]. Chromatin modifications, which are important for the preservation and expression of DNA functions, include phosphorylation, methylation, acetylation, ubiquitination, and others, and are part of the DDR. With aging, there are occurrences such as the loss of histones and the increase in "ambiguity" of nucleosomes, which are related to chromatin remodeling, imbalance of activation and inhibition of histone modifications, and transcriptional changes. DNA methylation in various tissues of humans and experimental animal models changes diversely with age, and is now referred to as the epigenetic clock, used as one of the lifespan clocks. Tens of thousands of DNA damage inductions and repairs are continuously performed every day, leaving epigenetic traces, which are thought to contribute to the epigenetic heterogeneity between cells with aging. In other words, DDR is a major cause of epigenetic changes that lead to deterioration in gene expression control, and this is thought to be one of the causes of somatic cell heterogeneity and overall functional decline with aging [4]. Furthermore, DNA damage is said to cause pathological protein production and aggregation, among other things, through a decrease in transcription speed and an increase in transcription noise, affecting the assembly and function of proteins and protein complexes. A series of research results support that DNA damage, mutations, and epimutations play a central role as the main causes of increased protein toxicity stress with aging.

16.4 Antiaging Attempts Against Genome Instability

In progeria model mice with DNA repair problems, endonuclease IRE1α (inositol-requiring protein 1α) and transcription factor XBP1 (X-box binding protein 1) (both major regulators of the unfolded protein response; UPR) are induced, unrepairable DNA damage promotes cellular aging, and the burden of UPR increases [5]. Caloric restriction is thought to have the effect of reducing protein toxicity stress caused by DNA damage by reducing transcriptional stress and alleviating UPR. The mechanism by which caloric restriction works to inhibit aging is thought to involve not only reducing the levels of reactive oxygen species and other reactive compounds, but also activating SIRT1 and AMPK4, promoting DNA damage repair, signaling as epigenetic regulatory factors, increasing NER ability, and reducing DNA damage levels. On the other hand, inhibition of mTOR (mammalian target of rapamycin) in vivo by rapamycin extends lifespan and DNA repair protein MGMT (O6-methylguanine DNA raises the level of methyltransferase. In antiaging research, AKT, which is upstream of daf-16 and FOXO (forkhead box O) and holds the key to insulin signaling, negatively regulates DNA repair and suppresses major DDR factors such as CHK1 and p53. This suggests that metformin may play a role in antiaging and cancer suppression. Reducing DNA damage and enhancing DNA repair to restore genome integrity is also important, and it has already been proven that risks can be reduced by reducing DNA damage from external factors such as UV protection and smoking cessation. Dietary intervention may be able to somewhat suppress the causes of endogenous DNA damage, but the majority of naturally occurring DNA damage is still unavoidable. The DNA repair process is very complex and cannot be explained by defects in antioxidant defenses, immune responses, and other physiological defense systems. There is a lot of evidence that only DNA repair dysfunction is associated with systemic premature aging, so to prevent genome instability, i.e., DNA damage and DDR, we should aim for antiaging interventions that can explain the entire range of aging phenotypes, from atrophy to inflammation to cancer.

References

1. Maezawa Y, Takemoto M, et al. Advances in the diagnosis and research of progeria. J Jpn Soc Inter Med. 2019;108:124–30.
2. Schumacher B, Pothol J, et al. The central role of DNA damage in the aging process. Nature. 2021;592:695–703.
3. Niedernhofer LJ, Gurkar AU, et al. Nuclear genomic instability and aging. Annu Rev Biochem. 2018;87:295–322.
4. Bahar R, Hartmann CH, et al. Increased cell-to-cell variation in gene expression in ageing mouse heart. Nature. 2006;441:1011–4.
5. Vermeij WP, Dollé MET, et al. Restricted diet delays accelerated aging and genomic stress in DNA-repair-deficient mice. Nature. 2016;537:427–31.

17 Genes and Antiaging Medicine: Telomere Shortening and Antiaging

Tsuyoshi Hachiya

17.1 What Are Telomeres?

The human genome is composed of autosomes numbered 1–22 and two types of sex chromosomes (X and Y). Each chromosome is a folded double-stranded DNA molecule, a sequence of four types of bases (adenine [A], thymine [T], guanine [G], and cytosine [C]). At the ends of the chromosomes, there is a characteristic repetitive sequence called telomere DNA (in mammals, a repetition of TTAGGG). Various proteins bind to telomere DNA, and it is believed to form a loop structure. The complex of telomere DNA and proteins is called a telomere. Telomeres prevent chromosomes from accidentally fusing and contribute to the stabilization of chromosome structure. During cell division, when DNA is replicated, the normal DNA replication mechanism cannot completely replicate the very end of the chromosome. Therefore, in most cells, telomere DNA becomes shorter with each replication. Telomerase, expressed in germ cells and stem cells, can replicate the ends of chromosomes, so these cells can maintain the length of telomere DNA.

T. Hachiya (✉)
Department of Advanced Informatics for Genetic Diseases, Graduate School of Medicine, Juntendo University, Tokyo, Japan
e-mail: hachiyam@medicalgenome.info; thachiya@iwate-med.ac.jp

17.2 Telomere Shortening Is One of the Typical Features of Aging

Carlos López-Otín and his team at the University of Oviedo's Institute of Oncology in Spain have defined the conditions that the typical features of aging (hallmarks of aging) should meet: (1) they appear during the normal aging process, (2) when experimentally worsened, aging accelerates, (3) when experimentally improved, the normal aging process slows down, and lifespan increases. They listed nine features that meet these conditions, and telomere shortening was included among them [1]. The following points were raised as evidence that telomere shortening is a typical feature of aging: In experiments using genetically modified animal models, mice with shortened telomeres had shorter lifespans, while mice with lengthened telomeres lived longer. In models of telomere DNA-binding protein function loss, tissue regenerative ability rapidly declined, and aging accelerated. Reactivating telomerase in telomerase-deficient mice led to recovery from the aging typical of telomerase-deficient mice. In wild-type adult mice, introducing telomerase throughout the body via a virus delayed normal physiological aging without increasing cancer incidence. In humans, recent meta-analyses have shown a strong relationship between short telomeres and mortality risk, particularly at younger ages. Human telomerase deficiency or mutations

H. Yamada, Y. Naito (eds.), *Anti-Aging Medicine*, https://doi.org/10.1007/978-981-97-8514-8_17

in telomere DNA-binding proteins have been reported to be associated with early onset of diseases accompanied by decreased tissue regenerative ability, such as pulmonary fibrosis, congenital keratosis, and aplastic anemia.

17.3 The Telomeres of Centenarians and their Children Are Maintained Long

Yasumichi Arai and his team at Keio University examined the leukocyte telomere length of 1554 research subjects ranging in age from 50 to 115 years old [2]. The breakdown of the research subjects was as follows: 684 were centenarians (100–104 years old) and supercentenarians (105 years old and above), 167 pairs were children of centenarians and their spouses, and 536 were extremely elderly residents (85–99 years old) in the community. In elderly men and women unrelated to centenarians (under 100 years old), there was a tendency for leukocyte telomere length to shorten with increasing age. On the other hand, in the children of centenarians or supercentenarians, leukocyte telomere length was maintained long regardless of age. Also, the leukocyte telomere length of supercentenarians tended to be longer than that of centenarians. These results suggest that the telomere length of centenarians and their children is maintained long.

17.4 Keeping Telomeres Long May Be the Secret to Healthy Longevity

In the book "The Telomere Effect: A Revolutionary Approach to Living Younger, Healthier, Longer" [3, 4] by Elizabeth Blackburn, who won the Nobel Prize in Medicine and Physiology for her research on telomeres, and Elissa Epel, an authority on health psychology, factors that affect telomere length, such as stress, meditation, depression, exercise, sleep, obesity and insulin resistance, diet, and social environment, are mentioned. Here are some excerpts and summaries.

17.4.1 Stress

Mothers who care for children with chronic illnesses at home are said to be under great stress among caregivers. Elizabeth Blackburn and Elissa Epel conducted research on such mothers. In their study, the longer the care, and the stronger the mother's perceived stress, the shorter the mother's telomeres were. The body has a stress response system, and when it is on alert, stress hormones such as cortisol and adrenaline are secreted in large amounts. In the subjects they surveyed, several physiological stress responses were observed, such as high stress hormone secretion during sleep, and these were correlated with short telomeres and decreased expression of telomerase.

17.4.2 Meditation

Experienced meditators gathered in the mountains of the Colorado Rockies for 3 months of intensive meditation training. A Buddhist teacher led the training, which aimed to firmly focus the mind and create a clear and calm mental state. Since starting meditation in the mountains, the participants' mental state improved. Anxiety eased, resilience strengthened, and empathy towards others increased. They became able to concentrate their consciousness for a long time and control their habitual reactions well. Those who meditated also began to think deeply about their life goals. The group that meditated had about 30% more telomerase expression than the control group that did not. It was also confirmed that the higher the sense of life purpose, the higher the amount of telomerase expression. In another study, people who practiced mindfulness stress reduction, developed for people with little meditation experience, for 3 months had a 17% increase in telomerase expression compared to the control group. In a study targeting caregivers of dementia patients, practicing Kirtan Kriya meditation (a traditional meditation method born from the principles of yoga) for 12 min a day for 2 months resulted in a 43% increase in telomerase expression from the initial level.

17.4.3 Exercise

It has been reported that the twin who moves their body a lot has longer telomeres than the twin who doesn't move much. Also, multiple studies have reported that people who sit all the time have shorter telomeres than those who do a little exercise. There are also reports that increasing the amount of aerobic exercise, regardless of the type of exercise, increases the activity of telomerase. On the other hand, a survey of people undergoing extreme training found that their muscle cell telomeres were shortened. To maintain long telomeres, it seems important to move your body moderately, such as stable aerobic exercise, high-intensity interval training, and a little relaxed interval training, within the range that does not become overtraining. Things to be careful about to keep telomeres long (for example, reducing stress and exercising moderately) often overlap with things important for healthy longevity. It can be said that keeping telomeres long is the secret to healthy longevity.

References

1. López-Otín C, Blasco MA, et al. The hallmarks of aging. Cell. 2013;153:1194–217.
2. Arai Y, Martin-Ruiz CM, et al. Inflammation, but not telomere length, predicts successful ageing at extreme old age: a longitudinal study of semi-supercentenarians. EBioMedicine. 2015;2:1549–58.
3. Blackburn E, Epel E. The telomere effect: a revolutionary approach to living younger, healthier, longer. London: Orion Spring; 2017.
4. Blackburn E, Epel E. (translated by Moriuchi K) Rejuvenate from cells! Telomere effect the strongest program for healthy longevity. Tokyo: NHK Publishing; 2017.

Genes and Antiaging Medicine: Epigenetics and Antiaging

18

Yoshimasa Saito

18.1 Control of Gene Expression by Epigenetic Changes

Epigenetics refers to the chemical modifications to DNA and histones that control gene expression without changing the DNA base sequence, and these changes are usually inherited after cell division. Typical examples of epigenetic changes include DNA methylation and acetylation and methylation of histones. Also, while the general term for the information of the DNA base sequence is called the "genome," the general term for the epigenetic information that modifies this genome is called the "epigenome." DNA methylation is the only physiological modification in the genome of vertebrates. DNA methylation is a reaction in which a methyl group is added to the 5-position carbon atom of cytosine (C) in a 2-base sequence (CpG) arranged from the 5′ side of DNA in the order of cytosine (C), guanine (G), and is catalyzed by DNA methyltransferase (DNMT). Many gene promoter regions have clusters of CpG sequences called CpG islands (Fig. 18.1). In the promoter regions of genes that generally show active expression, DNA methylation is not observed, and histones are acetylated. On the other hand, when the ninth lysine residue (H3K9) of histone H3 is methylated, it recruits DNMT and histone deacetylase (HDAC), DNA is methylated, and histones are deacetylated. These modifications cause the chromatin structure to aggregate and gene expression to be inactivated (Fig. 18.1). Thus, changes in the epigenome play a role like a switch that controls the ON and OFF of gene expression.

Y. Saito (✉)
Division of Pharmacotherapeutics, Keio University
Faculty of Pharmacy, Tokyo, Japan

H. Yamada, Y. Naito (eds.), *Anti-Aging Medicine*, https://doi.org/10.1007/978-981-97-8514-8_18

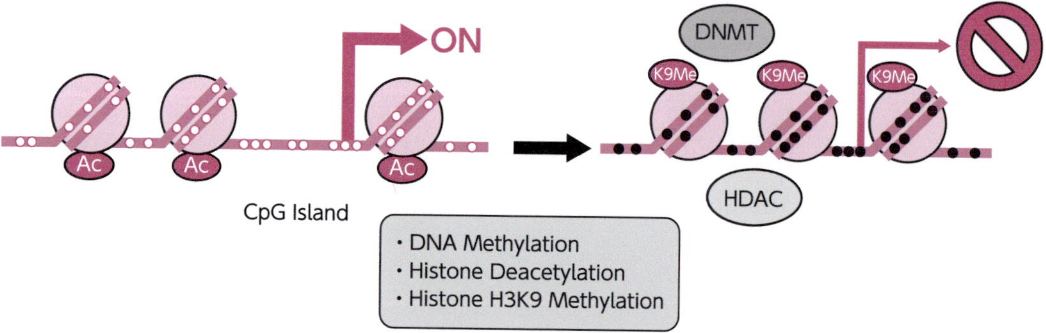

Fig. 18.1 Control of gene expression by epigenetic changes
When histone H3K9 is methylated, DNMT and HDAC are recruited, DNA is methylated, and histones are deacetylated. These modifications cause the chromatin structure to condense and gene expression to be inactivated. ●: Methylated DNA, ○: Non-methylated DNA, Ac: Histone acetylation, Me: Histone methylation

18.2 Epigenetic Changes with Aging

In recent years, it has become clear that epigenetics plays a significant role not only in diseases such as cancer, but also in aging and senescence. For example, identical twins have the same genome sequence, but due to different living environments and lifestyle habits, diseases such as diabetes and cancer may occur in only one of them, and their lifespans often differ. These changes are thought to be caused by epigenomic changes that occur with aging. In fact, a study comparing DNA methylation and histone acetylation in identical twins aged 3 and 50 reported that there is little difference in the epigenomic state of 3-year-old identical twins, but a clear difference is observed in 50-year-old identical twins [1]. It is thought that epigenomic changes accumulate as we age under different living environments and lifestyle habits, affecting the onset of diseases and lifespan. Also, the stem cell aging hypothesis has been proposed, which states that the function of stem cells, the source of all cells, declines with age, making it impossible to construct and maintain tissues. From previous studies, it has been pointed out that temporal epigenomic changes in tissue stem cells play an important role in the functional decline of individuals due to aging [2]. As shown in Fig. 18.2, the epigenomic state, such as DNA methylation and histone modification, is almost uniformly maintained in young stem cells. However, as we age and repeatedly self-replicate, and with the addition of exposure to environmental factors, chronic inflammation, and infections with viruses and bacteria, epigenomic changes accumulate. As aging progresses, epigenomic changes further accelerate, and it is thought that this ultimately leads to tissue dysfunction due to stem cell depletion, i.e., functional decline of individuals and proliferative abnormalities such as cancer [2]. In addition to the "chronological age" derived from the number of days since birth, the concept of "biological age," which is derived from the degree of aging of tissues and cells, has been proposed. It has been reported that there is a strong correlation between epigenetic modifications, especially DNA methylation levels, and biological age, and the biological age derived from DNA methylation levels (epigenetic age) is considered one of the reliable indicators of aging. In particular, the epigenetic age developed by Dr. Steve Horvath is a new indicator of biological age, also known as Horvath's aging clock, and is supported by many researchers [3]. Recent findings suggest a significant correlation between diseases caused by age changes and epigenetic age.

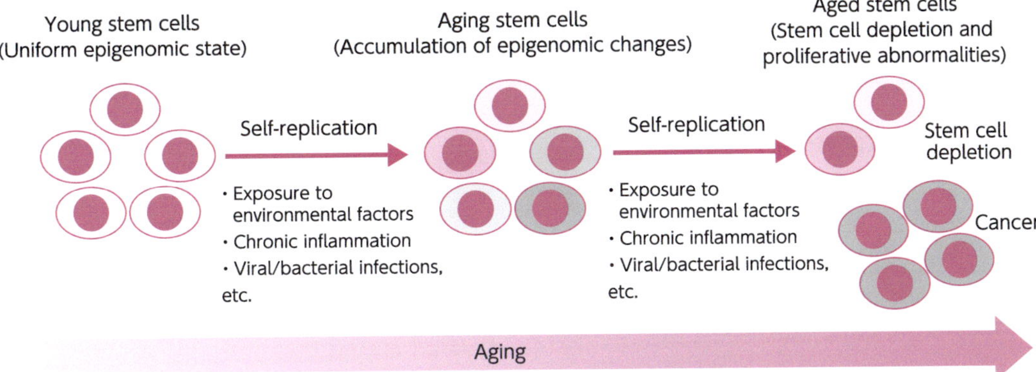

Fig. 18.2 Epigenome changes associated with aging in stem cells
In young stem cells, the epigenomic state, such as DNA methylation and histone modification, is almost uniformly maintained. As they age and repeatedly self-replicate, and with the addition of environmental factor exposure, chronic inflammation, and viral/bacterial infections, epigenomic changes accumulate. As aging progresses, these epigenomic changes intensify, and it is believed that they ultimately lead to proliferative abnormalities such as tissue dysfunction and cancer due to stem cell depletion

18.3 Possibility of Antiaging Intervention by Epigenetic Control

Epigenetic changes, being reversible by drugs such as DNA methylation inhibitors, histone deacetylase inhibitors, and histone methylation inhibitors, are attracting significant attention as potential new treatments for diseases such as cancer. In fact, the DNA methylation inhibitor azacitidine (Vidaza®) has been approved as a treatment for myelodysplastic syndromes and acute myeloid leukemia, and the histone deacetylase inhibitor vorinostat (Zolinza®) has been approved as a treatment for cutaneous T-cell lymphoma, both showing significant effects [4]. Aging, a phenomenon of life that accompanies aging, is expected to progress through complex interactions of multiple genetic changes. If these genes are regulated by epigenetic changes, it is conceivable that the expression of important aging-related genes could be controlled by artificially controlling epigenetic changes with new drugs. However, to apply these epigenome drugs to antiaging clinics, it will be necessary to identify key aging genes and develop gene-specific epigenetic controls. Recently, a groundbreaking epigenome editing technology has been developed that modifies the CRISPR-Cas9 technology to recruit methylation enzymes and demethylation enzymes region-specifically to control epigenetic changes [5]. It is thought that by correcting aging-specific epigenetic changes with epigenome editing technology, it will be possible to elucidate the molecular mechanisms of aging and develop new antiaging interventions. If successful in elucidating the molecular mechanisms of aging and developing new antiaging interventions, it is expected to lead to an extension of the healthy lifespan of the elderly and an improvement in their quality of life (QOL), as well as prevention of aging-related diseases such as cancer and an increase in the success rate of organ transplantation in the elderly. Ultimately, it is expected to contribute significantly to improving the medical economy, which is a major issue in our country, by reducing medical costs and care burdens for the elderly.

References

1. Fraga MF, Ballestar E, et al. Epigenetic differences arise during the lifetime of monozygotic twins. Proc Natl Acad Sci USA. 2005;102:10604–9.
2. Issa JP. Aging and epigenetic drift: a vicious cycle. J Clin Invest. 2014;124:24–9.
3. Horvath S. DNA methylation age of human tissues and cell types. Genome Biol. 2013;14:R115.

4. Campbell RM, Tummino PJ. Cancer epigenetics drug discovery and development: the challenge of hitting the mark. J Clin Invest. 2014;124: 64–9.
5. Morita S, Noguchi H, et al. Targeted DNA demethylation in vivo using dCas9-peptide repeat and scFv-TET1 catalytic domain fusions. Nat Biotechnol. 2016;34:1060–5.

Genes and Antiaging Medicine: Longevity Genes and Antiaging

19

Takashi Sasaki and Yasumichi Arai

A series of studies on C. elegans from the late 1980s revealed that mutations in the insulin/insulin-like growth factor signaling (IIS) genes can extend lifespan by more than double, demonstrating that aging is modifiable. Furthermore, it has been shown that IIS also plays a role in the lifespan control of fruit flies and mice, suggesting the existence of a pathway to longevity that transcends species, and the search for genes that contribute to human lifespan control is being vigorously pursued. The extent to which genetic factors contribute to human lifespan or longevity is reported to be 20–30% based on twin studies, with the remaining 70–80% attributed to the environment. On the other hand, Perls and his team, who lead the New England Centenarian Study, argue that the influence of genetics on longevity is stronger in the general population, as siblings of centenarians have a significantly higher probability of surviving past 80 years of age compared to siblings of those who died near the average lifespan, and there is a concentration of centenarian families. This section will provide an overview of the analysis of genetic factors related to longevity (hereinafter referred to as "longevity genes"), mainly targeting centenarians.

T. Sasaki · Y. Arai (✉)
Center for Supercentenarian Medical Research, Keio University School of Medicine, Tokyo, Japan
e-mail: yasumich@keio.jp

19.1 Apolipoprotein E (APOE) Gene

The Apolipoprotein E (APOE) gene is classified into ε2, ε3, and ε4 types due to a combination of two types of missense mutations, among which APOE ε4 type (APOE4) is the greatest genetic risk for Alzheimer's disease patients. In 1994, Schacter et al. first reported that in 338 French centenarians, the frequency of the APOE4 allele was lower and the frequency of the APOE2 allele was higher compared to a younger control group [1]. This sparked a surge in the search for longevity genes using a candidate gene approach, but the reproducibility was poor, and so far, only APOE and FOXO3 have been confirmed in follow-up studies across races. The authors conducted an allele frequency analysis and survival analysis using a cohort of elderly people aged 85–99 and a cohort of centenarians aged 100 or older to determine whether these reports are also true for elderly and centenarian Japanese people. As a result of the allele frequency analysis, the frequency of APOE4 and APOE2 carriers varied by age and gender, and the decrease in APOE4 allele frequency began from the elderly cohort, and a significant decrease in APOE4 allele frequency and an increase in APOE2 allele frequency were observed in centenarians (Fig. 19.1) [2]. From these results, it was clarified that APOE4 is a risk factor for longevity and APOE2 is a longevity factor in the Japanese population.

H. Yamada, Y. Naito (eds.), *Anti-Aging Medicine*, https://doi.org/10.1007/978-981-97-8514-8_19

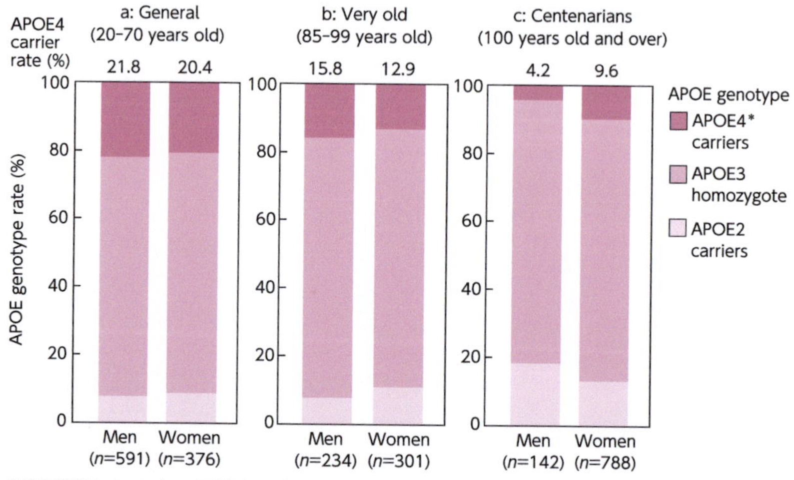

Fig. 19.1 APOE4 allele possession rate in elderly and centenarian Japanese [3]. (**a**) General (20–70 years old). (**b**) Super elderly (85–99 years old) (**c**) Centenarians (100 years old and over)

19.2 Longevity Genetic Factors Identified by Genome-Wide Association Analysis (GWAS)

In recent years, a method for searching for longevity genetic factors targeting the entire genome, namely, genome-wide association analysis (genome-wide association); GWAS, which compares the allele frequencies of genetic polymorphisms in a group of long-lived individuals and a nonlong-lived control group has been used. In 2009, the first GWAS targeting more than 1000 centenarians worldwide was conducted by an international consortium of centenarians by research teams in the United States, Italy, and the authors, and it was found that the ADAR gene, an RNA editing enzyme, is associated with longevity across races [3]. Subsequently, Zeng et al. identified rs2069837 in the IL6 region ($p = 1.08 \times 10–9$) and rs2440012 in the ANKRD20A9P region ($p = 3.73 \times 10–8$) through a GWAS analysis of the CLHLS cohort (case: 2178 centenarians, control: 2299 people aged 40–59) in the Chinese population [4]. Although the search for longevity genes by GWAS is being advanced in other countries, no reproducible results have been obtained due to differences in racial differences and the definition of the control group (such as centenarians and those over 90 years old). In addition, it has been pointed out that sampling of thousands of centenarians is practically difficult and there is a lack of detection power as a GWAS. Therefore, Deelen et al. conducted a meta-GWAS analysis using 20 cohorts including European, West Asian, and African American races, defining the longevity population as the 90th percentile (11,262 people) and 99th percentile (3484 people) of the population (young control group: 25,483 people below the 60th percentile) [5]. As a result, similar to previous studies, APOE4 (rs429358) was associated with a decrease in the odds of longevity ($p = 1.3 \times 10–56$), and APOE2 (rs7412) showed a genome-wide significant association with an increase in the odds of longevity ($2.4 \times 10–11$). They also found that a gene locus near GPR78 is significantly associated genome-wide ($4.3 \times 10–8$). GPR78 is a member of the G protein-coupled receptor family and primarily mediates physiological responses to various extracellular signals, including hormones and neurotransmitters, but little is known about the specific functions and clinical significance of GPR78. Furthermore, as a result of genetic correlation analysis by linkage disequilibrium (LD) score regression, it was shown that the longevity phenotype is genetically correlated with phenotypes related to the age of death of the father, coronary artery disease, and type 2 diabetes, suggesting that longevity may have a partially common genetic background with healthy aging (aging without disease).

19.3 Polygenic Risk Score and Longevity

GWAS has discovered thousands of genetic variations associated with various diseases, but many diseases, including type 2 diabetes, are overwhelmingly polygenic, and it is thought that multiple single nucleotide polymorphisms (SNPs) with small genetic effects accumulate to increase disease risk. The disease risk of individual SNPs is very small, so it is difficult to predict disease risk by a single SNP, but if information on thousands or millions of SNPs is integrated and summed, it can be beneficial for predicting disease risk. This is the idea behind the polygenic risk score. The Polygenic Risk Score (PRS) has been proposed. PRS is an index that scores combinations of tens to millions of genetic variations, evaluating an individual's genetic predisposition to specific diseases or complex genetic traits. So far, in the genome analysis of centenarians, it has been reported that the frequency of individual SNPs related to lifestyle diseases is not related to longevity. However, recently, the relationship between disease PRS such as coronary artery disease and longevity has begun to be verified. Gunn et al. calculated PRS for 54 diseases and phenotypes from the GWAS results of the New England Centenarian Study and Long Life Family Study, reporting that in centenarians, PRS for coronary artery disease and Alzheimer's disease is low, and PRS for parental longevity is high [6]. Since PRS needs to consider racial differences, we are also verifying various disease PRS in Japanese centenarians and examining the relationship between longevity and genetic factors of diseases.

19.4 Conclusion

Despite more than 30 years of searching for longevity genes, no genetic locus with a stronger effect on longevity than APOE has been identified, and it is believed that human longevity is unlikely to be controlled by a single gene like model animals. In order to detect susceptibility gene loci for multifactorial traits like longevity, further accumulation of longevity samples, including centenarians, is desired to identify gene polymorphisms with large effects on lifespan through rare variant analysis. Also, an approach to analyze cohorts of supercentenarians and supercentenarians, which are considered to have higher detection power, can be considered. Currently, the Keio University School of Medicine Centenarian Comprehensive Research Center is conducting a GWAS analysis using 960 Japanese centenarians, including supercentenarians over 110 years old, and its results are expected.

References

1. Schächter F, Faure-Delanef L, et al. Genetic associations with human longevity at the APOE and ACE loci. Nat Genet. 1994;6:29–32.
2. Sasaki T, Nishimoto Y, et al. Sex-specific effects of apolipoprotein ε4 allele on mortality in very old and centenarian Japanese men. J Gerontol A Biol Sci Med Sci. 2020;75:1874–9.
3. Sebastiani P, Montano M, et al. RNA editing genes associated with extreme old age in humans and with lifespan in C. elegans. PLoS One. 2009;4:e8210.
4. Zeng Y, Nie C, et al. Novel loci and pathways significantly associated with longevity. Sci Rep. 2016;6:21243.
5. Deelen J, Evans DS, et al. A meta-analysis of genome-wide association studies identifies multiple longevity genes. Nat Commun. 2019;10:3669.
6. Gunn S, Wainberg M, et al. Distribution of 54 polygenic risk scores for common diseases in long lived individuals and their offspring. Geroscience. 2022;44:719–29.

20 Genes and Antiaging Medicine: Aging Genes and Antiaging

Takahiko Shimizu

Aging genes refer to responsible genes that promote aging due to genetic abnormalities. Conversely, responsible genes that delay aging (life extension) due to genetic abnormalities are called longevity genes, but in a broad sense, they are all aging (lifespan) control genes. This section discusses the former, "aging genes" that promote aging due to genetic abnormalities.

20.1 Aging Genes Revealed from Progeria Research [1]

Genes that strongly influence the aging process of humans have been revealed from the search for the causative genes of hereditary progeria. These genes, unlike the "longevity genes" discovered from longevity mutants, are mostly genes that contribute to DNA repair or genome stability (Table 20.1). The most famous hereditary progeria is Werner syndrome. Werner syndrome is a genetic disease that develops after puberty and develops elderly symptoms such as cataracts, skin ulcers, arteriosclerosis, osteoporosis, diabetes, and cancer in the early 30 s. In 1996, it was discovered that there was a mutation in the WRN helicase gene from the genetic analysis of the disease family. This gene is thought to play an important role in chromosomal recombination repair, and therefore cells collected from patients with Werner syndrome exhibit abnormal chromosome formation and cellular aging. The RECQ helicase genes BLM (RECQL3) and RTS (RECQL4), which have similar structures to the WRN (RECQL2) gene, are also the causative genes for Bloom syndrome and Rothmund-Thomson syndrome, respectively, and the relationship between helicase activity and progeria onset is strongly suggested and is a research subject [2] (Table 20.1). Also, in Hutchinson-Gilford syndrome, where aging progresses at four times the speed from birth, patients develop adult diseases such as fractures, joint dislocations, and myocardial infarctions from childhood, and die from diseases such as strokes by around 20 years old. It is a very serious progeria. The responsible gene lamin A is a gene that codes for a backing protein of the nuclear membrane and plays an important role in maintaining the structure and function of the nuclear membrane. In lymphocytes collected from patients, the size of the nucleus increases and the chromatin structure protrudes from the nucleus, so the role of the nuclear membrane that stores the genome is being rerecognized. From the analysis of disease iPS cells derived from patients, when reprogramming, abnormal lamin A (progerin) does not express and aging phenotype does not appear, but it has been clarified that progerin expresses during differentiation and

T. Shimizu (✉)
Geroscience Research Center, National Center for Geriatrics and Gerontology, Aichi, Japan
e-mail: shimizut@ncgg.go.jp

H. Yamada, Y. Naito (eds.), *Anti-Aging Medicine*, https://doi.org/10.1007/978-981-97-8514-8_20

Table 20.1 Progeria and causative genes

Progeria	Causative gene	Gene function
Werner syndrome	WRN (RECQL2) helicase gene	Recombination repair
Hutchinson-Gilford syndrome	Lamin A gene	Nuclear envelope lining protein
Cockayne syndrome	CSA gene, CSB gene	Transcription-coupled DNA repair
Bloom syndrome	BLM (RECQL3) helicase gene	DNA repair enzyme
Rothmund-Thomson syndrome	RTS (RECQL4) helicase gene	DNA repair enzyme
Xeroderma pigmentosum	XP gene	DNA repair enzyme
Ataxia telangiectasia	ATM gene	Cell cycle checkpoint control
Down syndrome	Trisomy 21	Cause gene unknown

nuclear membrane abnormalities appear [3]. The aging control mechanism in the nuclear membrane is still unclear, and it is a research topic where progress is desired.

20.2 Aging Model Mice and Aging Genes [4]

So far, due to the advancement of gene modification technology, various types of aging model mice showing aging-like phenotypes have been reported, and "aging genes" have been identified in mice (Table 20.2). Telomeres are repeat structures that protect the DNA structure at the ends of chromosomes. It is also thought to be a clock that marks the aging of cells because the repeat structure becomes shorter with cell division. Telomerase is an enzyme that extends this repeat structure using RNA as a template, and high enzymatic activity is observed in cancer cells and germ cells. Mice completely lacking this telomerase (Tert or Terc) become infertile after six generations. The reproductive organs of these infertile mice are highly atrophied, and it has been found that the telomere length is also extremely shortened. Since the telomeres of mice are longer than those of humans, it is believed that six generations of breeding are necessary for telomere shortening. In cells that divide repeatedly, it has been confirmed that telomeres gradually shorten. Ultimately, it has been confirmed in model mice that telomeres are lost and cells cannot divide. These changes in telomeres are observed in rapidly dividing germ cells, hematopoietic cells, and skin cells during the wound healing process [5] (Table 20.2). Also, the SAMP series of mice, which are aging-promoting, are naturally occurring aging-promoting mice, such as the P1 series that develop amyloidosis, the P9 series that develop cataracts, and the P8 series that develop learning and memory disorders, and are widely used in basic aging research worldwide. If the responsible genes are identified from each of these series, the discovery of new aging genes is expected (Table 20.2). Klotho mice are premature mice with a significant growth disorder and a variety of aging symptoms in a few months, with an average lifespan of 60 days. Osteoporosis, arteriosclerosis, emphysema, skin aging, and gonadal atrophy are observed. The causative gene, the klotho gene, is highly expressed in the kidneys and is involved in calcium metabolism regulation. Due to gene mutations, calcium homeostasis cannot be maintained, and mechanisms such as causing bone calcium metabolism abnormalities and developing osteoporosis, and the activation of the protein-degrading enzyme m-calpain causing tissue damage have also been shown (Table 20.2). The imbalance of oxidation and reduction due to the deficiency of the antioxidant enzyme gene SOD is also a factor in accelerating aging. SOD, which decomposes superoxide into hydrogen peroxide and oxygen, constitutively expresses the cytoplasmic SOD1 and mitochondrial SOD2 genes in cells, and the deficiency of either gene presents tissue damage and frailty similar to aging. Genes involved in maintaining the body's oxidation-reduction balance are also considered aging genes (Table 20.2). The deficiency of genes known as longevity genes such as Sirt6, and clock genes such as Bmal1, has been shown to accelerate aging. As aging is controlled by various factors, it is expected that the relationship between

Table 20.2 Aging model mice and cause genes

Aging model mice	Aging traits	Function of cause gene
Telomerase-deficient mice	Infertility, hematopoietic disorders	Telomere extension
SAMP series mice	Accelerated aging, systemic changes	Polygenic system
Klotho mice	Osteoporosis, diverse systemic aging	Calcium metabolism-related gene
SOD1-deficient mice	Diverse systemic aging	Decomposition of reactive oxygen species
SOD2-deficient mice	Dilated cardiomyopathy, neurodegeneration, motor abnormalities	Decomposition of reactive oxygen species
Ercc1-deficient mice	Diverse systemic aging	DNA repair-related enzyme
Ku80-deficient mice	Diverse systemic aging	DNA repair-related enzyme
p53 mutant mice	Atrophy of systemic organs	Detection of gene mutations
Rad50 mutant mice	Cancer, hematopoietic disorders, diverse systemic aging	DNA repair-related enzyme
TTD mice (Xpd mutant mice)	Cancer, diverse systemic aging	DNA helicase (DNA repair enzyme)
Lamin A mutant mice	Diverse systemic aging	Nuclear envelope lining protein
Zmpste24-deficient mice	Systemic diverse aging	Prelamin A metalloprotease
Top3β-deficient mice	Nephritis, lymphoma hypertrophy	DNA topoisomerase III
SIRT6-deficient mice	Lymphocyte reduction, osteopenia	DNA repair
Bub1b mutant mice (Hypomorph)	Systemic diverse aging	Spindle formation checkpoint control
PolgA mutant mice	Systemic diverse aging	Mitochondrial DNA polymerase
Bmal1-deficient mice	Systemic diverse aging	Circadian rhythm

genes with unexpected functions and aging will be discovered in the future.

References

1. Schumacher B, Pothof J, et al. The central role of DNA damage in the ageing process. Nature. 2021;592:695–703.
2. Chu WK, Hickson ID. RecQ helicases: multifunctional genome caretakers. Nat Rev Cancer. 2009;9:644–54.
3. Liu GH, Barkho BZ, et al. Recapitulation of premature ageing with iPSCs from Hutchinson-Gilford progeria syndrome. Nature. 2011;472:221–5.
4. Takahiko Shimizu: Aging Genes. New gerontology. 3rd ed. Tokyo: University of Tokyo Press; 2010. p. 213–36.
5. Chakravarti D, LaBella KA, et al. Telomeres: history, health, and hallmarks of aging. Cell. 2021;184:306–22.

Genes and Antiaging Medicine: Applying Genome Editing Technology to Antiaging

21

Masaki Takasugi

21.1 Overview of Genome Editing

Genome editing refers to the technology that specifically modifies targeted sequences on the genome using artificial nucleases or nucleases that function in coordination with artificial ribonucleic acid (RNA). Genome editing can be performed relatively inexpensively and simply, has high sequence specificity, a wide range of applications, and unlike traditional genetic recombination technology, it basically does not leave sequences derived from foreign organisms in the genome. Its history is relatively short, with the first report of genome editing using zinc finger nuclease (ZFN) in 2005. In 2011, transcription activator-like effector nuclease (TALEN) was reported, and in 2012, genome editing using the now mainstream clustered regularly interspaced short palindromic repeats/CRISPR associated protein 9 (CRISPR/Cas9) was reported, and has rapidly spread since then. As can be seen from the fact that Dr. Emmanuelle Charpentier and Dr. Jennifer Doudna, who reported the latter, won the Nobel Prize in 2020, this technology has had a profound impact on the field of genetic engineering. Also, although it is highly controversial, it is said that in China, there are already humans born from embryos edited by this technology. Genome modification technology is a technology that can be applied to the treatment of progeria and age-related diseases, as will be described later, but it is important to note that the legal framework for controlling its practical use, as well as the formation of consensus for this, is not sufficiently established.

Genome editing technology that depends on the nuclease activity of site-specific nucleases including CRISPR/Cas9, although it has various important advantages over traditional genetic recombination technology as mentioned above, on the other hand, it has a serious problem of causing highly cytotoxic DNA double-strand breaks in the editing process. DNA double-strand breaks cause a strong stress response in cells, which can lead to cell aging or cell death, and it is also difficult to accurately control the degree of insertion or deletion of bases that occur during its repair. Therefore, a safer base editing technology that edits the genome without cutting it, based on CRISPR/Cas9, has been developed. By introducing a molecule into cells that has lost its nuclease activity due to mutation and is linked to an enzyme for editing bases, it is possible to edit bases instead of causing cuts in the target sequence. For example, in the case where a modified RNA adenine deaminase TadA, which recognizes DNA, is linked to Cas9 without nuclease activity, the adenine in the target sequence is deaminated, which is recognized as guanine by the polymerase, and after rep-

M. Takasugi (✉)
Department of Pathophysiology, Osaka Metropolitan University Graduate School of Medicine, Osaka, Japan
e-mail: w21343k@omu.ac.jp

H. Yamada, Y. Naito (eds.), *Anti-Aging Medicine*, https://doi.org/10.1007/978-981-97-8514-8_21

lication, the adenine is converted to guanine. This is called an adenine base editor.

A method called ABE has been developed to perform editing for other bases as well.

21.2 Treatment of Aging and Age-Related Diseases by Genome Editing

As of the time of writing this article (July 2022), the greatest discovery that genome editing technology has brought to the field of aging research is the 2021 report demonstrating that Hutchinson-Gilford Progeria Syndrome (HGPS) can be treated using ABE [1]. In approximately 90% of HGPS patients, a point mutation (G608G, GGC > GGT) is observed in the lamin A gene, which causes abnormal splicing and results in the production of abnormal lamin A (progerin) lacking the N-terminus, which is the cause of disease onset. This report shows that by using transgenic mice expressing human progerin and performing postnatal genome editing, the genetic mutation on progerin can be reversed, alleviating the symptoms. Lamin A underlies the nuclear envelope, maintaining its stability and controlling chromatin dynamics and gene expression, and the expression of progerin causes these to fail. Progerin is a dominant negative mutant. Transgenic mice expressing human progerin from both alleles exhibit premature aging and die around 215 days, whereas wild-type mice can survive for more than 2 years. When ABE is introduced into these transgenic mice using an adenovirus vector at 2 weeks postnatal, tissue-dependent 10. About 60% of the mutations are repaired, premature aging is improved, and the average lifespan becomes about 510 days, more than double the previous lifespan. These results strongly suggest that genome editing could be a highly effective treatment for patients with HGPS after birth. However, it should be noted that many reports on postnatal genome editing, at least at present, often involve the use of recombinant gene technology, such as the use of adenoviral vectors, as in this study. In a model mouse of Alzheimer's disease (5xFAD mouse), research on the development of treatment methods using genome editing technology has also been reported [2]. In this study, a protein that activates the transcription of neighboring genes, not an enzyme for editing bases, is linked to Cas9, which has lost its nuclease activity, to promote the expression of the target gene, in this case, the melatonin receptor type 1. It has been reported that the decrease in the expression of this gene, which plays an important role in circadian rhythm control and neuroprotection, is involved in the onset of Alzheimer's disease. When a mutant Cas9 is introduced into the brain of a 3-month-old 5xFAD mouse via a lentivirus, and the expression of the melatonin receptor type 1 is promoted, it has been shown that the pathology, including cognitive function, improves. However, in this case, since no changes occur in the DNA, it may not be accurate to call this genome editing. Also, it should be noted that the persistence of the effect basically requires continuous expression of the same factor. Finally, I would like to introduce a report on the treatment of muscle atrophy in mice by genome editing alone, without causing genetic recombination in the body [3]. It has been reported that the expression of miR-29b, which promotes muscle atrophy, increases in age-related muscle atrophy and denervation muscle atrophy, and attempts have been made to improve muscle atrophy by suppressing its expression. When Cas9 is expressed in skeletal muscle using an adenoviral vector and miR-29b is knocked out, muscle atrophy is suppressed. Furthermore, recently, it has been shown that muscle atrophy can be improved without causing genetic recombination in the body of a mouse by packing Cas9 and RNA that guides it to the miR-29b gene region into extracellular vesicles produced by cultured cells, collecting and producing extracellular vesicles in the culture medium, and then administering them to mice after inducing muscle atrophy, opening up another path to safe genome editing for the treatment of age-related diseases.

References

1. Koblan LW, Erdos MR, et al. In vivo base editing rescues Hutchinson-Gilford progeria syndrome in mice. Nature. 2021;589:608–14.
2. Park H, Kim J. Activation of melatonin receptor 1 by CRISPR-Cas9 activator ameliorates cognitive deficits in an Alzheimer's disease mouse model. J Pineal Res. 2022;72:e12787.
3. Chen R, Yuan W, et al. Delivery of engineered extracellular vesicles with miR-29b editing system for muscle atrophy therapy. J Nanobiotechnol. 2022;20:304.

Protein Homeostasis and Antiaging Medicine: Autophagy and Antiaging

22

Shuhei Nakamura and Tamotsu Yoshimori

22.1 Intracellular Degradation System Autophagy

Macroautophagy (hereafter referred to as autophagy) is a large-scale degradation system that exists universally in eukaryotes. Autophagy is always occurring at low levels within our body's cells, but it is strongly induced when cells detect starvation or various stresses. When autophagy is induced, a double-membrane structure called an autophagosome is created within the cell, which surrounds molecules and structures in the cytoplasm. The autophagosome then fuses with a lysosome, which contains a variety of hydrolytic enzymes, and the enclosed components are degraded and recycled (Fig. 22.1). Autophagy was traditionally thought of as a nonselective degradation system, but today it is known to exist as selective autophagy, which selectively removes organelles such as mitochondria, lysosomes, and endoplasmic reticulum (ER), aggregation-prone proteins, and invading bacteria. The membrane dynamics and molecular mechanisms of autophagy remained unclear for a long time, but the identification of a series of autophagy-related genes (ATG) through genetic analysis using yeast was a breakthrough, and the field has rapidly expanded over the past 30 years. Many reviews have been published on the detailed control mechanisms of autophagy, and readers are referred to these [1]. Through analysis so far, it has been found that autophagy not only secures necessary nutrients during starvation, but also plays a role in cellular metabolism and removal of harmful substances, and various physiological functions due to these actions are being revealed one after another. Below, I will introduce the involvement of autophagy in aging and lifespan control, which has become clear in recent years, and the potential for antiaging by autophagy.

S. Nakamura (✉)
Institute for Advanced Co-Creation Studies, Osaka University, Osaka, Japan

Department of Biochemistry, Nara Medical University, Nara, Japan
e-mail: shuhei.nakamura@naramed-u.ac.jp

T. Yoshimori
Graduate School of Frontier Biosciences, Osaka University, Osaka, Japan

Graduate School of Medicine, Osaka University, Osaka, Japan

H. Yamada, Y. Naito (eds.), *Anti-Aging Medicine*, https://doi.org/10.1007/978-981-97-8514-8_22

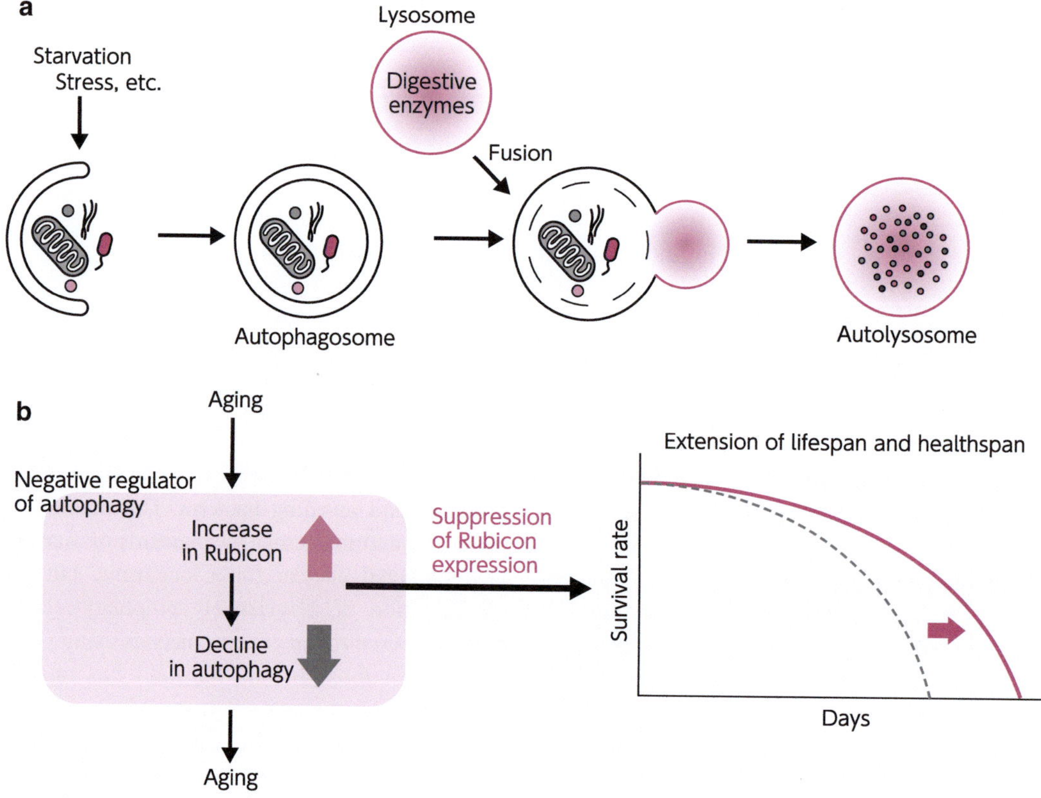

Fig. 22.1 Overview of autophagy (**a**) and mechanism of autophagy decline with aging (**b**)

22.2 Activation of Autophagy Leads to Suppression of Aging and Extension of Lifespan

In previous reports, it has been found that autophagy activity decreases with aging in various species. In the model organism nematode, a decrease in autophagy activity due to aging has been reported in various tissues. In fruit flies, a decrease in the expression of several autophagy-related genes such as Atg2 and Atg8a is observed with aging. In the thymus and liver of aged mice, a decrease in the number of autophagosomes is observed, and inhibition of fusion between autophagosomes and lysosomes is seen in neurons. In addition, it has been found that in humans, autophagy-related genes such as ATG5, ATG7, and BECN1 decrease with age. Experimentally, it has been reported that the inhibition of autophagy-related factors leads to a reduction in the lifespan of nematodes, fruit flies, and mice, and the progression of age-related diseases, suggesting that a decrease in autophagy is one of the factors in aging and disease onset. On the other hand, from analyses using model organisms, mainly nematodes, several pathways that extend the lifespan of animals have been revealed, and all of them commonly show activation of autophagy, and it has been found that this is essential for lifespan extension and aging suppression. The lifespan of a nematode mutant with a functional decline in the insulin/insulin-like growth factor 1 (IGF-1) signal, called daf-2, is known to be about twice as long, but if the activation of autophagy seen in this individual is prevented, the lifespan extension is not observed. The same has been reported for lifespan extension pathways due to calorie restriction, inhibition of the nutrient sensor mammalian target of rapamycin (mTOR), removal of germ cells, and inhibition of mitochondrial function. Furthermore, by using genetically modified animals to artificially induce autophagy, intracellular degradation system autophagy.

It has been shown that just activating autophagy can extend the lifespan of an organism. For example, in fruit flies, lifespan is extended by the overexpression of Atg1 or Atg8a, and in mice, by the overexpression of Atg5. Also, it was shown that the lifespan and healthspan of mice with a mutation in Beclin1, which inhibits its negative regulatory interaction with Bcl2, are extended due to the activation of systemic autophagy. Furthermore, it has been shown that the overexpression of HLH-30/TFEB, known as the master transcription factor of autophagy and lysosomal function, extends the lifespan of nematodes and improves various diseases, including neurodegenerative diseases, in mice. These series of results imply that not only does abnormal autophagy contribute to aging, but the activation of autophagy plays a central role in anti-aging and lifespan extension [1]. It was not well understood why autophagy activity decreases with aging, but the authors recently found that the increase of a negative regulator of autophagy, Rubicon, with aging is a factor. Rubicon is one of the few autophagy inhibitors that the authors previously identified, and it works as part of the class III PI3 kinase complex, inhibiting the fusion of autophagosomes and lysosomes. Interestingly, Rubicon increases commonly in nematodes, fruit flies, and mouse kidneys and livers with aging. On the other hand, in nematodes and fruit flies where Rubicon was inhibited, the activation of autophagy led to an extension of lifespan, improvement of motor function that decreases with aging, and reduction of protein aggregation that causes neurodegenerative diseases. Further experiments using mice suggested that Rubicon inhibition could potentially lead to the suppression of kidney fibrosis and Parkinson's disease progression due to aging, improvement of osteoporosis and age-related macular degeneration, and even inhibition of cellular aging. From these findings, it became clear that the increase in Rubicon is one of the factors for the decrease in autophagy and aging due to aging [2–5] (Fig. 22.1).

22.3 The Potential and Prospects of Antiaging by Autophagy

Considering this background, autophagy could potentially be a powerful target for anti-aging [1]. Some relatively immediate methods to activate autophagy include moderate exercise and calorie restriction. Recently, it has been shown in fruit flies that autophagy can be activated and lifespan and healthspan extended just by intermittent fasting at night without restricting total calories, and it will be interesting to see if the same happens in mammals. Furthermore, pharmacological methods to activate autophagy and inhibit aging are being actively explored. Autophagy is induced by the administration of rapamycin, an inhibitor of the nutrient sensor mTOR that suppresses autophagy, and extends lifespan in many animals. Also, resveratrol, a type of polyphenol found in red wine, is known to induce autophagy and extend the lifespan of yeast, nematodes, and mice. Spermidine, a type of polyamine, decreases with age, but its administration activates autophagy and extends the lifespan of yeast, nematodes, fruit flies, and mice. Recently, it has been shown that urolithin A, a component derived from pomegranates, induces mitophagy, a selective autophagy, extends the lifespan of nematodes, and improves memory impairment in a mouse model of Alzheimer's disease. However, these drugs have been reported to have effects other than autophagy, and further verification is needed for their application, as well as the development of more specific activators, including inhibitors of Rubicon. Also, it has been reported that activation of autophagy can have adverse effects on tissues and individuals depending on the cell type, tissue, and timing, so it is necessary to analyze in detail at the individual level where and when it is necessary to activate autophagy. For this purpose, it will be indispensable to establish a simple method to monitor the activity of autophagy in each individual in humans in the future.

References

1. Aman Y, Schmauck-Medina T, et al. Autophagy in healthy aging and disease. Nat Aging. 2021;1:634–50.
2. Nakamura S, Oba M, et al. Suppression of autophagic activity by Rubicon is a signature of aging. Nat Commun. 2019;10:847.
3. Yamamoto-Imoto H, Minami S, et al. Age-associated decline of MondoA drives cellular senescence through impaired autophagy and mitochondrial homeostasis. Cell Rep. 2022;38:110444.
4. Ando S, Hashida N, et al. Rubicon regulates A2E-induced autophagy impairment in the retinal pigment epithelium implicated in the pathology of age-related macular degeneration. Biochem Biophys Res Commun. 2021;551:148–54.
5. Yoshida G, Kawabata T, et al. Degradation of the NOTCH intracellular domain by elevated autophagy in osteoblasts promotes osteoblast differentiation and alleviates osteoporosis. Autophagy. 2022;18:2323–32.

23 Protein Homeostasis and Antiaging Medicine: Posttranslational Modification of Proteins and Antiaging

Masanori Itakura and Koji Uchida

23.1 Antiaging Effects of Antioxidants

It is known that various factors such as diet, lifestyle habits, and stress are complexly involved in the causes of aging and disease, and among them, the body's "rusting" or "oxidation" is considered one factor. The body generates energy using oxygen, but in the process, reactive oxygen species (ROS) are constantly produced as by-products. Normally, ROS are removed by antioxidant molecules in the body, but an increase in production or a decrease in removal ability can cause oxidative modifications of proteins, lipids, and DNA in the body, leading to many aging-related diseases such as cataracts, arteriosclerosis, and diabetes. In particular, oxidative modification of proteins not only reduces the original function of the protein, but also forms a new structure called an oxidation-specific epitope on the protein molecule, forming endogenous antigens called damage-associated molecular patterns (DAMPs). Excessive accumulation of DAMPs induces inflammation and is involved in diseases including aging and lifestyle-related diseases. As a defense system against such ROS-induced disruption of homeostasis (oxidative stress), food-derived antioxidants such as vitamin C, vitamin E, polyphenols, and carotenoids have long been attracting attention. In particular, the antioxidant properties of vitamin C have been reported to be important in antiaging. Unlike humans, mice can synthesize vitamin C in the body, but it has been reported that mice lacking the vitamin C synthesis enzyme show symptoms similar to aging and have a short lifespan when fed a low vitamin C diet [1]. On the other hand, natural antibodies play an important role in the removal of DAMPs. Natural antibodies are antibodies that are constantly produced mainly from B-1 cells even without specific antigen stimulation, and function as an existing immediate defense system to quickly counteract pathogens and viruses. Recently, the functionality of natural antibodies that recognize endogenous antigens, which are self-molecules, has been attracting attention, and it has been shown that the removal of DAMPs by natural antibodies improves inflammatory responses and arteriosclerosis [2].

23.2 Functionality Through Posttranslational Modification of Proteins by Antioxidants and Binding with Biomolecules

Antioxidants such as vitamin C and polyphenols, which are expected to work against aging, are oxidized after exhibiting their antioxidant ability.

M. Itakura · K. Uchida (✉)
Graduate School of Agricultural and Life Sciences, The University of Tokyo, Tokyo, Japan
e-mail: a-uchida@g.ecc.u-tokyo.ac.jp

H. Yamada, Y. Naito (eds.), *Anti-Aging Medicine*, https://doi.org/10.1007/978-981-97-8514-8_23

Many of the resulting oxidized intermediates are highly reactive and readily react with proteins that have abundant electron functional groups. In recent years, it has become clear that these antioxidant-modified proteins exhibit new functionalities related to antiaging, such as natural antibody production and antiinflammatory effects, by binding with specific biomolecules. For example, after vitamin C exhibits its antioxidant ability, it becomes oxidized vitamin C (oxidized vitamin C; oxVC) is oxidized, but oxVC and its decomposition products such as 2,3-diketogulonic acid and L-erythrose, due to their strong electrophilicity, glycate proteins by reacting with residues such as lysine. Furthermore, glycated proteins undergo reactions such as oxidation, dehydration, and condensation to produce heterogeneous modified proteins called advanced glycation end-products (AGEs). AGEs induce inflammatory responses through binding with AGE receptors and are involved in the onset of diseases. In addition, AGEs of extracellular matrix such as collagen fibers cause a decrease in skin elasticity, and AGEs of crystallins that make up the lens can cause cataracts, suggesting the involvement of protein glycation in aging. On the other hand, recent reports have revealed the existence of a homeostasis maintenance mechanism by AGEs. Kinzawa et al. reported an increase in B-1 cells producing natural antibodies by AGEs, in addition to the existence of natural antibodies recognizing AGEs derived from oxVC [3]. Furthermore, these anti-AGEs natural antibodies have been shown to cross react with modified proteins and apoptotic cells (DAMPs) produced by lipid oxidation. The pathophysiological significance of such multiple cross-reactive natural antibodies is still unknown, but it is believed that AGEs promote the production of natural antibodies, prevent the accumulation of DAMPs, and contribute to the maintenance of homeostasis. In the latest research, histones have been identified as biological binding molecules of AGEs derived from oxVC, and it has been revealed that AGEs bind to histone H2B expressed as a plasminogen receptor on monocytes and macrophages, suppressing inflammatory cell infiltration and acting antiinflammatory [4]. In this way, it was first proposed that vitamin C may show antiinflammatory properties through post-translational protein modification after exerting its antioxidant ability. Polyphenols are functional food components contained in vegetables and fruits, and they show various health-promoting effects such as antiinflammatory and antiaging effects, which are mainly thought to be due to antioxidant activities such as ROS removal. On the other hand, it is known that polyphenols that have exerted their antioxidant ability are converted into reactive intermediates through oxidation and chemically modify proteins. In particular, epigallocatechin gallate (epigallocatechin of Gallate (EGCG) and Piceatannol, a derivative of Resveratrol found in red wine, are notable for causing deamination of lysine residues on proteins and inducing protein cross-linking and polymerization. Interestingly, it has been reported that natural antibodies show cross reactivity to proteins modified by such polyphenol-modified proteins [5]. From these basic research results, it is expected that the essential function of antioxidants is to act on the immune system through protein modification. With this mechanism, even a small amount of antioxidant intake in the body would keep the body's defense system on standby by remembering the antigen (antioxidant-modified protein). This is a very reasonable hypothesis. Thus, in the antiaging and homeostasis maintenance mechanism by antioxidant molecules known for a long time, not only the antioxidant effect but also a new action mechanism through posttranslational protein modification is becoming clear (Fig. 23.1). In the future, by clarifying the details of the chemical modification structure pattern by antioxidant molecules and the body's reception and response mechanism, more effective antiaging is expected.

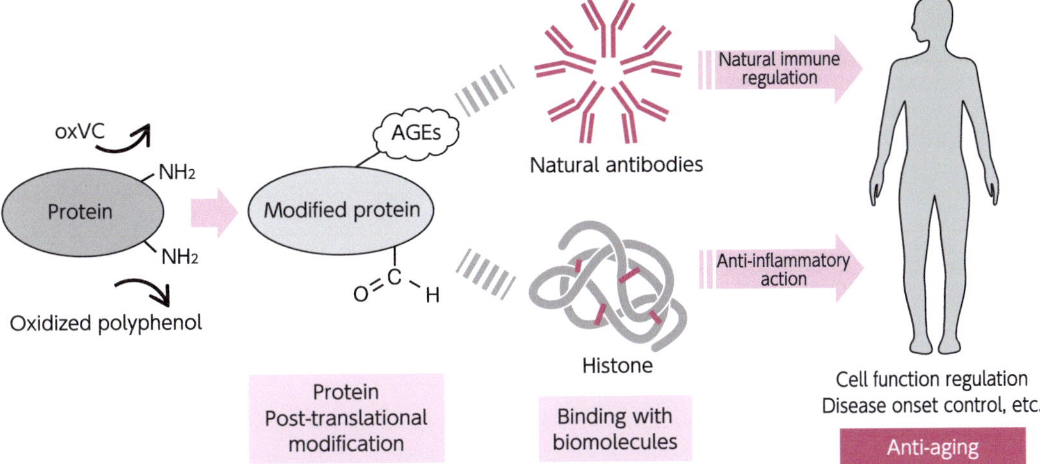

Fig. 23.1 Antiaging effect by antioxidant molecules through posttranslational protein modification

References

1. Ishigami A, Kondo Y, et al. SMP30 deficiency in mice causes an accumulation of neutral lipids and phospholipids in the liver and shortens the life span. Biochem Biophys Res Commun. 2004;315:575–80.
2. Que X, Hung MY, et al. Oxidized phospholipids are proinflammatory and proatherogenic in hypercholesterolaemic mice. Nature. 2018;558:301–6.
3. Chikazawa M, Otaki N, et al. Multispecificity of immunoglobulin M antibodies raised against advanced glycation end products: involvement of electronegative potential of antigens. J Biol Chem. 2013;288:13204–14.
4. Itakura M, Yamaguchi K, et al. Histone functions as a cell-surface receptor for AGEs. Nat Commun. 2022;13:2974.
5. Hatasa Y, Chikazawa M, et al. Oxidative deamination of serum albumins by (−)-epigallocatechin-3-O-Gallate: a potential mechanism for the formation of innate antigens by antioxidants. PLoS One. 2016;11:e0153002.

Senescent Cells and Antiaging Medicine: Senescent Cells and Senescence-Associated Secretory Phenomenon (SASP)

24

Naoko Ohtani

24.1 What Is SASP?

Cellular senescence is an irreversible cell proliferation stop induced when strong DNA damage occurs, and the cell cycle checkpoint mechanism is strongly activated. This can be considered an important cancer suppression mechanism, along with apoptosis, to prevent the increase of abnormal cells inherent in normal cells. However, unlike apoptosis, which is cell death, it is known that cells that have undergone senescence are resistant to apoptosis and continue to survive for a long time. From these long-lived senescent cells, it has been revealed that various secretory proteins such as inflammatory cytokines, chemokines, extracellular matrix remodeling factors, and growth factors are produced. This phenomenon is called the senescence-associated secretory phenotype (SASP) [1]. The effects of SASP factors are diverse. At the cellular level, SASP factors work paracrinally on surrounding cells, causing them to age, or autocrinally on the senescent cells themselves, further enhancing senescence [1]. At the individual level, it has been reported that immune cells that have migrated due to SASP factors secreted from senescent hepatic stellate cells help to remove accumulated senescent cells [2, 3]. Also, as an important physiological action of SASP, it is known that SASP that occurs transiently in fibroblasts during tissue damage works for tissue repair. On the other hand, adverse effects have also been suggested. SASP in aging individuals where senescent cells accumulate is thought to lead to age-related inflammatory diseases and functional decline. Even in cancer-associated fibroblasts (CAFs) in the microenvironment of cancer tissues, senescence and SASP are observed [1], and it is known to promote cancer. Although the action of SASP is complex, there is a tendency for it to become a disadvantageous pathological action when SASP does not converge and persists (Fig. 24.1).

N. Ohtani (✉)
Department of Pathophysiology, Graduate School of Medicine, Osaka Metropolitan University, Osaka, Japan
e-mail: naoko.ohtani@omu.ac.jp

H. Yamada, Y. Naito (eds.), *Anti-Aging Medicine*, https://doi.org/10.1007/978-981-97-8514-8_24

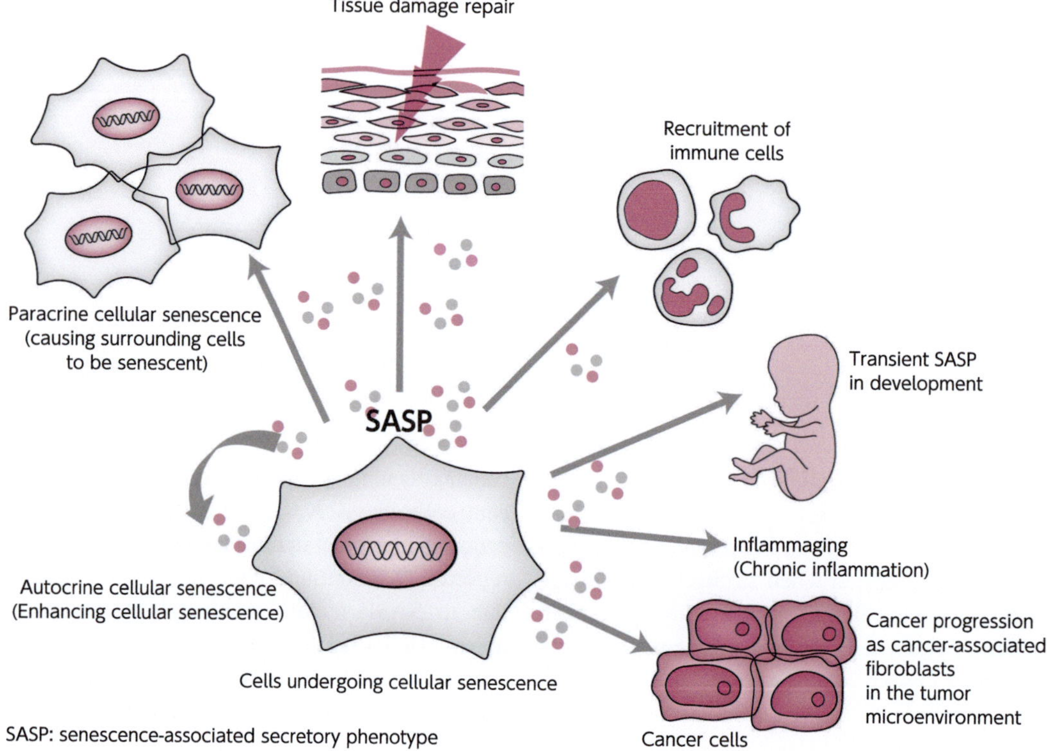

Fig. 24.1 Role of SASP
SASP factors are secreted from aging cells, inducing paracrine cellular aging or enhancing autocrine cellular aging. It is known that SASP factors contribute to tissue damage repair, migration of immune cells, development and reprogramming, inflammaging (chronic inflammation), and cancer progression as cancer-associated fibroblasts

24.2 Innate Immune Response via DNA Sensors (Endogenous SASP)

How is SASP induced in cells that have undergone senescence? As cellular senescence progresses, it is known that the expression of one of the nuclear membrane lining proteins, lamin B1, decreases. Therefore, chromatin flows out from the nucleus to the cytoplasm, and the DNA fragments that have flowed out into the cytoplasm can become ligands for various innate immune receptors, including DNA sensors, potentially triggering inflammatory signals. In fact, it has been revealed that such DNA fragments produced in cellular senescence act as ligands for the cGAS-STING pathway, one of the DNA sensors, and type I interferon is produced. Originally, cGAS (cyclic GMP-AMP Synthase is a molecule discovered as a natural immune receptor that recognizes DNA that has invaded the cytoplasm from viruses or bacteria. It has been shown in multiple papers that this is an important pathway that promotes cytokine production that occurs with cellular aging [4]. Furthermore, it has been reported that not only DNA fragments that have leaked from the nucleus, but also cDNA, which is a reverse transcription product of mRNA transcribed from retrotransposons, accumulates in the cytoplasm in aging cells, inducing SASP [5]. The genomes of organisms contain transposons and retrotransposons, which are repetitive DNA sequences of about 1 kb called transposons that can move freely on the genome. Among the retrotransposons, LINE1 (Long INterspersed Elements1, L1) is known to have high mobility. In the above report, it was reported that cDNA fragments reverse transcribed from L1 mRNA significantly increased in the cytoplasm of aged cells, and that this cDNA activated the cGAS-STING pathway, contributing to the induction of SASP [5]. Furthermore, it is expected that inhibitors of retrotransposon reverse transcriptase could be promising molecular targeted drugs for age-related inflammation, as inflammation improves when the reverse transcriptase activity of retrotransposons is inhibited [5]. DNA fragments in the cytoplasm are usually quickly degraded by enzymes that remove DNA [DNase2 and DNase3 (TREX1)], but when cellular aging occurs, the expression of these enzymes significantly decreases, and as a result, a part of the genome DNA fragments tends to accumulate in the cytoplasm [6]. Thus, it has been clarified that the induction of SASP in cellular aging is caused by the accumulation of unnecessary DNA fragments in the cytoplasm, activating the cGAS-STING pathway. These series of studies have shown that the cGAS-STING pathway is important as an "endogenous" factor in the induction of SASP.

24.3 Exogenous SASP Induction in the Microenvironment of Obesity-Induced Liver Cancer

The induction of SASP in individuals is not only due to the aforementioned endogenous factors, but it has also been found to be enhanced exogenously by the stimulation of various ligands circulating in the body. A notable example is a summary of our research results. In the microenvironment of high-fat diet-induced liver cancer, cellular aging occurs in hepatic stellate cells, which become CAF. When we examined the changes in the gut bacteria of these tumor-bearing mice, we found that in obese mice fed a high-fat diet, the proportion of Gram-positive bacteria significantly increased, and lipoteichoic acid (LTA), derived from the cell wall of Gram-positive bacteria, was prominently accumulated in the liver sinusoids. Hepatic stellate cells in the liver where lipoteichoic acid has accumulated were found to highly express toll-like receptor 2 (TLR2) compared to hepatic stellate cells in the normal liver, and were very sensitive to LTA. Through the LTA-TLR2 mediated signal, interleukin-1β (interleukin It has been found that the expression of various SASP factors, including interleukin-1 beta (IL-1β) and IL-33, and the rate-limiting enzyme cyclooxygenase-2 (COX-2), which produces prostaglandins, increases [7, 8]. Furthermore, in recent studies, the authors have revealed that signals through TLR2 promote the formation of pores on the cell membrane formed by the N-terminus of gasdermin D, and through these pores, SASP factors are released extracellularly [7]. Prostaglandin E2 (PGE2) excessively produced by COX-2 and IL-33 released extracellularly have been found to suppress antitumor immunity and promote the progression of liver cancer [7, 8]. Thus, LTA that has migrated to the liver is considered to be positioned as an "exogenous" inducer of SASP. So far, we have outlined SASP. In any case, in aging cells, the sensitivity to innate immune signals is increased, and when these signals are activated, it seems that secretory factors including various inflammatory cytokines are produced. It is expected that further elucidating the molecular mechanism of SASP and controlling cellular aging and SASP will lead to the prevention of cancer and age-related diseases and the realization of healthy longevity.

References

1. Takasugi M, Yoshida Y, et al. Cellular senescence and the tumour microenvironment. Mol Oncol. 2022;16:3333–51.
2. Krizhanovsky V, Yon M, et al. Senescence of activated stellate cells limits liver fibrosis. Cell. 2008;134:657–67.
3. Kang TW, Yevsa T, et al. Senescence surveillance of pre-malignant hepatocytes limits liver cancer development. Nature. 2011;479:547–51.
4. Li T, Chen ZJ. The cGAS-cGAMP-STING pathway connects DNA damage to inflammation, senescence, and cancer. J Exp Med. 2018;215:1287–99.
5. De Cecco M, Ito T, et al. L1 drives IFN in senescent cells and promotes age-associated inflammation. Nature. 2019;566:73–8.
6. Takahashi A, Loo TM, et al. Downregulation of cytoplasmic DNases is implicated in cytoplasmic DNA accumulation and SASP in senescent cells. Nat Commun. 2018;9:1249.
7. Yamagishi R, Kamachi F, et al. Gasdermin D-mediated release of IL-33 from senescent hepatic stellate cells promotes obesity-associated hepatocellular carcinoma. Sci Immunol. 2022;7:eabl7209.
8. Loo TM, Kamachi F, et al. Gut microbiota promotes obesity-associated liver cancer through PGE2-mediated suppression of antitumor immunity. Cancer Discov. 2017;7:522–38.

25 Senescent Cells and Antiaging Medicine: Cellular Aging Control Mechanism

Goro Katsuumi and Tohru Minamino

The mechanism of cellular aging has been extensively studied using cultured cells, and is now well understood. This section provides an overview of the key factors and signals that control cellular aging.

25.1 Modes of Cellular Aging

Normal body cells usually stop dividing and become unable to proliferate after a certain number of cell divisions when cultured. This state is defined as cellular aging. This concept was reported and proposed by Hayflick and others in the 1960s and has been widely supported to this day [1]. The cellular aging caused by repeated cell divisions is called "replicative senescence." In addition, cellular aging can be induced despite having sufficient telomere length due to various stresses such as radiation, reactive oxygen, exposure to anticancer drugs, acquisition of cancer genes, high blood sugar or specific lipid load, activation of sympathetic nerve signals or renin-angiotensin system signals. This mode of cellular aging is called "premature senescence." Both modes of cellular aging essentially occur when the accumulation of DNA damage exceeds a certain level, and cell division is stopped by the increased expression of negative cell division control proteins such as p53/p21, p16/Rb, due to the activation of the DNA damage response mechanism [2]. Of course, if these mechanisms fail, the cells will become immortal and cancerous, so cellular aging is also recognized as part of the cancer suppression mechanism.

G. Katsuumi (✉) · T. Minamino
Department of Cardiovascular Biology and Medicine, Juntendo University School of Medicine, Tokyo, Japan
e-mail: g.katsuumi.qv@juntendo.ac.jp

25.2 Cellular Aging Control Mechanism via p53

The famous cancer suppressor gene p53 is a major control protein for cellular aging. p53 is tightly controlled at the protein level by various kinases, acetylases, proteases, etc., and when the cell receives DNA damage, the checkpoint proteins ataxia telangiectasia and Rad3-related (ATR) and Chk1 pathway, and the ataxia-telangiectasia mutated (ATM) and Chk2 pathway are activated. Both proteins are serine/threonine selective kinases that drive phosphorylation cascades. p53 is mainly activated by phosphorylation via the ATM-Chk2 pathway. In addition, the MDM2 and MDM4 (MDMX) proteins that degrade p53 are inactivated by phosphorylation by ATM/ATR proteins, stabilizing p53 and indirectly increasing its expression. In addition, FOXO4 also binds to p53 and contributes to its stabilization, leading to increased expression of

H. Yamada, Y. Naito (eds.), *Anti-Aging Medicine*, https://doi.org/10.1007/978-981-97-8514-8_25

p21 and antiapoptotic factors [3]. In addition, during the induction of cellular aging, molecules such as Ras, Myc, and EIA are activated, and their downstream p14Arf. The activation of p19Arf (mouse equivalent) degrades Mdm2, leading to an increase in p53 expression. p53 promotes the inhibition or degradation of downstream cell division checkpoint proteins through the activation of p21, stopping cell division at the G1 or G2 phase. Even in the G2 phase, if the expression of p53 is sufficiently maintained and the G2/M checkpoint proteins are sufficiently suppressed, it is known that the M phase in cell division can be avoided, leading to irreversible cellular aging [4].

25.3 Cellular Aging Control Mechanism via p16

p16 is also well known as a major cellular aging control factor. It is known that the ATR-Chk1 pathway due to DNA damage response is the main factor in expression control. In addition, p16 is recognized as a more mature aging marker as its expression increases later than the aforementioned p53–p21 pathway. In this context, p16 is repressively controlled by p53, and during the DNA damage response, the F-box protein gene Fbxo22 is induced along with p53. Fbxo22 binds to the SCF protein degradation complex, ubiquitinates and induces the degradation of the increased p53, thereby inducing an increase in p16 expression (p53 rather decreases in the later stages of cellular aging) [5]. p16 controls irreversible cell division stoppage by inhibiting CDK (cyclin-dependent kinase) through the inactivation of Rb protein along with p21.

25.4 Survival and Protection Mechanisms of Aging Cells

Aging cells are supposed to be removed by cell death or immune mechanisms, but it is known that aging cells also try to prolong their life by activating various survival and protection mechanisms. In the process of cellular aging, an increase in the expression of apoptosis-inducing molecules such as Bax, NOXA, and PUMA through p53 is observed, but in reality, cell death does not easily occur. It is known that an antiapoptosis pathway called SCAP is also activated at the same time, and an increase in the expression of antiapoptosis factors such as BCL-2 and BCL-xL is known. The control mechanism is not yet clear, but it is suggested that the activation of the PI3K/Akt signal and the increase in the expression of FOXO4 are involved, and in fact, if FOXO4 is lost or PI3K/Akt is inhibited, cell death easily occurs. From the fact that apoptosis can be easily induced by BCL-2 inhibitors such as ABT-263, which is a type of senescent cell removal drug, it can be inferred that the activation of the antiapoptosis pathway is essential for the survival of senescent cells. Proteins such as GPNMB and GLS1 also increase in the process of cellular aging. Both are known to be proteins involved in the function of lysosomes. The former binds to v-ATPase on the lysosome and maintains its function, contributing to the long-term survival of senescent cells by keeping the lysosome normal [6], and the latter produces ammonia in the process of converting glutamine to glutamic acid, preventing the decrease in intracellular pH due to the leakage of acidic substances from the dysfunctional lysosome, and it is thought to work protectively in senescent cells by avoiding cell damage derived from lysosomes [7]. In addition, in order to avoid removal from endogenous immune mechanisms, it has been shown that some senescent cells increase the expression of HLA-E, an HLA that works suppressively on NK (natural killer) cells [8], and increase the expression of PD-L1, a known immune function suppressor molecule of T cells, and it is thought that these molecules help senescent cells avoid removal by immune mechanisms [9].

25.5 Conclusion

I mentioned the representative proteins and signals for the control mechanism of cellular aging, but it is shown that there are still many mechanisms involved in the control of cellular aging that cannot be written in this section due to space constraints. Cellular aging itself is not necessarily bad from the perspective of cancer suppression, and it is a bit difficult to target for treatment, but further elucidation of the control mechanism will likely lead to new treatment strategies such as senescent cell removal therapy discussed in other sections. There are still many things that are not known in detail, and further development of basic research is greatly expected in the future.

References

1. Hayflick L, Moorhead PS. The serial cultivation of human diploid cell strains. Exp Cell Res. 1961;25:585–621.
2. Di Micco R, Krizhanovsky V, et al. Cellular senescence in ageing: from mechanisms to therapeutic opportunities. Nat Rev Mol Cell Biol. 2021;22:75–95.
3. Baar MP, Brandt RMC, et al. Targeted apoptosis of senescent cells restores tissue homeostasis in response to chemotoxicity and aging. Cell. 2017;169:132–47.e16.
4. Johmura Y, Shimada M, et al. Necessary and sufficient role for a mitosis skip in senescence induction. Mol Cell. 2014;55:73–84.
5. Johmura Y, Sun J, et al. SCF(Fbxo22)-KDM4A targets methylated p53 for degradation and regulates senescence. Nat Commun. 2016;7:10574.
6. Suda M, Shimizu I, et al. Glycoprotein nonmetastatic melanoma protein B regulates lysosomal integrity and lifespan of senescent cells. Sci Rep. 2022;12:6522.
7. Johmura Y, Yamanaka T, et al. Senolysis by glutaminolysis inhibition ameliorates various age-associated disorders. Science. 2021;371:265–70.
8. Pereira BI, Devine OP, et al. Senescent cells evade immune clearance via HLA-E-mediated NK and CD8+ T cell inhibition. Nat Commun. 2019;10:2387.
9. Wang TW, Johmura Y, et al. Blocking PD-L1-PD-1 improves senescence surveillance and ageing phenotypes. Nature. 2022;611:358–64.

26

Senescent Cells and Antiaging Medicine: Development of Senolytics

Sae Aratani and Makoto Nakanishi

"Aging" is a phenomenon that no one can avoid. In recent years, "senolytic drugs" have been attracting attention as modern elixirs of youth. Senolytics are designed to specifically kill and remove senescent cells. Senescent cells are defined as those that have reached the limit of their division number and have permanently stopped proliferating. Senescent cells have a property of resistance to apoptosis, accumulate in the body with age, and some of them cause a phenomenon called senescent-associated secretory phenotype (SASP) [1]. SASP secretes cytokines, chemokines, and growth factors, and the chronic inflammation caused by these can become the breeding ground for geriatric diseases, highlighting the harmfulness of senescent cells. It is said that 90% of people over 65 have some kind of geriatric disease, and more than 70% have multiple diseases. Until now, individual geriatric diseases have been treated individually, but as it becomes clear that cellular aging is involved in the root cause of geriatric diseases and individual aging, senolytic drugs, which could be a fundamental solution to these problems, are being actively developed [2].

S. Aratani (✉)
Division of Cancer Cell Biology, Institute of Medical Science, University of Tokyo, Tokyo, Japan

Department of Nephrology, Graduate School of Medicine, Nippon Medical School, Tokyo, Japan
e-mail: sae-a@g.ecc.u-tokyo.ac.jp

M. Nakanishi
Division of Cancer Cell Biology, Institute of Medical Science, University of Tokyo, Tokyo, Japan

26.1 Development and Types of Senolytics (Fig. 26.1) [2, 3]

In 2011, a genetically modified mouse (INK-ATTAC mouse) was developed that could induce apoptosis in p16INK4A positive cells, a marker of aging, at any time depending on drug administration, and remove them. When this mouse was crossed with a premature aging mouse (BurR1H/H mouse) and senescent cells were removed, it was reported that the onset of various geriatric diseases improved, making the effectiveness of removing senescent cells clear. Based on this, bioinformatics based on the knowledge of the unique apoptosis resistance of senescent cells, SCAPs (senescent cell anti-apoptotic pathways), was used to analyze how to kill senescent cells by releasing the apoptosis suppression mechanism. As a result, the analysis of receptor tyrosine The kinase; RTK inhibitor dasatinib (D) and the flavonoid polyphenol quercetin (Q) have been identified as the first senolytic drugs. DQ combination therapy is currently being tested for efficacy in numerous age-related diseases, with reported effects in mice on frailty, osteoporosis, liver fibrosis, insulin resistance, neurodegenerative diseases, physical mobility, pulmonary fibrosis, chronic kidney disease, and even life

H. Yamada, Y. Naito (eds.), *Anti-Aging Medicine*, https://doi.org/10.1007/978-981-97-8514-8_26

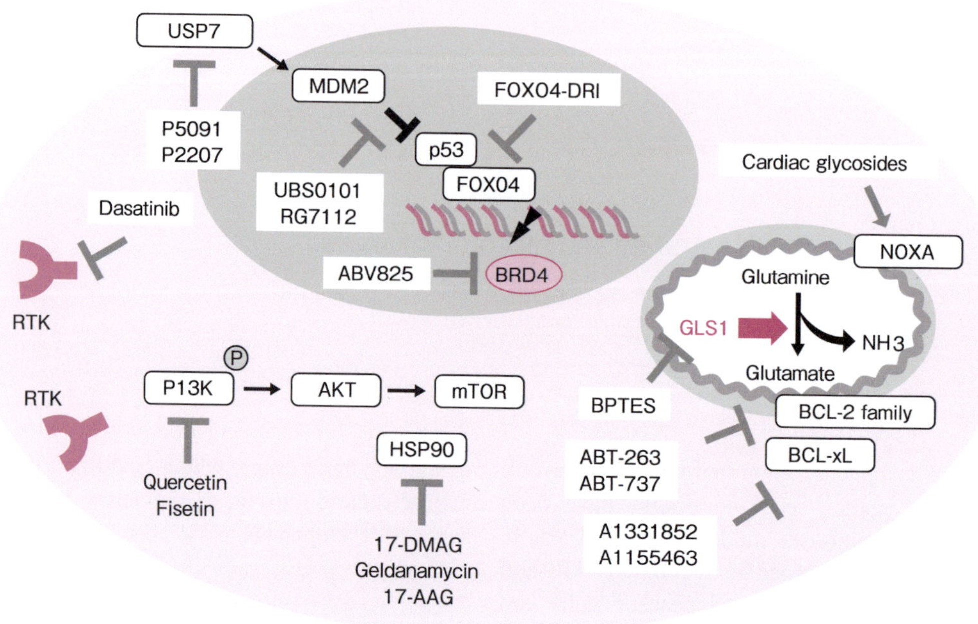

Fig. 26.1 Mechanism of action of senolytic drugs

extension. Similar effects have been reported for fisetin, a flavonoid natural compound similar to Q. The BCL-2 family (BCL-2, BCL-xL, BCL-W) is overexpressed in aging cells and plays a role in the mechanism of apoptosis inhibition. Therefore, ABT-263 (navitoclax), ABT-737, which inhibit BCL-2, BCL-xL, BCL-W, and A1331852, A1155463, which inhibit BCL-W, are known as senolytic drugs. ABT-263 has been reported to reduce cancer treatment-induced senescent cells and suppress SASP, thereby reducing cancer recurrence and metastasis. Initially, there were concerns about thrombocytopenia caused by ABT-263, but prodrugs (Nav-Gal) that are degraded by SA-β-gal, which is overexpressed in aging cells, and proteolysis-targeting chimera (PROTAC) technology that forcibly binds BCL-xL to E3 ligase and promotes ubiquitination and degradation of target cells have been developed, and side effects are being overcome. In aging cells, the expression of FOXO4 (forkhead box proteinO4) is increased, inhibiting the nuclear translocation of p53 and suppressing cell death. FOXO-DRI, which inhibits the interaction between FOXO4 and p53, and UBX0101 and RG7112, which inhibit the interaction with MDM2 that degrades p53, are known. Also, USP7 (ubiquitin-specific peptidase 7) Inhibitors such as P5091 and P2207, which activate p53 and induce apoptosis, are also mentioned as senolytic drugs. Heat shock proteins have the function of molecular chaperones and maintain proteostasis. From the screening search for senolytic drugs, inhibitors of heat shock protein 90, 17-DMAG, geldanamycin, 17-AAG, have been reported as senolytic drugs. As part of the mechanism of action, it is believed that 17-DMAG impairs the interaction between HSP-90 and AKT, thereby inhibiting AKT activation. On the other hand, in aging cells, the homeostasis of proteostasis decreases, and the production of aggregated proteins is enhanced due to the increased synthesis of RNA and proteins. Aggregated proteins are degraded in lysosomes, but excessive aggregated proteins puncture the lysosomal membrane, releasing protons from the lysosome into the cytoplasm and acidifying the cell. Intracellular acidification induces apoptosis from increased permeability of the mitochondrial membrane. However, in aging cells, it was found that the enzyme activity of GLS1 is enhanced, activating glutaminolysis, and the by-product ammonia is increased, and this ammonia neutralizes protons, maintaining survival. In our laboratory, we have reported the GLS1 inhibitor BPTES as a new senolytic drug [4]. In addition, the PROTAC drug (BETdegrader) of BET (bromodomain and extra-terminal family protein), ARV825, has been reported as a senolytic. This degrades BRD4, one of the BET proteins, by PROTAC action, inhibits the nonhomologous end joining repair mechanism of DNA double-strand breaks in aging cells via BRD4, and at the same time, promotes the expression of autophagy-related gene groups to induce cell death. Also, as next-generation senolytics, methods are being developed to identify small molecules and seno-antigens that are specifically expressed in aging cells, and to develop antibodies or vaccines against them. So far, vaccine therapy for GPNMB, CD153, and CAR-T therapy for uPAR have been reported.

26.2 Clinical Application of Senolytics and Conclusion

Expectations for the clinical application of senolytics are increasing, and several clinical studies have begun [5] (Table 26.1). However, the clinical trial of the aforementioned UBX0101 ended in failure without recognizing the effect, and there are many challenges. Rather than blindly removing aging cells, we must distinguish aging

Table 26.1 Clinical trials of senolytic drugs by the Translational Geroscience Network (TGN)

Diseases	Drug	Gender	Age	Phases	No.	NCT Number	Status	Start	End
HTSS	D, Q	All	18 and above	Not Applicable	10	2652052	Recruiting	From 16/02	Until 23/12
CKD	D + Q	All	40 and above	Phase 2	30	2848131	Enrolling by invitation	From 16/06	Until 23/04
IPF	D + Q	All	50 and above	Phase 1	26	2874989	Completed	From 16/12	Until 19/06
Frailty	Fisetin	F	70 and above	Phase 2	40	3430037	Recruiting	From 18/02	Until 23/06
DM and DKD	Fisetin	All	40 and above	Phase 2	30	3325322	Recruiting	From 18/02	Until 25/04
Frailty	Fisetin	All	70 and above	Phase 2	40	3675724	Recruiting	From 18/11	Until 23/06
Influenza vaccination	Entolimod	All	65 and above	Phase 2	61	4176133	Completed	From 19/10	Until 22/03
Alzheimer's Disease	D + Q	All	65 and above	Phase 1 Phase 2	5	4063124	Recruiting	From 20/02	Until 23/08
Osteoarthritis	Fisetin	All	40 and above	Phase 1 Phase 2	100	4210986	Active, not recruiting	From 20/06	Until 22/12
COVID-19	Fisetin	All	18 and above	Phase 2	80	4476953	Enrolling by invitation	From 20/08	Until 23/07
Skeletal health	D + Q, Fisetin	F	70 and above	Phase 2	120	4313634	Recruiting	From 20/09	Until 23/03
COVID-19	Fisetin	All	18 and above	Phase 2	150	4771611	Enrolling by invitation	From 21/01	Until 23/06
Alzheimer's Disease	D + Q	All	65 and above	Phase 2	48	4685590	Recruiting	From 21/12	Until 23/01
COVID-19	Fisetin	All	65 and above	Phase 2	150	4537299	Enrolling by Invitation	From 22/01	Until 23/12
Alzheimer's disease	D + Q	All	55 and above	Phase 1 Phase 2	20	4785300	Enrolling	From 22/05	Until 24/06
Frailty	D + Q, Fisetin	All	18 and above	Phase 2	60	4733534	Recruiting	From 22/05	Until 24/07

cells involved in chronic inflammation and consider differences in tissues and cell types. We want to deepen our understanding of the diversity of aging cells and strive for more effective and safe drug discovery.

References

1. Gorgoulis V, Adams PD, et al. Cellular senescence: defining a path forward. Cell. 2019;179:813–27.
2. Zhang L, Pitcher LE, et al. Targeting cellular senescence with senotherapeutics: senolytics and senomorphics. FEBS J. 2023;290:1362–83.
3. Kirkland JL, Tchkonia T. Senolytic drugs: from discovery to translation. J Intern Med. 2020;288: 518–36.
4. Johmura Y, Yamanaka T, et al. Senolysis by glutaminolysis inhibition ameliorates various age-associated disorders. Science. 2021;371:265–70.
5. Wissler Gerdes EO, Misra A, et al. Strategies for late phase preclinical and early clinical trials of senolytics. Mech Ageing Dev. 2021;200:111591.

Senescent Cells and Antiaging Medicine: Selective Removal of Senescent Cells

27

Hironori Nakagami

Aging is a phenomenon that progresses with age, interrelated at the levels of cells, organs, and individuals. Cellular senescence is an old-known phenomenon [1], but in recent years, it has been reported that a large amount of inflammatory cytokines are secreted from senescent cells, adversely affecting surrounding cells and promoting cellular senescence, a "senescence-associated secretory phenomenon" (SASP). This has clarified the relationship between cellular senescence and organ aging/individual aging (Fig. 27.1). Using genetically modified mice, a model mouse that removes these senescent cells from the body was created. As a result, it was possible to suppress various symptoms associated with aging and extend lifespan [2], proving the concept of anti-aging treatment by removing senescent cells (senolysis), and various methods are being developed worldwide for senescent cell removal therapy. Senescent cells do not proliferate, but they have the characteristic of not leading to cell death, and they accumulate little by little in the body with aging. Therefore, as a treatment method, attempts are being made to selectively lead senescent cells to cell death (apoptosis). This section provides an overview of the removal of senescent cells by immunity.

H. Nakagami (✉)
Department of Health Development and Medicine, Osaka University Graduate School of Medicine, Osaka, Japan
e-mail: nakagami@gts.med.osaka-u.ac.jp

H. Yamada, Y. Naito (eds.), *Anti-Aging Medicine*, https://doi.org/10.1007/978-981-97-8514-8_27

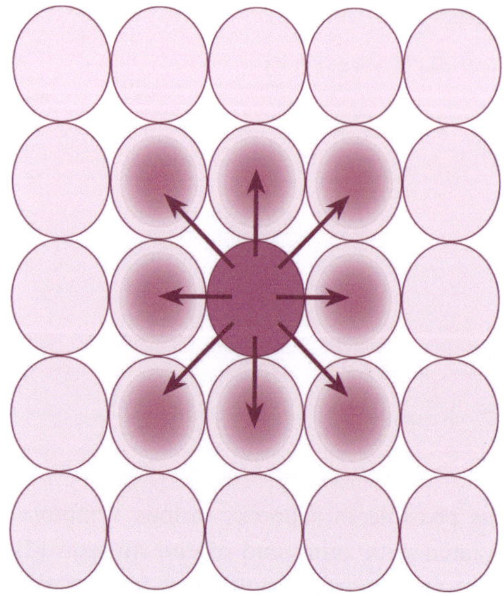

Fig. 27.1 Senescent cell removal therapy

27.1 Cell Removal by Immunity (CAR-T Therapy)

Immunity is a function that identifies and attacks things that should be eliminated from our bodies, such as pathogens (bacteria and viruses) or cancer cells. Immunity is broadly divided into innate immunity and acquired immunity. Innate immunity, which initially responds nonspecifically to viruses and bacteria that have invaded from the outside, plays an important role in activating the subsequent specific immune response, which is acquired immunity. White blood cells such as granulocytes, lymphocytes, and monocytes/macrophages are central to immune function, and among lymphocytes, there are T cells and B cells. Acquired immunity includes cellular immunity and humoral immunity. T cells are cells that directly attack cells to attack and eliminate abnormal cells infected with pathogens and cancer cells as cellular immunity, and helper T cells recognize antigens presented by dendritic cells, produce cytokines, and these cytokines activate macrophages, cytotoxic T cells, etc. On the other hand, the immune response centered on B cells and antibodies is humoral immunity, and when B cells are stimulated by cytokines produced by helper T cells, a large amount of antibodies are produced. In recent years, research on cancer cells has progressed, and it has become clear that cancer cells cannot be completely eliminated by immunity because cancer tissues construct a mechanism to escape from the mechanism of immunity. CAR-T cell therapy is a treatment developed as a treatment for refractory cancer, and it is a treatment using T cells, CAR-expressing T cells (CAR-T cells), which are modified to express genes that can proliferate while attacking cancer cells by taking out the patient's own T cells, designing a CAR (chimeric antigen receptor) that can recognize specific antigens expressed on the surface of cancer cells, etc. Using this technology, it has been reported that anti-aging effects were observed when senolytic CAR-T cells, which remove aging cells in the same mechanism as cancer cells, were administered to mice [3].

27.2 Cell Removal Vaccine

Attempts are also being made to develop cell-removal vaccines that induce therapeutic antibodies through humoral immune activation. Vaccines are therapies that utilize the body's immune function by activating natural and acquired immunity, and our immune function has the characteristic that once immunity is acquired, it is remembered by the body, and when the same immune stimulus is received later, a quicker response can be made than the first time, so the activation of acquired immunity by vaccines is effective in preventing infectious diseases. On the other hand, when considering a therapeutic vaccine against senescent cells, endogenous proteins become target molecules, and different strategies are required from infectious diseases, but it was found that antibody induction against endogenous proteins can be achieved by simultaneous administration of adjuvant [4]. A vaccine was designed to induce the production of antibodies that recognize the surface molecule CD153 of senescent T cells, and when administered to mice with an adjuvant, the production of anti-CD153 antibodies was induced, and the increase of senescent T cells was shown to be suppressed. Compared to mice fed a normal diet, in model mice fed a high-fat diet, senescent T cells in visceral fat increased from 6.8% to 16.0%, but when antibodies recognizing the surface molecule CD153 of senescent T cells were produced by vaccine administration, the increase of senescent T cells in high-fat diet mice was suppressed to 7.6%. Therefore, while administering the vaccine, mice were fed a high-fat diet, and when glucose tolerance was compared, it was shown that mice that had removed senescent cells could suppress blood glucose elevation in the glucose tolerance test. Also, the infiltration of macrophages, which are the starting point of inflammation in fat, was also suppressed [5]. Note: An adjuvant is a general term for substances or components used in combination with a drug to enhance or assist its effects. It is a nonspecific immune activator that activates the natural immune response and the antigen-antibody reaction. Known adjuvants include nucleic acid adjuvants (DNA, RNA, CpG) and alum (aluminum).

27.3 Future Expectations for Senescent Cell Removal Therapy

Tools for senescent cell removal therapy that are being developed worldwide, such as CAR-T therapy and cell removal therapy using vaccines, have already been reported in mice, demonstrating the potential for new applications of gene therapy and vaccine therapy that have been used in cancer treatment and other applications. Since the number of senescent cells is not as large as that of cancer cells or bacteria/viruses, and the progression is chronic, we expect that the development of cell removal therapy will evolve into more practical research.

References

1. Hayflick L, Moorhead PS. The serial cultivation of human diploid cell strains. Exp Cell Res. 1961;25:585–621.
2. Baker DJ, Childs BG, et al. Naturally occurring p16(Ink4a)-positive cells shorten healthy lifespan. Nature. 2016;530:184–9.
3. Amor C, Feucht J, et al. Senolytic CAR T cells reverse senescence-associated pathologies. Nature. 2020;583:127–32.
4. Nakagami H. Cellular senescence and senescence-associated T cells as a potential therapeutic target. Geriatr Gerontol Int. 2020;20:97–100.
5. Yoshida S, Nakagami H, et al. The CD153 vaccine is a senotherapeutic option for preventing the accumulation of senescent T cells in mice. Nat Commun. 2020;11:2482.

28 Mitochondria and Antiaging Medicine: Mitochondria and Antiaging

Jun Yoshino and Hiroshi Itoh

Mitochondria have long been known as intracellular organelles that play a central role in energy metabolism by producing adenosine triphosphate (ATP) through oxidative phosphorylation (OXPHOS), the TCA cycle, and beta-oxidation of fatty acids. Furthermore, mitochondria are thought to be involved in various biological aspects related to aging, such as the production of reactive oxygen species (ROS), genome stability, mitochondrial DNA (mtDNA) replication, mitophagy, telomere function, and cellular aging. These mitochondrial functions decline with aging and have been reported by various researchers to underlie the pathogenic mechanisms of age-related diseases such as diabetes, chronic kidney disease, heart failure, and Alzheimer's disease [1]. It has also been suggested that mitochondria mediate the benefits of calorie restriction, which delays the onset of age-related diseases and extends the lifespan of the individual, thus increasing the importance of mitochondrial research in anti-aging medicine.

J. Yoshino (✉)
Division of Endocrinology, Metabolism and Nephrology, Department of Internal Medicine, Keio University School of Medicine, Tokyo, Japan

Division of Nephrology, Department of Internal Medicine, Faculty of Medicine, Shimane University, Matsue, Shimane, Japan

The Center for Integrated Kidney Research and Advance (IKRA), Faculty of Medicine, Shimane University, Matsue, Shimane, Japan
e-mail: jyoshino@keio.jp

H. Itoh
Center for Preventive Medicine, Keio University, Tokyo, Japan

28.1 Lifespan Control Signals and Mitochondria

Recent research results suggest that evolutionarily conserved nutrient response signaling pathways mediate the antiaging and lifespan-extending effects of calorie restriction [1, 2]. These lifespan control signals play a crucial role in maintaining the homeostasis of mitochondrial function (Fig. 28.1). For example, the inhibition of insulin/insulin-like growth factor 1 (IGF1) signaling extends lifespan across a wide range of species through the activity of DAF-16/FOXO, enhances mitochondrial OXPHOS activity, beta-oxidation of fatty acids, and increases the expression of peroxisome proliferator-activated receptor gamma coactivator-1α (PGC1A), a major regulator of mitochondrial biogenesis. Also, 5-AMP-activated protein kinase (AMPK), which is activated by a decrease in cellular energy (increase in AMP/ATP ratio), positively regulates PGC1A expression and controls mitochondrial biogenesis, beta-oxidation of fatty acids, mitochondrial fusion and fission, and mitophagy. Similarly, the inhibition of the nutrient excess signal mammalian target of rapamycin (mTOR), a serine-threonine kinase that controls cell growth, activates various mito-

H. Yamada, Y. Naito (eds.), *Anti-Aging Medicine*, https://doi.org/10.1007/978-981-97-8514-8_28

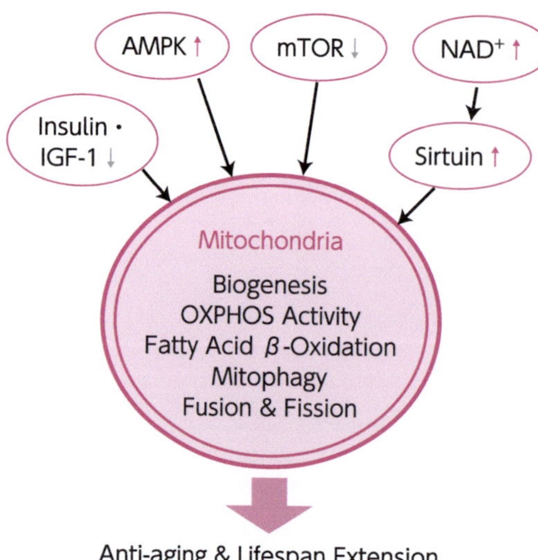

Fig. 28.1 Mitochondria and aging control signal

chondrial functions. Consistent with these findings, metformin, an AMPK activator, and rapamycin, an mTOR inhibitor, have been reported to improve glucose metabolism in mice and extend their lifespan.

28.2 Sirtuins and Mitochondria

Sirtuins have recently been in the spotlight for their role in the crosstalk at the molecular level between mitochondrial function regulation and aging/lifespan control. Sirtuins are a group of enzymes that deacetylate and deacylate substrate proteins in a nicotinamide adenine dinucleotide (NAD+)-dependent manner. In mammals, including humans, there are seven types of sirtuins, SIRT1 to SIRT7. Sirtuins have been reported to mediate the anti-aging effects of calorie restriction and play an important role in lifespan control. For example, it has been shown that mice with a systemic overexpression of SIRT6 and brain-specific overexpression of SIRT1 live longer [3]. Furthermore, the role of sirtuins in mitochondrial function control is being elucidated [4]. For instance, SIRT1 deacetylates and activates PGC1A in a NAD+ -dependent manner, controlling mitochondrial biogenesis, ROS production, fatty acid beta-oxidation, and mitophagy. Meanwhile, mitochondrial sirtuins (SIRT3 to SIRT5), located in the mitochondria, deacetylate and regulate the activity of major metabolic enzymes in the mitochondria, such as long-chain acyl coenzyme A dehydrogenase (LCAD), glutamate dehydrogenase (GDH), and carbamoylphosphate synthetase (CPS1). Moreover, other sirtuins (SIRT2, SIRT6, SIRT7) are also thought to be involved in mitochondrial function control through deacetylation and deacylation reactions, and the importance of sirtuins in mitochondrial biology is being established. Interestingly, it has been suggested that there is crosstalk between sirtuins and other lifespan control signals. For example, SIRT1 deacetylates and activates FOXO, a major mediator of lifespan extension associated with insulin/IGF1 signal inhibition. Also, while SIRT1 deacetylates and activates LKB1, an upstream activator of AMPK, AMPK activates SIRT1 through promoting NAD+ synthesis. Furthermore, the increase in SIRT1 activity and the enhancement of S6K1 deacetylation and phosphorylation increase the activity of mTOR complex 1 signaling. It is believed that multiple nutrient response signaling pathways form complex crosstalk and control mitochondrial function, aging, and lifespan in a multilayered manner, and more detailed examination is awaited in the future. In addition, research on NAD+, which controls sirtuin activity, has been rapidly expanding in recent years, and the importance of NAD+ metabolism in mitochondrial function control is also being established (see the next Chap. 29). Polyphenols that activate sirtuins (SIRT1), such as resveratrol, low molecular weight compounds like SRT2104, and NAD+ intermediary metabolites like nicotinamide mononucleotide (NMN), have been reported to improve mitochondrial function in many disease model mice, and their clinical application is being developed [5].

28.3 Outlook

As introduced in this article, it is no exaggeration to say that the importance of mitochondrial research in the mechanism of aging and lifespan control has been established by recent research results. In the future, mitochondrial biology is expected to be an attractive drug discovery target in antiaging medicine. It is thought that it will be important to further elucidate the molecular mechanism by which calorie restriction, exercise, etc. enhance mitochondrial function through basic research using animal models, and to verify the significance of mitochondria in human aging and lifespan control.

References

1. Amorim JA, Coppotelli G, et al. Mitochondrial and metabolic dysfunction in ageing and age-related diseases. Nat Rev Endocrinol. 2022;18:243–58.
2. Akbari M, Kirkwood TBL, et al. Mitochondria in the signaling pathways that control longevity and health span. Ageing Res Rev. 2019;54:100940.
3. Satoh A, Imai S, et al. The brain, sirtuins, and ageing. Nat Rev Neurosci. 2017;18:362–74.
4. van de Ven RAH, Santos D, et al. Mitochondrial sirtuins and molecular mechanisms of aging. Trends Mol Med. 2017;23:320–31.
5. Bonkowski MS, Sinclair DA. Slowing ageing by design: the rise of NAD+ and sirtuin-activating compounds. Nat Rev Mol Cell Biol. 2016;17:679–90.

29 Mitochondria and Antiaging Medicine: NAD+ Metabolic Intermediates and Antiaging

Jun Yoshino and Hiroshi Itoh

Nicotinamide adenine dinucleotide (NAD+) was discovered in 1906 by British biochemist Dr. Arthur Harden and has long been known as a coenzyme that plays an important role in redox reactions. In 2000, the NAD + -dependent deacetylase activity of the aging and lifespan control factor sirtuin was discovered by Imai and Guarente at the Massachusetts Institute of Technology, leading to a dramatic development in NAD+ biology research. As a result, it is suggested that a decrease in NAD+ levels contributes to the molecular mechanisms of aging through a decrease in sirtuin activity, and also plays a part in the pathogenesis of various aging-related diseases such as diabetes, chronic kidney disease, heart failure, and Alzheimer's disease. Currently, with about 115 years of history, NAD+ biology research is entering a second golden (renaissance) period centered on two NAD+ metabolic intermediates, nicotinamide mononucleotide (NMN) and nicotinamide riboside (NR), and it is no exaggeration to say that it is forming the mainstream of aging and lifespan research and antiaging medicine [1, 2].

29.1 Salvage NAD+ Synthesis Pathway

The materials for NAD+ synthesis are known to be tryptophan and water-soluble vitamin B3, col lectively referred to as nicotinamide (NAM) and nicotinic acid, and NR. In mammals, the salvage synthesis pathway starting from NAM plays a major role in intracellular NAD+ synthesis (Fig. 29.1). In this salvage synthesis pathway, NAM is converted to NMN by nicotinamide phosphoribosyltransferase (NAMPT), the rate-limiting enzyme for NAD+ synthesis, and then synthesized into NAD+ by the second synthesis enzyme, nicotinamide/nicotinic acid mononucleotide adenylyltransferase (NMNAT). Also, NR is converted to NMN by nicotinamide riboside kinase (NRK) and used for NAD+ synthesis. The synthesized NAD+ is used by NAD + -dependent enzymes such as redox reactions, sirtuins, poly ADP ribose polymerase (PARP), CD38, and plays an important role in various biological aspects such as metabolism, circadian rhythms, inflammation, etc. [1, 2].

J. Yoshino (✉)
Division of Endocrinology, Metabolism and Nephrology Department of Internal Medicine, Keio University School of Medicine, Tokyo, Japan

Division of Nephrology, Department of Internal Medicine, Faculty of Medicine, Shimane University, Matsue, Shimane, Japan

The Center for Integrated Kidney Research and Advance (IKRA), Faculty of Medicine, Shimane University, Matsue, Shimane, Japan
e-mail: jyoshino@keio.jp

H. Itoh
Center for Preventive Medicine, Keio University, Tokyo, Japan

H. Yamada, Y. Naito (eds.), *Anti-Aging Medicine*, https://doi.org/10.1007/978-981-97-8514-8_29

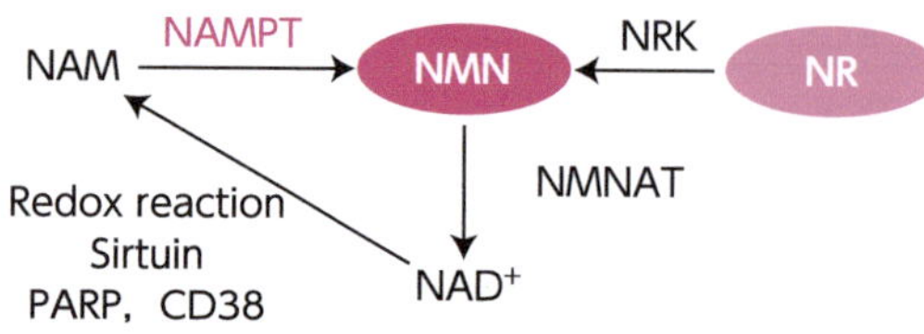

Fig. 29.1 Salvage NAD+ synthesis pathway

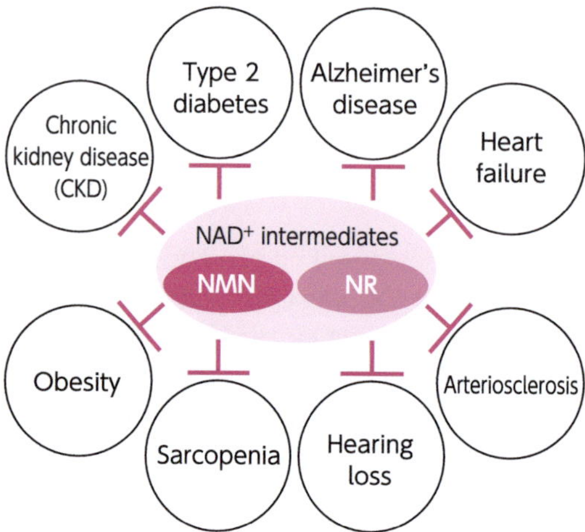

Fig. 29.2 Utility of NAD+ metabolic intermediates NAD+ metabolic intermediates

29.2 Aging and Aging-Related Diseases and NAD+ Metabolic Intermediates

Analyses using genetically modified animals of NAD+ synthesis rate-limiting enzyme NAMPT and NAD+ consuming enzymes CD38, PARP suggest that the decrease in NAD+ levels with aging is deeply involved in the pathophysiology of aging and age-related diseases through mechanisms such as the decrease in sirtuin activity. Furthermore, administration of NAD+ intermediate metabolites represented by NMN and NR has been reported to restore NAD+ levels in various organs, activate sirtuins, and exhibit significant therapeutic effects in animal models of various age-related diseases such as type 2 diabetes, obesity, chronic kidney disease, heart failure, arteriosclerosis, Alzheimer's disease, sarcopenia, and hearing loss [1, 2] (Fig. 29.2). Interestingly, long-term administration of NMN in drinking water for about a year in wild-type mice has been reported to prevent the onset of mitochondrial dysfunction in skeletal muscle, metabolic syndrome, insulin resistance, inflammation in adipose tissue, and osteoporosis associated with aging, and to exert various antiaging effects [3]. This has raised expectations for NAD+ intermediate metabolites as drug targets for antiaging.

29.3 Life Span Control and NAD+ Intermediate Metabolites

While there are reports that promoting NAD+ synthesis extends cell and individual lifespan in cell culture and experimental systems using yeast and nematodes, the effect of NAD+ intermediate metabolites on lifespan in mammals is not clear at this time. However, it has recently been revealed that calorie restriction, which has a lifespan-extending effect, increases NAMPT-NAD+ levels in various organs, and overexpression of NAD + -dependent deacetylase sirtuins (SIRT1, SIRT6) extends the lifespan of mice (see previous Chap. 28). These findings suggest that NAD+ intermediate metabolites may exert lifespan-extending effects through the activation of sirtuins. Furthermore, NAD+ intermediate metabolites have been reported by various researchers to restore mitochondrial functions such as oxidative phosphorylation and beta-oxidation of fatty acids, and improve disease conditions and prognosis in various disease model mice including mitochondrial disease mice and naturally aging mice [1, 2]. Considering the importance of mitochondrial function in aging and lifespan control (see previous Chap. 28), NAD+ intermediate metabolites may extend the lifespan of individuals through the enhancement of mitochondrial function, and further examination is awaited.

29.4 Outlook

The remarkable therapeutic effects and antiaging properties of NAD+ metabolic intermediates, represented by NMN and NR, are becoming increasingly clear in various disease model animals. Currently, NAD+ biology research is moving towards the stage of clinical application and societal implementation. For example, in Japan, NMN has been confirmed to be safe for oral administration in humans and its pharmacokinetics have been confirmed, leading the world [4]. More recently, it has been reported that oral administration of NMN improves skeletal muscle insulin signaling and insulin sensitivity (glucose uptake ability) in a double-blind randomized controlled trial targeting postmenopausal overweight and obese women [5]. These results are consistent with the results of animal models, suggesting the potential for NAD+ metabolic intermediates to become antiaging drugs in humans. Further development of NAD+ biology basic and clinical research in antiaging medicine is desired in the future.

References

1. Yamaguchi S, Ito H, et al. Significance and potential of translational research on NAD+ biology in super-aged society Japan. J Japan Geriatr Soc. 2020;57:213–23.
2. Yoshino J, Baur JA, et al. NAD+ Intermediates: the biology and therapeutic potential of NMN and NR. Cell Metab. 2018;27(3):513–28.
3. Mills KF, Yoshida S, et al. Long-term administration of nicotinamide mononucleotide mitigates age-associated physiological decline in mice. Cell Metab. 2016;24:795–806.
4. Irie J, Inagaki E, et al. Effect of oral administration of nicotinamide mononucleotide on clinical parameters and nicotinamide metabolite levels in healthy Japanese men. Endocr J. 2020;67:153–60.
5. Yoshino M, Yoshino J, et al. Nicotinamide mononucleotide increases muscle insulin sensitivity in prediabetic women. Science. 2021;372:1224–9.

Oxidative Stress and Antiaging Medicine: Oxidative Stress Response and Antiaging

30

Hiroshi Ichikawa

Aerobic organisms, including humans, can efficiently produce adenosine triphosphate (ATP) using oxygen in the air. However, 1–2% of the oxygen consumed becomes a reactive oxygen species (ROS) called superoxide ($O_{2}\cdot$), and mitochondria are constantly exposed to oxidative stress [1]. In addition, a large amount of ROS is produced from inflammatory cells such as neutrophils, and the body is exposed to strong oxidative stress. At the same time, the body is exposed to various stimuli, not only from within the body but also from what is called the exposome (ultraviolet rays, radiation, tobacco, alcohol, food additives, environmental hormones, etc.), and is exposed to oxidative stress from outside the body. Therefore, from the standpoint of antiaging, it is required to maintain a body and cells that can overcome ROS from both inside and outside the body. Now, when cells are exposed to oxidative stress, for example, when exposed to ultraviolet rays, many people seem to think that the stimulation by ROS [hydroxyl radicals ($OH^{\cdot}$) and singlet oxygen ($^{1}O_{2}\cdot$)] directly produced by ultraviolet rays leads to cell damage. However, when human keratinocytes are irradiated with ultraviolet rays of intensity that can cause cell death, the ROS directly produced by the ultraviolet rays are actually at the level of several μM when converted to hydrogen peroxide, and the irradiation time is also a few minutes. Considering that it is a ROS with a very short half-life, there is almost no possibility that HaCaT keratinocytes (human keratinocytes) will be damaged by the ROS directly produced by ultraviolet rays. Also, as is widely known, it has been found that applying sunscreen after sunburn suppresses skin inflammation. From the above, it can be considered that the response of cells when exposed to exogenous oxidative stress is not due to ROS from the outside, but to ROS produced secondarily inside the cells after stimulation from the outside. In other words, endogenous ROS become second messengers, and downstream transcription factors and gene expression are induced, leading to complex cellular responses. The mechanisms of cell death due to exogenous oxidative stress, transcription factors, gene expression control, etc., will be left to other sections, and in this section, we will focus on ROS produced secondarily inside cells due to exogenous stimuli.

H. Ichikawa (✉)
Major of Medical Life Systems, Graduate School of Life and Medical Sciences, Doshisha University, Kyoto, Japan
e-mail: hichikaw@mail.doshisha.ac.jp

30.1 Exposure to Exogenous Singlet Oxygen Secondarily Produces ROS Inside Cells

Singlet oxygen is known as one of the selective intermediate active oxygen species naturally generated in biological systems by light and nonlight

H. Yamada, Y. Naito (eds.), *Anti-Aging Medicine*, https://doi.org/10.1007/978-981-97-8514-8_30

mediated processes such as ultraviolet rays. Singlet oxygen plays an important role in cell signaling and homeostasis maintenance, and several in in vitro studies have been investigating the pathways through which singlet oxygen mediates the oxidation of biological molecules and potential pathogenesis. However, detailed reports on whether singlet oxygen causes cellular damage through the subsequent generation of ROS have not been observed. We established a cellular damage model using a hydrophobic endoperoxide (EP) solution as a singlet oxygen stimulant to elucidate the damage mechanism of singlet oxygen to HaCaT keratinocytes, and after confirming that singlet oxygen is spontaneously produced from EP at 35 °C or higher by chemiluminescence, we revealed that the amount of ROS production increases under the coexistence of HaCaT keratinocytes and EP [2, 3]. Next, we demonstrated that ROS is continuously produced from within the cell secondarily as a result of EP stimulation. Furthermore, to elucidate the mechanism of cellular damage caused by EP stimulation, we examined using various antioxidant enzymes and antioxidants, and found that platelet activating factor receptor antagonist (WEB2086), sodium azide (NaN_3), and histidine significantly inhibited cellular damage regardless of their addition after EP stimulation. In other words, it was revealed that hydrogen peroxide (H_2O_2) and singlet oxygen, which were secondarily produced in the cell after EP stimulation, are greatly involved in subsequent cellular damage. Furthermore, to identify the source of ROS that is secondarily produced in the cell, we attempted to detect singlet oxygen in the mitochondria using a confocal microscope with a Si-DMA fluorescent probe. As a result, we confirmed that singlet oxygen is continuously produced in the mitochondria even after the production of singlet oxygen from EP disappears after EP stimulation. Also, the intracellular singlet oxygen secondarily produced by EP stimulation disappeared by the addition of mitochondrial electron transport system inhibitors, intracellular superoxide scavengers (MnTMPyP), and WEB2086, suggesting that the singlet oxygen produced in the cell by EP stimulation is derived from mitochondria and platelet activating factor receptors (Fig. 30.1). We also examined the intracellular calcium level using the intracellular calcium assay method. As a result, it was revealed that the intracellular cal-

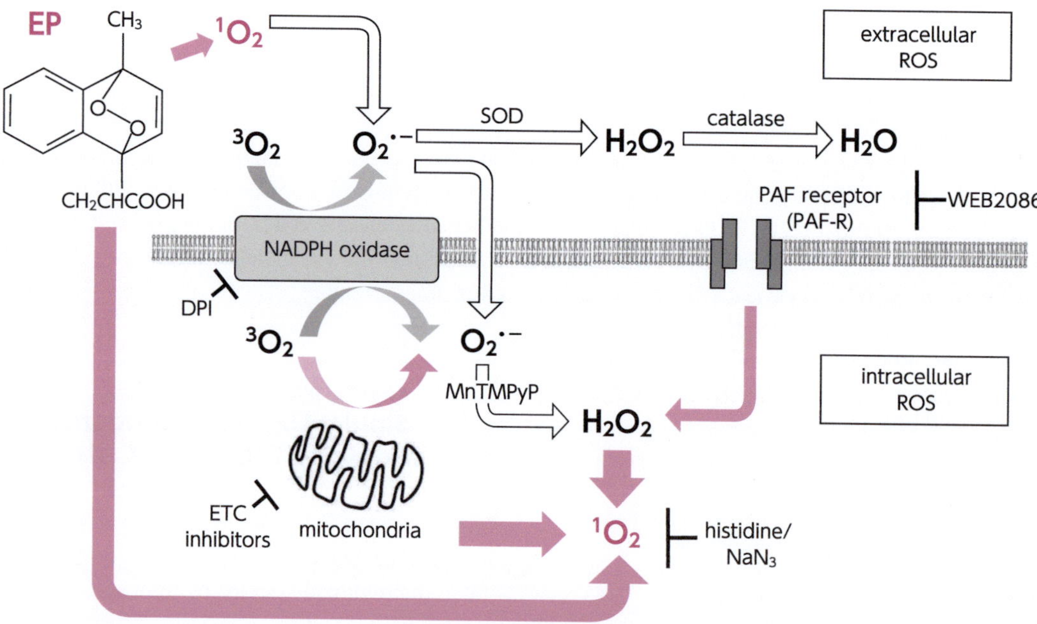

Fig. 30.1 ROS is produced secondarily in the cell by EP stimulation

cium level increases due to EP stimulation, and this increase is inhibited by NADPH oxidase inhibitor (DPI) and WEB2086. These results suggest the possibility that the indirect continuous release of intracellular ROS in response to singlet oxygen produced from EP is involved in cellular damage, and at the same time, these ROS are likely to be produced through the calcium-sensitive pathway via platelet activating factor receptors and mitochondria.

30.2 It Is Important to Control Intracellular ROS to Protect Cells from Exogenous Stimuli

The mechanism of cell damage caused by exogenous singlet oxygen is, as mentioned above, it has been clarified that it leads to the induction of transcription factors downstream and gene expression due to the persistent production of ROS such as hydrogen peroxide and singlet oxygen in the cell. In addition, the mechanism of secondary intracellular ROS production by external stimuli is suggested to be the same not only in this model but also in ultraviolet stimulation and other physical stimuli. That is, because strong damaging ROS (hydrogen peroxide and singlet oxygen) produced in the cell are involved in cell damage induction as second messengers, from the standpoint of antiaging, maintaining the elimination activity of hydrogen peroxide and singlet oxygen in the cell at a high level is expected to lead to the prevention of all diseases related to oxidative stress [4]. All mechanisms of cell damage caused by external stimuli are assumed to be due to ROS produced secondarily in the cell, so by correctly evaluating the active oxygen elimination activity of the body, various methods will be used to increase the antioxidant capacity of the body, which will lead to the realization of antiaging.

References

1. Majima HJ, Indo HP, et al. Chasing great paths of Helmut Sies "Oxidative Stress". Arch Biochem Biophys. 2016;595:54–60.
2. Mukai K. Antioxidant Activity of foods: development of singlet oxygen absorption capacity (SOAC) assay method. J Nutr Sci Vitaminol (Tokyo). 2019;65:285–302.
3. Elshafei ME, Minamiyama Y, et al. Singlet oxygen from endoperoxide initiates an intracellular reactive oxygen species release in HaCaT keratinocytes. J Clin Biochem Nutr. 2022;71:198–205.
4. Sies H. Oxidative eustress: On constant alert for redox homeostasis. Redox Biol. 2021;41:101867.

31 Oxidative Stress and Antiaging Medicine: Antiaging through Antioxidation

Yasuki Higashimura

31.1 Oxidative Stress and Aging

Reactive oxygen species include superoxide, which is oxygen reduced by one electron at a time, hydrogen peroxide, hydroxyl radicals, and singlet oxygen, which is excited oxygen. Also, hypochlorous acid produced from hydrogen peroxide and chlorine ions, peroxynitrite, a reaction product of nitric oxide and superoxide, lipid hydroperoxides and lipid peroxyl radicals produced by lipid oxidation, are recognized as reactive oxygen species in a broad sense. These reactive oxygen species do not necessarily harm the body, and it is known that an appropriate concentration of reactive oxygen species functions as a signal in the body. For example, it is known that reactive oxygen species produced in mitochondria cause epigenetic changes to nuclear DNA, which in turn act to extend lifespan. On the other hand, the body has an antioxidation mechanism to prevent oxidative damage by reactive oxygen species. Oxidative stress is a state where the balance between oxidative power and antioxidative power is disrupted, leaning towards the former, and can be understood as a state where "the production of reactive oxygen species increases, damage to the body accumulates to an irreparable extent, and damage to bodily functions occurs." The theory that supported the relationship between oxidative stress and aging is the "oxidative stress hypothesis of aging," proposed by Dr. Denham Harman in 1956. Unfortunately, the current situation is a mix of studies that support and do not support this theory, and a consensus on oxidative stress and aging in higher animals such as humans, monkeys, and mice has not yet been reached. However, in the case of Caenorhabditis elegans (C. elegans), which is frequently used as a model organism, it has been proven that oxidative stress is involved in aging [1]. Specifically, a mutant that shows a reduction in lifespan depending on the oxygen concentration was isolated, and the causative gene was identified as cyt-1, a subunit of mitochondrial electron transport complex II. This gene is conserved in humans as succinate dehydrogenase complex subunit C (SDHC). Interestingly, mice introduced with a mutant SDHC gene with a similar mutation show symptoms related to aging, such as neurological damage and infertility. In this section, we will discuss antioxidation and anti-aging, including research examples in model organisms such as C. elegans, and the current status and future prospects.

31.2 Antiaging Through Chain-Breaking Antioxidants

In biological components, lipids are the most susceptible to oxidative damage by reactive oxygen species. Hydroxyl radicals, which are one of the

Y. Higashimura (✉)
Department of Food Science, Ishikawa Prefectural University, Nonoichi City, Ishikawa, Japan
e-mail: yasuki@ishikawa-pu.ac.jp

H. Yamada, Y. Naito (eds.), *Anti-Aging Medicine*, https://doi.org/10.1007/978-981-97-8514-8_31

reactive oxygen species with strong oxidizing power, attack unsaturated fatty acids in cell membranes, causing cell membrane damage through chain lipid peroxidation reactions. Therefore, the accumulation of chronic oxidative cell membrane damage is considered to be one of the factors of aging. In order to stop this chain reaction, it is necessary to capture lipid peroxyl radicals. Substances with this ability are called chain-breaking antioxidants, such as vitamin E. The antiaging effects of these chain-breaking antioxidants have been demonstrated in studies using C. elegans and Drosophila melanogaster as model organisms, showing lifespan extension effects due to antioxidant action [2].

31.3 Antiaging by Preventive Antioxidants

Hydrogen peroxide, a typical reactive oxygen species, and lipid hydroperoxides, which are oxidation products of lipids, do not have particularly strong oxidizing power. However, when they are reduced by one electron by iron ions or copper ions, the former gives hydroxyl radicals and the latter gives lipid peroxyl radicals and lipid alkoxy radicals. These radicals, which are the products, exert strong oxidizing power, thereby exacerbating biological damage. Therefore, substances that can prevent radical generation by chelating metal ions or immediately eliminate generated radicals are called preventive antioxidants, and transferrin, metallothionein, and uric acid are included. Antiaging effects have also been demonstrated for these preventive antioxidants. In C. elegans administered with uric acid, a significant extension of lifespan was observed, and recent studies have revealed that the insulin/insulin-like growth factor signaling mechanism and the oxidative stress response mechanism starting from the transcription regulator nuclear factor-erythroid 2-related factor 2 (Nrf-2) are involved in its mechanism of action [3].

31.4 Antiaging by Antioxidant Enzymes

In order to prevent oxidative damage to biological components including lipids, it is necessary to eliminate reactive oxygen species and radicals that trigger it. In addition to the above-mentioned antioxidants, the body has a variety of antioxidant enzymes. For example, superoxide dismutase (SOD), which converts superoxide into oxygen and hydrogen peroxide, and catalase, which decomposes hydrogen peroxide, are included. The involvement of these antioxidant enzymes in aging is also supported in primates. For example, human and chimpanzee genes have 98.77% homology, but there is a two-fold difference in maximum lifespan. As a factor in this, antioxidant enzyme activity has been pointed out, and it is suggested that human SOD activity is twice as high as that of chimpanzees. Also, C. In studies using C. elegans, it has been confirmed that individuals with mutations in the gene coding for glutathione peroxidase-4, which reduces oxidized phospholipids in biological membranes, have a significantly shorter lifespan compared to the wild type [4].

31.5 Future Prospects

At present, there is no scientific evidence that oxidative stress accumulation correlates with lifespan in higher animals, including humans. However, in addition to the aforementioned C. elegans research, termites, known as long-lived insects, have been reported to achieve longevity by producing high amounts of uric acid, a preventive antioxidant, and maintaining high enzyme activity of SOD [5]. On the other hand, in naked mole-rats, which have a longer lifespan than mice or rats, it is known that the level of oxidative stress in cells is high and oxidative damage to biological components also accumulates. In other words, the view that the accumulation of oxidative stress causes aging varies greatly depending on the species, and there are conflicting research examples, so it is difficult to say that lifespan extension by antioxidants has been proven as a hypothesis. However, it is an undeniable fact that controlling the accumulation of oxidative stress, which has biological damage, leads to the prevention of the onset and progression of many diseases, and if the control of age-related diseases is achieved as a result, the author believes that there is a sufficient possibility of achieving healthy longevity starting from antioxidants.

References

1. Ishii N, Fujii M, et al. A mutation in succinate dehydrogenase cytochrome b causes oxidative stress and ageing in nematodes. Nature. 1998;394:694–7.
2. Ernst IMA, Pallauf K, et al. Vitamin E supplementation and lifespan in model organisms. Ageing Res Rev. 2013;12:365–75.
3. Wan QL, Fu X, et al. Uric acid induces stress resistance and extends the life span through activating the stress response factor DAF-16/FOXO and SKN-1/NRF2. Aging. 2020;12:2840–56.
4. Sakamoto T, Maebayashi K, et al. Inhibition of lipid peroxidation during the reproductive period extends the lifespan of Caenorhabditis elegans. J Clin Biochem Nutr. 2020;66:116–23.
5. Tasaki E, Sakurai H, et al. Uric acid, an important antioxidant contributing to survival in termites. PLoS One. 2017;12:e0179426.

Oxidative Stress and Antiaging Medicine: Nrf2 Signaling and Antiaging

32

Ken Itoh and Shuya Kasai

32.1 The Central Transcription Factor Nrf2, Which Is Responsible for Oxidative Stress Response in Higher Animals

Nuclear factor erythroid 2-related factor 2 (Nrf2) binds to the antioxidant response element (ARE), a gene expression control sequence, and uniformly controls the gene expression of the second phase detoxification enzymes, antioxidant enzymes, and the biosynthesis pathways of antioxidants [1]. Its activity is mainly regulated tissue-specifically by Nrf2 protein degradation and synthesis (mRNA and protein). The control of Nrf2 protein degradation involves the Keap1 (Kelch-like ECH-associated protein 1)/Cul3 (Cullin 3) system and the ubiquitin-proteasome control system mediated by GSK3β (glycogen synthase kinase 3 β)-SCF (Skp1-Cullin1-F-box protein) βTRCP (Fig. 32.1). Various electrophilic substances and reactive oxygen species/nitrogen species oxidatively modify the reactive cysteine of Keap1, thereby changing the conformation of Keap1 and inhibiting the degradation activity of Nrf2 by the Keap1/Cul3 system. The activation of Nrf2 by growth factors such as insulin-like growth factor 1 (IGF-1) is thought to be due to the inhibition of Nrf2 degradation by GSK3β-SCFβTRCP. In addition, Nrf2 gene expression is positively regulated by myocyte enhancer factor 2 (MEF2) [2] and positive feedback by Nrf2 itself [3]. The autophagy factor p62 is also a target gene of Nrf2, but p62 has an STGE motif similar to the ETGE motif, which is a Keap1 binding motif [4]. When the serine in the STGE motif is phosphorylated under autophagy induction, the STGE motif strongly competes with the binding of Nrf2 and Keap1, activating Nrf2. Phosphorylation of the STGE motif has been reported to be involved in the basal expression of Nrf2 mediated by TGF-β activated kinase 1 (TAK1) in the small intestinal epithelium and the activation of Nrf2 in white muscle [5, 6].

K. Itoh (✉) · S. Kasai
Department of Stress Response Science, Center for Advanced Medical Science, Hirosaki University Graduate School of Medicine, Aomori, Japan

Department of Stress Response Science, Biomedical Research Center, Hirosaki University Graduate School of Medicine, Aomori, Japan
e-mail: itohk@hirosaki-u.ac.jp

H. Yamada, Y. Naito (eds.), *Anti-Aging Medicine*, https://doi.org/10.1007/978-981-97-8514-8_32

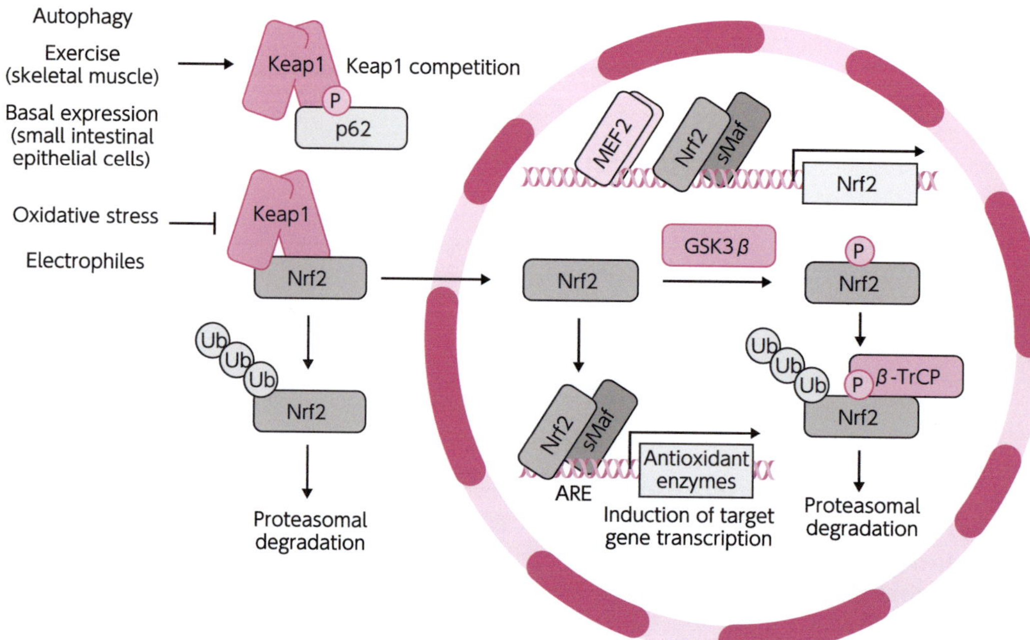

Fig. 32.1 Nrf2 activation control in the body

32.2 The Role of Nrf2 in Aging (Determining Lifespan)

The "Mitochondrial Free Radical Theory of Aging" proposed by Dr. Harman is widely accepted with some modifications. According to this theory, mitochondrial DNA mutations accumulate during the aging process, increasing the amount of reactive oxygen species and leading to cell death. Nrf2 expression and transcriptional response decrease with aging, so Nrf2 may play an important role in individual aging. In fact, in C57BL/6 male mice, the maximum lifespan is shortened due to the loss of Nrf2 [7]. Comparative biological analysis of 10 different rodent species with different lifespans has reported that Nrf2's ARE binding activity positively correlates with lifespan, while protein levels of Keap1 and SCFβTRCP negatively correlate [8]. It has also been reported that in cells of patients with Hutchinson-Gilford Progeria Syndrome, Nrf2 is inactivated due to mutations in nuclear lamin proteins [9]. The nematode Nrf2 homolog SKN-1 controls the maximum lifespan of nematodes [10]. Nrf2 is thought to be involved in controlling the maximum or average lifespan in various animals. Regarding lifespan extension by calorie restriction, there are reports that Nrf2 is not involved in mice [7, 11].

32.3 Network of Aging Control Responses via Nrf2—Centered on SIRT6, PGC1α—

Nrf2 forms a heterodimer with small Maf proteins and binds to ARE, but it also interacts with many other nuclear factors to form a transcription factor network [12]. Transcription co-factors include CBP, Baf chromatin remodeling complex, and mediator complex. It also cooperatively activates antioxidant genes with activating transcription factor 4 (ATF4) and heat shock factor 1 (HSF1). From the perspective of aging control, the interactions with peroxisome proliferator-activated receptor-γ co-activator-1α (PGC1α), known as the master control factor of mitochondrial biogenesis, and sirtuin 6 (SIRT6) are important [13, 14]. SIRT6 is a nuclear sirtuin that has attracted attention because its knockout mice exhibit premature aging, and it is known to interact with Nrf2 to activate transcription. Furthermore, it is known that it

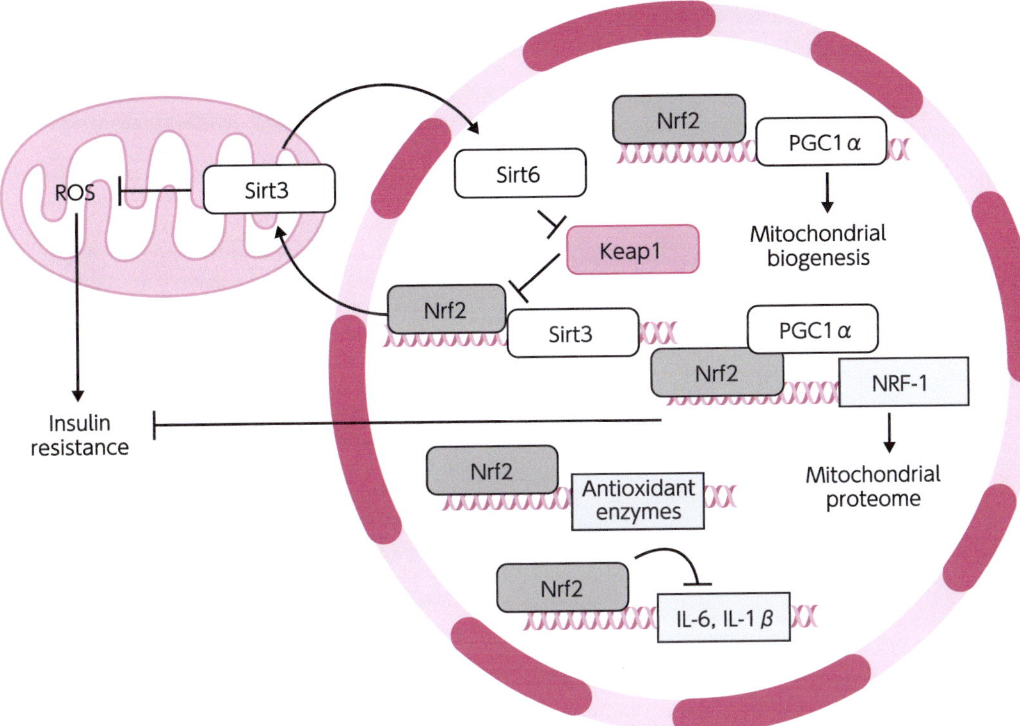

Fig. 32.2 Aging control response pathway by Nrf2 network

is controlled downstream of sirtuin 3 (SIRT3) and suppresses the gene expression of Keap1 (Fig. 32.2). PGC1α is known as a control factor for mitochondrial biogenesis in skeletal muscle, but it also works as an upstream, downstream, and co-activator of Nrf2, and the activation of both is important for healthy lifespan. Nrf2 controls the gene expression of PGC1α and nuclear respiratory factor1 (Nrf-1), contributing to the antioxidant action and mitochondrial activation of skeletal muscle and heart by exercise [15, 16], and it has been shown to contribute to the improvement of insulin resistance by exercise [17]. It has also been reported that NADPH oxidase 4 (NOX-4), which produces hydrogen peroxide, is involved in the activation of Nrf2 by exercise [7, 11].

32.4 The Role of Nrf2 in Cellular Aging (Senescence)

Aging cells secrete various inflammatory proteins, known as senescence-associated secretory Cellular aging promotes the aging phenomenon of an individual by causing a senescence-associated secretory phenotype (SASP). Moreover, cellular aging has been established as one of the good experimental systems for aging. In human fetal fibroblasts, Nrf2 has been shown to be defensive against replicative senescence by controlling the expression of proteasome subunits [18]. Also, in human mesenchymal stem cells, Nrf2 has been shown to be defensive against senescence induced by oxidative stress [19]. Activation of AMP-activated protein kinase (AMPK), which controls energy metabolism, reduces oxidative stress and endoplasmic reticulum stress and inhibits senescence, but a pathway involving Nrf2 activation has been reported [20].

32.5 Nrf2's Defense Mechanism Against Age-Related Diseases

If Nrf2 is resistant to aging, it should have a preventive effect against diseases that occur with aging. Consistent with this hypothesis, various

Nrf2 activators have been reported to exert preventive or therapeutic effects in preclinical trials in model mice of diabetes, neurodegenerative diseases, chronic kidney disease, etc [21]. The decline in the function of major organs plays a significant role in human lifespan control, but Nrf2 is important for maintaining the function of various stem cells [22]. Nrf2 activators are abundantly present as phytochemicals in the secondary metabolites of plants, represented by sulforaphane contained in cruciferous plants [23]. Today, as we face a super-aged society, preventing diseases through antiaging via Nrf2 activators has significant medical and social scientific significance.

References

1. Itoh K, Mimura J, et al. Discovery of the negative regulator of Nrf2, Keap1: a historical overview. Antioxid Redox Signal. 2010;13:1665–78.
2. Tsushima M, Liu J, et al. Emerging evidence for crosstalk between Nrf2 and mitochondria in physiological homeostasis and in heart disease. Arch Pharm Res. 2020;43:286–96.
3. Kwak MK, Itoh K, et al. Enhanced expression of the transcription factor Nrf2 by cancer chemopreventive agents: role of antioxidant response element-like sequences in the nrf2 promoter. Mol Cell Biol. 2002;22:2883–92.
4. Silva-Islas CA, Maldonado PD. canonical and non-canonical mechanisms of Nrf2 activation. Pharmacol Res. 2018;134:92–9.
5. Yamada M, Iwata M, et al. p62/SQSTM1 and Nrf2 are essential for exercise-mediated enhancement of antioxidant protein expression in oxidative muscle. FASEB J. 2019;33:8022–32.
6. Hashimoto K, Simmons AN, et al. TAK1 regulates the Nrf2 antioxidant system through modulating p62/SQSTM1. Antioxid Redox Signal. 2016;25:953–64.
7. Pomatto LCD, Dill T, et al. Deletion of Nrf2 shortens lifespan in C57BL6/J male mice but does not alter the health and survival benefits of caloric restriction. Free Radic Biol Med. 2020;152:650–8.
8. Lewis KN, Wason E, et al. Regulation of Nrf2 signaling and longevity in naturally long-lived rodents. Proc Natl Acad Sci USA. 2015;112:3722–7.
9. Kubben N, Zhang W, et al. Repression of the antioxidant NRF2 pathway in premature aging. Cell. 2016;165:1361–74.
10. Blackwell TK, Steinbaugh MJ, et al. SKN-1/Nrf, stress responses, and aging in Caenorhabditis elegans. Free Radic Biol Med. 2015;88:290–301.
11. Pearson KJ, Lewis KN, et al. Nrf2 mediates cancer protection but not prolongevity induced by caloric restriction. Proc Natl Acad Sci USA. 2008;105:2325–30.
12. Tonelli C, Chio IIC, et al. Transcriptional regulation by Nrf2. Antioxid Redox Signal. 2018;29:1727–45.
13. Kasai S, Shimizu S, et al. Regulation of Nrf2 by mitochondrial reactive oxygen species in physiology and pathology. Biomol Ther. 2020;10:320.
14. Gureev AP, Shaforostova EA, et al. Regulation of mitochondrial biogenesis as a way for active longevity: interaction between the Nrf2 and PGC-1α Signaling Pathways. Front Genet. 2019;10:435.
15. Hancock M, Hafstad AD, et al. Myocardial NADPH oxidase-4 regulates the physiological response to acute exercise. elife. 2018;7:e41044.
16. Merry TL, Ristow M. Nuclear factor erythroid-derived2-like 2 (NFE2L2, Nrf2) mediates exercise-induced mitochondrial biogenesis and the anti-oxidant response in mice. J Physiol. 2016;594:5195–207.
17. Xirouchaki CE, Jia Y, et al. Skeletal muscle NOX4 is required for adaptive responses that prevent insulin resistance. Sci Adv. 2021;7:eabl4988.
18. Kapeta S, Chondrogianni N, et al. Nuclear erythroid factor 2-mediated proteasome activation delays senescence in human fibroblasts. J Biol Chem. 2010;285:8171–84.
19. Shin JH, Jeon HJ, et al. Epigallocatechin-3-gallate prevents oxidative stress-induced cellular senescence in human mesenchymal stem cells via Nrf2. Int J Mol Med. 2016;38:1075–82.
20. Yuan H, Xu Y, et al. Role of Nrf2 in cell senescence regulation. Mol Cell Biochem. 2021;476:247–59.
21. Schmidlin CJ, Dodson MB, et al. Redox regulation by NRF2 in aging and disease. Free Radic Biol Med. 2019;134:702–7.
22. Dai X, Yan X, et al. Nrf2: redox and metabolic regulator of stem cell state and function. Trends Mol Med. 2020;26:185–200.
23. Eggler AL, Savinov SN. Chemical and biological mechanisms of phytochemical activation of Nrf2 and importance in disease prevention. Recent Adv Phytochem. 2013;43:121–55.

33 Immunology and Antiaging Medicine: What Is Immunosenescence?

Mitsuo Maruyama

33.1 Aging Changes in the Immune System

The immune system is a biological function with a complex network of cells and tissues essential for maintaining the health of our bodies. Characteristic changes with aging include ① increased susceptibility and severity due to decreased body defense ability against pathogens and response to vaccination, ② enhanced autoimmune reactions causing tissue damage, ③ accumulation of damage-associated molecular patterns (DAMPs) in the body, and decreased immune surveillance ability to monitor and remove aging cells and cancer cells, leading to disease onset and chronic inflammation. The decline in function with aging is more susceptible in the acquired immune system, which is centered on antigen-specific immune memory of T cells and B cells, compared to the innate immune system, which primarily eliminates pathogens through phagocytosis and associated inflammatory reactions. Immunosenescence is a phenomenon where multifaceted declines in immune function are observed with aging, and understanding and elucidating the mechanisms of its control are attracting attention due to its deep connection with extending healthy lifespan in an aging society. In fact, about 95% of annual deaths from seasonal influenza are elderly people aged 65 and over, and the severity and mortality rate of elderly people who have contracted infectious diseases caused by foreign pathogens, such as the recent COVID-19, are extremely high, which is a serious situation considering the decline in immune function against pathogens with aging. On the other hand, the function to distinguish self from non-self also decreases with aging, not only making the immune response fragile because foreign pathogens cannot be recognized as non-self, but also causing autoimmune diseases by attacking self-antigens with their own immune system, which is common in the elderly. Here, instead of summarizing the multifaceted aging changes that cause immunosenescence at the individual immune cell level, which have been clarified so far, we would like to discuss the contribution and role of cellular aging in immunosenescence at the immune tissue or individual level from the perspective of cellular aging and tissue aging research, which has been making remarkable progress in recent years.

33.1.1 Aging Research and Immunosenescence

When cells encounter stress or irreparable damage, they fall into an irreversible state of proliferative arrest accompanied by cell cycle arrest, a

M. Maruyama (✉)
Geroscience Research Center, National Center for Geriatrics and Gerontology, Aichi, Japan
e-mail: michan@ncgg.go.jp

H. Yamada, Y. Naito (eds.), *Anti-Aging Medicine*, https://doi.org/10.1007/978-981-97-8514-8_33

phenomenon of cellular aging that became clear in the mid-twentieth century. At that time, immunological research revealed the mechanism controlling the acquired immune system through the existence of T and B lymphocytes as two types of lymphocytes with different functions, and aging research and immunological research entered a new stage almost simultaneously. Regarding the correlation between the immune system and aging cells, if it is the immune system of a young individual, it is possible to quickly remove cancer cells and aging cells after recognition, but in aging individuals, this metabolic turnover is slow, leading to chronic accumulation. In fact, it has been reported that when this immune surveillance mechanism decreases, the accumulation of aging cells is promoted [1]. Analysis is also underway on the mechanism by which aging cells evade immune surveillance and accumulate, but there are still many unknowns. However, since it has been reported that aging cells cause disturbances in the production and function of stem cells, it cannot be denied that the exhaustion and dysfunction of hematopoietic stem cells (HSCs) in the bone marrow may adversely affect the differentiation and proliferation of immune cells, and may be one of the fundamental causes of immune aging due to aging. In any case, based on the knowledge gained so far, we will organize the fundamental aspects of aging that promote immune aging in each immune tissue, considering how aging cells (cellular aging) that show various functional declines affect immune aging.

33.2 Immune Aging and Aging Cells in Immune Tissues

33.2.1 Bone Marrow

Most immune cells are produced from HSCs that exist in the bone marrow and self-replicate. With aging, this bone marrow is also affected by aging changes, the number of HSCs decreases, and the decline in functions such as self-replication ability and clone formation ability is remarkable. This age-related bias in the differentiation of HSCs to the bone marrow system can also be confirmed in a system where HSCs derived from young mice are transplanted into aged individuals, suggesting that it is partly due to the bone marrow environment. As a factor related to this bias towards the bone marrow system, chemokine ligand 5 (C-C motif chemokine ligand 5; CCL5) is suggested because this bias is suppressed by knocking it out [2]. In addition, the accumulation of bone marrow adipocytes is partly due to the increased expression of receptor activator of nuclear factor kappa-B ligand (RANKL). In aged bone marrow, increases in p16INK4a positive cells, SASP (senescence-associated secretory phenotype) factors (CCL5 and RANKL), cells with DNA damage, and reactive oxygen species (ROS) are observed, suggesting a correlation between tissue aging of the bone marrow and accumulation of aging cells. Recent research suggests that, in humans, bone marrow-derived mesenchymal stem aging of bone marrow mesenchymal stem cells (BM-MSC) has been reported to be involved in a series of age-related hematopoietic stem cell (HSC) functional disorders [3].

33.2.2 Thymus

The thymus, as a tissue where T cells mature and differentiate, undergoes changes with age, where functional thymic cells transform into fibroblasts, adipocytes, and senescent cells, leading to atrophy. Thymic atrophy is associated with a decrease in turnover to new T cells, a reduction in the repertoire of T cell receptors, and the production of highly self-reactive T cells, which are characteristics of immunosenescence deeply related to the decline in T cell function. In adult thymic epithelial cells (TEC), senescence-associated β-galactosidase (SA-βGal) positive cells exist, and the high staining of oxidative DNA damage markers such as γH2AX in human thymic tissue and the thymus of aged mice suggests a possible correlation between thymic atrophy, DNA damage, and cellular aging. There are also interesting reports that increased oxidative stress and DNA damage in stromal cells and TEC promote thymic aging [4].

33.2.3 Spleen

The spleen, a secondary lymphoid organ that functions as the center of acquired immune activation, undergoes significant changes in the types and localization of immune cells with age. In addition to the decline in function of stromal cells and marginal zone macrophages, the distinction between T cell and B cell areas in the white pulp is not clear in aged individuals. The accumulation of senescent cells is confirmed by the presence of p16INK4a positive cells, the expression of SASP factors, and increased DNA damage. The increased expression of the SASP factor interleukin-6 (IL-6) in stromal cells that separate the follicular zone and the marginal zone suggests a decline in acquired immune functions such as appropriate T cell responses to pathogen invasion, and appropriate antigen presentation by B cells, macrophages, and dendritic cells, accompanied by the accumulation of age-dependent inflammation. On the other hand, there are reports that the phagocytic ability of aged marginal zone macrophages is inefficient in vivo, but is comparable to that of young mice in vitro [5], suggesting the existence of SASP-independent involvement in immunosenescence accompanied by structural changes in aged spleen tissue.

33.2.4 Lymph Nodes

Lymph nodes, which have a filtration function for various immune cells flowing into the blood vessels from surrounding tissues, also show significant age-related changes in number and function, such as changes in permeability, accumulation of senescent cells, and increased inflammation. In humans, an increase in the fat ratio and fibrosis of lymph nodes with age has also been reported. It is known that the replicative ability of stromal cells in aged lymph nodes decreases in response to stimulation. At the same time, these cells produce chemokine ligand 2 (C-C motif chemokine 2; The inhibitory function of CCL2 in germinal center formation is attracting attention as a decrease in immune response via lymph nodes. Despite being an important function to prevent unnecessary germinal center formation in the absence of antigens, CCL2 is a SASP factor, suggesting that aged stromal cells that constantly produce CCL2 may be involved in the immunosenescence of lymph nodes supporting the germinal center.

33.2.5 Mucosal Lymphoid Tissue

An immune system localized on the surface of mucosal tissues, classified into various categories such as nasopharynx, trachea, and conjunctiva, but most research has been done on gut-associated lymphoid tissue (GALT), where there is a lot of knowledge about the accumulation of aging cells and the decline of immune function with age. In GALT, the repertoire of naive T and B cells decreases and is replaced by memory cells. The number of dendritic cells also decreases, and their responsiveness decreases, suppressing the activation of the acquired immune system like the spleen. Considering that the surface of the mucosal tissue is exposed to many environmental stress factors compared to other tissues, it can be inferred that the chronic activation of the immune system at a low level and the accumulation of aging cells become apparent from an earlier stage, and the phenotype of immunosenescence may start relatively early compared to the immune system of other related tissues. The aging changes that cause immunosenescence at the immune tissue level are multifaceted, and individual changes are diverse. Therefore, we hope that understanding the link between the basic mechanism of cellular aging and immunosenescence will progress by capturing immunosenescence as a whole body function beyond solving aging seen in tissues and locally on a disease basis, and by using senolytics aimed at eliminating aging cells and senostatics targeting SASP, and that the possibility of improving function in immunosenescence by knowing aging cells will deepen.

References

1. Ovadya Y, Landsberger T, et al. Impaired immune surveillance accelerates accumulation of senescent cells and aging. Nat Commun. 2018;9:5435.

2. Ergen AV, Boles NC, et al. Rantes/Ccl5 influences hematopoietic stem cell subtypes and causes myeloid skewing. Blood. 2012;119:2500–9.
3. Gnani D, Crippa S, et al. An early-senescence state in aged mesenchymal stromal cells contributes to hematopoietic stem and progenitor cell clonogenic impairment through the activation of a pro-inflammatory program. Aging Cell. 2019;18:e12933.
4. Griffith AV, Venables T, et al. Metabolic Damage and Premature Thymus Aging Caused by Stromal Catalase Deficiency. Cell Rep. 2015;12:1071–9.
5. Birjandi SZ, Ippolito JA, et al. Alterations in marginal zone macrophages and marginal zone B cells in old mice. J Immunol. 2011;186:3441–51.

Immunology and Antiaging Medicine: Aging Immune Cells and Antiaging

34

Shinya Yamamoto and Motoko Yanagita

34.1 Mechanism of Immune Aging and Aging Immune Cells

The immune system recognizes self-tissues and foreign factors such as microbes, and is a biological defense mechanism that eliminates non-self. The immune system consists of innate immunity and acquired immunity, the former non-specifically targets foreign factors, while the latter recognizes and efficiently eliminates specific targets through immune memory by T cells and B cells antigen-specific receptors. Changes in immune function with aging are called immune aging, and both innate and acquired immunity change, but the decline in the latter's function is particularly noticeable. Under the conditions of immune aging, autoimmune risk and inflammatory factors increase. Clinically, issues such as increased susceptibility in the elderly, onset of autoimmune diseases, carcinogenesis, and decreased vaccine efficacy become problems. Although details will be given in the next section, systemic noninfectious chronic inflammation associated with individual aging is called inflammaging and is attracting attention as one cause of aging diseases such as hypertension, diabetes, and sarcopenia. In recent years, we will outline the cell populations and important factors involved in the onset of immune aging that are particularly noteworthy. With aging, it is known that especially acquired immunity decreases and T cells are more affected by aging than B cells. The thymus, which is the site of T cell differentiation, gradually atrophies with adolescence as the peak in humans, and is replaced by adipose tissue in old age. Therefore, in old age, the supply of naive T cells to the periphery decreases, while homeostatic proliferation of T cells suppresses the decrease. This reaction is regulated by factors such as interleukin-7 (IL-7) and interleukin-15 (IL-15). However, excessive homeostatic proliferation induces T cell aging and causes abnormalities in the immune system and immune aging [1]. The increase in inflammatory traits associated with T cell aging is called senescence-associated secretory phenotype (SASP), characterized by a significant increase in the production of inflammatory cytokines and angiogenesis inducers. Excessive SASP causes chronic inflammation, resulting in irreversible remodeling of tissue structure and decreased biological function. The involvement of regulatory T cells (Treg) has also been reported in the prolongation of inflammation with aging. In addition to becoming apoptosis-resistant with age, the supply of Tregs from the thymus decreases. This decrease in Treg diversity leads to a decrease in the suppression of abnormal activation of effector T

S. Yamamoto (✉) · M. Yanagita
Department of Nephrology, Kyoto University, Kyoto, Japan
e-mail: kidney2011@kuhp.kyoto-u.ac.jp

H. Yamada, Y. Naito (eds.), *Anti-Aging Medicine*, https://doi.org/10.1007/978-981-97-8514-8_34

cells, which is one of the causes of chronic inflammation. In recent research on immune aging, senescence-associated T cells (senescence-associate T cells) are being studied. Tissues; SA-T) have been identified and are attracting attention. SA-T cells, which are memory-type helper T cells, express the inhibitory receptor PD-1 (programmed cell death 1) and CD30L (CD153), which belongs to the tumor necrosis factor (TNF) superfamily. SA-T cells have distinct characteristics from regular helper T cells, such as high expression of bone marrow cell genes and halted proliferation. Furthermore, while they hardly produce typical cytokines in response to stimulation of the T cell receptor, they secrete large amounts of inflammatory cytokines and chemokines such as IFN-γ (interferon-γ), osteopontin, CCL3, CCL4, contributing to the increase of inflammatory factors. Studies in mice have reported that SA-T cells exist not only in lymphoid organs but also in various peripheral tissues such as the kidneys in systemic lupus erythematosus, and that they increase in visceral adipose tissue due to a high-fat diet, inducing insulin resistance through the activation of osteopontin [2]. So far, we have explained the mechanism of immunosenescence focusing on T cells, but recent reports suggest that changes in B cells with aging may also contribute to immunosenescence. In the elderly, B cell subsets are changing, and an increase in a group of cells called age-associated B cells (ABCs) has been reported [3]. These cells first accumulate in the spleen, then in the bone marrow, and increase significantly with aging. In addition to aging, they increase with various viral infections and vaccinations, and are suggested to be involved in the onset and progression of autoimmune diseases. ABCs are characterized by the expression of CD11b/CD11c and T-bet, unlike other B cells. Studies on the cell cycle in mice have shown that ABCs are in the resting phase and are not a subset of self-replicating cells. Their functional characteristics include strong antigen-presenting ability and the ability to produce inflammatory cytokines and chemokines compared to regular B cells, and their differentiation induction involves TLR7 or TLR9 signals, interleukin-21 (IL-21), and IFN-γ stimulation. Also, in a systemic lupus erythematosus (SLE) mouse model, it has been clarified that interferon-regulatory factor 5 promotes the formation of ABCs through IL-21 stimulation. In the future, it is expected that ABCs will become a new therapeutic target for autoimmune diseases. From these findings, the essence of immunosenescence is largely due to the collective and qualitative changes in immune cells such as T cells and B cells, and these changes are strongly associated with the onset and pathogenesis of age-related diseases. Next, we will introduce the possibility of developing therapies from the perspective of overcoming immunosenescence, including our research findings.

34.2 Potential New Therapeutic Targets in Immunosenescence

We have found that in aged kidneys, tertiary lymphoid cells; FDC) appear, and the characteristics and roles of TLT have been focused on. TLT is closely related to immunosenescence and is a lymphoid tissue that occurs in non-lymphoid tissues induced by chronic inflammatory stimuli. TLT is structurally and qualitatively completely different from mere infiltration of inflammatory cells. It is mainly composed of CD4 positive T cells and B cells, and the special fibroblasts inside show strong positivity for p75neurotrophin receptor, a neural crest marker. These fibroblasts produce homeostatic chemokines such as CCL19 and CXCL13, causing lymphocytes to migrate and cluster, maintaining the formation of TLT. When TLT matures, follicular dendritic Follicular dendritic cells (FDCs) are induced, forming B cell regions on the FDC scaffold, and germinal center reactions are observed in some B cell regions. In other words, tertiary lymphoid tissues (TLTs) can perform adaptive immune responses similar to lymph nodes, playing roles in antigen presentation, production of inflammatory cytokines, and lymphocyte proliferation and activation. The impact of TLTs in the kidney on the host organ varies depending on the clinical situation. In pathological conditions such as urinary tract infections and malignant tumors, they may play a beneficial role for the host through immune responses against pathogens and cancer cells. On the other hand, in conditions related to chronic inflammation and autoimmune abnormalities such as aging kidney, transplanted kidney, lupus nephritis, and IgA nephropathy, they have been reported to contribute to the progression of renal damage. In our study of renal TLTs in aged mice, the formation of TLTs was suppressed by anti-CD4 antibodies, and improvements in inflammation and renal tissue damage and fibrosis were observed, suggesting that CD4 positive T cells are important for the formation and maintenance of TLTs. Furthermore, after renal damage in aged mice, SA-T cells and ABCs induced with aging appear and accumulate specifically within the TLT [4]. Comprehensive gene analysis by single-cell analysis of immune cells in the kidney that induced TLT revealed that SA-T cells highly express cytokines that induce the differentiation of ABCs such as IFN-γ and IL-21, and that ABCs specifically express CD30, the receptor for CD153. The increase in SA-T cells and ABCs in correlation with the expansion of TLT suggests that their interaction plays an important role in the formation and expansion of TLT. In addition, receptor-ligand analysis between the two cell types identified the CD153-CD30 pathway as a novel signaling molecule connecting SA-T cells and ABCs, and in the renal damage model of aged CD153 knockout mice, not only the reduction of TLT and ABCs, but also the reduction of renal fibrosis and renal function impairment were observed. Similarly, in the renal damage model of aged CD30 knockout mice, the reduction of TLT and the decrease in the ability of SA-T cells to produce cytokines involved in the induction of ABC differentiation such as IFN-γ and IL-21 were observed. These results suggest that the CD153-CD30 pathway is important for the formation of renal TLT. In humans, it is known that CD153 positive cells are observed inside the renal TLT in the elderly, and it is considered to be a promising therapeutic target for kidney disease in the elderly. As detailed in the Sect. 26.1 the concept of anti-aging treatment by removing aging cells (senolysis) has been proposed as another approach, and the development of aging cell removal therapy is being attempted. Yoshida et al. developed a peptide vaccine targeting CD153 with the aim of removing SA-T cells, and when administered to mice fed a high-fat diet, it was found to suppress the increase of SA-T cells in visceral fat and improve glucose tolerance abnormalities [5]. Such new approaches to immune aging may have potential effects on various aging-related diseases, and the development of future research is expected.

References

1. King C, Ilic A, et al. Homeostatic expansion of T cells during immune insufficiency generates autoimmunity. Cell. 2004;117:265–77.
2. Shirakawa K, Yan X, et al. Obesity accelerates T cell senescence in murine visceral adipose tissue. J Clin Invest. 2016;126:4626–39.
3. Ma S, Wang C, et al. B cell dysfunction associated with aging and autoimmune diseases. Front Immunol. 2019;10:318.
4. Sato Y, Oguchi A, et al. CD153/CD30 signaling promotes age-dependent tertiary lymphoid tissue expansion and kidney injury. J Clin Invest. 2022;132:e146071.
5. Yoshida S, Nakagami H, et al. The CD153 vaccine is a senotherapeutic option for preventing the accumulation of senescent T cells in mice. Nat Commun. 2020;11:2482.

35 Immunology and Antiaging Medicine: Inflammaging due to Chronic Inflammation

Masafumi Takahashi

35.1 Chronic Inflammation

Inflammation is a biological response in tissues that occurs when harmful stimuli such as infection by pathogens, foreign bodies, or tissue damage are applied. The classical four signs of inflammation, as described in Latin, are rubor (redness), calor (heat), tumor (swelling), and dolor (pain). These signs represent reactions common to acute inflammation, consisting of vasodilation and increased permeability, migration and infiltration of inflammatory cells into tissues due to various inflammatory mediators, neutralization of inflammatory responses by immune reactions, and healing processes through tissue regeneration and repair. On the other hand, when the causative factors of inflammation persist and cannot be eliminated, or when the resolution mechanism of inflammation does not function properly, inflammation becomes chronic. Chronic inflammation causes functional impairment at the site of inflammation, which is considered the fifth sign of inflammation. In chronic inflammation, infiltration of inflammatory cells centered on macrophages, tissue destruction, angiogenesis, and tissue remodeling (structural changes) due to fibrosis occur, eventually leading to irreversible organ dysfunction. However, many aspects of the molecular mechanism are still unclear. The World Health Organization (WHO) defines chronic diseases such as cardiovascular diseases, cancer, metabolic diseases, respiratory diseases, and neuropsychiatric diseases as non-communicable diseases (NCDs), and positions them as important issues in medicine and public health. Chronic inflammation is a common underlying pathology in these NCDs.

35.2 Inflammaging

In the elderly, despite the absence of clear causes such as infection, the blood levels of the inflammation marker C-reactive protein (CRP) and inflammatory cytokines such as interleukin-6 (IL-6) and tumor necrosis factor-α (TNF-α) are increased, indicating the presence of chronic low-level sterile inflammation with aging [1]. Moreover, it has been shown that this chronic inflammation is associated with the onset of diseases that increase with aging, such as atherosclerotic diseases, metabolic diseases, neurodegenerative diseases, and cancer, which are included in NCDs. In 2000, the concept of inflammaging, a combination of inflammation and aging, was proposed by Franceschi et al. [2]. Furthermore, chronic inflammation due to aging leads to inflammaging (Fig. 35.1). The existence of inflammaging has been shown to promote

M. Takahashi (✉)
Division of Inflammation Research, Center for Molecular Medicine, Jichi Medical University, Tochigi, Japan
e-mail: masafumi2@jichi.ac.jp

H. Yamada, Y. Naito (eds.), *Anti-Aging Medicine*, https://doi.org/10.1007/978-981-97-8514-8_35

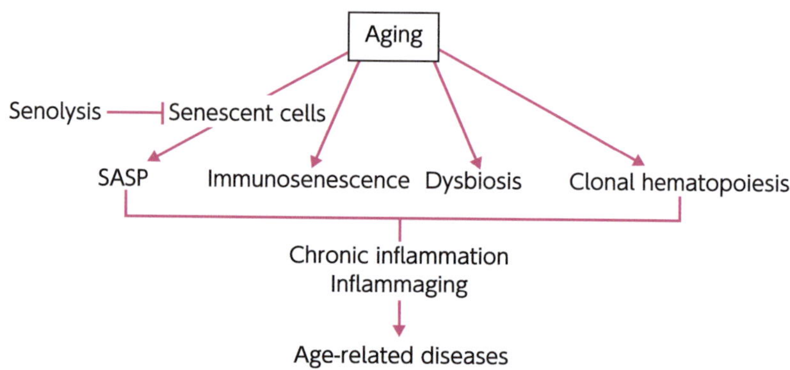

Fig. 35.1 Chronic inflammation and inflammaging SASP due to aging

aging itself, resulting in a vicious cycle of inflammation and aging. Several mechanisms are involved in the production of inflammatory factors that cause inflammaging [3]. One characteristic of inflammaging is the increased production of inflammatory cytokines/chemokines, growth factors, and proteases such as extracellular matrix-degrading enzymes in the blood. This phenomenon is called the senescence-associated secretory phenotype (SASP), and these SASP factors are mainly produced by aged cells. As DNA damage occurs with aging, apoptosis is induced via p53, while activation of the p21 or p16 pathway causes cell cycle arrest. This is a beneficial response to prevent cancer, but it results in the accumulation of senescent cells in the body with age. Furthermore, in these senescent cells, fragmentation of nuclear DNA, decreased expression of lamin B1, and decreased DNase2/3 occur, and nuclear DNA fragments leak into the cytoplasm, activating the cGAS/STING pathway, which induces the expression of inflammatory cytokines. Also, SASP factors can be induced by epigenetic changes such as DNA methylation and histone modification. Recently, from the perspective that SASP factors produced from senescent cells cause chronic inflammation and promote the onset of aging-related diseases and aging itself, the development and clinical trials of a new treatment method called senolysis (senotherapy), which selectively induces cell death in senescent cells and removes them, are underway. In fact, in animal experiments, it has been reported that removing senescent cells with drugs or vaccines that induce senolysis improves aging phenotypes, suppresses the onset of aging-related diseases, and even extends lifespan. Changes in the immune system, called immunosenescence, also contribute to inflammaging. In the elderly, thymic atrophy reduces acquired immune responses mediated by T cells, while shifts to the hematopoietic bone marrow system and decreased phagocytic ability of macrophages occur. Some senescent cells that have suffered DNA damage undergo apoptosis and are phagocytosed and processed (efferocytosis) by macrophages, but when this ability decreases with age, the debris of dead cells acts as a danger signal, activating innate immune pathways such as Toll-like receptors (TLRs) and NLRP3 inflammasomes, leading to inflammation. The oral and intestinal microbiota play an important role in maintaining homeostasis of the body, but age-related decline in barrier function leads to loss of diversity of the microbiota, changes in bacterial composition, i.e., dysbiosis, leading to inflammation. Furthermore, in elderly people without blood diseases, genetic mutations in hematopoietic stem cells lead to an increase in the population (clone) of their blood cells, and a state in which they occupy a certain proportion in peripheral blood is called clonal hematopoiesis (CH, also known as CH with indeterminate. This is referred to as CHIP (clonal hematopoiesis of indeterminate potential), and it has been reported that when there are mutations such as TET2 (ten–eleven translocation-2) and JAK2 (janus kinase 2) V617F, which are relatively frequently observed, inflammation is induced in macrophages through the inflammasome, promoting arteriosclerosis. This is considered one of the mechanisms of inflammaging (Masafumi Takahashi).

References

1. Shaw AC, Goldstein DR, et al. Age-dependent dysregulation of innate immunity. Nat Rev Immunol. 2013;13:875–87.
2. Franceschi C, Bonafè M, et al. Inflamm-aging. An evolutionary perspective on immunosenescence. Ann N Y Acad Sci. 2000;908:244–54.
3. Fulop T, Larbi A, et al. Immunology of aging: the birth of inflammaging. Clin Rev Allergy Immunol. 2021;18 https://doi.org/10.1007/s12016-021-08899-6.

Brain and Antiaging Medicine: Age-Related Changes in Cognitive Function

36

Naoyuki Sato

Cognitive function is defined as "the ability to know the higher properties of events based on reasoning and thinking, not on sensibility" (Kojien seventh Edition) and is divided into memory, orientation (time, place), executive function (social activities, housework), and activities of daily living (personal affairs). Age-related changes in cognitive function can be broadly divided into physiological and pathological (dementia). Dementia includes Alzheimer's disease (AD), Lewy body dementia, frontotemporal lobar degeneration, vascular dementia, neurofibrillary tangle dementia, argyrophilic grain dementia, and limbic-predominant age-related TDP-43 encephalopathy (LATE). Among these, AD is one of the most common types of dementia. In studies that accurately counted the number of neurons using AD autopsy brains, there is not much decrease in the number of neurons due to physiological aging in the olfactory cortex and hippocampus in the medial temporal lobe, but a significant decrease in neurons is observed in AD [1]. In fact, in brain imaging tests such as MRI, hippocampal atrophy due to aging is about 1% per year, but it rises to 5% in typical AD. An epidemiological survey targeting a wide age range of Japanese people reported that the score of the Mini-Mental State Examination (MMSE), a cognitive function test, begins to decline around the age of 55–64 [2]. Although rigorous studies using human autopsy brains are difficult, studies using primates have reported a decrease in the number of dendritic spine of neurons and changes in synapse morphology due to aging, suggesting that similar morphological and functional changes in synapses may occur in humans due to aging. Physiological cognitive decline includes a known decrease in logical memory (immediate and delayed reproduction of long sentences) due to normal aging. In AD, senile plaques, neurofibrillary changes, and neuronal death are the three major pathologies. Senile plaques are accumulations of beta-amyloid outside the cell. Neurofibrillary changes are accumulations of excessively phosphorylated tau inside neurons. Neuronal death in AD is observed mainly in the temporal and parietal lobes and is associated with neurofibrillary changes and dystrophic "Neuritic plaques" associated with neurites, that is, "accumulation of phosphorylated tau" correlates with diffuse senile plaques and "total senile plaques," that is, "accumulation of β-amyloid" does not correlate [1]. This suggests that neuronal cell death in Alzheimer's disease (AD) correlates with tau and not directly with β-amyloid. APOE genotype is a strong risk gene for AD, and β-amyloid accumulation is influenced by the APOE genotype. On the other hand, tau accumulation correlates with aging. Recently, it has been

N. Sato (✉)
Department of Aging Neurobiology, National Center for Geriatrics and Gerontology, Aichi, Japan
e-mail: nsato@ncgg.go.jp

H. Yamada, Y. Naito (eds.), *Anti-Aging Medicine*, https://doi.org/10.1007/978-981-97-8514-8_36

reported that the APOE genotype also affects neuronal cell death due to tau accumulation. In addition to AD, there are dementia with Lewy bodies (accumulation of α-synuclein), frontotemporal dementia (accumulation of tau and TDP-43), vascular dementia, and neurofibrillary tangle dementia and argyrophilic grain dementia, which are seen in older age, and in the latter two, it is known that tau accumulates [primary age-related tauopathy (PART)]. Furthermore, limbic-predominant TDP-43 encephalopathy (LATE), in which TDP-43 accumulates in the cerebral limbic system, is attracting attention in relation to hippocampal sclerosis (neuronal cell death in the hippocampus with hippocampal atrophy), and the pathological significance of TDP-43, which was originally found in frontotemporal dementia and amyotrophic lateral sclerosis, is expected to be elucidated. The combination of these proteins is also important in considering cognitive decline due to aging. In fact, the more you age, the more the combination of β-amyloid, tau, α-synuclein, and TDP-43 pathology increases. Even at the stage of mild cognitive impairment, which is a precursor to dementia, only β-amyloid + tau (2 types) is 40%, and β-amyloid + tau + α-synuclein (3 types) is 20%. Cognitive function is "the ability to know the higher properties of events based on reasoning and thinking, not on sensibility" (Kojien seventh edition) and is divided into memory, orientation (time, place), executive function (social activities, housework), and activities of daily living (personal matters). Changes in cognitive function due to aging are broadly divided into physiological and pathological (dementia). Dementia includes Alzheimer's disease (AD), dementia with Lewy bodies, frontotemporal dementia, vascular dementia, neurofibrillary tangle dementia, argyrophilic grain dementia, and limbic-predominant age-related TDP-43 (TAR DNA-binding protein-43) encephalopathy (Limbic-predominant age-related TDP-43). Among dementia, one of the most common types is Alzheimer's disease (AD). In studies that meticulously counted the number of neurons using AD autopsy brains, it was found that there is not much decrease in the number of neurons due to physiological aging in the olfactory cortex and hippocampus within the temporal lobe, but a significant decrease in neurons is observed in AD. In fact, in brain imaging tests such as MRI, hippocampal atrophy due to aging is about 1% per year, but it rises to 5% in typical AD. In an epidemiological survey targeting a wide age range of Japanese people, it has been reported that the score of the Mini-Mental State Examination (MMSE), a cognitive function test, begins to decline around the age of 55–64. Although rigorous studies using human autopsy brains are difficult, studies using primates have reported a decrease in the number of dendritic spines of neurons and changes in synapse morphology due to aging, suggesting that similar morphological and functional changes in synapses may occur in humans due to aging. As a physiological cognitive decline, it is known that logical memory (logical memory: immediate and delayed reproduction of long sentences) decreases due to normal aging. In AD, senile plaques, neurofibrillary changes, and neuronal death are the three major pathologies. Senile plaques are accumulations of beta-amyloid outside the cell. Neurofibrillary changes are accumulations of excessively phosphorylated tau inside neurons. Neuronal death in AD is observed mainly in the temporal and parietal lobes and correlates with neurofibrillary changes and "senile plaques with dystrophic neurites (neuritic plaques)", that is, "accumulation of phosphorylated tau", but not with diffuse senile plaques or "total senile plaques", that is, "accumulation of beta-amyloid". This suggests that neuronal death in AD correlates with tau and not so much directly with beta-amyloid. APOE genotype is a strong risk gene for AD, and beta-amyloid accumulation is affected by the APOE genotype. On the other hand, tau accumulation correlates with aging. In recent years, there have been reports that the APOE genotype also affects neuronal death due to tau accumulation. In addition to AD, as mentioned above, there are Lewy body dementia (accumulation of alpha-synuclein), frontotemporal lobar dementia (accumulation of tau and TDP-43), vascular dementia, and dementia observed in older age such as neurofibrillary tangle dementia and argyrophilic grain dementia, in the latter two of which it is known that tau accumulates, such as primary age-related tauopathy (primary Age-related tauopathy (PART)) is also drawing attention in relation to hippocampal sclerosis (characterized by neuronal cell death accompanied by hippocampal atrophy) where TDP-43 accumulates in the cerebral limbic system, known as Limbic-predominant Age-related TDP-43 Encephalopathy (LATE). In addition to the originally discovered frontotemporal dementia and amyotrophic lateral sclerosis, the pathological significance of TDP-43 is expected to be clarified. The combined pathology of these proteins is also important in considering age-related cognitive decline. In fact, the older one gets, the more the pathology of β-amyloid, tau, α-synuclein, and TDP-43 increases. Even at the stage of mild cognitive impairment, which is a precursor to dementia, 40% have only β-amyloid + tau (2 types), 20% have β-amyloid + tau + α-synuclein (3 types), and 10% have β-amyloid + tau + TDP-43 (3 types). In dementia, it has been reported that 20% have all four types: β-amyloid + tau + α-synuclein + TDP-43 [3]. In addition to the accumulation of these proteins, the increase in vascular lesions with age has a significant impact on cognitive function. Diabetes, hypertension, and dyslipidemia, known as vascular risk factors, each affects cognitive function. Interestingly, it has been reported that the impact of diabetes on cognitive decline depends on the APOE genotype, i.e., it is observed in APOE2 and APOE3 types, but not in APOE4 type [4]. The elucidation of this mechanism is awaited. Also, although the causal relationship is unclear, it has been reported that there is an inverse correlation between the body mass index (BMI) and brain amyloid. There is still much we do not know about the decline in cognitive function due to aging, but from what has been mentioned above, it is thought that the following equation can be established for "changes in cognitive function due to aging". That is, "changes in cognitive function due to aging" = "physiological aging changes + pathological changes in the brain + vascular damage + effects from the whole body (peripheral) tissues (including vascular risk factors) (including interactions with genetic factors)". Recently, it has been shown that cognitive functions such as execution function and processing speed improve by multi-domain (multifactor) interventions such as exercise, diet, cognitive training, and vascular risk management for people at risk of dementia (FINGER study) [5]. This research is spreading worldwide. Disease-modifying drugs such as antibody therapy against β-amyloid are still being developed for AD, and recently, the success of the phase III trial of Lecanemab was reported. It is expected that disease-modifying drugs targeting abnormally accumulating proteins will continue to be developed not only for AD, but also for dementia as a whole. Since the development of disease-modifying drugs is expected to take time, a combination of "non-drug therapies such as multi-domain interventions" and "development of disease-modifying drugs targeting key molecules of etiology

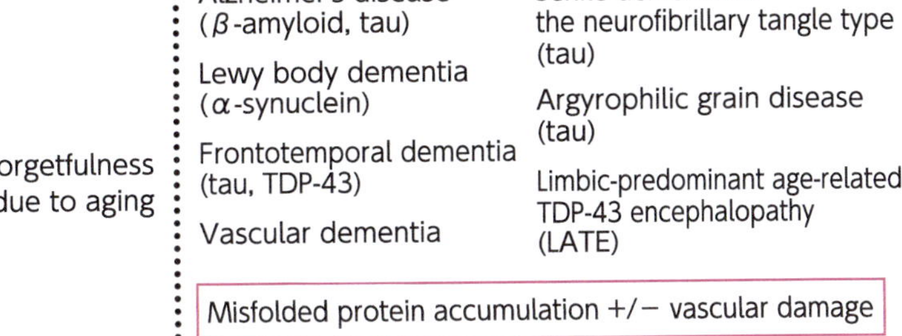

Fig. 36.1 Response to dementia combining "multi-domain intervention" and "accumulated protein-targeting drugs"

including accumulating proteins" is expected for dementia (Fig. 36.1).

References

1. Gómez-Isla T, Price JL, et al. Profound loss of layer II entorhinal cortex neurons occurs in very mild Alzheimer's disease. J Neurosci. 1996;16:4491–500.
2. Nakahata N, Nakamura T, et al. Age-related cognitive decline and prevalence of mild cognitive impairment in the Iwaki Health Promotion Project. J Alzheimers Dis. 2021;84:1233–45.
3. Karanth S, Nelson PT, et al. Prevalence and clinical phenotype of quadruple misfolded proteins in older adults. JAMA Neurol. 2020;77:1299–307.
4. Shinohara M, Tashiro Y, et al. Interaction between APOE genotype and diabetes in cognitive decline. Alzheimers Dement (Amst). 2020;12:e12006.
5. Ngandu T, Lehtisalo J, et al. A 2 year multidomain intervention of diet, exercise, cognitive training, and vascular risk monitoring versus control to prevent cognitive decline in at-risk elderly people (FINGER): a randomised controlled trial. Lancet. 2015;385:2255–63.

37 Brain and Antiaging Medicine: Aging Changes in Memory and Antiaging

Takeshi Tabira

37.1 Types of Memory

37.1.1 Short-Term and Long-Term Memory

Short-term memory is the memory used when temporarily remembering a phone number to make a call or doing mental arithmetic, and it disappears quickly. Long-term memory spans from several days to several decades, but memories lasting several hours to several days are also included in long-term memory. Memories from a few minutes to a few days are called recent memory, and memories from several months to several decades are called remote memory.

37.1.2 Declarative and Non-Declarative Memory

Declarative memory is the memory of experienced events (episodic memory) or the meaning of words and general knowledge (semantic memory), which can be expressed in words. Non-declarative memory is the memory that the body learns, such as how to ride a bicycle (procedural memory).

T. Tabira (✉)
Department of Diagnosis, Prevention and Treatment of Dementia, Graduate School of Medicine, Juntendo University, Tokyo, Japan
e-mail: spfe6he9@mild.ocn.ne.jp

37.1.3 Past and Future Memory

Memories of events, the meaning of words learned, and the names of people met are past memories. The memory of plans to be made is future memory, also known as prospective memory. This includes memories of plans such as what time to visit the dentist today, who to meet at what time, and to mail a letter if there is a mailbox on the way. It is said that the anterior part of the frontal lobe is involved in this.

37.2 Elements of Memory

When people who come to the memory clinic for a consultation are asked, "Do you forget things?", they often answer, "No, I can't remember," or "I can't recall people's names or the names of things." Memory has three elements: encoding (remembering), storage (preserving what is remembered), and retrieval (retrieving what is remembered). Any impairment can cause memory impairment, which is referred to as forgetfulness. Memory has a capacity for what can be remembered at once. In the case of adults, they can hold 7 ± 2 numbers for a short time (digit span). Memories of impressive events, etc., are easy to retain. It becomes easier to remember if you associate things, group-related things together, or use mnemonics (subjective organization).

H. Yamada, Y. Naito (eds.), *Anti-Aging Medicine*, https://doi.org/10.1007/978-981-97-8514-8_37

37.3 Aging Changes in Memory

Aging changes in memory first appear in recent memory lasting a few minutes to a few days, and prospective memory is also impaired, but remote memory is characterized by being retained. Immediate recall of three words such as "cherry blossom, cat, train" is maintained, but delayed recall after about 3–5 min is easily impaired. In the elderly, impairment of episodic memory is the strongest, and procedural memory is the least likely to be impaired. Semantic memory is also retained in the elderly, and vocabulary can be increased even in old age. Digit The span decreases with age, and it has been reported to be 5.4 in the 70 s. In the elderly, it can be difficult to recall the names of well-known people or famous entertainers on the spot, and the words they want to say may not come out quickly. This is called word-finding difficulty (word recall disorder), which is mainly caused by a decline in search function. In the elderly, providing cues (hints) at the recall stage can improve this.

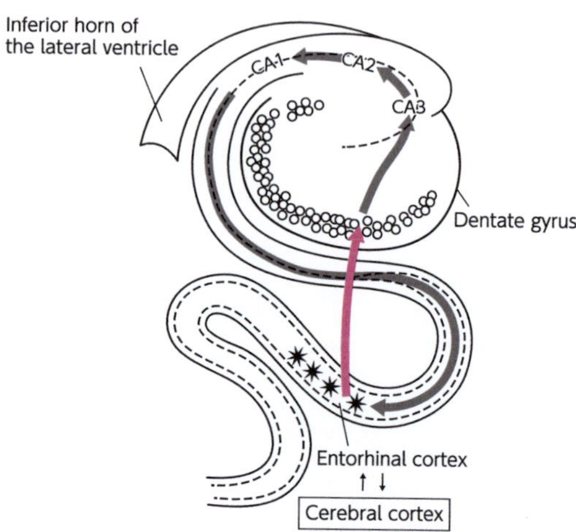

Fig. 37.1 Hippocampal memory circuit. The figure shows a cross-section of the hippocampus. Information from the cerebral cortex association area converges in the olfactory cortex and is transmitted to the dentate gyrus through perforant pathways. This information is passed on to CA3 → CA2 → CA1, processed, and returned to the olfactory cortex via the hippocampal fimbria. Short-term memory is temporarily stored in this circuit. Important information, impressive information, is sent to the cerebral cortex association area and stored as long-term memory. Age-related changes in nerve fibers appear earliest in the olfactory cortex (marked with an asterisk), followed by the hippocampus, and then spread to the cerebral limbic system

37.4 Memory Circuits and Age-Related Changes

Two circuits are known for declarative memory. One is a circuit in which information entered into the cerebral cortex association area converges in the entorhinal cortex, is sent to the hippocampal dentate gyrus by perforant fibers, and then returns to the entorhinal cortex via the hippocampal CA3, CA1, and hippocampal fimbria, and is projected to various parts of the cerebral cortex (Fig. 37.1). In this hippocampal circuit, memory is temporarily stored, and selected important ones are sent to the cerebral cortex and stored as long-term memory. In age-related changes in the hippocampus and Alzheimer's disease, hippocampal impairment appears early, and recent memory is impaired because temporary storage of memory in the hippocampal circuit is not possible. The other is a circuit from the hippocampus to the brain arch, mammillary body, mammillothalamic tract, anterior thalamic nucleus, cingulate gyrus, and hippocampus (Papez's circuit). This circuit can cause significant memory impairment in heavy alcohol drinkers and vitamin B1-deficient individuals, which is called Wernicke-Korsakov syndrome, causing severe recent memory impairment.

The volume of the hippocampus shrinks with age, which is due to the reduction of projection neuron protrusions and the dendritic protrusions and dendritic spine of the CA1 region. The decline in memory function due to aging is associated with a decrease in the number of synapses and a decrease in synaptic plasticity. Also, the number of nerve protrusions projecting from the entorhinal cortex to the hippocampal dentate gyrus decreases with age, which is thought to be due to age-related changes in the nerve fibers of the entorhinal cortex neurons. Furthermore, the decrease in neurogenesis due to aging of the granule cells of the hippocampal dentate gyrus is involved in the decline in memory function.

37.5 Antiaging of Memory

The N-methyl-D-aspartic acid (NMDA) receptor is involved in memory, and substances with NMDA receptor enhancing function have been shown to have memory-improving effects. Also, it has been shown in animal experiments that long-term memory is restored with histone deacetylase inhibitors. However, these have not yet been put into practical use. Passive and active immunotherapies targeting Alzheimer's disease amyloid are in clinical trials and are close to practical use. Memory function of mice raised in an enriched environment significantly improves. Strong stress can cause neuronal death in the hippocampal CA3 region, and in humans, hippocampal atrophy can be seen in proportion to the duration of stress exposure. In humans, hobbies and social interactions are considered beneficial, but proving this is difficult. In small-scale intervention trials in humans, aerobic exercise has been shown to improve memory function in the elderly and improve hippocampal functional imaging [1]. Brain training shows improvement in task performance, but it did not lead to improvement in overall cognitive function and memory function. In the large-scale multidomain intervention trial in Finland (The Finnish Geriatric Intervention Study to Prevent Cognitive Impairment and Disability; FINGER), significant suppression of decline in executive function was observed, but suppression of memory decline was not observed [2]. Acetylcholinesterase inhibitors such as donepezil hydrochloride increase acetylcholine in the brain and have a slight memory-improving effect. Huperzine A, contained in club moss, has been shown to have a memory-improving effect, but in February 2022, the Ministry of Health, Labour and Welfare classified club moss as a pharmaceutical, making it no longer available as a supplement. In a 3-year double-blind trial of early AD with food containing DHA, uridine-1-phosphate, folic acid, etc. (Souvenaid®), no improvement in memory function was observed, but it significantly suppressed the overall decline in cognitive function including memory [3]. Further research is needed for memory antiaging.

References

1. Won J, Callow DD, et al. Hippocampal functional connectivity and memory performance after exercise intervention in older adults with mild cognitive impairment. J Alzheimers Dis. 2021;82:1015–31.
2. Ngandu T, Lehtisalo J, et al. A 2 year multidomain intervention of diet, exercise, cognitive training, and vascular risk monitoring versus control to prevent cognitive decline in at-risk elderly people (FINGER): a randomised controlled trial. Lancet. 2015;385:2255–63.
3. Soininen H, Solomon A, et al. 36-month LipiDiDiet multinutrient clinical trial in prodromal Alzheimer's disease. Alzheimers Dement. 2021;17:29–40.

Brain and Antiaging Medicine: Antiaging of Cognitive Function: Exercise

38

Masahiro Okamoto and Hideaki Soya

38.1 Antiaging of Cognitive Function Through Exercise

The prefrontal cortex, located at the front of the cerebral cortex, and the hippocampus, a part of the limbic system, are both important brain regions for cognitive function. The prefrontal cortex is responsible for attention, concentration, selection, and decision-making (executive functions), while the hippocampus is responsible for memory and learning abilities. Brain atrophy associated with aging is notably observed in the cerebral cortex (especially the prefrontal cortex) and the hippocampus, but exercise can halt this atrophy and exert an antiaging effect on cognitive function. For example, older adults who walk more at the age of 65 have larger volumes of the prefrontal cortex and hippocampus nine years later and have a lower risk of transitioning to mild cognitive impairment (MCI) or dementia 13 years later [1]. In animal studies, making elderly rats exercise on a wheel for a month increases the number of new neurons (neurogenesis) and volume in the hippocampus and improves spatial learning ability. Antiaging of cognitive function is achieved not only by inhibiting atrophy in specific brain regions, but also by maintaining efficient neural activity and strengthening cooperation with other regions. Based on the concept of cognitive reserve, even if a certain brain region atrophies or is damaged, it is possible to prevent functional decline by, first, maintaining brain activity and neural networks in an efficient state (neural reserve), and second, compensating by mobilizing other brain regions that were not originally used (neural compensation). It has been widely confirmed that aerobic exercise training for a year strengthens the functional connections between brain regions. Also, making elderly people do moderate-intensity exercise for 10 min increases neural activity in areas other than those responsible for task performance, improving executive functions. In other words, exercise not only inhibits atrophy in specific brain regions, but also maintains normal brain networks and induces compensatory brain activity, thereby enhancing cognitive reserve.

38.2 Molecular Mechanism by Which Aerobic Exercise Enhances Cognitive Function

Various factors are assumed to be involved in the molecular mechanisms by which structural and functional changes occur due to aerobic exercise (Fig. 38.1). Recent studies have

The original version of the chapter has been revised. A correction to this chapter can be found at https://doi.org/10.1007/978-981-97-8514-8_165

M. Okamoto · H. Soya (✉)
Laboratory of Exercise Biochemistry and Neuroendocrinology, Sport Neuroscience Division, Advanced Research Initiative for Human High Performance (ARIHHP), Institute of Health and Sport Sciences, University of Tsukuba, Ibaraki, Japan
e-mail: okamoto.masahiro.ga@u.tsukuba.ac.jp; soya.hideaki.gt@u.tsukuba.ac.jp

H. Yamada, Y. Naito (eds.), *Anti-Aging Medicine*, https://doi.org/10.1007/978-981-97-8514-8_38

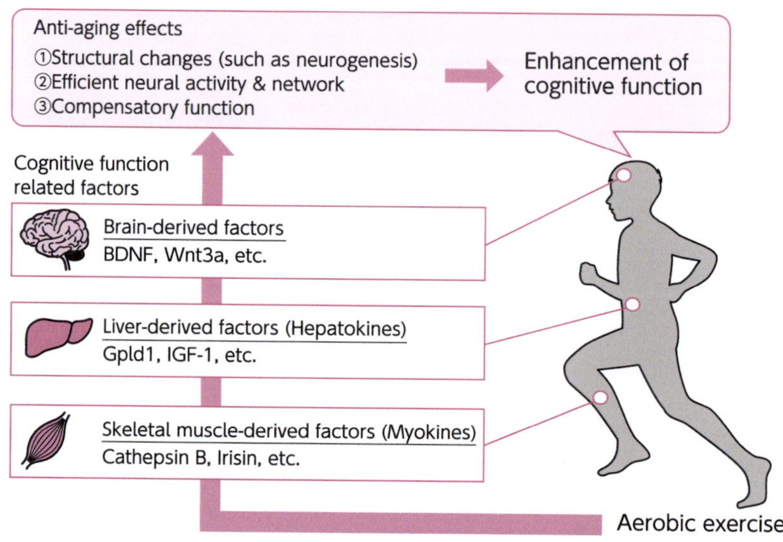

Fig. 38.1 Anticipated antiaging mechanism of cognitive function by exercise

reported that when elderly mice are made to exercise and their plasma is administered to sedentary elderly mice, the brain-derived neurotrophic factor (BDNF) and neurogenesis in the hippocampus increase, and memory learning ability improves. At this time, the blood level of glycosylphosphatidyl-inositol (GPI)-specific phospholipase D1 (Gpld1) shows strong correlation with improved memory, and in the blood of active rlderly people, Gpld1 is higher compared to sedentary elderly people, suggesting that Gpld1, a liver-derived factor, is attracting attention as a blood-borne factor responsible for the antiaging effects of exercise [2]. In addition, factors derived from skeletal muscles (myokines) such as cathepsin B and irisin, which increase with exercise, are also expected to be factors in preventing dementia, including Alzheimer's disease. Furthermore, the molecules that enhance cognitive function through exercise are not only transported through the blood, but also involve brain-derived factors such as BDNF, which have been pointed out for a long time. One of the factors contributing to the decline in cognitive function with aging is Wnt3a (Wnt family member 3A), which is secreted from a type of glial cell called astrocytes. The age-related decrease in Wnt3a secretion has been pointed out in relation to cognitive decline, and exercise reactivates this Wnt3a secretion and enhances neurogenesis. At this time, since the neural differentiation ability of stem cells itself is less affected by aging, it is assumed that exercise may exert antiaging effects by stimulating astrocytes around the neurons [3].

38.3 Even Low-Intensity Exercise Can Be Effective

When aiming to improve aerobic capacity, it is desirable to exercise at a moderate intensity of about 50–60% of the maximum oxygen uptake for more than 30 min at a time. Regarding cognitive function, research from meta-analysis shows that moderate- to high-intensity exercise for 45–60 min per session is associated with improved cognitive function in the elderly. Recently, it has been reported that high-intensity interval training (HIIT) also enhances cognitive function. HIIT, which is effective even in a short time compared to endurance exercise, is expected to have antiaging effects as an exercise with excellent time efficiency. However, it is difficult to continue exercising at or above moderate intensity if you are not good at exercise or if you have pain somewhere in your body. Therefore, attention is being paid to low-intensity exercise that makes your breath bounce lightly (about 30% of maximum oxygen uptake). In animal experiments, even low-intensity exercise training increases neurogenesis in the hippocampus and improves spatial learning ability. In addition, in human studies, it has been clarified that 10 min of low-intensity exercise increases the activity of the prefrontal cortex and hippocampus along with increased arousal, improving executive function and memory [4]. The effectiveness of low-intensity exercise has also been confirmed in a research project conducted in Tone Town, Ibaraki Prefecture. The experinets participants received exercise classes 2-6 times per month, and a low-intensity exercise program called "Furi Furi Gupper" was performed at home to music for 30 min every day. As a result of the two-year intervention, elderly people who exercised had suppressed atrophy of the prefrontal cortex, and their performance on memory and executive function tasks improved [5]. This suggests that aerobic exercise suppresses brain atrophy and enhances cognitive reserve, contributing to the antiaging of cognitive function. It is unclear why improving cardiovascular function leads to improved cognitive function, but even low-intensity exercise may be effective, and there are high hopes for future research.

References

1. Erickson KI, Raji CA, et al. Physical activity predicts gray matter volume in late adulthood: the Cardiovascular Health Study. Neurology. 2010;75:1415–22.
2. Okamoto M, Inoue K, et al. Reduction in paracrine Wnt3 factors during aging causes impaired adult neurogenesis. FASEB J. 2011;25:3570–82.
3. Horowitz AM, Fan X, et al. Blood factors transfer beneficial effects of exercise on neurogenesis and cognition to the aged brain. Science. 2020;369:167–73.
4. Suwabe K, Byun K, et al. Rapid stimulation of human dentate gyrus function with acute mild exercise. Proc Natl Acad Sci U S A. 2018;115:10487–92.
5. Tamura M, Nemoto K, et al. Long-term mild-intensity exercise regimen preserves prefrontal cortical volume against aging. Int J Geriatr Psychiatry. 2015;30:686–94.

39 Brain and Antiaging Medicine: Antiaging of Cognitive Function: Nutrition

Kazuya Yamashita

According to a report by the Lancet Commission on Dementia Prevention, Intervention, and Care, it is expected that by continuously avoiding 12 risk factors related to dementia (educational history, midlife hearing loss, high blood pressure, obesity, smoking, depression, social isolation, lack of exercise, diabetes in those over 65, excessive alcohol consumption, traumatic brain injury, air pollution), the onset can be prevented by about 40% [1]. The onset of Alzheimer's disease (AD), in particular, among dementia is considered to occur latently with the addition of environmental factors as mentioned above, against a genetic background, and can be prevented by correcting these environmental factors. However, there are few cases where various risk factors and protective factors pointed out in animal experiments and large-scale epidemiological studies have been proven to have a reliable preventive effect with reproducibility in intervention studies.

39.1 Antioxidants and Vitamins

Antioxidants such as vitamin B, vitamin C, vitamin E, carotene, and flavonoids (a type of polyphenol) weaken the damage to nerve cells caused by free radicals and amyloid beta protein (amyloid Beta; Aβ) weakens toxicity. In food, the intake of vegetables, fruits, green tea, coffee, wine, etc. is mentioned. Also, curcumin contained in curry (a yellow pigment of turmeric, a spice of curry, a type of polyphenol) also works as an antioxidant.

39.1.1 Vitamin B Group

Homocysteine is an amino acid produced as a metabolic by-product of the essential amino acid methionine, and with a deficiency of vitamin B6, vitamin B12, and folic acid, the conversion from homocysteine to methionine stagnates, and homocysteine migrates from inside cells to the blood, inducing the deposition of Aβ and tau proteins and neuronal cell death, and impairing cognitive function.

39.1.2 Vitamin D

Elderly people with vitamin D deficiency have an increased risk of developing AD. The reason for this is pointed out to be the strong affinity of vitamin D for neurons, glial cells, macrophages, and myelin in the brain.

K. Yamashita (✉)
The University of Shimane, Matsue, Shimane, Japan
e-mail: k-yamashita@u-shimane.ac.jp

H. Yamada, Y. Naito (eds.), *Anti-Aging Medicine*, https://doi.org/10.1007/978-981-97-8514-8_39

39.2 Fatty Acids

In many studies, the more fish you eat, the more memory loss in healthy participants is suppressed. That is, the usefulness of ω3 fatty acids, especially docosahexaenoic acid (docosahexaenoic acid; DHA), contained in fish (especially blue fish) for cognitive function has been pointed out, and in the sub-analysis of the Framingham study, a 16-year follow-up survey reported that the incidence of dementia was 47% lower in the group with high blood DHA concentration [2]. The intake of medium-chain fatty acids contained in coconut oil and the like may also suppress memory loss. On the other hand, it has been pointed out that the increase in blood trans fatty acids, which are abundant in fast food, may be involved in the onset of dementia.

39.3 Diet

The Mediterranean diet (Mediterranean diet) (Fig. 39.1), which includes a lot of oranges, apple juices, tomatoes, broccoli and other vegetables, peas and other legumes, bread, rice and other grains, olive oil, a lot of fish, less meat, and a moderate amount of wine, has established evidence as a dietary method for preventing cardiovascular diseases, but it is also recognized to be effective in preventing AD and cognitive dysfunction. Also, the DASH (Dietary Approaches to Stop Hypertension) diet is a dietary method to improve hypertension by firmly taking minerals (potassium, calcium, magnesium) with a salt excretion effect, and it is said to have a cognitive function improvement effect [3]. Furthermore, the MIND (Mediterranean-DASH Intervention for Neurodegenerative Delay) diet is a dietary therapy that combines the Mediterranean diet and the DASH diet, and it has been reported to have the potential to prevent AD and age-related cognitive decline [4]. However, in our country where the food culture is very different, it is difficult to introduce the Mediterranean diet, DASH diet, etc. into the diet, but in the Hisayama town study, it has been reported that a diet that adds milk and dairy products to ordinary Japanese food rich in soybeans and seaweed prevents dementia [5].

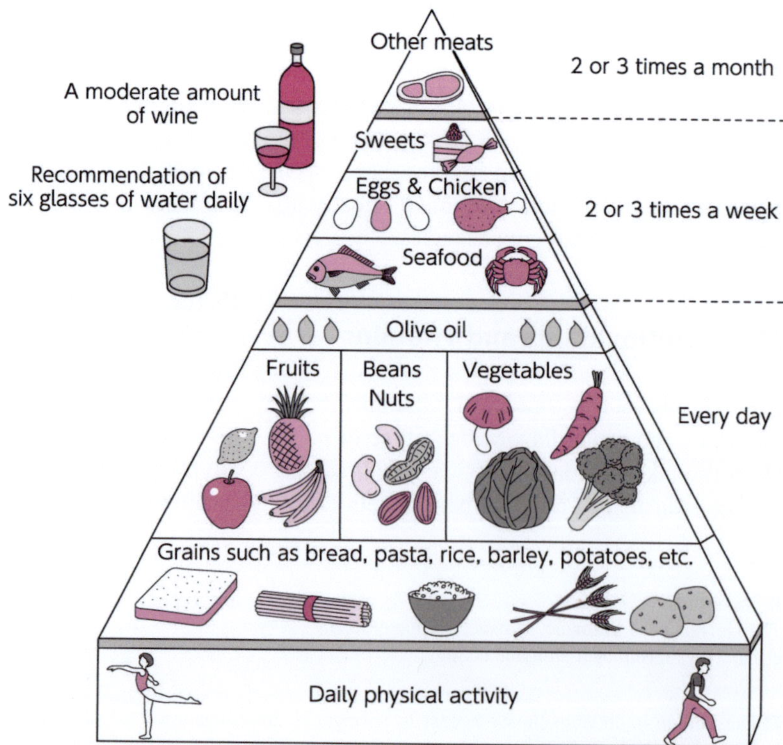

Fig. 39.1 Mediterranean diet pyramid

39.4 Metabolic Syndrome, Malnutrition, Frail

Metabolic syndrome is a definite risk factor for cerebrovascular dementia, and it has been reported that the more components such as obesity, low HDL cholesterol, high triglycerides, and high blood sugar one has, the more likely one is to develop dementia, although there are also negative reports. In the elderly, malnutrition increases the risk of dementia, so it is important to actively consume meat, fish, milk, etc., not just a rough diet, to maintain cognitive function. As people age, various functions decline and they become in need of care, and frailty due to aging is a cause of the need for care. Mild cognitive impairment (MCI) and physical frailty affect each other and are reversible, so the combined state of the two (cognitive frailty) is attracting attention.

39.5 Diabetes

Poor blood sugar management is associated with cognitive decline and deterioration, and not only is it associated with vascular dementia, but the frequency of AD and MCI is also high. Insulin resistance and hyperinsulinemia are speculated to cause a low insulin state in the central nervous system and promote the accumulation of Aβ in the brain. Gut microbiota: There are tens of trillions of bacteria in the gut, and the number of species is estimated to be nearly 1000. This gut microbiota and the gut form a complex gut system, and it has become clear that the biological functions of gut bacteria and the host are closely related. The gut microbiota is thought to affect diabetes, obesity, heart disease, Parkinson's disease, dementia, etc. The causal relationship between gut microbiota and dementia is unknown, but it has been pointed out that the composition ratio of gut bacteria is different in dementia patients compared to normal.

39.6 Conclusion

In the prevention of cognitive decline and dementia, nutritional epidemiological studies targeting Japanese people suggest that a diet with a good nutritional balance using various ingredients, rather than a single food, is effective. There is still a lack of scientific evidence for the preventive effects of AD and cognitive decline with reproducibility for supplements containing antioxidants and fatty acids.

References

1. Livingston G, Huntley J, et al. Dementia prevention, intervention, and care: 2020 report of the Lancet Commission. Lancet. 2020;396:413–46.
2. Schaefer EJ, Bongard V, et al. Plasma phosphatidylcholine docosahexaenoic acid content and risk of dementia and Alzheimer disease: the Framingham Heart Study. Arch Neurol. 2006;63:1545–50.
3. Smith PJ, Blumenthal JA, et al. Effects of the dietary approaches to stop hypertension diet, exercise, and caloric restriction on neurocognition in overweight adults with high blood pressure. Hypertension. 2010;55:1331–8.
4. Dhana K, James BD, et al. MIND Diet, common brain pathologies, and cognition in community-dwelling older adults. J Alzheimers Dis. 2021;83:683–92.
5. Ozawa M, Ninomiya T, et al. Dietary patterns and risk of dementia in an elderly Japanese population: the Hisayama Study. Am J Clin Nutr. 2013;97:1076–82.

40 Metabolism and Antiaging Medicine: Metabolic Control and Antiaging

Hirotsugu Suwanai and Ryo Suzuki

40.1 The Concept of Metabolism

Metabolism refers to the entire process in which a living organism takes in nutrients from the environment to form its biological components, and then excretes waste while producing energy. Metabolism can be broadly divided into material metabolism and energy metabolism. Material metabolism is divided into anabolism and catabolism. Anabolism refers to the process of absorbing nutrients into the body and converting them into biological components. On the other hand, catabolism refers to the process of breaking down biological components. Energy metabolism is divided into energy-absorbing reactions and energy-releasing reactions. The reaction that absorbs energy into the body during the conversion of nutrients into biological components is called an energy-absorbing reaction. The phenomenon of energy being produced when biological components are catabolized is called an energy-releasing reaction. The body excretes the biological components that have completed the energy-releasing reaction as waste (Fig. 40.1).

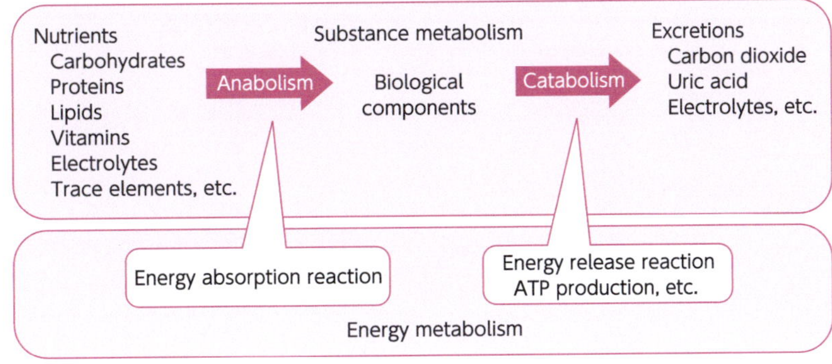

Fig. 40.1 Conceptual diagram of metabolism

Hirotsugu Suwanai and Ryo Suzuki contributed equally to this work.

H. Suwanai (✉) · R. Suzuki
Department of Diabetes, Metabolism, and Endocrinology, Tokyo Medical University, Tokyo, Japan
e-mail: suwanai-h@umin.ac.jp; suzukir@tokyo-med.ac.jp

H. Yamada, Y. Naito (eds.), *Anti-Aging Medicine*, https://doi.org/10.1007/978-981-97-8514-8_40

40.2 Metabolic Hypothesis of Nutrient Metabolism and Aging

The three major nutrients are carbohydrates, proteins, and fats. In addition, vitamins, electrolytes, and trace elements are absorbed and used to create biological components. After digestion and absorption, carbohydrates are broken down from simple sugars to pyruvic acid, proteins to amino acids, and fats to fatty acids and glycerol. These become acetyl-CoA, which produces adenosine triphosphate (ATP) from the citric acid cycle and electron transport chain in the mitochondria, and emits carbon dioxide. Importantly, during aerobic respiration in the mitochondria, which consumes oxygen, reactive oxygen species (ROS) are produced as by-products (Fig. 40.2). ROS react with biological macromolecules, leading to lipid peroxidation, DNA mutations, protein denaturation, and enzyme inactivation, and are thought to lead to aging. In ancient China, Laozi conveyed the saying, “The brighter the candle, the quicker it burns out,” which illustrates the characteristics of the metabolic theories of aging. In the 1920s, Raymond Pearl proposed the metabolic hypothesis, suggesting that lifespan is inversely correlated with basal metabolic rate. Subsequent research led to the calculation that the total amount of metabolic energy used up in a lifetime, when converted per body weight, is the same amount (7 × 105 J/g). Also, it is well known that the heartbeat hypothesis was proposed, stating that in mammals, smaller animals have faster resting heart rates, larger animals have slower heart rates, and lifespan ends with one billion heartbeats. These theories suggest that the accumulation of damage from ROS, a by-product of respiration resulting from heart rate and energy production, leads to aging. Calorie restriction (calorie Caloric restriction (CR) has been proven to extend the lifespan of animals, which may be partly due to a mechanism that reduces secondary ROS production and regulates and reprograms metabolic pathways in a more efficient way to achieve longevity [2].

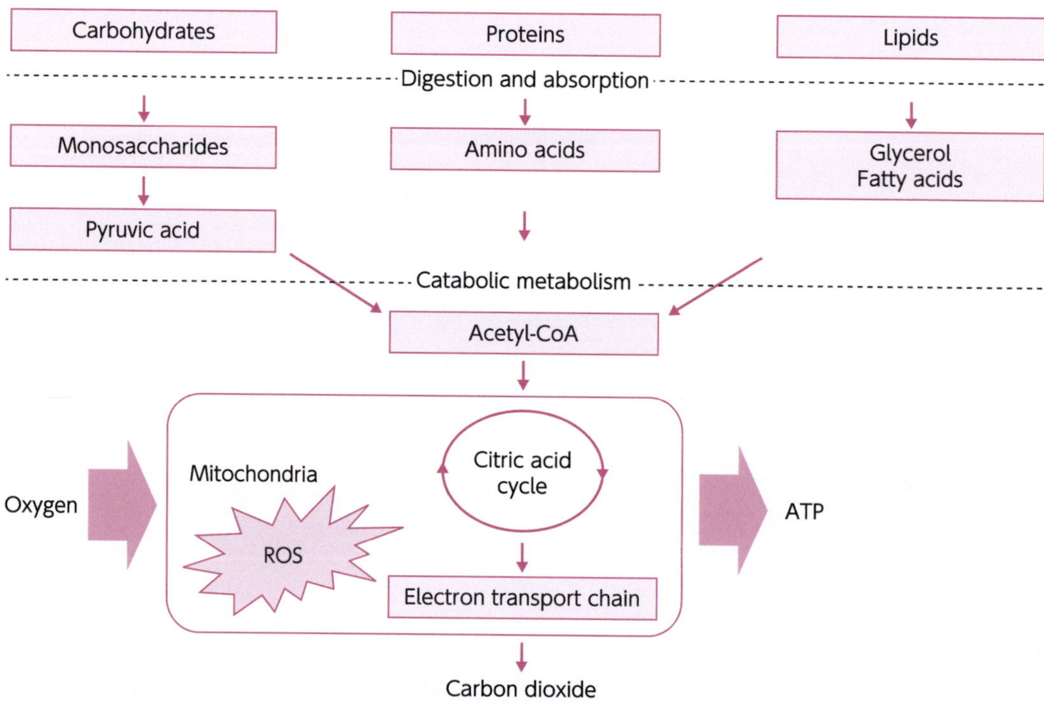

Fig. 40.2 Metabolic pathway of carbohydrates, proteins, and fats

40.3 Energy Metabolism Abnormalities and Aging

Diseases such as obesity, insulin resistance, glucose tolerance disorders, and dyslipidemia are considered to be associated with energy metabolism abnormalities and have been referred to as metabolic syndrome. They are related to lifestyle habits such as excessive high-energy food intake and insufficient exercise, and in particular, type 2 diabetes, which is mainly caused by insulin resistance and glucose tolerance disorders, is showing an increasing prevalence worldwide. Also, the global increase in obesity is a major problem. Diabetes causes microvascular complications such as neuropathy, retinopathy, and nephropathy. Moreover, obesity, dyslipidemia, and diabetes increase the risk of macrovascular diseases such as heart disease and cerebrovascular disease, greatly affecting lifespan.

40.4 Insulin Resistance and Metabolic Control

Insulin resistance is a condition where the blood glucose-lowering effect of insulin is not sufficiently exerted. Insulin promotes Na reabsorption in the kidneys, so hypertension often coexists in patients with insulin resistance, making it easier for atherosclerotic diseases to progress. Also, insulin stimulates cell proliferation in addition to metabolic control, and there are research results that cancer occurrence is higher in type 2 diabetics and obese individuals. This is thought to be due to the possibility that hyperinsulinemia associated with insulin resistance promotes cell proliferation and induces carcinogenesis [3]. Physical exercise enhances insulin sensitivity in muscles. The effects of exercise are acute and chronic. The acute effect enhances glucose uptake into skeletal muscles, similar to insulin. Chronic exercise enhances insulin sensitivity over the long term. Continuous exercise increases the expression of genes related to glucose and lipid metabolism in skeletal muscles, enhances the function of mitochondria abundant in slow-twitch muscles, and contributes to antiaging. Recent large-scale clinical studies have shown that metabolic control is important, as it has been shown that life prognosis can be improved by treating diabetes, hypertension, and dyslipidemia in various ways, such as drug therapy, diet therapy, and exercise therapy. Many large-scale clinical studies improve prognosis by drug intervention for diseases such as diabetes, dyslipidemia, and hypertension, but metformin, in the MILES (Metformin in Longevity Study) and TAME (Targeting Aging with Research aimed at antiaging, such as Metformin, is being conducted, and further research and reports are awaited [4].

References

1. Baynes JW, Dominiczak MH. (Naoyuki Taniguchi, Kazuhiro Iwai, et al. Supervised translation): Baynes & Dominiczak biochemistry original 4th edition. Tokyo: Maruzen Publishing; 2018.
2. Rodwell VW, Bender DA, et al. (Takao Shimizu Supervised translation): illustrated Harper's Biochemistry Original. 30th ed. Tokyo: Maruzen Publishing; 2016.
3. Kahn CR, et al. Edited (Yasunori Kanazawa, Masato Kasuga, et al. Supervised translation): Joslin diabetes. 2nd ed. Tokyo: Medical Science International; 2007.
4. Mohammed I, Hollenberg MD, et al. A critical review of the evidence that metformin is a putative anti-aging drug that enhances healthspan and extends lifespan. Front Endocrinol (Lausanne). 2021;12:718942.

41 Metabolism and Antiaging Medicine: Sarcopenic Obesity and Antiaging

Masahide Hamaguchi

Sarcopenia is a term that means a decrease in skeletal muscle mass and a decrease in muscle strength with aging. Sarcopenic obesity is a pathological condition in which obesity is combined with sarcopenia, and mechanisms such as insulin resistance, inflammation, and oxidative stress are considered to link obesity and sarcopenia [1]. In obesity, studies have been accumulated focusing on the accumulation of neutral fat in white adipocytes in subcutaneous adipose tissue and visceral adipose tissue. In recent years, in addition to these orthotopic fat accumulations, the elucidation of the pathological conditions of ectopic fat accumulation in fatty muscles, fatty liver, fatty pancreas, and pericardial adipose tissue has been attracting attention. In sarcopenic obesity, where sarcopenia and obesity coexist, fatty muscle is an important element.

41.1 Sarcopenia and the Quantity and Quality of Skeletal Muscles

The prevalence of sarcopenia in community-dwelling elderly people is thought to be about 15%. Various research groups have proposed criteria for determining sarcopenia, but it is often determined by three indicators: grip strength, walking speed, and skeletal muscle mass, and the criteria reported from the European and Asian Sarcopenia Working Groups are often used. Skeletal muscles are affected by aging not only in quantity but also in quality, such as muscle strength, endurance, and muscle function, and both quantity and quality gradually decrease from around the age of 40. Among skeletal muscles, large muscle groups on the surface that contain a lot of type II fibers (fast muscle fibers) are particularly susceptible to aging. Many of these muscles are called antigravity muscles and are important muscle groups for maintaining posture and daily life. Sarcopenia is a physical risk of frailty, which is closely related to nursing care. Furthermore, it has been reported that it is also a risk of all-cause mortality.

41.2 Orthotopic Fat Accumulation and Ectopic Fat Accumulation

Fat cells include white fat cells that store neutral fat, brown fat cells that consume neutral fat, and beige fat cells. Brown fat cells take in neutral fat, metabolize it in mitochondria to produce heat, and help maintain body temperature. Brown fat cells are distributed around the neck, collarbone, and shoulder blade areas. It is believed that brown fat

M. Hamaguchi (✉)
Department of Endocrinology and Metabolism, Kyoto Prefectural University of Medicine, Kyoto, Japan
e-mail: mhama@koto.kpu-m.ac.jp

H. Yamada, Y. Naito (eds.), *Anti-Aging Medicine*, https://doi.org/10.1007/978-981-97-8514-8_41

Table 41.1 Ectopic and orthotopic fat accumulation

Orthotopic fat accumulation	Subcutaneous fat tissue	
	Visceral fat tissue	
Ectopic fat accumulation	Fat muscle	Intramyocellular lipids (IMCL) Accumulation
		Excessive accumulation to Intramuscular adipose tissue (IMAT)
	Fatty liver	
	Fatty pancreas	
	Pericardial fat accumulation	

cells decrease with age. In recent years, in addition to brown and white fat cells, beige fat cells have been discovered and have attracted attention. Beige fat cells are thought to be cells that have differentiated within white adipose tissue in response to some stimulus and have functions similar to brown fat cells. Unlike brown fat cells, it is known that beige fat cells increase with age, and in the elderly, beige fat cells play a major role in heat production for maintaining body temperature. Beige fat cells are distributed in subcutaneous adipose tissue and are fat cells with characteristics intermediate between white and brown fat cells, but their feature is that they metabolize neutral fat like brown fat cells and produce heat in mitochondria. Cold, exercise, and some food nutrients are known as external factors that stimulate brown and beige fat cells. If the calories consumed exceed the calories burned, they are basically stored as neutral fat in white fat cells. White fat cells are distributed subcutaneously and in the abdominal cavity and are respectively called subcutaneous adipose tissue and visceral adipose tissue (Table 41.1) [1]. Until now, the increase in insulin resistance associated with the enlargement of visceral adipose tissue has been a concern. Visceral adipose tissue is also considered an endocrine organ that produces adipokines, hormones, and cytokines. When visceral fat cells enlarge due to excessive calorie intake, the production of adipokines from visceral adipose tissue is impaired, leading to an increase in systemic insulin resistance. Particularly in our country, the enlargement of visceral adipose tissue is considered an essential component of metabolic syndrome. In recent years, the concept has been proposed that the storage of triglycerides in subcutaneous adipose tissue and visceral adipose tissue is normal fat accumulation, and the accumulation of triglycerides in tissues other than these is ectopic fat accumulation [2]. Currently, fatty liver, fatty muscle, fatty pancreas, and pericardial fat accumulation are mentioned as ectopic fat accumulations. The liver and skeletal muscles are the two main targets of insulin, and ectopic fat accumulation in these organs increases insulin resistance. Fatty liver is basically fat accumulation in liver cells. On the other hand, in fatty muscles, there are intramyocellular lipids (IMCL) deposition, which is fat accumulation in muscle tube cells, and excessive fat accumulation in intramuscular adipose tissue (IMAT). IMCL is evaluated histologically by MRI and tissue collection from muscle biopsy and Oil-red O staining, and IMAT is mainly evaluated by CT, so reports where IMCL and IMAT are evaluated simultaneously are rare. When the term "fatty muscle" is used, it is necessary to pay attention to whether the increase in IMCL is being discussed or whether excessive fat accumulation in IMAT is being discussed.

41.3 Quality of Skeletal Muscle and Intramuscular Fat/ Ectopic Fat Accumulation

Recently, fatty muscles have been attracting attention as a pathological condition that reduces the quality of skeletal muscles [3]. It has been reported that the accumulation of neutral fats in IMAT increases with age, and that the increase in neutral fat accumulation in IMAT is associated with decreased muscle strength, falls, and ability disorders [4]. Recently, it is believed that sarcopenia progresses in various pathological conditions that deviate from the decline due to aging. The decrease in the amount of skeletal muscle, the decrease in quality, and both have various adverse effects depending on the pathological condition, but in sarcopenic obesity, it is considered that the quality of skeletal muscle decreases from fatty muscle [3]. In addition, Tamura et al.

reported that the accumulation of IMCL causes insulin resistance [5]. The BMI of Japanese people with type 2 diabetes is known to be low compared to other countries, and one of the reasons is believed to be that the amount of skeletal muscle decreases due to sarcopenic obesity.

41.4 Fatty Muscle and Exercise

A meta-analysis has been conducted on the suppression of IMAT accumulation and exercise, and it has been shown that moderate-intensity aerobic training and a combination of resistance training and aerobic training suppress the accumulation of IMAT [3]. Tamura and others compared diet therapy and exercise therapy in people with type 2 diabetes and found that diet therapy with calorie restriction alone did not reduce IMCL much, but adding exercise therapy to the diet reduced IMCL by 19% and improved insulin resistance by 57%. It has been reported that not only exercise but also daily activities consume calories, and increasing daily activities can also reduce IMCL. It is known that the reduction of IMCL improves insulin sensitivity in skeletal muscles. Thus, both IMAT and IMCL, which are fat muscles, are the cause of insulin resistance, but it has been shown that they can be suppressed by increasing exercise and daily activities, and insulin resistance can be improved.

41.5 Conclusion

Sarcopenic obesity is not only associated with physical dysfunction, but also with metabolic disorders and progression of arteriosclerosis, and is considered to have a high cardiovascular risk. The development of clinical research on sarcopenic obesity requires the establishment of diagnostic criteria. Among them, fat muscle is one of the important indicators in considering sarcopenic obesity, and the organization of concepts and the establishment of diagnostic criteria centered on IMCL and IMAT are required.

References

1. Obara K. Sarcopenic obesity. Jpn J Geriatr Med. 2014;51:99–108.
2. Shimabukuro M, Yamakawa K, et al. Obesity and ectopic fat, lipotoxicity. Jpn J Intern Med. 2011;100:983–8.
3. Tuñón-Suárez M, Reyes-Ponce A, et al. Exercise training to decrease ectopic intermuscular adipose tissue in individuals with chronic diseases: a systematic review and meta-analysis. Phys Ther. 2021;101:pzab162.
4. Akima H, Yoshiko A, et al. Relationship between quadriceps echo intensity and functional and morphological characteristics in older men and women. Arch Gerontol Geriatr. 2017;70:105–11.
5. Tamura Y, Tanaka Y, et al. Effects of diet and exercise on muscle and liver intracellular lipid contents and insulin sensitivity in type 2 diabetic patients. J Clin Endocrinol Metab. 2005;90:3191–6.

42 Metabolism and Antiaging Medicine: Anti-glycation and Antiaging

Sho-ichi Yamagishi

42.1 Diabetes Accelerates Aging

Currently, there are 36.4 million elderly people aged 65 and over in our country, accounting for an estimated 29.1% of the total population. The proportion of the elderly population is unparalleled worldwide, especially with over 20.57 million women aged 65 and over, it is reported that one in four Japanese women is over 70 years old. Furthermore, due to the spread of sedentary lifestyles and Western eating habits, the number of patients with metabolic diseases such as obesity, metabolic syndrome, and diabetes is also increasing. In fact, according to the "National Health and Nutrition Survey" in the first year of Reiwa, the proportion of "people strongly suspected of having diabetes" is 19.7% for adult men and 10.8% for adult women, and if limited to those over 70 years old, the proportion is 26.4% for men and 19.6% for women. On the other hand, our country is also a representative longevity country in the world, with an average life expectancy of 81.6 years for men and 87.7 years for women, and for women, it feels like we have entered the era of living to 90 years old. However, there is a 9 to 12-year gap between the so-called healthy life expectancy, the period in which one can live independently without relying on continuous medical care or nursing care, and the average life expectancy. And diabetes is a major obstacle to extending healthy life expectancy. In fact, in diabetes, various levels of blood vessels are damaged, and the risks of dialysis, blindness, myocardial infarction, heart failure, cerebrovascular disorders, and lower limb amputation increase, significantly impairing "healthy life expectancy". In addition, it has been clarified that in diabetes, the aging process progresses, and the risk of developing various geriatric diseases such as Alzheimer's disease, cancer, osteoporosis, and osteoarthritis also increases [1–5].

42.2 Memory of Hyperglycemia (Metabolic Memory)

According to the EDIC-DCCT study, which followed up on the DCCT (Diabetes Control and Complications Trial), if the blood sugar control of type 1 diabetes patients is poor for the initial 6.5 years, even if blood sugar control is improved thereafter, it does not necessarily sufficiently suppress the progression of vascular complications [5]. In fact, in the conventional therapy group where blood sugar control was insufficient during the DCCT period, compared to the intensive therapy group where blood sugar control was attempted from the beginning, the risk of progression of microvascular diseases such as nephropa-

S.-i. Yamagishi (✉)
Department of Medicine, Showa University School of Medicine, Tokyo, Japan
e-mail: shoichi@med.showa-u.ac.jp

H. Yamada, Y. Naito (eds.), *Anti-Aging Medicine*, https://doi.org/10.1007/978-981-97-8514-8_42

thy, retinopathy, peripheral neuropathy, and cardiac autonomic neuropathy is high for 14–18 years after the end of DCCT, and the risk of cardiovascular events and death is more than double 11 years later, and this phenomenon has been reported to continue for 27 years [5]. Also, in the United Kingdom Prospective Diabetes Study conducted on patients with type 2 diabetes, in the follow-up study of the United Kingdom Prospective Diabetes Study (UKPDS), UKPDS80, it has been reported that strict blood glucose control from the beginning can have a suppressive effect on vascular complications over the long term and can exert what is called a legacy effect [5]. These facts indicate that, like in experimental animals, a phenomenon called "metabolic memory" of hyperglycemia exists in the onset and progression of human diabetic vascular complications.

42.3 Advanced Glycation End Products (AGE)

Monosaccharides such as glucose, glycerol aldehyde, and fructose react nonenzymatically with the amino groups of proteins, lipids, and nucleic acids to form Schiff bases and Amadori compounds, eventually forming advanced glycation end products (AGEs). AGEs are irreversibly formed and accumulated in the body depending on the degree and duration of blood glucose control, and once formed, they are metabolized very slowly, which best explains the phenomenon of "metabolic memory" [1–5]. It has also been clarified that the more AGEs were accumulated in the skin at the end of the DCCT, the more pronounced the progression of retinopathy and nephropathy during the EDIC-DCCT period. Furthermore, it has been clarified that the expression of the receptor for AGE (RAGE), a cell surface receptor that recognizes AGEs, is induced by AGEs themselves, and the local production of AGEs is also enhanced [1–5]. These facts suggest that once formed in the body, AGEs, which are hardly metabolized and remain in the tissue for a long time, and the interaction between AGEs and RAGE, whose expression is induced by AGEs, continuously activate the AGE-RAGE system, shaping the "metabolic memory". Conversely, if blood glucose control is strict from the beginning, the formation of AGEs can be kept low, a brake can be applied to the progression of subsequent vascular complications, and a legacy effect can be observed. It has been reported that the AGE value of the skin measured by the AGE Reader (manufactured by Diagnoptics Technologies B.V., imported and sold by Serista Co., Ltd.) can be a marker predicting various vascular complications and total mortality in diabetic patients, including Japanese [1].

42.4 AGE-RAGE System Inhibiting Aptamer

The authors have successfully developed a DNA aptamer that specifically binds to AGEs [4]. The AGEs bound by the aptamer are efficiently degraded by macrophages, so the aptamer can effectively remove AGEs that have accumulated in the body. In addition, the binding of AGEs and RAGEs is also blocked by the aptamer, so the AGE-RAGE system can be strongly inhibited by the AGE aptamer. In fact, it has been shown that the progression of nephropathy in type 2 diabetes model animals is suppressed, the onset of early retinopathy in type 1 diabetes model animals is suppressed, and abnormalities such as vascular restenosis and remodeling after balloon injury, enlargement of visceral fat, and abnormalities of adipocytokines due to excessive fructose load are corrected by administering the AGE aptamer [4]. Furthermore, it has been found that by continuously administering the AGE aptamer to tumor-bearing nude mice using an osmotic pump, angiogenesis and macrophage infiltration are suppressed, and melanoma proliferation is inhibited. Recently, the authors have also been the first in the world to successfully develop a functional RAGE antagonistic aptamer that binds to RAGE and inhibits the AGE-RAGE system [4]. It has been shown that the onset and progression of nephropathy in type 1 diabetes model animals is suppressed by the administration of the RAGE aptamer, and renal damage in hypertensive model animals is suppressed. In addition, it has been shown that by administering the RAGE aptamer to tumor-bearing animals, the AGE-RAGE-oxidative stress system is blocked in melanoma, angiogenesis and macrophage infiltration are suppressed, and melanoma proliferation and liver metastasis are suppressed, and long-term local administration of the RAGE aptamer suppresses melanoma proliferation. In addition, it has been found that the mortality rate of sepsis model animals is dramatically reduced by the RAGE aptamer.

References

1. Yamagishi S, Fukami K, et al. Evaluation of tissue accumulation levels of advanced glycation end products by skin autofluorescence: a novel marker of vascular complications in high-risk patients for cardiovascular disease. Int J Cardiol. 2015;185:263–8.
2. Yamagishi S. Potential clinical utility of advanced glycation end product cross-link breakers in age- and diabetes-associated disorders. Rejuvenation Res. 2012;15:564–72.
3. Yamagishi S, Matsui T, et al. Role of receptor for advanced glycation end products (RAGE) and its ligands in cancer risk. Rejuvenation Res. 2015;18:48–56.
4. Koga Y, Sotokawauchi A, et al. DNA-Aptamer raised against receptor for advanced glycation end products improves survival rate in septic mice. Oxidative Med Cell Longev. 2021;2021:9932311.
5. Yamagishi S, Nakamura N, et al. Glycation and cardiovascular disease in diabetes: a perspective on the concept of metabolic memory. J Diabetes. 2017;9:141–8.

43 Metabolism and Antiaging Medicine: Phosphorus Metabolism and Antiaging

Makoto Kuro-o

Until now, aging research has pursued universal mechanisms that are preserved across species. As a result, it has been clarified that caloric restriction (CR) suppresses aging and the accumulation of cellular aging promotes individual aging. CR mimetics and senolytic drugs have been developed and are attracting attention as therapeutic drugs for age-related diseases, and clinical trials have begun. However, the process of aging is greatly influenced by the biological characteristics of the species and the living environment. Therefore, it is necessary to pursue the mechanism of species-specific aging from a new perspective that is not an extension of previous aging research.

43.1 Phosphorus and Aging

In 1997, the causative gene of a mutant mouse exhibiting premature aging was identified [1]. This gene, named Klotho after the Greek mythological "goddess who spins the thread of life", causes a variety of aging-like symptoms such as vascular calcification, cardiac hypertrophy, sarcopenia, pulmonary emphysema, hearing loss, dementia, frailty, and shortened lifespan when deficient in mice. Subsequent research revealed that the Klotho gene codes for the receptor of fibroblast growth factor-23 (FGF23). FGF23 is a hormone secreted by bone cells that sense phosphorus intake by some mechanism. When it binds to the Klotho receptor expressed in the kidney, it inhibits phosphorus reabsorption in the renal tubules and increases the amount of phosphorus excreted per nephron. In other words, FGF23 functions as a phosphorus diuretic hormone that increases urinary phosphorus excretion in response to phosphorus intake. In Klotho-deficient mice, FGF23 cannot function, so phosphorus diuresis does not occur even if phosphorus is ingested, and phosphorus accumulates in the body. Furthermore, it was shown that premature aging improves if Klotho-deficient mice are fed a low-phosphorus diet, giving birth to the concept that "phosphorus accelerates aging" [2]. How does phosphorus accelerate aging?

43.2 Acceleration of Renal Aging by Excessive Phosphorus Intake

Modern dietary habits involve excessive phosphorus intake. In addition to the increased intake of foods rich in phosphorus, such as meat and dairy products, food additives may also contain large amounts of phosphorus, and we are regu-

M. Kuro-o (✉)
Division of Anti-aging Medicine, Center for Molecular Medicine, Jichi Medical University, Tochigi, Japan
e-mail: mkuroo@jichi.ac.jp

H. Yamada, Y. Naito (eds.), *Anti-Aging Medicine*, https://doi.org/10.1007/978-981-97-8514-8_43

larly consuming several times the necessary amount of phosphorus. On the other hand, the number of nephrons decreases with age. This is part of the aging phenomenon and its cause is unknown, but if lifestyle diseases that cause renal complications such as diabetes and hypertension are added, the decrease in the number of nephrons accelerates, leading to chronic kidney disease. It progresses to chronic kidney disease (CKD). When people with a low number of nephrons, such as middle-aged and elderly people or CKD patients, consume too much phosphorus, they are forced to increase the amount of phosphorus excreted per nephron, causing FGF23 to rise. FGF23 inhibits phosphorus reabsorption in the renal tubules, causing the concentration of phosphorus in the primary urine to increase. When the concentration of phosphorus in the primary urine exceeds the solubility, calcium phosphate precipitates, but the Fetuin-A protein promptly adsorbs it, inhibiting the growth of the crystals. As a result, microparticles composed of calcium phosphate and Fetuin-A protein are formed. These microparticles are called calciprotein particles (CPP). CPP is phagocytosed through the Toll-like receptor (TLR4) expressed on the luminal side of the proximal renal tubules, but renal tubular cells that have phagocytosed CPP beyond their processing capacity undergo cell death. In other words, if the intake of phosphorus is too high for the remaining number of nephrons, FGF23 increases and CPP is formed in the primary urine, causing renal tubular damage. As a result, the number of nephrons decreases further, and FGF23 increases further, creating a vicious cycle that accelerates kidney aging [3].

43.3 From Kidney Aging to Systemic Aging

CPP also appears in the blood. When phosphorus is consumed, even if kidney function is normal, blood phosphorus concentration transiently increases (postprandial hyperphosphatemia), and with it, the level of CPP in the blood also increases. As kidney aging or CKD progresses and the number of nephrons decreases, postprandial hyperphosphatemia may be enhanced and prolonged, and the increase in blood CPP may also be enhanced and prolonged. In experiments using cultured cells, CPP has been shown to induce calcification in vascular smooth muscle cells and innate immune responses in macrophages. Clinical studies have shown that blood CPP levels strongly correlate with blood phosphorus levels and are associated with coronary artery calcification score, aortic pulse wave velocity, cardiovascular events, and inflammatory markers such as hs-CRP. From these facts, it has been pointed out that CPP may be a causative substance of arteriosclerosis and chronic inflammation [2]. In other words, when the number of nephrons decreases due to kidney aging, blood CPP increases, which may accelerate systemic aging. An extreme example of this is patients with renal failure.

43.4 Renal Failure: A Clinical Model of Accelerated Aging

When the number of nephrons decreases drastically, no matter how much FGF23 is increased, the ingested phosphorus cannot be fully excreted in the urine, leading to a breakdown in phosphorus homeostasis and a persistent increase in blood phosphorus concentration. Along with this, blood CPP also increases, and vascular calcification, cardiac hypertrophy, sarcopenia, frailty, etc. progress rapidly, and the overall mortality rate increases. The expression of Klotho in the kidneys is extinguished. This pathological condition of renal failure patients is similar to the "animal model of accelerated aging," the Klotho-deficient mouse, and the idea of considering renal failure patients as a "clinical model of accelerated aging" was born [4]. However, while the breakdown of phosphorus homeostasis in renal failure patients is due to the extinction of overall renal function, the breakdown of phosphorus homeostasis in Klotho-deficient mice is due to a single defect in urinary phosphorus excretion, which is a crucial difference. Klotho-deficient mice cannot excrete phosphorus, but other renal functions are normal, and blood creatinine levels do not increase, so they are not in renal failure. Nevertheless, they exhibit symptoms similar to those of renal failure patients. The various symptoms of renal failure patients are called "uremia," and it has been believed that they are caused by various "uremic toxins" that should be excreted in the urine. However, the fact that a condition similar to uremia occurs due to a single defect in urinary phosphorus excretion suggests that phosphorus may be the most important uremic toxin.

43.5 Phosphorus: A Unique Aging Accelerator for Modern Humans

About 400 million years ago, the first organisms with bones made of calcium phosphate evolved. These were osteichthyes [5]. Bones made of calcium phosphate are harder than the cartilage or calcium carbonate skeletons that previous organisms had, and they made it possible for terrestrial vertebrates, which need to support their body weight and move around on land without the help of buoyancy in water, to evolve. These organisms use a strategy of keeping the concentration of phosphorus and calcium in the extracellular fluid supersaturated and giving cues for deposition when and where they want to make bones. For this strategy to work, they needed to acquire a system to keep the concentration of phosphorus in the extracellular fluid constant. That is the FGF23-Klotho endocrine system. In fact, the Klotho gene is a gene that only exists in organisms after osteichthyes, and it can be said to be a gene that evolved for phosphorus. Furthermore, if calcium phosphate is deposited in tissues other than bones, they also need to acquire a defense system to suppress its growth. That is the formation of CPP. The FGF23-Klotho endocrine system and CPP formation are essential systems for acquiring bones made of calcium phosphate, but as mentioned above, they can also be a double-edged sword that can accelerate aging. Especially for modern humans, they are the first in the evolutionary process of organisms to face chronic excessive phosphorus intake. Organisms that have acquired bones made of calcium phosphate may be destined to age due to phosphorus.

References

1. Kuro-o M, Matsumura Y, et al. Mutation of the mouse klotho gene leads to a syndrome resembling ageing. Nature. 1997;390:45–51.
2. Kuro-o M. The Klotho proteins in health and disease. Nat Rev Nephrol. 2019;15:27–44.
3. Shiizaki K, Tsubouchi A, et al. Calcium phosphate microcrystals in the renal tubular fluid accelerate chronic kidney disease progression. J Clin Invest. 2021;131:e145693.
4. Stenvinkel P, Larsson TE. Chronic kidney disease: a clinical model of premature aging. Am J Kidney Dis. 2013;62:339–51.
5. Kuro-o M, Moe OW. FGF23-αKlotho as a paradigm for a kidney-bone network. Bone. 2017;100:4–18.

44 Hormones and Antiaging Medicine: Hypothalamic-Pituitary Hormones and Antiaging

Yutaka Takahashi

The hypothalamus is the central part of the endocrine system and the autonomic nervous system. In the hypothalamus, the central nervous system and neuroendocrine are interconnected through a network of neurons, regulating instinctive behavior and accompanying visceral functions to maintain homeostasis. It also houses centers for appetite, energy consumption, temperature regulation, thirst, and sexual behavior, integrating internal and external information and adapting to various stresses. It regulates the function of the pituitary gland, which is located at the top of all endocrine organs in the body. These functions are closely related to aging and lifespan regulation. In this section, I would like to discuss the role of the hypothalamus and pituitary hormones, mainly the GH (growth hormone)-IGF (insulin-like growth factor) system, which is involved in lifespan regulation, and the recently reported role of the hypothalamus.

Y. Takahashi (✉)
Department of Endocrinology and Metabolism, Nara Medical University, Nara, Japan
e-mail: takahash@naramed-u.ac.jp

44.1 GH-IGF System and Lifespan in Animal Models

Table 44.1 shows that most of the hypothalamic-pituitary hormone-related gene modifications or mutant mice related to lifespan regulation are in the GH-IGF system, and the importance of this system is clear [1]. All mice with reduced GH-IGF system signals live longer. Ames dwarf mice with natural mutations in the pituitary transcription factor Prop1, Snell dwarf mice with natural mutations in Pit1, lit/lit mice with natural mutations in the Ghrh receptor, Gh receptor-deficient mice, and Igf-I receptor heterozygous knockout mice [2] each shows a lifespan extension effect of 24–70%, and they all have common features of reduced IGF-I action and dwarfism. Igf-I receptor heterozygous knockout mice and their embryonic fibroblasts show resistance to oxidative stress, but lifespan extension and oxidative stress resistance are only observed in females, indicating a gender difference. Another mechanism is that IGF-I/insulin regulates the expression of FOXO1, which plays an important role in stress resistance, suggesting the involvement of FOXO1. In fact, in nematodes, the lifespan extension effect of insulin/Igf-I receptor homolog mutants is not observed in Foxo1 homolog mutants. In addition, although not listed in Table 44.1, knockout mice of molecules downstream of insulin/IGF-I signaling also show longevity, suggesting the involvement of the Akt-mTOR pathway.

H. Yamada, Y. Naito (eds.), *Anti-Aging Medicine*, https://doi.org/10.1007/978-981-97-8514-8_44

Table 44.1 Lifespan of modified or mutant mice in the neuroendocrine system (Modified with permission from [1])

Gene	Modification or mutation	Lifespan	Characteristics
Igf-I receptor	Brain-specific hetero knockout	Extended	Growth disorder, obesity
Irs2	Brain-specific hetero, homo knockout	Extended	Obesity, glucose tolerance abnormality
Prop1	Natural mutation (Ames)	Extension	Decrease in GH, PRL, TSH
Pit1	Natural mutation (Snell)	Extension	Decrease in GH, PRL, TSH
Gh receptor	Knockout	Extension	Decrease in serum IGF-I, growth disorder
Ghrh	Knockout	Extension	Decrease in serum GH, IGF-I, growth disorder
Ghrh receptor	Natural mutation (lit/lit)	Extension	Decrease in serum GH, IGF-I, growth disorder
Sirt1	Brain-specific transgenic	Extension	Delayed aging, decrease in cancer
Iκbα	Hypothalamus-specific transgenic	Extension	Delayed aging
Ikkβ	Hypothalamus-specific transgenic	Extension	Delayed aging

44.2 GH-IGF System and Aging, Lifespan in Humans

In humans, acromegaly, a hyperactive model of the GH-IGF system, shortens lifespan if left untreated. On the other hand, adult GH deficiency, a decrease in the GH-IGF system, unlike in lower animals such as mice, also shortens lifespan, a phenomenon known as the GH paradox. Furthermore, it has been suggested that at least some of the aging process may occur due to a decrease in the GH-IGF system. For example, GH secretion decreases with age, a phenomenon known as somatopause following menopause [3]. General changes observed with aging include changes in body composition (increased body fat, decreased muscle mass and bone salt), lipid metabolism disorders, increased cardiovascular disease due to arteriosclerosis, and decreased mental activity. Interestingly, these are similar to the symptoms of adult GH deficiency. In fact, in adult GH deficiency, there is an increase in visceral fat, fatty liver, nonalcoholic steatohepatitis (NASH), lipid metabolism disorders, and a high incidence and mortality rate of cardiovascular complications. Furthermore, GH replacement therapy has been suggested to not only improve complications, but also improve prognosis. However, the effects of GH replacement therapy on the general elderly population are limited and side effects such as edema, joint pain, and glucose tolerance abnormalities are common, so its use for antiaging purposes is prohibited. On the other hand, in acromegaly, which causes an excess of the GH-IGF-I system, complications such as hypertension, diabetes, lipid abnormalities, and heart failure occur, worsening the life prognosis, consistent with the results of animal models. And when GH-IGF-I levels normalize with treatment, the prognosis also improves. The mechanism of worsening prognosis is suggested to involve not only the impact of each complication but also the essential involvement of GH-IGF-I system excess. Although there are still many unknowns, increased oxidative stress [4] and telomere shortening have been reported in patients with acromegaly, and mechanisms such as increased oxidative stress, telomere shortening, and induction of cellular aging by IGF-I have been suggested [5].

44.3 The Role of the Hypothalamus in Aging

The hypothalamus not only regulates pituitary hormones through hypothalamic hormones, but it has also been revealed that it regulates aging and lifespan through various mechanisms. I will introduce a newly reported pathway.

44.3.1 Aging Due to Inflammation and Decreased GnRH

A high-fat diet causes inflammation in the hypothalamus and is associated with obesity, but with aging, inflammation increases in the hypothalamus, especially in the medial basal part, and is associated with systemic aging. Interestingly, inflammation in the medial basal part leads to a decrease in GnRH, which regulates gonadotropins, while supplementation with GnRH improves aging phenotypes independently of sex hormone action, highlighting a new role for GnRH as a mediator of aging [6].

44.3.2 The Hypothalamus as a Control Center for Aging

SIRT1, one of the longevity genes sirtuins, is a nicotinamide adenine dinucleotide-(NAD) dependent deacetylase. Dr. Shinichiro Imai of Washington University proposes an aging regulation mechanism by the NAD-SIRT1 system with the hypothalamus as the control center; that is, through the regulation of the autonomic nervous system and the endocrine system from the hypothalamus, in adipose tissue, eNAMPT secretion, which is the NAD synthesis rate-limiting enzyme, secretion of fibroblast growth factor-21 (FGF21) from the liver, and Interleukin-6 (IL-6), FNDC5 secretion from muscles to regulate NAD synthesis, and control sirtuins in the hypothalamus to regulate aging and lifespan. This is a whole-body NAD regulation system (NAD World). In the NAD World, the hypothalamus as the control center, skeletal muscles as mediators that mediate signals from there to other tissues, and adipose tissue as modulators that regulate the function of the control center interact to control the process of aging and lifespan [7].

44.3.3 The Role of Hypothalamic Stem Cells in Aging

It has been shown that hypothalamic stem cells play an important role in aging and lifespan regulation. In a mouse model, transplanting healthy hypothalamic stem cells slowed the progression of aging, while losing stem cells accelerated the progression of aging. The mechanism involves exosome miRNA (106a-5p, 20a-5p and 466 m-5p) secreted by hypothalamic neural stem cells into the cerebrospinal fluid [8].

44.4 Conclusion

It has become clear that the hypothalamus is involved in aging and lifespan regulation not only through the pathway involving hypothalamic hormones and pituitary hormones, but also through various mechanisms. The role and intervention methods in humans will be important themes in the future.

References

1. Tang Y, Purkayastha S, et al. Hypothalamic microinflammation: a common basis of metabolic syndrome and aging. Trends Neurosci. 2015;38:36–44.
2. Bartke A, Darcy J. GH and ageing: Pitfalls and new insights. Best Pract Res Clin Endocrinol Metab. 2017;31:113–25.
3. Lamberts SW, van den Beld AW, et al. The endocrinology of aging. Science. 1997;278:419–24.
4. Nishizawa H, Handayaningsih AE, et al. Enhanced oxidative stress in GH-transgenic rat and acromegaly in humans. Growth Hormon IGF Res. 2012;22:64–8.
5. Matsumoto R, Fukuoka H, et al. Accelerated telomere shortening in acromegaly: IGF-I induces telomere shortening and cellular senescence. PLoS One. 2015;10:e0140189.
6. Zhang G, Li J, et al. Hypothalamic programming of systemic ageing involving IKK-β, NF-κB and GnRH. Nature. 2013;497:211–6.
7. Satoh A, Imai S, et al. The brain, sirtuins, and ageing. Nat Rev Neurosci. 2017;18:362–74.
8. Zhang Y, Kim MS, et al. Hypothalamic stem cells control ageing speed partly through exosomal miRNAs. Nature. 2017;548:52–7.

45 Hormones and Antiaging Medicine: Thyroid Hormones and Antiaging

Masahide Hamaguchi

Metabolic changes in accordance with growth and aging are controlled by various hormones. Thyroid hormones play an important role in controlling the metabolic functions of the whole body, and they become the target of treatment for both functional renewal and decline. There is a condition called subclinical thyroid dysfunction, which requires careful handling as to whether to treat it, even though the thyroid function is not within the normal range. This section describes in particular the subclinical hypothyroidism, where the relationship with aging is important.

45.1 Thyroid Hormones and Whole Body Metabolism

The thyroid is an endocrine organ located just below the cricoid cartilage in the anterior neck, consisting of two lobes connected by an isthmus, resembling a butterfly spreading its wings. Iodine taken in from food reaches the thyroid follicular cells via the bloodstream, is actively taken up by the thyroid follicular cells from the sodium iodide symporter (NIS), and is converted to organic iodine by being organified by thyroid peroxidase. Follicular cells surround a space filled with colloid, and organic iodine is released into the colloid. The colloid consists of thyroglobulin, a glycoprotein with tyrosine residues in its matrix. Tyrosine residues in contact with the cell membrane of follicular cells are iodinated at one site (monoiodotyrosine) or two sites (diiodotyrosine) and combine with each other to form two types of thyroid hormones, triiodothyronine (T3) and thyroxine (T4). T3 and T4 are taken up again by the thyroid follicular cells and secreted from the thyroid follicular cells into the blood vessels. Thyroid hormones act on cells throughout the body, activating metabolism. As a result, oxygen consumption increases, basal metabolism and heat production increase, body temperature rises, and sweating increases. They act on fat cells, promoting the breakdown of neutral fats and cholesterol. They act on cardiac muscle cells, increasing contractility, heart rate, and cardiac output. They act on skeletal muscles, increasing muscle contractility and promoting the synthesis of muscle proteins. During the growth period, overall growth and development are promoted. They also act on brain and nerve cells, leading to maturation of the nervous system and acceleration of thought. They promote bone turnover and promote bone growth and development during the growth period. They promote the absorption of blood sugar from the digestive tract and the use of absorbed sugar due to increased basal metabolism. Thus, thyroid hormones, which are indispensable for metabolism

M. Hamaguchi (✉)
Department of Endocrinology and Metabolism, Kyoto Prefectural University of Medicine, Kyoto, Japan
e-mail: mhama@koto.kpu-m.ac.jp

H. Yamada, Y. Naito (eds.), *Anti-Aging Medicine*, https://doi.org/10.1007/978-981-97-8514-8_45

throughout the body, are regulated to be within an appropriate range by a negative feedback mechanism. The secretion of thyroid hormones is controlled by thyroid stimulating hormone (TSH). TSH is produced in the anterior pituitary gland, reaches the thyroid follicular cells via the bloodstream, and promotes the production of thyroid hormones by binding to the TSH receptors of the thyroid follicular cells. The production of TSH in the pituitary gland is controlled by thyrotropin-releasing hormone (TRH) secreted from the hypothalamus. When the concentrations of T3 and T4 increase, the secretion of TRH from the hypothalamus and the secretion of TSH from the pituitary gland are both suppressed.

45.2 Thyroid Dysfunction and Malaise

Thus, symptoms due to thyroid hormone abnormalities appear throughout the body. While thyroid dysfunction is diagnosed by measuring thyroid function, if thyroid dysfunction is not measured, it may be overlooked because it is difficult to detect abnormalities in general internal medicine examinations. In fact, there may be cases where the diagnosis of thyroid dysfunction is delayed because mental disorders such as depression, menopausal disorders, and indefinite complaints are suspected in cases who visited the hospital with the main complaint of malaise. Symptoms of thyroid dysfunction include symptoms of hyperfunction, hypofunction, and common symptoms (Fig. 45.1). If these symptoms are suspected, measurement of thyroid function is recommended. The diagnosis of thyroid disease is published and updated as needed from the Japan Thyroid Association, so please refer to it [1]. In addition to these overt thyroid dysfunctions, there is latent thyroid dysfunction. Latent thyroid dysfunction refers to a mild excess or deficiency of thyroid hormones that does not appear in symptoms or findings. Even though the free thyroid hormone in the blood (free T4; FT4) is within the reference range, only the simultaneously measured TSH is higher than the normal value, it is called latent hypothyroidism (a state where the thyroid hormone is slightly low and requires a lot of TSH stimulation to maintain normal), and if it is low, it is called latent thyrotoxicosis (a state where the thyroid hormone is slightly high and TSH stimulation is suppressed to maintain normal), and together they are called latent thyroid dysfunction.

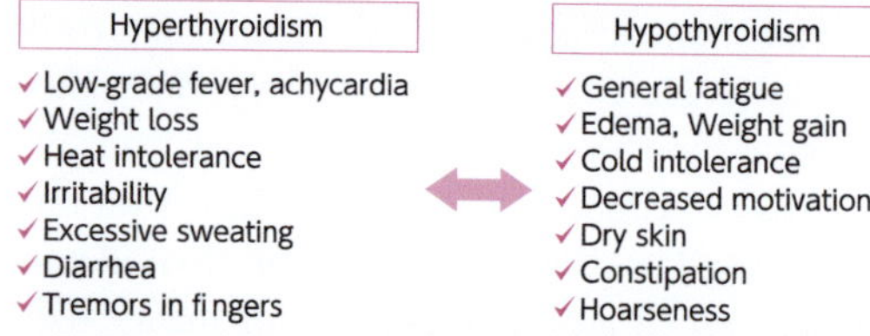

Fig. 45.1 Symptoms of hyperthyroidism and hypothyroidism

45.3 Latent Hypothyroidism and Aging

Hypothyroidism includes overt hypothyroidism with high TSH and latent hypothyroidism with no clear increase in TSH. Latent hypothyroidism basically occurs for the same reasons as overt hypothyroidism. Therefore, Hashimoto's disease, iodine excess (eating a lot of seaweed, drugs, contrast agents, etc.), after-thyroid surgery, and after-isotope treatment are the causes of latent hypothyroidism. Other causes include after-neck radiation exposure, recovery period of destructive thyroiditis, etc. When there is a lack of thyroid hormone, symptoms such as fatigue, lack of energy, and forgetfulness appear. Furthermore, in women, a relationship with infertility and miscarriage has been pointed out. Latent hypothyroidism is reported to be seen in 0.7–0.9% of women, 0.2–0.4% of men, overt hypothyroidism in 2.1–10.4% of women, and 0.4–9.6% of men.

45.4 Screening and Replacement Therapy for Hypothyroidism in the Elderly

It is known that the frequency of hypothyroidism increases with age, and overt hypothyroidism is reported to be found in 15% of women and 10% of men in their 70s. Furthermore, the prevalence of overt hypothyroidism in those aged 65 and over is reported to be increasing year by year [2]. In 2007, it was 8.4% in women and 2.7% in men, but by 2015, it had increased to 12.6% in women and 4.5% in men [2]. Even in the elderly, overt hypothyroidism may be thought to be due to aging and symptoms may not be recognized. Also, symptoms and signs may be atypical. On the other hand, hypothyroidism increases the relative risk of total mortality by 1.26 (95% CI 1.15–1.37). Therefore, it is recommended to evaluate thyroid function by screening in the elderly. On the other hand, the risk of death is between 60 and 79 years old, and it is not a clear risk in those over 80. There are reports that subclinical hypothyroidism is a risk of death, and those with a risk of cardiovascular disease are a risk of death, and those that are not. There are reports that levothyroxine supplementation for hypothyroidism in those aged 65 and over improves survival, but the effect was not clear in subclinical hypothyroidism [3]. The risk of heart failure also increases in subclinical hypothyroidism with TSH levels above 10 mIU/L, but it was not clear below 10 mIU/L. Hypothyroidism did not show a significant association with cognitive function, depressive symptoms, physical function, or frailty [4]. Levothyroxine supplementation for subclinical hypothyroidism significantly reduces TSH levels, but does not lead to improvement in clinical symptoms. Based on these results, "The Lancet Diabetes & Endocrinology" recommends levothyroxine supplementation for those aged 65 and over with TSH levels above 10 mIU/L and consideration of replacement therapy for those aged 65–84 with TSH levels between 7 and 9.9 mIU/L, but excessive treatment should be avoided [5]. If TSH is suppressed to below 0.1 mIU/L with replacement therapy, the risk of atrial fibrillation and fractures increases.

45.5 Conclusion

Metabolism associated with aging is regulated by thyroid hormones. Abnormal thyroid function is easily diagnosed by measuring thyroid function, but if thyroid function is not measured, diagnosis may be delayed. Be aware that symptoms associated with abnormal thyroid function, especially hypothyroidism, may be mistaken for aging. On the other hand, be aware that replacement therapy in subclinical hypothyroidism does not necessarily lead to improvement in clinical symptoms, and it is necessary to check the latest evidence and respond accordingly. Also, abnormal thyroid function requires caution as it has been pointed out in relation to female infertility and miscarriage.

References

1. Japan Thyroid Association. Thyroid disease diagnostic guidelines 2021. https://www.japanthyroid.jp/doctor/guideline/japanese.html. Revised 2 Jun 2022.
2. Lage MJ, Espaillat R, et al. Hypothyroidism treatment among older adults: evidence from a claims database. Adv Ther. 2020;37:2275–87.
3. Peng CC, Huang HK, et al. Association of thyroid hormone therapy with mortality in subclinical hypothyroidism: a systematic review and meta-analysis. J Clin Endocrinol Metab. 2021;106:292–303.
4. Du Puy RS, Poortvliet RKE, et al. Outcomes of thyroid dysfunction in people aged eighty years and older: an individual patient data meta-analysis of four prospective studies (towards understanding longitudinal international older people studies consortium). Thyroid. 2021;31:552–62.
5. Biondi B, Cappola AR, et al. Subclinical hypothyroidism in older individuals. Lancet Diabetes Endocrinol. 2022;10:129–41.

Hormones and Antiaging Medicine: DHEA and Antiaging

46

Michiaki Fukui

46.1 What Is DHEA?

Dehydroepiandrosterone (DHEA) exists in the body as DHEA sulfate (DHEA-S) and can be converted to each other. It is mainly produced in the reticular layer of the adrenal glands and the gonads, but is also produced in some peripheral tissues. The fluctuations in blood DHEA levels are almost in sync with the fluctuations in blood DHEA-S levels, and it is common to measure DHEA-S levels in general medical practice. Blood DHEA-S levels begin to increase around the age of 6–7, peak around the age of 12–13, maintain high levels until around the age of 13–25, and then decrease linearly with age (Fig. 46.1). In recent years, DHEA has been proven, mainly at the level of animal experiments, to have various beneficial biological effects on the body, such as immune activation, antidiabetic, anti-osteoporosis, anti-atherosclerosis, anti-obesity, antitumor, and central nervous system effects. Based on these results, DHEA supplementation therapy is being attempted in humans, mainly in Europe and the United States.

M. Fukui (✉)
Department of Endocrinology and Metabolism, Graduate School of Medical Science, Kyoto Prefectural University of Medicine, Kyoto, Japan
e-mail: michiaki@koto.kpu-m.ac.jp

H. Yamada, Y. Naito (eds.), *Anti-Aging Medicine*, https://doi.org/10.1007/978-981-97-8514-8_46

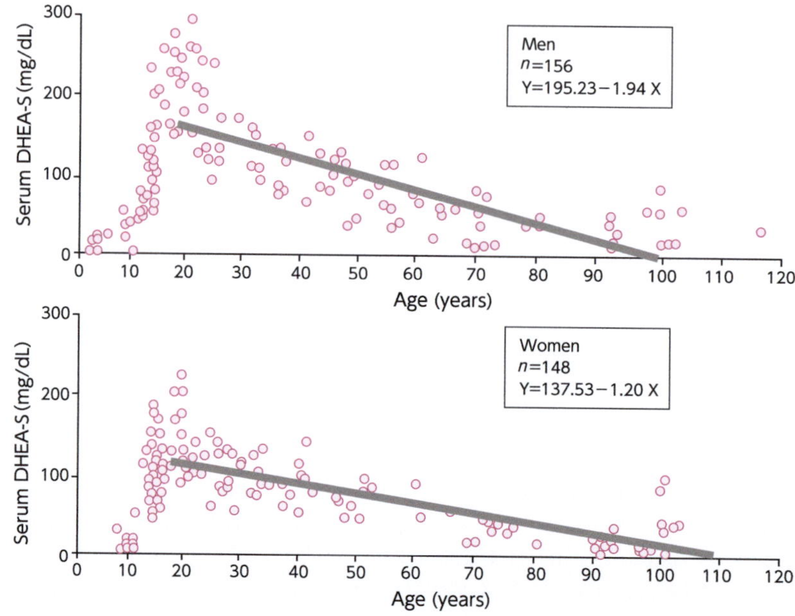

Fig. 46.1 Changes in serum DHEA-S concentration with aging. (Reprinted from [6] with permission)

46.2 Changes in Hormone Levels due to DHEA Supplementation

Tanaka et al. have investigated the fluctuations in blood DHEA-S levels 24 h after oral administration of DHEA (10–25 mg) in those who have been administered DHEA for more than a year and found that the blood concentration increases to about 2–3 times the pre-dose level 1–6 h after administration, returns to the pre-dose level 24 h later, and the average blood concentration of DHEA administered for 24 h is about twice the pre-dose level [1]. However, if DHEA is suddenly discontinued after more than a year of administration, the DHEA-S level significantly decreases 2 weeks after discontinuation, and it is lower than before DHEA administration, so it is thought that it needs to be gradually reduced when discontinuing. DHEA administration not only increases the blood concentration of DHEA itself, but also increases other sex hormones such as androstenedione, testosterone, and estradiol, acting as a prodrug and increasing the blood concentration of all sex hormones. It has been reported that there was no difference in the change in blood hormone concentration between 50 and 25 mg/day administration, so it is considered that a supplement of 25 mg/day could be effective for Japanese people.

46.3 Multifaceted Effects of DHEA Supplementation

Recent review articles have shown the effects of DHEA supplementation therapy on various diseases and conditions [2]. The effect of DHEA on obesity is being targeted… Administering 50 mg/day or placebo for 6 months, it has been reported that both visceral fat area and subcutaneous fat area evaluated by MRI decreased only in the DHEA group, and insulin resistance improved only in the DHEA group. It has also been reported that when DHEA is supplemented in men in their 60s, improvements in endothelial function and insulin sensitivity are observed. Several sex hormones, including DHEA, have been shown to act like neurotransmitters in the brain, and it has been reported that they may be related to memory and cognitive function as neurosteroids. In humans, it has been reported that there is a significant correlation between cognitive function and blood DHEA concentration in patients with Alzheimer's disease and schizophrenia. It has been recognized that DHEA supplementation improves the decrease in bone density in postmenopausal women. From the results in vitro, it has been reported that DHEA directly promotes the differentiation and proliferation of osteoblasts and inhibits bone resorption by osteoclasts, and it is expected to have an osteoporosis inhibitory effect by a mechanism different from bisphosphonates and estrogen. DHEA may prevent the onset of osteoporosis in postmenopausal women by being locally converted to estrogen, that is, by intracrine action.

46.4 Effectiveness of DHEA Supplementation

In a trial of 50 mg of DHEA/day for 24 female adrenal insufficiency patients conducted by Arlt et al., not only the blood concentration of DHEA but also DHEA-S, androstenedione, and testosterone levels increased after 4 months of DHEA administration, and it was shown that the sense of health and sexual thoughts and interests increased, and the usefulness of DHEA supplementation therapy was reported [3]. On the other hand, in the results reported by Nair et al., in a double-blind trial using DHEA for 27 people and placebo for 30 people among 57 elderly women for 2 years, although an increase in blood DHEA concentration and bone density was observed in the DHEA administration group, no changes were observed in body fat mass or insulin sensitivity, and no improvement was observed in quality of life (QOL) [4]. Nair et al. also conducted a trial comparing DHEA or testosterone with placebo for 87 elderly men and reported that neither DHEA nor testosterone supplementation therapy affected body composition, physical ability, QOL, or insulin sensitivity in men. In a randomized controlled trial conducted by Corona et al., the results of a meta-analysis of randomized controlled trials (RCTs) showed that administering DHEA for an average of 36 weeks to 1353 elderly men significantly reduced body fat. However, this significance disappeared when corrected for other sex hormones such as testosterone. It also reported that it did not improve glucose metabolism, bone metabolism, sexual function, or QOL [5]. Many other administration trials have been conducted, and the effectiveness has been judged from various perspectives such as body composition, glucose metabolism, and cognitive function, but the results that DHEA supplementation is effective are limited to cases of adrenal insufficiency and clear androgen deficiency. While DHEA supplementation therapy is expected to be a new treatment in antiaging medicine, so far, the results have not been as good as expected. Even if similar studies are conducted, the results often differ from trial to trial, and the problem is that stable results have not been obtained. One of the factors is that each trial is small in scale and short in duration, resulting in large variations in results, and a consistent view has not been reached. Hormones have an optimal concentration to work effectively, and any hormone can have harmful effects on the body due to excess or deficiency. Considering these problems, clinical studies with expanded trial scales and durations are being attempted overseas, but it can be said that it is necessary to conduct our own large-scale clinical trials in our country. When using DHEA in hormone replacement therapy, it may be necessary to do so as part of personalized medicine, taking into account various genetic and environmental factors, including racial differences. For this, it is necessary not only for clinical trials but also for the molecular mechanism to be elucidated. Whether DHEA supplementation therapy will be a blessing for an aging society will be brought about by the mutual development of basic medicine and clinical medicine in the future, and it is hoped that more interest will be gathered in this field in the future.

References

1. Tanaka T, Nagata Y, et al. Study on long-term DHEA supplementation therapy. Hormone therapy for the elderly and elderly. Mod Physician. 2007;27:1145–8.
2. Sahu P, Gidwani B, et al. Pharmacological activities of dehydroepiandrosterone: a review. Steroids. 2020;153:108507.
3. Arlt W, Callies F, et al. Dehydroepiandrosterone replacement in women with adrenal insufficiency. N Engl J Med. 1999;341:1013–20.
4. Nair KS, Rizza RA, et al. DHEA in elderly women and DHEA or testosterone in elderly men. N Engl J Med. 2006;3:1647–59.
5. Corona G, Rastrelli G, et al. Dehydroepiandrosterone supplementation in elderly men: a meta-analysis study of placebo controlled trials. J Clin Endocrinol Metab. 2013;98:3615–26.
6. Yanase T. III. Basics of anti-aging medicine. F. Hormones and anti-aging medicine. 3. Adrenal hormones. In: Basics and clinical of anti-aging medicine. 3rd ed. Tokyo: Medical View Co., Ltd.; 2015. p. 109.

47 Hormones and Antiaging Medicine: Adipokines and Antiaging

Nobuhiro Shojima and Toshimasa Yamauchi

While adipose tissue stores triglycerides in preparation for starvation, it also produces and secretes various physiologically active substances known as adipokines. Adipokines not only act on adipose tissue but also function as hormones, affecting distant organs and regulating glucose and lipid metabolism as well as vascular function. Abnormal expression of adipokines plays a significant role in the onset of metabolic syndrome. It has become clear that with aging, subcutaneous fat decreases, visceral fat increases, brown fat decreases, and adipokines change. These changes in adipokines are thought to promote aging. This section provides an overview of the relationship between aging, antiaging, and the representative adipokines, adiponectin and leptin.

47.1 Adiponectin

Adiponectin is a physiologically active substance that is specifically expressed and secreted by adipocytes and has been independently discovered by four groups worldwide and named by Matsuzawa and his team at Osaka University [1]. As obesity increases, the blood level of adiponectin decreases and animal experiments have shown that replenishing it by injection improves glucose tolerance and lipid metabolism abnormalities. The decrease in adiponectin levels due to obesity is at least partly responsible for the onset of metabolic syndrome.

47.2 Enhancement of Insulin Sensitivity by Adiponectin

Adiponectin reduces the tissue neutral fat content, which is a cause of insulin resistance, by promoting the burning of fatty acids in the liver. The AMP-activated protein kinase (AMPK) in skeletal muscle, which is reported to be activated by exercise, promotes insulin-independent glucose uptake and fatty acid burning, thereby supplying the energy needed for exercise. Adiponectin activates AMPK. In skeletal muscle, insulin sensitivity is enhanced through AMPK signaling, activating mitochondrial function and fatty acid oxidation. The action of adiponectin in skeletal muscle was found to improve the quantity and function of mitochondria, enhancing metabolism and exercise endurance, and producing effects similar to those of exercise [1].

47.3 Adiponectin Receptor and Exercise Mimicry Effect

The molecule believed to transmit the insulin-sensitizing effect of adiponectin is called the adiponectin receptor. There are two molecules of the

N. Shojima (✉) · T. Yamauchi
Department of Diabetes and Metabolism, The University of Tokyo, Tokyo, Japan
e-mail: nshojima-tky@umin.ac.jp

H. Yamada, Y. Naito (eds.), *Anti-Aging Medicine*, https://doi.org/10.1007/978-981-97-8514-8_47

adiponectin receptor, AdipoR1 and AdipoR2. AdipoR1 is highly expressed in skeletal muscles and the liver, while AdipoR2 is abundantly expressed in the liver [2]. Activation of AdipoR1 inhibits hepatic gluconeogenesis and fatty acid synthesis, while activation of AdipoR2 burns fatty acids. It was discovered that adiponectin, through AdipoR1 in skeletal muscles, not only increases intracellular Ca^{2+} concentration, but also activates the longevity gene Sirtuin 1, mimicking the antiaging effects of exercise [1, 3].

47.4 Adiponectin and Anti-atherosclerotic Effects

It is suggested that adiponectin may exert anti-atherosclerotic effects by directly acting on vascular endothelial cells and infiltrating inflammatory monocytes/macrophages involved in atherosclerotic lesion formation, inhibiting lipid accumulation and exerting anti-inflammatory effects.

47.5 Adiponectin Receptor Agonist AdipoRon

Enhancing the action of adiponectin could potentially be a fundamental treatment strategy for diabetes and metabolic syndrome through its antiaging effects. The authors found that a low molecular weight compound of about 430, like adiponectin, binds to the adiponectin receptor and increases the amount of mitochondria in skeletal muscle cells [4]. Furthermore, oral administration to mice improved insulin resistance caused by a high-fat diet through the adiponectin receptor. It has been reported that diabetes gene homozygous mice (db/db), which develop diabetes from overeating, become significantly short-lived when a high-fat diet is added as an environmental factor. Even when consuming a high-fat diet, simultaneous administration of AdipoRon did not significantly affect body weight, but improved diabetes and significantly extended lifespan. It was also reported that AdipoRon promotes hair growth in mice.

47.6 Antiaging Strategy Targeting Adiponectin Receptors

The decrease in adiponectin action associated with obesity has been clarified as a major cause and molecular mechanism of lifestyle-related diseases, such as insulin resistance, metabolic syndrome, diabetes, cardiovascular disease, and cancer, which are at increased risk with obesity. Drug development targeting adiponectin receptors can be an antiaging strategy against these diseases and contribute to the realization of healthy longevity. The elucidation of the three-dimensional structure of the human adiponectin receptor is expected to accelerate clinical application to humans [5]. In addition, the metabolic improvement effect has been confirmed in mice expressing the human adiponectin receptor.

47.7 Leptin

Leptin is a factor identified as a causative gene in overeating obese mice, secreted from fat cells and acting on the feeding center in the hypothalamus to suppress appetite. In a model of lipoatrophic diabetes in mice, there is a deficiency of white adipose tissue, severe insulin resistance and diabetes, severe fatty liver, and hypertriglyceridemia, but these are improved by the administration of leptin. Patients with lipoatrophic diabetes in humans also have similar conditions, and improvements can be seen when leptin is administered.

47.8 Leptin and Leptin Resistance

Leptin not only acts centrally, but also improves glucose and lipid metabolism by acting on the liver and skeletal muscles. In simple obesity, the expression of leptin in adipose tissue is high and increases with the degree of obesity. In obesity, leptin becomes resistant to leptin, and metabolic abnormalities occur. It has been reported that leptin resistance appears with aging.

References

1. Yamauchi T, Kadowaki T. Adiponectin receptor as a key player in healthy longevity and obesity-related diseases. Cell Metab. 2013;17:185–96.
2. Yamauchi T, Kamon J, et al. Cloning of adiponectin receptors that mediate antidiabetic metabolic effects. Nature. 2003;423:762–9.
3. Iwabu M, Yamauchi T, et al. Adiponectin and AdipoR1 regulate PGC-1α and mitochondria by Ca2+ and AMPK/SIRT1. Nature. 2010;464:1313–9.
4. Okada-Iwabu M, Yamauchi T, et al. A small-molecule AdipoR agonist for type 2 diabetes and short life in obesity. Nature. 2013;503:493–9.
5. Tanabe H, Fujii Y, et al. Crystal structures of the human adiponectin receptors. Nature. 2015;520:312–6.

48 Hormones and Antiaging Medicine: Endothelial Cells as Endocrine Cells

Yuichi Oike

48.1 Endothelial Cells Are Endocrine Organs

Maintaining the morphology of the circulatory system and its normal functioning is the most basic and essential for life support. The length of blood vessels extends to about 100,000 km in total, and except for a few exceptions such as the cornea and epithelium, they are distributed in all parts of the body, and like tree branches, make nonoverlapping branches to enjoy sunlight and blood vessels are spread out so that oxygen and nutrients can be supplied to peripheral tissues without excess or deficiency. Therefore, as can be easily imagined from the famous quote by Dr. William Osler (1849–1919), "Humans age with blood vessels;" the failure of the vascular system as a route for oxygen and nutrient supply leads to organ aging, which appears as changes and failures in various tissue functions. Endothelial cells, which exist as the innermost layer of cells on the inner side of the vascular lumen anatomically, produce and secrete various physiologically active factors such as nitric oxide (NO) and endothelin and play an important role in maintaining the homeostasis of vascular functions such as contraction and relaxation (dilation) regulation, inflammation and permeability regulation, and coagulation and fibrinolytic system regulation. The total area covered by endothelial cells is 7000 km^2, the total weight is 1 kg, and it is the largest endocrine organ in the human body, and its dysfunction is the basis of various brain and cardiovascular diseases whose onset risk increases with age.

48.2 Discovery of NO as a Vascular Physiologically Active Factor

In 1980, Dr. Robert F. Furchgott found that the presence of endothelial cells is essential for the vasodilatory action shown by the neurotransmitter acetylcholine and speculated that a vasodilatory substance (endothelium-derived relaxing factor; EDRF) is produced from endothelial cells by acetylcholine [1]. Dr. Louis J. Ignarro clarified that the main body of EDRF is NO, and on the other hand, nitroglycerin, which is also an explosive and was used as a drug for heart diseases such as angina and hypertension without knowing the mechanism of action, is metabolized in the body to become NO, and Dr. Ferid Murad found that NO is involved in vasodilation. For these achievements, "Discovery of nitric oxide as a signaling molecule in the cardiovascular system," the 1998 Nobel Prize in Physiology or Medicine was awarded to the three.

Y. Oike (✉)
Department of Molecular Genetics, Graduate School of Medical Sciences, Kumamoto University, Kumamoto, Japan
e-mail: oike@gpo.kumamoto-u.ac.jp

H. Yamada, Y. Naito (eds.), *Anti-Aging Medicine*, https://doi.org/10.1007/978-981-97-8514-8_48

48.3 NO Production Mechanism in Endothelial Cells

In endothelial cells, endothelial nitric oxide synthase (eNOS) produces NO using the essential amino acid L-arginine as a substrate (Fig. 48.1). The activation of eNOS, which is key in this process, is mainly regulated by two mechanisms (Fig. 48.2). First, agonists present in the blood, such as acetylcholine and serotonin, bind to receptors expressed on vascular endothelial cells, activating inositol trisphosphate (IP3) through the G protein/PLC signaling pathway, causing the release of Ca^{2+} from the endoplasmic reticulum and an increase in intracellular Ca^{2+} concentration. This forms a complex with calmodulin, activating eNOS. Another mechanism is the activation of the PI3K/Akt signaling pathway by "shear stress" based on blood flow in the vessels, which also activates eNOS. Certain drugs such as statins, estrogen, and "shear stress" also contribute to the transcriptional activation of eNOS. Given that "shear stress" is constantly occurring in the vessels, it is clear that NO plays an important role in maintaining vascular physiological functions such as vasodilation (expansion), anti-inflammatory, and anticoagulant effects by being constantly produced and secreted from vascular endothelial cells. In mice with eNOS gene deficiency, in addition to vascular endothelial dysfunction, obesity, hypertension, lipid metabolism abnormalities, and glucose metabolism abnormalities (insulin resistance) are observed [2], demonstrating the importance of NO produced and secreted from vascular endothelial cells in disease prevention.

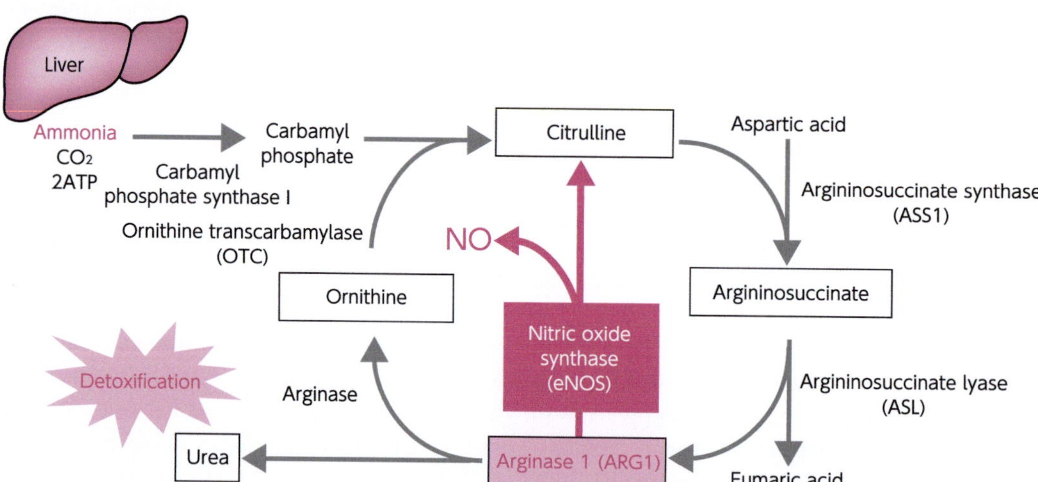

Fig. 48.1 Arginine metabolism and NO synthesis system. It is important to take in enough arginine from food (soy, chicken, sesame, brown rice, cocoa, etc.) or supplements for detoxification and NO synthesis

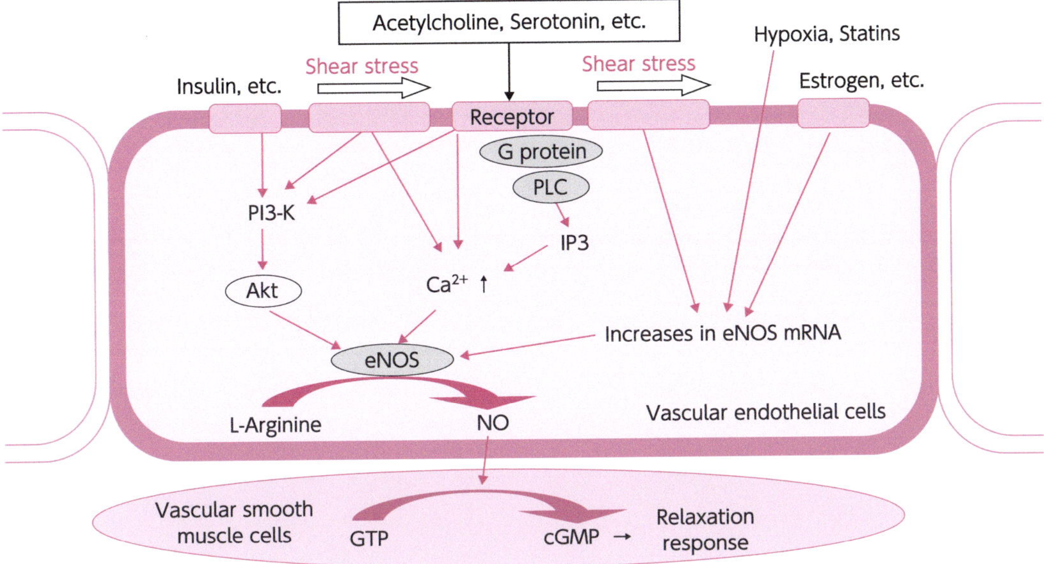

Fig. 48.2 NO production mechanism and action in vascular endothelial cells

48.4 NO and Aging, Age-Related Vascular Diseases

With aging, NO production and blood levels decrease, dropping sharply in men in their 40s and women in their 50s [3]. Given that nitroglycerin, an angina drug, is metabolized into NO in the body, it is easy to imagine that NO is beneficial in the prevention and treatment of cardiovascular diseases from the perspective of vascular protection. Indeed, people with low blood NO levels have an increased incidence of cardiovascular diseases, and improvements in cardiovascular diseases have been shown by increasing blood NO levels. Since a decrease in NO production is involved in the increased risk of various age-related diseases, including cardiovascular diseases, maintaining blood NO levels is important for health maintenance. Unlike nitroglycerin intake, NO produced and secreted from vascular endothelial cells is regulated by the enzymatic reaction of eNOS and is not overproduced, so it is recommended to consume foods and supplements rich in arginine, a substrate (Fig. 48.2). Also, aerobic exercises such as walking and jogging are expected to increase NO production from vascular endothelial cells by improving blood flow and increasing "shear stress" stimulation. When considering antiaging from the perspective of maintaining vascular function, it is important to practice a lifestyle that is conscious of NO levels derived from vascular endothelial cells, such as dietary habits and exercise habits.

48.5 Other Physiologically Active Factors Derived from Vascular Endothelial Cells

Due to space constraints, this section focuses on NO and provides an overview. The angiopoietin-like factor 2 (ANGPTL2) that we identify and analyze is also produced and secreted from vascular endothelial cells. In addition, the ANGPTL2 signal, which is enhanced with aging, acts on vascular endothelial cells in an autocrine manner, causing vascular inflammation, leading to a decrease in vascular endothelial dilation function, promoting arteriosclerosis, and accelerating the onset of cardiovascular diseases [4]. Many physiologically active factors derived from vascular endothelial cells that are involved in the onset of aging-related diseases due to overproduction or deficiency have been identified, suggesting the importance of maintaining the normal function of vascular endothelial cells, the largest endocrine organ in the human body, for antiaging.

References

1. Furchgott RF, Zawadzki JV. The obligatory role of endothelial cells in the relaxation of arterial smooth muscle by acetylcholine. Nature. 1980;288:373–6.
2. Duplain H, Burcelin R, et al. Insulin resistance, hyperlipidemia, and hypertension in mice lacking endothelial nitric oxide synthase. Circulation. 2001;104:342–5.
3. Celermajer DS, Sorensen KE, et al. Aging is associated with endothelial dysfunction in healthy men years before the age-related decline in women. J Am Coll Cardiol. 1994;24:471–6.
4. Horio E, Kadomatsu T, et al. Role of endothelial cell-derived angptl2 in vascular inflammation leading to endothelial dysfunction and atherosclerosis progression. Arterioscler Thromb Vasc Biol. 2014;34:790–800.

49 Hormones and Antiaging Medicine: Melatonin and Antiaging

Atsuhiko Hattori

49.1 The Synthesis Pathway of Melatonin

Melatonin is a hormone synthesized from tryptophan via serotonin. How the clock information of the suprachiasmatic nucleus (the main clock) is transmitted to the pineal gland is shown in Fig. 49.1 [1]. The oscillations produced by the transcription/translation feedback mechanism of the so-called "clock genes" such as Clock, Bmal1, Per, Cry, etc., expressed in the suprachiasmatic nucleus, are not exactly 24 h, but a 24 ± 4 h cycle, hence they are called circadian clocks, and each species shows its own unique cycle. This clock mechanism is precisely synchronized to a 24-h cycle by stimuli from the outside world that follow a 24-h cycle (synchronizing factors). In other words, the 24-h light information received by the retina is transmitted to the suprachiasmatic nucleus, aligning the phase of the body clock with the outside world and synchronizing the clock cycle to a 24-h cycle. Next, the time information generated in the suprachiasmatic nucleus stimulates the pineal gland cells only at night via norepinephrine (NE), resulting in the activation of arylalkylamine *N*-acetyltransferase (AANAT), the rate-limiting enzyme for melatonin synthesis. On the other hand, during the day, NE is not secreted, so AANAT is hardly synthesized, and as a result, the concentration of melatonin in the blood shows a clear diurnal variation, being low during the day and high at night. This clear diurnal rhythm of melatonin is observed not only in the blood but also in all body fluids such as cerebrospinal fluid, saliva, aqueous humor, follicular fluid, and even breast milk. In other words, melatonin is a messenger that transmits the "night" time information to the whole body under the command of the body clock.

A. Hattori (✉)
Department of Biology, College of Liberal Arts and Sciences, Tokyo Medical and Dental University, Chiba, Japan

Department of Sport and Wellness, College of Sport and Wellness, Rikkyo University, Saitama, Japan
e-mail: ahattori1005@rikkyo.ac.jp

H. Yamada, Y. Naito (eds.), *Anti-Aging Medicine*, https://doi.org/10.1007/978-981-97-8514-8_49

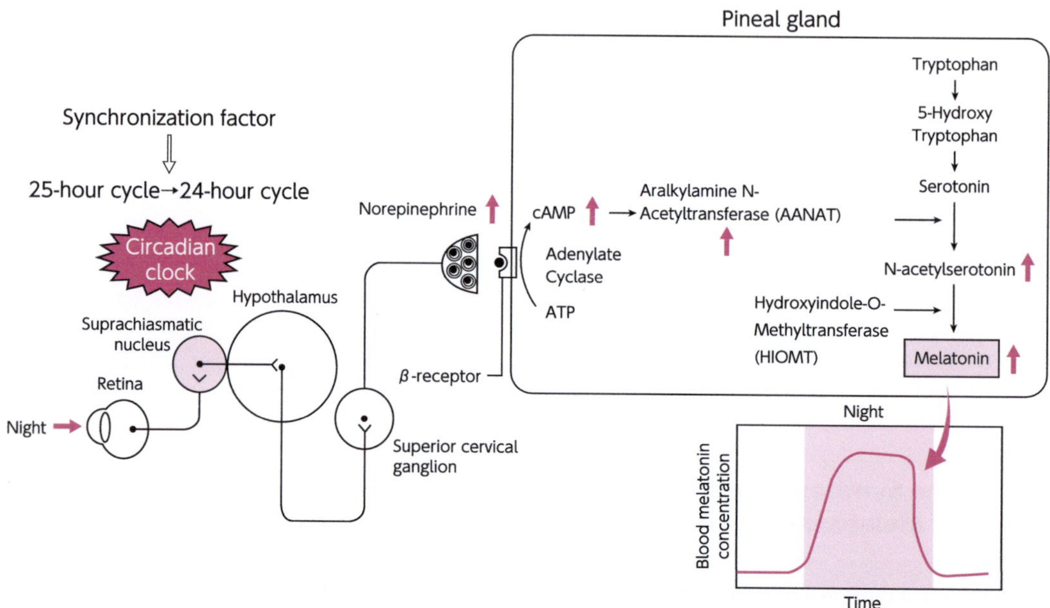

Fig. 49.1 Melatonin synthesis pathway in the human pineal gland.
Melatonin is produced only at night in the pineal gland. The arrows (↑) in the figure indicate the sites that are activated (increased) at night.

(Quoted from Junhiko Hattori: Melatonin. Anti-Aging Medicine: Basics and Clinical Practice. Revised 2nd Edition, Medical View Co., Ltd., Tokyo, 2008, p. 103–6)

49.2 Melatonin and Disease

However, the amount of melatonin secreted at night decreases dramatically with age. On the other hand, even among young people, there are those with low melatonin levels at night. The biggest factor is "exposure to light at night". Especially, blue light has a strong inhibitory effect. It is well known that the use of smartphones and PCs at night causes insomnia, and one of the causes is believed to be a decrease in melatonin [2]. Since melatonin can easily pass through the blood-brain barrier, it is effective even when administered orally. With an intake of about 0.1–0.3 mg, the concentration of melatonin in the blood reaches almost nighttime levels after about 30 min. Many of the reports on the effectiveness of melatonin for insomnia have targeted the elderly. One reason for this is that as people age, there is no difference in melatonin levels between day and night, resulting in a decrease in the contrast between day and night. This is thought to be the main cause of the common "irregular sleep-wake type sleep disorder" (often waking up at night and feeling sleepy during the day). There are many reports that taking melatonin is effective for elderly people with such insomnia. In addition, many reports have been made on the correlation between various diseases and the decrease in melatonin secretion at night. Associations have been reported with the onset risk of diseases such as breast cancer, prostate cancer, cardiovascular diseases, Alzheimer's disease, and more recently, type 2 diabetes. On the other hand, known receptors for melatonin are the G protein-coupled membrane receptors MT1 and MT2, but melatonin is also a scavenger of free radicals. The characteristics of melatonin as an antioxidant include (1) the ability to easily pass through the blood-brain barrier, (2) not only is it a scavenger of free radicals, but the substances produced after scavenging, *N*1-acetyl-*N*2-formyl-5-methoxykynuramine (AFMK) and *N*1-acetyl-5-methoxykynuramine (AMK), also have the ability to scavenge radicals, and (3) melatonin bound to membrane receptors enhances the expression of antioxidant enzymes in the body (such as superoxide dismutase; SOD). Recently, it has been reported that AMK metabolized in the brain has an effect that induces long-term memory

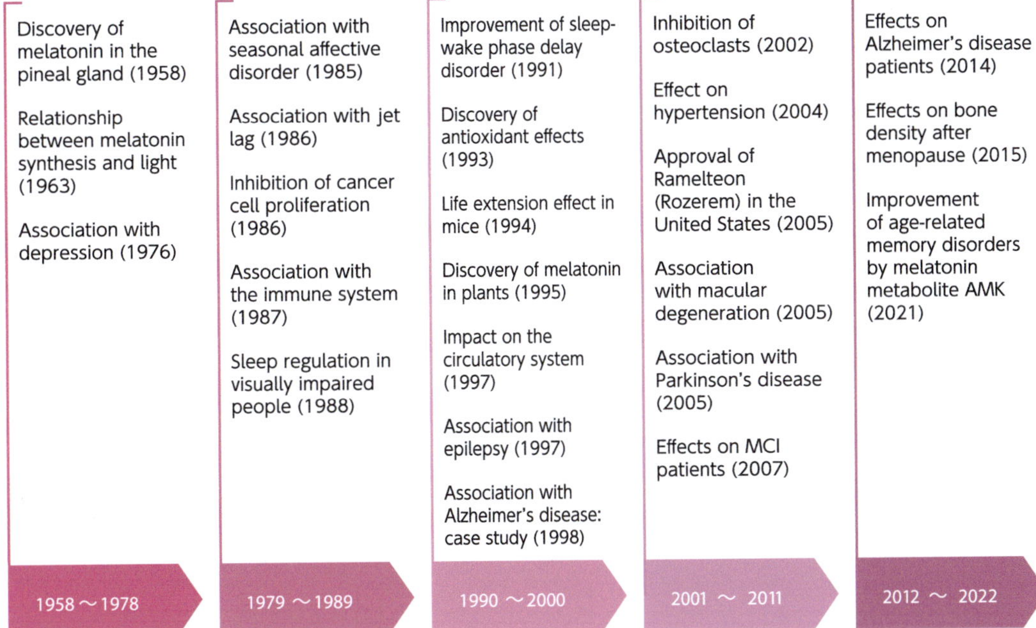

Fig. 49.2 The main historical discoveries about melatonin, mainly related to humans, are listed

(memory consolidation effect) [3]. It has been said that memory is consolidated during sleep, and this can be considered a study that clarifies part of the mechanism. MT1/MT2 are present not only in the suprachiasmatic nucleus, but also in various organs throughout the body (cerebrum, retina, spinal cord, spleen, thymus, pituitary gland, adrenal gland, kidney, liver, small intestine, heart, skin, lung, testis, ovary, blood vessels, bone). Figure 49.2 shows the history of major discoveries related to melatonin, focusing on humans. Recently, reports have also been made on the relationship between polymorphisms of melatonin receptors and various diseases. It is not hard to imagine a relationship between insomnia and MT1 polymorphisms, but reports have been made on the relationship between a certain MT2 polymorphism and an increased risk of developing type 2 diabetes, and between MT1 polymorphisms and acute myocardial infarction, rheumatoid arthritis, and kidney stones. In addition, drugs targeting MT1/MT2 receptors (such as ramelteon) have been developed. Melatonin is an evolutionarily old substance and exists not only in animals but also in plants [4]. It has recently been reported that people who continue to consume foods rich in melatonin have a lower mortality rate [5]. In the future, melatonin is expected to become an increasingly useful substance for antiaging, not only from the perspective of pharmaceuticals but also from the perspective of health foods.

References

1. Hattori J. Melatonin. Anti-aging medicine: basics and clinical practice. Revised 2nd ed. Tokyo: Medical View Co.; 2008. p. 103–6.
2. Ayaki M, Hattori A, et al. Protective effect of blue-light shield eyewear for adults against light pollution from self-luminous devices used at night. Chronobiol Int. 2016;33:134–9.
3. Iwashita H, Matsumoto Y, et al. The melatonin metabolite N1-acetyl-5-methoxykynuramine facilitates long-term object memory in young and aging mice. J Pineal Res. 2021;70:e12703.
4. Hattori A, Migitaka H, et al. Identification of melatonin in plants and its effects on plasma melatonin levels and binding to melatonin receptors in vertebrates. Biochem Mol Biol Int. 1995;35:627–34.
5. Nagata C, Wada K, et al. Associations between dietary melatonin intake and total and cause-specific mortality among Japanese adults in the Takayama study. Am J Epidemiol. 2021;190:2639–46.

Hormones and Antiaging Medicine: Insulin and Antiaging

50

Kohjiro Ueki

50.1 Insulin Signaling

The signaling of insulin or its similar IGF-1 (insulin-like growth factor-1) is a positive regulator of cell proliferation, reproduction, and metabolism, and the signaling pathway is conserved from nematodes to mammals (Fig. 50.1). In nematodes and flies, it is thought to suppress stress resistance through various mechanisms. In nematodes and flies, although there are multiple insulin-like peptides (ILPs), orthologs of insulin or IGF-1, including those that show inhibitory effects on receptors, it is thought that a single receptor mediates both metabolic and proliferative effects, unlike in mammals. However, it is not entirely clear how these are regulated. On the other hand, in mammals, insulin mainly has a metabolic effect, and IGF-1 has a cell proliferative effect, but the two are somewhat complementary. This is thought to be because each hormone can bind not only to its own receptor, such as the insulin receptor (IR) or the IGF-1 receptor (IGF-1R), but also to different receptors with low affinity, and the two types of receptors can form hybrid receptors.

K. Ueki (✉)
Diabetes Research Center, Research Institute, National Center for Global Health and Medicine, Tokyo, Japan
e-mail: uekik@ri.ncgm.go.jp

H. Yamada, Y. Naito (eds.), *Anti-Aging Medicine*, https://doi.org/10.1007/978-981-97-8514-8_50

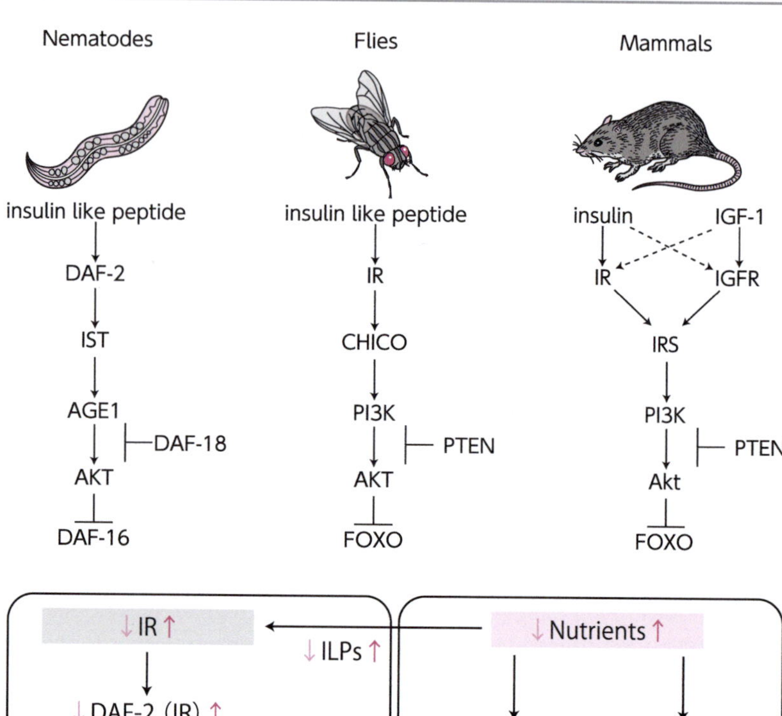

Fig. 50.1 Insulin/IGF-1 signaling in nematodes, flies, and mammals. From nematodes to mammals, the basic signaling molecules are conserved, but in mammals, in addition to the existence of IR and IGF-1R, there are multiple isoforms of IRS and PI3K, forming more complex signaling pathways

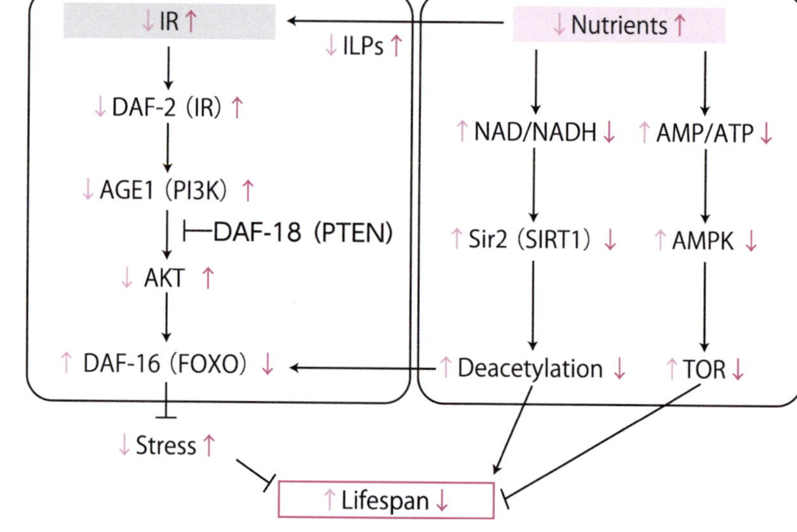

Fig. 50.2 Aging/lifespan regulation by insulin/nutrient signaling in nematodes. Insulin signaling promotes aging by inhibiting the activity of the transcription factor FOXO. Nutrients increase the production of ILP, activating insulin signaling, and at the same time, inhibit the activity of Sir2, which activates FOXO

50.2 Insulin Signaling and Lifespan in Animal Models

In nematodes and flies, it has been shown that the insulin signal of insulin receptor → PI3 kinase (phosphoinositide-3 kinase; PI3K) → AKT negatively regulates lifespan. In fact, lifespan is extended by the action of reduced mutants of DAF2, AGE1, AKT1/2, etc., which are orthologs of these molecules in nematodes. This is thought to be because AKT, which is activated by the insulin signal, phosphorylates the transcription factor FOXO (DAF-16 in nematodes), which suppresses stress signals such as oxidative stress and expels it from the nucleus, thereby inhibiting its activity [1–3]. In addition, this pathway is activated by increased production of insulin-like peptides due to nutrition, but it is also controlled by the decrease in activity of Sir2 due to nutritional status and the increase in activity of the TOR pathway [4] (Fig. 50.2). On the other hand, unlike nematodes and flies, in humans, the inactive type mutation of the insulin receptor, which is recognized to extend lifespan in lower animals, often results in severe diabetes and a short lifespan. In mice, a systemic knockout of the insulin receptor results in death due to ketosis immediately after birth [5]. Also, homozygous deficiency of the IGF-1 receptor or knockout of the p110α or p110β subunits of PI3K in mice results in fetal death, and double deficiency of Akt1/2 is also lethal [6]. In lower animals, knockout of Akt1 and Akt2, which extend lifespan, results in sarcopenia and osteoporosis, characteristics of aging, and a shortened lifespan in mice with skeletal muscle-specific knockout [7]. On the other hand, lifespan extension is observed in mice with heterozygous deficiency of IGF-1R [8], and knockout mice and fibroblasts derived from them were resistant to oxidative stress.

50.3 Insulin Signaling and Antiaging

The differences between lower animals and mammals are not clear. In mammals, the effects of insulin are diversified in each organ, and the impact of metabolic disorders due to decreased insulin action on aging may be greater. In humans, hyperinsulinemia caused by obesity is known to promote the onset of arteriosclerosis and malignant tumors, so in antiaging, it is considered important to ensure sufficient insulin action in each organ, and it is thought that maintaining blood insulin levels within a very narrow range may be appropriate.

References

1. Kimura KD, Tissenbaum HA, et al. daf-2, an insulin receptor-like gene that regulates longevity and diapause in Caenorhabditis elegans. Science. 1997;277:942–6.
2. Ogg S, Paradis S, et al. The Fork head transcription factor DAF-16 transduces insulin-like metabolic and longevity signals in C. elegans. Nature. 1997;389:994–9.
3. Kenyon C, Chang J, et al. A C. elegans mutant that lives twice as long as wild type. Nature. 1993;366:461–4.
4. Kenyon CJ. The genetics of ageing. Nature. 2010;464:504–12.
5. Accili D, Drago J, et al. Early neonatal death in mice homozygous for a null allele of the insulin receptor gene. Nat Genet. 1996;12:106–9.
6. Nandi A, Kitamura Y, et al. Mouse models of insulin resistance. Physiol Rev. 2004;84:623–47.
7. Sasako T, Umehara T, et al. Deletion of skeletal muscle Akt1/2 causes osteosarcopenia and reduces lifespan in mice. Nat Commun. 2022;13:5655.
8. Holzenberger M, Dupont J, et al. IGF-1 receptor regulates lifespan and resistance to oxidative stress in mice. Nature. 2003;421:182–7.

51 Hormones and Antiaging Medicine: Male Hormones and Antiaging

Hisamitsu Ide

51.1 Production and Physiological Action of Testosterone

Male hormones (androgens) include testosterone (T), which is mainly produced in the Leydig cells in the interstitium of the testes, and dehydroepiandrosterone (DHEA) derived from the adrenal glands. T is also produced in the hippocampus, muscles, and fat. About 98% of T in the blood is bound to sex hormone-binding globulin (SHBG) and albumin, and the remaining 1–2% exists as free (unbound) T [1]. Total T includes both bound and free forms. In target tissues, T is converted to dihydrotestosterone (DHT), which has high activity, by the action of 5α-reductase. It is believed that adrenal-derived DHEA also acts by being converted to T or DHT. The androgen receptor (AR), a receptor for T and DHT, is a nuclear receptor. When it binds to T or DHT, it acts as a transcription factor for target genes, showing various biological activities. T has diurnal variations and also varies with stress and lifestyle habits. T plays many important physiological roles, including differentiation of male reproductive organs, manifestation of secondary sexual characteristics, sperm formation, and effects on muscles, bones, central nervous system, prostate, bone marrow, skin, and sexual function.

51.2 Changes in Testosterone with Aging

The increase in T during adolescence is essential for physical development such as spermatogenesis. T continues to rise until around the age of 20 and then begins to decrease gradually. DHEA, which is secreted from the adrenal glands, also decreases with age. T decreases by about 1–3% per year, and DHEA decreases by up to 4% [2]. These continue to decrease until around the age of 90, and functional disorders occur due to the decrease in sex hormones. T is associated with dependency in life, quality of daily life, falls, balance maintenance, and walking. Observational studies on the association between T levels and life prognosis have been reported, but a stronger correlation is seen with mortality in elderly men. The decrease in T causes various health problems such as decreased muscle strength, obesity, diabetes, depression, decreased motivation, decreased sexual desire, and lowered quality of life (QOL).

H. Ide (✉)
Department of Urology, Dokkyo University School of Medicine, Saitama, Japan

Department of Urology, Juntendo University Graduate School of Medicine, Tokyo, Japan
e-mail: h.ide.me@juntendo.ac.jp

H. Yamada, Y. Naito (eds.), *Anti-Aging Medicine*, https://doi.org/10.1007/978-981-97-8514-8_51

51.3 Late-Onset Hypogonadism (LOH) Syndrome in Aging Men

Among hypogonadism conditions, the syndrome caused mainly by the decrease in T levels due to aging or stress is referred to as LOH (late-onset hypogonadism). Symptoms of LOH syndrome include general fatigue, decreased libido, lack of motivation, erectile dysfunction (ED), decreased concentration, insomnia, irritability, shoulder stiffness, heaviness in the head, and a decrease in early morning erections, among others. In addition, a decrease in T in aging men is associated with depressive states, decreased sexual function, decreased cognitive function, osteoporosis, cardiovascular disease, increased visceral fat, worsening insulin resistance, and increases in total cholesterol and low-density lipoprotein (LDL) cholesterol, which are risk factors for cardiovascular disease and diabetes. Conversely, diseases such as metabolic syndrome, arteriosclerosis, diabetes, hypertension, and hyperlipidemia must be considered in the background of LOH syndrome. The Aging Males' Symptoms (AMS) score (Table 51.1) is used internationally as a symptom score for LOH syndrome. Men who present with mood disorders may also require psychological testing for depression. If psychological symptoms such as suicidal ideation are strong, referral to a psychiatrist should be recommended promptly. Blood tests include measurements of total T and free T in the blood. Since T varies throughout the day, it is recommended that blood be drawn in the morning. If you are taking antidepressants such as sulpiride, T may be low, so medication confirmation is essential. In the guidelines for LOH syndrome treatment, the diagnostic criteria are a total T value of 250 ng/dL or less or a free T value of 7.5 pg/mL or less [3].

Table 51.1 Aging Males' Symptoms (AMS) score

1	Overall, I don't feel well (health status, how I feel about myself)
2	Joint and muscle pain (back pain, joint pain, pain in hands and feet, back pain)
3	Severe sweating (sudden unexpected sweating, feeling hot unrelated to tension or exercise)
4	Sleep problems (difficulty falling asleep, not being able to sleep deeply, etc.)
5	Often feel sleepy, frequently feel tired
6	Irritability (snapping at people, getting angry over small things, becoming grumpy)
7	Became neurotic (easily tense, mentally restless, etc.)
8	Feeling of anxiety (panic state)
9	Body fatigue and decrease in activity (general decrease in activity, lack of interest in leisure activities, etc.)
10	Decrease in muscle strength
11	Depressed mood (feeling down, sad, tearful, lack of motivation, etc.)
12	Feeling that "the peak of life has passed"
13	Feeling "exhausted" or "at rock bottom"
14	Beard growth has slowed down
15	Decline in sexual ability
16	Decrease in the number of morning erections
17	Decrease in sexual desire (sex is not enjoyable, no desire for intercourse)

Each item is scored as follows: none: 1 point, mild: 2 points, moderate: 3 points, severe: 4 points, extremely severe: 5 points

The total score looks at the severity of male menopause symptoms. 17–26 points: none, 27–36 points: mild, 37–49 points: moderate, 50 points and above: severe

51.4 Treatment of LOH Syndrome

The decision to start treatment is made when LOH syndrome is suspected from symptoms. T values vary greatly due to individual differences and stress, and it is important to note that a single measurement of T values does not serve as an indicator for diagnosing LOH syndrome, understanding disease progression, or assessing improvement with treatment. Overseas guidelines recommend measuring at least twice. Also, in the guidelines for LOH syndrome treatment, T values are not a mandatory item in determining treatment eligibility. This is because the CAG repeat sequence in the N-terminal region of the AR gene can affect the activity efficiency of the androgen receptor. In Asians, who tend to have a long CAG repeat sequence and a low activity efficiency of the androgen receptor, testosterone replacement therapy may be effective even if the T value is above the standard value [4]. Currently available testosterone preparations in our country include enantate testosterone (Enarmon Depot®) as an injection and male hormone ointment (Glowmin®) as a general medicine. The usual T replacement therapy is Enarmon Depot® 125–250 mg every 2 weeks. Administer every ~4 weeks. The dosage and interval of administration should be adjusted according to the patient's symptoms and age. Side effects of T replacement therapy include polycythemia, cardiovascular disease, sleep apnea syndrome, oily skin, acne, breast pain, liver dysfunction, infertility, etc. T replacement therapy for patients with heart disease should be carefully considered. Many observational studies deny the increased risk of prostate cancer due to T replacement therapy. In T replacement therapy for LOH syndrome, regular blood collection, evaluation of treatment effects, risk management of polycythemia and cardiovascular disease, and follow-up of PSA (prostate-specific antigen) values are performed. As a future issue, new T formulations and management tools may be needed to adjust the T dose and dosage according to individual and symptoms.

References

1. Södergård R, Bäckström T, et al. Calculation of free and bound fractions of testosterone and estradiol-17 beta to human plasma proteins at body temperature. J Steroid Biochem. 1982;16:801–10.
2. Walther A, Philipp M, et al. The rate of change in declining steroid hormones: a new parameter of healthy aging in men? Oncotarget. 2016;7:60844–57.
3. Japanese Urological Association, Japanese Men's Health Medical Association, et al., editor. Guide to the treatment of LOH syndrome (aging male/gonadal dysfunction). Tokyo: Medical Book Publishing; 2022.
4. Tirabassi G, Cignarelli A, et al. Influence of CAG repeat polymorphism on the targets of testosterone action. Int J Endocrinol. 2015;2015:298107.

Hormones and Antiaging Medicine: Female Hormones and Antiaging

52

Osamu Hiraike

52.1 Basic Concept of the Antiaging Effects of Estrogen

Estrogen is a general term for substances with female hormone activity, with 17β-estradiol (E2) being the representative estrogen. Since the discovery of E2 in 1923, other natural estrogens such as estrone and estriol have been identified. Estetrol is a natural estrogen produced only in the fetal liver [1], and there are four types of natural estrogens. The main estrogen-producing organs in women are the ovarian granulosa cells, but estrogen is also produced in the adrenal cortex and in men's testes. While men's estrogen secretion does not change significantly throughout life, women's estrogen secretion peaks during puberty and significantly decreases after menopause due to the depletion of primordial follicles, falling below the level of men's secretion (Fig. 52.1) [2]. Women experience an increase in the frequency of degenerative diseases during the same period as this decrease in estrogen, suggesting that changes in estrogen secretion signifi-

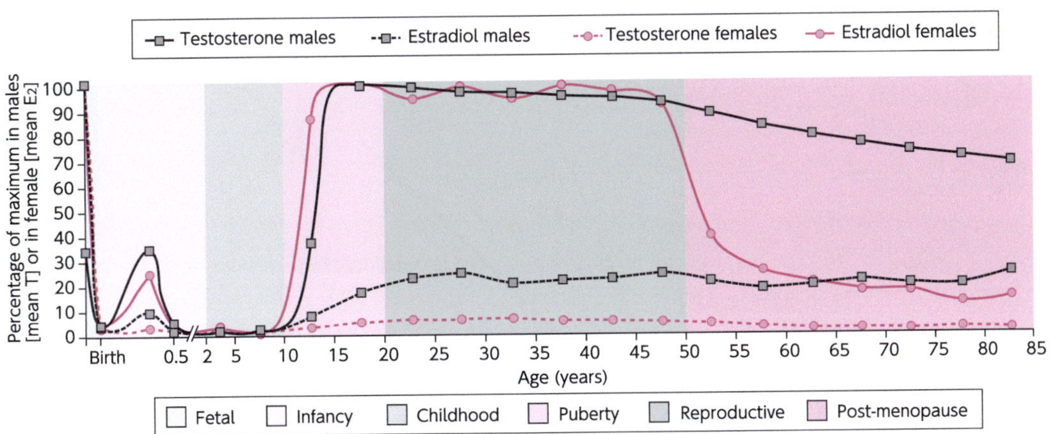

Fig. 52.1 Conceptual diagram of age-related and temporal changes in estrogen production in men and women

O. Hiraike (✉)
Department of Obstetrics and Gynecology, The University of Tokyo, Tokyo, Japan

H. Yamada, Y. Naito (eds.), *Anti-Aging Medicine*, https://doi.org/10.1007/978-981-97-8514-8_52

cantly affect aging. Specific diseases include auditory abnormalities, cancer, cardiovascular diseases, muscle atrophy, neuropsychiatric disorders, visual degeneration, osteoporosis, age-related skin changes, and urinary tract disorders, which appear after menopause. Some of these diseases improve with estrogen administration, suggesting that estrogen can be considered an antiaging substance.

52.2 Mechanism of Estrogen Action

52.2.1 Estrogen Biosynthesis Pathway

After the onset of secondary sexual characteristics, estrogen secretion changes in an amplitude pattern, peaking during ovulation. Estrogen is synthesized from cholesterol through the action of gonadotropins released from the pituitary gland and the collaborative work of ovarian granulosa cells and surrounding theca cells, resulting in the formation of estrone, E2, and estriol [3].

52.2.2 Mechanism of Estrogen Action

The biological effects of estrogen are mediated by the estrogen receptor (ER). There are two subtypes of ER, ERα (ESR1) and ERβ (ESR2), both of which have estrogen-dependent transcriptional activation ability. The central mechanism is thought to be that the ER, when bound to estrogen, acts on the basic transcription apparatus complex to promote transcriptional activation (genomic function). In addition to this, there are four mechanisms: the direct action of estrogen itself (non-genomic function), the function of ERs present in the membrane, and the modification of ER by various signaling factors (Fig. 52.2). The ER activation effect of estrogen varies depending on the binding mode of various estrogens and ER, and the expression distribution of

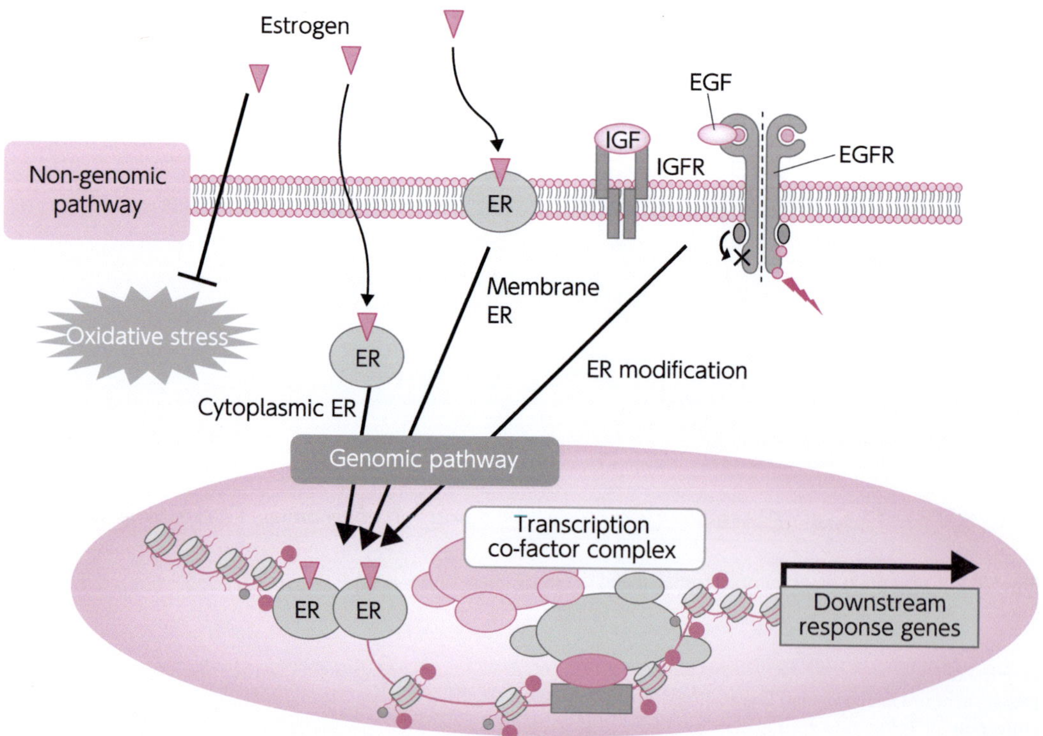

Fig. 52.2 Diagram explaining the estrogen action pathway. It is known that there are four pathways

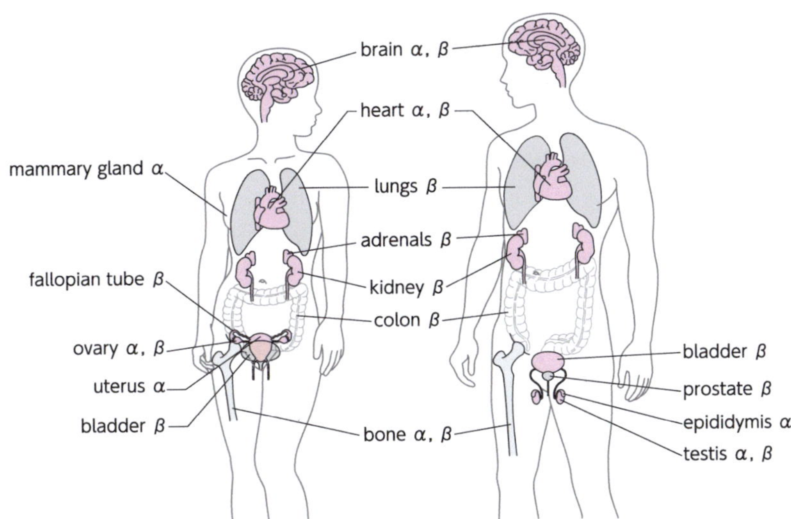

Fig. 52.3 Whole body distribution of ERα (ESR1) and ERβ (ESR2). The difference in distribution is important for understanding tissue specificity

ER in the human body is tissue- (organ) specific (Fig. 52.3) [4]. Therefore, there are tissue-specific mechanisms for the biological effects of estrogen.

52.3 Antiaging Effects of Estrogen

52.3.1 Antioxidant Effects

It is known that there are sex differences in antioxidant effects across species, and the activities of Mn-superoxide dismutase (Mn-SOD) and glutathione peroxidase (GPx) are higher in females than in males, leading to the theory that females live longer. The antioxidant effect of estrogen in the body is strong, and in ovariectomized rats, there is a decrease in GPx activity, lipid oxidation, mitochondrial damage, and an increase in H_2O_2 production, but these oxidative stress accumulation states are improved with estrogen supplementation. Cell death is also suppressed in cultured skin fibroblasts. The mechanism by which estrogen increases the expression of Mn-SOD and GPx is known [5].

52.3.2 Anti-inflammatory Effects

In postmenopausal women, not only aging but also a decrease in estrogen increases inflammatory markers such as interleukin-1β (IL-1β), interleukin-6 (IL-6), and tumor necrosis factor-α (TNF-α) and decreases CD4 + T cells and B cells, but it promotes the secretion of anti-inflammatory cytokines such as interleukin-10 (IL-10) and transforming growth factor-β (TGF-β). It is also known that interferon-γ (INF-γ), a Th1 cytokine, is subject to gene expression control by estrogen's genomic function, and estrogen and ER signals have a significant impact on the immune system.

52.3.3 Telomerase Inhibitory Effects

Human somatic cells have a limit to the number of divisions, and telomere (chromosome end) DNA replication shortens with each DNA replication. Telomere length shortens with the progression of aging, but it has been pointed out that estrogen promotes the expression of human telomerase reverse transcriptase (hTERT) and may act inhibitory against telomere length shortening [6].

52.4 Systemic Antiaging Effects of Estrogen

The following effects are also known.

52.4.1 Uterus

The action of estrogen to proliferate uterine cells is important for the interaction between endometrial epithelial cells and stromal cells. ERα expressed in stromal cells is important, and endometrial epithelial cells proliferate through various growth factors [epidermal growth factor (EGF), insulin-like growth factor 1 (IGF-1), hepatocyte growth factor (HGF), keratinocyte growth factor (KGF), fibroblast growth factor (FGF), transforming growth factor-α (TGF-α)] mediated by ERα. In ERβ knockout mice, the cell proliferation-promoting effect of ERα becomes prominent, leading to hypertrophy and edema of the uterus [7].

52.4.2 Skin

As humans age, the degeneration of collagen and elastin, sagging of subcutaneous tissue, and a decrease in skin elasticity lead to an increase in wrinkles and thinning of both the epidermis and dermis. However, these can be improved by estrogen supplementation. Hormone replacement therapy increases the number of skin papillae, enhancing the fineness of the skin texture, and has positive effects even in areas where estrogen is not applied. Aging delays skin wound healing, but there are reports that appropriate ERβ stimulation, which significantly decreases after the age of 70, brings anti-inflammatory effects and promotes wound healing [8]. In fact, there are reports that leg ulcers decrease with hormone replacement therapy.

52.4.3 Urinary and Reproductive Organs

After the perimenopausal period, the urinary and reproductive systems undergo significant atrophic changes. The symptoms of atrophy are a major disease state that greatly impairs quality of life, and in recent years, it has been attracting attention as the genitourinary syndrome of menopause (GSM). In the vagina, ERα is dominant, and without ERα, excessive proliferation occurs, causing keratinization abnormalities in the vaginal mucosa. Estrogen administration significantly improves vaginal atrophy symptoms. Recently, the selective estrogen receptor modulator ospemifene has been used for GSM treatment. Estrogen is known to improve the nutritional status of the urethral mucosa and increase the thickness of the urethral wall, showing a certain effect on stress urinary incontinence in postmenopausal women, even with local administration.

References

1. Reame NK. Estetrol for menopause symptoms: the Cinderella of estrogens or just another fairy tale? Menopause. 2020;27:841–3.
2. Ober C, Loisel DA, et al. Sex-specific genetic architecture of human disease. Nat Rev Genet. 2008;9:911–22.
3. Sluijmer AV, Heineman MJ, et al. Relationship between ovarian production of estrone, estradiol, testosterone, and androstenedione and the ovarian degree of stromal hyperplasia in postmenopausal women. Menopause. 1998;5:207–10.
4. Drummond AE, Fuller PJ. The importance of ERbeta signalling in the ovary. J Endocrinol. 2010;205:15–23.
5. Viña J, Sastre J, et al. Modulation of longevity-associated genes by estrogens or phytoestrogens. Biol Chem. 2008;389:273–7.
6. Bayne S, Li H, et al. Estrogen deficiency reversibly induces telomere shortening in mouse granulosa cells and ovarian aging in vivo. Protein Cell. 2011;2:333–46.
7. Wada-Hiraike O, Hiraike H, et al. Role of estrogen receptor β in uterine stroma and epithelium: insights from estrogen receptor β−/− mice. Proc Natl Acad Sci USA. 2006;103:18350–5.
8. Campbell L, Emmerson E, et al. Estrogen promotes cutaneous wound healing via estrogen receptor beta independent of its antiinflammatory activities. J Exp Med. 2010;207:1825–33.

Organ Correlation and Antiaging Medicine: Gut-Brain Correlation and Antiaging

53

Shin Fukudo

53.1 Concept of Gut-Brain Correlation

The functional relationship between the central nervous system and the digestive tract is called the brain-gut interaction [1]. The functional relationship between the central nervous system and the digestive tract can be broadly divided into (1) signals from the central nervous system to the digestive tract, and (2) signals from the digestive tract to the central nervous system (Fig. 53.1). The brain-gut interaction assumes a bidirectional relationship of (1) and (2). On the other hand, when the order of the words "gut" and "brain" is reversed to "gut-brain interaction", it often implies an emphasis on the (2) signal [2]. When analyzing the effects of intestinal bacteria on the body, the components that intestinal bacteria have on the (2) signal become the main focus.

S. Fukudo (✉)
Department of Behavioral Medicine, Tohoku University Graduate School of Medicine, Sendai, Miyagi, Japan

Department of Psychosomatic Medicine, Japanese Red Cross Ishinomaki Hospital, Ishinomaki, Miyagi, Japan
e-mail: sfukudo@med.tohoku.ac.jp

H. Yamada, Y. Naito (eds.), *Anti-Aging Medicine*, https://doi.org/10.1007/978-981-97-8514-8_53

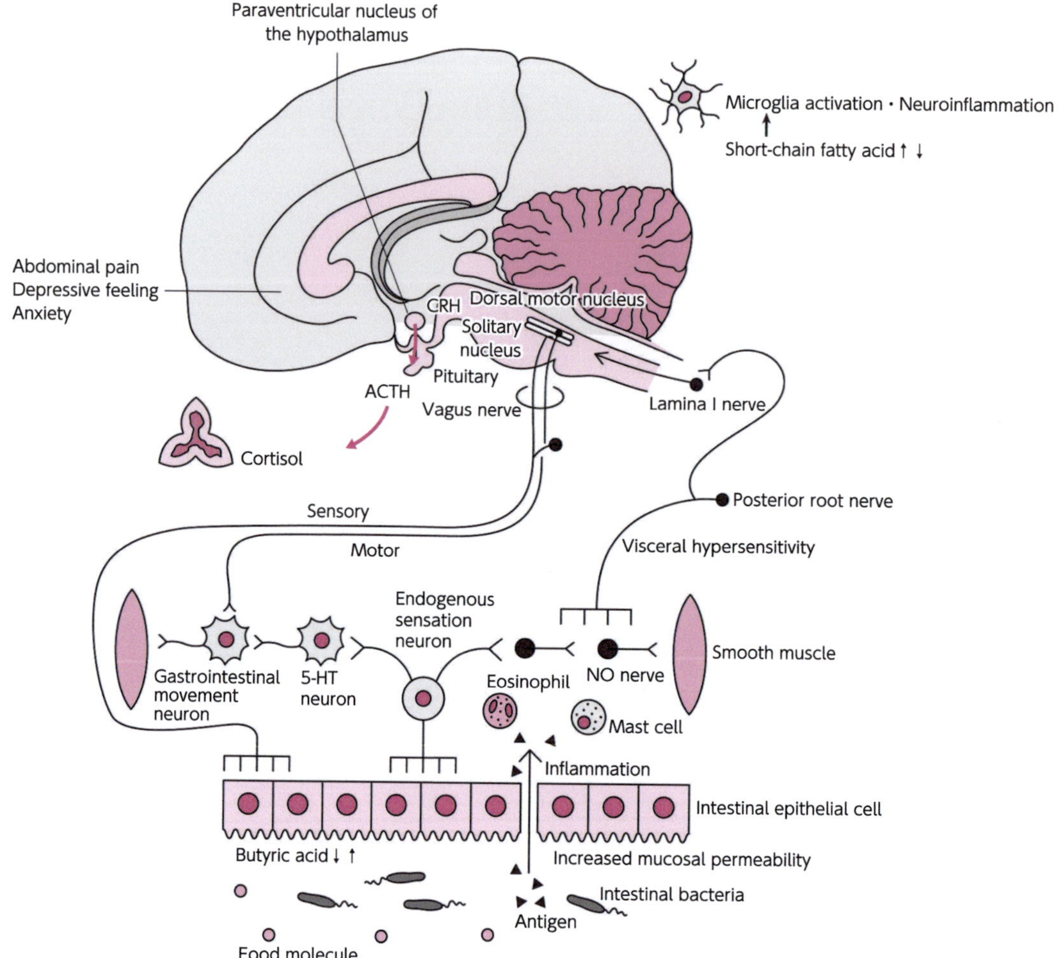

Fig. 53.1 Pathway of gut-brain correlation. When food molecules enter the intestinal lumen, various metabolisms occur due to the action of intestinal bacteria. There are bacteria that produce molecules with anti-inflammatory, antidepressant, and antitumor effects, such as butyric acid, a short-chain fatty acid, and bacteria that enhance mucosal permeability, facilitate the invasion of antigens into the body, and cause minor inflammation. When stress is loaded on the body, CRH is released from the paraventricular nucleus of the hypothalamus, ACTH is secreted from the anterior pituitary, and cortisol is secreted from the adrenal cortex. Mast cells degranulate due to nerve stimulation and CRH stimulation, and together with eosinophils, they exacerbate inflammation. The resulting inflammation sensitizes nerves, and the ascending of signals in sensory nerves causes central excitation. Products of intestinal bacteria, including short-chain fatty acids, affect the activation of microglia in the brain, thereby affecting brain function

53.2 Importance of Gut-Brain Correlation Disorders

Gut-brain correlation disorders (disorders of gut-brain interaction; DGBI) are a collective term for a group of diseases where the gut-brain correlation is considered the main pathology. These were previously referred to as functional gastrointestinal disorders, among which irritable bowel syndrome (IBS), functional constipation, and functional dyspepsia are representative. The global prevalence based on the current international diagnostic criteria for DGBI, Rome IV, is 40.3% for DGBI, 4.1% for IBS, 11.7% for functional constipation, and 7.2% for functional dyspepsia [3]. Under the more lenient Rome III criteria, the prevalence of IBS is 10.1% [3]. In other words, these are highly frequent. In developed countries, the negative impact of IBS on medical costs is enormous. Furthermore, the burden on patients is extremely high due to the impairment of quality of life (QOL) caused by IBS.

53.3 Aging and Gut-Brain Correlation Disorders

IBS, functional constipation, and functional dyspepsia, which are representative disorders of DGBI, all decrease with age in epidemiological studies based on cross-sectional and random sampling designs [3]. Also, all of these have a higher prevalence in women than in men [3]. In Japan's age-specific prevalence survey by the Ministry of Health, Labour and Welfare, the prevalence of chronic constipation increases with age, but this is thought to be due to an increase in drug-induced, symptomatic, and organic constipations, which are different from functional constipation and constipation-predominant IBS.

53.4 Intestinal Bacteria and Gut-Brain Correlation

It has been known that IBS can develop after infectious enteritis. When observing the onset of IBS after all inflammation has subsided in a group of patients with acute enteritis, IBS develops at a high rate of 29.3%. Such IBS is postinfectious (PI)-IBS. PI-IBS is more likely to develop if there are psychological abnormalities such as depression, anxiety, or somatization at the time of acute enteritis. In IBS, the composition of intestinal bacteria is also different from that of healthy individuals [4]. Although there are some differences in the research, generally, an increase in the Firmicutes/Bacteroides ratio, an increase in *Ruminococcus torques*, *Roseburia*, *Blautia*, *Veillonella*, Gammaproteobacteria, a decrease in *Bifidobacterium*, and a decrease in *Clostridiales* are observed [4]. These changes are related to bacterial metabolic products, gastrointestinal symptoms, and changes in psychological behavior. In IBS patients, the stress response is enhanced, and there is an abnormality in the corticotropin-releasing hormone (CRH) system, and the influence of intestinal bacteria on this response has been suggested [4]. Data showing the specific relationship between intestinal bacteria and neurotransmitters in IBS have also been obtained. Brain function imaging has clarified the disease image of brain-gut correlation in IBS, and abnormalities in the sensory motor network, salience network, and emotional arousal network derived from visceral sensation, which is an internal sensation, have been detected [4].

53.5 Changes in Neural Function due to Intestinal Bacteria

Research that neural function changes due to changes in intestinal bacteria is attracting attention. The quality of life (QOL) of human groups is good in individuals dominated by Prevotella and bad in individuals dominated by Bacteroides enterotype 2. When comparing healthy individuals and those with depression, the composition ratio of intestinal bacteria was Prevotella > Bacteroides enterotype 2 in healthy individuals, but Prevotella < Bacteroides enterotype 2 in those with depression. Here, it is shown that Coprococcus as an intestinal bacterium is involved in the production of the dopamine-related substance 3,4-dihydroxyphenylacetic acid, and individuals rich in Coprococcus maintain a good QOL. Constipation is important as an initial symptom of Parkinson's disease, and there are patients who show abnormalities in dopamine transporter scintigraphy (DaT Scan) and ^{123}I-meta-iodobenzyl-guanidine (MIBG) myocardial scintigraphy early on [5]. Epidemiologically, there are reports that IBS increases the risk of Parkinson's disease or Alzheimer's disease [4]. These are thought to be due to the fact that abnormalities in intestinal bacteria lead to increased mucosal permeability and minor inflammation of the gastrointestinal mucosa, and at the same time, a decrease in the concentration of butyric acid in the gastrointestinal lumen leads to activation of microglia in the brain and neuroinflammation.

53.6 Intervention Trials of Brain Function by Intestinal Bacteria

In a randomized controlled trial of placebo versus *Bifidobacterium longum* in IBS patients, *Bifidobacterium longum* administration was observed to improve gastrointestinal symptoms and depression scales more than placebo administration [4]. At the same time, it was shown that when shown the expression of fear in others, the degree of amygdala activation measured using functional magnetic resonance imaging was suppressed [4]. By transforming the intestinal bacteria of IBS, it is possible to improve brain function and mental state through gut-brain correlation.

References

1. Koseki T, Muratsubaki T, et al. Impact of mindfulness tendency and physical activity on brain-gut interactions. J Gastroenterol. 2023;58:158–70.
2. Drossman DA, Hasler WL. Rome IV-functional GI disorders: disorders of gut-brain interaction. Gastroenterology. 2016;150:1257–61.
3. Sperber AD, Bangdiwala SI, et al. Worldwide prevalence and burden of functional gastrointestinal disorders, results of Rome Foundation Global Study. Gastroenterology. 2021;160:99–114.e3.
4. Fukudo S, Okumura T, et al. Evidence-based clinical practice guidelines for irritable bowel syndrome 2020. J Gastroenterol. 2021;56:193–217.
5. Sakakibara R, Doi H, et al. Lewy body constipation. J Anus Rectum Colon. 2019;3:10–7.

Organ Correlation and Antiaging Medicine: Exercise and Organ Correlation

54

Wataru Aoi

Habitual exercise is known to contribute widely to the improvement of physical strength, prevention of metabolic and cardiovascular diseases, cancer prevention, and even the prevention and improvement of physical and mental frailty. Various organ functions, such as nutrient and oxygen supply through blood flow and excitation-contraction coupling through motor nerves, are necessary for skeletal muscle contraction. Furthermore, humoral factors such as hormones, cytokines, and metabolites affect the function of skeletal muscles. In addition, skeletal muscles themselves secrete physiologically active factors called myokines, which communicate intricately with all organs of the body, bringing about the benefits of exercise.

54.1 Humoral Factors Controlling Skeletal Muscle Contraction

The main energy substrates for skeletal muscle contraction are carbohydrates and lipids. In addition to the supply of substrates from meals, the liver, and body fat, the ability of skeletal muscles to metabolize these substrates is regulated by various humoral factors. If the ability to metabolize carbohydrates and lipids is high, it leads to improved endurance and management of blood glucose levels and body fat mass, contributing to the prevention of lifestyle-related diseases. In particular, the ability to take up glucose and the oxygen consumption ability of mitochondria are known as major functions that determine the metabolic capacity of skeletal muscles. The metabolic capacity of skeletal muscles rapidly increases due to hormones such as catecholamines and thyroid hormones. In addition, adiponectin secreted from fat cells, lactate, bile acids, and other metabolites also activates glucose uptake and aerobic metabolism. Many of these increase in the blood with single or habitual exercise, activating metabolism. On the other hand, the decrease in sex hormone secretion due to aging weakens metabolic capacity. Also, among polyphenols and carotenoids contained in food, some enhance mitochondrial activity. Many adipokines derived from visceral fat decrease insulin sensitivity and mitochondrial activity [1]. When a decrease in intestinal barrier function occurs due to dysbiosis, the concentration of endotoxins and inflammatory cytokines in the circulation increases, which is suggested to impair skeletal muscle metabolism [2]. On the other hand, habitual exercise is thought to suppress the decrease in metabolic capacity by reducing body fat and improving the intestinal environment.

W. Aoi (✉)
Laboratory of Nutrition Science, Graduate School of Life and Environmental Sciences, Kyoto Prefectural University, Kyoto, Japan
e-mail: waoi@kpu.ac.jp

H. Yamada, Y. Naito (eds.), *Anti-Aging Medicine*, https://doi.org/10.1007/978-981-97-8514-8_54

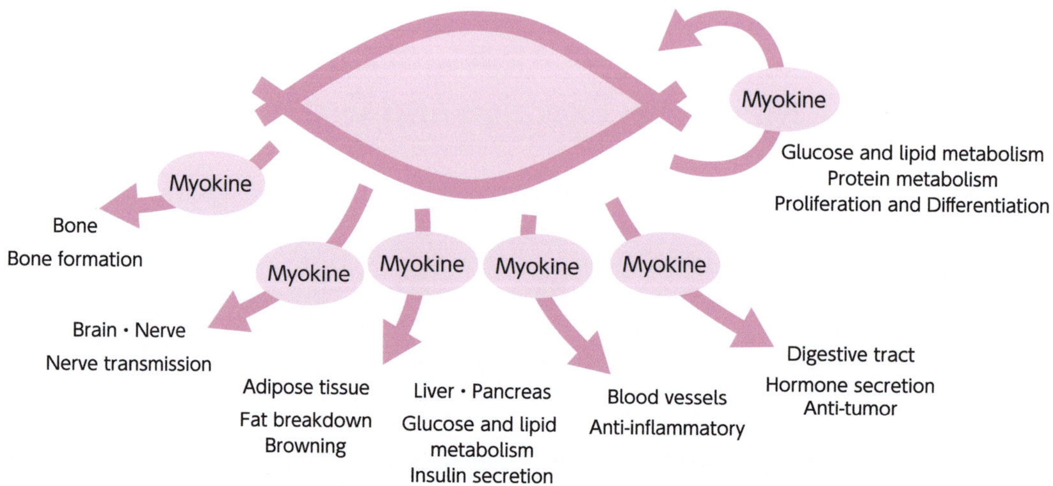

Fig. 54.1 Secretory function of skeletal muscle and the action of myokines. (Modified with permission from [3])

54.2 Physiologically Active Substance Myokine

Skeletal muscle is also recognized as a secretory organ that produces the physiologically active protein myokine. The secreted myokines regulate the function of various organs through autocrine, paracrine, and endocrine pathways (Fig. 54.1) [3]. Many myokines are secreted in response to muscle contractions and are deeply involved in the acute physiological responses obtained through exercise, as well as adaptations from habitual exercise. They also affect the weakening of the musculoskeletal system due to disease risk and aging and are attracting attention as target factors for antiaging.

54.3 The Diverse Functions of Myokines

Many myokines are involved in the regulation of nutrient metabolism. Interleukin-6 (IL-6) is a typical myokine whose concentration in the blood increases with a single bout of exercise, enhancing the metabolism of carbohydrates and lipids in skeletal muscles, as well as promoting the breakdown of neutral fats in adipose tissue and glycogen in the liver. This is thought to contribute to blood glucose regulation and the supply of energy substrates needed during exercise. PPAR-γ coactivator-1α (PGC-1α) is a transcription coactivator that regulates the transcriptional activity of various factors involved in aerobic metabolism. Habitual exercise increases PGC-1α expression in skeletal muscles, under whose control several secretory factors exist. Irisin, a representative factor, contributes to the browning of white adipocytes, bone metabolism, and the regulation of neural functions. β-Aminoisobutyric acid and meteorin-like are also known to improve the metabolism of adipose tissue and the liver. Some myokines have anti-inflammatory effects and may contribute to the prevention of cardiovascular and metabolic diseases by suppressing chronic inflammation. IL-6 enhances the production of anti-inflammatory factors such as interleukin-10 (IL-10) and interleukin-1 (IL-1) receptor agonists in neutrophils and the liver. However, IL-6 is also known as an inflammatory cytokine, and it is necessary to distinguish between transient/moderate increases due to exercise and transient/high secretion or persistent/low increases during chronic inflammation or infection. Secreted protein acidic and rich in cysteine and oncostatin M, which increase with exercise and decrease with aging, contribute to cancer prevention by controlling apoptosis and growth factors. It has also been suggested that myokines are involved in exercise-induced bone metabolism and improvements in depression and cognitive function. Many myokines also interfere with the weakening of the musculoskeletal system due to aging [4]. Myostatin, a member of the Transforming growth factor-β (TGF-β) superfamily, is widely known as a myokine that promotes muscle atrophy. On the other hand, several myokines, such as follistatin and decorin, directly or indirectly inhibit the action of myostatin and suppress muscle atrophy. Although the blood concentration of these myokines does not necessarily correlate significantly with the progression of sarcopenia, it has been reported that exercise intervention reduces myostatin and increases follistatin and decorin concentrations. Insulin-like growth factor-1 (IGF-1), which is known to activate protein synthesis and promote hypertrophy of muscle fibers, is a well-known secretarory factor. In cross-sectional studies, it has been reported that the blood concentration of IGF-1 is lower in the sarcopenia group than in the non-sarcopenia group. Furthermore, the blood concentration increases with exercise intervention, and a correlation is observed between muscle mass and muscle strength. Although the use of these myokines for the treatment of sarcopenia is expected, the effectiveness and application methods are not yet clear.

54.4 Future Prospects

The antiaging effects of exercise habits are diverse, and evidence is accumulating year by year. The existence of myokines supports the mechanism by which physical activity affects not only skeletal muscles, but also overall body function. In recent years, factors secreted from organs due to exercise have come to be called exerkines [5]. The stimulation of muscle contraction associated with exercise spreads throughout the body, and factors secreted from various organs, exerkines, interact with each other to form the effects of exercise. The development of biomarkers to estimate physical strength improvement and antiaging effects, functional materials that mimic the effects of exercise, and the expansion into drug development using factors related to exercise and organ correlation are expected.

References

1. Nicholson T, Church C, et al. The role of adipokines in skeletal muscle inflammation and insulin sensitivity. J Inflamm (Lond). 2018;15:9.
2. Cavallari JF, Schertzer JD. Intestinal microbiota contributes to energy balance, metabolic inflammation, and insulin resistance in obesity. J Obes Metab Syndr. 2017;26:161–71.
3. Aoi W. Characteristics of skeletal muscle as a secretory organ. In: The plasticity of skeletal muscle: from molecular mechanism to clinical applications. Berlin: Springer; 2017. p. 195–210.
4. Aoi W. Myokines: a potential key factor in development, treatment, and biomarker of sarcopenia. In: Sarcopenia: molecular mechanism and treatment strategies. Amsterdam: Elsevier; 2021. p. 171–85.
5. Chow LS, Gerszten RE, et al. Exerkines in health, resilience and disease. Nat Rev Endocrinol. 2022;18:273–89.

Organ Correlation and Antiaging Medicine: Intestinal-Kidney Correlation and Antiaging

55

Takehiro Suzuki and Takaaki Abe

55.1 Progression of Aging-Related Diseases due to Deterioration of the Intestinal Environment and Chronic Inflammation in Patients with Chronic Kidney Disease

Nutrients ingested by humans are metabolized by intestinal bacteria into short-chain fatty acids with anti-inflammatory and metabolic regulatory effects, antioxidants, amino acids, vitamins, etc., which are involved in maintaining human health. In symbiosis where beneficial resident bacteria coexist, intestinal bacteria break down dietary fiber to produce short-chain fatty acids such as propionic acid, butyric acid, and acetic acid. These short-chain fatty acids serve as nutrients for the intestinal epithelium, maintaining the intestinal barrier function, inducing regulatory T cells (Treg) that suppress inflammation, promoting the secretion of GLP-1, an intestinal hormone with a blood sugar lowering effect, and controlling glucose metabolism and lipid metabolism, thus having a preventive effect on diabetes and hypertension [1] (Fig. 55.1a). Also, the intestine forms an intestinal barrier composed of intestinal epithelial cells, the intestinal mucus layer, antimicrobial peptides secreted from intestinal epithelial cells, and secretory dimeric IgA secreted from IgA-producing plasma cells under the mucosa, preventing the invasion of pathogenic bacteria and bacterial lipopolysaccharides (LPS) into the body [1] (Fig. 55.1a). On the other hand, indoxyl sulfate, *p*-cresyl sulfate, and phenyl sulfate, which are uremic toxins accumulated in the body of patients with chronic kidney disease (CKD), are intestinal-derived uremic toxins produced by the metabolism of tryptophan and tyrosine, amino acids derived from dietary protein, by intestinal bacteria into precursor substances such as indole, *p*-cresol, and phenol, which are then sulfated in the liver [1, 2] (Fig. 55.1b). In CKD patients, dysbiosis, a state of deterioration of the intestinal bacterial environment due to changes in the intestinal microbiota caused by constipation, dietary fiber intake restriction, frequent use of antibiotics, impairment of the intestinal barrier, and uremic toxins, exists [3]. In the intestinal microbiota of CKD patients, the number of beneficial bacteria such as *Bifidobacterium*, *Lactobacillus*, and *Roseburia* spp., which produce butyric acid, decreases, while the number of bacteria involved in the production of uremic toxins such as indoxyl sulfate and *p*-cresyl sulfate increases [3]. In addition, in CKD, intestinal bacteria and LPS penetrate the intestinal barrier due to impairment of the intestinal barrier, infiltrating the submucosal tissue of the intestine and the bloodstream, inducing T helper 17 cells (Th17).

T. Suzuki (✉) · T. Abe
Division of Medical Science, Graduate School of Biomedical Engineering, Tohoku University, Sendai, Miyagi, Japan
e-mail: suzuki2i@med.tohoku.ac.jp

H. Yamada, Y. Naito (eds.), *Anti-Aging Medicine*, https://doi.org/10.1007/978-981-97-8514-8_55

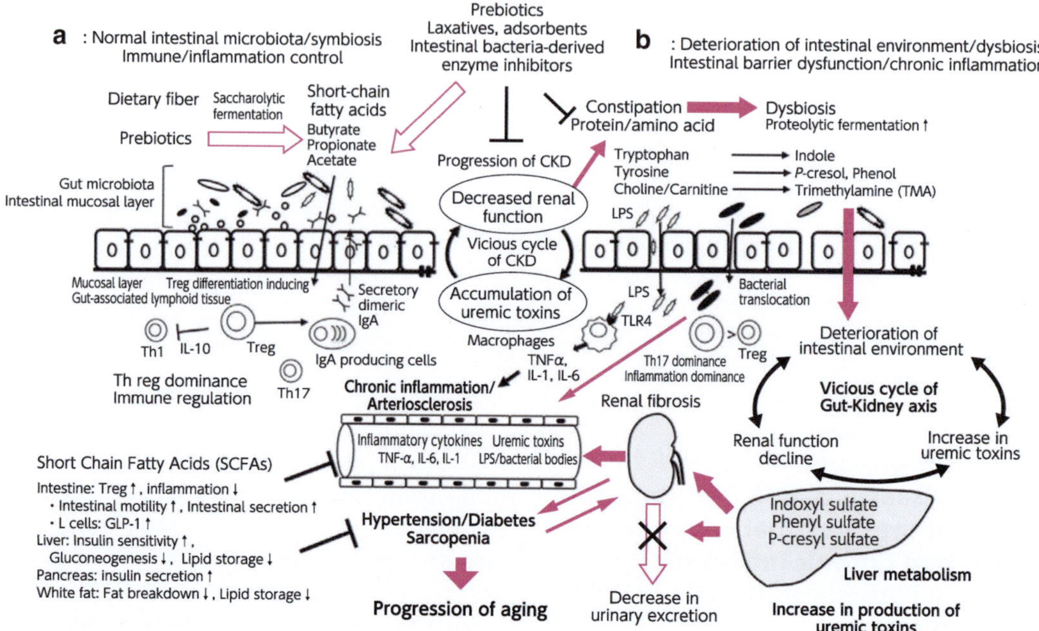

Fig. 55.1 Chronic kidney disease treatment and antiaging through intestinal-kidney correlation. (**a**) Normal intestinal microbiota/symbiosis/immune and inflammation control; (**b**) Deterioration of intestinal environment/dysbiosis/intestinal barrier dysfunction/chronic inflammation

This further activates macrophages and induces Tumor Necrosis Factor-alpha (TNF-α), Interleukin-6 (IL-6), Interleukin-1 (IL-1), and other inflammatory cytokines' increase, causing oxidative stress and chronic inflammation throughout the body, leading to the progression of arteriosclerosis, hypertension, diabetes, sarcopenia, and other aging-related diseases, thereby further exacerbating Chronic Kidney Disease (CKD) itself, creating a "vicious cycle" [1] (Fig. 55.1b).

55.2 Intestinal-Renal Correlation in Chronic Kidney Disease Treatment and Antiaging

The progression of CKD and the deterioration of the intestinal environment, along with dysbiosis, cause oxidative stress and chronic inflammation, leading to epigenomic changes, metabolic abnormalities, and cellular damage. This results in the decline, degeneration, and fibrosis of organs and tissues throughout the body and also triggers a decrease in the function and number of stem cells responsible for peripheral tissue regeneration [4]. These changes promote the progression of aging and underlying diseases such as arteriosclerosis, hypertension, diabetes, and sarcopenia [4]. Conversely, correcting the negative intestinal-renal correlation, where CKD and intestinal environment deterioration mutually form a vicious cycle, could be a treatment strategy for antiaging (Fig. 55.1). In CKD patients, the prevalence of constipation is high, and the dysbiosis associated with constipation further progresses CKD (Fig. 55.1b). Therefore, proper management of bowel movements in CKD patients could be a potential new treatment. We administered a novel laxative, lubiprostone (Amitiza®), to adenine renal failure model mice [5]. Lubiprostone administration increased *Lactobacillus* and *Prevotella*, which are short-chain fatty acid-producing bacteria, and decreased intestinal uremic toxins such as indoxyl sulfate, improving renal function and reducing fibrosis and inflammatory cell infiltration in the kidneys of renal failure mice [5]. We recently revealed that phenylsulfate (PS) is a causative substance for the progression of diabetic kidney disease (DKD), the most common disease leading to dialysis initiation in our country, and is a prognostic marker [2]. Administering PS to a diabetes model induced podocyte damage and basement membrane thickening in the renal glomeruli, increasing albuminuria. Because PS is derived from phenol which is converted from tyrosine in the diet in the human body by the enzyme tyrosine phenol-lyase (TPL), only found in gut microbiota, the production of PS decreases and albuminuria significantly decreases when a TPL inhibitor is administered to a diabetes model [2]. Furthermore, when this drug was administered to mice with renal failure, not only was there a decrease in PS, but a protective effect on renal function was also observed [2]. Next, with the cooperation of Okayama University, we clarified how PS is involved with DKD in the clinical setting through the analysis of a diabetic patient cohort (U-CARE study, $n = 362$). PS significantly correlates with albuminuria, and in patients with microalbuminuria, PS was the only factor predicting worsening albuminuria 2 years later [2]. From these studies, it is thought that laxatives, intestinal bacteria-derived enzyme inhibitors, or short-chain fatty acid production promoters and intestinal bacteria (probiotics) or food/supplements (prebiotics) that improve the intestinal environment can be effective as treatments for CKD, hypertension, diabetes, arteriosclerosis, sarcopenia, etc., which accelerate aging, by improving chronic inflammation and metabolic abnormalities throughout the body and are expected to be a new antiaging treatment strategy in future research [2, 4, 5] (Fig. 55.1).

References

1. Meijers B, Evenepoel P, et al. Intestinal microbiome and fitness in kidney disease. Nat Rev Nephrol. 2019;15:531–45.
2. Kikuchi K, Saigusa D, et al. Gut microbiome-derived phenyl sulfate contributes to albuminuria in diabetic kidney disease. Nat Commun. 2019;10:1835.
3. Chung S, Barnes JL, et al. Gastrointestinal microbiota in patients with chronic kidney disease: a systematic review. Adv Nutr. 2019;10:888–901.
4. Franceschi C, Garagnani P, et al. Inflammaging: a new immune-metabolic viewpoint for age-related diseases. Nat Rev Endocrinol. 2018;14:576–90.
5. Mishima E, Fukuda S, et al. Alteration of the intestinal environment by lubiprostone is associated with amelioration of adenine-induced CKD. J Am Soc Nephrol. 2015;26:1787–94.

56 Microbiome and Antiaging Medicine: Predicting Longevity from the Microbiome

Yuji Naito

56.1 Is Gut Bacterial Flora Information Useful for Predicting Lifespan?

If it becomes possible to use an individual's gut microbiota, bacterial flora information as an indicator of biological age, and propose specific measures, it could contribute to lifespan extension (healthy longevity). For this, analysis based on long-term cohort study data and information from intervention trials during relatively chronic periods, not acute, would be useful. Wilmanski et al. [1] conducted a study on the diversity of gut bacterial flora and proposed a unique indicator, Bray–Curtis uniqueness, a new indicator of β diversity at the genus level or ASV (amplicon sequence variant) level. This uniqueness index increases with age and showed the highest correlation with chronological age compared to various clinical biomarkers. Further analysis of gut bacteria correlated with this uniqueness revealed a correlation between increased uniqueness and a decrease in the genus Bacteroides, and groups with low uniqueness or a high occupancy rate of the genus Bacteroides had a lower survival rate 4 years later, as revealed by a cohort study. Salosensaari et al. [2] tracked the relationship between gut bacterial flora and mortality in a population cohort of 7211 people for 15 years and revealed the characteristics of gut bacterial flora associated with mortality. They found that α diversity (Shannon, Observed) was not related to mortality, and among β diversity, the third component (PC3) of principal component analysis had a positive correlation with mortality, and they proceeded with the analysis of factors affecting PC3. As a result, the correlation between PC3 and mortality was not affected by age, sex, BMI, smoking, diabetes, anticancer drugs, immunomodulators, hypertension, and antihypertensive drugs and was observed similarly in the western and eastern districts of Finland with different lifestyles, and it was revealed that it was strongly influenced by the gut bacterial family (Enterobacteriaceae family). Gut bacterial family bacteria are dominant bacteria of the Proteobacteria phylum, facultative anaerobes, and ferment glucose, and many have nitrate reductase activity, including *Escherichia*, *Klebsiella*, *Proteus*, *Salmonella*, *Shigella*, *Yersinia*, etc. Furthermore, groups with a high occupancy rate of the gut bacterial family have a high mortality rate from cancer, especially gastrointestinal cancer (HR 1.80, 95% CI 1.35–2.41) was also shown. These two cohort studies indicate that there is no single gut bacterium that predicts healthy longevity, and that the balance of gut microbiota found in diversity indices is useful. When considering occupancy rates, it seems

Y. Naito (✉)
Department of Human Immunology and Nutrition Science, Kyoto Prefectural University of Medicine, Kyoto, Japan
e-mail: ynaito@koto.kpu-m.ac.jp

H. Yamada, Y. Naito (eds.), *Anti-Aging Medicine*, https://doi.org/10.1007/978-981-97-8514-8_56

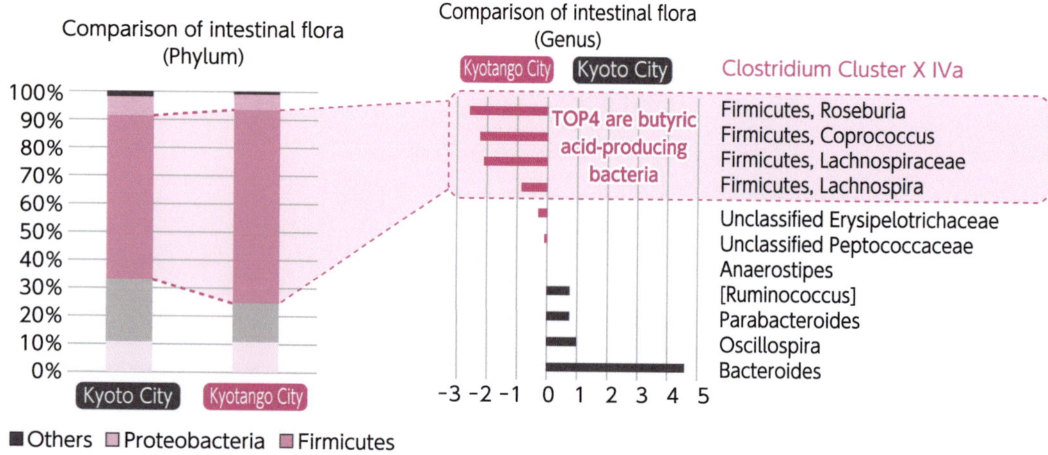

Fig. 56.1 Comparison of gut microbiota in Kyoto City and Kyotango City targeting elderly people aged 65 and over. In Kyotango, compared to Kyoto City, Proteobacteria and Bacteroidetes decreased, Firmicutes increased, and at the genus level, the occupancy rate of butyric acid-producing bacteria such as Roseburia was high. (Modified with permission from [3])

that a low rate of Bacteroides and gut bacteria is associated with longevity, but it is also known that the gut microbiota of Japanese people has low similarity with Westerners, so careful analysis is needed to determine whether it can be applied to Japanese people. At least in the analysis of gut microbiota data from healthy elderly people in the Kyotango Longevity Cohort Study, a low rate of Bacteroides and Proteobacteria and a high rate of Firmicutes have been observed [3] (Fig. 56.1).

56.2 Life Extension Targeting Gut Microbiota

Research results using nematodes, fruit flies, medaka, and mice have been reported, suggesting that changing the gut microbiota may lead to life extension. There are also some attempts at life extension through probiotics. *Streptococcus thermophilus* induces antioxidant enzymes, extending the lifespan of nematodes [4] (Fig. 56.2). *Bifidobacterium animalis* subsp. *lactis* LKM512 induces polyamine production, has skin aging inhibitory effects, and extends mouse lifespan [5]. For LKM512 bacteria, a randomized placebo-controlled parallel-group double-blind trial in healthy people revealed that intake of LKM512 strain increases polyamines such as spermidine in feces and blood, suppresses gut bacteria-derived toxins such as trimethylamine and inflammatory cytokine TNF-α (tumor necrosis factor α) in the blood, and maintains vascular endothelial function [6]. Spermidine has been reported to be involved in T cell differentiation, but recently, it has been possible to rejuvenate aged T cells (immunosenescent T cells), and it has been reported that the therapeutic effect of immune checkpoint inhibitors in cancer animal models is restored, attracting attention [7].

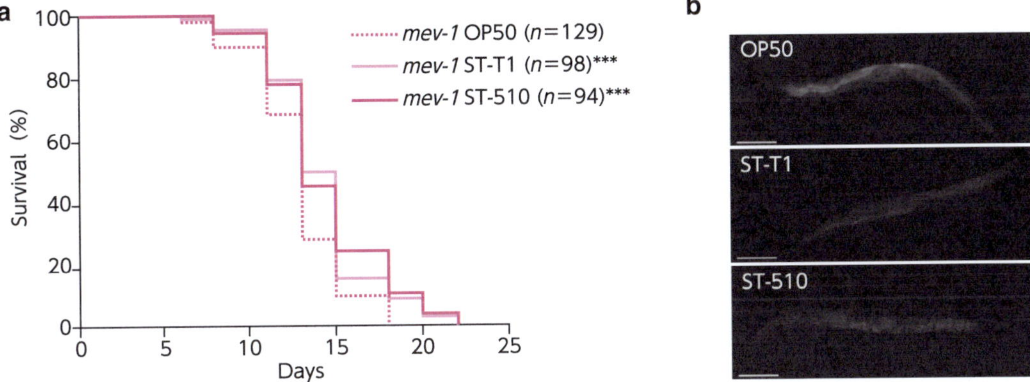

Fig. 56.2 *Streptococcus thermophilus* extends the lifespan of nematodes and reduces oxidative stress. *Streptococcus thermophilus* (ST-T1 strain, ST-510 strain) significantly (***$p < 0.01$) extended lifespan (**a**) and suppressed intracellular hydrogen peroxide levels (BES-H2O2-Ac) on Day 14 (**b**). (Modified with permission from [4])

56.3 Life Extension by Fecal Transplantation

Bárcena et al. [8] used two different mouse models of progeria to find dysbiosis characterized by an increase in the occupancy rate of Proteobacteria and Cyanobacteria, a decrease in the occupancy rate of Verrucomicrobia, and a decrease in diversity in early life. They also reported that fecal transplantation therapy from wild-type mice alleviated findings such as decreased body temperature, hypoglycemia, and perirenal vascular calcification associated with aging in the two progeria model mice and restored lifespan. Furthermore, the same effects has been confirmed by bacterial transplantation of *Akkermansia muciniphila*. Metabolome analysis of ileal contents revealed that progeria mice, as a result of dysbiosis, had impaired secondary bile acid production pathways, which were restored by transplantation of *Akkermansia muciniphila*. From these results, it was revealed that by transplanting the feces of normal mice into a progeria model, the aging phenomenon is suppressed along with the recovery of bile acid metabolism, and both health span and lifespan are extended. Recently, intestinal mucosal gene expression was analyzed 36 weeks after administering *Akkermansia muciniphila* to aged mice of 60 weeks old. As a result, the expression of gene groups involved in tight junction proteins, Muc2 type mucus, intestinal stem cells, and cell rotation was renewed, and intestinal permeability evaluated by lipopolysaccharide was also suppressed [9]. The rejuvenating effect of *Akkermansia muciniphila* on the intestine is attracting attention.

56.4 Conclusion

Involvement of gut microbiota in the pathophysiology of aging has been clarified, and important results have been reported that suggest that controlling aging of the gut (aging gut) may lead to the prevention of lifestyle-related diseases and the extension of healthy longevity. We cannot clearly answer the question, "What gut bacteria are necessary for the extension of lifespan (healthy longevity) of organisms including humans?" However, information obtained from gut microbiota analysis may be useful as an indicator of biological age, not calendar age. In addition, with the advancement of analytical techniques centered on mass spectrometry, the existence of many bacterial metabolites that affect the host's immunity, inflammation, and metabolism is becoming clear.

References

1. Wilmanski T, Diener C, et al. Gut microbiome pattern reflects healthy ageing and predicts survival in humans. Nat Metab. 2021;3:274–86.
2. Salosensaari A, Laitinen V, et al. Taxonomic signatures of cause-specific mortality risk in human gut microbiome. Nat Commun. 2021;12:2671.
3. Naito Y, Takagi T, et al. Gut microbiota differences in elderly subjects between rural city Kyotango and urban city Kyoto: an age-gender-matched study. J Clin Biochem Nutr. 2019;65:125–31.
4. Desaka N, Ota C, et al. Streptococcus thermophilus extends lifespan through activation of DAF-16-mediated antioxidant pathway in Caenorhabditis elegans. J Clin Biochem Nutr. 2022;70:7–13.
5. Matsumoto M, Kurihara S, et al. Longevity in mice is promoted by probiotic-induced suppression of colonic senescence dependent on upregulation of gut bacterial polyamine production. PLoS One. 2011;6:e23652.
6. Matsumoto M, Kitada Y, et al. Endothelial function is improved by inducing microbial polyamine production in the gut: a randomized placebo-controlled trial. Nutrients. 2019;11:1188.
7. Al-Habsi M, Chamoto K, et al. Spermidine activates mitochondrial trifunctional protein and improves antitumor immunity in mice. Science. 2022;378:eabj3510.
8. Bárcena C, Valdés-Mas R, et al. Healthspan and lifespan extension by fecal microbiota transplantation into progeroid mice. Nat Med. 2019;25:1234–42.
9. Shin J, Noh JR, et al. Ageing and rejuvenation models reveal changes in key microbial communities associated with healthy ageing. Microbiome. 2021;9:240.

57 Microbiome and Antiaging Medicine: Gut Bacterial Metabolites and Antiaging

Ryuji Ohue-Kitano and Ikuo Kimura

57.1 Aging and Changes in Gut Bacterial Metabolites

With aging, there is a decrease in basal metabolic rate and immune function, and changes in diet, among other factors, have been reported to reduce the diversity of gut bacteria in the elderly compared to younger individuals and to alter the composition of bacterial species. Furthermore, with the recent advancements in omics analysis, attention is now being focused on gut bacterial metabolites as substantial molecular entities, in addition to the identification of individual species related to host aging control. In fact, from cohort studies and animal experiments, it has become clear that some gut bacterial metabolites fluctuate with aging. For example, it is suggested that the decrease in typical gut bacterial metabolites such as short-chain fatty acids, polyamines, and secondary bile acids with aging is involved in age-related diseases [1]. On the other hand, interesting findings have been reported that these gut bacterial metabolites are at the same level in centenarians as in younger individuals. It is suggested that gut bacterial metabolites are closely related to host aging and age-related diseases, and the control of the production of gut bacterial metabolites is expected to be an important target for the realization of antiaging (Fig. 57.1).

R. Ohue-Kitano (✉)
Graduate School of Biostudies, Kyoto University, Kyoto, Japan

Department of Biological and Environmental Chemistry, Kindai University, Fukuoka, Japan
e-mail: ohue.ryuji.8e@kyoto-u.ac.jp

I. Kimura
Graduate School of Biostudies, Kyoto University, Kyoto, Japan
e-mail: kimura.ikuo.7x@kyoto-u.ac.jp

H. Yamada, Y. Naito (eds.), *Anti-Aging Medicine*, https://doi.org/10.1007/978-981-97-8514-8_57

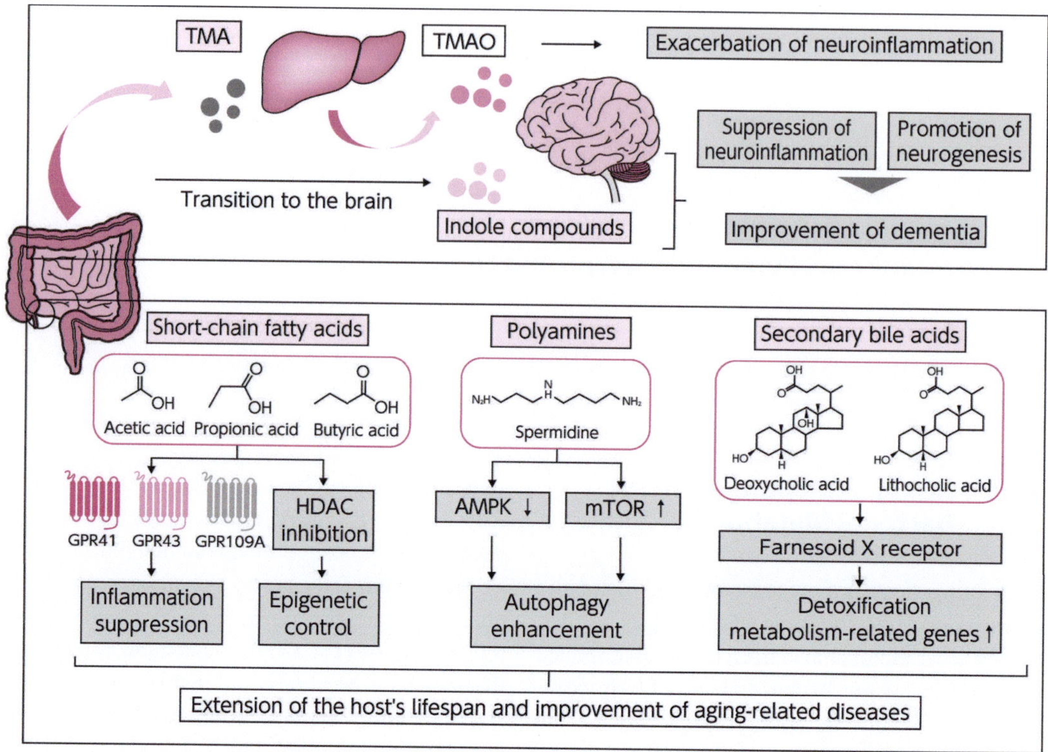

AMPK: AMP-activated protein kinase, HDAC: histone deacetylase, mTOR: mammalian target of rapamycin, TMA: trimethylamine, TMAO : trimethylamine *N*-oxide

Fig. 57.1 The potential for antiaging by intestinal bacterial metabolites. Intestinal bacterial metabolites contribute to the improvement of aging-related diseases and dementia, and even to the extension of the host's lifespan

57.2 Mechanism of Antiaging Action by Gut Bacterial Metabolites

57.2.1 Short-Chain Fatty Acids

Short-chain fatty acids such as acetic acid, propionic acid, and butyric acid are produced by the fermentation of indigestible polysaccharides by intestinal bacteria and are used as an energy source for the host. To date, GPR41, GPR43, GPR109A, and Olfr78 have been identified as short-chain fatty acid receptors, and it has become clear that short-chain fatty acids contribute to the regulation of host metabolic functions, immune regulation, and even brain function improvement through these receptors [2]. On the other hand, it has been pointed out that the amount of short-chain fatty acids in the human intestine decreases with age, leading to age-related diseases such as dementia and chronic inflammation associated with aging. In fact, it has been revealed from studies using mammals such as mice and monkeys that the administration of short-chain fatty acid-producing bacteria contributes to the improvement of Alzheimer's disease by inhibiting the accumulation of amyloid β in nerve cells. Furthermore, it has been reported that short-chain fatty acids may contribute to inflammation suppression through immune regulation actions such as induction of regulatory T cells and inhibition of neutrophil migration and regulation of IgA production via GPR43 and GPR109A. Interestingly, it has been suggested that short-chain fatty acids are directly involved in extending the lifespan of organisms. As a mechanism of action, epigenetic control through the inhibition of class I and II histone deacetylases by short-chain fatty acids has been pointed out, contributing not only to anticancer effects but also to the maintenance of host homeostasis by suppressing chronic inflammation and regulating the expression of metabolism-related genes [3].

57.2.2 Polyamines

Polyamines, including putrescine, spermidine, and spermine, are growth factors involved in cell division and protein synthesis, universally present in the cells of organisms. However, it is known that the concentration in the body decreases due to the decrease in the activity of ornithine decarboxylase, the rate-limiting enzyme of polyamine biosynthesis, with aging. In recent years, it has been suggested that intestinal bacteria are one of the important sources of polyamines, and it has been reported that polyamine-producing bacteria are involved in extending healthy lifespan. For example, as a result of administering *Bifidobacterium animalis* subsp. *lactis* LKM512 (LKM512) to mice, an increase in the concentration of polyamines in the intestine and a lifespan extension effect have been observed. Furthermore, it has been reported that the administration of arginine, a precursor of polyamines, in addition to LKM512, significantly increases the concentration of putrescine in the intestine and the concentration of spermidine and spermine in the blood, resulting in the suppression of chronic inflammation and the improvement of dementia. As a mechanism of action, the enhancement of autophagy through the activation of mammalian target of rapamycin (mTOR) signaling and the inhibition of AMP-activated protein kinase (AMPK) signaling by polyamines has been pointed out [4].

57.2.3 Secondary Bile Acids

Secondary bile acids are intestinal bacterial metabolites that are converted from cholic acid and chenodeoxycholic acid, which are synthesized in the liver as primary bile acids, by being deconjugated by intestinal bacteria in the intestine, into deoxycholic acid and lithocholic acid. The relationship between bile acids and longevity was first observed in growth hormone-deficient mice. It has been shown that the loss of function of growth hormone-releasing hormone is associated with the longevity of mice, and it has been revealed that growth hormone-deficient mice live about 20% longer than wild-type mice by activating detoxification metabolism-related genes through the farnesoid X receptor due to an increase in bile acid levels. Furthermore, a comparison of bile acids in young and old mice revealed that the ratio of secondary bile acids to primary bile acids decreases with age. This is due to changes in the composition of the gut microbiota in old mice, and it has been shown that the ratio of secondary bile acids to primary bile acids increases in old mice when they live with young mice. Furthermore, in humans, it has been reported that a decrease in secondary bile acids is correlated with obesity, metabolic abnormalities, cardiovascular diseases, and other aging-related diseases [5].

57.2.4 Metabolites Derived from Choline and Amino Acids

Other intestinal bacterial metabolites that have been pointed out in relation to aging include trimethylamine derived from choline, indole compounds derived from tryptophan, and phenylacetylglutamine derived from phenylalanine. Trimethylamine is absorbed from the intestine to the liver and metabolically converted to trimethylamine-*N*-oxide. Trimethylamine-*N*-oxide is implicated in the onset of cardiovascular diseases such as atherosclerosis and the worsening of neuroinflammation and dementia with aging due to its transition to the brain. On the other hand, indole compounds are suggested to contribute to the improvement of dementia by improving the intestinal barrier function that decreases with age through the aryl hydrocarbon receptor and by directly affecting the brain, promoting neurogenesis, and suppressing neuroinflammation. Phenylacetylglutamine is considered one of the metabolic markers of longevity because its concentration in the body increases with age, and it shows anti-inflammatory effects. On the other hand, it has been reported to show high values in patients with dementia and diabetes, and further clarification of its pathophysiological significance is needed.

References

1. Zhou Y, Hu G, et al. Host and microbiota metabolic signals in aging and longevity. Nat Chem Biol. 2021;17:1027–36.
2. Kimura I, Ichimura A, et al. Free fatty acid receptors in health and disease. Physiol Rev. 2020;100:171–210.
3. Fellows R, Denizot J, et al. Microbiota derived short chain fatty acids promote histone crotonylation in the colon through histone deacetylases. Nat Commun. 2018;9:105.
4. Eisenberg T, Knauer H, et al. Induction of autophagy by spermidine promotes longevity. Nat Cell Biol. 2009;11:1305–14.
5. Ma J, Hong Y, et al. Gut microbiota remodeling reverses aging-associated inflammation and dysregulation of systemic bile acid homeostasis in mice sex-specifically. Gut Microbes. 2020;11:1450–74.

Microbiome and Antiaging Medicine: Vaginal Microbiome and Antiaging

58

Remi Yoshikata

58.1 Ideal Vaginal Microbiome

The ideal vaginal environment is a state where the vaginal microbiome, i.e., the bacterial flora in the vagina, is predominantly composed of lactic acid bacteria, mainly of the *Lactobacillus* genus. The vagina maintains its self-cleaning function under the presence of lactic acid bacteria, and a correlation with health benefits such as reduced risk of gynecological diseases has been shown. The *Lactobacillus* genus is classified as the largest genus among lactic acid bacteria and is widely distributed in nature, and is also included in fermented foods. Lactobacilli are primarily classified into five communities (community-state type; CST) in molecular studies based on the cluster state of the bacterial species [1]. CST I is *Lactobacillus crispatus*, CST II is *Lactobacillus gasseri*, CST III is *Lactobacillus iners*, CST V is *Lactobacillus jensenii*, and the diverse group with fewer lactobacilli is classified as CST IV (subtypes A, B). CST I is the most common species in the *Lactobacillus* genus, with a pH of 4.4 or less. CST IV lacks lactic acid bacteria and is dominated by diverse anaerobic bacteria, with a pH of 5.3 or higher [1]. CST IV is a state with many bacterial vaginosis-related organisms in the vaginal microbiome, and it has been shown to increase the risk of sexually transmitted diseases and human papillomavirus (HPV) infection, decrease fertility, promote inflammation, affect immune regulation, and increase the risk of gynecological cancer [2–5]. Also, the proportion of each CST varies by race. In a cohort study targeting sexually mature women, the overall prevalence of lactic acid bacteria-dominated CST I, II, III, and V was 73% on average for all races, 80% for Asians, and when viewed by CST, the most common CST for Asians was CST III at 41%. The prevalence of the diverse group CST IV was 27% on average for all races and 19% for Asians [1].

58.2 The Relationship Between Estrogen and the Vaginal Microbiome

In the vagina, lactobacilli are produced by estrogen. Estrogen activates the production of glycogen in the vaginal mucosal epithelium, and lactobacilli, a lactic acid bacterium, is produced by the metabolism of the sloughed vaginal mucosal epithelium, glycosidase, and vaginal bacteria. The main function of lactic acid bacteria is known to be the destruction of bacteria by acidifying vaginal cells, cell membrane destruction, and stimulating host immunity, playing a role similar to antibiotics. The vaginal pH and self-cleaning function vary depending on each type of lactic acid bacteria-dominated CST I, II, III, and V,

R. Yoshikata (✉)
Hamasite Clinic, Tokyo, Japan
e-mail: r-yoshikata@tokyomidtown-mc.jp

H. Yamada, Y. Naito (eds.), *Anti-Aging Medicine*, https://doi.org/10.1007/978-981-97-8514-8_58

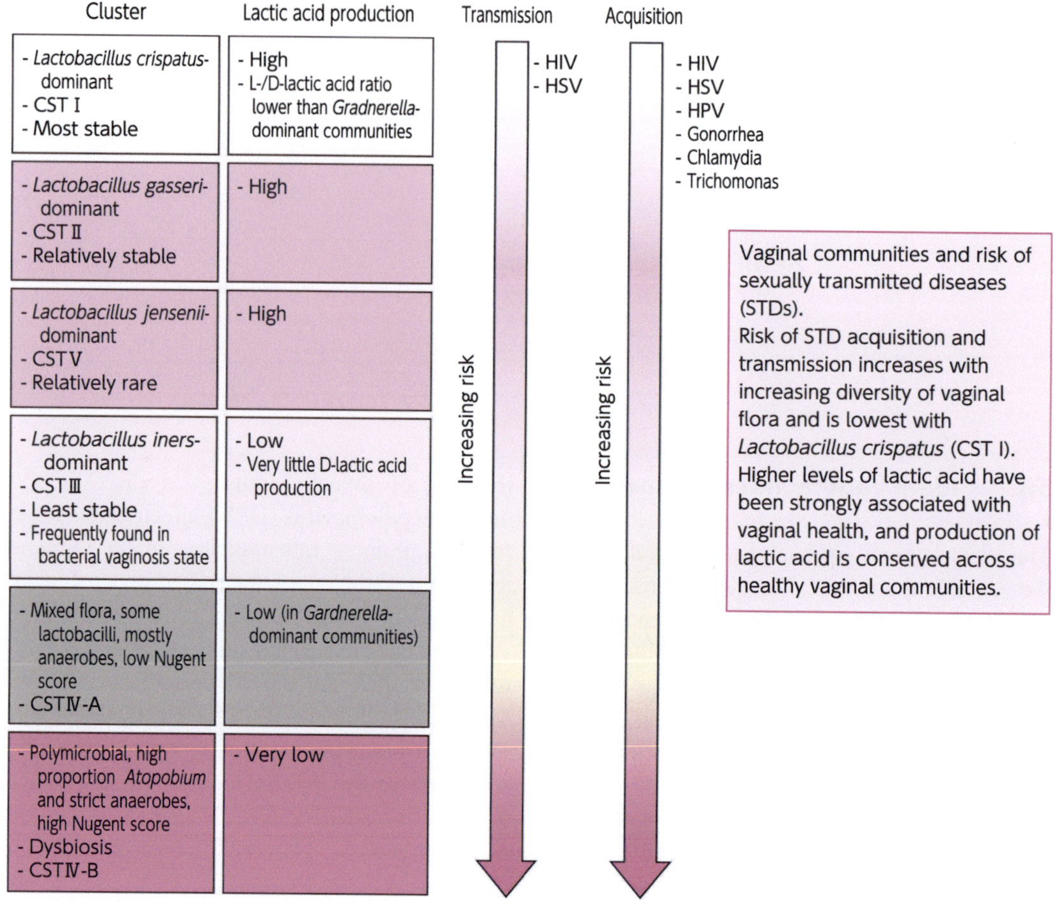

Fig. 58.1 Vaginal microbiome and women's health [5]

with CST I having a higher level of self-cleaning function and being strongly associated with vaginal health [5] (Fig. 58.1).

58.3 Changes in the Vaginal Microbiome due to Aging

Due to aging and estrogen deficiency, the ability of lactic acid bacteria in the vagina to produce decreases. The acidity in the vagina decreases, and the development and metabolic activity of the vaginal mucosa decrease, as do collagen proliferation and moisture retention. The vaginal mucosa becomes dry and thin, falling into a state of chronic inflammation and atrophic vaginitis (Fig. 58.2). In our own case, a next-generation sequencer analysis of the vaginal microbiome was conducted on 69 women aged 30–70 (average 47.6 years old), and a significant difference was observed before and after menopause. In premenopausal cases, about 65% were dominated by lactobacilli, and CST III showed the highest proportion of 38.2%, similar to known reports in Asians. The proportion of pathogenic bacteria was 23%. In postmenopausal cases, the proportion of lactobacillus-dominant cases decreased to 5.7%, while the proportion of pathogenic bacteria increased to 37% [6] (Fig. 58.3).

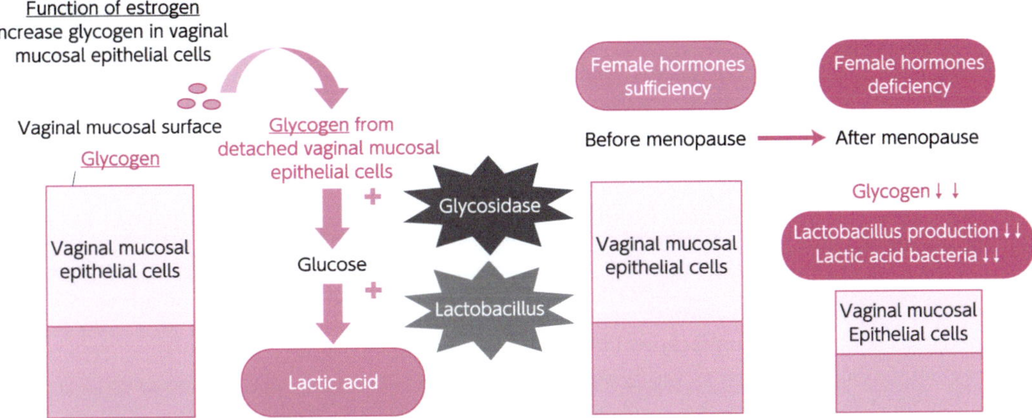

Fig. 58.2 Changes in the vaginal mucosa due to aging and atrophic vaginitis

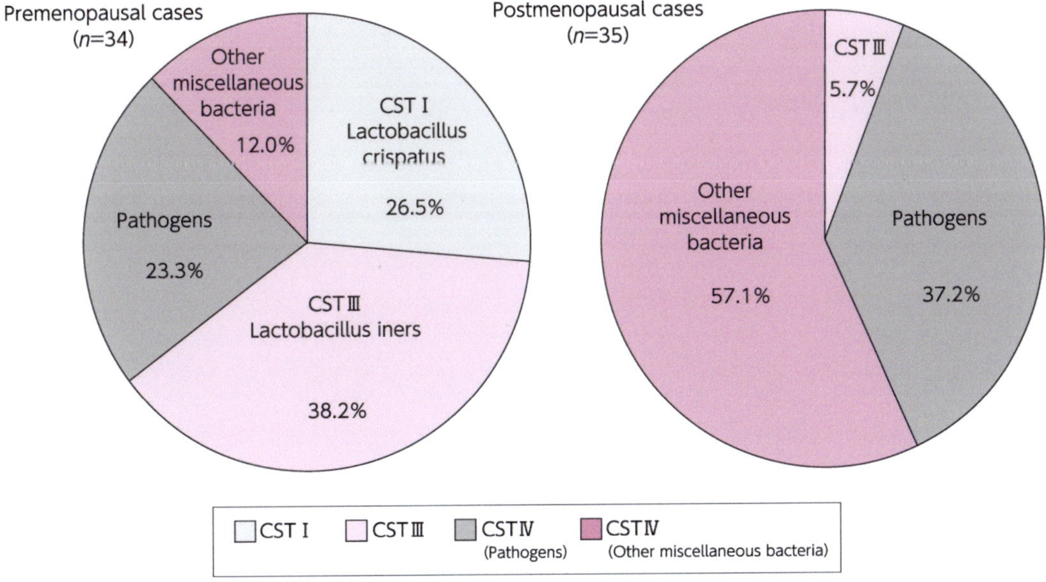

Fig. 58.3 Vaginal microbiome community type (CST): comparison of premenopausal and postmenopausal cases [6]

58.4 Antiaging of the Vaginal Microbiome

For the treatment of genitourinary symptoms of menopause (GSM), including atrophic vaginitis, the 2020 position statement of The North American Menopause Society (NAMS) lists the regular use of Femzone moisturizers and lubricants as the first line, followed by low-dose estrogen and dehydroepiandrosterone (DHEA) vaginal agents and hormone replacement therapy if menopausal symptoms such as hot flashes are present. In our own case, we have observed a decrease in vaginal pathogenic bacteria and an increase in lactobacilli with the use of a moisturizer containing lactobacilli and vaginal gel for Femzone care [7], and the antiaging effect of the vaginal microbiome by self-care is also expected.

References

1. Ravel J, Gajer P, et al. Vaginal microbiome of reproductive-age women. Proc Natl Acad Sci U S A. 2011;108(Suppl 1):4680–7.
2. Haahr T, Jensen JS, et al. Abnormal vaginal microbiota may be associated with poor reproductive outcomes: a prospective study in IVF patients. Hum Reprod. 2016;31:795–803.
3. Mitra A, MacIntyre DA, et al. Cervical intraepithelial neoplasia disease progression is associated with increased vaginal microbiome diversity. Sci Rep. 2015;5:16865.
4. Audirac-Chalifour A, Torres-Poveda K, et al. Cervical microbiome and cytokine profile at various stages of cervical cancer: a pilot study. PLoS One. 2016;11:e0153274.
5. Lewis FMT, Bernstein KT, et al. Vaginal microbiome and its relationship to behavior, sexual health, and sexually transmitted diseases. Obstet Gynecol. 2017;129:643–54.
6. Yoshikata R, Ohta H, et al. Age-related changes, influencing factors, and crosstalk between vaginal and gut microbiota: a cross-sectional comparative study of pre- and postmenopausal women. J Womens Health (Larchmt). 2022;31:1763–72.
7. Yoshikata R, Ohta H, et al. Evaluation of the efficacy of Lactobacillus-containing feminine hygiene products on vaginal microbiome and genitourinary symptoms in pre- and postmenopausal women: a pilot randomized controlled trial. PLoS One. 2022;17:e0270242.

59 Sleep, Circadian Rhythm, and Antiaging Medicine: Circadian Rhythm and Antiaging

Kazuhiro Yagita

The circadian rhythm is a physiological function that maintains the homeostasis of living organisms through adaptation to environmental cycles. The circadian clock, which controls the circadian rhythm, is also commonly referred to as the biological clock and is widely present in vertebrates including humans, as well as in insects, plants, and even photosynthetic bacteria such as cyanobacteria. The origin of the circadian clock is thought to begin with the assimilation of the Earth's rotation period, i.e., the relationship with the sun, into the basic order of life phenomena within cells. It is a function that was created as one of the adaptation mechanisms to the Earth's environment. This is an important basic concept when considering the relationship between the circadian rhythm and aging or antiaging. In other words, when there is a mismatch between the day-night cycle of the natural environment and one's own life rhythm, a failure of the circadian clock inherent in each cell occurs, which disrupts the order of cellular functions, including metabolism. In mammals, the suprachiasmatic nucleus (SCN) in the hypothalamus synchronizes the circadian clocks throughout the body and controls the individual function rhythm as the central circadian rhythm. In recent years, the impact of this circadian rhythm control system on aging and lifespan has been suggested, and attention is being paid to the relationship between the circadian rhythm and antiaging.

59.1 Age-Related Changes in Circadian Rhythm

The relationship between the circadian rhythm and aging can be discussed from two perspectives. One is "changes in the circadian rhythm with aging," and the other is "acceleration of aging due to circadian rhythm mismatch." Here, we will first discuss changes in the circadian rhythm with aging. The decrease in total activity with aging is one of the most well-known indicators of changes in the circadian rhythm due to aging in studies using animals such as mice [1]. Along with the decrease in activity with aging, a blurring of the behavioral rhythm is observed. This is thought to be due to a desynchronization between nerve cells in the suprachiasmatic nucleus with aging, resulting in a weakening of the rhythm of the suprachiasmatic nucleus as a whole. In fact, electrophysiological analysis of the suprachiasmatic nucleus shows that the amplitude of the firing rhythm of nerve cells is large in youth, but decreases in old age. On the other hand, in humans, there is a large individual difference, and it has not been generalized as an indicator of physiological aging. However,

K. Yagita (✉)
Department of Physiology and Systems Bioscience, Kyoto Prefectural University of Medicine, Kyoto, Japan
e-mail: kyagita@koto.kpu-m.ac.jp

H. Yamada, Y. Naito (eds.), *Anti-Aging Medicine*, https://doi.org/10.1007/978-981-97-8514-8_59

changes in spontaneous activity need to be paid attention to as indicators of aging. Also, one of the characteristics of age-related changes in behavioral rhythm is changes in the period of the circadian rhythm. In humans, this is recognized as a change in chronotype [2]. It is well known that humans tend to be nocturnal in youth and become morning-type as they age, but this is due to a shortening of the period of the circadian clock with aging. The period of the human circadian clock is approximately 24.5. The human circadian rhythm is thought to be approximately 25 h, and it is believed to approach 24 h with aging, leading to earlier bedtimes and wake times. On the other hand, changes in the circadian rhythm of behavior associated with aging have been reported in animals such as mice. Interestingly, in mice, the period becomes longer with age, contrary to humans. This is due to the fact that the mouse's circadian clock has an endogenous period of about 23.5 h, which is shorter than 24 h. In other words, it can be said that the aging-related changes in the behavioral rhythm are due to the endogenous circadian clock's period approaching exactly 24 h with age. In addition, it is known that the response to light-dark cycle shifts and time differences changes with age. In experiments exposing rodents to light-dark shift environments, the mortality rate of aged mice significantly increased compared to young mice. Considering that the number of days required for re-synchronization after a phase shift of the circadian rhythm is longer in aged mice than in young mice, it is suggested that the ability to adapt to phase changes in the light-dark cycle and time differences decreases with age. There is also a clear circadian rhythm control in deep body temperature, and a daily fluctuation of about 1.5 °C in body temperature can be seen in young adults. On the other hand, in the elderly, the decrease in body temperature during sleep is weakened, and the amplitude of the circadian rhythm of deep body temperature significantly weakens [2]. Also, clear circadian rhythms can be seen in the secretion of melatonin and cortisol, but a decrease in hormone secretion rhythm is observed in the elderly. In particular, melatonin is a hormone involved in sleep and body temperature regulation and is thought to play a role in "rhythm regulation" from its action on the suprachiasmatic nucleus. Therefore, the decrease in melatonin secretion at night in the elderly suggests the possibility of falling into a vicious cycle that exacerbates circadian rhythm disorders by promoting circadian rhythm control failure. On the other hand, there are reports that exposure to bright light during the day improves melatonin secretion at night, which is noteworthy from the perspective of antiaging using the circadian rhythm.

59.2 Aging Promotion by Circadian Misalignment

As molecular mechanisms controlling aging, the involvement of the insulin/insulin-like growth factor-1 (IGF-1) signaling pathway, the sirtuin family, and the mechanistic/mammalian target of rapamycin (mTOR) signaling pathway has been clarified so far, but all of these gene groups are known to be controlled by the circadian clock. Therefore, in the state of circadian clock control failure, or "circadian misalignment," caused by jet lag syndrome or shift work, it naturally affects the aging control mechanism. In fact, shift work including night shift (night-shift work), it has been clarified that the risk of developing type 2 diabetes, metabolic syndrome, and even some cancers significantly increases [3], supporting the association with age-related pathologies. In animal experiments using mice, it has been confirmed that long-term disruption of the light-dark cycle and circadian rhythm mismatch can reproduce some of the pathological mechanisms in humans, such as chronic inflammation in the liver and the occurrence of liver cell cancer, and significantly shorten lifespan [4, 5]. Furthermore, in long-term circadian rhythm mismatches, there is an increase in PD-1 (programmed death receptor-1) positive CD4+ T cells and Th17 cells, and an acceleration of immune aging (immune senescence), which is an age-related change in immune function [5]. These results suggest that circadian rhythm mismatch induces cellular aging through aging control mechanisms, which is manifested as a failure in the homeostasis of individual functions, and ultimately as individual aging. There are still many unknowns about the relationship between circadian rhythms and antiaging, especially how circadian rhythm mismatch affects aging control mechanisms, and how various physiological functions fail to maintain homeostasis. In particular, in the study of aging promotion mechanisms by circadian rhythm mismatch, it is expected that the importance of long-term observational studies for understanding human lifestyle diseases and carcinogenesis will increase.

References

1. Nakamura TJ, Takasu NN, et al. The suprachiasmatic nucleus: age-related decline in biological rhythms. J Physiol Sci. 2016;66:367–74.
2. Hood S, Amir S. The aging clock: circadian rhythms and later life. J Clin Invest. 2017;127:437–46.
3. Acosta-Rodríguez VA, Rijo-Ferreira F, et al. Importance of circadian timing for aging and longevity. Nat Commun. 2021;12:2862.
4. Kettner NM, Voicu H, et al. Circadian homeostasis of liver metabolism suppress hepatocarcinogenesis. Cancer Cell. 2016;30:909–24.
5. Inokawa H, Umemura Y, et al. Chronic circadian misalignment accelerates immune senescence and abbreviates lifespan in mice. Sci Rep. 2020;10:2569.

60 Sleep, Circadian Rhythm, and Antiaging Medicine: Sleep Disorders and Antiaging

Hiroshi Kadotani

Human sleep is classified into two qualitatively different stages: non-REM sleep and REM sleep. Non-REM sleep is further divided into three stages, N1 to N3 (light to deep), based on the depth of sleep (brain wave activity). Deep non-REM sleep (stage N3) is determined when slow waves (large brain waves that oscillate slowly at 0.5–2.0 Hz with an amplitude of 75 μV) are present more than 20% of the time, hence stage N3 is also referred to as slow wave sleep (SWS). The non-REM-REM sleep cycle lasts approximately 60–120 min (average about 90 min), and as sleep progresses into the latter half, the duration of non-REM sleep shortens, and the appearance of sleep slow waves decreases. Sleep also changes with age. Due to changes in the body clock with age, older people tend to go to bed and wake up earlier than younger people. Also, sleep time decreases with age, especially slow wave sleep decreases, and the time of awakening in the middle of sleep increases [1] (Fig. 60.1). People who are sleep-deprived appear less attractive to others. In a study involving healthy subjects, photographs of their faces without makeup were taken in a state of sufficient sleep and in a state of sleep deprivation after limiting sleep time to 4 h for two consecutive nights. When these photos were presented to other people, the sleep-deprived photos were perceived as less attractive, less healthy, and more sleepy, and people did not want to interact with them, but there was no difference in trustworthiness [2]. Thus, we are sensitive to changes in appearance related to sleep, and securing sleep time is necessary for good human relationships.

H. Kadotani (✉)
Department of Psychiatry, Shiga University of Medical Science, Otsu, Shiga, Japan
e-mail: kadotani@belle.shiga-med.ac.jp

H. Yamada, Y. Naito (eds.), *Anti-Aging Medicine*, https://doi.org/10.1007/978-981-97-8514-8_60

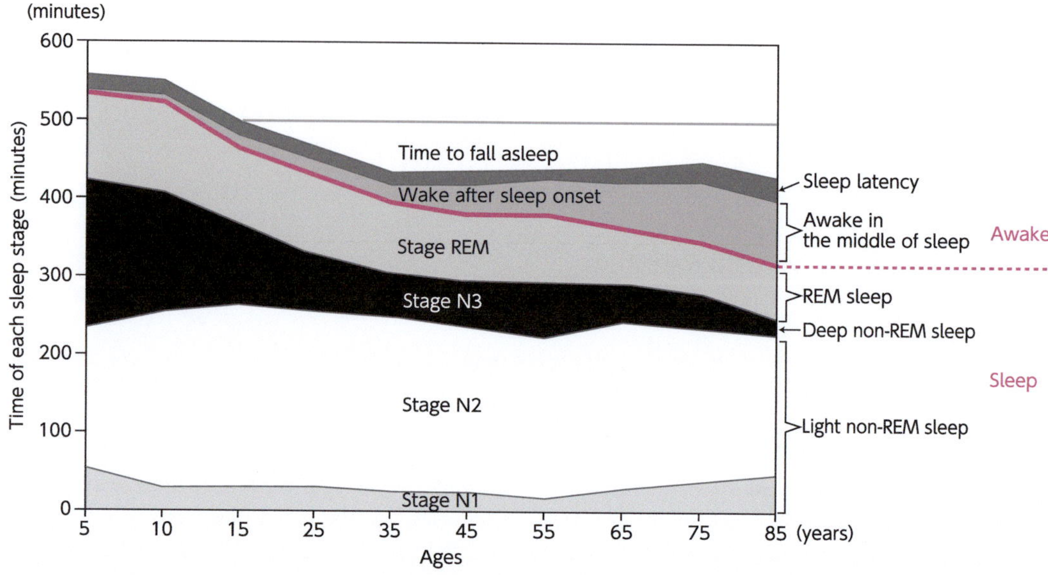

Fig. 60.1 Age and sleep stages (Adapted with permission from [1])

60.1 Sleep and Growth Hormone

The secretion of growth hormone is high in childhood, peaks in adolescence, and then decreases with age, but even after the age of 60, about 20% of the amount during adolescence is secreted. Growth hormone is an indispensable hormone for children's growth, but it also plays an important role in maintaining health in adults by regulating metabolism. Growth hormone is secreted intensively during sleep, especially during stage N3 (slow wave sleep) regardless of the time [3] (Fig. 60.2). The time when growth hormone, which promotes skin regeneration, is easily secreted is sometimes referred to as the "golden time for skin" from 10 p.m. to 2 a.m., but this is incorrect. Growth hormone is secreted even when you sleep during the day after a night shift, and conversely, even if you sleep at night, if you have little slow wave sleep due to sleep disorders such as sleep apnea, the amount of growth hormone secreted decreases [4]. It is important to get enough stage N3 sleep for the secretion of growth hormone.

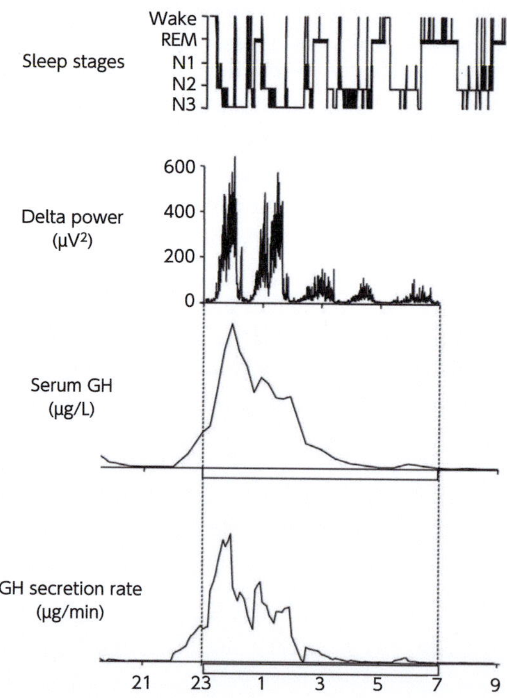

Fig. 60.2 Sleep stages and growth hormone (GH) secretion. Delta power (brain waves less than 4 Hz) is a measure of slow wave activity and is high during deep non-REM sleep (N3). (Modified with permission from [3])

60.2 Sleep and the Glymphatic System

In recent years, the glymphatic system (a portmanteau of "glia" and "lymphatic system"), which is responsible for the excretion of waste products in the brain (such as amyloid beta and tau proteins), has been proposed as a clearance system of the brain. It is believed to be involved in the onset of various brain diseases, including Alzheimer's disease. When sleep is reduced or disrupted, the activity of the glymphatic system decreases, causing the concentration of amyloid beta and tau proteins in the cerebrospinal fluid to increase [5]. It is believed that the glymphatic system is particularly active during sleep, especially during non-REM sleep [5]. When considering the antiaging of brain cells, it is necessary to consider this glymphatic system.

60.3 Conclusion

As such, sleep affects antiaging, and it is believed that disrupted sleep has a negative impact on antiaging.

References

1. Ohayon MM, Carskadon MA, et al. Meta-analysis of quantitative sleep parameters from childhood to old age in healthy individuals: developing normative sleep values across the human lifespan. Sleep. 2004;27:1255–73.
2. Sundelin T, Lekander M, et al. Negative effects of restricted sleep on facial appearance and social appeal. R Soc Open Sci. 2017;4:160918.
3. Van Cauter E, Latta F, et al. Reciprocal interactions between the GH axis and sleep. Growth Horm IGF Res. 2004;14(Suppl A):S10–7.
4. Mishima K. 89th Double sinful "golden time of skin" legend. Web National Geographic serial: cutting urban legends of sleep. 2018. https://natgeo.nikkeibp.co.jp/atcl/web/15/403964/032000083/.
5. Chong PLH, Garic D, et al. Sleep, cerebrospinal fluid, and the glymphatic system: a systematic review. Sleep Med Rev. 2022;61:101572.

Sleep, Circadian Rhythms, and Antiaging Medicine: Mild Disorders and Sleep Disorders

61

Jun Nishihira

61.1 What Is a Mild Disorder?

In recent years, there has been concern about the decline in labor productivity during work hours due to physical and mental disorders such as excessive stress (presenteeism) and the accompanying economic loss. On the other hand, it has been revealed that workers who are free from physical and mental disorders and have a high degree of happiness have high productivity and creativity [1]. Against this social background, it is expected that early detection of mild physical and mental disorders that appear before workers' health deteriorates, and prevention of excessive stress through improvement of lifestyle habits such as eating habits, will be effective in mental health measures. Most physical and mental disorders are often evaluated by questionnaire methods based on subjective responses. In our country, the "Occupational Stress Brief Survey Form" created by the Ministry of Health, Labor and Welfare's research team on the prevention of work-related diseases is widely used as a simple and valid survey form [2]. Among physical and mental disorders, mild disorders do not require medical treatment and can be prevented from becoming serious by self-management. It is believed that early detection of this mild disorder (mild disorder) can lead to avoidance of serious stress conditions. With the aim of early detection of mild disorders, we will explain the method of evaluating mild disorders simply using the "Occupational Stress Brief Survey Form" and the factors necessary for understanding the state of the disorder, based on the results of previous research [3].

61.2 Exploration of Mild Disorder States

The "Occupational Stress Brief Survey Form" evaluates six psychosomatic reactions (vitality, irritability, fatigue, anxiety, depression, physical complaints). The method of determining the degree of stress is in accordance with the "Stress Check System Implementation Manual" (Ministry of Health, Labor and Welfare), and an exploratory factor analysis was conducted using 29 questions to explore the psychosomatic disorder state lurking in the "working subject group" [4]. After confirming the sample validity and considering the optimal number of factors, we concluded that a five-factor solution is valid compared to the currently widely used six factors. That is, the two scales of "anxiety" and "depression," which were evaluated as individual reaction scores, were integrated into one scale as "anxiety/depression" (Fig. 61.1). Also, the two questions "no appetite" and "can't sleep well"

J. Nishihira (✉)
Health Information Science Center, Hokkaido Information University, Ebetsu, Japan
e-mail: nishihira@do-johodai.ac.jp

H. Yamada, Y. Naito (eds.), *Anti-Aging Medicine*, https://doi.org/10.1007/978-981-97-8514-8_61

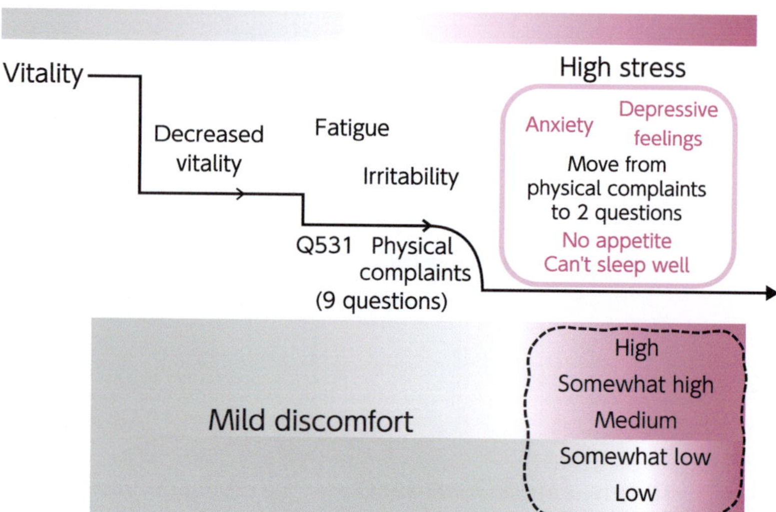

Fig. 61.1 Definition of mild discomfort. Improvements can be made by revising dietary habits and lifestyle

included in the "physical complaints" group were integrated into the same group of "fatigue/depression," and "physical complaints" were reduced to two questions. Based on these results, we concluded that it is appropriate to use the five factors of "decreased vitality," "irritability," "fatigue," "physical complaints," and "anxiety/depression" as indicators to define mild disorders.

61.3 Stress Characteristics of Mild Disorders

Using item response theory, we calculated the average values for each factor of stress characteristics for each group, showing the item response curves (Fig. 61.2). The factors used here are "anxiety/depression," "decreased vitality," "irritability," "fatigue," and "physical complaints." Of these five factors, the first four, excluding "physical complaints," showed high slopes and high discriminative power. Among the five factors, "decreased vitality" appears first at the far left threshold in the early stages, followed by "fatigue" and "irritability," and finally "anxiety/depression" under high stress. From these results, it was concluded that "decreased vitality" is a symptom recognized in the early stages when the degree of stress felt by workers is relatively low. Next, "irritability" and "fatigue" appear, followed by "anxiety/depression," which is the symptom with the highest degree of stress.

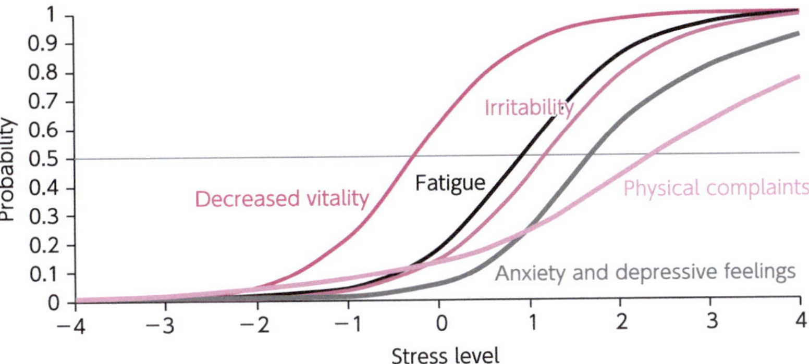

Fig. 61.2 Item response characteristics. Shows the stress level as a latent characteristic. Convert each item's (1: almost none, 2: sometimes, 3: often, 4: always) to (0-0-1-1)

61.4 Relationship Between Sleep Disorders and Mild Discomfort

Sleep disorders (insomnia syndrome) are defined as difficulty in initiating or maintaining sleep [5]. Twenty to thirty-five percent of the general population complain of insomnia symptoms, and 10–20% have severe symptoms, and sleep disorders are associated with depression and loss of appetite. In the results of the research survey on mild discomfort, it was suggested that there is a close relationship between "anxiety/depression" and "unable to sleep well/no appetite," and these questions were integrated into one scale (Fig. 61.2). In dealing with workers' stress problems, it is necessary to keep in mind sleep disorders and focus on the evaluation of "anxiety/depression" in high-stress states. In cases with sleep disorders, it is necessary to first rule out depression and aim to improve the state of mild discomfort in nondepressed individuals.

61.5 Concept of Mild Discomfort and Future Development

From the research on mild discomfort, there will be no objection to evaluate the four factors of "decreased vitality" that appear in the early stages of stress, "irritability" and "fatigue" that appear next, and "physical complaints" that gradually appear from the early stages as targets for evaluation of mild discomfort. On the other hand, careful response is required to exclude the diagnosis of depression when including the factor of "anxiety/depression," which is classified as a high-stress state, in mild discomfort. In conclusion, understanding the state of mild discomfort can lead to early recovery from workers' discomfort and improvement of presenteeism and can contribute to productivity improvement.

References

1. Proto E. Are happy workers more productive? Firms' concerns about the well-being of their employees are largely supported by the evidence. IZA World of Labor; 2016. p. 315.
2. Ministry of Health, Labour and Welfare. Occupational stress simple survey form (57 items). 2015. https://www.mhlw.go.jp/bunya/roudoukijun/anzeneisei12/dl/stress-check_j.pdf. Accessed 8 Feb 2023.
3. Kagami-Katsuyama H, Sato-Ueshima M, et al. The relationship between mental and physical minor health complaints and the intake of dietary nutrients. Nutrients. 2023;15:865.
4. Ministry of Health, Labour and Welfare. Manual for implementing the stress check system based on the occupational safety and health act. 2021. https://www.mhlw.go.jp/content/_000533925.pdf. Accessed 8 Feb 2023.
5. Li L, Wu C, et al. Insomnia and the risk of depression: a meta-analysis of prospective cohort studies. BMC Psychiatry. 2016;16:375.

62 Caloric Restriction, Starvation, and Antiaging Medicine: Caloric Restriction Research and Antiaging

Takuya Chiba

62.1 Molecular Mechanisms of Antiaging Effects by Caloric Restriction

The antiaging effects of caloric restriction, which involves reducing the food given to experimental animals by about 30% and keeping them on this diet for a long period, have been confirmed across various species, from nematodes and fruit flies to mice and monkeys. Therefore, it is considered to be one of the traits that organisms have acquired during the process of evolution. Organisms sense their nutritional status and choose whether to produce the necessary energy using the nutrients they have ingested or to store them for future energy shortages. This is thought to have maximized both the chance of survival of the organism itself and the chance of survival of the species through reproduction. The antiaging effects of caloric restriction are suggested to be closely related to this nutrient-sensing function of organisms, and the center that controls this is thought to be a group of neurons in the arcuate nucleus in the hypothalamus of the brain. Neurons expressing neuropeptide Y (NPY) are known to be important for the antiaging effects of caloric restriction. Activation of NPY enhances feeding behavior and suppresses energy consumption. Caloric restriction enhances the expression of NPY in the arcuate nucleus. On the other hand, it has been reported that when caloric restriction is applied to NPY knockout mice, their lifespan does not extend, and the antiaging effects of caloric restriction, such as enhanced oxidative stress resistance and reduced tumor incidence, disappear or weaken [1]. Insulin, leptin, and ghrelin are known to be important factors in the regulation of NPY expression. Insulin is a hormone-like factor secreted from pancreatic β cells, leptin from fat cells, and ghrelin from the stomach. Insulin and leptin negatively regulate the expression of NPY, while ghrelin positively regulates it. It has been reported that animals subjected to caloric restriction have decreased body weight and blood glucose levels, along with decreased blood insulin and leptin concentrations and increased ghrelin concentration.

62.2 Nutrient-Sensing Signals and Caloric Restriction

The growth hormone (GH)/insulin-like growth factor 1 (IGF-1)/insulin signal usually regulates processes such as enhanced glucose metabolism and fat accumulation. GH is secreted from the anterior pituitary gland, binds the growth hormone receptor (GHR), and activates JAK2, thereby increasing the expression of IGF-1 and promoting its secretion. Insulin is secreted in

T. Chiba (✉)
Biomedical Gerontology Laboratory, Faculty of Human Sciences, Waseda University, Saitama, Japan
e-mail: takuya@waseda.jp

H. Yamada, Y. Naito (eds.), *Anti-Aging Medicine*, https://doi.org/10.1007/978-981-97-8514-8_62

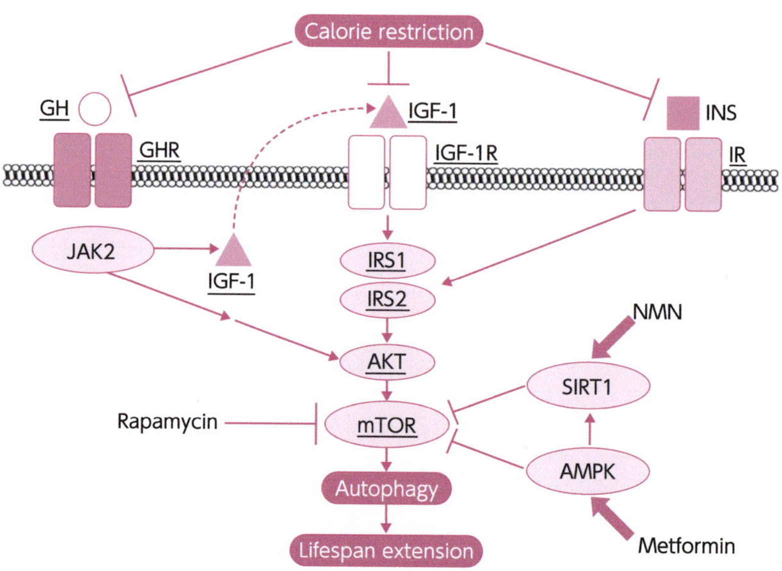

Fig. 62.1 Signaling systems related to the antiaging effect of calorie restriction. Calorie restriction is thought to exert lifespan extension and antiaging effects by inhibiting mTOR signaling through the suppression of GH/IGF-1 and insulin signaling. The inhibition of mTOR regulates various cellular functions such as the reduction of protein synthesis, but the activation of autophagy is suggested to be involved in lifespan extension. Also, it is suggested that factors that show lifespan extension effects by loss (underlined) due to genetic mutations or genetic modifications are important as development targets for substances that mimic the antiaging effects of calorie restriction, as they are present in many of these signaling systems. *AMPK* AMP-activated kinase, *GH* growth hormone, *GHR* growth hormone receptor, *IGF-1* insulin-like growth factor 1, *IGF-1R* insulin-like growth factor 1 receptor, *INS* insulin, *IR* insulin receptor, *IRS1* insulin receptor substrate 1, *IRS2* insulin receptor substrate 2, *JAK2* Janus kinase 2, *mTOR* mammalian target of rapamycin, *NMN* nicotinamide mononucleotide

response to elevated blood glucose levels and binds to the insulin receptor (IR) expressed in organs such as the liver and adipose tissue, activating downstream signals. Similarly, IGF-1 is secreted from the liver in response to meals and binds to the insulin-like growth factor 1 receptor (IGF-1R) in the liver and adipose tissue, activating downstream insulin receptor substrate 1 (IRS1) and insulin receptor substrate 2 (IRS2)/AKT via the mammalian target of rapamycin (mTOR) signal, playing a crucial role in nutrient sensing. These intracellular signaling pathways are highly conserved across species and are thought to be important as antiaging signals regulated by calorie restriction. When the nutritional status is negative, as in calorie restriction, molecules such as AMP-activated kinase (AMPK) are activated and mechanisms to produce the deficient energy are activated, along with autophagy. From these findings, it has become clear from various genetically modified animal studies that the inhibition of mTOR signaling is key to the lifespan-extending effects of calorie restriction. Molecules present in these signaling pathways are also attracting attention as targets for substances such as rapamycin, metformin, and nicotinamide mononucleotide (NMN), which are expected to mimic the effects of calorie restriction (Fig. 62.1).

62.3 The Effect of Calorie Restriction on Humans

In humans, it is often difficult to reduce calorie intake by about 30% long-term, as in experimental animals, because it causes discomfort to the subjects. Among the calorie restriction studies conducted on humans so far, I will discuss the main studies where randomized comparative trials were conducted. In the Diet, Obesity and Genes (DiOGenes) project, calorie restriction was implemented for 8 weeks. In the calorie-restricted group compared to the ad libitum group, the indicator of insulin resistance significantly decreased, suggesting an improvement in insulin resistance due to calorie restriction [2]. However, in the Comprehensive Assessment of Long-term Effects of Reducing Intake of Energy (CALERIE) study, no clear results were obtained regarding the improvement of indicators showing insulin resistance. The CALERIE study aimed to implement a 25% calorie restriction, and effects such as anti-obesity effects, anti-inflammatory effects, and enhanced oxidative stress resistance were observed [3]. In this 2-year clinical trial, it is not clear to what extent there is an extension effect on lifespan. Although a milder calorie restriction than planned was implemented in the CALERIE study, it is an important finding that an effect leading to a certain antiaging was obtained. However, research using experimental organisms suggests that induction of a very strong stress state, close to a state of starvation, is necessary to exert the antiaging effect of calorie restriction. Therefore, it is suggested that the mechanism is essentially different from the physical changes due to mere weight loss to the extent that humans do so-called dieting. To demonstrate the antiaging effect of calorie restriction in humans, it is suggested that long-term studies using calorie restriction mimetics are necessary.

References

1. Chiba T, Tamashiro Y, et al. A key role for neuropeptide Y in lifespan extension and cancer suppression via dietary restriction. Sci Rep. 2014;4:4517.
2. Montastier E, Déjean S, et al. Adipose tissue CIDEA is associated, independently of weight variation, to change in insulin resistance during a longitudinal weight control dietary program in obese individuals. PLoS One. 2014;9:e98707.
3. Kraus WE, Bhapkar M, et al. 2 Years of calorie restriction and cardiometabolic risk (CALERIE): exploratory outcomes of a multicentre, phase 2, randomised controlled trial. Lancet Diabetes Endocrinol. 2019;7:673–83.

63 Caloric Restriction, Fasting, and Antiaging Medicine: Fasting, the Biological Signals of Starvation, and Antiaging

Masaharu Uno and Eisuke Nishida

As represented by phrases like "No need for a doctor if you eat until you're eight parts full," "No need for a doctor with persimmons," and "An apple a day keeps the doctor away," there exist sayings related to health and diet. We have accumulated knowledge since ancient times that diet and health are closely related. We, as living organisms, need to intake and utilize energy from the outside world to maintain our life systems. Therefore, the pathways for energy assimilation and metabolism are often preserved across species. How we handle the energy taken in from the outside world is a very important issue for us as living organisms. Indeed, appetite is listed as one of the three major desires of us, humans. As a result of this desire and the industrial revolution, we have come to enjoy overeating. Consequently, the amount of energy we intake has dramatically increased, leading to obesity and lifestyle-related diseases becoming social issues. On the other hand, simple meals and fasting were once part of our lives. For example, fasting was practiced as part of training under various religious sects. A famous example is "Ramadan" in Islam, where fasting is practiced during the ninth month of the Hijri calendar. Recent studies have been scientifically verifying that simple meals and fasting are related to health and even life extension. As a result, simple meals and fasting are attracting attention as choices in an era of overeating. In this section, we will introduce findings on the effects of caloric restriction (simple meals) and fasting on health and lifespan. In 1935, McCay and others found that lifespan was extended in rats with restricted diets [1]. Since then, it has been reported that restricting the diet in various animals, including yeast, nematodes, fruit flies, mice, and even monkeys, extends lifespan and also has beneficial effects on health. There are various methods of dietary restriction. Some methods restrict specific nutrients, but this section will mainly focus on two methods. These two methods are reducing caloric intake and incorporating periods of starvation (Fig. 63.1). Caloric restriction is, as the name suggests, a method of restricting the calories consumed daily (calorie restriction; CR) (Fig. 63.1 left). On the other hand, there are various methods of incorporating periods of starvation, depending on the length and frequency of these periods (Fig. 63.1 right). There are three main types: (1) alternative day fasting (ADF), where free eating and fasting are alternated every other day, (2) time-restricted feeding (TRF), where a fasting period is set within a day, (3) periodic fasting, where a fasting period is set regularly. There are three methods of fasting (PF). In studies using model organisms, CR, ADF, and TRF are often researched, while in human studies, CR and PF are more common.

M. Uno (✉) · E. Nishida
RIKEN Center for Biosystems Dynamics Research (BDR), Hyogo, Japan
e-mail: masaharu.uno@riken.jp; eisuke.nishida@riken.jp

H. Yamada, Y. Naito (eds.), *Anti-Aging Medicine*, https://doi.org/10.1007/978-981-97-8514-8_63

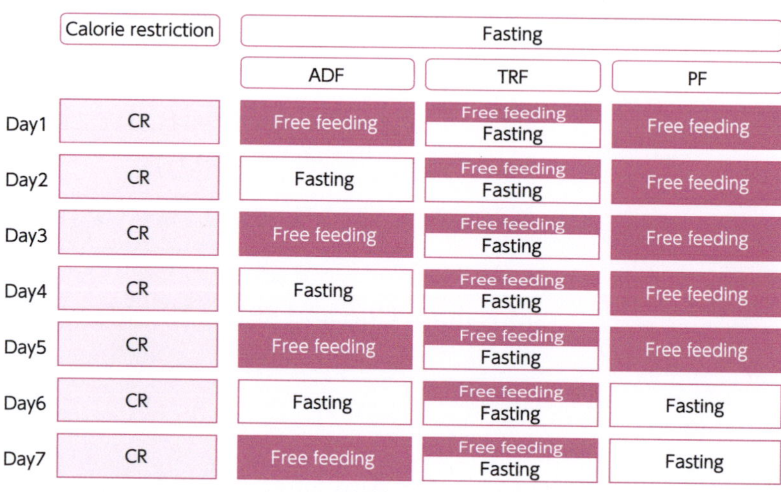

Fig. 63.1 Methods of dietary restriction. Calorie restriction reduces daily calorie intake. On the other hand, fasting can be broadly divided into ADF, TRF, and PF, depending on the frequency

However, these four methods have been reported to have effects on lifespan and health in both humans and mice. However, there are few studies directly comparing these four methods, and our understanding of the differences in their antiaging mechanisms is still limited. Organisms exposed to calorie restriction or fasting need to maintain homeostasis based on less nutrition, so various nutrition-related signaling pathways work during these dietary restrictions [2]. As a result of dietary restriction, adenosine triphosphate (ATP) and NADH (reduced form of nicotinamide adenine dinucleotide), which are used as cellular energy, as well as amino acids, are depleted in cells. As a result, amino acid depletion inhibits the activity of the mechanistic target of rapamycin (mTOR), ATP depletion (increase in adenosine monophosphate; AMP) activates AMP-activated protein kinase (AMPK), and NADH depletion (increase in NAD^+; oxidized form of nicotinamide adenine dinucleotide) induces the activation of sirtuin (SIRT). It is known that the inhibition of mTOR activity and the activation of AMPK suppress protein synthesis and ribosome biogenesis and promote the activation of autophagy. SIRT1 is known to adapt cells to a state of low nutrition by activating various transcription factors. Also, a decrease in carbohydrate intake is known to reduce the activity of the insulin/insulin-like growth factor 1 (IGF-I) signaling pathway, resulting in the upregulation of stress response genes and autophagy via the downstream transcription factor forkhead box O (FoxO). Thus, various intracellular signaling pathways respond to a state of low nutrition, changing the state of the cell (Fig. 63.2). And these cells adapted to a state of low nutrition have been found to work advantageously against various stresses and cellular aging. Various transcription factors are involved in the changes that adapt to a state of low nutrition at the cellular level. These transcription factors include forkhead transcription factor FoxO, nuclear factor erythroid 2-related factor 2 (NRF2), peroxisome proliferator-activated receptor-α (PPAR-α), and peroxisome proliferators-activated. Receptor-γco-activator-1α (PGC-1α) is known among others. FoxO controls gene groups related to gluconeogenesis and cell cycle, in addition to anti-stress gene groups. NRF2 is known to control the expression of a very large number of genes involved in cell protection. PPAR-α is known for its involvement in liver lipid metabolism and ketone body production, and PGC-1α is known to regulate mitochondrial function. Thus, various transcription factors are known to function to adapt cells to a state of malnutrition. It is thought to be important that autophagy is enhanced when cells are exposed to a state of malnutrition [3]. Autophagy is thought to have the function of recycling dysfunctional organelles and proteins. The function of autophagy has been observed to decline with aging, and it has been reported that the decline in autophagy func-

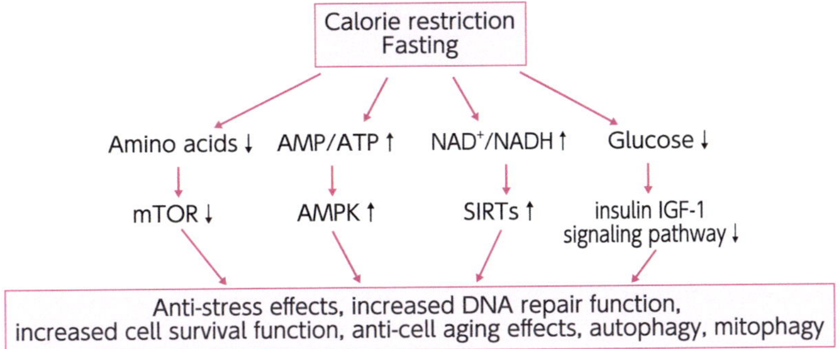

Fig. 63.2 Signal transduction activated by calorie restriction and fasting. The mTOR pathway is triggered by amino acid deficiency, AMPK and SIRTs are activated by cellular energy depletion, and insulin signal transduction pathway is inhibited by glucose reduction. As a result, antiaging effects such as antioxidant stress action and autophagy are triggered

tion is involved in the onset of various diseases. Therefore, it is expected that the enhancement of autophagy by dietary restriction or drugs plays an important role in extending healthy lifespan and average lifespan. The control of autophagy enhancement involves energy sensing signaling pathways such as AMPK, SIRT, and mTOR. From these, it is thought that autophagy plays an important role in the antiaging effects and promotion of health by dietary restriction. Individual-level responses to dietary restriction include changes in metabolic substances and fluid factors in the blood [4]. Blood molecules, which play important roles in signal transmission and energy supply for maintaining homeostasis of the individual, also play important roles in aging and lifespan control. Individuals exposed to starvation or severe calorie restriction respond by changing their energy utilization. Specifically, they shift from glucose utilization to lipid utilization. When a state of malnutrition occurs due to dietary restriction, first, the use of glycogen stored in the liver and gluconeogenesis from glucogenic amino acids occurs. Subsequently, an increase in the concentration of free fatty acids and ketone bodies in the blood is observed. Ketone bodies are taken up by cells and enter the TCA cycle (tricarboxylic acid) and contribute to the production of ATP, the energy within cells, through the cycle. Thus, it is known that ketone bodies, whose concentration increases in the blood, function as an energy source for muscles and the brain, which consumes a lot of energy, and the transition of the center of energy during energy depletion is becoming understood. Interestingly, it has also been found that when given as a diet, ketone bodies also show antiaging effects. It is thought that the change in the means of energy utilization of organisms may have an antiaging effect. In a state of malnutrition, if there are things like free fatty acids and ketone bodies whose concentration increases in the blood, there are also things whose blood concentration decreases. It is known that insulin, IGF-1, glucose, gluconeogenic amino acids, and neutral fatty acids decrease. In addition, an increase in ghrelin, a hormone that stimulates appetite, and an increase in adiponectin, which promotes fatty acid oxidation in adipose tissue, have also been observed. In addition to these, fibroblast growth factor 21 (FGF21) has been reported to be released of from the liver into the bloodstream in that condition. Overexpression of FGF21 is known to have antiaging effects and is expected to be a factor involved in lifespan extension due to dietary restriction. Changes in intracellular energy status in response to dietary restriction are thought to exert antiaging effects by triggering various signal transduction pathways and metabolic products, stress responses, autophagy, and even enhancement of mitochondrial function. However, there are still many unknowns, such as the spatiotemporal order in which these effects occur and the extent of dietary restriction required to exert these effects. Several studies have been conducted on the difference in longevity effects between calorie restriction and starvation stimuli in model organisms. In experiments using nematodes as model organisms, it has been reported that the lifespan extension effect of starvation stimuli is greater than that of calorie restriction [5, 6]. It has also been reported that the factors involved in lifespan extension due to calorie restriction and starvation stimuli are different, suggesting a difference. In experiments using mice, it has been reported that periodically providing starvation stimuli is more effective in extending lifespan than calorie restriction. From these findings, it is expected that starvation stimuli will also bring about more effective antiaging effects in humans. However, the knowledge about the lifespan extension effect of starvation stimuli is lacking, as the lifespan extension effect of calorie restriction has been elucidated earlier [7]. Thus, the antiaging effects of calorie restriction and starvation stimuli are expected antiaging operations because many things have been understood about their mechanisms and their implementation is easy. However, it has also been reported that the lifespan extension effect varies greatly among individuals in mice, so in order to obtain its antiaging effect more effectively, it is necessary to accumulate more extensive knowledge in humans.

References

1. McCay CM, Crowell MF, et al. The effect of retarded growth upon the length of life span and upon the ultimate body size. 1935. Nutrition. 1989;5:155–71; discussion 172.
2. Kenyon CJ. The genetics of ageing. Nature. 2010;464:504–12.
3. Rubinsztein DC, Mariño G, et al. Autophagy and aging. Cell. 2011;146:682–95.
4. García-Flores LA, Green CL. Of mice and men: impacts of calorie restriction on metabolomics of the cerebellum. J Gerontol A Biol Sci Med Sci. 2021;76:547–51.
5. Honjoh S, Yamamoto T, et al. Signalling through RHEB-1 mediates intermittent fasting-induced longevity in C. elegans. Nature. 2009;457:726–30.
6. Uno M, Honjoh S, et al. A fasting-responsive signaling pathway that extends life span in C. elegans. Cell Rep. 2013;3:79–91.
7. Acosta-Rodríguez V, Rijo-Ferreira F, et al. Circadian alignment of early onset caloric restriction promotes longevity in male C57BL/6J mice. Science. 2022;376:1192–202.

64 Caloric Restriction and Starvation in Antiaging Medicine: Caloric Restriction Mimetics

Ken Shinmura

Caloric restriction (CR) has a wealth of evidence to be considered the golden standard in antiaging medicine. However, due to various difficulties and unresolved issues associated with long-term implementation of CR, there has been a growing interest in the development of drugs and compounds that mimic the effects of CR, known as CR mimetics or CR mimicry [1]. Table 64.1 shows the currently proposed CR mimetics. In the manifestation of CR effects, four intracellular signals play a crucial role: insulin/IGF-1 (insulin-like growth factor 1), AMPK (AMP-activated protein kinase), sirtuin, and mTOR (mammalian target of rapamycin) [1, 2] (Fig. 64.1). Furthermore, these signals are not independently effective, but construct a network of cellular function regulation while forming close interrelationships [1, 2]. Therefore, it was hypothesized that the desirable effects of CR could be reproduced by activating or deactivating these four signaling pathways. For instance, the CR mimetics listed in the first half of Table 64.1 mainly target the suppression of insulin/IGF-1. However, metformin, an antidiabetic drug, is known to have the ability to activate AMPK and sirtuin and partially inhibit mitochondria [3]. Resveratrol, a polyphenol found in red wine, has attracted attention for its sirtuin-activating properties. However, the sirtuin activation by resveratrol is indirect, and its antioxidant and anti-inflammatory effects are now considered to be the main causes of its clinical effects [4]. As research progresses, it has become clear that promising CR mimetics have complex mechanisms of action. Conversely, it has been found that if one specializes in a single cellular signal like sirtuin-activating compounds (STACs), it does not necessarily lead to a stronger CR effect [1, 2]. Therefore, when considering the use of CR mimetics, it is necessary to carefully select between (1) types with broad effects and (2) selective types targeting specific intracellular signals, according to the intended use. Among these, metformin, resveratrol, NAD^+ (nicotinamide adenine dinucleotide) precursors, and spermidine have been the focus of attention for many years as CR mimetics. This is mimicry, and clinical research has advanced considerably. In the United States, with the financial support of the National Institute on Aging, large-scale trials are underway to investigate whether metformin can slow the progression of chronic diseases associated with aging, such as heart disease, cancer, and dementia, in the elderly [3]. The latter three are available as supplements in our country and can be used in daily practice. However, we believe that more evidence is needed for the dosage and duration of administration when expecting antiaging effects clinically. Recently, SGLT2 (sodium-glucose cotransporter 2) inhibitors and β-hydroxybutyrate have been attracting attention

K. Shinmura (✉)
Department of General Internal Medicine, Hyogo Medical University, Nishinomiya, Hyogo, Japan
e-mail: ke-shimmura@hyo-med.ac.jp

H. Yamada, Y. Naito (eds.), *Anti-Aging Medicine*, https://doi.org/10.1007/978-981-97-8514-8_64

Table 64.1 CR mimicry candidates

Main action	Drug/compound name	Other actions, etc.
Glycolysis inhibition	2-Deoxyglucose	Clinical applications other than cancer treatment are insufficiently studied
	Mannoheptulose	Natural substance derived from avocado, hexokinase inhibitor
	3-Bromopyruvate	GAPDH inhibitor
	Dichloroacetate	PDK inhibitor
AMPK activation	Metoformin	Improves insulin sensitivity, activates sirtuins, inhibits mTOR, has mitochondrial inhibitory effects
	AICAR	More effective in muscle
	Aspirin, salicylate	Has EP300, COX, mTOR, NF-κB inhibitory effects
Adiponectin increase	Pioglitazone	PPARγ agonist, improves insulin sensitivity
	AdipoRon	AdipoR1, R2 agonist
	Ozmotin	Phytochemical
Sirtuin activation	Resveratrol	Red wine polyphenol, activates AMPK, eNOS, inhibits NF-κB, antioxidant, promotes autophagy, and has various other effects
	SRT1720, SRT2104	Produced as so-called STACs
	NAD^+ precursors (NMN, NR)	Converts to NAD^+ in the body, available as a supplement
	Quercetin	Polyphenol
	Berberine	Contained in goldenseal
mTOR inhibition	Rapamycin	Difficult to use clinically due to immunosuppressive effects
	Rapalogs	Rapamycin derivatives
Activation of autophagy	Spermidine	Polyamine
	Epigallocatechin-3-gallate	A type of green tea polyphenol
	4,4′-Methoxychalcone	Polyphenol
	Curcumin	Turmeric polyphenol
	Hydroxycinnamic acid	Inhibits acetyl CoA synthesis
Increase in ketone bodies	β-Hydroxybutyric acid (ketone salts and esters)	Not only used as an energy substrate, but also has intracellular signaling functions such as anti-inflammatory effects
	Medium-chain fatty acids (ketogenic diet)	Coconut oil and palm kernel oil
	SGLT2 inhibitors	Has CR-like effects such as improved insulin sensitivity
RAS inhibition	ACE inhibitors	
	ARB	
	Anti-aldosterone drugs	
Inhibition of vascular aging	Statins	

GAPDH glyceraldehyde-3-phosphate dehydrogenase, *PDK* pyruvate dehydrogenase kinase, *AMPK* AMP (5′ adenosine monophosphate)-activated protein kinase, *AICAR* 5-aminoimidazole-4-carboxamide-1-β-D-ribofuranoside, *COX* cyclo-oxygenase, *mTOR* mammalian target of rapamycin, *NF-κB* nuclear factor-κB, *PAAR γ* peroxisome proliferator-activated receptorγ, *eNOS* endothelial nitric oxide synthase, *STACs* sirtuin-activating compounds, *NAD^+* nicotinamide adenine dinucleotide, *NMN* nicotinamide mononucleotide, *NR* nicotinamide riboside, *SGLT2* sodium-glucose cotransporter 2, *RAS* renin-angiotensin-aldosterone, *ACE* angiotensin converting enzyme, *ARB* angiotensin receptor blocker

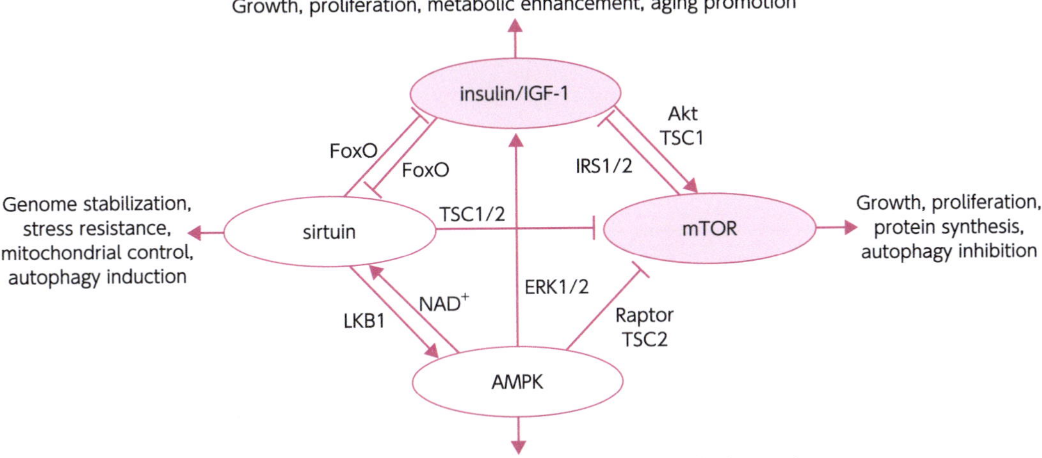

Fig. 64.1 Four intracellular signals targeted by calorie restriction. Calorie restriction acts inhibitory on insulin/IGF and mTOR, and stimulatory on sirtuins and AMPK. *IGF-1* insulin-like growth factor 1, *FoxO* forkhead box protein O, *IRS1/2* insulin receptor substrate 1/2, *TSC1/2* tuberous sclerosis complex 1/2, *LKB1* liver kinase B1, *NAD*+ nicotinamide adenine dinucleotide, *ERK1/2* extracellular signal-regulated kinase 1/2, *AMPK* AMP (5′ adenosine monophosphate)-activated protein kinase

as CR mimicry. SGLT2 inhibitors are antidiabetic drugs, but their use as heart failure and chronic kidney disease treatments is also increasing. The main action of SGLT2 inhibitors, which excrete carbohydrates from the kidneys and reduce the total amount of calories used in the body, can be said to be a concept similar to CR. In fact, the reactions caused throughout the body by the administration of SGLT2 inhibitors have many common points with the reactions caused by CR [5]. Furthermore, it is noteworthy that ketosis is caused by SGLT2 inhibitors. It is said that the total blood ketone body concentration reaches 1 mmol/L or more due to intense exercise or 24-h fasting, and the total blood ketone body concentration can rise up to about 1 mmol/L even with the administration of SGLT2 inhibitors [5]. Furthermore, the total blood ketone body concentration can reach up to 3–4 mmol/L at most with a ketogenic diet or β-butyrate administration. If ketone bodies are mediators of the antiaging effects of CR and intermittent fasting, these may be promising CR mimicry. In recent years, the antitumor effects of the fasting-mimicking diet have also been actively researched, and a combined discussion is necessary. However, there are many issues left regarding the specific treatment strategy for aging control by ketone bodies, such as what method and how much ketosis to induce and whether chronic or intermittent ketosis is preferable [5]. Therefore, their use in anticipation of antiaging effects is not recommended at this time.

References

1. Shinmura K. Nutritional interventions for cardiovascular aging and age-related cardiovascular diseases. In: Nutrition, exercise and epigenetics: ageing interventions. Cham: Springer; 2015. p. 179–209.
2. Madeo F, Carmona-Gutierrez D, et al. Caloric restriction mimetics against age-associated disease: targets, mechanisms, and therapeutic potential. Cell Metab. 2019;29:592–610.
3. Kulkarni AS, Gubbi S, et al. Benefits of metformin in attenuating the hallmarks of aging. Cell Metab. 2020;32:15–30.
4. Xia N, Daiber A, et al. Antioxidant effects of resveratrol in the cardiovascular system. Br J Pharmacol. 2017;174:1633–46.
5. Shinmura K. Cardiac aging and anti-aging medicine. Anti-Aging Medicine/Japanese Society of Anti-Aging Medicine. 2022;18:186–91.

65 Blood Component Factors and Antiaging Medicine: Blood Component Factors and Antiaging

Hiroshi Kondoh

65.1 What Is a Blood Metabolite?

Blood is a liquid factor that circulates throughout the body, reflecting genetic, epigenetic, homeostatic maintenance, disease influences, physiological responses, lifestyle, nutritional intake, and other environmental factors. Therefore, it deeply involves the aging of individuals, organs, and cells. Blood contains cellular and noncellular components. Here, we focus on blood metabolites, referring to other items for cellular components of blood (such as immune cells), hormones, and inflammatory factors. "Metabolites" are very small low-molecular metabolites/components synthesized and metabolized while cells are alive, reflecting the metabolic and vital activities of cells, tissues, and individuals. With the latest equipment (such as liquid chromatography mass spectrometers LC-MS), comprehensive analysis of many metabolites (metabolome) is possible. Metabolomics is one of the new scientific methods that detect these low-molecular metabolites and analyze them by statistical methods. Metabolomics is used for elucidating the pathophysiology of diseases, searching for diagnostic markers and biomarkers. Blood is an easily obtainable biological sample and an important analysis target for metabolomics. Many human blood metabolites (about 80%) are evolutionarily conserved in other species metabolites (such as *E. coli* and yeast) and can play an important role in understanding fundamental life phenomena, including aging. Many studies targeting blood have traditionally targeted plasma or serum, which are noncellular components. One reason for this is the difficulty in stably handling cell-derived metabolites. In recent years, metabolome analysis techniques that perform nontargeted and comprehensive analysis in whole blood (containing both cellular and noncellular components) have also been reported. About half of blood metabolites have been found to be derived from cellular components, and in whole blood metabolome, both (cellular and noncellular components) can be analyzed, and there are advantages such as maintaining the stability of metabolites [1].

H. Kondoh (✉)
Geriatric Unit, Graduate School of Medicine, Kyoto Univeristy, Kyoto, Japan
e-mail: hkondoh@kuhp.kyoto-u.ac.jp

65.2 Aging Metabolites

As an aging indicator, in addition to "calendar age," "biological age (degree of aging)" is also important. Recently, it has become possible to measure the telomeres of human blood cell components, and it is attracting attention as a new aging indicator. Alternatively, it has been reported that the decrease of nicotinamide adenine dinucleotide (NAD^+), an important metabolite for the activation of sirtuins, is reported in blood and internal organs with human aging. In

H. Yamada, Y. Naito (eds.), *Anti-Aging Medicine*, https://doi.org/10.1007/978-981-97-8514-8_65

Table 65.1 Summary of whole blood metabolome

	Decrease	Increase
Aging metabolites (14)	Antioxidant system (ophthalmic acid, carnosine, and acetylcarnosine), muscle system (leucine, isoleucine), etc.	Urea system (dimethylguanosine, *N*-acetylarginine, etc.)
Starvation metabolites (44)		Antioxidant system (uric acid, carnosine, ophthalmic acid, ergothioneine, etc.), TCA system, signal system (3-hydroxytyrosine, etc.), ketone bodies, carnitine, BCAA
Frailty metabolites (15)	Antioxidant system (acetylcarnosine, uric acid, ophthalmic acid, ergothioneine, etc.)	
Sarcopenia metabolites (22)	Uremic toxins TCA system Muscle system Methylation system, etc.	

a comprehensive whole blood metabolomics analysis (126 metabolites), 14 aging-related metabolites (11%) were found [1]. Among them, 9 metabolites, including NAD^+, were found to decrease in the elderly. In the elderly, antioxidants (ophthalmic acid, carnosine, and acetyl carnosine) and oxidation-reduction (nicotinamide adeninedinucleotide phosphates; $NADP^+$ or NAD^+), metabolites related to muscles (leucine, isoleucine) decrease, while metabolites involved in the urea cycle (dimethylguanosine, *N*-acetylarginine, etc.) increase. We observed various and complex life phenomena in human aging (Table 65.1). The blood metabolome may reflect the aspect or both aspects of "unwanted retention or deficiency of metabolites in aging."

65.3 Starvation Metabolites

The "calorie restriction hypothesis" is one of the aging hypotheses focusing on metabolism. Calorie restriction activates sirtuins, AMP-activated protein kinase (AMPK), FOXO transcription factors, etc., leading to lifespan extension. One of the important targets of the FOXO transcription factor is the radical scavenger gene group, and indeed, a reduction in oxidative stress is observed with calorie restriction. However, the healthy verification in humans is very difficult. On the other hand, humans are resistant to starvation environments, while mice are very vulnerable. Traditionally, during human starvation, it was thought that the increase in metabolites such as ketone bodies, carnitine, and branched-chain amino acids was important for energy replenishment. Comprehensive whole blood metabolomics analysis in 58-h fasting in humans revealed that as many as 44 metabolites (about 30%) increased, indicating that metabolic activation occurs much more than previously expected [2]. This included 14 well-known starvation markers such as ketone bodies, carnitine, and branched-chain amino acids. In addition, (1) TCA cycle system, (2) antioxidant system, and (3) purine-pyrimidine system metabolites also increased (Table 65.1). Antioxidant metabolites include uric acid, carnosine, ophthalmic acid, ergothioneine, pentose phosphate pathway (PPP) (a major production pathway for NADPH essential for redox control), metabolites, etc. which are consistent with the finding that antioxidant power increases with calorie restriction. Also, 3-hydroxybutyric acid has been reported to have a histone deacetylase (HDAC) inhibitory effect. Metabolites that increase in response to starvation may have antiaging effects due to increased antioxidant power and signal response modification.

65.4 Frailty and Sarcopenia-Related Metabolites

Frailty is a state in which physiological reserve capacity decreases in old age, leading to increased vulnerability to stress, resulting in functional impairments, need for care, and death. It is a precursor to bedridden states. Frailty is "reversible", and there is room for improvement through exercise and nutrition intake. In frailty, physiological functions such as walking and cognitive functions age and decline in a complex manner, while sarcopenia is an indicator of muscle aging. A comprehensive whole blood metabolome analysis has recently reported differences between frailty and sarcopenia. Of the 15 frailty markers, 7 were reductions in antioxidant-related metabolites (acetylcarnosine, uric acid, ophthalmic acid, ergothioneine, etc.) [3]. This suggests that antioxidants may be reduced in a wide range of systems in frailty. On the other hand, of the 22 sarcopenia metabolites found in the same patient group, many (20) were metabolites related to renal failure, but they did not overlap at all with the above-mentioned frailty metabolites [4]. Therefore, it was inferred that the involvement of metabolite markers in frailty and sarcopenia is significantly different (Table 65.1). From the above, it is believed that the identification of aging-related metabolites will continue to be useful in antiaging medicine [5].

References

1. Chaleckis R, Murakami I, et al. Individual variability in human blood metabolites identifies age-related differences. Proc Natl Acad Sci USA. 2016;113:4252–9.
2. Teruya T, Chaleckis R, et al. Diverse metabolic reactions activated during 58-hr fasting are revealed by non-targeted metabolomic analysis of human blood. Sci Rep. 2019;9:854.
3. Kameda M, Teruya T, et al. Frailty markers comprise blood metabolites involved in antioxidation, cognition, and mobility. Proc Natl Acad Sci USA. 2020;117:9483–9.
4. Kameda M, Teruya T, et al. Reduced uremic metabolites are prominent feature of sarcopenia, distinct from antioxidative markers for frailty. Aging (Albany NY). 2021;13:20915–34.
5. Kondoh H, Kameda M, et al. Whole blood metabolomics in aging research. Int J Mol Sci. 2020;22:175.

Blood Component Factors and Antiaging Medicine: Parabiosis and Antiaging

66

Ichiro Manabe

66.1 What Is Parabiosis?

Parabiosis is derived from the Greek words para (next, close) and bios (life) and refers to a state where two living animals are surgically connected and share a single circulatory system. It is often likened to conjoined twins that naturally occur as parabiosis. The experimental technique of parabiosis was established in rats by French physiologist Paul Bert in the 1860s. It was centrally used for the functional analysis of liquid factors exchanged between connected animals, contributing to the development of endocrine physiology research [1]. For example, it was widely used for hormone function analysis. However, the number of papers rapidly decreased with the peak in the 1960s and 80s, and it was temporarily forgotten as a research method. However, since the beginning of this century, the number of applications in fields such as immunology, stem cells, organ interrelationships, aging, and regeneration has rapidly increased. It is widely known not only for distinguishing the contribution of the circulatory and nervous systems by comparing cells and molecules transmitted through the bloodstream with local cells and molecules in tissues, but also as an experimental method that brings innovative hypotheses. Currently, mice are mainly used for parabiosis experiments. An incision is made in the skin of the flank, and the skin flaps of the two animals are sutured together [2]. Nowadays, it is common to make a long incision on the entire side of the body and connect it, and sometimes an incision is made in the peritoneum to connect the abdominal cavity. With the angiogenesis associated with wound healing at the connection site, the blood vessels of the two animals connect and they start to share a circulatory system. When inbred animals are used, rejection does not occur, and they can survive for a long time as a pair. An example of the presence of physiologically active substances in the blood suggested by parabiosis is the identification of leptin. Coleman found that by connecting a db/db mouse, a model of diabetes and obesity, with a wild-type mouse, the wild-type mouse stopped eating and starved to death [3]. From this result, Coleman thought that there might be a factor in the blood of the db/db mouse that conveys a sense of fullness and strongly suppresses the appetite of the wild-type mouse. Later, Friedman and others identified leptin.

I. Manabe (✉)
Department of Systems Medicine, Graduate School of Medicine, Chiba University, Chiba, Japan
e-mail: imanabe@chiba-u.jp

66.2 Application of Parabiosis to Aging Research

In parabiosis, the circulatory system is shared, so if two individuals with different conditions are connected, it becomes possible to examine the

H. Yamada, Y. Naito (eds.), *Anti-Aging Medicine*, https://doi.org/10.1007/978-981-97-8514-8_66

effects of the differences in their overall internal environment on each other. Specifically, for example, it enables the examination of differences in molecules and cells flowing in the blood, using physiological readouts as indicators without identifying these molecules. If parabiosis (heterochronic parabiosis) is performed between aged and young animals, it is possible to examine the overall factors and organ function contributions in various functional changes (such as tissue function decline) and pathological conditions (such as age-related diseases) due to aging. Specifically, it has been hypothesized that the decline in function with aging can be improved by performing parabiosis with a healthy young individual, and this hypothesis has been examined. As early as 1972, it was revealed that heterochronic parabiosis between aged and young rats has a life-extending effect [2]. Conboy et al. used heterochronic parabiosis to show that a young systemic environment improves the proliferation and regenerative ability of skeletal muscle stem cells (satellite cells) and liver precursor cells in aged individuals [2]. The results of this paper suggested that environmental factors where stem cells are located greatly affect the decline in regenerative ability with aging [2]. In response to these results of parabiosis, the identification of blood mediators that stimulate or inhibit stem cell function is being conducted. For example, it was reported that growth differentiation factor 11 (GDF11) promotes skeletal muscle cell regeneration and nerve regeneration [4]. However, further verification is required for the model that GDF11 is an antiaging factor. Thus, due to the recovery of the aging phenotype by parabiosis with young mice, attention has been focused on the identification of blood mediators. However, it should be noted that the results of parabiosis do not only indicate the action of blood mediators. In parabiosis, substances and cells flowing in the blood are shared. The shared substances include various essential substances for survival, such as glucose, lipids, and oxygen, in addition to mediators such as hormones (Fig. 66.1). Indeed, what is shared is substances flowing in the blood, but this includes

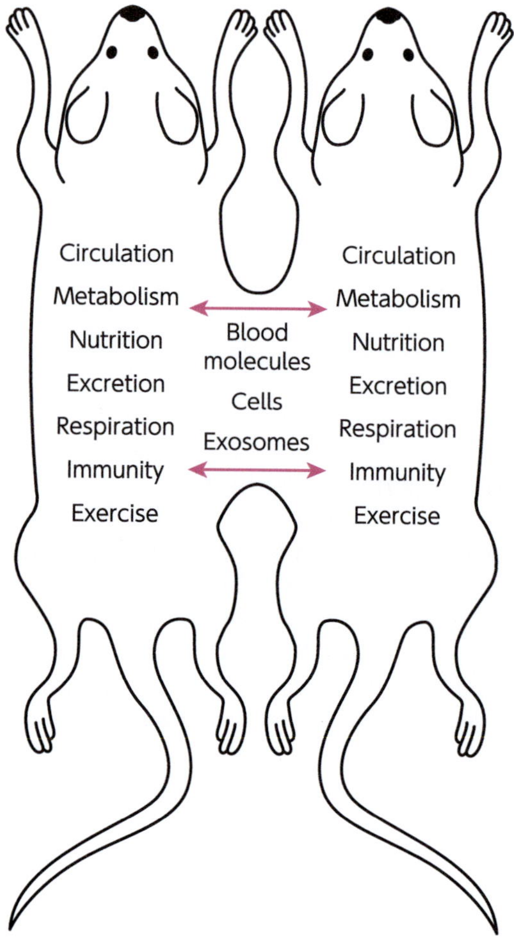

Fig. 66.1 Parabiosis shares organ function through blood flow

waste and harmful substances, and the metabolic function of healthy young mouse organs contributes to their excretion and elimination. In other words, metabolism and elimination also contribute to the effects of parabiosis, and by sharing the circulatory system, the internal environment or organs of the whole body are shared. Also, the contribution of cells flowing in the blood is important. Changes in immune function due to aging (immunosenescence) are likely to contribute to organ dysfunction with aging. For example, a decrease in phagocytic function due to aging is suggested to cause the accumulation of damage-associated molecular patterns in tissues and cause chronic inflammation (inflammaging). Chronic inflammation is a common underlying pathology in various age-related diseases and can also induce a decrease in stem cell function. Therefore, parabiosis may act on the aging phenotype by improving immune cell function. Furthermore, aged mice paired with young mice increase their amount of exercise to match the movements of the young mice. Thus, factors other than blood mediators must be considered in interpreting the results of parabiosis.

References

1. Finerty JC. Parabiosis in physiological studies. Physiol Rev. 1952;32:277–302.
2. Conboy MJ, Conboy IM, et al. Heterochronic parabiosis: historical perspective and methodological considerations for studies of aging and longevity. Aging Cell. 2013;12:525–30.
3. Coleman DL. A historical perspective on leptin. Nat Med. 2010;16:1097–9.
4. Sinha M, Jang YC, et al. Restoring systemic GDF11 levels reverses age-related dysfunction in mouse skeletal muscle. Science. 2014;344:649–52.

Part III

Clinical Anti-aging Medicine

Clinical Antiaging Medicine: Antiaging Medicine from the Perspective of Centenarian Research

67

Yasumichi Arai

As global aging progresses, achieving healthy longevity is becoming increasingly important for both individuals and society. Many centenarians live independently until their 90s and are considered models of healthy longevity. Our research group began a medical survey of centenarians in the Tokyo area in 1992, and in 2002, we started a nationwide survey of those aged 105 and over (supercentenarians). Here, we will explain the population dynamics and medical biological characteristics of centenarians (for genetic characteristics, please refer Chap. 19).

67.1 Purpose of Centenarian Research

The purpose of centenarian research is, first and foremost, to elucidate the mechanism of human healthy longevity. Centenarians, especially supercentenarians over 110 years old, not only have long lifespans but also long healthy lifespans. It is most important to clarify the factors that determine the healthy longevity of centenarians from comprehensive perspectives such as medical biology, genetics, and psychosocial studies and to apply them to extend the healthy lifespan of the general elderly population. The second purpose is to bridge basic aging science and human longevity research. Aging is a universal phenomenon observed in all organisms. Recent advances in basic aging research are beginning to elucidate the molecular mechanisms essentially involved in aging. It is important to clarify whether the slow progression of such aging processes in centenarians leads to healthy longevity, as this could provide important implications for considering the application of therapeutic interventions targeting aging delay in humans. Third, while centenarians have long healthy lifespans, they are also aging. Observing the ultimate aging phenomenon in humans and clarifying how centenarians adapt to age-related functional decline and maintain their quality of life comprise an important objective that cannot be substituted by research using model animals.

67.2 Population Statistics of Centenarians

Figure 67.1 shows the transition in the number of centenarians in Japan from 1963 to 2021. The number of centenarians based on the 2021 resident register is reported to be 86,510, which is about 20 times the number when we started our survey in 1992. It is a worldwide phenomenon that women live longer, and in Japan, women make up about 88% of centenarians. While the United States boasts the largest population of

Y. Arai (✉)
Center for Supercentenarian Medical Research, Keio University School of Medicine, Tokyo, Japan
e-mail: yasumich@keio.jp

H. Yamada, Y. Naito (eds.), *Anti-Aging Medicine*, https://doi.org/10.1007/978-981-97-8514-8_67

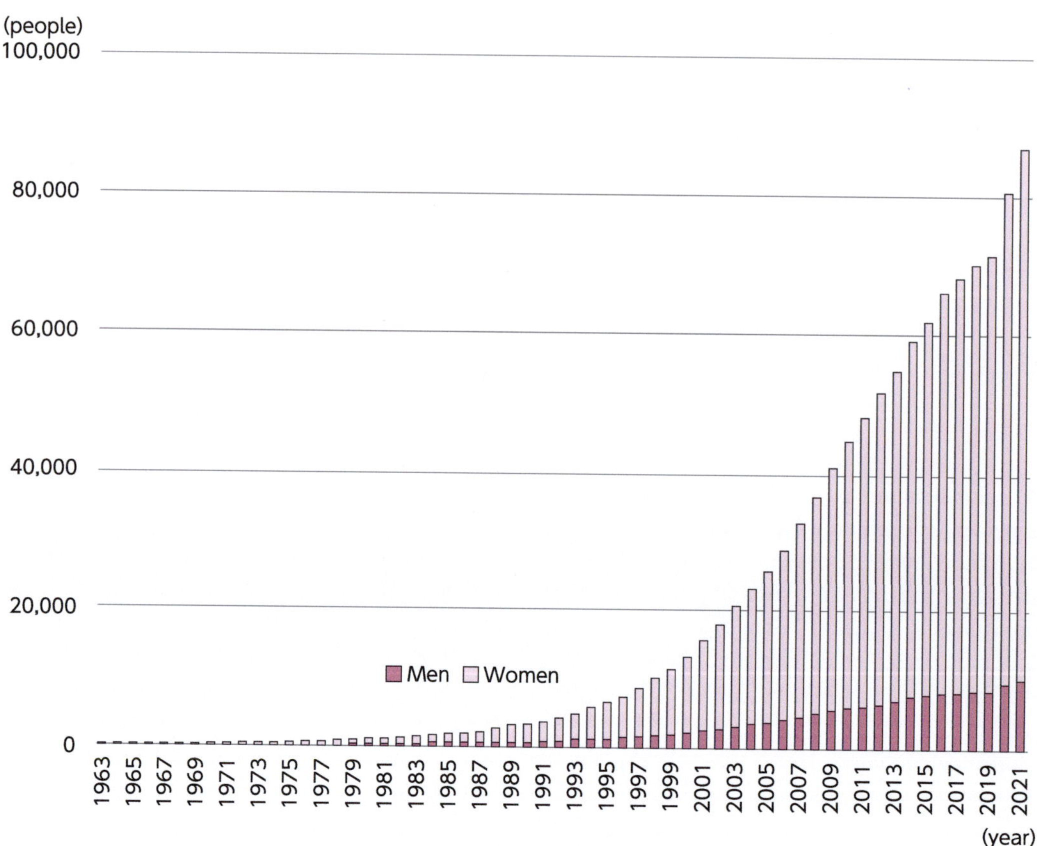

Fig. 67.1 Transition of the centenarian population in Japan (1963–2021)

centenarians in the world, when comparing the number of centenarians per 100,000 population, Japan is at the top, followed by Spain, Italy, and other European countries.

67.3 Medical Characteristics of Centenarians

From the results of centenarian research so far, it has been found that slow cognitive decline with aging, low risk of cardiovascular disease, and prevention of frailty are important for achieving healthy longevity.

67.3.1 Cognitive Function

Among 642 elderly people aged 100 and over, our study was classified individuals into three groups: those aged 104 (narrowly defined as "centenarians"), those aged 105–109 ("semi-supercentenarians"), and those who reached 110 years old ("supercentenarians"), and compared their cognitive function (Mini Mental State Examination; MMSE) and daily living activities (activities of daily living; ADL) [1]. The results showed that cognitive function and ADL were higher in supercentenarians, supercentenarians, and centenarians in that order at the age of 100, indicating that maintaining cognitive function is important for survival and maintaining independent living after the age of 100. It is important to understand the molecular mechanisms by which centenarians delay cognitive decline with aging and prevent the onset of dementia, and neuropathological research is important for this. Takao and others analyzed the autopsy of brains of four supercentenarians and showed that overall atrophy was mild, Alzheimer's neuropathology and Lewy body pathology were mild, and arteriosclerosis was mild, indicating that age-related neuropathology was relatively mild [2]. In the future, it is expected that the elucidation of defensive factors against dementia will progress through molecular and cellular level analysis.

67.3.2 Cardiovascular Disease Risk

As people age, the mortality rate from heart disease increases, and in people over 90, it is the leading cause of death. Under the hypothesis that centenarians, especially supercentenarians, can prolong their healthy lifespan by preventing or overcoming cardiovascular diseases, a prospective study was conducted to verify the relationship between cardiovascular disease biomarkers and survival time [3]. First, from the medical history, both centenarians and supercentenarians had significantly lower prevalence of diabetes, hypertension, and lipid metabolism disorders than the very elderly (85–99 years old), confirming that low cardiovascular disease risk is a characteristic of healthy longevity. Next, the relationship between blood biomarkers related to cardiovascular disease and total mortality in each age group was examined. High levels of N-terminal pro-brain natriuretic peptide (NT-proBNP), interleukin 6, cystatin C, and low levels of albumin were associated with increased all-cause mortality risk in the elderly. In particular, NT-proBNP was strongly associated with survival rate over 105 years old. Furthermore, when the group over 100 years old was divided into three groups by age at death and the relationship between biomarkers and age at measurement was examined, supercentenarians had significantly lower NT-proBNP concentrations than centenarians who died at a younger age when compared at the same measurement age. NT-proBNP is clinically applied as a biomarker indicating the severity of heart failure, and it also shows a high correlation with kidney function. In our analysis, NT-proBNP and cystatin C were correlated, and we proposed a working hypothesis that the breakdown of circulatory homeostasis due to age-related decline in heart and kidney function may be behind the high NT-proBNP blood levels in supercentenarians [4]. Future verification by echocardiography or pathological tissue analysis is desired.

67.3.3 Frailty

Frailty, along with dementia, is a significant factor in the loss of independence and need for care in the elderly. The J-CHS criteria and the deficit index proposed by Rockwood et al. are frequently used as diagnostic criteria for frailty. In centenarian surveys, it is often difficult to evaluate physical functions such as walking speed, and Rockwood's deficit index is often used. The Chinese Longitudinal Healthy Longevity Survey (CLHLS) evaluated frailty using the deficit index in 4434 centenarians and 14,051 elderly people in the age groups of 65–79, 80–89, and 90–99 [4]. As a result, the average value of the deficit index increased with age, and the frailty index was highest in centenarians. The frailty index was significantly associated with total mortality, ADL disability, and cognitive impairment in all age groups, including centenarians, indicating that it is a hindrance to healthy longevity. We also compared the deficit index at the ages of 100–104 in three centenarian groups: (narrowly defined) centenarians, supercentenarians, and supercentenarians. The results suggested that delaying the onset of frailty could lead to an extension of healthy life expectancy [5]. These results indicate that, like the elderly, preventing frailty is important for achieving healthy longevity in centenarians.

67.4 Conclusion

The medical and biological characteristics of centenarians and supercentenarians have been revealed through research. To further apply these findings to interventions for age-related diseases, it is necessary to understand how the molecular-level characteristics of aging, or Hallmarks of aging, identified from research on aging model organisms, are controlled in supercentenarians, how delayed aging leads to the avoidance or delay of age-related diseases, and to elucidate the key pathways. The development of translational research, where basic research and clinical research are integrated, is important in the future.

References

1. Arai Y, Inagaki H, et al. Physical independence and mortality at the extreme limit of life span: supercentenarians study in Japan. J Gerontol A Biol Sci Med Sci. 2014;69:486–94.
2. Takao M, Hirose N, et al. Neuropathology of supercentenarians—four autopsy case studies. Acta Neuropathol Commun. 2016;4:97.
3. Hirata T, Arai Y, et al. Associations of cardiovascular biomarkers and plasma albumin with exceptional survival to the highest ages. Nat Commun. 2020;11:3820.
4. Gu D, Feng Q, et al. Frailty still matters to health and survival in centenarians: the case of China. BMC Geriatr. 2015;15:159.
5. Arai Y, Hirose N. Frailty/Sarcopenia and longevity. J Jpn Soci Intern Med. 2020;109:427–33.

68 Clinical Antiaging Medicine: Obesity, Metabolic Syndrome, and Antiaging Medicine

Hiroaki Masuzaki

68.1 The Concept of Metabolic Syndrome and Antiaging

Metabolic syndrome is a preventive medical disease concept that positions the excessive accumulation of "visceral fat tissue" located in an anatomical position that can directly affect the nutrition and energy metabolism of the liver, such as mesenteric fat tissue and omental fat tissue, whose venous return merges into the portal vein, as the most upstream pathological condition, aiming to prevent the onset and progression of atherosclerotic diseases, including coronary artery disease. The formation of metabolic syndrome involves a variety of combinations for each individual, such as decreased physical activity, overeating or unbalanced diet, broad stress, biological rhythm disorders, aging, racial/ethnic differences, genetic background, and endocrine system dysfunction, and the decline in organ and vascular function with aging has a significant impact on the progression and worsening of metabolic syndrome [1]. In metabolic syndrome, insulin resistance, hyperglycemia, lipid metabolism abnormalities (especially hypertriglyceridemia, low HDL-C), hypertension, etc., a group of risk factors for atherosclerotic diseases and type 2 diabetes accumulate in various combinations in the same individual, and it is also important as an intervention target for the prevention of the onset of type 2 diabetes and hypertension, whose risk of onset increases with age. High LDL-C is an established risk factor for atherosclerotic diseases, and there is no direct causal relationship with excessive accumulation of visceral fat, so it is not included in the diagnostic criteria for metabolic syndrome, but of course, in metabolic syndrome with high LDL-C, it is desirable to further strengthen the guidance and management to improve lifestyle habits and reduce visceral fat [2]. The risk of developing type 2 diabetes in metabolic syndrome is about 3–6 times higher than in non-metabolic syndrome. Also, the risk of cardiovascular disease onset and cardiovascular disease death in metabolic syndrome is about 1.5–2 times higher than that in non-metabolic syndrome. The purpose of guidance and management for those with metabolic syndrome is particularly to prevent the onset and progression of type 2 diabetes and cardiovascular disease, and the key is to improve lifestyle habits, including diet therapy, exercise therapy, smoking cessation, and reduce visceral fat [2].

H. Masuzaki (✉)
Division of Endocrinology, Diabetes and Metabolism, Hematology, Rheumatology, (Second Department of Medicine), Graduate School of Medicine, University of the Ryukyus, Okinawa, Japan
e-mail: hiroaki@med.u-ryukyu.ac.jp

H. Yamada, Y. Naito (eds.), *Anti-Aging Medicine*, https://doi.org/10.1007/978-981-97-8514-8_68

68.2 Visceral Fat Increases with Age

In 2005 (Heisei 17), eight associations, including the Japan Obesity Association, jointly established the diagnostic criteria for metabolic syndrome in our country [2]. If the waist circumference exceeds 85 cm in men and 90 cm in women, and if they have two or more of hypertension, hyperglycemia, and dyslipidemia, they are considered to have metabolic syndrome. The cutoff value for waist circumference corresponds to a visceral fat area of 100 cm^2 evaluated at the umbilical level by CT. The levels of fatty acids, adiponectin, leptin, inflammatory cytokines, chemokines, and other liquid physiologically active substances collectively referred to as "adipokines" secreted from visceral fat tissue are involved in the formation of pathological conditions. International data comparisons have revealed that East Asians, including Japanese, are prone to developing lifestyle-related diseases such as type 2 diabetes, even with mild obesity. From an epidemiological survey of about 150,000 Japanese people, the frequency of people with glucose tolerance abnormalities, hypertension, and dyslipidemia was compared with those with a BMI of 22, which is considered to have the lowest morbidity and mortality rate. The prevalence of hypertension, hypertriglyceridemia, and low HDL-C was approximately doubled at a BMI of 25 for hypertension, hypertriglyceridemia, and low HDL-C, and at a BMI of 27 for glucose tolerance abnormalities, providing academic evidence for defining obesity as a BMI of 25 or more in our country [3]. If the BMI is 25 or more but there are no health impairments and no excessive accumulation of visceral fat tissue, it is considered "healthy obesity" and is not a target for preventive medicine. Also, since the definition of obesity is an excessive increase in fat tissue weight, athletes with little fat tissue accumulation but a high skeletal muscle weight that increases the BMI value, pregnant women, and those with an increase in BMI due to edema from renal dysfunction are not considered obese. Men, despite having a BMI value of less than 25 and no health impairments such as glucose tolerance abnormalities, hypertension, and dyslipidemia, have a high proportion of nonobese metabolic syndrome with a visceral fat area exceeding 100 cm^2, raising an alarm for overlooking "hidden metabolic syndrome". Not only in men, but also in postmenopausal women, visceral fat tends to increase with age compared to subcutaneous fat, and it is significant to evaluate metabolic syndrome from an antiaging perspective.

68.3 Challenges of Metabolic Syndrome and Obesity in the Super-Aged Society

In our country, the aging population and declining birth rate began to progress around 1985, and it is estimated that by 2025, social security benefits will reach 150 trillion yen. By 2060, one in every 2.5 people will be over the age of 65, marking the arrival of an unprecedentedly aged society in human history. Currently, about 30% of national medical expenses are related to lifestyle diseases, and about 60% of the mortality rate by cause of death is occupied by cancer, cardiovascular disease, diabetes, etc. As an initiative to protect the people from lifestyle diseases and aim for a true healthy longevity society, specific health checkups (hereinafter, specific checkups) and specific health guidance targeting those from 40 to 74 years old started in 2008 (Heisei 20), and it is popular as "Metabo checkup", but the low rate of medical examination and regional differences remain a major problem, and it is noteworthy that the number of patients with metabolic syndrome and obesity is increasing again under Corona. Currently, about one in four examinees falls under metabolic syndrome, but especially in the elderly, it is not uncommon to develop "sarcopenic obesity" that combines an increase in visceral fat tissue and a decrease in skeletal muscle mass (sarcopenia) due to a decrease in physical activity and changes in dietary habits and hormone balance. It has been clarified that the risk of cardiovascular events and fractures increases further in sarcopenic obesity [4]. Recent large-scale epidemiological studies have reported that if there is no sarcopenia and skeletal muscle mass is relatively maintained, it works favorably for the life prognosis of mildly obese and type 2 diabetes patients (obesity paradox in type 2 diabetes) [5]. The group with the best life prognosis for type 2 diabetes patients is BMI 25.0–29.9, and even overweight type 2 diabetes patients who have exercise habits, have less visceral fat, and have excellent exercise tolerance tend to have fewer cardiovascular accidents [5]. Especially in the elderly, mild overweight can function as a reserve (metabolic reserve) against sarcopenia, frailty, osteoporosis, malnutrition, and immune deficiency associated with aging, and it is suggested that it may contribute to the improvement of life prognosis. From the perspective of antiaging, it is required to guide and treat not only evaluating metabolic syndrome and obesity based on BMI, but also paying attention to the "quality of obesity" at the whole body level, including body fat distribution, skeletal muscle mass, bone density, cardiopulmonary function, exercise tolerance, and brain function.

References

1. Masaki Y, Shimabukuro M. Obesity and metabolic syndrome: recent findings and prospects. J Jpn Soc Intern Med. 2017;106:477–83.
2. Metabolic Syndrome Diagnostic Criteria Study Committee. Definition and diagnostic criteria of metabolic syndrome. J Jpn Soc Intern Med. 2005;94:794–809.
3. Yoshiike N, Nishi N, et al. Relationship between the degree of obesity based on body mass index and risk factors for diabetes, hypertension, and hyperlipidemia: an epidemiological study based on multi-institutional collaborative research. Obes Res. 2000;6:4–17.
4. Batsis JA, Villareal DT. Sarcopenic obesity in older adults: aetiology, epidemiology and treatment strategies. Nat Rev Endocrinol. 2018;14:513–37.
5. Costanzo O, Cleland JG, et al. The obesity paradox in type 2 diabetes mellitus: relationship of body mass index to prognosis: a cohort study. Ann Intern Med. 2015;162:610–8.

69 Clinical Antiaging Medicine: Antiaging Medicine from the Perspective of the Heart, Blood Vessels, and Circulatory System

Hiroshi Akazawa

69.1 Antiaging Medicine for Blood Vessels

The words of William Osler, "Man ages with his blood vessels," are all too famous. Blood vessels are constantly exposed to metabolic substances in the blood and hemodynamic loads. Arteriosclerosis progresses with age and is further accelerated and worsened by lifestyle-related diseases and risk factors, causing functional disorders and diseases in important organs such as the brain, heart, blood vessels, and kidneys. It is no exaggeration to say that blood vessels are the most important target for antiaging.

69.2 Mechanism of Vascular Aging

The structure and function of blood vessels change with age. The vascular endothelium plays an important role in various vascular functions, such as regulating vascular tone, proliferation of vascular smooth muscle, coagulation fibrinolysis, inflammation, and control of angiogenesis. The vascular endothelium produces nitric oxide (NO), which relaxes blood vessels and inhibits atheroma formation, but vascular endothelial function, including NO production, decreases with age. In addition, lifestyle-related diseases such as hypertension, diabetes, and dyslipidemia, as well as lifestyle factors such as obesity, overeating, excessive salt intake, smoking, lack of exercise, and menopause, lead to increased cellular aging and oxidative stress and progression of vascular endothelial dysfunction [1].

69.3 Clinical Indicators for Vascular Antiaging

The ankle brachial index (ABI), brachial-ankle pulse wave velocity (baPWV), and cardio-ankle vascular index (CAVI) are widely used as indicators of arteriosclerosis. The vascular age is calculated by comparing the CAVI with the average value of the same sex and age. In addition, flow-mediated dilatation (FMD) is the most commonly performed test for evaluating vascular endothelial function in clinical practice [2]. This is a noninvasive and simple test that evaluates how much the arterial diameter expands due to NO derived from vascular endothelial cells by releasing the arm after compression using an ultrasound diagnostic device.

H. Akazawa (✉)
Department of Cardiovascular Medicine, Graduate School of Medicine, The University of Tokyo, Tokyo, Japan
e-mail: akzw@g.ecc.u-tokyo.ac.jp; akazawah-tky@umin.ac.jp

H. Yamada, Y. Naito (eds.), *Anti-Aging Medicine*, https://doi.org/10.1007/978-981-97-8514-8_69

69.4 Antiaging Medicine for the Heart

The heart also undergoes irreversible degenerative changes with age. The risk of developing heart disease increases due to the accumulation of hypertension, diabetes, and arteriosclerosis, but even in the absence of these risk factors, the incidence of heart hypertrophy increases with age, and while contractile ability is maintained, diastolic ability decreases, as reported in epidemiological studies [3]. Heart failure, especially heart failure with preserved Heart diseases such as heart failure with preserved ejection fraction (HFpEF), valvular disease, and atrial fibrillation often occur in the elderly, and overcoming these diseases is an important issue for extending healthy life expectancy in our aging society. There is a demand to keep the heart young through antiaging (antiaging) medical approaches and to prevent the onset of heart disease associated with aging.

69.5 Mechanism of Cardiac Aging

It is not yet clear how cardiac aging occurs, how important it is as a basis for the onset of heart disease, and what its mechanisms and roles are. It is well known that the activation of the renin-angiotensin system and the sympathetic nervous system promotes cardiac remodeling, such as myocardial cell hypertrophy, interstitial fibrosis, and enlargement of the ventricular cavity, which are major factors in worsening myocardial dysfunction and heart failure. Recently, our understanding of the molecular mechanisms of individual aging has been deepening. It is suggested that oxidative stress, mitochondrial dysfunction, cellular aging, and inflammation are involved in cardiac aging, and the roles of newly identified aging-promoting or aging-inhibiting factors are attracting attention. It is also suggested that antiaging genes such as sirtuins may be target molecules for cardiac antiaging.

69.6 Clinical Indicators for Cardiac Antiaging

Various imaging diagnoses and biomarker methods can be used to evaluate the morphology and function of the heart. In echocardiography, in addition to cardiac hypertrophy, aortic valve and mitral valve sclerosis and calcification are observed as morphological changes with aging. It has also been shown by speckle tracking that age-related functional changes occur earlier in the left atrium than in the left ventricle [4]. Blood B-type natriuretic peptide (BNP) or N-terminal pro-brain natriuretic peptide (NT-proBNP), which are commonly used as biomarkers of cardiac load, are known to significantly increase from the 60s even in the general population without cardiovascular disease [4]. It is also possible to estimate age using artificial intelligence (AI) from electrocardiograms and chest X-rays. Interestingly, patients whose estimated age from chest X-rays is older than their actual age have a higher risk of cardiovascular events, suggesting that chest X-ray age may be a new indicator related to the onset and prognosis of heart disease [5]. In addition to conventional clinical indicators, the cardiac age estimated from diagnostic modalities such as electrocardiograms, chest X-rays, echocardiograms, and MRIs using AI is expected to be a useful indicator for antiaging as a marker of cardiac aging (Fig. 69.1).

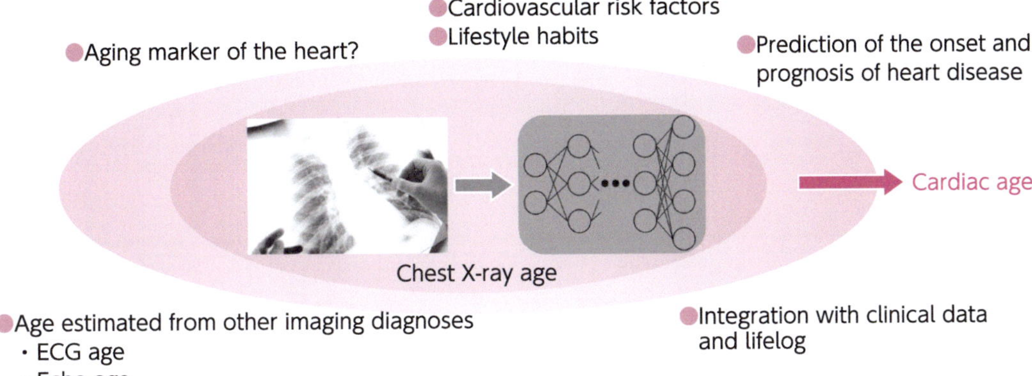

Fig. 69.1 Estimation of cardiac age using AI

69.7 Antiaging of the Heart, Blood Vessels, and Circulatory System

Recent research has deepened our understanding of the molecular mechanisms of cardiovascular aging, but we have only caught a glimpse of a small part of it. Drugs such as renin-angiotensin-aldosterone system inhibitors, statins, and incretin-related drugs have been reported to have vascular protective effects. While we cannot directly intervene in the aging process at present, the prevention of lifestyle diseases such as hypertension, diabetes, and dyslipidemia, as well as improvements in lifestyle such as diet, salt reduction, exercise, sleep, and smoking cessation, and the practice of a stress-free living environment, will likely reduce the risk of cardiovascular aging, delay the onset of cardiovascular diseases, and contribute to the extension of healthy life expectancy.

References

1. Minamino T, Komuro I. Vascular aging: insights from studies on cellular senescence, stem cell aging, and progeroid syndromes. Nat Clin Pract Cardiovasc Med. 2008;5:637–48.
2. Alexander Y, Osto E, et al. Endothelial function in cardiovascular medicine: a consensus paper of the European Society of Cardiology Working Groups on Atherosclerosis and Vascular Biology, Aorta and Peripheral Vascular Diseases, Coronary Pathophysiology and Microcirculation, and Thrombosis. Cardiovasc Res. 2021;117:29–42.
3. Lakatta EG, Levy D. Arterial and cardiac aging: major shareholders in cardiovascular disease enterprises: part I: aging arteries: a "set up" for vascular disease. Circulation. 2003;107:139–46.
4. Yoshida Y, Nakanishi K, et al. Alteration of cardiac performance and serum B-type natriuretic peptide level in healthy aging. J Am Coll Cardiol. 2019;74:1789–800.
5. Ieki H, Ito K, et al. Deep learning-based age estimation from chest X-rays indicates cardiovascular prognosis. Commun Med (Lond). 2022;2:159.

70 Clinical Antiaging Medicine: Antiaging Medicine from the Perspective of the Brain and Nervous System

Toru Yamashita

70.1 Outline

To practice antiaging for the brain, it is necessary to deliver oxygen and nutrients to brain cells and maintain the brain blood vessels that remove waste in a youthful and healthy state.

Lifestyle diseases (hypertension, diabetes, dyslipidemia) control and smoking cessation are important for preventing the onset of cerebrovascular disorders.

Depending on the stage of lifestyle diseases, it is recommended to intervene with exercise therapy, diet therapy, antihypertensive drugs, and statin drugs.

70.2 The Relationship Between Cerebrovascular Disorders and Dementia

Recent research results have revealed that lifestyle diseases such as hypertension, diabetes, dyslipidemia, obesity, and the subsequent cerebrovascular disorders are significantly involved in the onset and worsening of mild cognitive impairment and dementia such as Alzheimer's disease. For example, in a study of 1110 consecutive autopsy cases of dementia, it has been reported that 25.7% had a combination of vascular damage and Alzheimer's pathology [1]. The mechanism by which vascular damage causes Alzheimer's type dementia is thought to be due to the dysfunction of the perivascular drainage pathway that pumps out amyloid beta protein produced in the brain to the cerebrospinal fluid side due to vascular damage. Since oxidative stress around the blood vessels and the chronic inflammation caused downstream are largely involved in the pathogenesis of vascular damage, their control is important for preventing the progression of dementia and antiaging of the brain. In a 5-year observational study of 837 patients with mild cognitive impairment (MCI), 42.1% transitioned to Alzheimer's disease, and those who transitioned had significantly more lifestyle diseases (diabetes, hypertension, dyslipidemia) than those who did not transition (2.33 vs 1.15, $p < 0.001$) [2]. Also, in about 300 patients with Alzheimer's disease, among those with the above-mentioned vascular risk factors, the group that treated and managed all vascular risk factors (89 people) had a Mini Mental State Examination (MMSE) of about 21 after 30 months, which was almost flat, while the group that did not treat or manage at all (72 people) had a clear worsening of MMSE from 21 to 15, and the group that partially treated had a moderate worsening [3]. From the above, it can be said that it is important to control lifestyle diseases (hypertension, diabetes,

T. Yamashita (✉)
Department of Neurology, Okayama University Graduate School of Medicine, Dentistry and Pharmaceutical Sciences, Okayama, Japan
e-mail: toruyamashita@okayama-u.ac.jp

H. Yamada, Y. Naito (eds.), *Anti-Aging Medicine*, https://doi.org/10.1007/978-981-97-8514-8_70

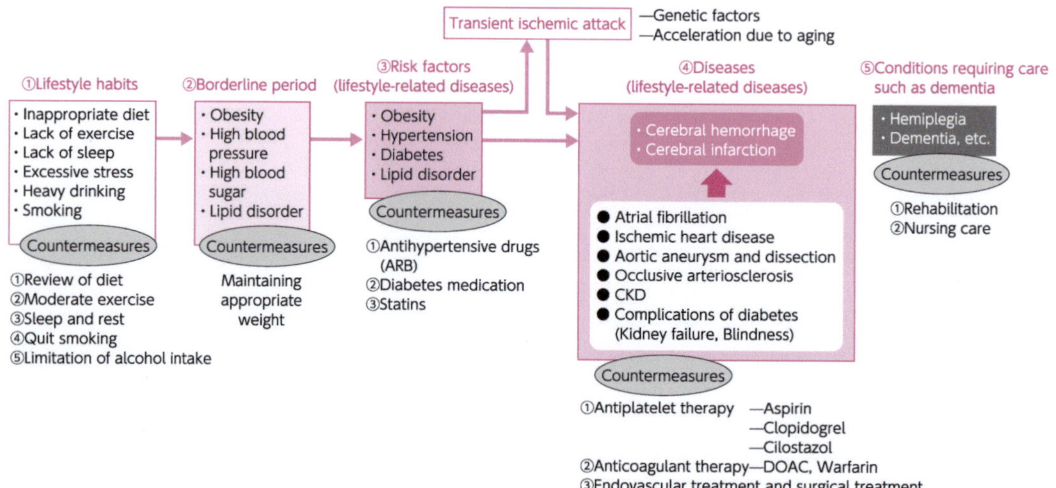

Fig. 70.1 Relationship between lifestyle-related diseases and cerebrovascular disorders. It is most important to suppress the progression of lifestyle-related diseases for the prevention of cerebrovascular disorders

dyslipidemia) that cause cerebrovascular disorders and to quit smoking in order to prevent the progression to dementia (Fig. 70.1).

70.3 Therapeutic Effects of Exercise Therapy, Diet Therapy, and Supplements

Based on previous research, it is assumed that exercise therapy has brain function improvement effects through various mechanisms. It is expected that exercise will increase cerebral blood flow and improve synaptic function due to the increase in neurotrophic factors [brain-derived neurotrophic factor (BDNF) and insulin-like growth factor 1 (IGF-1)]. From the accumulation of a vast amount of observational research, it has been shown that consuming too much high-calorie food mainly composed of carbohydrates, low-protein food, and low-fat food increases the risk of cerebrovascular disorders and dementia. A nutritional supplement containing three functional components, choline, uridine, and DHA, called Souvenaid®, was initially announced to have cognitive function improvement effects in randomized controlled trials, but a Cochrane review incorporating three trials with 1097 cases did not obtain evidence of dementia prevention effects or cognitive function improvement effects. On the other hand, Abe et al. conducted a randomized controlled trial for MCI using the antioxidant compound TwendeeX and reported a significant improvement in cognitive function. It is attracting attention for demonstrating in humans that oxidative stress is involved in the pathophysiology of cognitive dysfunction [4].

70.4 Treatment Strategies for Lifestyle-Related Diseases

Calcium channel blockers (CCB) and angiotensin II receptor blockers (ARB), which are antihypertensive drugs, have been reported to have not only antihypertensive effects but also anti-inflammatory effects. For example, in an administration experiment of telmisartan, one of the ARBs, for spontaneous hypertensive rats (SHRSR) [5], glial fibrillary acidic protein (GFAP) positive astrocytes were dissociated from NAGO positive vascular endothelial cells in the untreated group, but this dissociation was suppressed in the telmisartan administration group. From these results, it was suggested that the neurovascular unit (NVU), which is composed of vascular endothelial cells-basal membrane-astrocyte endfoot, is protected by ARB administration (Fig. 70.2). Telmisartan has not only the antihypertensive effect mediated by the angiotensin II type 1 receptor (AT1R), but also the anti-arteriosclerosis effect and the effect

Fig. 70.2 Blood-brain barrier (BBB) and neurovascular unit (NVU) in cerebrovascular disorders [7]. (**a**) Normal vascular structure. (**b**) Oxidative stress and inflammatory responses around the brain vessels cause vascular damage. (**c**) Calcium channel blockers (CCB), angiotensin II receptor blockers (ARB), statins, etc. are believed to have an effect in suppressing inflammation around the vessels

of suppressing the onset of cardiovascular diseases by activating the peroxisome proliferator-activated receptor γ (PPARγ), so a stronger vascular protective effect is expected. Also, statin drugs, which are treatments for dyslipidemia, showed memory function improvement effects in the statin administration group in administration experiments to Alzheimer's disease mouse models and significantly suppressed amyloid β positive senile plaques compared to the untreated group. In addition, the same NVU protective effect as in the telmisartan administration experiment has been confirmed (Fig. 70.2). It is suggested that the perivascular drainage pathway, which mainly passes through the basal membrane, was also protected, and the improvement of amyloid protein clearance may have been involved in the memory function improvement effect [6].

70.5 What is brain antiaging that can be done from now on?

The brain antiaging that should be practiced from now on is important to prevent the progression of lifestyle-related diseases, and depending on the stage, it is recommended to perform (1) exercise therapy (for example, aerobic exercise more than 3 times a week), (2) diet therapy and supplements, (3) interventions with ARB and statins.

References

1. Jellinger KA, Attems J. Is there pure vascular dementia in old age? J Neurol Sci. 2010;299:150–4.
2. Li J, Wang YJ, et al. Vascular risk factors promote conversion from mild cognitive impairment to Alzheimer disease. Neurology. 2011;76:1485–91.
3. Deschaintre Y, Richard F, et al. Treatment of vascular risk factors is associated with slower decline in Alzheimer disease. Neurology. 2009;73:674–80.
4. Tadokoro K, Morihara R, et al. Clinical benefits of antioxidative supplement Twendee X for mild cognitive impairment: a multicenter, randomized, double-blind, and placebo-controlled prospective interventional study. J Alzheimers Dis. 2019;71:1063–9.
5. Liu W, Yamashita T, et al. Protective effect of telmisartan on neurovascular unit and inflammasome in stroke-resistant spontaneously hypertensive rats. Neurol Res. 2015;37:491–501.
6. Kurata T, Kawai H, et al. Statins have therapeutic potential for the treatment of Alzheimer's disease, likely via protection of the neurovascular unit in the AD brain. J Neurol Sci. 2012;322:59–6.
7. Abe K. Shinkei Chiryou (in Japanese). 2018;35:380–384.

71 Clinical Antiaging Medicine: Antiaging Medicine from the Perspective of the Kidneys

Ryo Yamada and Motoko Yanagita

71.1 Aging and Kidney Function

The kidneys are the main organs that eliminate unnecessary by-products of metabolism and excess body fluids. About 20–25% of cardiac output flows into the kidneys, and 150 L of blood is filtered daily, 99% of which is reabsorbed, producing 1.5 L of urine. Therefore, under physiological conditions, the kidneys are highly metabolically active organs, prone to oxidative stress, and highly susceptible to the effects of aging. The kidney volume does not change relatively until the age of 50, but it is known to decrease with age thereafter. Such changes are mainly due to a decrease in the volume of the renal cortex, and histologically, they present various characteristic pathological findings such as arteriosclerosis, glomerulosclerosis, tubular atrophy, and interstitial fibrosis. Against the background of such pathological changes, the glomerular filtration rate (GFR), a functional indicator of the kidneys, decreases with age, leading to chronic kidney disease (CKD), which is a year-on-year decrease in GFR. Currently, Japan is one of the most aged societies in the world, with people aged 65 and over accounting for more than 25% of the total population, and CKD itself is a risk factor that determines the prognosis of life, including cardiovascular diseases, attracting social attention as an important issue.

R. Yamada (✉) · M. Yanagita
Department of Nephrology, Graduate School of Medicine, Kyoto University, Kyoto, Japan
e-mail: kidney2011@kuhp.kyoto-u.ac.jp

71.2 Chronic Kidney Disease and Cellular Aging

Various factors are involved in the pathogenesis of CKD, but in recent years, the involvement of cellular aging has been strongly suggested. First, cellular aging refers to a state of cells defined by irreversible cell cycle arrest, apoptosis resistance, and acquisition of an inflammatory promoting phenotype (senescence-associated secretory phenotype; SASP). This is a phenomenon discovered by Hayflick et al., where human fetal fibroblasts stop proliferating after a certain number of cell divisions, but do not die and maintain metabolic activity. It is believed to be caused by the shortening of telomere length due to repeated DNA replication. On the other hand, various internal and external stimuli, such as DNA damage response, oxidative stress, mitochondrial dysfunction, and overexpression of cancer genes, can lead to a state of cellular aging. Aging cells show an increase and activity change in lysosomal contents, as well as staining of senescence-associated beta-galactosidase (SA-βgal), which causes cell cycle arrest, and cyclin-dependent The presence of senescent cells is confirmed based on several findings, such as the increased expression of

H. Yamada, Y. Naito (eds.), *Anti-Aging Medicine*, https://doi.org/10.1007/978-981-97-8514-8_71

cyclin-dependent kinase (CDK) inhibitors like $p21^{Cip1}$ and $p16^{Ink4a}$ [1]. Senescent cells, identified by combining these markers, are not only confirmed in culture conditions but also in vivo, and it has been reported in various animal species that they accumulate in each organ with aging, leading to organ function decline and potentially determining life prognosis. They are one of the most focused research targets. As mentioned earlier, the kidney is considered one of the organs where senescent cells are likely to accumulate because it is easily exposed to various external factors such as oxidative stress and drug metabolism. In fact, in humans, the accumulation of senescent cells in the kidney has been shown not only with aging, but also in patients with chronic kidney disease (CKD). Similar reports exist for experimental animals, and the main site of senescent cell accumulation in the kidney is believed to be the proximal tubular epithelial cells, primarily composed of renal tubular epithelial cells. This is thought to be because proximal tubular epithelial cells in the kidney are the most metabolically active cells, including drug metabolism, and are vulnerable to hypoxic stimuli. Generally, senescent cells in renal damage appear in the acute phase and can remain and accumulate in the chronic phase. First, senescent cells that appear in the acute phase are CDK. It is believed that the expression of inhibitors not only temporarily halts the cell cycle, but also contributes to tissue regeneration by secreting tissue repair factors as part of the SASP. This includes factors involved in immune surveillance and functions to remove senescent cells after the acute phase to prevent their remnants. Furthermore, analysis using knockout mice of $p21^{Cip1}$ and $p16^{Ink4a}$ suggests that senescent cells appearing in the acute phase of kidney injury models such as ischemia-reperfusion injury and unilateral ureteral obstruction work to suppress fibrosis. These events suggest the possibility that senescent cells are beneficial in the acute phase of kidney injury. On the other hand, if the emerged senescent cells are not removed and remain until the chronic phase, they can have various effects. For example, cellular senescence is considered a cancer suppression mechanism in epithelial cells. As an example, in mice where the cancer suppressor gene Apc was specifically knocked out in renal tubular epithelial cells, the renal tubular epithelial cells undergo cellular senescence and proliferation inhibition, but it has been reported that early renal tumors are formed when $p21^{Cip1}$ and $p16^{Ink4a}$/$p19^{Arf}$ are further knocked out [2]. On the other hand, cellular senescence in interstitial fibroblasts is considered to be a hotbed for promoting cancer by presenting SASP that suppresses immune surveillance. In addition, the accumulation of senescent cells in epithelial cells causes inflammation of the microenvironment due to SASP, mimicking organ aging with age, even in the young, and plays an important role in the transition to or progression of CKD. This is important not only after acute kidney injury, but also in the process leading to CKD common to general kidney diseases such as IgA nephropathy, diabetic nephropathy, and hereditary kidney diseases (polycystic kidney disease, nephron tuberculosis, etc.). In kidney transplantation as well, it has been suggested that senescent cells cause local chronic inflammation through SASP and promote allograft rejection [3].

71.3 Removal of Senescent Cells as a Treatment for Chronic Kidney Disease and Its Prospects

From the above, recent research has been actively conducted under the idea that the removal of senescent cells (senolysis) may be able to suppress the progression of CKD. At the experimental animal level, it was suggested that senescent cells could become antiaging targets because the removal of $p16^{Ink4a}$-expressing cells using INK-ATTAC mice in 2004 resulted in the extension of individual lifespan and a decrease in the proportion of glomerulosclerosis in the kidneys [4]. Recently, it has been reported that senescent cells specifically express the glutamine metabolic enzyme Gls1, and the proportion of glomerulosclerosis in aged mice decreases with its

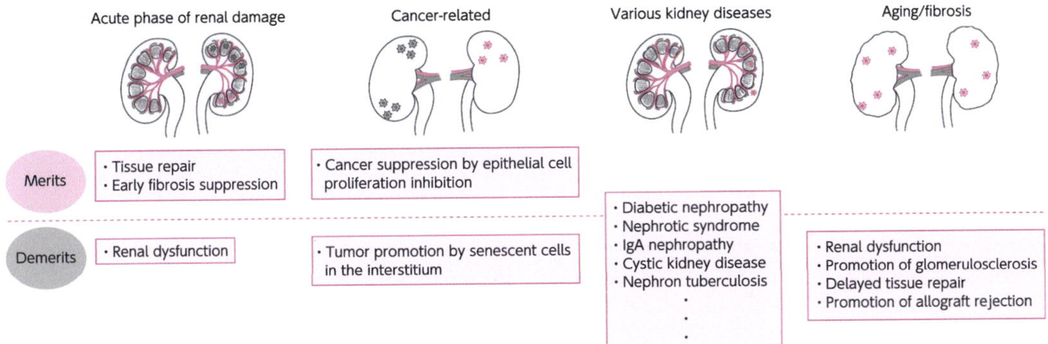

Fig. 71.1 Relationship between renal damage, individual aging, and cellular aging [4]. Senescent cells that appear in the acute phase of renal damage can have beneficial effects such as promoting tissue repair, but on the other hand, senescent cells that accumulate with chronic renal damage and aging are thought to be harmful, causing renal function decline and delayed tissue repair

inhibitor [5]. At the clinical research level, the main strategy is to remove senescent cells with drugs targeting the anti-apoptosis mechanism. Among them, treatments using a combination of dasatinib (D), a Srk kinase inhibitor, and quercetin (Q), a flavonoid, known as D + Q, or fisetin, a flavonoid with a shorter half-life, which have senescent cell removal effects in any cell type and fewer off-target effects, are currently undergoing clinical trials for various conditions. In the kidneys, trials of D + Q (NCT02848131) and fisetin (NCT03325322) targeting patients with diabetic nephropathy and CKD are underway, and the results are eagerly awaited (Fig. 71.1).

References

1. Gorgoulis V, Adams PD, et al. Cellular senescence: defining a path forward. Cell. 2019;179:813–27.
2. Cole AM, Ridgway RA, et al. p21 loss blocks senescence following Apc loss and provokes tumourigenesis in the renal but not the intestinal epithelium. EMBO Mol Med. 2010;2:472–86.
3. Sturmlechner I, Durik M, et al. Cellular senescence in renal ageing and disease. Nat Rev Nephrol. 2017;13:77–89.
4. Baker DJ, Wijshake T, et al. Clearance of $p16^{Ink4a}$-positive senescent cells delays ageing-associated disorders. Nature. 2011;479:232–6.
5. Johmura Y, Yamanaka T, et al. Senolysis by glutaminolysis inhibition ameliorates various age-associated disorders. Science. 2021;371:265–70.

72 Clinical Antiaging Medicine: Antiaging Medicine from the Perspective of the Digestive System

Tomohisa Takagi and Yuji Naito

72.1 Antiaging Medicine from the Perspective of the Digestive System

The digestive system, which includes the digestive tract from the mouth to the anus, as well as the liver, pancreas, and other solid organs connected to it, significantly affects healthy longevity as its function declines with age, a time-dependent natural physiological process, and various diseases develop and progress. Particularly in Japan, where super-aging is advancing, it has been clearly shown that the onset and progression of several digestive diseases, which are significantly increasing, can be suppressed by appropriate antiaging interventions centered on lifestyle corrections, contributing to healthy longevity. Therefore, in this section, I would like to touch on representative digestive diseases whose incidence increases with age.

72.2 Lifestyle and Digestive Tract Cancer

Digestive cancers account for about half of all cancer morbidity and mortality in Japan. Smoking and alcohol are common risk factors for all digestive cancers, and obesity is a common risk factor for colon, liver, and pancreatic cancers. Specific risk factors for digestive cancers at specific sites, such as salt (stomach), red meat/processed meat/physical inactivity (colon), and insufficient intake of vegetables and fruits (esophagus, stomach), have also been identified [1] (Table 72.1). In esophageal cancer, individuals with a deficiency of aldehyde dehydrogenase 2 (ALDH2), an alcohol metabolizing enzyme that causes facial flushing with one beer (flusher), have a significantly increased risk of esophageal cancer due to prolonged exposure to high concentrations of acetaldehyde, as alcohol metabolism is not smoothly performed [2]. The greatest risk factor for stomach cancer is persistent infection with *Helicobacter pylori* (*H. pylori*), which is detected in 99% of stomach cancer cases. It has been clearly shown that smoking, intake of salt and salt-preserved foods, low intake of vegetables and fruits, and excessive alcohol intake increase the risk in *H. pylori*-infected individuals. Many lifestyle habits

T. Takagi (✉)
Molecular Gastroenterology and Hepatology, Graduate School of Medical Science, Kyoto Prefectural University of Medicine, Kyoto, Japan

Department for Medical Innovation and Translational Medical Science, Graduate School of Medical Science, Kyoto Prefectural University of Medicine, Kyoto, Japan
e-mail: takatomo@koto.kpu-m.ac.jp

Y. Naito
Department of Human Immunology and Nutrition Science, Graduate School of Medical Science, Kyoto Prefectural University of Medicine, Kyoto, Japan

H. Yamada, Y. Naito (eds.), *Anti-Aging Medicine*, https://doi.org/10.1007/978-981-97-8514-8_72

Table 72.1 Certain/almost certain risk/prevention factors for gastrointestinal cancer [1]. Either "certain" or "almost certain" factors determined in any of the evaluations [6, 7]

	Risk factors	Prevention factors
Esophageal cancer (squamous cell carcinoma)	Smoking, alcohol, acetaldehyde associated with alcohol, hot food, and drinks[a]	Vegetables, fruits
Gastric cancer	*Helicobacter pylori* infection (non-cardia), smoking, alcohol[a], obesity (cardia)[a], salt/salted foods[a], processed meat (non-cardia[a])	
Colorectal cancer	Smoking, drinking, obesity, processed meat, tall height, red meat[a]	Physical activity (colon), whole grains[a], dietary fiber from food[a], dairy products[a], calcium supplements[a]
Liver cancer	Hepatitis B and C viruses, smoking, drinking, aflatoxin, obesity[a], diabetes[a]	Coffee consumption[a]
Pancreatic cancer	Smoking, obesity, drinking[a], tall height[a], diabetes[a]	

[a] Almost certain risk

are also involved in colon cancer. Smoking, drinking, obesity (especially in men), and processed meat are established risk factors, while high physical activity is a certain preventive factor, and intake of dietary fiber, dairy products, and calcium is almost certainly a preventive factor. A family history of colon cancer is also an established risk factor, with a particularly strong association in familial adenomatous polyposis and hereditary nonpolyposis colorectal cancer families.

72.3 Lifestyle and Liver and Pancreatic Diseases

The leading cause of liver cancer is persistent infection with hepatitis viruses caused by hepatitis B and C viruses, followed by those caused by alcoholic liver disease. In recent years, the progress of antiviral treatment for hepatitis viruses has been remarkable, and we have entered an era where hepatitis viruses can be eradicated with oral medication. On the other hand, the prevalence of fatty liver, where triglycerides, one of the visceral fats, accumulate in the liver, is increasing. Nonalcoholic steatohepatitis (NASH) is attracting attention as it can progress from chronic hepatitis-like changes to cirrhosis, and further, there is a risk of developing liver cancer, and its prevention and treatment are important. Fatty liver is greatly influenced by lifestyle habits, as it is called a liver disease of metabolic syndrome. Lifestyle habits related to fatty liver include diet, exercise, sleep, and preferences. Also, physical activity and the prevalence of fatty liver are inversely correlated, and lack of sleep induces disruption of the biological clock, becoming a worsening factor for fatty liver [3]. Alcohol, a preference, can be a cause and worsening factor of fatty liver, and it is important to correct lifestyle habits. The pancreas secretes digestive fluid and insulin, which is indispensable for glucose metabolism, from β cells, so it is considered significant to prevent pathological aging of the pancreas. Evidence has accumulated that genetic factors, smoking, drinking, and lifestyle-related diseases such as obesity and diabetes are involved in the onset of pancreatitis and pancreatic cancer, and again, it is important to correct lifestyle habits.

72.4 Lifestyle Habits and Lower Gastrointestinal Diseases

Hippocrates, an ancient Greek physician, said, "All disease begins in the gut." The famous saying, "All diseases begin in the gut," emphasizes the importance of maintaining a healthy digestive tract. In particular, the lower digestive tract, which governs the fundamental life phenomena of "digestion and absorption," is crucial to prevent its pathological aging. With the advent of an aging society, chronic constipation, which affects not only quality of life (QOL) but also life prognosis, is increasing. It is necessary to recognize this and correct lifestyle habits and appropriate treatment interventions. Reviewing the dietary and exercise habits of patients with chronic constipation is the first step in managing chronic constipation symptoms. In particular, it is well known that bowel habits are closely related to dietary habits, and reviewing these lifestyle habits is an important issue in clinical practice. The appearance of a bowel movement due to the gastroenteric reflex associated with meals is known to be most strongly induced after breakfast, and the habit of eating breakfast is important for establishing bowel habits. Also, it has been shown that intake of dietary fiber and water leads to improvement in stool conditions, and it is important to correct these lifestyle habits [4]. In addition, although the details will be left to another section, the gut microbiota is also becoming an important element of antiaging. The gut microbiota changes with age, and when the balance of the gut microbiota is disrupted (dysbiosis), it leads to many pathological conditions and aging. By combining a diet rich in dietary fiber and low-fat foods, regular exercise habits, and probiotics, it is expected to improve the aging process and age-related diseases in the field of antiaging medicine by adjusting the homeostasis of the gut microbiota and promoting a healthy gut environment where dysbiosis has been corrected [5].

References

1. Tsugane S. Trends in gastrointestinal cancer and its association with lifestyle. J Jpn Soc Intern Med. 2020;117:359–64.
2. Abiko S, Shimizu Y, et al. Risk assessment of metachronous squamous cell carcinoma after endoscopic resection for esophageal carcinoma based on the genetic polymorphisms of alcohol dehydrogenase-1B aldehyde dehydrogenase-2: temperance reduces the risk. J Gastroenterol. 2018;53:1120–30.
3. Shetty A, Hsu JW, et al. Role of the circadian clock in the metabolic syndrome and nonalcoholic fatty liver disease. Dig Dis Sci. 2018;63:3187–206.
4. Takagi T, Kamada K, et al. Chapter 4. Treatment of constipation. Improvement of dietary habits. Supervised by Norifumi Hibi, Hidekazu Suzuki: practical clinical manual for constipation. Tokyo: Kyowa Planning; 2020. p. 114–8.
5. Naito Y, Takagi T, et al. Obesity and intestinal bacteria: lower gastrointestinal diseases from the perspective of the intestinal environment. J Jpn Soc Intern Med. 2021;118:525–31.
6. International Agency for Research on Cancer. IARC monographs on the identification of carcinogenic hazards to humans (http://monographs.iarc.fr/), World Cancer Research Fund International. Diet and cancer report (https://www.wcrf.org/dietandcancer).
7. National Cancer Center Research and Development Fund. "Research on the evaluation of carcinogenicity and cancer prevention effects based on scientific evidence and recommendations for cancer prevention guidelines" (http://epi.ncc.go.jp/can_prev/)

Clinical Antiaging Medicine: Antiaging Medicine from the Perspective of the Respiratory System

73

Hisamitsu Omori

The physiological functions of all body organs decline with age. Among them, the respiratory system is one of the organs that most significantly exhibits the effects of aging and senescence [1]. This section provides an overview of normal aging of the lungs, chronic obstructive pulmonary disease (COPD), which is a state of accelerated lung aging, and factors believed to be related to the acceleration of lung aging.

73.1 Normal Aging of the Lungs

Normal aging of the lungs is believed to occur due to factors such as a decrease in respiratory muscle strength, stiffening of the chest wall, and a decrease in lung elastic recoil [1]. Respiratory function is an important indicator of lung aging. Respiratory function tests using spirometry show that vital capacity (VC), forced expiratory volume in one second (FEV_1), FEV_1/FVC ratio, and mid-expiratory flow (MEF) decrease with age. In contrast, residual volume (RV) and functional residual capacity (FRC) increase with age (Fig. 73.1). Among these, the decrease in FEV_1 is particularly useful as an indicator of annual changes [1–3]. The decrease in lung elastic recoil with age is attributed to degeneration and reduction of elastic tissue in the alveolar wall and parenchyma, expansion of the alveolar duct, and an increase in Kohn's pores [1]. These findings suggest that structural changes in the lungs affect function, a state referred to as "senile lungs" [1].

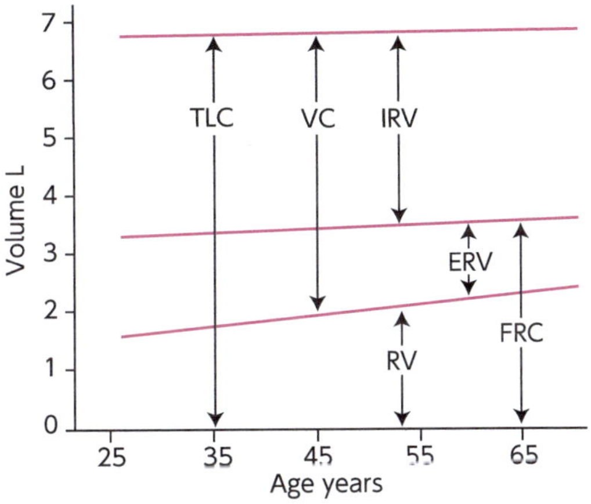

Fig. 73.1 Changes in respiratory function due to aging [2]. *TLC* total lung capacity, *VC* vital capacity, *IRV* inspiratory reserve volume, *ERV* expiratory reserve volume, *RV* residual volume, *FRC* functional residual capacity

73.2 Accelerated Aging of the Lungs

Chronic obstructive pulmonary disease (COPD) is a lung disease caused by long-term exposure to harmful substances, mainly tobacco smoke,

H. Omori (✉)
Department of Biomedical Laboratory Sciences, Faculty of Life Sciences, Kumamoto University, Kumamoto, Japan
e-mail: omorih@gpo.kumamoto-u.ac.jp

H. Yamada, Y. Naito (eds.), *Anti-Aging Medicine*, https://doi.org/10.1007/978-981-97-8514-8_73

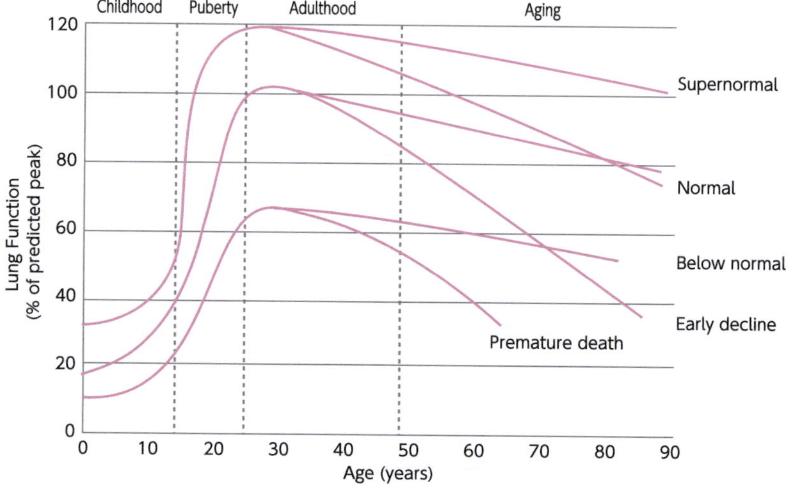

Fig. 73.2 Natural history of respiratory function throughout life. (Modified and reprinted with permission from [4])

and is characterized by airflow obstruction in respiratory function tests. Airflow obstruction occurs due to a combination of peripheral airway lesions and emphysematous lesions in various proportions [3]. The elastic fibers in the alveolar wall form a three-dimensional network structure, and the more the number of their partial destruction sites increases, the more it leads to the loss of lung elastic recoil. As a result, lung overinflation, increased residual volume, and obstructive ventilation disorder occur. The decrease in FEV_1 in spirometry reflects the natural history of COPD. Initially, patients are asymptomatic, but as the decrease in FEV_1 progresses, abnormalities become clearly visible in imaging diagnoses, and clinical symptoms such as dyspnea become severe [3]. As shown in Fig. 73.2, COPD can be described as an accelerated state of lung aging. Factors affecting the annual change in one-second volume in early adulthood include smoking, asthma, obesity, secondhand smoke, indoor air pollution including biomass, outdoor air pollution, occupational exposure, etc. have been reported [4]. Molecular, biological, and cellular biological factors related to the acceleration of lung aging, as cellular stress responses, include genetic instability, changes in telomere length, epigenetic changes, loss of proteostasis, deregulated nutrient-sensing, mitochondrial dysfunction, cellular senescence, stem cell exhaustion, and altered intercellular communication. These nine factors are mentioned [2, 5] (Table 73.1). An overview of the nine factors is shown [2, 5].

Table 73.1 Factors related to accelerated lung aging. (Adopted from [4] with permission)

Genetic instability	Primary hallmarks: Causes of cellular damage
Changes in telomere length	
Epigenetic alterations	
Loss of proteostasis	
Deregulated nutrient-sensing	Antagonistic hallmarks: Responses to damage
Mitochondrial dysfunction	
Cellular senescence	
Stem cell exhaustion	Integrative hallmarks: Culprits of the phenotype
Altered intercellular communication	

Genetic instability: Genome DNA is constantly damaged by various factors such as ultraviolet rays, radiation, various compounds, etc., but maintains stability by removing damage through DNA repair mechanisms and DNA damage response mechanisms. However, if there is a problem with the DNA repair mechanism or DNA damage response mechanism, it becomes impossible to properly remove the damage, which induces genome DNA instability and is thought to cause functional abnormalities in the body.

Telomere shortening (replicative aging): With each DNA replication, the telomere length gradually shortens, eventually reaching a critical length. Then replication is hindered, and cell division stops. DNA damage response is initiated by very short telomeres without a cap structure, and cellular aging is induced. Telomere shortening is thought to be the cause of replicative aging.

Epigenetic alterations: This is a chemical modification that occurs to DNA to control the on and off of genes.

Loss of proteostasis: As cells age, stress from the environment accumulates, and the mechanism to maintain the appropriate protein quantity ratio and distribution in the cell begins to weaken. Proteins lose stability, the process of autophagy begins to malfunction, and misfolded proteins accumulate.

Deregulated nutrient-sensing: Metabolic activity stresses cells. Excessive changes in metabolic activity, nutritional conditions, and composition are causes of faster cell aging.

Mitochondrial dysfunction: Mitochondrial energy production via acetyl coenzyme A (acetyl CoA) regulates cellular homeostasis. Mitochondrial dysfunction can contribute to aging (Aging) due to increased production of reactive oxygen species (ROS). Mitochondrial DNA (mtDNA) is susceptible to oxidative stress, and an association has been suggested between the accumulation of mtDNA mutations and aging (Aging). Smoking has been reported to affect the structure and function of mitochondria.

Cellular senescence: Aging and cellular senescence are not the same phenomenon. While aging is a progressive phenomenon that occurs over time, cellular senescence can occur at any point during the lifespan, even during embryonic development.

Stem cell exhaustion: With aging, human stem cells ultimately lose their ability to divide. Furthermore, humans cannot replenish stem cells that have migrated, differentiated, or died.

Altered intercellular communication: With aging, cells show an increase in self-preservation signals that can cause damage to other cells. Changes in intercellular communication with aging can contribute to a decline in tissue integrity.

References

1. Japanese Respiratory Society Pulmonary Physiology Committee. Clinical respiratory function test. 8th ed. Tokyo: Medical Review Co.; 2016.
2. MacNee W. Is chronic obstructive pulmonary disease an accelerated aging disease? Ann Am Thorac Soc. 2016;13_Suppl_5:S429–37.
3. Japanese Respiratory Society COPD Guideline 6th Edition Committee. Guidelines for the diagnosis and treatment of COPD (chronic obstructive pulmonary disease), 6th ed. Tokyo: Medical Review Co.; 2022.
4. Lange P, Ahmed E, et al. Natural history and mechanisms of COPD. Respirology. 2021;26:298–321.
5. Schneider JL, Rowe JH, et al. The aging lung: physiology, disease, and immunity. Cell. 2021;184:1990–2019.

74 Clinical Antiaging Medicine: Antiaging Medicine from the Perspective of the Locomotor System

Kenji Takahashi and Naoki Okubo

In a super-aged society, it is necessary to extend the healthy lifespan beyond the average lifespan. Healthy lifespan is defined as the period during which one can live without being restricted by health problems. On the other hand, according to the National Livelihood Survey, musculoskeletal disorders, including fractures and joint diseases, are the most common cause of needing support and care, accounting for about 23%. This article provides an overview of osteoporosis, which is the cause of fractures in the elderly, and knee osteoarthritis (knee OA), the most common joint disease that restricts activities of daily living (ADL), from the perspective of antiaging.

74.1 Knee Osteoarthritis

Knee osteoarthritis is a disease caused by excessive mechanical stress on the vulnerable knee structure due to aging. Not only aging, but also genetics, being female, past knee injuries (sports, overuse), and obesity increase the risk of knee OA progression. In addition, local progression factors include joint surface mismatch, lower limb alignment abnormalities, muscle weakness, and changes in the knee structure. Meniscal and ligament injuries are important changes in the knee structure. Pathologically, it causes progressive cartilage loss, bone modification under the cartilage, osteophyte formation, and synovitis [1].

74.1.1 Diagnosis

The Kellgren and Lawrence grade (K-L classification) [2] is often used based on standing simple X-ray frontal images. However, since it is difficult to regenerate joint cartilage once it degenerates, there is a growing idea to diagnose cases with a high risk of progression as early knee OA when knee joint function begins to decline and to intervene. Early knee OA is often considered to be cases where there are no OA findings on X-rays, but there are restrictions on knee pain and activities of daily living, and there are physical findings such as knee tenderness. In recent years, it has been reported that abnormal images such as deviation of the medial meniscus to the medial side of the medial joint space by ultrasound diagnosis, meniscal abnormalities, bone marrow lesions, cartilage damage, etc. by MRI are risk factors for progression [3].

74.1.2 Treatment

In the end stage, surgical treatments such as osteotomy and artificial joint surgery are performed.

K. Takahashi (✉) · N. Okubo
Department of Orthopaedics, Kyoto Prefectural University of Medicine, Kyoto, Japan
e-mail: kenji-am@koto.kpu-m.ac.jp

H. Yamada, Y. Naito (eds.), *Anti-Aging Medicine*, https://doi.org/10.1007/978-981-97-8514-8_74

Table 74.1 Recommendations for conservative treatment of knee OA [4]

	Without complications	Gastrointestinal complications	Cardiovascular complications	Frail	Total body pain and depression
Core treatment	Patient education, weight loss, exercise				
Level 1A Strong	Topical NSAID	Topical NSAID		Topical NSAID	
Level 1B Conditional	Nonselective NSAID	COX-2 inhibitors	Steroid joint injection	Steroid joint injection	Nonselective NSAID
	COX-2 inhibitors	Steroid joint injection	Hyaluronic acid joint injection	Hyaluronic acid joint injection	COX-2 inhibitors
	Steroid joint injection	Hyaluronic acid joint injection			
Level 2 in favor	Hyaluronic acid joint injection	Nonselective NSAID			Topical NSAID
					Steroid/ Hyaluronic acid joint injection
					Duloxetine

However, even if surgery is performed, it is not possible to expect functional recovery that allows activities such as sports and heavy work. Therefore, it is necessary to intervene with treatment at an early stage (Table 74.1). The basics of progression prevention are improvements in daily life and exercise centered on strength training. However, while exercise improves symptoms, there is no evidence that it inhibits the progression of cartilage degeneration. Low to moderate intensity.

There is a need to efficiently maintain muscle strength through regular exercise. In recent years, exercises aimed at improving knee kinematics have also been studied. No drug therapy to inhibit cartilage degeneration has been developed yet. Regenerative medicine using adipose-derived stem cells has shown potential and is expected. However, it is considered insufficient to improve the natural course of knee OA with drug therapy alone, and it is necessary to combine it with improvements in the mechanical environment (such as orthotics, exercise, weight loss, etc.).

74.2 Osteoporosis

The World Health Organization (WHO) defines osteoporosis as a "disease that weakens bone strength and increases the risk of fractures". Bone strength is defined by "bone density" and "bone quality", and imbalances in bone metabolism leading to decreased bone density and oxidative stress causing deterioration in bone quality result in decreased bone strength (Fig. 74.1). The estimated number of patients in Japan is about 12.8 million (three million men, 9.8 million women).

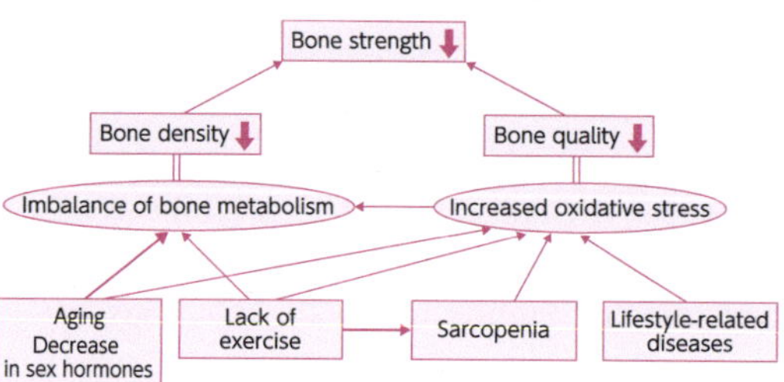

Fig. 74.1 Pathology of decreased bone strength

74.2.1 Diagnosis

In the "Guidelines for Prevention and Treatment of Osteoporosis 2015 Edition" [5], the diagnosis of osteoporosis consists of "history of fragile fractures" and "decreased bone density" and does not include bone quality. However, even if the bone density is the same, bone strength can decrease due to aging and lifestyle-related diseases that affect bone quality, so bone quality is considered an important indicator. Methods to detect deterioration in bone quality, such as measuring pentosidine in urine and homocysteine in blood, have been reported, but they are currently not covered by insurance, and further research is expected.

74.2.2 Prevention and Treatment

The choice of treatment drugs takes into account the patient's age, the severity and condition of osteoporosis, and comorbidities. The guidelines suggest that it is desirable to differentiate between bone formation promoting drugs and bone resorption inhibiting drugs according to bone metabolism. Bone resorption inhibitors such as bisphosphonates and anti-RANKL antibodies (denosumab) have many evidences of not only bone density increasing effect but also fracture inhibiting effect and are widely used. On the other hand, there is also a concern that excessive bone metabolism inhibition increases the risk of long bone fractures called "atypical fractures". This is thought to be because the accumulation of reduced repair ability such as microfractures and advanced glycation end products (AGEs) deteriorates bone quality. Bone formation-promoting drugs have a usage limit of 1 or 2 years, but they do not cause excessive bone metabolism inhibition, so they can be expected to have a positive effect on not only bone density but also bone quality. In addition to drug therapy, exercise therapy is important from the perspective of prevention and treatment. Immobility causes a decrease in bone density and deterioration of bone quality, while light dynamic load exercises such as walking and tai chi have been shown to increase lumbar spine bone density, and strong dynamic exercises such as jumping and jogging increase femur bone density. In recent years, it is noteworthy that reports have increased urinary pentosidine in patients with decreased grip strength and walking ability [6]. Future osteoporosis treatment will require not only focusing on bone density, but also evaluating bone quality and sarcopenia, and treating the entire musculoskeletal system.

References

1. Katz JN, Arant KR, et al. Diagnosis and treatment of hip and knee osteoarthritis: a review. JAMA. 2021;325:568–78.
2. Kellgren JH, Lawrence JS. Radiological assessment of osteo-arthrosis. Ann Rheum Dis. 1957;16:494–502.
3. Sharma L, Hochberg M, et al. Knee tissue lesions and prediction of incident knee osteoarthritis over 7 years in a cohort of persons at higher risk. Osteoarthritis Cartilage. 2017;25:1068–75.
4. OARSI. Guidelines for the non-surgical management of knee, hip and polyarticular osteoarthritis; 2019.
5. Osteoporosis Prevention and Treatment Guidelines Committee (Japan Osteoporosis Society, Japan Bone Metabolism Society, Osteoporosis Foundation). Guidelines for prevention and treatment of osteoporosis, 2015 ed. Tokyo: Life Science Publishing; 2015.
6. Moriwaki K, Matsumoto H, et al. Urinary pentosidine level is associated with grip strength and gait speed in community-dwelling adults: a cross-sectional study. BMC Musculoskelet Disord. 2021;22:392.

75 Clinical Antiaging Medicine: Antiaging Medicine from the Perspective of Sensory Organs (Eyes)

Nobuyoshi Kitaichi

Humans obtain 80% of information from the outside world as visual information, and the importance of antiaging of the eyes is high among sensory organs. In the field of ophthalmology, there are cases where the effectiveness of food factors has been shown in clinical intervention trials for the prevention and progression suppression of age-related diseases and eye fatigue, and we will explain some representative ones.

75.1 Antioxidant Vitamins and AREDS Trials

A large-scale clinical study on the intake of antioxidants targeting healthy individuals was conducted under the leadership of the National Eye Institute (NEI) in the United States [Age-Related Eye Diseases Study (AREDS) trial] [1]. The effects of vitamins A, C, E, zinc, copper, etc. were verified in 11 facilities and a total of 3640 people. No significant effect was obtained for cataracts, but a preventive effect on the progression of severe cases was observed in age-related macular degeneration, one of the typical age-related eye diseases. Specifically, in the case of the presence of large/medium drusen or age-related macular degeneration in one eye, the progression rate to age-related macular degeneration of the test eye was reduced by 25% in the antioxidant vitamin intake group (vitamin C 500 mg, vitamin E 400 IU) + trace mineral group (zinc 80 mg, copper 2 mg) compared to the placebo group. Vitamin A is essential for visual function as it has an antioxidant effect and is also a precursor of rhodopsin. However, during this trial, the incidence of lung cancer was significantly higher in the vitamin A intake group among smokers. Smokers should not be advised to take vitamin A. The AREDS trial is an important study that proved the medical effectiveness of so-called supplements such as vitamins and minerals in a large-scale randomized comparative trial.

75.2 Polyunsaturated Fatty Acids (PUFA) and AREDS2 Trials

Polyunsaturated fatty acids (PUFA) are classified into ω3 and ω6 lineages from their chemical structure. Arachidonic acid is synthesized from linoleic acid, which is ω6PUFA, and inflammatory substances such as prostaglandin E2 are synthesized by the arachidonic acid cascade. On the other hand, the inflammatory inducing action of ω3PUFA represented by α-linolenic acid, eicosapentaenoic acid (EPA), and docosahexaenoic acid (DHA) is very weak and can be obtained from fish oil by consuming fish. ω3 and ω6

N. Kitaichi (✉)
Department of Ophthalmology, Health Sciences University of Hokkaido Hospital, Sapporo, Hokkaido, Japan
e-mail: nobukita@hoku-iryo-u.ac.jp

H. Yamada, Y. Naito (eds.), *Anti-Aging Medicine*, https://doi.org/10.1007/978-981-97-8514-8_75

PUFAs do not convert to each other in the body, and if you consume a lot of ω3PUFA, ω6PUFA will decrease relatively. In animal experiments, the intake of EPA reduces arachidonic acid in the body and suppresses choroidal neovascularization, but in the large-scale clinical trial AREDS2 in humans conducted in response to the results of the aforementioned AREDS trial, even if DHA + EPA is added to antioxidant supplements, the preventive effect on age-related macular degeneration did not increase. However, in twin studies in the United States, the habit of eating fish rather than meat reduces the risk of developing age-related macular degeneration [2].

75.3 Lutein and Age-Related Macular Degeneration

Lutein is a type of carotenoid and, along with its optical isomer zeaxanthin, belongs to the xanthophylls. Although about 40 types of carotenoids exist in the human body, only lutein and zeaxanthin are selectively absorbed in the macula, the most sensitive part of the retina. Therefore, these two carotenoids are referred to as "macular pigments". Their maximum absorption wavelength is 446 nm, very close to blue light, and they absorb harmful blue light while also having a strong antioxidant effect. In other words, they protect the macula of the retina through both a filter effect and an antioxidant effect. In groups that consume a diet rich in carotenoids, the risk of age-related macular degeneration decreased by 43%, and particularly, the intake of 6 mg/day of lutein/zeaxanthin was found to be the most effective [3].

75.4 Anthocyanins and Glaucoma

Anthocyanins, a type of flavonoid, have long been said to have beneficial effects on the eyes. Anthocyanins are abundant in berries such as blueberries, blackberries, chokeberries, raspberries, and haskap, as well as in red cabbage. Recently, it has been reported that the intake of blackcurrant fruit extract anthocyanins can inhibit the progression of glaucoma visual field defects [4]. However, compared to the general public's perception that "blueberries are good for the eyes", large-scale clinical trials are insufficient, and it remains an important research topic.

75.5 Astaxanthin and Eye Fatigue

Astaxanthin is an orange pigment found in salmon, roe, and crustacean shells and has various physiological functions such as antioxidant and antitumor effects, and strong anti-inflammatory and tissue protective effects have been observed in animal experiments. In a clinical study targeting healthy Japanese adults who often work on computers and are aware of eye fatigue, significant improvements were observed in subjective symptoms such as "eyes get tired easily", "dry eyes", "stiff shoulders", and in the quasi-subjective regulation function in the objective eye fatigue evaluation [5]. It also improves the retinal blood flow speed in healthy individuals [5]. Today, its excellent effects are widely recognized not only in the field of ophthalmology but also in sports medicine and cosmetics.

75.6 Catechins and Adjustment Power

Green tea contains a large amount of functional ingredients such as catechins, polyphenols, and caffeine. In fact, antioxidant effects, antitumor effects, lipid metabolism improvement effects, and blood pressure rise suppression effects have been reported so far. Therefore, we examined eye fatigue and adjustment ability by intake of "Sun Rouge Tea" containing tea catechins and anthocyanins, "Yabukita Tea" containing tea catechins, and "Barley Tea" containing no active ingredients in a prospective randomized placebo-controlled double-blind trial. In the "Sun Rouge Tea" group, subjective symptoms significantly improved compared to the placebo group. The objective adjustment ability also significantly improved immediately after near work and after rest in the "Sun Rouge Tea" group [6]. Since the effectiveness was significantly demonstrated in Sun Rouge Tea, the main cause of the effect is considered to be anthocyanin. However, since anthocyanin was effective in a smaller amount than previous reports, it is speculated that the effect of improving eye fatigue is enhanced by taking tea catechins and anthocyanins at the same time.

75.7 Conclusion

The recent work environment has more management work than physical work, and the spread of computers and smartphones is making the burden on the eyes increasingly severe. The nutrients and their clinical effects demonstrated in the human clinical trials discussed in this section are summarized in the table (Table 75.1). We would like to take these appropriately and aim for antiaging of the eyes.

Table 75.1 Various nutrients and clinical effects in humans

Eye disease/ symptom	Subjects	Nutrient	Clinical effect
Age-related macular degeneration	Americans	Antioxidant vitamins[a] + trace minerals	Progression suppression
		Fish diet (ω3 polyunsaturated fatty acids)	Onset suppression
		Lutein	Onset suppression, symptom improvement
Glaucoma	Japanese	Anthocyanin	Progression suppression
Eye fatigue	Japanese	Astaxanthin	Improvement in self-awareness symptoms, improvement in eye adjustment function, improvement in fundus blood flow
Eye fatigue	Japanese	Anthocyanin + Catechin	Improvement in self-awareness symptoms, improvement in eye adjustment function

[a] However, intake of vitamin A in smokers increased the incidence of lung cancer

References

1. Age-Related Eye Disease Study Research Group. A randomized, placebo-controlled, clinical trial of high-dose supplementation with vitamins C and E, beta carotene, and zinc for age-related macular degeneration and vision loss: AREDS report no. 8. Arch Ophthalmol. 2001;119:1417–36.
2. Seddon JM, George S, et al. Cigarette smoking, fish consumption, omega-3 fatty acid intake, and associations with age-related macular degeneration: the US twin study of age-related macular degeneration. Arch Ophthalmol. 2006;124:995–1001.
3. Seddon JM, Ajani UA, et al. Dietary carotenoids, vitamins A, C, and E, and advanced age-related macular degeneration. Eye Disease Case-Control Study Group. JAMA. 1994;272:1413–20.
4. Ohguro H, Ohguro I, et al. Two-year randomized, placebo-controlled study of black currant anthocyanins on visual field in glaucoma. Ophthalmologica. 2012;228:26–35.
5. Nobuyoshi K, Susumu I. Application of astaxanthin to eye diseases. Supervised by Yoshikawa Toshiichi, Naito Yuji: function and application of astaxanthin. Tokyo: CMC Publishing; 2019. p. 138–45.
6. Maeda-Yamamoto M, Nishimura M, et al. A randomized, placebo-controlled study on the safety and efficacy of daily ingestion of green tea (Camellia sinensis L.) cv. "Yabukita" and "Sunrouge" on eyestrain and blood pressure in healthy adults. Nutrients. 2018;10:569.

76 Clinical Anti-Aging Medicine: Gender Medicine and Anti-Aging Medicine

Keiko Amano

Gender medicine aims to advance research on conditions where the male–female ratio is overwhelmingly skewed towards one gender, conditions where the incidence rate is almost the same but there are clinical differences between men and women, conditions where physiological and biological understanding is still lagging in men or women, and the relationship between social status and health of men and women. The goal is to reflect the results in disease diagnosis, treatment methods, and preventive measures—a medical reform known as Gender-specific Medicine (GSM) [1]. By providing "gender-considerate medical care," it aims to lead a healthier life, which is the same goal as "anti-aging medicine," a field of medicine that intervenes in the biological process of aging, aims to reduce the incidence of age-related diseases such as arteriosclerosis and cancer, and aims for healthy longevity. Since the 1990s, gender medicine research originating in the United States has actively sought factors that cause gender differences in biological processes and age-related diseases. As a result, the factors that cause gender differences have been clarified as the "Sex biological differences" and "Gender psychological or cultural differences" shown in Tables 76.1 and 76.2 [2].

Table 76.1 Biological differences in sex [2]

Differences related to genetics
Differences in gene transcription
Differences in gene expression
Differences in mutation frequency
Differences in the transmission of genetic information
Differences related to sex hormone levels and their actions
Differences in sex hormone levels within individuals, between individuals, and between sexes
Differences in information transmission
Differences in receptor expression increase
Differences related to Epigenetics
Differences related to the expression of receptors, enzymes, and binding proteins
Differences in enzyme expression and binding activity
Differences in binding protein and binding activity
Differences in receptor expression and its degree
Differences in enzyme induction
Anatomical differences, differences in organ size
Differences in organ function
Differences in the expression of functional units in organs
Differences in susceptibility to injury, aging, and regeneration in organs
Differences related to metabolism
Differences in metabolic function
Differences in metabolic induction
Differences in metabolic cofactors
Differences in metabolism due to reproductive stage, age

K. Amano (✉)
Department of Internal Medicine, Seifuso Hospital, Niiza, Japan
e-mail: k.amn@jcom.home.ne.jp

H. Yamada, Y. Naito (eds.), *Anti-Aging Medicine*, https://doi.org/10.1007/978-981-97-8514-8_76

Table 76.2 Gender psychological and cultural differences [2]

Differences in individual or societal perceptions of male and female roles
Role as a labor force
Role as a spouse and parent in a family
Role as a member of a household, including servants
Role as a caregiver
Differences in individual or societal perceptions and coping with health and disease
Differences in risk behaviors, risk factors
Differences in general perceptions of health and disease
Differences in recognition and confirmation of symptoms caused by illness
Differences in expressing about disease, actions taken for health
Differences in adherence to treatment
Differences in acceptance of disease
Differences due to individual or societal stereotypical thinking about disease and perception of disease frequency
Differences due to patients' and doctors' attributive thinking about disease
Differences in presenting diagnosis and treatment
Association between gender and stereotypical understanding of disease
Differences in awareness
Differences in information dissemination to public health agencies
Differences in non-medical issues affecting access to healthcare
Differences in literacy rates
Differences in awareness of individual rights
Differences in self-deprecating behaviors
Differences in time allocation for daily work and activities
Differences in economic affluence
Differences in the extent of family and social assistance

76.1 Consider Gender and Age Differences in Health Checkup Diagnostic Criteria and Guidance [3]

In 2008, metabolic syndrome checkups began, and depending on the high risk of developing lifestyle-related diseases, people were divided into two groups, "motivation support" and "active support," and diet and exercise guidance was strengthened by public health nurses and dietitians. Unfortunately, the diagnostic criteria for metabolic syndrome do not take into account gender and age differences. Health guidance that does not take into account variations due to age and gender can sometimes be lenient for young people and strict for the elderly. As a result, abnormalities in young people may be overlooked, and in the elderly, unreasonable lifestyle guidance and treatment may be conducted according to conventional criteria, potentially reducing their quality of life. For dyslipidemia, in recent years, treatment guidelines that take into account gender and age differences have been proposed by the Japan Atherosclerosis Society [4]. In our country, clinical guidelines have been established in many areas, but gender and age differences should be considered in clinical guidelines.

76.2 Sex Hormones and the Onset of Aging

Aging and the accompanying decrease in hormone secretion can lead to menopausal disorders, muscle weakness, decreased bone density, obesity, increased blood pressure and blood sugar levels, arteriosclerosis, and cognitive decline. Noticing these physiological changes early and taking action is the beginning of anti-aging. Estrogen exerts its effects by binding to estrogen receptors (ER). There are two types of ER, α and β, which are distributed differently in tissues and have different functions. ER-α is mainly distributed in female reproductive organs such as the mammary glands and uterus. ER-β is found in bones, brain, liver, prostate, blood vessel walls, lungs, thyroid, bladder, synovial membranes of joints, tendons, ligaments, etc., regardless of gender, and has a wider physiological significance. Women's estrogen levels begin to decrease around the age of 40, and menopause occurs on average at 50–51 years of age. Menopausal symptoms appear around menopause, aging progresses rapidly after menopause, and 10–15 years later, the incidence of osteoporotic fractures and cardiovascular diseases such as myocardial infarction and stroke increases. Treatments include advice on lifestyle habits related to menopausal symptoms and physiological changes due to aging; hormone replacement therapy for women; Oriental medicine such as Chinese herbal medicine, acupuncture, and qigong; thermal therapies such as hot spring cures and Japanese warm therapies; counseling, antidepressants, anti-anxiety drugs, and sleeping pills for mental symptoms. Sometimes collaboration with a psychiatrist is necessary. In men, it has become clear that whole-body fatigue, decreased motivation, erectile dysfunction (ED), and other physical and mental problems specific to middle-aged and older men are caused by a decrease in male hormones (testosterone) due to aging, which is called "late-onset hypogonadism (LOH syndrome)" or male menopausal disorder [5]. Male hormones are known as substances that promote secondary sexual characteristics, but they also promote various physiological activities throughout the body, such as muscle and bone formation, hematopoietic function, maintenance of sexual function, and lipid and carbohydrate metabolism. They are also involved in cognitive and psychological functions in the brain. There are several factors that decrease testosterone, the most typical of which are aging and stress. Testosterone is mainly produced in the testes under the command of the hypothalamus in the cerebrum, but it is thought that if psychological stress is prolonged, the state of sympathetic dominance occurs, and the cerebrum inhibits testosterone production. The decrease in male hormones is gradual, and male menopausal disorders can occur at any age after the decrease begins at 40. The peak age of onset is in the 50s and 60s. There are also men in their 70s and 80s who complain of symptoms. The treatment is testosterone replacement therapy, which has been shown to improve muscle mass, strength, bone density, serum lipids, insulin sensitivity, mood and libido, and sense of health, and the action of PDE5 inhibitors is enhanced for erectile dysfunction.

76.3 Intervention in Lifestyle-Related Diseases Is Lifelong

There are many diseases in the elderly that have clear gender differences based on lifestyle habits such as diet, exercise, rest, smoking, and drinking. Obesity, hyperlipidemia, and hypertension, which are risk factors for arteriosclerotic diseases, are higher in young men and gradually increase in women after menopause, eventually surpassing men. The incidence of cardiovascular diseases is overwhelmingly higher in young men, and in women, it begins to increase with the decrease in estrogen, and there is almost no gender difference in the late elderly. Osteoporosis, which is more common in women, is also a disease with clear gender differences. Cerebrovascular dementia is more common in men, and Alzheimer's disease is more than twice as common in women as in men, but it is also clear that lifestyle-related diseases increase the risk of developing dementia, and good lifestyle habits are the first step in anti-aging. To put it extremely, education is needed from childhood for this purpose (Keiko Amano).

References

1. Amano K. Gender medicine—General. Thrombosis and Circulation. 2015;23:234–238.
2. Oertelt-Prigione S. Sex and gender in medical literature. In: Oertelt-Prigione S, Regitz-Zagrosek V, editors. Sex and gender aspects in clinical medicine. London: Springer; 2012. p. 9–15.
3. Amano K. Problems of medical examination from the perspective of gender differences. Health Eval Promot 2013; 40: 613–622.
4. Japan Atherosclerosis Society. Guidelines for the prevention of atherosclerotic diseases, 2017 edition. Tokyo: Japan Atherosclerosis Society; 2017.
5. Department of Urology, Juntendo University Hospital. LOH syndrome (late-onset hypogonadism). https://juntendo-urology.jp/disease/loh/.

Clinical Anti-Aging Medicine: Cancer Treatment and Anti-Aging Medicine

77

Takuro Kobayashi and Shigeo Horie

Cancer is one of the age-related diseases that increase with age, and about half of the Japanese population is diagnosed with cancer in their lifetime. In recent years, research has been advancing on how the content of research in the anti-aging field affects cancer treatment. In this section, we will introduce some recent trends in anti-aging medicine in cancer treatment.

77.1 Testosterone and Cancer Cachexia

Testosterone is a sex steroid hormone, about 95% of which is produced in the testes in men, and the remaining 5% is produced in the adrenal glands. On the other hand, in women, it is produced in the ovaries and adrenal glands, but the amount produced is very small compared to men. Low levels of testosterone in men are associated with a decline in physical and mental function and quality of life, and maintaining testosterone may contribute to extending healthy life expectancy. In advanced cancer, various mental and physical symptoms appear when it becomes a state of cancer cachexia. It has been recently reported that supplementing testosterone in male patients with advanced cancer reduces the blood concentration of TNF-α (tumor necrosis factor α), which reflects the state of cancer cachexia [1]. In addition, it was clarified that the "sense of unhappiness" improved by administering testosterone, and that the decline in gonadal function is one of the causes of cancer cachexia. It is expected that the symptoms of patients with advanced cancer will be treated by testosterone replacement therapy in the future.

77.2 Frailty, Sarcopenia, and Surgery

Frailty is a concept that indicates a "state between a healthy state and a state requiring care (a state where support is needed in daily life)." Many elderly people are thought to go from a healthy state to a state of "sarcopenia" where muscle strength declines, then to a state of "frailty" where life functions decline in general, and finally to a state requiring care. In a cohort study of more than 400,000 people, it was reported that patients who were frail before surgery had a significantly higher mortality rate within 6 months after surgical examination or treatment compared to patients who were not frail [2]. Furthermore, in the results of subgroup analysis by invasiveness, even in low-invasive surgical procedures like cystoscopy, the 30-day mortality rate in frail patients was 1.55% (95% CI, 1.20%–1.97%), which is higher than the general 30-day mortality rate of

T. Kobayashi (✉) · S. Horie
Department of Urology, Graduate School of Medicine, Juntendo University, Tokyo, Japan
e-mail: ta-kobayashi@juntendo.ac.jp

H. Yamada, Y. Naito (eds.), *Anti-Aging Medicine*, https://doi.org/10.1007/978-981-97-8514-8_77

1% in high-risk surgery. Since treatments that are low to moderate in load may be high risk in frail patients, it is necessary to universally apply screening for frailty.

77.3 Gut Bacteria and Cancer Immunotherapy/Endocrine Therapy

It has been suggested that gut bacteria can affect the response to immune checkpoint inhibitors in cancer patients. In one study, 30 patients with metastatic renal cell carcinoma were randomly assigned to receive nivolumab and ipilimumab with or without the probiotic (CBM588) [3]. As a result, the group receiving nivolumab and ipilimumab with CBM588 had a significantly longer progression-free survival (PFS) than the group without it (12.7 months vs. 2.5 months, HR = 0.15, 95% CI 0.05–0.47, P = 0.001). Although there was no statistically significant difference, the response rate was also higher in the CBM588 group (58% vs. 20%, P = 0.06). Changes in gut bacteria due to CBM588 may improve clinical outcomes in cancer immunotherapy, and further research is expected. Also, hormone therapy for prostate and breast cancer has been associated with risks of decreased bone density and obesity. Recently, it has been revealed that the diversity of gut microbiota in patients significantly decreases due to the decrease in serum testosterone concentration caused by endocrine therapy for prostate cancer [4].

77.4 Genomic Instability and Radiation Therapy

It has been observed that the Y chromosome in male leukocytes is partially lost with age, a phenomenon known as mosaic loss of chromosome Y (mLOY). mLOY, which is also increased by smoking, has been attracting attention as an indicator of genomic instability from recent genome-wide association studies [5]. Cancer patients have a higher future risk of mLOY. In a study involving more than 200,000 people in the UK Biobank, patients with solid tumors had a higher future risk of mLOY, and this trend was also observed in bladder cancer. In our study involving 348 prostate cancer patients, we found a significant association between mLOY and patients who had received local radiation therapy (OR 2.55, 95% CI 1.08–6.50, P = 0.04).

77.5 Pre-treatment Geriatric Assessment in Cancer Treatment

There is growing interest in the comprehensive geriatric assessment (CGA) as a tool to comprehensively evaluate the health status of elderly cancer patients. CGA is a method to improve QOL by organizing and evaluating problems comprehensively from six items such as medical evaluation, life function, mental function, and social/environment (living environment, care environment, relationships with family and friends) for the elderly (Table 77.1). Full-scale CGA requires evaluation by experts and time, and it is difficult to use in actual clinical practice, so more convenient screening tools such as

Table 77.1 Comprehensive Geriatric Assessment (CGA)

Evaluation item	Evaluation method (Example)
Functional and physical performance	Activities of daily living (ADL) Instrumental activities of daily living (IADL) History of falls
Comorbidity/medication status	Charlson comorbidity index (CCI) Medication count
Cognitive function	Mini-mental state examination (MMSE) Hasegawa dementia scale-revised (HDS-R)
Psychological state	Geriatric depression scale 15 (GDS 15) Hospital anxiety and depression scale (HADS)
Nutritional status	Mini-nutritional assessment (MNA) BMI
Social support	Economic status, mobility, care provider situation, etc.

CGA7 and G8 have been developed, and the validity of these tools is being evaluated in Japan.

77.6 Conclusion

Anti-aging, in a word, means "youthful." Even if the calendar age is the same, the so-called biological age varies from individual to individual. In recent years, there has been a growing interest in measuring biological age, and how to evaluate and control the invisible aging process within the body has become an important issue. The application of anti-aging to cancer treatment is expected to lead to the prevention and early detection of cancer, as well as the suppression of its progression.

References

1. Izumi K, Iwamoto H, et al. Androgen replacement therapy for cancer-related symptoms in male: result of prospective randomized trial (ARTFORM study). J Cachexia Sarcopenia Muscle. 2021;12:831–42.
2. Shinall MC Jr, Arya S, et al. Association of preoperative patient frailty and operative stress with postoperative mortality. JAMA Surg. 2020;155:e194620.
3. Dizman N, Meza L, et al. Nivolumab plus ipilimumab with or without live bacterial supplementation in metastatic renal cell carcinoma: a randomized phase 1 trial. Nat Med. 2022;28:704–12.
4. Kure A, Tsukimi T, et al. Gut environment changes due to androgen deprivation therapy in patients with prostate cancer. Prostate Cancer Prostatic Dis. 2023;26(2):323–30. (Epub ahead of print.)
5. Kobayashi T, Yatsuya T, et al. Loss of somatic Y chromosome. Urology. 2022;15:418–23.

78 Clinical Anti-Aging Medicine: Appearance and Anti-Aging Medicine

Hidekazu Yamada

78.1 Appearance

The phenotype referred to as "appearance" [1] is important in anti-aging medicine for aging evaluation. This is because it is one of the most easily understood phenotypes. Currently, standards for aging evaluation for intervention are being established, and it is an important item in the aging clock. When looking at the correlation with chronological age, recent advancements in image evaluation through machine learning have made it one of the most accurate items (with an error of 2.4 years). The area treated as "appearance" covers a wide range, including not only morphology but also function (Table 78.1). Therefore, the appearance area that people can recognize is evaluated by dividing it into skin, facial features, and body shape (Fig. 78.1). Examples of appearance evaluations when aging include wrinkles, spots, sagging, white hair, hair loss, excessive hair growth in certain areas (such as ear hair or long eyebrows), decrease in height, and changes in body shape. The relationship with chronological age has been clarified for each. (From conventional quantification, image analysis using machine learning is becoming mainstream and is also being used for pre-disease judgment.)

Table 78.1 Appearance (Modified with permission from [2])

Skin	• Skin tone • Texture, wrinkles, dullness • Hair quality • Body odor, fragrance • Capillary structure
Appearance	• Facial and other shapes • Eyebrows, eyelashes, hairstyle • Deep wrinkles, sagging • Genital area (second FACE) • Poise, behavior
Body shape	• Size • Body image • Muscle distribution, fat distribution, shape of chest and buttocks (BWH ratio with differences in insulin resistance) • Skeletal structure • Body shape

The area treated as appearance is expanding. The morphology of capillaries, in the beauty field, is divided into skin, appearance, and body shape. The genital area is included in the appearance. Scent and movement are also important. It was summarized at the Antiaging Research Society for Appearance. It is expected to expand further in the future

H. Yamada (✉)
Anti-Aging Center, Kindai University, Osaka, Japan
e-mail: yamada@mitameonline.com

H. Yamada, Y. Naito (eds.), *Anti-Aging Medicine*, https://doi.org/10.1007/978-981-97-8514-8_78

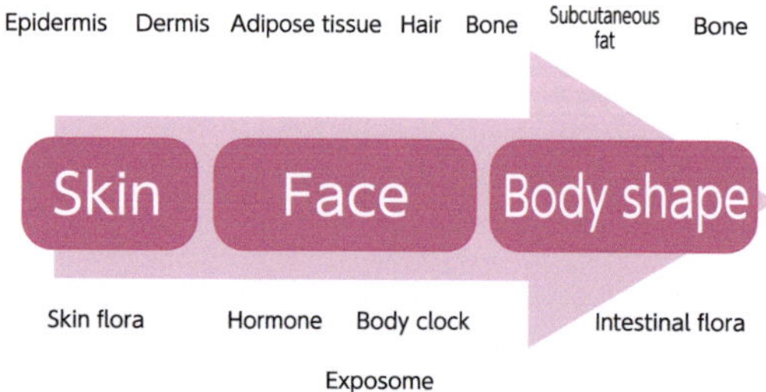

Fig. 78.1 Conceptual diagram of appearance. The area to be addressed. Basically, it is necessary to look at both the relationship between the genome and the environment. Therefore, the exposome is considered an important concept. In aging, epigenetic changes are important

78.2 Changes in Appearance and Epigenetics

Even in identical twins, it has been found that environmental factors can cause significant differences in appearance, indicating that environmental factors are more important than genetic background. Furthermore, advances in DOHaD (developmental origins of health and disease; "future health and susceptibility to specific diseases are strongly determined by the influence of the fetal and early postnatal environment") research and the understanding that social factors such as educational background, human relationships, and income are related to aging have been made.

78.3 Internal Aging and External Aging

Traditionally, aging has been explained as being divided into internal aging and external aging. Internal aging is the process of aging that progresses due to the original aging mechanism, but even in this case, wrinkles, sagging, and changes in body shape occur. Therefore, it is understood that preventive interventions regarding exercise, nutrition, and mental health (brain, sleep) are important even in internal aging. As for behavior, it is said that genetic factors play a large role.

Extrinsic aging refers to aging caused by environmental exposure factors such as ultraviolet rays. Recently, the concept of the exposome (the cumulative sum of environmental exposures in life) has been considered important in terms of appearance. The total amount of ultraviolet exposure, chemical exposure, and PM2.5 exposure since birth is related. In photoaging, wrinkles are related to UVA (mainly dermis), UVB (mainly epidermis), and near-infrared rays. It has been found that the function of skin flora and gut flora as an external environment affects the appearance. In addition, gut flora and skin flora are considered external factors on the exposome. From the standpoint of epigenetics, it has been found that even if DNA repair is complete, aging can be promoted through DNA methylation. Aging is attracting attention to the "information theory of aging" [3] that assumes it is "due to the loss of temporal transcription networks and epigenetic information." In the Dunedin study, a correlation between brain function, cognitive function, and apparent age has been shown. Although a causal relationship between the appearance of aging and biological age, including blood tests, cannot be proven, they have a strong correlation.

78.4 The Meaning of Appearance

From the idea that appearance is the fourth asset [4], in a capitalist economy, it may have meaning in values such as youthfulness. This is probably the reason for the interest in the beauty field. Furthermore, it has been suggested that beauty is involved in human evolution along with time, fire, and words [5], and it will be necessary to deepen the discussion in relation to appearance and neuroaesthetics (the center of beauty and the medial orbitofrontal cortex) [6].

References

1. Yamada H. General theory of anti-aging for appearance. Aesthet Dermatol. 2020;30:98.
2. Yang JH, Hayano M, et al. Loss of epigenetic information as a cause of mammalian aging. Cell. 2023;186:305–26.e27.
3. Hakim C. Erotic capital. New York: Basic Books; 2011.
4. Vince G. Transcendence: how humans evolved through fire, language, beauty, and time. New York: Basic Books; 2020.
5. Kirk U, Skov M, et al. Modulation of aesthetic value by semantic context: an fMRI study. Neuroimage. 2009;44:1125–32.
6. Yamada H. Appearance. Anti-Aging Study Group on Appearance 2015 (Differentiation Meeting of the Japan Anti-Aging Medical Association).

Clinical Anti-Aging Medicine: Anti-Aging Medicine from the Perspective of Oral Function

79

Takayoshi Sakai and Kenji Mishima

79.1 The Importance of Oral Function and Aging

Anti-aging medicine aims for homogeneous aging and requires a cross-sectional approach aimed at overall health, as well as comprehensive knowledge that encompasses each area. The oral cavity is the entrance to the digestive tract and is positioned as an organ that has not only life science aspects necessary for maintaining life, such as respiration, occlusion and mastication, eating and swallowing, and saliva secretion, but also cultural aspects necessary for human life, such as speech and language, taste, and emotional expression. If the eating and swallowing function decreases, nutritional intake is impaired, and overall activity decreases immediately. If saliva secretion and respiratory function decrease, the risk of aspiration pneumonia and suffocation increases. Also, a decrease in taste and speech/language function greatly affects not only the quality of life (QOL) but also the maintenance of cognitive function. Mastication is programmed in the brainstem and operates based on sensory information from the periphery. Appetite is enhanced through smell and vision, and taste determines whether food is edible or not. The cerebrum works perceptually and cognitively, considering the appropriate intake method for the food. Eating affects overall physiological functions such as hypnotic effects and increased body temperature, and it is known that these changes are transmitted to the eating and satiety centers in the hypothalamus [1]. Humans chew food beautifully in coordination with the tongue, teeth, gums, and cheek mucosa under the presence of saliva. After forming a food bolus of the appropriate nature, the food is sent to the entrance of the esophagus by the swallowing reflex, which is a coordinated movement of the swallowing-related muscles, and is transferred to the esophageal sphincter by peristaltic movement and gravity, where it is digested. It is known that when the parts related to this eating and swallowing function weaken due to aging, eating and swallowing disorders occur, leading to malnutrition (Fig. 79.1). Approximately 70% of the muscles in the maxillofacial region exist around the oral cavity, and if the masticatory ability decreases with age, it leads to a decrease in QOL due to oral function disorders. The development of various coping methods is expected, and it is an issue that must be addressed.

T. Sakai (✉)
Department of Rehabilitation for Orofacial Disorders, Osaka University Graduate School of Dentistry, Osaka, Japan
e-mail: sakai.takayoshi.dent@osaka-u.ac.jp

K. Mishima
Division of Pathology, Department of Oral Diagnostic Sciences, Showa University School of Dentistry, Tokyo, Japan

H. Yamada, Y. Naito (eds.), *Anti-Aging Medicine*, https://doi.org/10.1007/978-981-97-8514-8_79

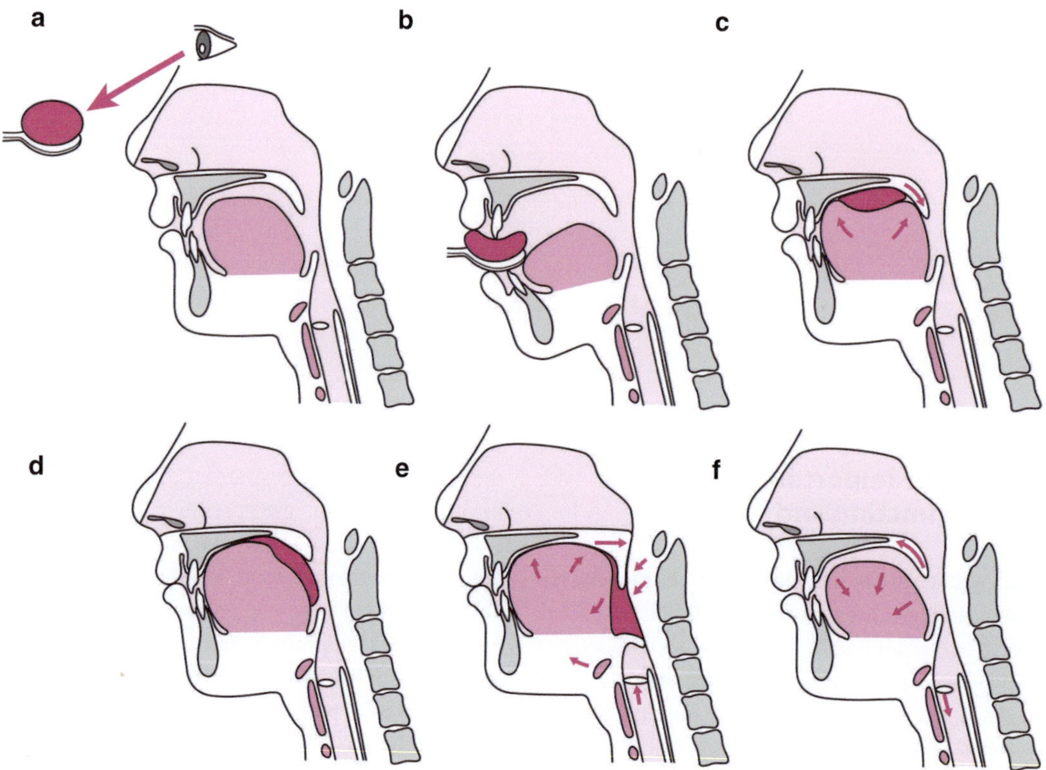

Fig. 79.1 Anatomy and mechanism of eating and swallowing. Modified with permission from [1]). (**a**) Preceding period (confirmation of food), (**b**) Preparation period (1) (intake into the mouth), (**c**) Preparation period (2) (mastication and bolus formation) (a: nasal cavity, b: hard palate, c: soft palate, d: bolus, e: tongue, f: pharynx, g: laryngeal lid, h: larynx, i: trachea, j: esophagus), (**d**) Oral period (tongue root, sending to the pharynx), (**e**) Pharyngeal period (passing through the pharynx, sending to the esophagus), (**f**) Esophageal period (sending from the entrance of the esophagus to the stomach)

79.2 The Function and Role of Saliva

Chewing has been reported to stimulate the motor cortex of the cerebral cortex, and there exists a powerful neural network that links the periphery and the central nervous system. The increase in cerebral blood flow due to chewing and occlusion is effective in activating brain function and saliva secretion, and has been shown to promote metabolism throughout the body. Saliva contains a variety of important growth factors, bioactive substances, antimicrobial substances, immunoglobulins, etc., that are important for homeostasis of the organism. In addition, it has digestive, self-cleaning, neutralizing, antimicrobial, and mucosal protective functions, making the promotion of saliva secretion an indispensable issue for anti-aging medicine. Furthermore, it is useful as a test material for evaluating various hormones, stress substances, antioxidants, etc. Non-invasive tests using saliva are expected to be introduced in future aging tests, and the National Institutes of Health (NIH) in the United States has reported an information site called the Human Salivary Proteome Wiki (HSP Wiki) as a search tool for human salivary proteome analysis. It is expected to be used in the diagnosis of aging in the future [2]. The causes of dry mouth (oral dryness) are diverse, and most of them are due to aging-related diseases that are the target of anti-aging medicine, such as lifestyle-related diseases, menopausal disorders, and stress. Since many cases develop from daily lifestyle habits, it is important to propose a lifestyle that includes exercise and dietary guidance, and an approach that understands anti-aging medicine is necessary [3]. Aspiration pneumonia in the elderly is a leading cause of death, and it has been reported that the cause is due to bacteria in the mouth and pharynx. Aspiration pneumonia is a disease that occurs when food and saliva are allowed to flow into the lungs due to eating and swallowing disorders, and the body's immune defense mechanism does not work properly due to a decrease in overall immunity. It is said that many types of resident bacteria live in the oral cavity. It is clear that the environment in the oral cavity deteriorates due to a decrease in saliva secretion and resident bacteria proliferate, and it is possible to prevent aspiration pneumonia by working on oral care and sufficient saliva secretion as a bacterial exclusion mechanism. The new coronavirus infection that caused a global pandemic at the end of 2019 also tends to become severe in the elderly. It is predicted that not only immune function but also aging changes in oral function are affecting, and research and development including prevention are expected [4].

79.3 Anti-Aging from the Mouth

In Japan, which is a super-aged society, the "food" of the elderly is being re-recognized. The media is discussing the need for gastric fistulas, but it is important to correctly recognize whether the function of eating has really been lost, and whether appropriate evaluations and responses to swallowing function are being implemented. From functional evaluations such as swallowing endoscopy and swallowing radiography to guidance on oral care and swallowing training, collaboration with other departments is necessary. As an approach from the mouth, it is also necessary to actively incorporate dentures and tongue feeding aids that consider swallowing function. Not only functional tests and training, but also anti-aging to maintain the function of eating will become important. Obstructive sleep apnea syndrome (OSAS) is a disease in which breathing weakens or stops during sleep, the oxygen concentration in the body decreases, and sleep is disturbed. The prevalence in Japan is about 5% of the adult population, and the estimated number of patients is said to be over five million [5]. Sleep apnea occurs when the upper airway narrows or closes during sleep. The decrease in muscle tension involved in airway maintenance and the anatomical balance of the pharynx are involved in its occurrence. It is known that muscle tension decreases during sleep, and this decrease in muscle tension also occurs in muscles involved in airway maintenance, such as masticatory muscles and muscles around the pharynx. The decrease in muscle tension of the masticatory muscles eliminates the contact between the upper and lower teeth, resulting in a mild open state, and especially in the supine position, the lower jaw sinks towards the pharynx, narrowing the airway. In addition, the activity of the muscles around the pharynx involved in airway maintenance, especially the styloglossus muscle, decreases, and the airway narrows. When inhaling, the soft palate is drawn to the narrowed airway and vibrates, causing a snoring sound, and when the airway is completely blocked, it becomes apnea. Snoring is a sign that the airway is narrowing, and in severe snoring, it becomes a state of hypoventilation [5]. When apnea or hypoventilation occurs during sleep, it not only affects the respiratory system, but also accelerates aging throughout the body, including the circulatory system, needless to say. We would like to continue to pay attention to this as an important theme for anti-aging medicine from the perspective of the oral cavity.

References

1. Shinichi A. Decline and reactivation of eating and swallowing functions. Anti-Aging Med. 2015;11:221–9.
2. Lau WW, Hardt M, et al. The human salivary proteome wiki: a community-driven research platform. J Dent Res. 2021;100:1510–9.
3. Okuyoshi S, Ichiro S, et al. Special feature: anti-aging of oral function. Anti-Aging Med. 2015;11:187–234.
4. Usami Y, Hirose K, et al. Brief communication: immunohistochemical detection of ACE2 in human salivary gland. Oral Sci Int. 2021;18:101–4.
5. Okuyoshi S. Oral appliance treatment for sleep apnea syndrome by dentists for dentists. Tokyo: Ishiyaku Publishers; 2014.

Clinical Anti-Aging Medicine: Sarcopenia, Frailty, and Anti-Aging Medicine

80

Hiroaki Ohta

80.1 The Challenge of "Frailty" Towards a 100-Year Life

Amid growing attention to "frailty" that impairs the healthy lifespan of the elderly, the Ministry of Health, Labour and Welfare has decided to tackle the understanding of frailty in the medical check-up of the late-stage elderly medical system for those aged 75 and over, taking the fiscal year 2020, when the number of late-stage elderly surpasses the early-stage elderly for the first time, as an opportunity. "Frailty" is the ultimate medical and nursing concept that captures the multifaceted weakening due to aging, including physical factors, mental and psychological factors, and social factors, without missing signs of aging, supports the independence of the elderly in a multifaceted way, and revitalizes the elderly. Until now, it has been overlooked as an "aging phenomenon," but by using "frailty" as a unified name, we aimed to reform the awareness of medical care and nursing. In other words, frailty was defined as "a state where physiological reserve capacity decreases in old age, vulnerability to stress increases, and it is easy to fall into outcomes such as functional disability, need for care, death, etc." Not only life prognosis and disease measures for the elderly, but also dealing with "frailty" with life function as an endpoint through integrated understanding will be a new mission in geriatric medicine. Frailty is accelerated by malnutrition and sarcopenia, but the "Frailty Health Check Sheet" [1] consisting of 10 categories and 15 questions includes malnutrition, but it is not enough with just walking speed and falls for sarcopenia. It is thought that items related to sarcopenia, such as "difficulty in opening the cap of a pet bottle," which indicates a decrease in grip strength, are also necessary. It might be good to add checks that do not miss signs of aging due to sarcopenia, such as "stand up, walk" and "climb stairs," which are basic daily activities for the elderly, from the sarcopenia screening method SARC-F/SARC-Calf [2].

80.2 Clinical Prospects of Frailty Starting from Sarcopenia

As we age, the function of all our organs decreases, and one of these changes is in the skeletal muscles, which make up the largest proportion of our body weight. The amount of skeletal muscle peaks between the ages of 20 and 30, and gradually decreases in middle age, with the rate of decrease accelerating in old age. Factors that determine the decrease in skeletal muscle mass from middle age to old age include lifestyle habits such as nutrition and physical activity, as well as the impact of various diseases and

H. Ohta (✉)
Department of Obstetrics and Gynecology, Kawasaki Medical Center, Okayama, Japan
e-mail: ohta-h@feel.ocn.ne.jp

H. Yamada, Y. Naito (eds.), *Anti-Aging Medicine*, https://doi.org/10.1007/978-981-97-8514-8_80

genetic factors. On the other hand, the decrease in skeletal muscle mass can lead to fatigue from exercise and a corresponding decrease in appetite, which raises concerns about the impact on other organ functions. In the frailty cycle, the decrease in skeletal muscle mass not only affects physical activity, but also various metabolisms, and if proactive preventive interventions are not taken, the aging changes throughout the body will accelerate. That is, the decrease in skeletal muscle mass leads to a decrease in physical function, weight loss, fatigue, and a decrease in physical activity, leading to physical frailty. This physical frailty can affect cognitive function and social activity, increasing the risk of needing care and death, and thus affecting healthy life expectancy. In other words, it is extremely important to prevent the decrease in skeletal muscle mass from middle age in order to extend healthy life expectancy, and this is also of great significance in disease treatment. Therefore, the onset of sarcopenia, which is an age-related decrease in muscle mass, is the starting point for various frailties. Clinically, the presence of osteoporosis increases the risk of future sarcopenia, and the presence of sarcopenia also tends to increase the risk of osteoporosis and fractures. Furthermore, epidemiological studies [3] have shown that the presence of osteoporosis is somewhat involved in the onset of sarcopenia and frailty. Extending the healthy life expectancy of the elderly is attracting attention to drug research targeting muscle–bone interactions, and it is also expected that simultaneous exercise intervention for sarcopenia and osteoporosis will lead to exercise benefits for both muscles and bones.

80.3 Incidence of Frailty and Pre-Frailty

The "Frailty Health Check" is expected to be used in various situations, not only during specific health checks, but also in community care prevention and daily life support comprehensive projects, and medical institutions, with the aim of increasing interest in frailty in the elderly and promoting lifestyle improvements. This health check form has been created considering the evidence related to the elderly so far and the burden on the health business and respondents. It is a future task to evaluate the multifaceted frailty in a limited time on site, and to link the measures from the evaluation to the results. There is no established standard for evaluating frailty, but the "physical frailty" by Fried's physical function phenotype is usually used [4] (Table 80.1). Frailty is defined as applying to three or more of the five items, and 1 ~In item 2, it is pre-frail. A systematic review and meta-analysis of over 120,000 elderly people aged 60 and over in 28 countries, including Japan, quantified the risk of frailty in the elderly for the first time on a global scale [5]. As a result, the incidence per 1000 people per year was 43.4 people (4.3% annually) for frailty and 150.6 people (15.1% annually) for pre-frailty (Fig. 80.1). From this study, (1) The incidence of frailty was higher in low- and middle-income countries compared to high-income countries. (2) The risk of frailty in the elderly is a global issue. (3) It showed that it is an issue faced by countries where the aging of the population is progressing. (4) It pointed out that strengthening muscle training and increasing protein intake may be useful in preventing and suppressing the progression of frailty. (5) It made five recommendations to evaluate the risk of frailty through regular check-ups and intervene appropriately at an early stage.

Table 80.1 General criteria for evaluating frailty [4]

1.	Decreased physical activity
2.	Decreased grip strength
3.	Low energy state (easily tired)
4.	Decreased walking speed
5.	Unintentional weight loss

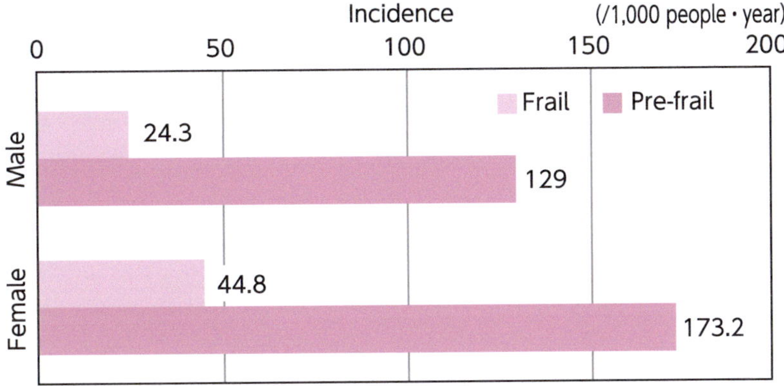

Fig. 80.1 Incidence of frailty and pre-frailty by gender [5]

References

1. Material for the 34th study on health checkups and health guidance by insurers. Ministry of Health, Labor and Welfare; 2019. https://www.mhlw.go.jp/stf/shingi2/_0000200938_00005.html.
2. Sarcopenia Clinical Practice Guide Creation Committee. Sarcopenia clinical practice guide. Tokyo: Japan Sarcopenia and Frailty Society; 2019.
3. Yoshimura N, Muraki S, et al. Do sarcopenia and/or osteoporosis increase the risk of frailty? A 4-year observation of the second and third ROAD study surveys. Osteoporos Int. 2018;29:2181–90.
4. Fried LP, Tangen CM, et al. Frailty in older adults: evidence for a phenotype. J Gerontol A Biol Sci Med Sci. 2001;56:146–56.
5. Ofori-Asenso R, Chin KL, et al. Global incidence of frailty and prefrailty among community-dwelling older adults: a systematic review and meta-analysis. JAMA Netw Open. 2019;2:e198398.

81 Clinical Anti-Aging Medicine: Anti-Aging Medicine for Female Reproductive Organs

Kuniaki Ota

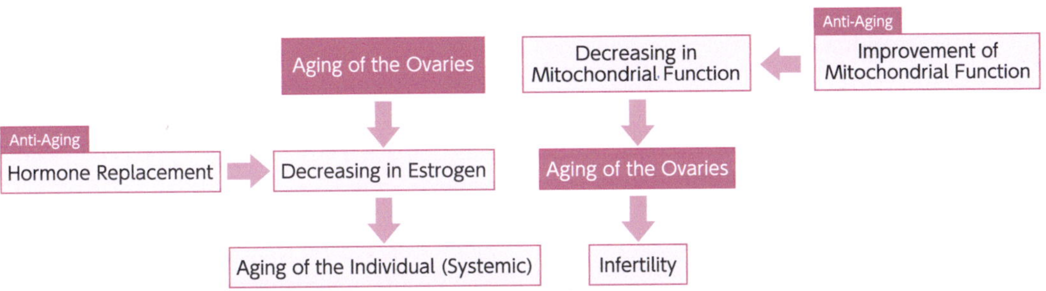

Fig. 81.1 Aging of the female reproductive organs: Two trends

It is known that the aging (senescence) speed of female reproductive organs is faster than other organs in the body. The aging of female reproductive organs (Fig. 81.1) exists in two aspects: (1) aging as an individual and (2) aging in terms of fertility. The former is aging due to estrogen deficiency, and the latter is aging of the egg due to age.

81.1 Aging as an Individual

81.1.1 Estrogen and Aging

The main estrogen-producing organs in women are the ovarian granulosa cells, but estrogen is also produced in the adrenal cortex and testes in men. However, while men's estrogen secretion does not change significantly throughout their lives, women's estrogen secretion peaks during puberty, but significantly decreases with age as the primordial follicles deplete. By the age of 55 after menopause, the blood estrogen levels in men and women become equivalent, and after the age of 60, women's estrogen levels are about half of men's (Fig. 81.2). Therefore, compared to men, estrogen deficiency directly has a significant impact on aging, and anti-aging (anti-senescence) by estrogen is practicable. The anti-aging effects of estrogen are not only seen in vasomotor symptoms (such as hot flashes) in menopause, but also in cardiovascular protection, prevention of osteoporosis, improvement of skin degenerative changes, and improvement of urogenital organs. All of these are assumed to work through estrogen receptors, but it is also assumed that the anti-aging effects of estrogen involve

K. Ota (✉)
Department of Obstetrics and Gynecology, Kawasaki Medical School, Okayama, Japan

H. Yamada, Y. Naito (eds.), *Anti-Aging Medicine*, https://doi.org/10.1007/978-981-97-8514-8_81

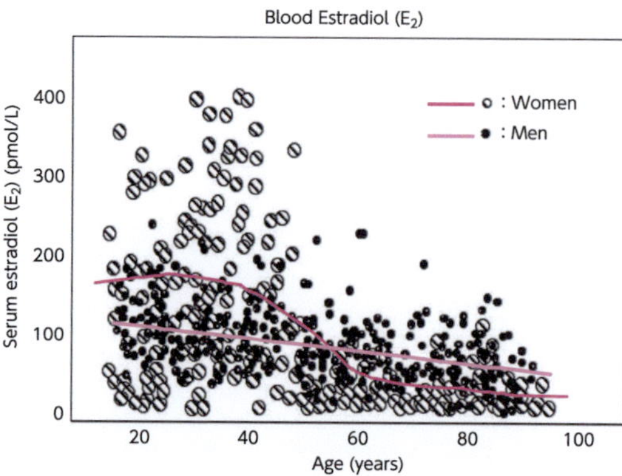

Fig. 81.2 Age-related changes in blood estrogen levels in men and women. (Modified with permission from [1])

antioxidant, anti-inflammatory, and telomerase activities.

81.1.2 Organ (1): Anti-Aging for the Musculoskeletal System

Treatment for postmenopausal osteoporosis is diversifying with the advent of biologics, but it is widely known that hormone replacement therapy containing estrogen can prevent bone loss [2]. In addition to postmenopausal estrogen, estrogen in skeletal muscles decreases with age, and it is suggested that a decrease in estrogen concentration is related to muscle mass reduction and fall risk. It has been shown that hormone replacement therapy in postmenopausal women can inhibit the decrease in skeletal muscle mass [3]. At present, hormone replacement therapy (HRT) is not recommended for anti-aging medicine to maintain skeletal muscle, but it is thought that the day is not far off when HRT will be used to prevent the increasing population of locomotive syndrome.

81.1.3 Organ (2): Anti-Aging for Lipid Metabolism (Atherosclerosis)

Estrogen, one of the female hormones, has multifaceted anti-atherosclerotic effects, and thus, the incidence of atherosclerotic diseases increases in women after menopause. The Framingham study in the United States has found a significant difference in the incidence of coronary artery disease between premenopausal women in their 40s and 50s and postmenopausal women [4]. Although atherosclerotic diseases such as coronary artery disease and cerebral infarction are rare in young women, they rapidly increase from the age of 50 and eventually reach the same level as men [5]. Therefore, hormone replacement therapy (HRT) for postmenopausal women is thought to reduce atherosclerotic diseases, and it has been reported that the onset of atherosclerotic diseases is less in women who have not undergone HRT. However, in 2002, both the Women's Health Initiative (WHI) [6], a study on the primary prevention effect of atherosclerotic diseases by HRT, and the Heart and Estrogen/Progestin Replacement Study (HERS) [7], a secondary prevention study, reported that they could not suppress the onset of the diseases. However, since the cohorts of these studies were elderly women (average age 63) and more than two-thirds were obese, it is questionable whether these results can be directly applied to Japanese women in their 50s who undergo HRT. On the other hand, other prospective cohort studies have reported that starting HRT immediately after menopause significantly reduces the incidence of heart disease [8]. However, at present, there is still no solid data to recommend HRT for improving lipid abnormalities and preventing atherosclerotic diseases as anti-aging medicine, and further examination of HRT is needed in the future.

81.1.4 Organ (3): Anti-aging for the Nervous System

Estrogen affects the nervous system in various ways through estrogen receptors in the brain. Specifically, estrogen increases the concentration of acetylcholine in cholinergic synapses and inhibits apoptosis associated with the neurotoxicity of glutamate and amyloid beta protein. It is known that β-protein (Aβ) deposition inhibits apoptosis, glycogen synthase kinase 3β activity, and tau phosphorylation [9]. Therefore, by performing Hormone Replacement Therapy (HRT), blood vessels are dilated through prostacyclin and nitric oxide of vascular endothelial cells, and cerebral blood flow increases, among other things, there are many reports suggesting the neuroprotective effects of estrogen [10]. Also, women's estrogen secretion decreases rapidly with menopause, which is thought to adversely affect brain function and increase the risk of developing Alzheimer's disease. Based on this hypothesis, the preventive effect of HRT on Alzheimer's disease has been verified so far, and it has been reported that it is desirable to start HRT in the menopausal transition to early postmenopause and use it for a long time to prevent Alzheimer's disease [11, 12]. On the other hand, in a meta-analysis of previous reports, short-term use of HRT could prevent dementia, but there was no preventive effect on the onset of Alzheimer's disease with long-term administration [13]. Therefore, the preventive effect of HRT on Alzheimer's disease may reduce the risk of onset, but there is not enough evidence, and there are many issues to be clarified such as the timing and duration of administration for clinical use, and accumulation of future data is essential.

81.1.5 Organ (4): Anti-Aging for the Skin

Skin aging is accelerated not only by external factors such as sunlight but also by internal factors such as aging. Furthermore, in women, skin abnormalities such as "wrinkles and spots have become noticeable," "severely dry," "feel an itch that has never been experienced," "cosmetics that were regularly used suddenly do not fit" appear as one of the symptoms of estrogen deficiency due to menopause. Estrogen is thought to have an effect of preventing skin deterioration by promoting the production of collagen and elastin that give skin elasticity and lipid components that are one of the moisturizing components, and maintaining the amount of water [14]. In fact, when estrogen is administered to postmenopausal women, it has been confirmed that there are rejuvenating effects not only from various menopausal disorders but also from various skin indicators [15].

81.1.6 Organ (5): Anti-Aging for the Genitals

The vagina has estrogen receptors, and when estrogen action is insufficient, keratinization of the skin does not occur normally, so vaginal atrophy begins, it dries, and it becomes a state without resistance, causing atrophic vaginitis and senile vaginitis from histological mucosal thinning [16]. For the same disease, the North American Menopause Society (NAMS) and the International Society for the Study of Women's Sexual Health (ISSWSH) in 2014 included histological and functional disorders of the vagina and vulva as a disease concept called genitourinary syndrome of The concept of Genitourinary Syndrome of Menopause (GSM) was proposed by Palacios et al. [17] In Japan, the Japan Society of Obstetrics and Gynecology adopted the official Japanese translation of GSM as "menopausal genitourinary syndrome" in 2019, and the concept of the disease is gradually being established. As mentioned earlier, the symptoms are due to a lack of estrogen, and the main treatment is local administration of estrogen to the vagina. However, oral administration of ospemifene, a selective estrogen receptor modulator approved by the U.S. Food and Drug Administration (FDA), and treatment with fractional carbon dioxide laser or erbium YAG laser are also practiced, but in Japan, they are performed within the scope of private practice because they are not covered by insurance. In the context of improving women's quality of life as a right, it can be considered as anti-aging medicine that is attracting attention. From these facts, it is necessary to accumulate more evidence to reach a consensus on the anti-aging effects of estrogen, and it is thought that it will take time to reach a conclusion. On the other hand, Salpeter et al. used Bayesian analysis to determine whether HRT ultimately contributes to a reduction in individual mortality, and found from 19 randomized trials

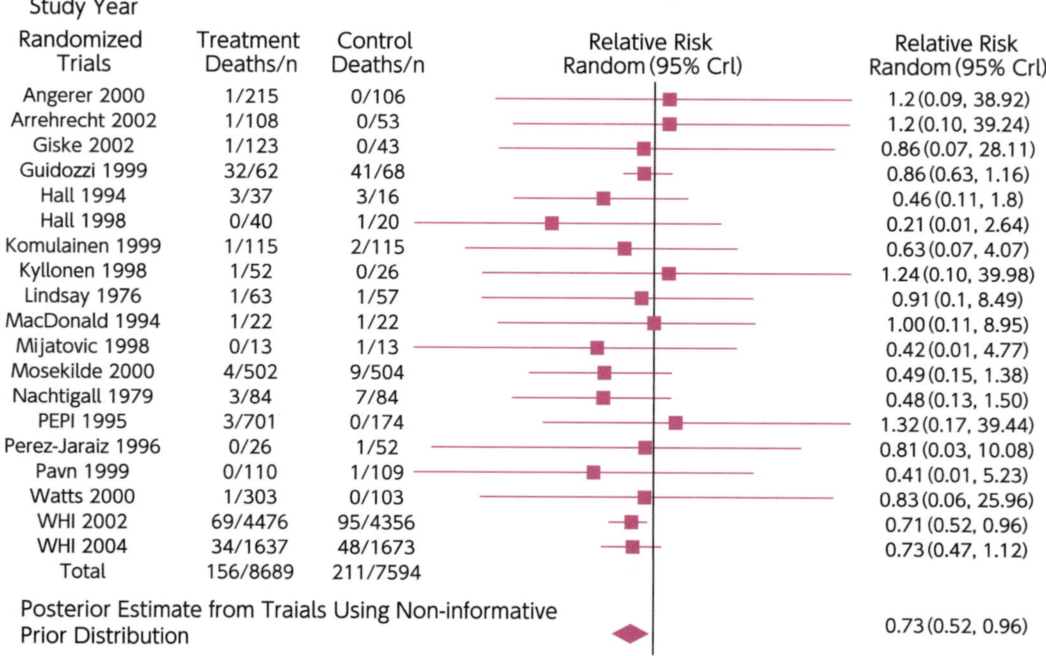

Fig. 81.3 Meta-analysis of mortality related to HRT [18]

following 16,000 women (average age 55) for 83,000 person-years that the mortality rate for HRT users was a relative risk of 0.73 (95% confidence interval 0.52–0.96) [18] (Fig. 81.3). This suggests that HRT, centered on estrogen, is indeed a comprehensive anti-aging medicine for women. In the future, we hope to see evidence of the anti-aging effects of HRT on various organs as introduced in this section.

81.2 Anti-Aging as Fertility

Women's fertility decreases with age [19]. In treatments using assisted reproductive technology (ART) with donor eggs from young women, the production rate decreased even as the recipient's age increased, so the main cause of decreased fertility is "decreased egg quality," or "egg aging." The cause of egg aging is thought to be a decrease in mitochondrial function, and anti-aging of fertility requires an attempt to improve mitochondrial function. In basic research, there have been reports of successful anti-aging of eggs with mitochondrial transplantation and vitamin-like substances involved in lipid metabolism such as L-carnitine and coenzyme Q10 (CoQ10), which are essential factors in the electron transport system in mitochondria, but in humans, none of these attempts have been established. Therefore, at present, there are no established treatments or preventive measures for egg anti-aging, and realistic measures such as freezing one's own young eggs before aging and the development of anti-aging medical alternatives using artificial ovaries or egg stem cells as regenerative medicine are expected [20].

References

1. Khosla S, Melton LJ 3rd, et al. Relationship of serum sex steroid levels and bone turnover markers with bone mineral density in men and women: a key role for bioavailable estrogen. J Clin Endocrinol Metab. 1998;83:2266–74.
2. Nash Z, Al-Wattar BH, et al. Bone and heart health in menopause. Best Pract Res Clin Obstet Gynaecol. 2022;81:61–8.
3. Tiidus PM, Lowe DA, et al. Estrogen replacement and skeletal muscle: mechanisms and population health. J Appl Physiol. 2013;115:569–78.

4. Kannel WB, Hjortland MC, et al. Menopause and risk of cardiovascular disease: the Framingham study. Ann Intern Med. 1976;85:447–52.
5. Kannel WB, Wilson PW. Risk factors that attenuate the female coronary disease advantage. Arch Intern Med. 1995;155:57–61.
6. Rossouw JE, Anderson GL, et al. Risks and benefits of estrogen plus progestin in healthy postmenopausal women: principal results from the Women's Health Initiative randomized controlled trial. JAMA. 2002;288:321–33.
7. Grady D, Herrington D, et al. Cardiovascular disease outcomes during 6.8 years of hormone therapy: heart and estrogen/progestin replacement study follow-up (HERS II). JAMA. 2002;288:49–57.
8. Schierbeck LL, Rejnmark L, et al. Effect of hormone replacement therapy on cardiovascular events in recently postmenopausal women: randomised trial. BMJ. 2012;345:e6409.
9. Manly JJ, Merchant CA, et al. Endogenous estrogen levels and Alzheimer's disease among postmenopausal women. Neurology. 2000;54:833–7.
10. Ohkura T, Teshima Y, et al. Estrogen increases cerebral and cerebellar blood flows in postmenopausal women. Menopause. 1995;2:13–8.
11. Zandi PP, Carlson MC, et al. Hormone replacement therapy and incidence of Alzheimer disease in older women: the Cache County study. JAMA. 2002;288:2123–9.
12. Henderson VW, Benke KS, et al. Postmenopausal hormone therapy and Alzheimer's disease risk: interaction with age. J Neurol Neurosurg Psychiatry. 2005;76:103–5.
13. Hogervorst E, Bandelow S. Sex steroids to maintain cognitive function in women after the menopause: a meta-analyses of treatment trials. Maturitas. 2010;66:56–71.
14. Brincat MP, Baron YM, et al. Estrogens and the skin. Climacteric. 2005;8:110–23.
15. Hall G, Phillips TJ. Estrogen and skin: the effects of estrogen, menopause, and hormone replacement therapy on the skin. J Am Acad Dermatol. 2005;53:555–68; quiz 69–72.
16. Portman DJ, Gass MLS. Genitourinary syndrome of menopause: new terminology for vulvovaginal atrophy from the International Society for the Study of Women's Sexual Health and the North American Menopause Society. Menopause. 2014;21:1063–8.
17. Palacios S, Mejía A, et al. Treatment of the genitourinary syndrome of menopause. Climacteric. 2015;18(Suppl 1):23–9.
18. Salpeter SR, Cheng J, et al. Bayesian meta-analysis of hormone therapy and mortality in younger postmenopausal women. Am J Med. 2009;122:1016–22.e1.
19. Wood JW. Fecundity and natural fertility in humans. Oxf Rev Reprod Biol. 1989;11:61–109.
20. Fisch B, Abir R. Female fertility preservation: past, present and future. Reproduction. 2018;156:F11–27.

Part IV

Anti-ageing Practice and Diagnostics

Let's Start Anti-Aging Medical Practice

82

Ko Tanaka

82.1 The Treatment Goals of Our Clinic that Has Incorporated Anti-Aging Medical Practice

Our clinic is a regular insurance medical institution, and we conduct medical treatment incorporating anti-aging concepts within our regular practice. Of course, as part of our free medical services, we also offer various anti-aging checkups, limited intravenous therapies, sales of supplements, etc. In this article, we would like to present the current situation of our clinic and hope it will serve as a gateway for many doctors to enter the field of anti-aging medicine [1]. Our clinic is located in a residential area of Shizuoka City and is a specialized insurance medical institution for gastrointestinal diseases, conducting about 7000 endoscopic examinations annually. Currently, not only patients with gastrointestinal diseases but also many patients with lifestyle-related diseases, including dementia, visit our clinic. The ultimate needs of patients are cancer prevention, adult disease prevention, and further, dementia prevention. The basics of lifestyle, nutrition, and exercise therapy for these diseases are common and are essentially anti-aging medicine. Even if we tell patients that "doing this will prevent disease" in their daily lives, they often don't listen. However, when we explain that "doing this will lead to anti-aging," they become motivated. In this way, the term "anti-aging" itself has a "charisma," and we are convinced that promoting anti-aging medical practice will lead to the prevention of all diseases. As we approach a longevity society, we have listed our clinic's treatment goals in Table 82.1.

K. Tanaka (✉)
Tanaka G.I. Clinic, Shizuoka, Japan
e-mail: tanaka-c@mtj.biglobe.ne.jp

Table 82.1 Treatment goals for a longevity society

Extend healthy life expectancy and end life in a lively and sudden manner
Reduce the period of care to the limit
Do not think, say, or let others say you are old until you are 90
Aim for holistic medical care unique to each patient
Detect the buds of disease at a very early stage and intervene appropriately

82.2 Basic Concept of Anti-Aging Medicine

The basics of anti-aging medicine are summarized in nutritional therapy and exercise therapy. Furthermore, it's wonderful when high-quality sleep and positive, bright thinking are added. We would like to introduce the basics of our patient guidance in line with anti-aging in Table 82.2. When giving lifestyle guidance, it is doubtful how effective it is to instruct abstractly to "have a balanced diet and exercise." After deepening the understanding of the patient's lifestyle, it is desirable to provide as concrete guidance as pos-

H. Yamada, Y. Naito (eds.), *Anti-Aging Medicine*, https://doi.org/10.1007/978-981-97-8514-8_82

sible that can be implemented and continued. It goes without saying that encouraging a regular, nutritionally balanced diet is the ideal form of nutritional guidance, but for various reasons such as work, it is often difficult to continue an ideal diet. In the current general dietary lifestyle, while carbohydrates, proteins, and fats are partially excessively ingested, there is a tendency for micronutrients such as vitamins and minerals to be generally deficient. Also, considering that the amount of micronutrients contained in food has decreased compared to half a century ago, the use of supplements (nutritional supplements) can be an option as a realistic measure in terms of nutrition. Figure 82.1 shows the nutrient intake compared to the dietary intake standards of Japanese in their 20s. The basic principles of our clinic's nutritional guidance are Table 82.3 is shown. We recommend eating rice, especially brown rice, as a staple food as much as possible. We advise to refrain from wheat-based foods, namely bread, ramen, soba, udon, pasta, etc. Wheat has many problems, such as allergies and excessive gluten due to genetic manipulation, and we want to cherish rice, which is a traditional culture of Japan. On the other hand, exercise therapy is essential for anti-aging medicine. As we enter an aging society, many people lament that they cannot exercise enough, but on the other hand, a 95-year-old woman who is an active tennis player visits our clinic. She still goes to the tennis club three times a week and enjoys playing for 1–2 h. She has won twice in the Grand Master division (85 years and older) of the Japan Tennis Association. Her lumbar spine and femur bone density maintain a level of

Table 82.2 Basics of anti-aging guidance

1.	Understand that work, play, and alcohol should be limited to 80% fullness for those over 40
2.	Absolutely no smoking!!
3.	Reduce carbohydrates!! From gourmet to simple meals
4.	Increase muscle!! 40 min of walking to sweat every day
5.	7 h of sound sleep per day
6.	Always have a positive smile

Table 82.3 Basic principles of nutritional guidance

The basic intake calories are standard weight × 30–35 kcal
Carbohydrates should be meals with a low glycemic index
Brown rice is full of nutrients compared to white rice
The first step in carbohydrate restriction is from 40% carbohydrates
Lipids are obtained from fish, such as EPA and DHA
Proteins are obtained from soybeans, natto, tofu, fish, and a little bit of meat at the end
People over 75 should eat meat to prevent locomotive syndrome
Vegetables, especially root vegetables, are full of fiber and help prevent cancer with good bowel movements
Intake of nutrients that prevent oxidation
Natural foods should be eaten in their seasonal prime
It's better to limit salt intake, as cancer loves salt

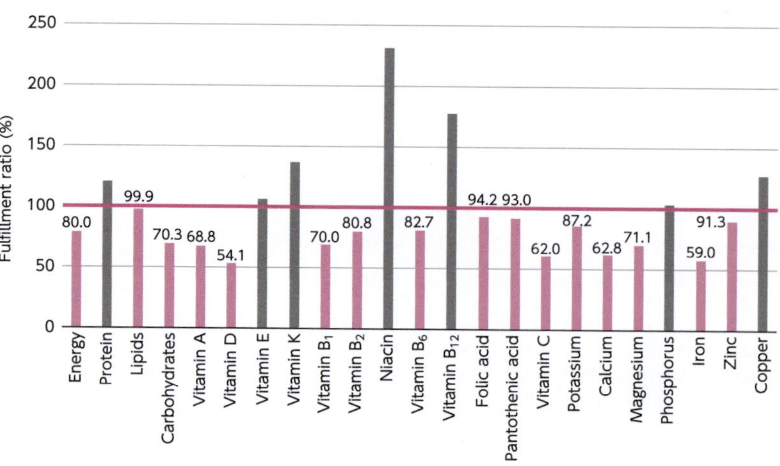

Fig. 82.1 Nutrient intake compared to dietary intake standards for women aged 20–29. (Reprinted with permission from [3])

over 100% compared to young people. As a side note, before the Corona disaster, she was so energetic that she booked her own plane tickets and traveled alone all over Japan. We explain her current situation to the elderly who visit our clinic, and tell them that being old is not the cause of decreased bone density and muscle strength, but the problem is the feeling of neglecting effort because they think they are old. A 15-min brisk walk every day can extend your healthy life expectancy by 3 years [2]. Exercise therapy can also reduce the mortality rate of cardiovascular diseases, cancer mortality rate, and the incidence rate of dementia. The exercise time that increases the healthy life expectancy the most is considered to be 90 min of walking every day and 40 min of jogging. However, in the case of the elderly, many people have individual diseases such as low back pain and knee pain and cannot exercise enough. Even in such cases, we recommend exercises such as stepping, muscle training using pet bottles, and squats, depending on each person's situation. On the other hand, for patients who are energetic but say they don't have time, we recommend walking during work, climbing stairs, and squats. In this way, it is important to incorporate exercise therapy into daily life according to the patient's specific situation.

82.3 About Supplements

In our clinic's survey, many patients are taking supplements voluntarily, and about 80% of them are taking supplements without understanding how they benefit them. This is judged to be a reflection of the fact that the market is flooded with many products and consumers are confused by irresponsible information. Consumers tend to choose products with attractive catchphrases, and often take special supplements without sufficient consideration for the five major nutrients that should be prioritized. General practitioners should provide guidance to avoid unnecessary supplements and actively provide appropriate information so that they can choose supplements with clear and reliable evidence. In providing nutritional guidance, it is correct to promote a balanced diet of the three major nutrients and to recommend the intake of so-called multivitamins and minerals, which are properly combined with trace nutrients, as the first choice for necessary patients. It is also effective to add antioxidant nutrients (beta-carotene, vitamin E, vitamin C, selenium, coenzyme Q10, catechins, etc.) for patients whose oxidative stress is increasing as a result of medical checkups [4]. The principles of supplement administration in our clinic are presented in Table 82.4. We have introduced our anti-aging medical care, and we hope that many doctors will join us in this effort. Each clinic's unique approach is desired, and each unique approach is important [5].

Table 82.4 Principles of supplement administration

Choose the appropriate supplement from the anti-aging dock results
Multivitamins and minerals for patients with poor health, patients with a lot of stress
Multivitamins and minerals for malnutrition due to gastrectomy, surgery, etc.
Heme iron for those with iron deficiency anemia who have difficulty taking inorganic iron
For early dementia, omega-3 fatty acids, vitamins, water-soluble choline, etc.
Choose the appropriate supplement according to individual circumstances

References

1. Takashi T, Yoshiko N, et al. Anti-aging medical practice for general practitioners (special feature: anti-aging and lifestyle). Clin Pract. 2007;26:551–7.
2. Wen CP, Wai JP, et al. Minimum amount of physical activity for reduced mortality and extended life expectancy: a prospective cohort study. Lancet. 2011;378:1244–53.
3. Report of the National Health and Nutrition Examination Survey of Japan. Dietary reference intakes for Japanese (2020 Edition). Healthy Pass Inc.
4. Tadashi T. If you eat this, you don't need supplements. Tokyo: Toyo Keizai Shinposha; 2015.
5. Imura Y, Editor G. Medical care to open up Japan's future: from therapeutic medicine to preventive medicine. Tokyo: Diagnosis and Treatment Company; 2012.

Physical Activity and Anti-Aging: Prevention of Metabolic and Cardiovascular Diseases Through Exercise

83

Yutaka Kimura

Lack of exercise is recognized as a clear factor in the risk of metabolic and cardiovascular diseases. Recently, not only lack of exercise but also an increase in sedentary time has been reported to be associated with obesity-related metabolic and cardiovascular diseases [1]. Therefore, not only conscious physical activity in daily life but also physical activity through daily activities such as housework, shopping, commuting, etc., should be actively considered as a broad definition of exercise. In addition to traditional low-intensity activities such as walking, moderate- to high-intensity exercise has also been recognized as beneficial for the prevention of metabolic and cardiovascular diseases due to increases in type II fiber in peripheral muscle tissue, increases in capillary beds, and increases in peripheral circulation blood volume [2]. The effects of exercise in the metabolic and cardiovascular areas, such as hemoglobin A1c (HbA1c), HDL cholesterol, blood pressure values, and specific numbers for weight loss in obesity cases, have also been calculated by meta-analysis [3] (Table 83.1). The lowering effect of exercise on resting systolic/diastolic blood pressure is 3.4/2.5 mmHg, and it is said that about 60–70% have a blood pressure-lowering effect with exercise therapy. This effect is generally comparable to the effect of the first dose of a blood pressure-lowering diuretic. Regarding blood sugar, a 0.8% reduction in HbA1c with 30 min of moderate exercise twice or more per week, and even greater effects with the addition of resistance exercise, have been reported. These numbers are the effects of short-term exercise, and it is difficult to calculate the numbers for long-term exercise continuation, and epidemiological analysis is considered useful for long-term exercise effects. These effects can potentially be obtained simultaneously through exercise, which is interesting. The mechanism of the blood pressure-lowering effect of exercise can also be compared with drugs. In addition to the effect as a blood pressure-lowering diuretic, the inhibition of sympathetic nerve function can be seen as the action of a β-blocker, and the improvement of endothelial function can be seen as part of the effect of Ca antagonists or ACE inhibitors, ARB drugs. Various mechanisms are considered for the specific mechanism of the

Table 83.1 Effects of appropriate exercise obtained by meta-analysis [3]

Target disease	Item	Improvement
Diabetes	A1c	0.80%
Dyslipidemia	HDL	2.5 mg/dL
Hypertension	Blood pressure value (systolic/diastolic)	3.4/2.4 mmHg
Obesity	Body weight	6.7 kg/year

Y. Kimura (✉)
Health Science Center, Kansai Medical University, Osaka, Japan
e-mail: kimuray@hirakata.kmu.ac.jp; nishigur@hirakata.kmu.ac.jp

H. Yamada, Y. Naito (eds.), *Anti-Aging Medicine*, https://doi.org/10.1007/978-981-97-8514-8_83

exercise effect (Fig. 83.1). Recently, static stretching as a vascular stretch has been reported to activate vascular endothelial function and show anti-atherosclerotic effects [4]. These reports indicate the usefulness of stretching as a new exercise therapy in the cardiovascular area and atherosclerosis, different from traditional aerobic exercise and resistance training. The improvement of joint and tendon mobility by stretching is thought to contribute to the increase in exercise volume, and it is necessary to actively perceive flexibility and stretching as the third exercise. Skeletal muscles secrete hormones and cytokines collectively referred to as myokines, which have been reported to affect insulin sensitivity and various metabolisms, indicating the importance of skeletal muscles in metabolic diseases [5] (Fig. 83.2). Irisin, which was initially noticed, is related to fat metabolism, obesity, and atherosclerosis. Brain-derived neurotropic factor (BDNF) is said to be involved in neural function and cognitive function, and is also useful for preventing dementia in the elderly. Recently, myonectin, which is involved in the suppression of skeletal muscle metabolism, has also been noticed. With the elucidation of these myokines, the importance of resistance training as a stimulus to the skeletal muscle system along with traditional aerobic exercise is suggested. Also, resistance exercise as a muscle stimulus without skeletal muscle hypertrophy can be expected to have sufficient effects on metabolic diseases. Furthermore, skeletal muscle mass (strength) itself has been reported to be useful for the pre-

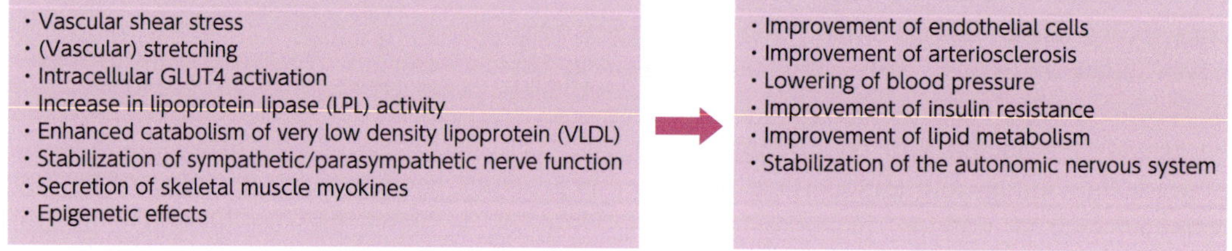

Fig. 83.1 Mechanism and its effects of exercise

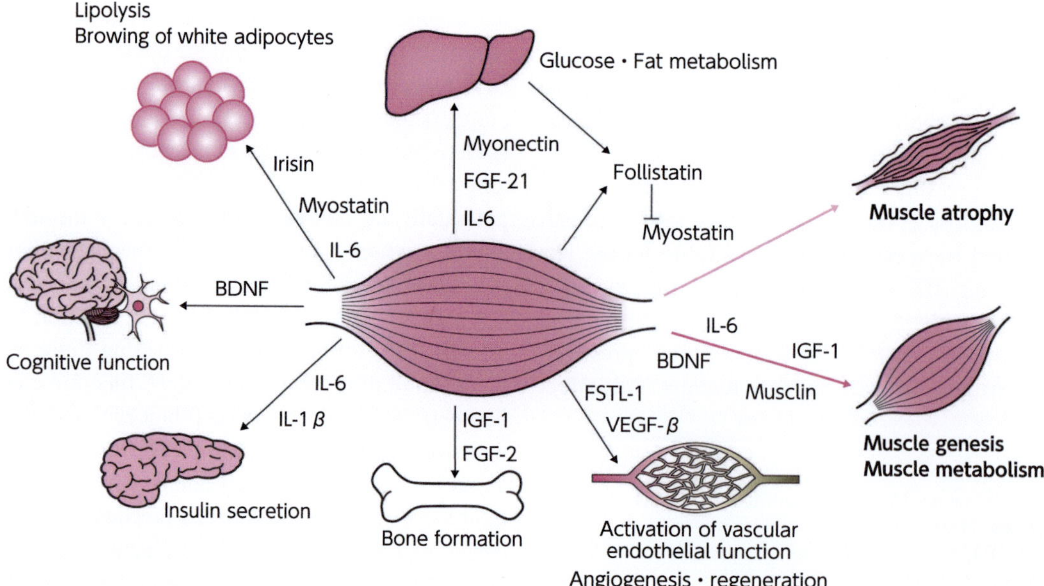

Fig. 83.2 Various effects of myokines [6]

vention and improvement of lifestyle-related diseases. Therefore, when considering resistance exercise, it is necessary to broadly perceive it from high-load intensity aimed at increasing muscle mass and strength to low intensity aimed only at muscle stimulation. Even 5 min of exercise can have an effect, so it is important to actively exercise even for a short time. Regarding exercise frequency, there are reports that even once a week is effective for primary prevention of cardiovascular diseases and other mortality rates. Considering the effects of the aforementioned myokines, if you only look at the simple energy consumption in muscles due to exercise, the diet effect of exercise is small, but if you consider the metabolic effects of myokines other than energy consumption, exercise can be more useful for dieting than imagined.

References

1. Bull FC, Al-Ansari SS, et al. World Health Organization 2020 guidelines on physical activity and sedentary behavior. Br J Sports Med. 2020;54:1451–62.
2. Zheng C, Liu Z. Vascular function, insulin action, and exercise: an intricate interplay. Trends Endocrinol Metab. 2015;26:297–304.
3. Metkus TS Jr, Baughman KL, et al. Exercise prescription and primary prevention of cardiovascular disease. Circulation. 2010;121:2601–4.
4. Kruse NT, Scheuermann BW. Cardiovascular responses to skeletal muscle stretching: "stretching" the truth or a new exercise paradigm for cardiovascular medicine? Sports Med. 2017;47:2507–20.
5. Hoffmann C, Weigert C. Skeletal muscle as an endocrine organ: the role of myokines in exercise adaptations. Cold Spring Harb Perspect Med. 2017;7:a029793.
6. Pedersen BK. Compr Physiol. 2013; Demntis F, et al. Aging Cell. 2013; Ost M, et al. Free Radical Biol. 2016.

84 Physical Activity and Anti-Aging: Cancer Prevention Through Exercise

Wataru Aoi

According to the 2020 Population Dynamics Statistics [1], the proportion of deaths due to malignant neoplasms (cancer) is 27.6%, making it the leading cause of death since 1981. When looking at the breakdown of cancer by organ, lung cancer, stomach cancer, and colorectal cancer are the most common in men, while colorectal cancer, lung cancer, and pancreatic cancer are the most common in women. The mortality and incidence rates of several organ cancers, including colorectal cancer, are still increasing year by year. Normal cells can become cancerous due to a complex interplay of genetic factors, environmental factors, lifestyle habits, and aging. However, this mechanism is extremely complex and not yet fully understood, limiting the development and establishment of treatment methods. It is important to improve lifestyle habits and actively prevent cancer to reduce the risk of cancer.

W. Aoi (✉)
Laboratory of Nutrition Science, Graduate School of Life and Environmental Sciences, Kyoto Prefectural University, Kyoto, Japan
e-mail: waoi@kpu.ac.jp

84.1 The Cancer Prevention Effects of Exercise in Epidemiological Studies

Regular exercise is well known to prevent obesity and insulin resistance, contributing to the prevention of metabolic and cardiovascular diseases. Moreover, exercise and physical activity are supported by many epidemiological studies as lifestyle habits that reduce the risk of cancer. The "Physical Activity Guidelines for Health Promotion 2013" also includes the concept of physical activity and exercise for cancer prevention. According to the report "Diet, Nutrition, Physical Activity and Cancer: a Global Perspective" [2] compiled by the World Cancer Research Fund/American Institute for Cancer Research, cancers that can be prevented by physical activity include colorectal cancer, breast cancer, uterine cancer, esophageal cancer, lung cancer, and liver cancer (Table 84.1). Among these, physical activity was the only lifestyle habit that was definitively effective in reducing the risk of colorectal (especially colon) cancer (Table 84.1). In a prospective cohort study of Japanese people, it was observed that the incidence of colorectal cancer was lower depending on the amount of physical activity [3]. The cancer prevention effect of physical activity and exercise depends on the amount of physical activity, and both occupational physical activity and leisure-time physical activity are effective. There is also

H. Yamada, Y. Naito (eds.), *Anti-Aging Medicine*, https://doi.org/10.1007/978-981-97-8514-8_84

Table 84.1 Cancer prevention effects of exercise [2]

Judgment	Factors that reduce the risk of colorectal cancer	Cancer sites where risk is reduced by exercise
Certain	Physical activity	Colon
Highly likely	Whole grain fiber dairy products calcium	Breast (pre/post menopause) endometrium
Possible	Vitamin C-containing foods fish vitamin D multivitamins	Esophagus lung liver

a suggested relationship with exercise intensity, for example, people who walk at a fast pace or occasionally engage in high-intensity sports have a lower risk. Furthermore, the effect varies depending on the frequency of exercise and various exercise conditions.

84.2 The Mechanism of Cancer Prevention Through Exercise

The cancer prevention effects of physical activity and exercise are suggested to involve several mechanisms [4]. The decline in immune function contributes to cellular carcinogenesis, but habitual exercise revitalizes the immune function. In particular, it has long been pointed out that natural killer cells are activated by exercise. Also, excessive oxidative stress causes DNA mutations and promotes cellular carcinogenesis. However, exercise reduces oxidative stress by inducing antioxidant enzymes not only in skeletal muscles but also in visceral tissues, the nervous system, and circulation. Other suggested effects include anti-inflammatory actions through cytokine secretion regulation, hormone secretion regulation, improvement of energy metabolism to suppress excessive insulin secretion, improvement of bile acid quantity and composition, and improvement of bowel movements through activation of intestinal peristalsis. These factors are thought to work together to prevent cancer. It has been found that myokines secreted from skeletal muscles mediate the effects of exercise. Myokines are attracting attention as they explain changes in physiological functions and adaptive phenomena during exercise [5]. So far, several reports have been made, mainly on those that regulate energy metabolism, and they have been studied in relation to the supply of energy substrates during exercise and the regulation of metabolic functions in peripheral tissues. In addition, it has been suggested that there are myokines that affect brain function, pancreatic insulin secretion ability, and bone metabolism, and their functions are gradually becoming clear. Furthermore, it has been found that there are myokines that suppress tumor formation and contribute to cancer prevention. Secreted protein acidic and rich in cysteine (SPARC) is a type of matrix cell protein that exists around cells and plays an important role in maintaining and developing various tissue morphologies. In SPARC-deficient mice, phenotypes such as cataracts, bone weakening, skin disorders, and fat accumulation are observed, indicating that it is a multifunctional protein. It has been suggested that SPARC, which is secreted from skeletal muscles into the blood by exercise, acts on colon cells and leads cancer buds to apoptosis. Also, it has also been shown that myokine oncostatin M contributes to the inhibition of breast cancer cell proliferation. It is unclear whether these myokines can have the same effect on organ cancers other than colorectal and breast cancer, but it can be said that they support the evidence that exercise is effective in cancer prevention. Cancer cells secrete various inflammatory cytokines, induce cachexia, and cause various systemic symptoms. In particular, in skeletal muscles, protein degradation signaling systems, including the ubiquitin–proteasome system, are enhanced, causing muscle atrophy. In addition, loss of appetite often occurs, resulting in insufficient energy and protein intake from meals, often leading to malnutrition. Therefore, as cancer progresses, physical activity and quality of life decrease, and endurance to surgery, chemotherapy, and radiation therapy also decreases, narrowing the treatment options. In contrast, it is believed that maintaining physical strength by incorporating appropriate exercise therapy during cancer treatment is effective.

References

1. Table 6 Number of deaths, mortality rate (per 100,000 population), and proportion by cause of death rank (up to 10th) by gender. Overview of vital statistics (final figures) for 2020. Ministry of Health, Labour and Welfare; 2022. https://www.mhlw.go.jp/toukei/saikin/hw/jinkou/kakutei20/dl/10_h6.pdf.
2. World Cancer Research Fund, American Institute for Cancer Research. Diet, nutrition, physical activity and cancer: a global perspective. 2018.
3. Lee KJ, Inoue M, et al. Physical activity and risk of colorectal cancer in Japanese men and women: the Japan Public Health Center-based prospective study. Cancer Causes Control. 2007;18:199–209.
4. Rogers CJ, Colbert LH, et al. Physical activity and cancer prevention: pathways and targets for intervention. Sports Med. 2008;38:271–96.
5. Aoi W. Characteristics of skeletal muscle as a secretory organ. The plasticity of skeletal muscle: from molecular mechanism to clinical applications. Berlin: Springer; 2017. p. 195–210.

Physical Activity and Anti-Aging: Prevention of Sarcopenia and Frailty Through Exercise

85

Sumito Ogawa

The usefulness of exercise has been demonstrated in many clinical studies aimed at improving the physical function of the elderly. In particular, sufficient resistance exercise (strength training) is considered necessary to increase the skeletal muscle mass of the elderly. Regarding the exercise effect on sarcopenia and frailty, the "Sarcopenia Treatment Guidelines 2017 Edition" [1] recognizes the improvement effect of exercise, but its evidence level is low, and the recommendation level is not necessarily high. On the other hand, in the "Frailty Treatment Guidelines 2018 Edition" [2], exercise is effective in preventing the onset and progression of frailty, and both its evidence level and recommendation level are high.

85.1 Exercise for the Prevention of Sarcopenia

In the systematic review conducted in the "Sarcopenia Treatment Guidelines 2017 Edition," a cross-sectional study (case-control study) targeting elderly Japanese people found a correlation between midlife exercise habits and the onset of sarcopenia. In other cross-sectional studies, a positive relationship was found between physical activity and skeletal muscle mass. In a 10-year observational study of 3608 Japanese people, it was found that the more steps and physical activity, the lower the onset of sarcopenia. In addition, a systematic review of randomized controlled trials (RCTs) diagnosing sarcopenia based on decreased skeletal muscle mass and decreased muscle strength or physical function related to exercise therapy for sarcopenia has been reported. In the three trials extracted in the systematic review, a comprehensive exercise training program that added balance exercises and walking to resistance exercises was implemented for 60 min once, twice a week, for 3 months, and as a result, significant improvements were observed in limb skeletal muscle mass, normal walking speed, and knee extension strength. In the systematic review, it cannot necessarily be said that the subjects who met the latest sarcopenia diagnostic criteria [3–5] were selected at the time of intervention, and the evidence level is low. In the statement of the above treatment guidelines, the evidence level of exercise for the prevention and suppression of the onset of sarcopenia is "low," and the exercise therapy as a treatment for sarcopenia is "very low." In addition, the recommendation level of exercise for the prevention and suppression of sarcopenia is "strong," and the recommendation level of exercise therapy as a treatment for sarcopenia is "weak" [1] (see Table 85.1).

S. Ogawa (✉)
Department of Geriatric Medicine, The University of Tokyo, Tokyo, Japan
e-mail: suogawa@m.u-tokyo.ac.jp

H. Yamada, Y. Naito (eds.), *Anti-Aging Medicine*, https://doi.org/10.1007/978-981-97-8514-8_85

Table 85.1 Evidence and recommendation levels in the "Sarcopenia Medical Guide 2017 Edition"

CQ Can exercise prevent or suppress the onset of sarcopenia?
Statement Regular exercise and abundant physical activity may prevent the onset of sarcopenia, and an active lifestyle is recommended (Evidence level: Low, Recommendation level: Strong)
CQ Is exercise therapy effective as a treatment for sarcopenia?
Statement Exercise intervention for people with sarcopenia has the effect of improving limb skeletal muscle mass, knee extension strength, normal walking speed, and maximum walking speed, and is recommended (Evidence level: Very low, Recommendation level: Weak)

Table 85.2 Evidence and recommendation levels in the "Frailty Medical Guide 2018 Edition"

CQ Is exercise intervention effective in preventing the onset and progression of frailty?
Statement Exercise intervention for frailty is recommended as it can improve walking, muscle strength, physical mobility, and daily activity levels, and potentially prevent the progression of frailty (Evidence level: 1, Recommendation level: A)
CQ What kind of exercise is recommended to prevent the onset and progression of frailty?
Statement • As an exercise program to prevent the onset and progression of frailty, a multifactorial exercise program combining resistance exercise, balance training, and functional training is recommended (Evidence level: 1+, Recommendation level: A) • The exercise program is recommended to be of moderate to high intensity, with a gradual increase in exercise intensity (Evidence level: 1+, Recommendation level: A)

85.2 Exercise for the Prevention of Frailty

In the "Frailty Medical Guide 2018 Edition" [2], exercise intervention for the prevention of the onset and progression of frailty is recommended as it can improve walking, muscle strength, physical mobility, and daily activity levels, and potentially prevent the progression of frailty (Evidence level: 1, Recommendation level: A). According to a meta-analysis of RCTs investigating the effects of exercise intervention on physical ability in frailty, improvements were observed in walking and the short physical performance battery (SPPB). On the other hand, the effectiveness of exercise was not recognized for overall balance ability, physical function scale, TUG (timed up & go test), or QOL. Furthermore, subgroup analysis revealed that exercise intervention was effective for mild to moderate frailty, but not for severe frailty. A multifactorial exercise program consisting of resistance exercise, endurance exercise, stretching, and balance training is recommended as an exercise program for the prevention of the onset and progression of frailty from the early stages of frailty [2] (Table 85.2). Also, for the exercise program, it is recommended to gradually increase the exercise intensity from moderate to high intensity.

References

1. Sarcopenia Treatment Guideline Committee. Sarcopenia treatment guidelines, 2017 edition. Tokyo: Life Science Publishing. 2017. p. 36–7, p. 46–9.
2. Arai H, Chief Editor. Frailty treatment guide, 2018 edition. Tokyo: Life Science Publishing. 2018. p. 33–6.
3. Cruz-Jentoft AJ, Baeyens JP, et al. Sarcopenia: European consensus on definition and diagnosis: report of the European Working Group on Sarcopenia in Older People. Age Ageing. 2010;39:412–23.
4. Chen LK, Liu LK, et al. Sarcopenia in Asia: consensus report of the Asian Working Group for Sarcopenia. J Am Med Dir Assoc. 2014;15:95–101.
5. Chen LK, Woo J, et al. Asian Working Group for Sarcopenia: 2019 consensus update on sarcopenia diagnosis and treatment. J Am Med Dir Assoc. 2020;21:300–7.e2.

Physical Activity and Anti-Aging: Improvement of Brain Function (Cognitive Impairment, Depression) by Exercise

86

Yudai Yamazaki and Hideaki Soya

86.1 Brain Decline and Cognitive Impairment, Depression Due to Aging

Structural and functional decline of the brain due to aging is prominent in the prefrontal cortex involved in higher cognitive functions and the hippocampus involved in learning and memory. Therefore, in old age, cognitive impairments such as executive function and memory decline, which these areas are responsible for, are likely to occur. Furthermore, the prefrontal cortex and the hippocampus are areas deeply associated with stress and depression. Many patients with depression also show cognitive impairment, and chronic stress and depression in old age can be risk factors that accelerate cognitive impairment due to aging. Therefore, the prefrontal cortex and the hippocampus are important targets for the prevention and improvement of cognitive impairment and depression due to aging.

86.2 Actual Effects of Exercise on Cognitive Impairment, Depression

Physical activity and exercise are expected to be effective as non-pharmacological therapies for improving cognitive impairment and depression. A meta-meta-analysis using six meta-analyses showed that exercise has a moderate effect size for improving depression and a small effect size for improving anxiety symptoms [1]. Demurtas et al. examined the effects of exercise on mild cognitive impairment (MCI)/dementia by integrating multiple meta-analyses and systematic reviews in an umbrella review [2]. As a result, exercise significantly improved comprehensive cognitive function in patients with MCI and was effective for attention, executive function, and memory, although the certainty of the evidence was low to moderate. For the effects of exercise on patients with dementia, it was found to improve comprehensive cognitive function in all types of dementia. However, the certainty of the evidence was low, and no effect was observed on attention, executive function, or memory. These integrated findings suggest that exercise may be effective in improving depression and cognitive impairment. However, many of the studies to date have potential bias risks such as subject selection and small sample sizes, and more sophisticated and controlled trials are needed to strengthen the effectiveness of exercise.

Y. Yamazaki (✉) · H. Soya
Laboratory of Exercise Biochemistry and Neuroscience, Faculty of Health and Sport Sciences, University of Tsukuba, Ibaraki, Japan
e-mail: soya.hideaki.gt@u.tsukuba.ac.jp

H. Yamada, Y. Naito (eds.), *Anti-Aging Medicine*, https://doi.org/10.1007/978-981-97-8514-8_86

86.3 Presumed Mechanism of Action of Exercise

The effects of exercise on the brain are diverse (Fig. 86.1), and long-term exercise practice and the accompanying improvement in aerobic work capacity bring about structural and functional improvements in brain regions, including the prefrontal cortex and the hippocampus. For example, it is well known that wheel running enhances neurogenesis in the dentate gyrus of the hippocampus [3]. Behind such structural and functional improvements in the brain by exercise, there are increases in monoamines, brain-derived neurotrophic factor (BDNF), and vascular endothelial growth factor due to exercise. Increased secretion of growth factors such as Vascular Endothelial Growth Factor (VEGF) and Insulin-like Growth Factor-1 (IGF-1) is expected. Brain-Derived Neurotrophic Factor (BDNF), a major nutrient factor in the brain, plays a crucial role in neuronal survival, differentiation, and neural plasticity. Monoamines such as dopamine, serotonin, and norepinephrine are essential for various brain functions, including cognition, movement, and emotion. Many patients with depression or cognitive impairment often exhibit decreased secretion or dysfunction of these substances. Exercise may contribute to the improvement of depression and cognitive impairment by promoting their secretion and normalizing brain function [4].

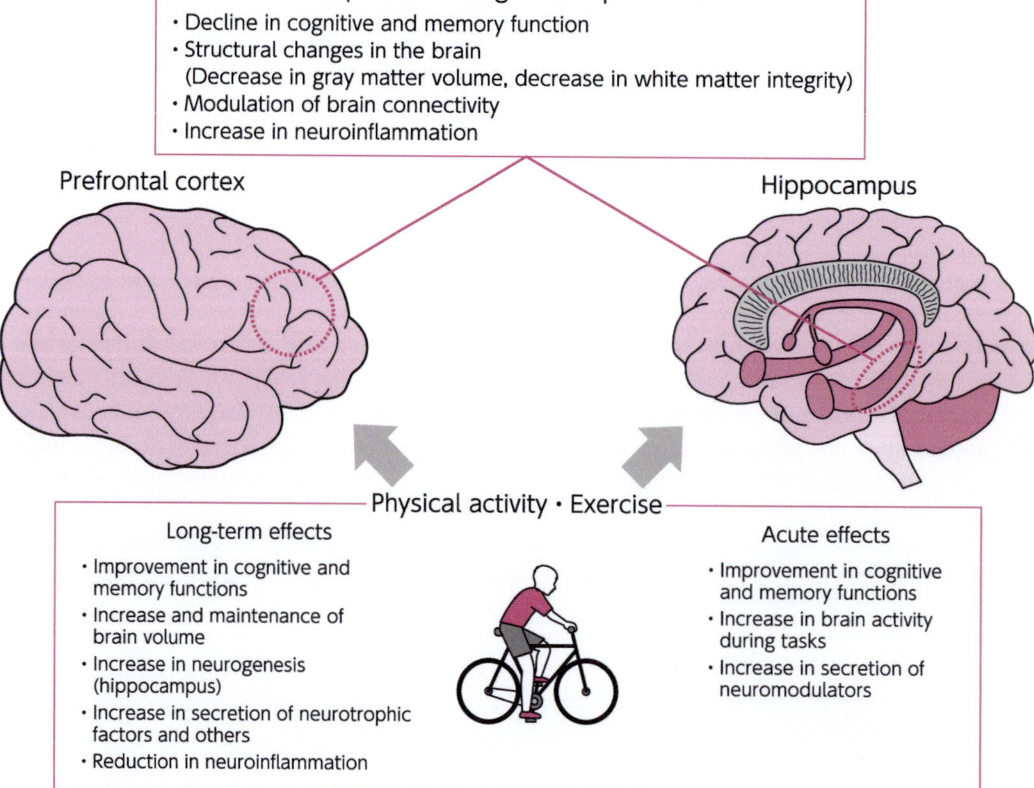

Fig. 86.1 Brain regions associated with cognitive impairment, depression, and targets of exercise effects

86.4 Neuroinflammation Hypothesis and Antidepressant Effects of Exercise

Chronic stress-induced inflammation and associated neuroinflammation have been pointed out to be related to depression. Long-term exercise may improve depression and associated cognitive impairment by reducing neuroinflammation. In particular, the kynurenine pathway has been attracting attention as a mechanism for exercise-induced neuroinflammation suppression. Kynurenine (KYN), a metabolite of tryptophan, is mostly produced and metabolized in the liver. However, stress and inflammation increase KYN production in immune cells, increasing peripheral KYN concentration. Since KYN can cross the blood–brain barrier, an increase in peripheral KYN concentration also increases the influx of KYN into the brain. Some of the KYN that enters the brain becomes quinolinic acid (QA), which acts as an agonist for N-methyl-D-aspartic acid (NMDA)-type glutamate receptors. Excessive brain QA concentration can lead to neurotoxicity due to QA's glutamate excitatory effect. On the other hand, peroxisome proliferator-activated receptor γ coactivator-1α (PGC-1α), produced from skeletal muscles during exercise, increases the production of kynurenine aminotransferase (KAT), an enzyme involved in the metabolic pathway from KYN to kynurenic acid (KA) [5] (Fig. 86.2). Since KA cannot cross the blood–brain barrier, the increased KAT due to exercise converts peripheral KYN to KA, suppressing its influx into the brain, preventing neuroinflammation in the brain, and potentially providing an antidepressant effect. This effect is expected as a new mechanism following the secretion of BDNF and monoamines, which have been assumed as mechanisms of the antidepressant action of exercise.

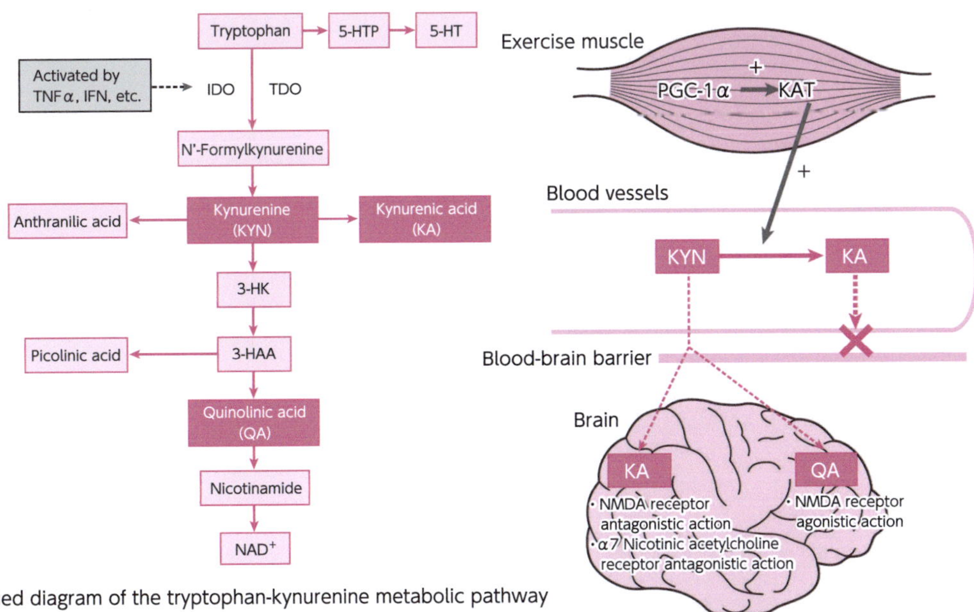

Fig. 86.2 Hypothesized mechanism of depression and the action mechanism of exercise effects, assuming neuroinflammation

References

1. Rebar AL, Stanton R, et al. A meta-meta-analysis of the effect of physical activity on depression and anxiety in non-clinical adult populations. Health Psychol Rev. 2015;9:366–78.
2. Demurtas J, Schoene D, et al. Physical activity and exercise in mild cognitive impairment and dementia: an umbrella review of intervention and observational studies. J Am Med Dir Assoc. 2020;21:1415–22.e6.
3. van Praag H, Kempermann G, et al. Running increases cell proliferation and neurogenesis in the adult mouse dentate gyrus. Nat Neurosci. 1999;2:266–70.
4. Cotman CW, Berchtold NC. Exercise: a behavioral intervention to enhance brain health and plasticity. Trends Neurosci. 2002;25:295–301.
5. Agudelo LZ, Femenía T, et al. Skeletal muscle PGC-1α1 modulates kynurenine metabolism and mediates resilience to stress-induced depression. Cell. 2014;159:33–45.

Physical Activity and Anti-aging: The Significance of Exercise at each Life Stage

87

Takumi Aoki and Kojiro Ishii

87.1 The Significance of Exercise for Children

Sports activities in sports clubs and extracurricular activities are the main places where children exercise, playing an important role in their healthy mental and physical development. In addition, it is characteristic that children have time for play by their own free will, not just sports activities. It has been reported that by engaging in "risky outdoor play" represented by play at high places and play with speed, increases in physical activity, decreases in sedentary behavior time, and improvements in self-esteem and concentration occur. In the physical activity guidelines formulated by WHO [1], it is recommended that children aged 5–17 "perform moderate to vigorous physical activity for 60 minutes or more per day," so it is desirable to accumulate activities mainly through sports and play based on the guidelines. It is also important to understand the negative effects that sports activities can have on children. Table 87.1 summarizes the views of the American Academy of Pediatrics on youth competitors [2]. While many benefits can be gained from youth sports, there are facts such

Table 87.1 The American Academy of Pediatrics' view on youth athletes [2]. (Quoted from Hideaki Soya, Mitsuyama K, et al. (eds): More Understandable and Usable Sports Science. Kodansha, Tokyo, 2017)

What can be gained from youth sports	Leadership, enjoyment, self-esteem, teamwork, athletic ability, sociability
Numerical facts	• 70% quit organized sports by the age of 13. • At least 50% of sports injuries were caused by overuse. • Only about 1% of high school athletes are at a level to receive scholarships. • Only 3–11% continue to compete in college. • Only about 0.03–0.5% make it to the professional level
Guidance	• To recover physically and mentally, set aside 3 months a year when no sports are played. • To gain lifelong skills and enjoyment in sports, participate in various sports. • Participating in multiple sports can reduce the risk of injuries, stress, and burnout. • To prevent injuries, take at least 1–2 days off from sports each week. • Avoiding specialization in a particular sport until late adolescence can increase the potential for higher-level performance. • Getting involved in many sports from an early age and delaying specialization contributes to both lifelong physical health and the chance to become an elite athlete.

T. Aoki (✉)
Faculty of Education, Miyagi Gakuin Women's University, Miyagi, Japan
e-mail: kishii@mail.doshisha.ac.jp

K. Ishii
Faculty of Health and Sports Science, Doshisha University, Kyoto, Japan

H. Yamada, Y. Naito (eds.), *Anti-Aging Medicine*, https://doi.org/10.1007/978-981-97-8514-8_87

as that 70% of children quit sports activities by the age of 13, 50% of sports injuries are caused by overuse, and only about 0.03–0.5% can advance to the professional level. In addition, there are disadvantages brought about by early specialization in a particular sport, so it is recommended to specialize in a sport after experiencing a variety of play and sports. Sports organizations and coaches involved in youth sports, as well as parents, are required to act based on scientific evidence and not limit the future possibilities of children.

87.2 The Significance of Exercise for Adults

According to a report from the 2019 National Health and Nutrition Survey announced by the Ministry of Health, Labour and Welfare in 2020, the proportion of adults who have a habit of exercising (those who exercise for more than 30 min at least twice a week and have continued to do so for more than a year) is 33.4% for men and 25.1% for women. However, this proportion varies by age group, with the lowest rates being in the 40 s for men (18.5%) and in the 30 s for women (9.4%). This indicates that building exercise habits during adulthood is a challenge. Looking at the obesity situation in our country in the same survey, the peak of the proportion of obese people with a BMI over 25.0 is in the 40 s for men (39.7%) and in the 60 s for women (28.1%). Considering the above, it is particularly important for men to engage in physical activity during adulthood to prevent and improve obesity and metabolic syndrome. The physical activity guidelines formulated by the WHO recommend that adults without chronic diseases or disabilities should "perform at least 150–300 minutes per week of moderate-intensity aerobic physical activity" or "perform at least 75–150 minutes per week of high-intensity aerobic physical activity." However, while the target values of these guidelines are strongly recommended based on evidence, they may not always be realistic depending on the state of exercise implementation. In the physical activity guidelines "Active Guide for Health Promotion" presented by the Ministry of Health, Labour and Welfare in 2013, the phrase "+10 (Plus Ten)," which calls for moving the body for an additional 10 min, is used. For those who do not have a habit of exercising, it would be good to use +10 as a stepping stone to form an exercise habit.

87.3 The Significance of Exercise in the Elderly

A decrease in muscle mass and strength can be cited as a physical change unique to old age. Figure 87.1 is a schematic diagram of the annual changes in muscle strength [3], and it is thought that the goal should be to maximize muscle strength in youth, maintain it in adulthood, and minimize its decline in old age. The decline in muscle mass, muscle strength, and physical function with aging is called sarcopenia (for details on the definition and diagnostic criteria, please

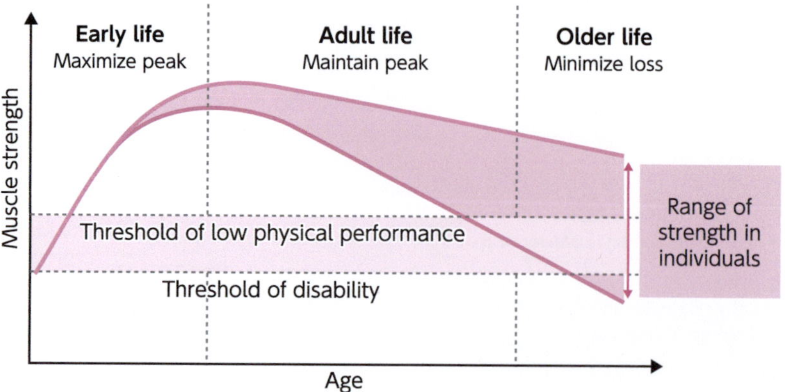

Fig. 87.1 Muscle strength and life course [3]

refer to other articles), and exercise plays an important role in addressing sarcopenia. The sarcopenia treatment guidelines [4] recommend ensuring a habit of exercise and a large amount of physical activity for the prevention and suppression of sarcopenia, and exercise intervention for those with sarcopenia is also recommended. In old age, it is easy to develop sarcopenic obesity, which is a combination of sarcopenia and obesity. Sarcopenic obesity, compared to each symptom alone, tends to affect instrumental activities of daily Decreased activities of daily living (ADL), falls, and death are particularly problematic. Although there is currently no consensus on the treatment of sarcopenic obesity, according to the "Obesity Treatment Guidelines for the Elderly 2018," [5] it is recommended to limit energy intake, consume sufficient protein, and perform resistance exercises. In other words, habitual exercise is indispensable for minimizing the decline in muscle mass and strength in old age, increasing decreased muscle mass and strength, and treating sarcopenic obesity for maintaining and promoting health in old age.

References

1. World Health Organization. WHO guidelines on physical activity and sedentary behaviour. 2020. https://apps.who.int/iris/handle/10665/336656.
2. Soya H, Mitsuyama K, et al., editors. More convincing, usable sports science. Tokyo: Kodansha; 2017.
3. Cruz-Jentoft AJ, Bahat G, et al. Sarcopenia: revised European consensus on definition and diagnosis. Age Ageing. 2019;48:16–31.
4. Sarcopenia Treatment Guidelines Committee. Sarcopenia treatment guidelines 2017 edition, partial revision. Tokyo: Life Science Publishing; 2020.
5. Japanese Geriatrics Society. Obesity in the elderly treatment guidelines 2018. https://www.jpn-geriat-soc.or.jp/tool/pdf/guideline2018_01.pdf.

Physical Activity and Anti-aging: Nutrients That Enhance the Anti-aging Effects of Exercise

88

Yoshikazu Takanami and Yukari Kawai

Numerous epidemiological and experimental studies conducted so far have shown that high levels of daily physical activity and regular exercise training improve various physiological functions and metabolic indicators, and also reduce the risk of various chronic diseases and premature death, contributing to anti-aging. On the other hand, these beneficial effects of exercise are not necessarily fully recognized if nutrient intake is not appropriate. The anti-aging effects of exercise are also manifested under conditions where appropriate nutritional conditions are met, and considering the deep complementary relationship between exercise and nutrition for health, the combined effect of nutrient intake is expected to further enhance the anti-aging effects of exercise (Table 88.1). In this section, we will overview the evidence on nutrients and dietary methods that have been examined from this perspective, including proteins and amino acids, vitamin D, omega-3 fatty acids, ketogenic diet, and medium-chain fatty acids.

Table 88.1 Nutrients and dietary methods that may enhance the anti-aging effects of exercise

Type of exercise	Nutrients/dietary methods to combine	Enhanced anti-aging effects
Resistance exercise	Protein/amino acids (Leucine)	Increase in muscle mass/strength[a]
Resistance exercise	Vitamin D	Increase in muscle strength[a]
Resistance exercise	Omega-3 fatty acids	Increase in muscle strength[b], improvement in physical function[b]
Resistance exercise	Ketogenic diet	Reduction in body fat[a]
Aerobic exercise	Ketogenic diet	Reduction in body fat[a]
Aerobic exercise	Medium-chain fatty acids	Increase in blood ketone concentration[a]
Resistance exercise	Medium-chain fatty acids	Increase in muscle strength/mass[b], reduction in body fat[b]

[a] Items reported with human trial evidence that enhances the anti-aging effects of exercise

[b] Items without direct evidence of enhancing the anti-aging effects of exercise, but with potential expected

88.1 Proteins and Amino Acids

When considering the anti-aging effects of a combination of exercise and protein/amino acids, the main target is skeletal muscle. Resistance

Y. Takanami (✉)
Department of Food Science, Otsuma Women's University, Tokyo, Japan
e-mail: yoshi.takanami@otsuma.ac.jp

Y. Kawai
Laboratory of Health and Sports, Louis Pasteur Center for Medical Research, Kyoto, Japan

H. Yamada, Y. Naito (eds.), *Anti-Aging Medicine*, https://doi.org/10.1007/978-981-97-8514-8_88

exercise and protein/amino acid intake are effective in preventing "sarcopenia," a condition where skeletal muscle mass and strength excessively decrease with age. Resistance exercise is an important factor that stimulates protein synthesis in skeletal muscles, and the rate of muscle protein synthesis significantly increases after exercise compared to at rest. Also, the intake of proteins and amino acids, mainly due to the action of essential amino acids and branched-chain amino acids such as leucine, promotes muscle protein synthesis. It has been reported many times that by taking proteins and essential amino acids in combination with resistance training (RT), a higher anabolic effect can be obtained compared to the effect of each alone. Proteins that can be expected to have a synergistic effect include whey and casein, which are milk proteins rich in essential amino acids, egg white protein, and soy protein [1]. As an amino acid supplement, essential amino acids, especially branched-chain amino acids, are effective, and it is presumed that the activation of leucine's muscle protein synthesis signal contributes. When taking proteins and amino acids as supplements, it is possible to enhance the effect by considering the timing of intake and the nutrients taken at the same time. In a 10-week intervention study, it was shown that if taken before and after RT, fat-free mass and muscle strength significantly increased compared to when taken at other times [2]. Also, when young people take amino acids and carbohydrates at the same time, the anabolic effect due to insulin secretion is added, and muscle protein synthesis greatly increases compared to amino acids alone. On the other hand, in the elderly, this additional effect is not seen due to insulin resistance, but it has been reported that aerobic exercise improves resistance and improves muscle protein synthesis ability stimulated by insulin [2]. In the case of the elderly, it is conceivable that a combination of aerobic exercise and RT and simultaneous intake of amino acids and carbohydrates could enhance muscle protein synthesis.

88.2 Vitamin D

Vitamin D is a vitamin that exhibits hormone-like effects involved in maintaining calcium homeostasis and regulating bone metabolism. Recently, its effects on immune function regulation and skeletal muscles are becoming clearer, attracting attention. Previous studies have reported that in the elderly, the lower the blood vitamin D concentration, the higher the risk of falling, and the easier it is for muscle strength, muscle mass, and physical function to decline, suggesting a link between vitamin D deficiency and muscle atrophy and sarcopenia. In a prospective cohort study targeting elderly Japanese, it was shown that muscle strength declined and the number of sarcopenia cases significantly increased in the vitamin D deficiency group compared to the vitamin D sufficient group [3]. There have been several studies on whether vitamin D enhances the effects of resistance training (RT), and their meta-analysis has been reported [4]. This meta-analysis showed that when RT and vitamin D intake were combined in the elderly, the muscle strengthening effect was significantly greater than RT alone, demonstrating the effectiveness of vitamin D. However, no significant additional effects were observed in muscle mass, bone density, or improvement in physical exercise function.

88.3 Omega-3 Fatty Acids

The intake of omega-3 fatty acids, which are abundant in fish oil, has been supported by many epidemiological studies and intervention studies to reduce the risk of cardiovascular disease. Recently, the functionality of the metabolic products of omega-3 fatty acids has also been elucidated, and anti-inflammatory effects, anti-allergic effects, and heart failure inhibitory effects have been reported. The effects on skeletal muscles have been examined from various aspects, and effects on muscle mass, muscle strength, physical exercise function, and metabolic aspects have been reported. Basic research has also shown the muscle atrophy inhibitory effect of omega-3 fatty acids and the effect of promoting muscle protein synthesis by exercise stimulation, and a synergistic effect with exercise is expected. A meta-analysis on the effects of omega-3 fatty acid intake alone and in combination with RT on the skeletal muscles of the elderly has been reported [5]. As a result of the analysis, omega-3 fatty acid intake may enhance muscle strength and physical exercise function, but no result was obtained that omega-3 fatty acid intake enhances the effect of RT. Also, no significant effect of omega-3 fatty acid intake on muscle mass was observed. However, since the number of previous studies adopted for the analysis is small, and many of them are studies with a small number of subjects, it is expected that more clear conclusions will be obtained by conducting high-quality randomized controlled trials with many subjects in the future.

88.4 Ketogenic Diet (KD) and Medium-Chain Fatty Acids (MCT)

The ketogenic diet (KD) generally refers to a dietary method that restricts carbohydrates and consumes fats as an alternative energy source. A low-carbohydrate, high-fat diet suppresses the rise in blood insulin levels, and the production of ketone bodies increases due to the increased use of free fatty acids in the liver, increasing the concentration of ketone bodies in the blood. Many studies support that this KD has the effect of reducing body weight and body fat, and it is expected to have anti-aging effects. However, in KD, insulin, which has an anabolic effect, may decrease, and the expression of AMP kinase in cells may increase, and the activity of mTORC1 (mechanistic/mammalian target of rapamycin complex 1) may decrease, which may reduce muscle protein synthesis and potentially decrease muscle mass. A review of studies examining the impact of combining KD with exercise on body composition changes has been reported [6]. According to this, when KD was combined with endurance training (ET), many studies reported that the decrease in body weight and body fat was greater than when ET was combined with a normal diet. This is thought to be the result of KD further enhancing the fat utilization promoted by ET. On the other hand, there was no significant decrease in lean body mass, an indicator of muscle mass, with the use of KD, and no significant changes were reported in a 12-week intervention trial. On the other hand, when KD was combined with resistance training (RT), many studies reported a decrease in lean body mass. This may be due to the low insulin levels caused by dietary restrictions and extremely low carbohydrate intake [7]. As a KD, there is a diet that uses medium-chain triglycerides (MCT). In a short-term intervention study targeting elderly women, it was shown that the increase in blood ketone body concentration was greater when MCT and aerobic exercise programs were combined than when used alone. The anti-aging effect of combining MCT and exercise is expected, but no human intervention studies have been reported yet. However, in an intervention study examining the frailty prevention effect of MCT on the elderly, it was shown that muscle mass and strength increased in the MCT group compared to the control group [8]. Caprylic acid, a medium-chain fatty acid, is an activator of the hormone ghrelin, which has a growth hormone secretion-promoting effect and an appetite-enhancing effect, so it may negate the negative effect of ketone bodies on muscle protein synthesis. High-quality human intervention studies are awaited in the future.

References

1. Ato S, Fujita S. Regulation of muscle protein metabolism by nutrition and exercise. J Phys Fitness Sports Med. 2017;6:119–24.
2. Fujita S. Exercise and nutritional intervention for muscle atrophy: latest evidence from basic research and application in the field. J Basic Phys Therapy. 2015;18:3–8.

3. Mizuno T, Hosoyama T, et al. Influence of vitamin D on sarcopenia pathophysiology: a longitudinal study in humans and basic research in knockout mice. J Cachexia Sarcopenia Muscle. 2022;13:2961–73.
4. Antoniak AE, Greig CA. The effect of combined resistance exercise training and vitamin D3 supplementation on musculoskeletal health and function in older adults: a systematic review and meta-analysis. BMJ Open. 2017;7:e014619.
5. Cornish SM, Cordingley DM, et al. Effects of omega-3 supplementation alone and combined with resistance exercise on skeletal muscle in older adults: a systematic review and meta-analysis. Nutrients. 2022;14:2221.
6. Ashtary-Larky D, Bagheri R, et al. Ketogenic diets, physical activity and body composition: a review. Br J Nutr. 2022;127:1898–920.
7. Vandenberghe C, Castellano CA, et al. A short-term intervention combining aerobic exercise with medium-chain triglycerides (MCT) is more ketogenic than either MCT or aerobic exercise alone: a comparison of normoglycemic and prediabetic older women. Appl Physiol Nutr Metab. 2019;44:66–73.
8. Abe S, Ezaki O, et al. Medium-chain triglycerides in combination with leucine and vitamin D increase muscle strength and function in frail elderly adults in a randomized controlled trial. J Nutr. 2016;146:1017–26.

Nutrition, Diet, and Anti-aging: The History of Nutrition Science in Japan and Anti-aging

89

Shaw Watanabe

89.1 The National Institute of Nutrition and the Shoku-Yo-Kai

Japan adopted German medicine during the Meiji era and ousted traditional Chinese medicine [1]. The study of physiology and biochemistry, such as energy metabolism, protein amino acid metabolism, and digestion, greatly advanced at the end of the nineteenth century. Tadasu Saeki, who studied at Yale University, proposed the concept of "nutrition science" based on his experiences upon returning to Japan, and in 1920, he established the National Institute of Nutrition. The term Nutrition originates from the Greek word meaning "to nourish," sharing the same root as the word nurse. Saeki's nutrition science made significant contributions to improving the nutrition of the Japanese people, but it was not recognized as a discipline and was taught in home economics departments as a subject for women. Even now, there is no nutrition science in medical education, and the training of dietitians is based in home economics departments. On the other hand, Japan has had the concept of dietary health since the Heian period. The "Yojokun" written by Ekiken Kaibara during the Edo period became a long-read classic. In the Meiji era, Sagyen Ishizuka, who became the army's pharmaceutical supervisor, advocated brown rice vegetarianism and the concept of "Shindo-Fuji," and formed the Shoku-Yo-Kai. This trend was passed on to Kenzo Futaki and Joichi Sakurazawa and is active as the Japan Comprehensive Medical Association and Macrobiotics. Brown rice vegetarianism became a national movement before the war. With the longevity of the population in the 2000s, the limitations of Western medicine have become apparent. Home medical care has become important in the medical system, and individualized medicine and integrated medicine that emphasize dietary habits to maintain high patient QOL are increasing within the Anti-Aging Medical Society [2].

89.2 Keio University's Institute of Dietary Health and Renal Diet Therapy

Hospital treatment food in Japan started with the Institute of Dietary Health established by Saburo Kitasato at Keio University's School of Medicine in 1924. At that time, it was understood that the cause of beriberi, a national disease, was due to white rice, and the idea of therapeutic food led to the concept of nutritional therapy. At the 33rd Japan Internal Medicine Society General Meeting (1936), "diet therapy" for heart disease, kidney disease, hypertension, gastrointestinal disease, infectious diseases, etc., was presented. Among

S. Watanabe (✉)
Medical Rice Association, Tokyo University of Agriculture, Tokyo, Japan

H. Yamada, Y. Naito (eds.), *Anti-Aging Medicine*, https://doi.org/10.1007/978-981-97-8514-8_89

them, low-protein diets were devised for kidney disease until the invention of hemodialysis in the late 1960s. From around the 1980s, drug treatments such as diuretics, antihypertensive drugs, and hypoglycemic drugs have become the main treatment. However, the increase in dialysis patients has not stopped, currently exceeding 340,000, with 40,000 new patients each year. Chronic kidney disease (CKD) patients often experience uremic dysbiosis, where the gut microbiota is often disturbed. As a result, desmosomes and others become loose, causing a leaky gut syndrome that can lead to the invasion of toxins into the liver, promoting the secretion of interleukin-6 (IL-6) and other substances, which can exacerbate minor inflammation and have negative effects on the kidneys and cardiovascular system. By improving the intestinal environment, the negative spiral of intestinal–kidney interaction can be improved. Low-protein processed brown rice JAS0027, which has obtained the certification of Manufacturing Process Management JAS, is a government-certified product that achieves packed rice through a series of technologies including the selection of raw rice, surface treatment with high-pressure steam, protein removal with lactic acid bacteria and enzyme solution, and steam cooking and packing. This packed rice has the advantages of low-protein white rice, such as (1) energy equivalent to white rice, (2) protein content less than one-tenth, (3) almost zero potassium, (4) phosphorus less than one-fourth, and also retains the goodness of brown rice, such as (5) dietary fiber, (6) γ-oryzanol, and (7) antioxidant capacity [3, 4]. Dietary therapy is fundamental for patients with renal failure, but compliance is not easy. By simply replacing the main meal with this packed rice, daily protein intake can be reduced by 10 g. Since the side dishes can be the same as those eaten by the family, it is easy to continue for people with decreased kidney function. Even dialysis patients can extend the interval between dialysis with appropriate low-protein diet. The fact that phosphorus is also low has many advantages for the bone-kidney circuit mediated by FGF23-Klotho gene product. It has been shown in diabetes and other chronic diseases that medication therapy alone is insufficient. Reevaluation of dietary therapy is necessary [5].

89.3 The Era of Individualized Nutrition and Future Anti-aging Medicine

The enactment of the Health Promotion Act (2004) has advanced measures against metabolic syndrome. From the nutrition of the group to the pre-disease state of the individual (Fig. 89.1), nutritional guidance to correct this

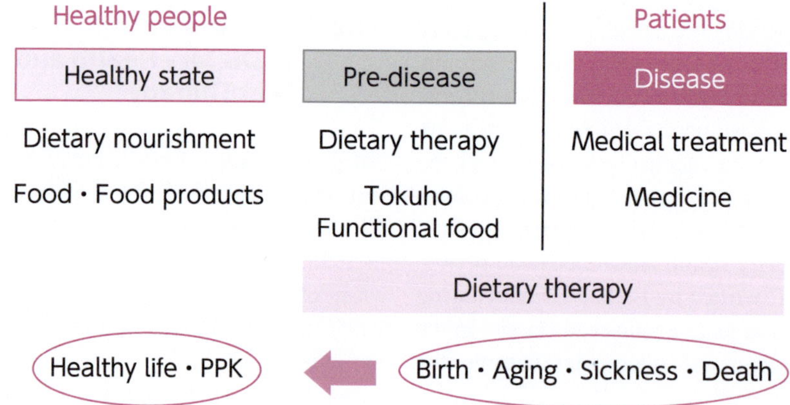

Fig. 89.1 Scope of dietary therapy. What is pre-disease? In chronic diseases, there are many continuous changes from health to disease. The current situation is to divide artificially at the boundary and respond to patients with medicine. Originally, diet is related to everything from health to disease. What is pre-disease? People who are healthy but have abnormalities in tests (e.g., metabolic syndrome), people who feel very bad but have normal test values (e.g., mild depression), and people who can return to a healthy state with diet, exercise, and mindset. We aim for a dignified healthy longevity by adjusting "food, mind, and body," aiming for a life with a purpose, a life that values life. Is birth, aging, sickness, and death inevitable in life? There is also a life where you stay healthy throughout your life and die suddenly. There are many such people among brown rice eaters. We aim for PPK (Pinpin Korori)

has begun. There are over a thousand types and hundreds of trillions of intestinal bacteria in the human gut, which have a unique composition depending on the food we eat. The metabolic products of intestinal bacteria can be beneficial or toxic to the body. To increase the number of intestinal bacteria that produce short-chain fatty acids such as butyric acid, it is essential to consume dietary fiber, which is abundant in brown rice. The intake of dietary fiber by Japanese people is decreasing year by year, but if brown rice is the staple food, the deficiency can be significantly compensated. Brown rice is also rich in vitamins and minerals and has high antioxidant capacity. Eating brown rice for a month can change the intestinal flora, relieve constipation, improve skin elasticity, and other changes in physical condition can be felt. The Medical Rice Association is working with the keyword "pre-disease," not only for health promotion through brown rice diet, but also for the development and maintenance of agriculture through organic rice, and for the preservation of the national environment, expecting a triple effect. In the unprecedented chaos of the post-corona and Ukraine war, how to live, how to choose a way of life, it is impossible to live a life with a purpose without choosing a way of life. The aging of farmers, the abandonment of cultivated land, the decrease in imports, and the overlapping of climate change are raising issues of food security. How to live independently in old age, especially in the 80s and 90s, is an issue that the Japan Anti-Aging Medical Association must also consider.

References

1. Watanabe S. Principles of nutrition. Tokyo: Nansei-do; 2009.
2. Watanabe S. New integrated medical science for regional comprehensive medical care, nursing, and medical field. Tokyo: Life Science Promotion Association; 2014.
3. Watanabe S, Ohtsubo K. Low-protein diet: history and use of processed low-protein rice for the treatment of chronic kidney disease. Food Secur. 2021;10:2255.
4. Watanabe S, Adachi K, et al. Dietary therapy with low protein genmai (brown rice) to improve the gut-kidney axis and reduce CKD progression. Asia Pac J Clin Nutr. 2022;31:341–7.
5. Watanabe S, Hirakawa A, et al. Food as medicine: the new concept of "medical rice". Adv Food Technol Nutr Sci Open J. 2016;2:38–50.

90 Nutrition and Diet and Anti-aging: Japanese Cuisine and Anti-aging

Naoki Saji

90.1 Background

As society ages, the impact of dementia on society, such as the driving of cars by elderly dementia patients, has become apparent. There is a demand for the development of new treatments for dementia and the timely and appropriate provision of medical care and nursing care, and a multi-institutional collaborative research system called the Orange Registry Study [1] has been established in our country. Subsequently, the Dementia Measures Promotion Outline was established in 2019, advocating for "coexistence" where people can live as themselves in familiar areas even if they have dementia, and working towards "prevention" of dementia. "Dementia prevention" can be said to be an important element from the perspective of anti-aging. In this section, we will focus on dementia prevention and diet.

90.2 Anti-aging of Cognitive Function

Age is a risk factor for dementia. The risk of dementia increases with age, but in reality, there are super-elderly people who still retain cognitive function. Even in old age, there is a possibility that the decline in cognitive function can be delayed by devising daily life. A research survey by the World Health Organization (WHO) Lancet Commission was published in 2020 [2]. According to the paper, there are about 40% of modifiable risk factors for dementia, and nutritional intervention, weight, hypertension, diabetes, dyslipidemia, etc., are recommended. Also, in the "Dementia Disease Treatment Guidelines 2017," exercise habits, dietary factors, leisure activities, social participation, mental activities, and cognitive training are listed as defensive factors for dementia. In recent years, new findings have been found from various clinical studies on dementia prevention. In the FINGER study [3], a clinical study from Northern Europe, the preventive effect of dementia was proven by multifactorial intervention through exercise and diet management, and verification of the preventive effect of dementia is being promoted by multiple countries. Age is an unavoidable disease risk for humans, but it is understood that a well-thought-out daily life contributes to the anti-aging of cognitive function.

90.3 Dementia Prevention from the Perspective of Food

Dementia prevention from the perspective of food has been attracting attention overseas as well. The Mediterranean diet has a preventive

N. Saji (✉)
Center for Comprehensive Care and Research on Memory Disorders, Hospital, National Center for Geriatrics and Gerontology, Aichi, Japan
e-mail: sajink@ncgg.go.jp

H. Yamada, Y. Naito (eds.), *Anti-Aging Medicine*, https://doi.org/10.1007/978-981-97-8514-8_90

effect on heart disease and lifestyle-related diseases, and also contributes to the prevention of dementia. The characteristics of the Mediterranean diet are (1) abundant intake of fruits and vegetables, (2) more fish intake than dairy products and red meat, (3) intake of olive oil, nuts, beans, and whole grains, (4) moderate intake of red wine, etc. Recently, dietary styles such as the DASH diet and the MIND diet, which are expected to have even more preventive effects on dementia, have been proposed. The DASH diet (Dietary Approaches to Stop Hypertension) is characterized by (1) appropriate total calorie intake, (2) intake of vegetables, fruits, and dietary fiber, and (3) reduction in fat intake and a shift from red meat to fish oil, and its blood pressure-lowering effect has been proven in randomized controlled trials. The MIND diet (Mediterranean-DASH Intervention for Neurodegenerative Delay diet) emphasizes the intake of whole grains, fruits, vegetables, legumes, and berries, and can be described as a mix of the Mediterranean and DASH diets. The "Dementia Disease Treatment Guidelines 2017" in our country also states, "High-calorie diets mainly composed of carbohydrates, low-protein diets, and low-fat diets increase the risk of mild cognitive impairment and dementia. Moderate drinking and fish intake reduce the risk of dementia." It is expected that improving dietary habits can reduce the risk of dementia.

90.4 Perspective from Japanese Cuisine

There is research data suggesting that the Japanese dietary pattern contributes to reducing the risk of dementia. The authors also conducted a food intake questionnaire for clinical research participants on intestinal bacteria, and investigated the relationship between diet, nutrition, dementia, and intestinal bacteria. From the food intake status of rice, miso soup, seaweed, pickles, vegetables, fish, soybeans, fruits, mushrooms, green tea, beef and pork, etc., we calculated the Japanese food score and also investigated the frequency of coffee intake. In previous studies, coffee, which is not a traditional Japanese food, was subtracted one point to calculate the Japanese food score, but due to its antioxidant effects, coffee is also proposed to have a preventive effect on dementia, so in this study, one point was added to calculate the modern Japanese food score. When comparing the Japanese food score for the presence or absence of dementia, the modern Japanese food score, including coffee intake, was significantly lower in the dementia group (Fig. 90.1).

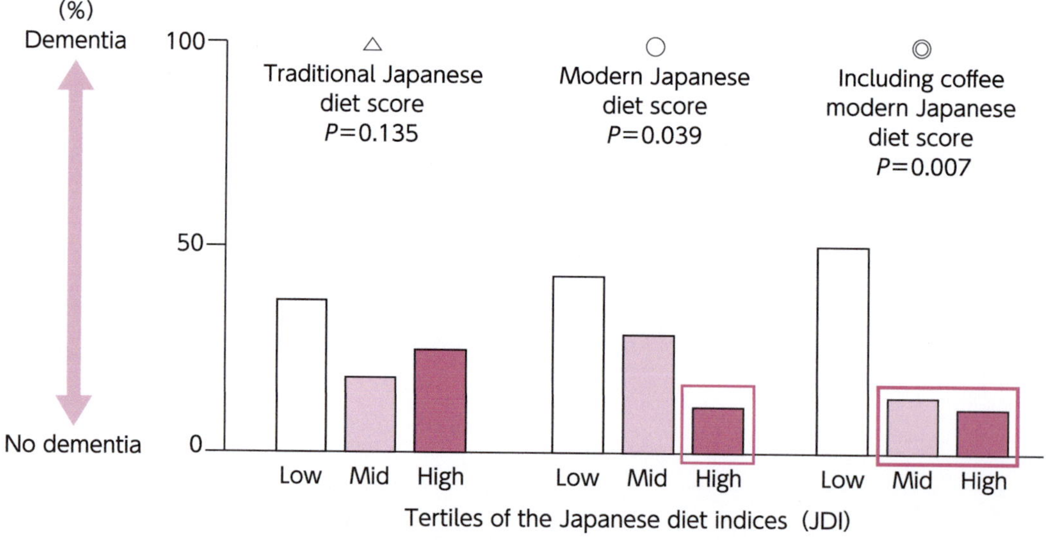

Fig. 90.1 Comparison of prevalence of dementia with Japanese diet score [4]. The higher the modern Japanese diet score including coffee, the lower the proportion of dementia

When analyzing the metabolic products of intestinal bacteria, the concentrations of P-cresol and indole (carcinogenic promoters produced by harmful intestinal bacteria, indicators of intestinal putrefactive fermentation) tended to be lower in groups that consumed more mushrooms and coffee, but no statistically significant difference was observed.

90.5 Relationship Between Intestinal Bacteria and Cognitive Function

The relationship between the intestinal environment and brain function is called the brain–gut correlation, and its importance is being recognized. The intervening mechanisms between the gut and the brain are speculated to be neurological mechanisms (stimulation of the autonomic nervous system), circulatory mechanisms (hormones and metabolic products, etc.), and immunological mechanisms (inflammation, immune response). In our previous studies, the imbalance of intestinal bacteria (dysbiosis) was independently associated with dementia and mild cognitive impairment, and the metabolic products of intestinal bacteria were also significantly associated with cognitive impairment (Fig. 90.2). Ammonia and some organic acids significantly increased in dementia groups (odds ratio 1.6 times), while lactate levels were reduced (odds ratio 0.28 times). These correlations remained similar even after adjusting for known risk factors through multivariate analysis, revealing that metabolic products are associated with cognitive function independently of gut microbiota composition. As the diet consumed can also affect the metabolic products of gut bacteria, a mechanism linking diet, gut bacteria, and brain function is suggested. Further scientific evi-

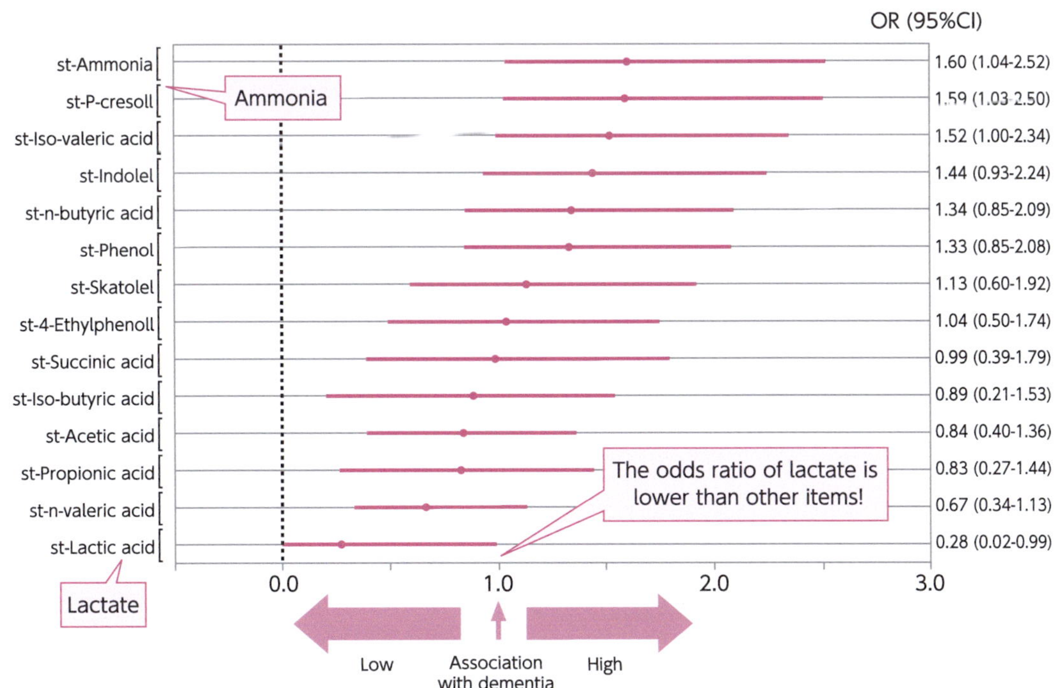

Fig. 90.2 Intestinal bacterial metabolites and dementia. Reprinted with modifications from [5]. Odds ratio (95% confidence interval) when the concentration of gut microbiota metabolic products and dementia metabolic products increases by 1SD. The stronger the association with dementia, the larger the red line is displayed above 1.0, and the range below 1.0 reduces the risk associated with dementia. High levels of organic acids such as ammonia were significantly associated with dementia, but lactate was significantly lower with an odds ratio of 0.28

dence from the perspective of gut bacteria contributing to good dietary intake patterns, such as the Mediterranean diet and Japanese diet, is expected for dementia prevention.

90.6 Summary

This section has outlined anti-aging of cognitive function, dementia prevention from the perspective of diet, and brain–gut correlations. Research on gut bacteria, an important factor involved in anti-aging, is expected to progress further. Dementia prevention is an important issue for the aging society in the future, and new evidence from the research theme of diet and gut bacteria is expected.

References

1. Saji N, Sakurai T, et al. ORANGE's challenge: developing wide-ranging dementia research in Japan. Lancet Neurol. 2016;15:661–2.
2. Livingston G, Huntley J, et al. Dementia prevention, intervention, and care: 2020 report of the lancet commission. Lancet. 2020;396:413–46.
3. Ngandu T, Lehtisalo J, et al. A 2 year multidomain intervention of diet, exercise, cognitive training, and vascular risk monitoring versus control to prevent cognitive decline in at-risk elderly people (FINGER): a randomised controlled trial. Lancet. 2015;385:2255–63.
4. Saji N, Tsuduki T, et al. Relationship between the Japanese-style diet, gut microbiota, and dementia: a cross-sectional study. Nutrition. 2022;94:111524.
5. Saji N, Murotani K, et al. Relationship between dementia and gut microbiome-associated metabolites: a cross-sectional study in Japan. Sci Rep. 2020;10:8088.

91 Nutrition, Diet, and Anti-aging: Mediterranean Diet and Anti-aging

Norie Sawada

91.1 What Is the Mediterranean Diet?

In relation to diet and aging or disease, people do not eat just one type of food; so, in recent years, research has been conducted to evaluate dietary patterns that categorize the overall diet to see the impact of dietary habits. Dietary patterns are thought to be more related to actual meals where various foods are consumed, and research is being conducted to clarify the relationship between dietary patterns and diseases. The evaluation methods of dietary patterns can be divided into two major categories. The first is to statistically identify and classify common patterns (such as Western type, healthy type, etc.) by conducting factor analysis, principal component analysis, etc., based on dietary data directly obtained from the subjects of the study. The second is to classify in advance based on previous reports on the effects of each food on health (such as the Mediterranean diet, glycemic index, etc.). The Mediterranean diet is a representative of dietary patterns that are classified in advance based on previous reports on the effects of each food on health. The Mediterranean diet recommends the intake of vegetables, fruits, nuts, legumes, whole grains, dairy products, olive oil, and seafood, and moderate drinking, while it recommends limiting the intake of meat, processed meat, and unsaturated fatty acids. Based on previous reports on the effects of each food on health, it is classified in advance, and if there are many recommended foods, one point is given, and if there are few restricted foods, one point is given, and more the points, the more the Mediterranean diet is being followed. The characteristics of the Mediterranean diet, such as vegetables, fruits, and whole grains, contain a lot of dietary fiber, and it has been reported that the vitamins contained in vegetables and fruits have antioxidant and anti-inflammatory effects. It has been reported that in groups with a high Mediterranean diet score, the telomere length, which is an indicator of aging, is longer [1], and it is believed to be related to anti-aging.

91.2 The Mediterranean Diet and Cardiovascular Disease

The Mediterranean diet has been reported to be associated with a reduction in the risk of cardiovascular disease due to improvements in blood pressure, lipid, and glucose metabolism abnormalities, endothelial function, and anti-inflammatory effects. In the case of the Mediterranean diet and cardiovascular disease, the results of a meta-analysis of 11 studies showed that the combined risk (relative risk; RR) and 95% confidence interval (95% CI) of all

N. Sawada (✉)
Division of Cohort Research, National Cancer Center Institute for Cancer Control, Tokyo, Japan
e-mail: nsawada@ncc.go.jp

H. Yamada, Y. Naito (eds.), *Anti-Aging Medicine*, https://doi.org/10.1007/978-981-97-8514-8_91

cardiovascular diseases in the highest group compared to the lowest group of the Mediterranean diet score were RR = 0.81 (95% CI = 0.74–0.88), and the combined risk of ischemic heart disease/acute myocardial infarction was RR = 0.70 (95% CI = 0.62–0.80), reporting a reduction in risk [2].

91.3 The Mediterranean Diet and Cognitive Function

It is known that age, being female, and the apolipoprotein E (ApoE) gene are risks for Alzheimer's type dementia, but lifestyle habits, including diet, have been reported as modifiable factors. The Mediterranean diet has been reported to reduce the risk of cardiovascular disease risk factors, such as blood pressure and lipids, which are intermediate factors in the risk of Alzheimer's type dementia, and to improve cognitive function and reduce the risk of dementia due to the neuroprotective effects of glucose metabolism in the brain and anti-inflammatory actions. In a meta-analysis of 11 studies on the Mediterranean diet and Alzheimer's type dementia and mild cognitive impairment, the combined risk of Alzheimer's type dementia in the highest group compared to the lowest group of the Mediterranean diet score was RR = 0.89 (95% CI = 0.84–0.93), and the combined risk of mild cognitive impairment was RR = 0.91 (95% CI = 0.85–0.97) [3].

91.4 Mediterranean Diet and Cancer

The anti-inflammatory and antioxidant effects of the Mediterranean diet have also been reported to reduce the risk of cancer. In a meta-analysis of 18 cohort studies on the relationship between the Mediterranean diet and cancer mortality, the combined risk in the highest group compared to the lowest group of the Mediterranean diet score was RR = 0.87 (95% CI = 0.82–0.92). By site, the combined risk in the highest group compared to the lowest group of the Mediterranean diet score was RR = 0.56 (95% CI = 0.44–0.72) for head and neck cancer (nine studies), RR = 0.70 (95% CI = 0.61–0.80) for stomach cancer (seven studies), RR = 0.83 (95% CI = 0.76–0.90) for colorectal cancer (17 studies), RR = 0.64 (95% CI = 0.54–0.75) for liver cancer (four studies), RR = 0.84 (95% CI = 0.76–0.94) for respiratory cancer (five studies), and RR = 0.87 (95% CI = 0.76–0.98) for bladder cancer (four studies), showing a reduction in risk. On the other hand, no association was observed for endometrial cancer, esophageal cancer, pancreatic cancer, and prostate cancer [4].

91.5 Conclusion

The Mediterranean diet is associated with improvements in antioxidant and anti-inflammatory actions, and has been reported to be associated with the risk of many diseases, including cancer, cardiovascular disease, and dementia (Fig. 91.1). However, as the name suggests, much of the research has been conducted in Western countries, including those along the Mediterranean coast, and may not fully apply to Japanese people. Therefore, there is a lack of research targeting Japanese and other Asian populations. On the other hand, in Japanese population, it has been reported that a higher Japanese food score, characterized by eight foods such as rice, miso soup, vegetables, fish, and green tea, reduces the risk of death [5]. The foods that make up the Japanese diet are thought to have antioxidant and anti-inflammatory effects, which may be similar to the dietary patterns associated with anti-aging. Further research on dietary patterns related to anti-aging in Japanese people is necessary.

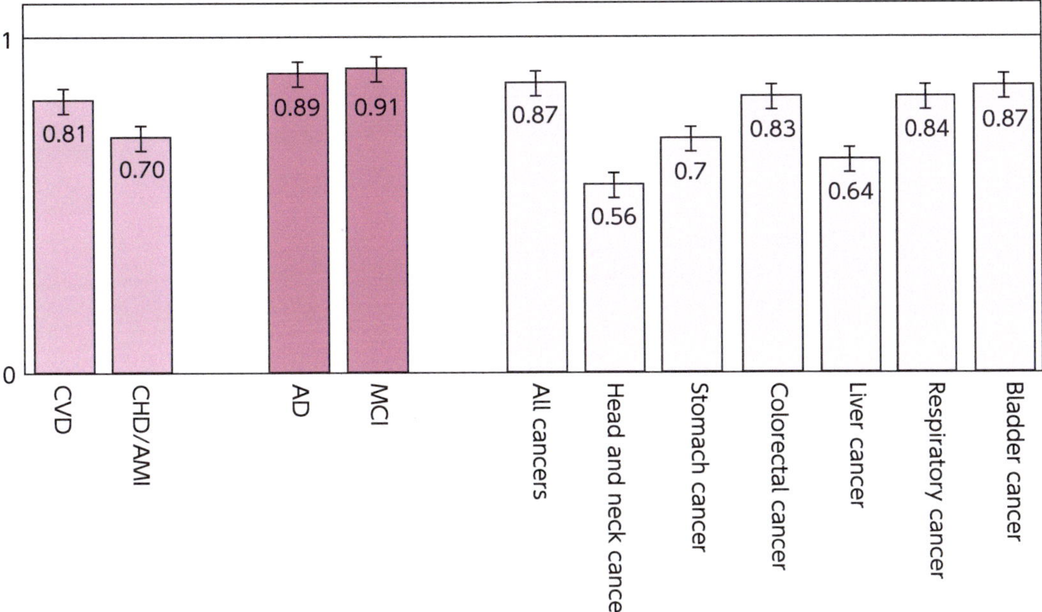

Fig. 91.1 Association between Mediterranean diet and disease [1–4]. Shows the risk of the highest group compared to the lowest group in the Mediterranean diet score. *CVD* cardiovascular disease (all circulatory diseases); *CHD/AMI* coronary heart disease/acute myocardial infarction; *AD* Alzheimer's dementia; *MCI* mild cognitive impairment

References

1. Canudas S, Becerra-Tomás N, et al. Mediterranean diet and telomere length: a systematic review and meta-analysis. Adv Nutr. 2020;11:1544–54.
2. Rosato V, Temple NJ, et al. Mediterranean diet and cardiovascular disease: a systematic review and meta-analysis of observational studies. Eur J Nutr. 2019;58:173–91.
3. García-Casares N, Gallego Fuentes P, et al. Alzheimer's disease, mild cognitive impairment and Mediterranean diet. A systematic review and dose-response meta-analysis. J Clin Med. 2021;10:4642.
4. Morze J, Danielewicz A, et al. An updated systematic review and meta-analysis on adherence to mediterranean diet and risk of cancer. Eur J Nutr. 2021;60:1561–86.
5. Matsuyama S, Sawada N, et al. Association between adherence to the Japanese diet and all-cause and cause-specific mortality: the Japan Public Health Center-based prospective study. Eur J Nutr. 2021;60:1327–36.

92 Nutrition, Diet, and Anti-aging: Dietary Inflammatory Index (DII) and Anti-aging

Norie Sawada

92.1 What Is the Dietary Inflammatory Index (DII)?

Chronic inflammation in vivo has been reported to be a risk factor for lifestyle-related diseases such as cancer and cardiovascular disease. There are various lifestyle factors that can contribute to an inflammatory state, and diet is considered one of them. There are many dietary nutrients that can affect inflammation, and food contains both inflammation-suppressing and promoting components. Since people do not eat single food, it is necessary to evaluate the overall impact of diet on inflammation. In recent years, the Dietary Inflammatory Index (DII), which evaluates the overall impact of diet on inflammation, has been developed from about 2000 previous studies on 45 types of nutrients and foods. It is used internationally, mainly in Western countries, and its association with diseases has been reported. That is, the lower the DII score (negative value), the more the diet is considered to suppress inflammation, and the higher the score (positive value), the more the diet is considered to promote inflammation (Table 92.1). Also, it has been shown in multiple studies that there is a correlation between the DII score and inflammatory biomarkers such as C-reactive protein (CRP), interleukin-6 (IL-6), and TNF-α (tumor necrosis factor-α). Examples of nutrients and foods that make up the DII score and have proven anti-inflammatory effects include polyphenols, ω3 unsaturated fatty acids, and dietary fiber. The higher the intake of polyphenols, the more it is associated with a decrease in inflammation markers through antioxidant pathways. Also, dietary fibers are suggested to bring about anti-inflammatory effects through short-chain fatty acids mediated by gut bacteria. It is thought that a high DII score is associated with anti-aging, as there are reports that it leads

Table 92.1 Foods and nutrients classified by inflammatory action in the Dietary Inflammatory Index (DII) [1]

Foods and nutrients classified as having anti-inflammatory action	Foods and nutrients classified as having a nutrient inflammatory effect
Vitamin A, Vitamin C, Vitamin D, Vitamin E, Vitamin B6, Niacin, Riboflavin, Thiamine, β-Carotene, Dietary Fiber, Folic Acid, Magnesium, Selenium, Zinc, Monounsaturated Fatty Acids, n-3 Unsaturated Fatty Acids, n-6 Unsaturated Fatty Acids, Polyunsaturated Fatty Acids, Isoflavones, Flavan-3-ols, Flavones, Flavonols, Flavonones, Anthocyanins, Pepper, Thyme/Oregano, Rosemary, Garlic, Ginger, Onion, Saffron, Eugenol, Turmeric, Green Tea/Black Tea, Caffeine, and Alcohol	Carbohydrates, Cholesterol, Energy, Total Fat, Iron, Protein, Saturated Fatty Acids, Trans Fatty Acids, and Vitamin B12

N. Sawada (✉)
Division of Cohort Research, National Cancer Center Institute for Cancer Control, Tokyo, Japan
e-mail: nsawada@ncc.go.jp

H. Yamada, Y. Naito (eds.), *Anti-Aging Medicine*, https://doi.org/10.1007/978-981-97-8514-8_92

to a decrease in the blood concentration of klotho protein [2], which is considered an anti-aging hormone.

92.2 Dietary Inflammatory Index (DII) Score and Diabetes

Long-term inflammation can potentially lead to metabolic abnormalities such as diabetes. Diets that promote inflammatory responses can potentially lead to the onset of diabetes by causing insulin resistance due to the increase in inflammatory cytokines (interferon γ, IL-1, IL-6, CRP, etc.). In the results of a meta-analysis of 48 studies that reported a relationship between the DII score and type 2 diabetes, no association was found between the DII score and type 2 diabetes, and the combined odds ratio (OR) and 95% confidence interval (confidence interval; CI) was OR (95% CI) = 1.03 (0.91–1.15). However, the quality of studies that met the inclusion criteria was evaluated using the Newcastle-Ottawa scale, and the integrated OR (95% CI) from eight high-quality studies was 1.58 (1.15–2.17), and it was reported that a high DII score was associated with a higher risk of diabetes [3].

92.3 Dietary Inflammatory Index (DII) Score and Osteoporosis

Chronic inflammation can lead to bone erosion and decreased bone density due to decreased osteoblast function and increased osteoclast activity, leading to osteoporosis and fractures. In a meta-analysis of 11 studies reporting the association between DII score and bone density, osteoporosis, and fractures (bone density: three studies, osteoporosis: three studies, fractures: five studies), it was reported that a high DII score was associated with lower bone density in the lumbar spine and proximal femur, and that the risk of osteoporosis was 31% higher and the risk of fractures was 28% higher in the group with the highest DII score compared to the group with the lowest score [4].

92.4 Dietary Inflammatory Index (DII) Score and Cancer, Cardiovascular Disease, and Depression

Chronic inflammation is associated with many diseases, including diabetes and osteoporosis mentioned above. In 2021, an umbrella review of 15 meta-analyses on the DII score and 38 diseases was reported [5]. Some of the 38 diseases targeted were death, cancer, cardiovascular disease, depression, and pathologies that are intermediate factors (hypertension, hyperglycemia, etc.) in these diseases. The results of this umbrella review were evaluated in five stages: convincing (certain), highly suggestive (almost certain), suggestive (possible association), weak (possible weak association), and no evidence (no association), based on the quality of the sample size and the strength of the association. In this umbrella review, it was reported that a high DII score was certainly associated with an increased risk of myocardial infarction, and that the association with the risk of all-cause mortality, all cancers, colorectal cancer, pancreatic cancer, head and neck/esophageal cancer, and oral cancer was almost certain (Table 92.2).

Table 92.2 Association between Dietary Inflammatory Index (DII) and disease [5]

Grade of certainty	Definition of grade	Disease
Convincing (Class I)	Number of cases >1000, statistically significant P value $<1 \times 10^{-6}$, $I2 < 50\%$, 95% confidence interval is statistically significant, the largest included study shows statistically significant results (P value ≤ 0.05), no influence of small-scale studies, no excessive significance bias	Myocardial infarction
Highly suggestive (Class II)	Number of cases >1000, statistically significant P value $<1 \times 10^{-6}$, the largest included study shows statistically significant results (P value ≤ 0.05), does not meet the definition of Class I	All deaths, all cancers, colorectal cancer, pancreatic cancer, head and neck/esophageal cancer, oral cancer
Suggestive (Class III)	Number of cases >1000, statistically significant P value $<1 \times 10^{-3}$, does not meet the definitions of Class I, II	Esophageal cancer, lung cancer, breast cancer, ovarian cancer, pharyngeal cancer, prostate cancer, depression, HbA1c, waist circumference
Weak (Class IV)	Statistically significant P value $\leqq 0.05$, does not conform to the definition of Class I–III	Cancer death, cardiovascular disease death, heart disease death, stomach cancer, kidney cancer, hypertension, cardiovascular disease, fasting blood glucose, insulin, HOMA-IR, systolic blood pressure
No evidence (Class V)	Statistically insignificant P value >0.05	Laryngeal cancer, endometrial cancer, urothelial cancer, heart disease, stroke, angina, metabolic syndrome, hyperglycemia, central obesity, waist-hip ratio, diastolic blood pressure

92.5 Conclusion

A diet with a high Dietary Inflammatory Index DII score is associated with an inflammatory state and is reported to be associated with the risk of many diseases, including cancer, cardiovascular disease, diabetes, and osteoporosis. A diet with a low DII score may contribute to a healthy longevity by reducing the incidence of age-related diseases such as arteriosclerosis and cancer.

References

1. Shivappa N, Steck SE, et al. Designing and developing a literature-derived, population-based dietary inflammatory index. Public Health Nutr. 2014;17:1689–96.
2. Zhang C, Zhang Z, et al. Association between dietary inflammatory index and serum klotho concentration among adults in the United States. BMC Geriatr. 2022;22:528.
3. Motamedi A, Askari M, et al. Dietary inflammatory index in relation to type 2 diabetes: a meta-analysis. Int J Clin Pract. 2022;2022:9953115.
4. Fang Y, Zhu J, et al. Dietary inflammatory index in relation to bone mineral density, osteoporosis risk and fracture risk: a systematic review and meta-analysis. Osteoporos Int. 2021;32:633–43.
5. Liu FH, Liu C, et al. Dietary inflammatory index and health outcomes: an umbrella review of systematic review and meta-analyses of observational studies. Front Nutr. 2021;8:647122.

93 Nutrition, Diet, and Anti-aging: Fermented Foods and Anti-aging

Yuji Naito

Various fermented foods are used in diets around the world. There is a wide variety of fermented foods, and the microbial communities involved in their fermentation are also diverse (Fig. 93.1). Not only lactic acid bacteria, but also yeasts, koji molds, acetic acid bacteria, and other diverse microorganisms have been used in fermented foods around the world. It is difficult to explain everything with the term "fermented foods." However, fermentation by microorganisms brings benefits to humans, and it is a fact that we eat fermented foods at our daily meals, and their

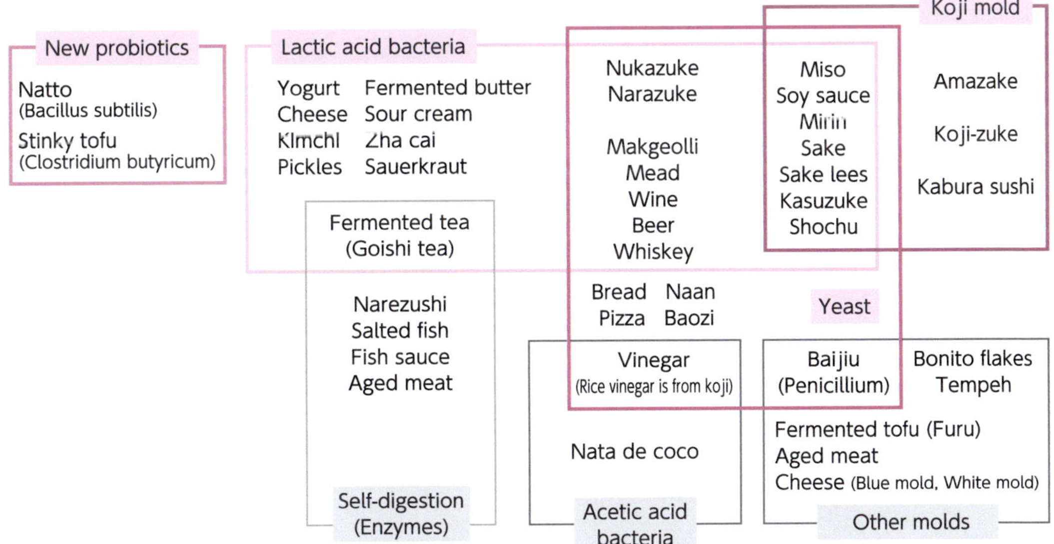

Fig. 93.1 Fermented foods in Japan and the world

Y. Naito (✉)
Department of Human Immunology and Nutrition Science, Kyoto Prefectural University of Medicine, Kyoto, Japan
e-mail: ynaito@koto.kpu-m.ac.jp

H. Yamada, Y. Naito (eds.), *Anti-Aging Medicine*, https://doi.org/10.1007/978-981-97-8514-8_93

safety is guaranteed. With the innovative progress of genetic analysis technology for these microbial communities, analysis technology for their metabolites, and analysis methods for host immunity, inflammation, and metabolism, the mystery of fermented foods is about to be unraveled. The current status and latest information on research on the functionality of fermented foods, especially their anti-aging effects, are explained.

93.1 Fermented Foods and Anti-aging Effects

Various health benefits of fermented foods have been shown, and basic and clinical research has been conducted [1]. Table 93.1 shows a summary of the analysis results of prospective cohort studies on fermented foods, milk, cheese, yogurt intake, and disease prevention, and the results of intervention trials. In the meta-analysis of prospective cohort studies, although a risk reduction effect for hypertension, metabolic syndrome, and type 2 diabetes is expected, in intervention trials aimed at elucidating the mechanism, no clear effectiveness has been proven for each item used as a biomarker. The anti-aging effect of fermented foods remains a mystery. The National Cancer Center is conducting a multipurpose cohort study (Japan Public Health Center-based Prospective, JPHC study). The JPHC study is a research project aimed at clarifying the relationship between various lifestyle habits and diseases such as cancer, stroke, and myocardial infarction, and contributing to the prevention of lifestyle-related diseases and the extension of healthy life expectancy in Japanese people. It reveals the relationship between a diet-centered lifestyle and lifestyle-related diseases. Using the results of a questionnaire survey, the intake of total soy products, fermented/non-fermented soy products, and each soy product (natto, miso, tofu) was calculated, divided into five equal groups, and then the relationship with mortality (total mortality, can-

Table 93.1 Health effects of fermented foods [1]

	Total fermented foods	Milk	Cheese	Yogurt
Prospective cohort study				
CVD	N	U	N	N
CAD/CHD	N	N	N	N
Stroke	F	N	F	N
Hypertension	F	F	N	N
MetS	F	F	U	U
T2DM	F	N	F	F
Intervention trial				
LDL cholesterol	NE	NE	NE	NE
HDL cholesterol	NE	U	U	NE
Fasting TGs	NE	U	NE	NE
Postprandial TGs	UD	NE	NE	UD
LDL size	UD	NE	UD	UD
apoB	UD	NE	NE	UD
Non-HDL cholesterol	UD	UD	UD	UD
Cholesterol ratios	UD	NE	NE	F
Inflammation	NE	NE	UD	NE
Insulin resistance	U	NE	NE	NE
Blood pressure	NE	NE	UD	NE
Vascular function	NE	NE	UD	NE

apoB apolipoprotein B; *CAD* coronary artery disease; *CHD* coronary heart disease; *CVD* cardiovascular disease; *HDL* high-density lipoprotein; *LDL* low-density lipoprotein; *MetS* metabolic syndrome; *T2DM* diabetes mellitus type 2; *TG* triglyceride; *N* neutral; *U* uncertain; *F* favorable; *NE* no effect; *UD* undetermined

cer mortality, cardiovascular disease mortality, heart disease mortality, and cerebrovascular disease mortality) over an average of about 15 years was investigated separately for men and women [2]. As a result, although no clear relationship was observed between the intake of total soy products and mortality, it was observed that the more fermented soy products were consumed, the lower the risk of total mortality (total mortality) for both men and women. Furthermore, when the relationship between the mortality risk and each of the fermented soy products, natto, miso, and tofu, was examined, it was found that the more natto and miso were consumed, the lower the mortality risk in women, but this trend was not observed in men. The fact that there was no trend of decrease in tofu for both men and women suggests that the health effects of fermented soy products depend on natto and miso.

93.2 Microbial Flora and Their Metabolites Contained in Fermented Foods

Fermented foods contain not only a variety of microbial flora but also their metabolites and nutrients as food, and a very complex analysis is required to understand their functionality when consumed. The most studied among the diverse microbial flora is lactic acid bacteria. A large-scale genome-wide analysis comparing lactic acid bacteria in fermented foods and human intestinal bacteria [3] has shown that closely related lactic acid bacteria strains are detected in both food and intestinal environments, and there is also evidence that fermented foods can be considered as a possible source of lactic acid bacteria in the intestinal bacteria. In other words, lactic acid bacteria as commensal bacteria, which are believed to play an important role in inflammation and immune responses, have settled in humans as commensal bacteria derived from fermented foods over a long history, mainly moving from the mother immediately after birth. Fermented foods contain a variety of microbial flora, and it is believed that there are many genes of enzymes that humans do not have, but this has not been fully elucidated. *Lactobacillus Plantarum* is a gram-positive rod bacterium mainly isolated from plant-derived fermentations. It contributes to the fermentation of high-salt pickles such as Japanese pickles, funazushi from Lake Biwa, Korean kimchi, and German sauerkraut. Although relatively rare, it can also be detected as a human resident bacterium. This bacterium possesses enzyme genes that saturate unsaturated fatty acids such as linoleic acid and α-linolenic acid, which are dietary lipids, through complex pathways. Furthermore, it is becoming clear that a variety of enzyme genes produce beneficial functional metabolites for the host. For example, the initial metabolite of linoleic acid, hydroxy fatty acid HYA (10-hydroxy-cis-12-octadecenoic acid), suppresses the production of inflammatory cytokines through the long-chain fatty acid receptor GPR40 expressed in intestinal epithelial cells, and has been reported to maintain intestinal barrier function, antimicrobial activity, and metabolic improvement [4]. The presence of HYA-related compounds has also been found, and they seem to be included in traditional Japanese fermented foods such as cheese, suguki, and miso. HYA is industrially produced from edible oils and is also being developed as a supplement. *Lactobacillus brevis* is frequently isolated from fermented pickles, not only producing lactic acid and acetic acid, but also secreting bacteriocin with antimicrobial activity, contributing to the preservation of fermented foods. A screening system using the human monocyte-like cell line KG-1 has been established for the exploration of new functionalities of L. brevis, and it has been shown that *Lactobacillus brevis* KB290 strain (*L. brevis* KB290) exerts anti-inflammatory effects by reducing CD103-dendritic cells in the presence of retinoic acid, resulting in a relatively dominant CD11c+ macrophages. Furthermore, in experimental models, colitis was suppressed [5], and a randomized double-blind placebo-controlled trial has been conducted for adults presenting symptoms of diarrhea-predominant irritable bowel syndrome with abdominal pain and discomfort [6]. As a result, it was shown that the intake of L. brevis KB290 and β-carotene reduced the intensity of abdominal pain and the frequency of bowel movements, improving labor productivity and daily life disruptions due to abdominal symptoms. Furthermore, in the examination of serum cytokine concentrations, the concentration of the anti-inflammatory cytokine IL-10 (interleukin10) was significantly higher, suggesting a possible involvement in anti-inflammatory action.

93.3 Prebiotics and Nutrients Contained in Fermented Foods

Traditional Japanese fermented foods such as miso and salt koji contain glycosylceramides (N-2'- hydroxyoctadecanoyl-1-O-b-D-glucopryranosyl-9-methyl-4,8-seramide) with unique structures specific to koji made with *Aspergillus oryzae* [7]. Koji glycosylceramides is thought to reach the large intestine without being broken down or absorbed by digestive enzymes in the small intestine, and serve as food for intestinal bacteria. It has been reported that administration of koji glycosylceramide to mice increases the number of *Blautia coccoides*, *Bacteroides sartorii*, *Hathewaya histolytica*, etc [8]. In other words, koji glycosylceramide is presumed to have a functionality similar to dietary fiber. It has also been suggested that acidic protease derived from *Aspergillus oryzae* affects the intestinal microbiota. Experimental data suggesting that enzyme activity derived from fermented foods is involved in the characteristic intestinal microbiota of Japanese people have also been reported. Yang et al. [9] administered purified acidic protease derived from *Aspergillus oryzae* to rats and reported a significant increase in the concentration of lactic acid, a metabolite of *Bifidobacterium*, in the cecum and feces. *Bifidobacterium* has been reported to be a dominant bacterium in the intestinal microbiota of Japanese people, and this result suggests that continuous intake of traditional fermented foods may be influencing this, in addition to genetic factors of the host and breast milk intake. Japanese people consume a relatively large amount of fermented foods using koji mold, such as miso, soy sauce, and sake, and a correlation with the extension of Japanese people's lifespan has also been suggested. Analysis of amino acid compounds and peptides in fermented foods including miso has led to the discovery of hydrophobic and hydrophilic pyroglutamyl peptides. It has been reported that hydrophilic peptides have a savory taste and hydrophobic peptides affect the intestinal microbiota [10]. Wada et al. [11] administered pyroglutamyl leucine (Pyroglu-Leu) to experimental colitis and reported a significant improvement in colitis. The concentration of Pyroglu-Leu significantly increased in the small intestine and colon 24 h after oral administration, indicating an improvement in dysbiosis caused by colitis. It has also been suggested that antimicrobial peptides secreted from the small intestinal epithelium are involved in this effect on the intestinal microbiota [12]. New insights have also been gained about the anti-inflammatory effects of miso. Okada et al. [13] found that the yeast *Zygosaccharomyces sapae* derived from miso induces IL-10 producing regulatory T cells and reported its inhibitory effect on DSS (dextran sulfate sodium) colitis. The intake of miso and miso-derived foods by Japanese people is on the decline, and it is attracting attention as a food ingredient that can be expected to have anti-inflammatory effects.

93.4 Impact of Fermented Foods on the Intestinal Microbiota

The impact of fermented foods on gut microbiota and immunity has been clarified through human intervention trials. Wastyk et al. [14] conducted a randomized controlled trial with 36 participants over 10 weeks, dividing them into a group that consumed a diet rich in fermented foods (fermented food group) and a group that consumed a diet high in dietary fiber (fiber group). The fermented food group consumed a lot of fermented foods and drinks such as yogurt, cottage cheese, kombucha, kefir, kimchi, sauerkraut, etc., while the fiber group consumed a lot of tofu, seeds, fruits, vegetables, whole grains, and nuts. As a result, in the fermented food group, the diversity of gut microbiota (Observed ASVs, Shannon Diversity) significantly increased, and gut bacteria such as *Lachnospira, Roseburia, Ruminococcus, Faecalibacterium prausnitzii,* which have anti-inflammatory effects, increased. However, no significant changes were observed in the fiber group. According to measurements using the Olink® method, which is attracting attention as a highly sensitive plasma proteome quantification method, the blood concentration of 19 types of proteins involved in promoting inflammation (IL6, IL12B, MCP-2, etc.) decreased in the fermented food group. Furthermore, inflammatory signals of classical monocytes, CD8+T cells, CD4+T cells, and B cells were also suppressed in the fermented food group by multi-parameter protein analysis (CyTOF® analysis) using a mass cytometer. It is extremely important that human prospective comparative trials have proven that fermented foods affect the gut microbiota and suppress inflammatory immune responses, but it is necessary to verify whether the same effects can be obtained for Japanese people who have different gut microbiota and consume a lot of fermented foods. Also, caution is required because the fermented foods recommended for consumption in this trial are quite different from the fermented foods that Japanese people usually consume.

93.5 Conclusion

As the association between gut microbiota and various diseases is becoming clear, the functionality of fermented foods is attracting attention. We introduced the research field where it is gradually becoming clear that the diverse microbial communities contained in fermented foods, their metabolites, nutrients, etc., are involved in the host's immunity and inflammatory response. Recently, the anti-inflammatory effects of foods have been attracting attention, and approaches to inflammatory bowel disease have also begun. Now that effective biologic treatments are advancing, we believe it is time to challenge the maintenance of remission and prevention of recurrence with anti-inflammatory foods.

References

1. Gille D, Schmid A, et al. Fermented food and non-communicable chronic diseases: a review. Nutrients. 2018;10:448.
2. Katagiri R, Sawada N, et al. Association of soy and fermented soy product intake with total and cause specific mortality: prospective cohort study. BMJ. 2020;368:m34.
3. Pasolli E, De Filippis F, et al. Large-scale genome-wide analysis links lactic acid bacteria from food with the gut microbiome. Nat Commun. 2020;11:2610.
4. Miyamoto J, Mizukure T, et al. A gut microbial metabolite of linoleic acid, 10-hydroxy-cis-12-octadecenoic acid, ameliorates intestinal epithelial barrier impairment partially via GPR40-MEK-ERK pathway. J Biol Chem. 2015;290:2902–18.
5. Fuke N, Takagi T, et al. Lactobacillus brevis KB290 with vitamin A ameliorates murine intestinal inflammation associated with the increase of CD11c+ macrophage/CD103$^-$ dendritic cell ratio. Inflamm Bowel Dis. 2018;24:317–31.
6. Fuke N, Aizawa K, et al. Effect of combined consumption of Lactobacillus brevis KB290 and β-carotene on minor diarrhoea-predominant irritable bowel syndrome-like symptoms in healthy subjects: a randomised, double-blind, placebo-controlled, parallel-group trial. Int J Food Sci Nutr. 2017;68:973–86.
7. Kitagaki H. Medical application of substances derived from non-pathogenic fungi aspergillus oryzae and A. Luchuensis-Containing Koji. J Fungi (Basel). 2021;7:243.
8. Hamajima H, Matsunaga H, et al. Japanese traditional dietary fungus koji aspergillus oryzae functions as a prebiotic for Blautia coccoides through glycosylceramide: Japanese dietary fungus koji is a new prebiotic. Springerplus. 2016;5:1321.
9. Yang Y, Iwamoto A, et al. Consumption of an acid protease derived from Aspergillus oryzae causes bifidogenic effect in rats. Nutr Res. 2017;44:60–6.
10. Hirai S, Horii S, et al. Anti-inflammatory effect of pyroglutamyl-leucine on lipopolysaccharide-stimulated RAW 264.7 macrophages. Life Sci. 2014;117:1–6.
11. Wada S, Sato K, et al. Ingestion of low dose Pyroglutamyl leucine improves dextran sulfate sodium-induced colitis and intestinal microbiota in mice. J Agric Food Chem. 2013;61:8807–13.
12. Shirako S, Kojima Y, et al. Pyroglutamyl leucine, a peptide in fermented foods, attenuates dysbiosis by increasing host antimicrobial peptide. NPJ Sci Food. 2019;3:18.
13. Okada Y, Tsuzuki Y, et al. Novel probiotic yeast from miso promotes regulatory dendritic cell IL-10 production and attenuates DSS-induced colitis in mice. J Gastroenterol. 2021;56:829–42.
14. Wastyk HC, Fragiadakis GK, et al. Gut-microbiota-targeted diets modulate human immune status. Cell. 2021;184:4137–53.e14.

Nutrition, Diet, and Anti-aging: High Fermentation Dietary Fiber and Anti-aging

94

Yuji Naito

94.1 Dietary Fiber Intake and Risk of Lifestyle Diseases

The intake of dietary fiber among Japanese people has been decreasing over the years, and it has been epidemiologically proven to be a risk factor for various diseases, including lifestyle diseases. According to a meta-analysis on the intake of dietary fiber and the risk of lifestyle diseases, in observational studies, the group with the highest intake of dietary fiber had a 15–30% lower risk of all-cause mortality, cardiovascular disease mortality, coronary artery disease onset, stroke onset and death, type 2 diabetes onset, and colon cancer onset compared to the group with the lowest intake [1]. Furthermore, in a meta-analysis of prospective clinical trials, the group with a higher intake of dietary fiber had significantly lower body weight, systolic blood pressure, and total cholesterol compared to the group with a lower intake. The association between these important outcomes and the risk reduction by dietary fiber is maximized when the daily intake of dietary fiber is between 24 and 29 g (Fig. 94.1). Moreover, it was revealed that an increase of 8 g/day in dietary fiber intake from food reduces the all-cause mortality rate by 7% (95% CI 0.90–0.95), the incidence of coronary heart disease by 19% (95% CI 0.73–0.90), the incidence of type 2 diabetes by 15% (95% CI 0.82–0.89), and the incidence of colon cancer by 8% (95% CI 0.89–0.95). It was suggested that a higher intake of dietary fiber could be even more beneficial in preventing cardiovascular diseases, type 2 diabetes, colon cancer, and breast cancer. The "Dietary Guidelines for Japanese People (2020 Edition)" [2] set by the Ministry of Health, Labor, and Welfare also indicates that it is effective in preventing lifestyle diseases as shown in Table 94.1.

Y. Naito (✉)
Department of Human Immunology and Nutrition Science, Kyoto Prefectural University of Medicine, Kyoto, Japan
e-mail: ynaito@koto.kpu-m.ac.jp

H. Yamada, Y. Naito (eds.), *Anti-Aging Medicine*, https://doi.org/10.1007/978-981-97-8514-8_94

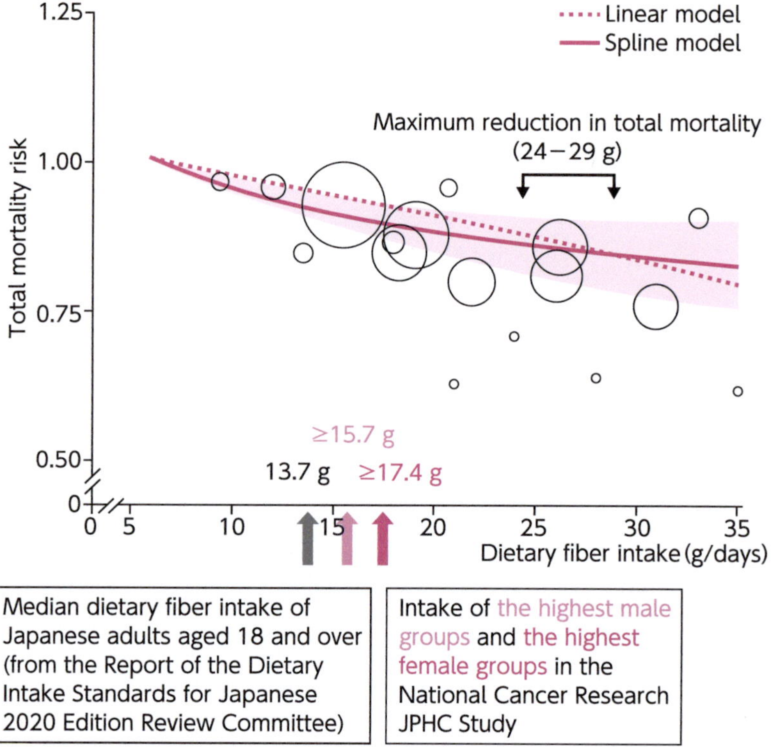

Fig. 94.1 Dietary fiber intake and total mortality risk, current situation in Japan and the world. (Modified with permission from [1])

Table 94.1 Dietary fiber intake and prevention of lifestyle-related diseases [2]. The intake of dietary fiber has been studied in relation to the incidence or mortality rate of numerous lifestyle-related diseases, and meta-analysis has reported a significant negative correlation with many diseases, making it a rare nutrient

• Total mortality
• Incidence and death from myocardial infarction
• Stroke onset
• Incidence and death from cardiovascular diseases
• Onset of type 2 diabetes
• Breast cancer onset
• Stomach cancer onset
• Colorectal cancer onset

[Quoted from "Dietary Intake Standards for Japanese (2020 Edition)", supervised by Sadayoshi Ito and Satoshi Sasaki, Daiichi Publishing, Tokyo, 2020]

94.2 Changes in Dietary Fiber Intake among Japanese People

The average intake of dietary fiber among Japanese people is reported to be 14 g/day, which means that about 10 g or more is lacking. The intake of dietary fiber exceeded 20 g/day in 1995, but by 2015, it had fallen to less than 15 g. The National Cancer Center's research team conducted a cohort study on the risk of diseases such as cancer, stroke, and myocardial infarction related to diet, which is called the Japan Public Health Center-based Prospective Study (JPHC study). The relationship between dietary fiber intake and disease risk has been examined by a 17-year follow-up study conducted since 1995, which analyzed the dietary surveys of approximately 90,000 men and women aged from 45 to 74 [3]. The intake of dietary fiber was divided into five groups per daily amount. As a result, it has become clear that the more dietary fiber Japanese people consume, the lower their risk of death. In Western studies, it is known that the intake of whole grains as dietary fiber has a significant impact, but in this dietary survey, there is a tendency that the more dietary fiber derived from beans, vegetables, and fruits is consumed, the lower the overall mortality risk. Although this survey revealed that there is little correlation between grain intake and overall mortality risk, it may be due to the fact that the intake of dietary fiber from grains is decreasing among the Japanese population as a whole. Looking at Fig. 94.1, we can see the current situation in Japan. The median intake of dietary fiber among Japanese people over 18 years is 13.7 g. Furthermore, in the JPHC study, when divided into five groups based on the daily intake of dietary fiber, the intake of the highest group (the high dietary fiber intake group that falls within the top 20% overall) is ≥15.7 g for men and ≥17.4 g for women. The intake of dietary fiber by the Japanese is considerably lower than international standards. The "Japanese Dietary Intake Standards (2020 Edition)" set by the Ministry of Health, Labor and Welfare states that the "target amount" per day (the intake amount that the current Japanese should aim for in order to prevent lifestyle-related diseases) is 21 g or more for men and 18 g or more for women aged 18–64. However, it is noted in the addendum that "the ideal target amount of dietary fiber for adults is considered to be 24g/day or more, but considering the current intake situation of the Japanese and the feasibility of achieving this, it has been set lower." This revision significantly revised the amount of dietary fiber in each ingredient according to a new international measurement method, and caution is required when comparing with previous data.

94.3 What Is High Fermentable Dietary Fiber?

Dietary fiber is not zero calories, and its caloric value is converted according to the degree of fermentation as shown below.

1. Those that reach the large intestine and are completely fermented are considered 2 kcal/g.
2. Dietary fiber that does not undergo fermentation is considered 0 kcal/g as a rule.
3. For dietary fiber with a clear fermentation decomposition rate, the following applies.

Those with a fermentation decomposition rate of less than 25%--------------0 kcal/g

Those with a fermentation decomposition rate of 25% or more, but less than 75%--------------1 kcal/g

Those with a fermentation decomposition rate of 75% or more--------------2 kcal/g

According to this classification, dietary fiber with a fermentation decomposition rate of 75% or more is calculated as 2 kcal/g and is also referred to as "high" fermentable dietary fiber (Table 94.2).

- Guar Gum Enzyme Hydrolysate: Guar gum is made from the seeds of the guar bean, an annual legume cultivated for food in India and Pakistan. The main component of guar gum is galactomannan, a type of polysaccharide, which is hydrolyzed and depolymerized by enzymes to produce guar gum enzyme hydrolysate (partially hydrolyzed guar gum; PHGG). It may also be referred to by its English abbreviation, PHGG. PHGG is a high molecular weight, difficult-to-digest polysaccharide with an average molecular weight of about 20,000, and a mannose to galactose ratio of 2:1. PHGG is fully fermented and

Table 94.2 Energy conversion factor of dietary fiber (kcal/g) [4]

Dietary fiber material name	Energy conversion factor (kcal/g)
Agar Xanthan gum Psyllium husk Gellan gum Cellulose Low molecular weight sodium alginate Polydextrose	0 (non-fermented)
Arabic gum Indigestible dextrin Beet fiber	1 (low fermentation)
Guar gum (guar flower, guar gum) Enzymatically hydrolyzed guar gum Wheat germ Heat-moisture treated starch (resistant starch) Water-soluble soy dietary fiber (WSSF) Tamarind seed gum Pullulan	2 (high fermentation)

decomposed when it reaches the large intestine, so its energy conversion factor is calculated as 2 kcal. A characteristic of PHGG is that it is tasteless and odorless, and can dissolve in various substances as a water-soluble dietary fiber. It is used in a variety of foods, not only solid foods but also beverages, jellies, liquid foods, soups, health foods, etc. There are stick-shaped supplements, each containing 5 g, which are easy to use and convenient. The most characteristic effect of PHGG on the intestinal environment is that it works to improve both constipation and diarrhea. While the effect on constipation is common for dietary fibers, the effectiveness of PHGG for healthy individuals with a tendency toward diarrhea has also been confirmed [5]. Its effect on human intestinal bacteria has been studied, and its most characteristic effect is the increase in bifidobacteria. Based on clinical experience to date, there are no diseases that would require withholding or contraindicating its use.

- Arabinoxylan (Wheat Bran)

  Arabinoxylan is a polysaccharide composed of arabinose and xylose, and is the main component of hemicellulose found in the cell walls of the outer husks of grains such as rice, wheat, corn, rye, oats, and barley. Although hemicellulose is representative of insoluble dietary fiber, arabinoxylan is a water-soluble dietary fiber. It not only contains a lot of insoluble dietary fiber, but also contains a lot of water-soluble dietary fiber arabinoxylan, minerals, vitamins, etc., and its health effects are expected. Arabinoxylan is a net-like structure that is not digested in the human small intestine, but reaches the large intestine, dissolves in water, and is utilized by intestinal bacteria. Human trials of arabinoxylan have shown effects such as an increase in bifidobacteria, an increase in short-chain fatty acids, and an improvement in bowel movements. In basic research, immune activation activity is also attracting attention. Since it is a component derived from normal food ingredients, no particular health damage or side effects are known.

- β-Glucan (Super Barley)

  Beta-glucan is a water-soluble dietary fiber found in barley. When barley beta-glucan dissolves in water, it absorbs moisture and solidifies into a jelly-like state. This jelly-like barley beta-glucan wraps around food and slowly moves through the digestive tract, which has the effect of suppressing the absorption of carbohydrates. Overseas, it is recommended to consume 3000 mg of barley beta-glucan per day, and in the United States, it is recognized to display the effect of "reducing the risk of coronary heart disease." Among foods, hulled barley contains a lot of dietary fiber, with 6.0 g of water-soluble dietary fiber and 3.6 g of insoluble dietary fiber per 100 g. Barley can be divided into "hulled barley" with less sticky and popping texture, and "glutinous barley" with a strong sticky and chewy texture. This glutinous barley is a water-soluble dietary fiber material that contains a lot of beta-glucan and is attracting attention. The best way to consume it is as "barley rice," especially for "breakfast." It has been proven that the function of "barley beta-glucan" to suppress the absorption of carbohydrates continues until the next meal. This function is called the "second meal effect." Recently, Australian barley, known as super barley, has also been attracting attention. Super barley contains not only beta-glucan, but also arabinoxylan, fructan, and the resistant starch mentioned next, and 21.6% is considered to be dietary fiber.

- Resistant Starch

  Resistant starch is "indigestible starch," a starch that is difficult to digest. It is thought to have a similar effect to dietary fiber because it reaches the large intestine without being digested much and is utilized by the intestinal flora. Resistant starch is found in foods such as snap beans, corn, barley, white rice, whole wheat (wholemeal flour), and potatoes. Also, resistant starch decreases significantly when heated and increases again when cooled. This is because the molecules of starch in unheated ingredients are in a state that is difficult to digest, or because starch that has been heated once and gelatinized, like cold rice, recrystallizes and becomes difficult to digest when cooled. Slices of yam and rice balls are considered to be the best sources of resistant starch. Unlike water-soluble dietary fiber, resistant starch is less likely to ferment by the intestinal flora, and it is known that it is not immediately utilized in the proximal colon, but is gradually utilized in the distal colon, increasing the production of butyric acid.

94.4 Conclusion

High fermentable dietary fiber shows an effect of improving the intestinal environment, and in addition to the intestinal regulation effect due to the improvement of bowel movements, immunological or nutritional effects through fermentative metabolic products are attracting attention.

References

1. Reynolds A, Mann J, et al. Carbohydrate quality and human health: a series of systematic reviews and meta-analyses. Lancet. 2019;393:434–45.
2. Supervised by Sadayoshi Ito, Toshi Sasaki. Dietary intake standards for Japanese (2020 edition). Tokyo: First Publishing; 2020.
3. Katagiri R, Goto A, et al. Dietary fiber intake and total and cause-specific mortality: the Japan public health center-based prospective study. Am J Clini Nutr. 2020;111:1027–35.
4. Japan Ministry of Health, Labour and Welfare. Shokushinhatsu No. 0217002.
5. Yasukawa Z, Inoue R, et al. Effect of repeated consumption of partially hydrolyzed guar gum on fecal characteristics and gut microbiota: a randomized, double-blind, placebo-controlled, and parallel-group clinical trial. Nutrients. 2019;11:2170.

95 Nutrition, Diet, and Anti-aging: Low-Salt Diet and Anti-aging

Tomohiro Katsuya

95.1 Salt Intake and Disease Risk

It has long been known that an increase in salt intake raises blood pressure and increases the risk of cerebrovascular and cardiovascular diseases. According to the 2019 National Health and Nutrition Survey, the average salt intake was 10.9 g for men and 9.3 g for women, which is far from the revised "Dietary Intake Standards for Japanese" in 2020, which recommends less than 7.5 g for men and less than 6.5 g for women. An analysis of the NIPPON DATA, which tracked 8072 Japanese participants in the National Nutrition Survey for 24 years, showed that for every 2 g/1000 kcal increase in household salt intake density, total mortality increased by 1.07 times, cardiovascular mortality by 1.11 times, coronary artery disease mortality by 1.25 times, and stroke mortality by 1.12 times, all significantly increased (Fig. 95.1) [1]. According to the data from the JPCH study, a multi-purpose cohort study of Japanese people, which followed about 40,000 people for 10 years, the risk of developing stomach cancer was 2.23 times higher in men and 1.32 times higher in women when comparing the lowest and highest quintiles of salt intake. It is often found in people who eat salty foods, such as salted fish and sea urchin paste. There are also reports that the risk is even higher in those infected with Helicobacter pylori. The Japanese Society of Hypertension's Hypertension Treatment Guidelines 2019 (JSH2019) recommends an intake of less than 6 g/day for hypertensive patients. Sodium is the most important extracellular cation in the body, necessary for many physiological processes, and is strictly regulated in many processes (kidney, biochemistry, endocrinology, immunity, nerves) to maintain serum sodium within the normal range [2]. Therefore, it is not recommended to drastically reduce sodium intake in the short term, and in the elderly and those with reduced kidney function, it is necessary to carefully monitor electrolytes and kidney function while reducing salt intake.

T. Katsuya (✉)
Katsuya Clinic, Hyogo, Japan
e-mail: tkatsuya@iris.eonet.ne.jp

H. Yamada, Y. Naito (eds.), *Anti-Aging Medicine*, https://doi.org/10.1007/978-981-97-8514-8_95

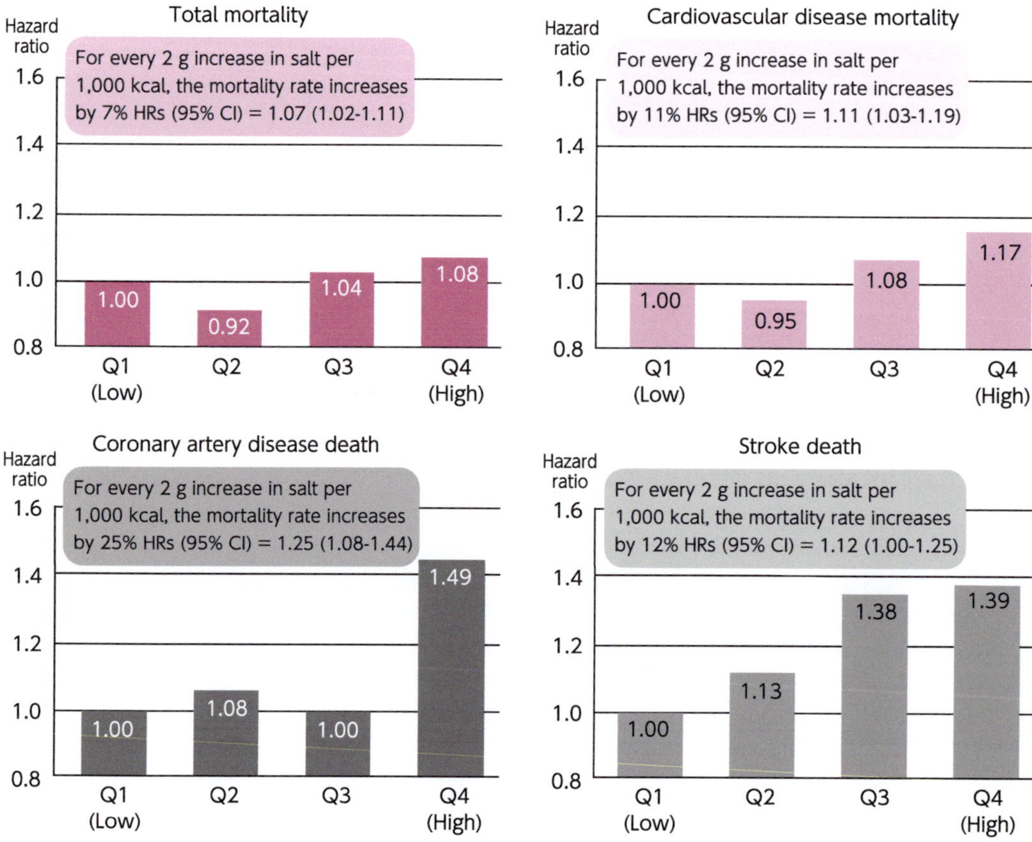

Fig. 95.1 Association between salt intake density in Japanese households and total mortality and various disease mortality. Reused with permission from [1]. Shows the association between quartiles of household salt intake density (Q1~Q4), and household salt intake density of 2 g/1000 kcal and each mortality. HRs (95% CI): Hazard ratios (95% confidence interval). The hazard ratio indicates an estimate of the risk size. The 95% confidence interval indicates the range in which the true hazard ratio exists with a 95% probability

95.2 Verification and Practice of Salt Reduction Effects

The estimation of average salt intake in a population is more feasible using a formula validated from a single fasting urine sample (converted using Kawasaki's formula). For example, in the PURE study, the sodium intake of over 100,000 people was investigated based on the estimated value of 24-h urinary sodium excretion from a single fasting urine sample, and a positive association with systolic blood pressure of 2.11 mmHg per 1 g of sodium (equivalent to 2.4 g of salt) was reported [3]. In a meta-analysis of 185 intervention trials, including intervention studies such as DASH (dietary approaches to stop)-Sodium) and TONE, it was found that reducing salt intake from an average of 11.8 g/day to 3.9 g/day in urine salt excretion resulted in a decrease in blood pressure of 5.51/2.88 mmHg in hypertensive patients, and a decrease of 7.75/2.68 mmHg was confirmed in a sub-analysis of Asians [4]. In JSH2019, it is recommended to evaluate salt intake using 24-h urine collection or weighing by a dietitian, or 24-h recall dietary survey in hypertension specialty facilities, and sodium and creatinine measurement in random urine or second urine after waking up (calculated by Kawasaki method), or salt check sheet in general facilities. It is also important to guide appropriate salt intake according to body size, nutritional status, physical activity level, and overall condition in frail elderly people and chronic dialysis patients. The Salt and Nutrition Committee of the Japanese

Society of Hypertension has published a list of low-salt foods available in stores (https://www.jpnsh.jp/data/salt_foodlist.pdf), and the National Cardiovascular Research Center is also promoting "Karushio Certified Products" (https://www.ncvc.go.jp/karushio/mark/mark_4577/) and Karushio certification system.

95.3 Effective Use and Effects of Alternative Salt (KCl)

In cooking, it has been investigated whether the taste and texture change when part of the regular salt is replaced with alternative salt (KCl). When the proportion of KCl is high, it tastes metallic, so attempts have been made to improve the flavor by combining it with taste enhancers (e.g., food acids, amino acids, umami substances) or using other processing techniques. When less than 30% of table salt is replaced with KCl, it has been reported that most people cannot distinguish it from food cooked with regular salt [5]. The major advantage of alternative salt is that it can reduce sodium while maintaining the salty taste of food, and it is thought that a long-term effort is needed to comprehensively reduce the total amount of salt in a wide food environment. In a prospective cohort study of 20,995 residents with an average age of 65.4 in 600 villages in the rural part of China, followed for 4.74 years [6], the incidence of stroke was significantly reduced by 14%, major cardiovascular events by 13%, and total mortality by 12% in the group using alternative salt (NaCl 75%, KCl 25%) compared to regular salt (NaCl 100%). On the other hand, the incidence of serious adverse events due to hyperkalemia was not significantly different between the alternative salt group and the regular salt group. From the results of this trial, it was shown that the use of alternative salt is useful in reducing stroke, major cardiovascular events, and total mortality in people with a history of stroke and hypertensive patients over 60 years old, and its impact is considered to be comparable to the SPRINT trial that rewrote hypertension treatment guidelines worldwide.

95.4 Salt Sensitivity and Anti-aging

Excessive salt intake increases the risk of hypertension, but the reactivity of blood pressure, or salt sensitivity, varies among individuals. Salt sensitivity is determined by the interaction of environmental factors such as aging, malnutrition, and stress, and genetic factors, and in adults, obesity is often associated with salt-sensitive hypertension. Children of women who experienced malnutrition during pregnancy have a higher risk of developing obesity, diabetes, and salt-sensitive hypertension in adulthood, and offspring of mice given a low-protein diet during pregnancy develop salt-sensitive hypertension through abnormal DNA methylation of the angiotensin II type 1 receptor gene (AT1AR) in the hypothalamus, causing overexpression of hypothalamic AT1AR and increased sympathetic nerve hyperactivity in the kidneys. On the other hand, in aging model mice, the production of the anti-aging factor klotho in the kidneys decreases due to promoter methylation, and circulating soluble klotho also decreases. In a state of klotho deficiency, salt-sensitive hypertension is easily induced with aging due to the activation of the Wnt5a-RhoA pathway in blood vessels by salt, leading to decreased renal blood flow and increased peripheral vascular resistance. It has become clear that excessive salt intake from fetal to old age not only increases traditional risks in metabolism, cardiovascular and kidneys, but also causes abnormal DNA methylation of specific genes [7]. In considering anti-aging through salt reduction, we believe that appropriate salt intake (we call this "appropriate salting") that takes into account the epigenetic memory at each stage of life, such as fetal nutrition, obesity, and aging, is necessary.

References

1. Shima A, Miyamatsu N, et al. Relationship of household salt intake level with long-term all-cause and cardiovascular disease mortality in Japan: NIPPON DATA 80. Hypertens Res. 2020;43:132–9.

2. Mente A, O'Donnell M, et al. Sodium intake and health: what should we recommend based on the current evidence? Nutrients. 2021;13:3232.
3. Mente A, O'Donnell M, et al. Associations of urinary sodium excretion with cardiovascular events in individuals with and without hypertension: a pooled analysis of data from four studies. Lancet. 2016;388:465–75.
4. Dobashi T. How much can Japanese reduce salt? J Nutr. 2020;78:49–56.
5. Greer RC, Marklund M, et al. Potassium-enriched salt substitutes as a means to lower blood pressure: benefits and risks. Hypertension. 2020;75:266–74.
6. Neal B, Wu Y, et al. Effect of salt substitution on cardiovascular events and death. N Engl J Med. 2021;385:1067–77.
7. Kawarazaki W, Fujita T. Kidney and epigenetic mechanisms of salt-sensitive hypertension. Nat Rev Nephrol. 2021;17:350–63.

Nutrition, Diet, and Anti-aging: Ketogenic Diet and Anti-aging

96

Kohjiro Ueki

96.1 What Is a Ketogenic Diet?

A ketogenic diet is one where the daily intake of protein is ensured, carbohydrate intake is usually kept below 50 g/day, and about 60–80% of the total energy intake comes from fats. The main energy source shifts from carbohydrates to fats, leading to an increase in the production of ketone bodies (acetoacetic acid, acetone, and β-hydroxybutyric acid (βOHB)). This is why it is generally referred to as a ketogenic diet. Under a ketogenic diet, the increased dietary fat and decreased carbohydrate intake lead to low insulin levels, which in turn increase fat breakdown (lipolysis) in fat cells. The resulting fatty acids flow into the liver, undergo β-oxidation, and are converted to acetyl-CoA. This is then converted to HMG-CoA by mitochondrial hydroxymethylglutaryl (HMG)-CoA synthase 2 (HMGCS2), and further to β-OHB via acetoacetic acid. The βOHB produced in the liver is released into the circulating blood by the specific transporter SLC16A (Fig. 96.1). Therefore, most of the ketone bodies in circulation are βOHB. The ketogenic diet is thought to work through the effects of this increased βOHB and the effects based on low insulin levels due to carbohydrate restriction. In a state where a balanced diet is being consumed, the blood concentration of ketone bodies is several to tens of μM, but it is said to increase to 200–300 μM after about half a day of fasting, up to 6–8 mM after about 2 days of fasting, and up to 2–3 mM in a ketogenic diet.

K. Ueki (✉)
Diabetes Research Center, Research Institute,
National Center for Global Health and Medicine,
Tokyo, Japan
e-mail: uekik@ri.ncgm.go.jp

H. Yamada, Y. Naito (eds.), *Anti-Aging Medicine*, https://doi.org/10.1007/978-981-97-8514-8_96

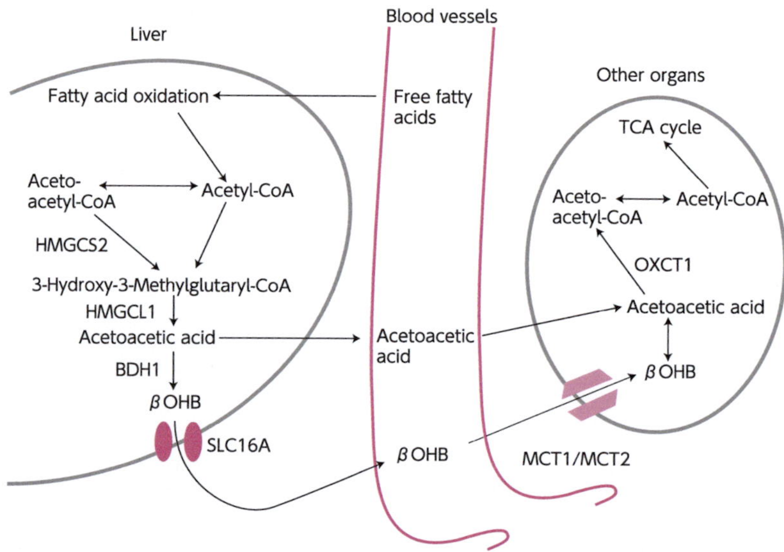

Fig. 96.1 Production of ketone bodies in the liver and tissue migration

96.2 The Role of Ketone Bodies in the Body

- Ketone bodies as energy substrates

  The βOHB released into the circulating blood is taken up into cells via two cell surface transporters, MCT (monocarboxylate transporter) 1/MCT2. The taken up βOHB is converted back to acetoacetic acid via the reverse pathway of synthesis, and is loaded with Co-A from succinyl-CoA by OXCT1 (3-oxoacid CoA-transferase 1) to be converted to acetoacetyl-CoA. Acetoacetyl-CoA becomes two molecules of acetyl-CoA, enters the TCA cycle, and is consumed for ATP production (Fig. 96.1).
- Ketone bodies as signaling molecules

  βOHB binds to at least two G-protein-coupled receptors on the cell surface, HCAR2 (hydroxycarboxylic acid receptor 2 or GPR109) and FFAR3 (free fatty acid receptor 3 or GPR41). HCAR2 is activated by the binding of βOHB, which inhibits fat breakdown in fat cells. FFAR3, on the other hand, is expressed in the sympathetic nerves and enhances their tension, but is inhibited by the binding of βOHB, leading to a decrease in the individual's energy consumption (Fig. 96.2).
- Ketone bodies as transcription factors

  Histone deacetylase (HDAC) deacetylates the lysine residues of histones and suppresses transcription, but it has been revealed that ketone bodies bind to HDAC and inhibit its activity. βOHB is thought to inhibit the activity of HDAC1, HDAC3, and HDAC4, induce high acetylation of histone H3K9 and H3K14, and regulate the expression of stress resistance genes through epigenomic changes [1]. Furthermore, βOHB is also thought to indirectly induce high acetylation of histones. βOHB, which increases with fasting, is taken up into cells and acetyl-CoA accumulates in cells through the aforementioned pathway. It is thought that this leads to an increase in protein acetylation.

96.3 Known Effects of a Ketogenic Diet

In addition to weight loss effects through mechanisms such as increased fat breakdown due to hypoinsulinemia, the ketogenic diet is known to have therapeutic effects for refractory epilepsy. The brain uses glucose as its main energy source after eating, and ketone bodies when fasting continues for a long time. The brain is composed of neurons and glial cells (astrocytes), and glucose not only becomes energy when taken up into neurons and glial cells via GLUT1 and GLUT3, but also when glucose taken up into glial cells is metabolized to lactate, which is then excreted

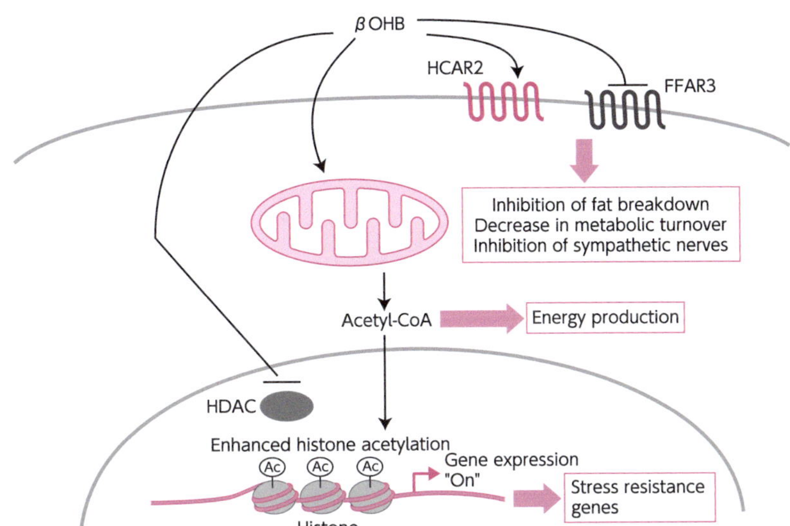

Fig. 96.2 Action of ketone bodies

outside the cell and flows into neurons where it is converted to pyruvate and enters the TCA cycle to be converted to ATP. Recently, it has been clarified that the "astrocyte-neuron lactate shuttle," where glucose taken up into glial cells is metabolized to lactate and transported to neurons, regulates neuronal electrical activity, and inhibition of lactate dehydrogenase (LDH) reduces lactate transport and shows antiepileptic effects. In a ketogenic diet, it is thought that the reduction of glucose influx into glial cells and the use of ketone bodies as an energy source by neurons exert antiepileptic and neuroprotective effects. Furthermore, ketone bodies are said to cause neuronal inhibition through changes in the activity of neuronal ion channels and glutamate transporters. Also, a ketogenic diet has been reported to improve cognitive function in Alzheimer's disease (in the case of ApoE4 negative) [2].

96.4 Ketogenic Diet and Anti-aging

There is consistent evidence from yeast to mammals that calorie restriction has some effects on lifespan extension and anti-aging. On the other hand, while the neuroprotective and weight loss effects of a ketogenic diet are evident in both rodents and humans, conclusions about its lifespan extension and anti-aging effects have not been reached [3]. In mice, a ketogenic diet with extremely reduced carbohydrates can extend average lifespan, improve cognitive function, and reduce age-related diseases such as cancer, but its effects are smaller compared to calorie restriction [4, 5]. Furthermore, it is unclear whether the anti-aging effects of a ketogenic diet are due to the action of ketone bodies as mentioned above, or whether the anti-obesity effects contribute to a certain extent (Fig. 96.3). Nowadays, not only the classical ketogenic diet but also diets that significantly increase protein intake like the Atkins diet are often referred to as ketogenic diets, but it must be considered that practicing and maintaining a ketogenic diet in humans involves considerable difficulty. In any case, it is considered difficult to verify the anti-aging effects of a ketogenic diet in humans through randomized controlled trials, and it is expected that the identification of biomarkers reflecting some anti-aging effects and the verification of effects by drugs that mimic the in vivo action of a ketogenic diet will be needed.

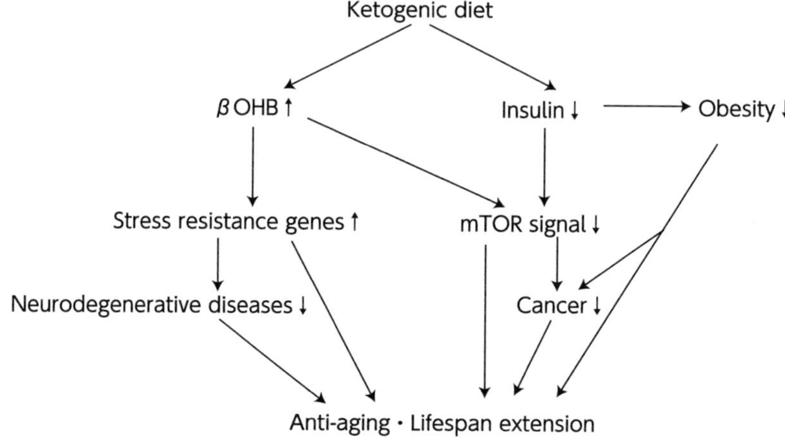

Fig. 96.3 Anti-aging effects of a ketogenic diet (hypothesis). *mTOR* mammalian target of rapamycin

References

1. Shimazu T, Hirschey MD, et al. Suppression of oxidative stress by β-hydroxybutyrate, an endogenous histone deacetylase inhibitor. Science. 2013;339:211–4.
2. Hertz L, Chen Y, et al. Effects of ketone bodies in Alzheimer's disease in relation to neural hypometabolism, β-amyloid toxicity, and astrocyte function. J Neurochem. 2015;134:7–20.
3. Lee MB, Hill CM, et al. Antiaging diets: separating fact from fiction. Science. 2021;374:eabe7365.
4. Newman JC, Covarrubias AJ, et al. Ketogenic diet reduces midlife mortality and improves memory in aging mice. Cell Metab. 2017;26:547–57.e8.
5. Roberts MN, Wallace MA, et al. A ketogenic diet extends longevity and Healthspan in adult mice. Cell Metab. 2017;26:539–46.e5.

Nutrition, Diet, and Anti-aging: Low AGE Diet and Anti-aging

97

AGEs Derived from Dietary Habits

Sho-ichi Yamagishi

Monosaccharides such as glucose and fructose react non-enzymatically with the amino groups of proteins, lipids, and nucleic acids to form Schiff bases and Amadori products. This reaction, although slow, leads to the irreversible formation of a yellow-brown substance with a unique fluorescence, known as advanced glycation end products (AGEs), through repeated dehydration and condensation reactions. In recent years, it has been revealed that AGEs are not only produced endogenously under oxidative stress and hyperglycemia, but are also ingested exogenously from food and smoking, and about 7% of AGEs derived from diet and tobacco remain in the body for a certain period of time [1]. It is known that there is a positive correlation between the estimated intake of AGEs from dietary records and interviews and serum AGE levels, and exogenous AGEs are said to account for about one-third of the AGEs present in the body. Some AGEs, such as pentosidine and crossline, emit specific fluorescence due to their structure [2]. Utilizing this property, a method has been developed in recent years to estimate the accumulation of AGEs in the body by non-invasively quantifying the skin autofluorescence (SAF) value using the AGE Reader (manufactured by Diagnoptics Technologies B.V., imported and sold by Serista Co., Ltd.). Although the fluorescent substances in the skin include not only AGEs but also other biological substances, it has been reported that the SAF value correlates with the amount of fluorescent and non-fluorescent AGEs in plasma, and 76% of the SAF value is derived from AGEs accumulated in the skin [3]. From a survey of Japanese residents, it has been revealed that the accumulation of AGEs estimated from the SAF value is accelerated by distortions in lifestyle habits (smoking, lack of exercise, mental stress, lack of sleep, heavy drinking, skipping breakfast, consuming sweets, fried foods, and processed foods) independent of age [4].

S.-i. Yamagishi (✉)
Department of Medicine, Showa University School of Medicine, Tokyo, Japan
e-mail: shoichi@med.showa-u.ac.jp

97.1 Dietary AGEs and Aging Face

It has been reported that perceived age can predict future cardiovascular disease risk and lifespan, and serve as an indicator of aging. In the general Japanese population, it has been found that the SAF value significantly correlates with the decrease in grip strength of the dominant hand and "perceived age," or the degree of an aging face. The decrease in grip strength of the dominant hand can be a marker predicting future total mortality and cardiovascular death [5], suggesting that dietary AGEs are involved in "perceived age," or signs of aging in appearance, and are related to aging phenomena inside and outside the body, including muscle weakness.

H. Yamada, Y. Naito (eds.), *Anti-Aging Medicine*, https://doi.org/10.1007/978-981-97-8514-8_97

97.2 Dietary AGEs and Cancer

People who consume a high amount of AGEs derived from food have a higher risk of developing pancreatic and breast cancer, and it has been reported that an increase in AGE intake is involved in the increased risk of pancreatic cancer associated with increased red meat intake. Furthermore, in our joint research with the "European Prospective Investigation into Cancer and Nutrition (EPIC)," we have found that regardless of the presence or absence of diabetes, a high blood AGE level increases the risk of developing rectal cancer 4 years later.

97.3 Dietary AGEs and Osteoporosis, Cognitive Function

In the Cardiovascular Health Study, which followed 3373 people for an average of about 9 years, it was reported that the higher the blood AGE level, the higher the risk of proximal femur fractures, and even after adjusting for other factors such as body mass index, smoking, and kidney disease, the risk of proximal femur fractures increases by 1.17 times for each 1SD increase in blood AGE levels. In addition, regardless of the presence or absence of diabetes, it has been clarified that the higher the urinary AGE excretion level, the more likely cognitive function will decline after 9 years.

97.4 AGE as a Biomarker of Lifespan

In the CHIANTI study, which targeted about 1000 elderly people aged 65 and over, and the Baltimore Longitudinal Study of Aging, which targeted about 450–750 community-dwelling elderly people, it has been revealed that the blood level of carboxymethyllysine, one of the AGEs, correlates with decreased kidney function, anemia, and slow walking speed, and can be a biomarker predicting all-cause and cardiovascular death independent of classical risk factors. In a follow-up survey of 72,880 general residents without diabetes or cardiovascular disease for 4 years, it was reported that for each unit increase in SAF value measured by AGEReader, the risk of developing type 2 diabetes or cardiovascular disease triples, and the risk of death increases fivefold. And for people over 36 years old, the SAF value is the strongest predictor of death.

97.5 Evidence of Low AGE Diet

According to the results of a meta-analysis of randomized trials that intervened in the amount of dietary AGEs, a low AGE diet reduces blood inflammation, oxidative stress, vascular damage markers, cholesterol, AGE levels, increases adiponectin and longevity genes, and improves insulin resistance in diabetic patients. Furthermore, it has been reported that feeding a diet restricted in AGEs extends the lifespan of mice without calorie restriction, and that even if calorie restriction is performed, if AGE restriction is not performed at the same time, no lifespan extension effect is observed. This suggests that the amount of dietary AGEs derived from different dietary habits may affect age-related diseases and lifespan.

97.6 Ageless Life with AGE-Less Food

Generally, it is known that a large amount of AGEs are produced when meat products, fatty foods, and processed foods rich in fat are fried or grilled at high temperatures. On the other hand, cooking methods that use a lot of water and take a long time, such as steaming or boiling slowly, are less likely to produce AGEs. Fast food, which is high in calories and fat and often involves high-temperature cooking, should be avoided as much as possible. Also, the AGE reaction has long been known as the Maillard reaction or browning reaction. The browning of food can be a rough guide to the amount of AGEs it contains. In addition, fructose is about ten times more likely to form AGEs than glucose. We should advise people to avoid consuming large amounts of beverages containing high-fructose corn syrup, excessive intake of fructose, and highly processed foods with significant browning. Also, it is known that lemon and vinegar can suppress the production of AGEs during the cooking process, such as frying. Furthermore, medically activated charcoal such as Kremezin, high-purity cellulose charcoal (Dietary Carbon® pure charcoal powder, Dustech Co., Ltd.), chitin and chitosan contained in mushrooms and sakura shrimp, are suggested to have the potential to inhibit the absorption of AGEs. A diet rich in vegetables and seaweed, which are abundant in dietary fiber and phytochemicals, is recommended. We have clarified that the amount of AGEs in food can be estimated from the ingredients and cooking methods ("AGE Data Book: Understanding Ageless Meals in Numbers" supervised by Shoichi Yamagishi, Banraisha, 2019). There is a strong positive correlation between the actual measured values of AGEs in food reported by Uribarri et al. and the predicted values using our estimation formula. First of all, it may be useful to practice dietary guidance so that the total daily intake of AGEs is 15,000 kilo units or less, preferably 10,000 kilo units or less.

References

1. Yamagishi S, Matsui T. Pathological role of dietary advanced glycation end products in cardiometabolic disorders, and therapeutic intervention. Nutrition. 2016;32:157–65.
2. Yamagishi S, Fukami K, et al. Evaluation of tissue accumulation levels of advanced glycation end products by skin autofluorescence: a novel marker of vascular complications in high-risk patients for cardiovascular disease. Int J Cardiol. 2015;185:263–8.
3. Yamagishi S. Are finger skin fluorophores other than advanced glycation end products (AGEs) associated with impaired musculoskeletal properties? J Gerontol A Biol Sci Med Sci. 2020;75:401–2.
4. Isami F, West BJ, et al. Association of advanced glycation end products, evaluated by skin autofluorescence, with lifestyle habits in a general Japanese population. J Int Med Res. 2018;46:1043–51.
5. Leong DP, Teo KK, et al. Prognostic value of grip strength: findings from the Prospective Urban Rural Epidemiology (PURE) study. Lancet. 2015;386:266–73.

98 Nutrition, Diet, and Anti-aging: The Reality of Anti-aging Menus

Tomohisa Takagi and Yuji Naito

Kyoto Prefecture has regional characteristics in population dynamics, with the aging rate increasing as you move north from the southern part of the prefecture. The northernmost region of Kyoto Prefecture, the Kyotango area (Kyotango City, Miyazu City, Yosano Town, Ine Town), has become a super-aged region with an aging rate of 37.6% (as of March 2020, according to the Basic Resident Register). Given that the aging rate in Kyoto City during the same period is reported to be 26.5%, the Kyotango region is notably aging. However, this region is also known for its longevity, having produced Jiroemon Kimura, who is listed in the Guinness Book of Records as the world's oldest man (he died on June 12, 2013, at the age of 116). In fact, in Kyotango City, Kyoto Prefecture, the number of elderly people over 100 years old (centenarians) per 100,000 population is 160, nearly three times the national average, making it one of Japan's leading longevity regions. Therefore, in 2017, Kyoto Prefectural University of Medicine established the Longevity and Regional Epidemiology Course (led by Professor Seimei Matoba) and began research to uncover the secrets of healthy longevity in this region. So far, it has been revealed that in this region, the incidence of colon cancer is low, the onset rate of dementia is low, there are many people with young vascular age, and the frequency of sarcopenia onset is low. In the analysis of intestinal bacteria, many butyric acid-producing bacteria are detected, which is a characteristic feature. It is believed that the food culture rooted in this region plays an important role in this longevity, and the traditional Japanese food culture that is being inherited is attracting attention as a potential source of healthy longevity. Therefore, in this section, I would like to introduce the food culture in this region as an anti-aging menu. In a dietary survey in this region, it was interesting to find that a higher proportion of elderly people in Kyotango City consume food ingredients rich in dietary fiber, such as whole grains, seaweeds, vegetables, and beans, every day compared to urban elderly residents (Fig. 98.1) [1]. In a dietary lifestyle survey conducted earlier, it was revealed that they consume a variety of beans from childhood, consume seaweeds in various cooking methods, consume a lot of wild vegetables, consume a lot of seeds (especially sesame), and use dried sardines in almost all dishes. These unique local foods are being eaten against this

T. Takagi (✉)
Molecular Gastroenterology and Hepatology, Kyoto Prefectural University of Medicine, Kyoto, Japan

Department for Medical Innovation and Translational Medical Science, Graduate School of Medical Science, Kyoto Prefectural University of Medicine, Kyoto, Japan
e-mail: takatomo@koto.kpu-m.ac.jp

Y. Naito
Department of Human Immunology and Nutrition Science, Kyoto Prefectural University of Medicine, Kyoto, Japan
e-mail: ynaito@koto.kpu-m.ac.jp

H. Yamada, Y. Naito (eds.), *Anti-Aging Medicine*, https://doi.org/10.1007/978-981-97-8514-8_98

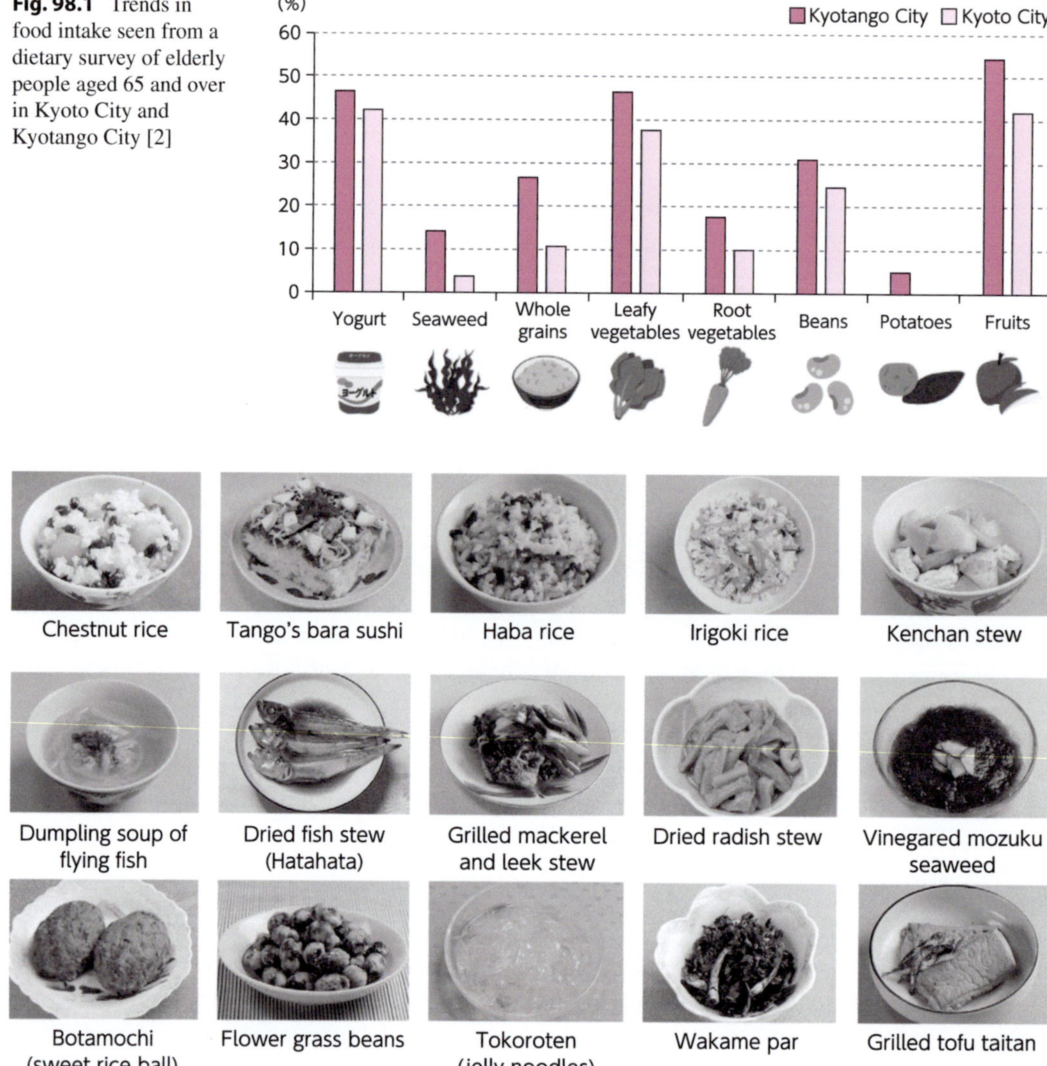

Fig. 98.1 Trends in food intake seen from a dietary survey of elderly people aged 65 and over in Kyoto City and Kyotango City [2]

Fig. 98.2 Representative local foods of the Kyotango region [3]

background and are still being passed down (Fig. 98.2). Interestingly, when checking the protein intake status of centenarians in this region, it is characteristic that they have been consuming a lot of protein from fish and beans continuously from childhood (Fig. 98.3). It is inferred that the health and longevity of the region are maintained by the local food culture. For more details, please refer to the examples listed in "Living Now - The Recipe for a Long Life in Kyotango" (4th Edition) (Produced by the Health Promotion Division of the Health and Longevity Welfare Department of Kyotango City, 2022). If you are interested, please read it (Fig. 98.4).

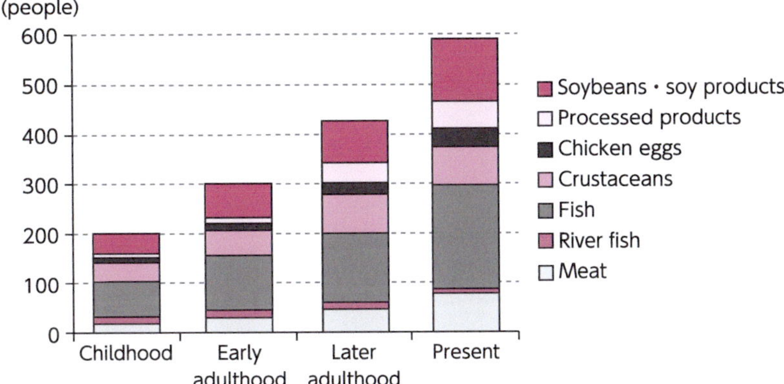

Fig. 98.3 Protein sources that the elderly in Kyotango City have been consuming. Modified with permission from [4]

Fig. 98.4 "Living Now - The Recipe for a Long Life in Kyotango City" [4]

References

1. Naito Y, Takagi T, et al. Gut microbiota differences in elderly subjects between rural city Kyotango and urban city Kyoto: an age-gender-matched study. J Clin Biochem Nutr. 2019;65:125–31.
2. Nikkei Health 2022 Spring, Nikkei.
3. Kyotango City HP. https://www.city.kyotango.lg.jp/top/kosodate_kyoiku/shokuiku/1/1/index.html.
4. Kyotango City Health Promotion Division. Living now—the recipe for a long life in Kyotango. 2013.

99 Nutrition, Diet, and Anti-aging: Cancer Prevention and Nutrition/Diet

Shoichiro Tsugane

99.1 Evidence-Based Cancer Prevention

While the concept of evidence-based medicine is well-established, prevention also needs to be based on evidence. When health care professionals intervene in the unique culture and preferences of individuals and ethnic groups in the name of dietary guidance, there is a need for solid evidence that the probability of getting cancer decreases in return for implementing it. Simply stating that "it has antioxidant effects" or "it suppressed cancer development in mice" is insufficient as evidence. The most reliable evidence comes from intervention studies with random allocation (randomized controlled trials). However, when the incidence of cancer is the endpoint, large-scale long-term studies under strict control are necessary, and the number of studies is limited, so there is not much evidence available. So far, antioxidant nutrients such as β-carotene, vitamin E, and vitamin C have not shown the expected cancer prevention effects theoretically, and it has been shown that high doses of β-carotene and vitamin E increase the risk of certain cancers and cerebrovascular diseases. Also, it has been clearly shown that reducing the intake of fat (fat energy ratio) in adulthood does not prevent breast cancer or colorectal cancer. The next most reliable evidence is provided by cohort studies. This is an observational study that sets up a large target group, first understands factors such as dietary habits through questionnaires, and then tracks the occurrence of cancer over a long period. It is less likely to be biased and relatively reliable. However, there is a limitation that the relationship seen between a factor and cancer incidence may actually be due to a third factor (confounding factor), and the research results may be apparent. When developing cancer prevention methods based mainly on cohort studies, it is necessary to have backing for the mechanism through experiments in animals or in test tubes. In addition, evidence from research methods such as case-control studies and cross-sectional studies can be used, but while they can be conducted in a relatively short time and at low cost, biases and reversals of causality from the time axis are likely to occur. Regardless of the type of research method, the relationship obtained in a single study cannot deny the possibility of chance, bias, and confounding, so it is necessary to evaluate the presence or absence of a causal relationship based on multiple studies.

S. Tsugane (✉)
International University of Health and Welfare Graduate School of Public Health, Tokyo, Japan
e-mail: stsugane@iuhw.ac.jp

H. Yamada, Y. Naito (eds.), *Anti-Aging Medicine*, https://doi.org/10.1007/978-981-97-8514-8_99

Table 99.1 Association between food/nutrition and cancer risk—current status of international causal relationship evaluation [1]

Factors that increase risk		Factors that decrease risk	
Factor	Cancer site	Factor	Site
Salted food	Stomach	Vegetables	Oral, pharyngeal, laryngeal, esophagus
Processed meat	<u>Colon</u>, non-cardia stomach	Fruits	Oral, pharyngeal, laryngeal, esophagus
Red meat	Colon	Dietary fiber	Colon
Glycemic load	Endometrium	Whole grains	Liver, endometrium
Arsenic in drinks	Bladder, <u>lung</u>, and skin	Coffee	Colon
Cantonese-style salted fish	Nasopharynx	Dairy products	Colon
Mate tea	Esophagus	Calcium supplement	Colon
Aflatoxin	<u>Liver</u>		
β-Carotene supplement	<u>Lung</u>		

Underlined convincing; *No underline* probable

99.2 International Current Status of the Evaluation of the Causal Relationship Between Nutrition/Diet-Related Factors and Cancer Risk

The World Cancer Research Fund continuously conducts causal evaluations on the relationship between dietary factors and cancer (http://www.dietandcancerreport.org/). A summary of the dietary factors (excluding alcohol, obesity, physical activity, etc.) that have been determined to be "certain" or "probably" related to cancer sites is shown in Table 99.1. The criteria for a "certain" judgment include evidence from two or more types of epidemiological studies, evidence from at least two cohort studies, no conflicting study results, supported by high-quality studies that can deny the possibility of chance, bias, and confounding, a dose-response relationship that can be biologically explained, and data that it can occur in a normal diet. If there is not enough evidence, it is judged as "probably," and if there is even less, it is judged as "suggestive." Based on the current status of causal evaluation, the following dietary advice for cancer prevention is proposed: (1) Maintain a healthy weight, (2) be physically active, (3) eat whole grains, vegetables, fruits, and legumes, (4) limit fast food, (5) limit red and processed meat, (6) limit sugary drinks, (7) limit alcohol intake, and (8) do not use supplements for cancer prevention. Also, it is recommended that nursing women, if possible, breastfeed, and that people who have had cancer follow this advice.

99.3 Cancer Prevention Methods That Can Be Recommended Based on the Causal Relationship Evaluation and Current Situation Based on Evidence from Japanese People

In Japan as well, several cohort studies involving tens to hundreds of thousands of people have been conducted since around 1990, and a large number of evidences are being reported. The research group at the National Cancer Center is continuously conducting causal evaluations based on evidence from Japanese people (http://epi.ncc.go.jp/can_prev/). The summary is shown in Table 99.2. Since the number of evidences is limited, the judgment is more conservative than the international evaluation. It is "almost certain" that salted foods (stomach) and hot foods and drinks (esophagus) increase the risk, and vegetables and fruits (esophagus) and coffee (liver) decrease the risk. The factors mentioned as "possible" are similar to the international evaluation, such as red and processed meat, dietary fiber, calcium, coffee, etc., but dietary factors characteristic of Japanese people, such as green tea and soybeans/isoflavones, are also shown. Based on

Table 99.2 Association between food/nutrition and cancer risk—current status of causal relationship evaluation by domestic research groups [2]

Factors that increase risk		Factors that decrease risk	
Factor	Site of cancer	Factor	Site
Salted foods	Stomach	Vegetables	Esophagus, stomach
Processed meat/red meat	Colon (female)	Fruits	Esophagus, stomach, lung
Hot food and drinks	Esophagus	Dietary fiber	Colon
Grains	Stomach	Soy	Breast, prostate
		Isoflavones	Breast, prostate
		Calcium	Colon
		Coffee	Liver, uterus, colon
		Green tea	(female)
		Fish	Stomach (female)
		Omega-3 unsaturated fatty	Cervix
		acids from seafood	Colon

Underlined almost certain; *Not underlined* possible

the international evaluation and the causal relationship evaluation based on evidence from Japanese people, the following dietary advice formulated by the Cancer Prevention Research Team is shown. In addition to dietary advice, smoking cessation, moderate drinking, maintaining body shape, physical activity, and infection control are also recommended.

- Diet: Balanced and not biased

  Specifically, the following three points are mentioned. (1) Minimize the intake of salted foods and salt. As a specific goal, consume less than 7.5 g of salt per day for men and 6.5 g for women, and particularly limit high-salt foods (such as salted fish, sea urchin paste, etc.) to once a week. (2) Do not lack vegetables and fruits. As a specific goal, consume 400 g of vegetables and fruits per day (for example, about five small bowls of vegetables and one plate of fruit). (3) Do not consume food and drinks while they are hot. In the future, as evidence accumulates, the content of causal relationship evaluations and dietary advice may change.

References

1. World Cancer Research Fund International. Diet, activity and cancer. http://www.dietandcancerreport.org/. Accessed June 2022.
2. National Cancer Center Cancer Research and Development Fund. Research on evaluation of carcinogenicity and cancer prevention effects based on scientific evidence and proposal of cancer prevention guidelines. http://epi.ncc.go.jp/can_prev/. Accessed June 2022.

100 Stress Management and Anti-aging: Understanding Stress

Sumito Ogawa

100.1 The Situation and Background of Stress in Our Country

According to the Labor Safety and Health Survey (Actual Situation Survey: Reiwa second year) conducted by the Ministry of Health, Labor, and Welfare, 54.2% (58.0% in the Heisei 30th year survey) of workers have strong anxiety or stress related to their current job or professional life, indicating that more than half of the people feel stress [1]. In addition, by age group, 58.3% of people in the 50–59 age group feel stress, which is higher than other age groups. Factors of stress in workers include the quality and quantity of work, failures at work, the occurrence of responsibilities, interpersonal relationships (including sexual harassment and power harassment), the future prospects of the company, complaints from customers and business partners, changes in roles and positions (promotions, upgrades, reassignments, etc.), and employment stability. In addition, in the home, factors such as nursing and caregiving for family members and bereavement are also mentioned, and a wide range of stress measures that take into account psychosocial factors are necessary. Among physical diseases, those in which psychosocial stress is closely involved in the onset and course, and functional and organic disorders are recognized are understood as psychosomatic diseases, and understanding of such psychosomatic correlations has been advancing. With the increasing complexity of social structures in recent years, a variety of psychosocial stresses are likely to occur, and stress-related diseases and psychosomatic diseases are also increasing. In such cases, there are diseases associated with acute stress, and it is not uncommon for lifestyle habits to be greatly disturbed under chronic stress conditions, leading to further diseases. Therefore, the evaluation of stress and its prevention and management measures are important. In 2015, the Labor Safety and Health Law was partially revised, and it became mandatory to conduct stress checks at least once a year, analyze the results, and improve the workplace environment during general health checkups. The importance of early screening of high-risk individuals and prevention of stress-related diseases through workplace environment improvement is shown there.

S. Ogawa (✉)
Department of Geriatric Medicine, The University of Tokyo, Tokyo, Japan
e-mail: suogawa@m.u-tokyo.ac.jp

H. Yamada, Y. Naito (eds.), *Anti-Aging Medicine*, https://doi.org/10.1007/978-981-97-8514-8_100

100.2 What Is Stress?

Stress is understood as a reaction that occurs in the mind and body due to external stimuli, and it is sometimes called stress, which includes external stimuli that cause stress (stressors) and the mind and body's reactions to them (stress responses). According to Dr. Hans Selye, who proposed the "stress theory," it is expressed as a general adaptation syndrome caused by non-specific harmful factors, and a series of non-specific reactions occur physically due to various stressors [2]. Stressors can be broadly divided into physical stressors (heat, cold, noise, congestion, etc.), chemical stressors (harmful substances, air pollution, drugs, alcohol, tobacco, carbon monoxide, etc.), psychological stressors (anxiety, dissatisfaction, anger, sadness, etc.), social stressors (conflicts in human relationships, problems at work and at home, etc.), and biological stressors (bacteria, mold, viruses, pollen, etc.), and various factors can become stressors. In fact, stress is broadly divided into two types: pleasant stress and unpleasant stress, and not all stress has a bad effect. Of these, pleasant stress consists of moderate tension and pressure that increase motivation, loads during exercise, etc., and is thought to contribute to the activation of the sympathetic nervous system and resistance. On the other hand, unpleasant stress refers to excessive stress and chronic stress, which are likely to lead to poor physical condition. Considering that there are large individual differences in overload and coping ability, it is important to notice your own stress condition at an early stage and strive to relieve stress and take sufficient rest.

100.3 The Stress Response and Coping with Them

The stress response caused by stressors is broadly divided into psychological changes (decreased vitality, interest, and concern, anxiety, depression, impatience, confusion, etc.), physical changes (headache, stiff shoulders, lower back pain, eye fatigue, palpitations, shortness of breath, stomach pain, decreased appetite, constipation or diarrhea, insomnia, trembling, etc.), and behavioral changes (increased drinking and smoking, overeating, venting anger, mistakes or accidents at work, increased near-miss incidents, etc.). There may be certain tendencies in the appearance of these changes due to individual differences in constitution, personality, and coping ability, and there may be people who are characterized as being easy to recognize psychological changes, physical changes, and behavioral changes. If these stress responses persist for a long period of time, it is important to review the content of daily life and introduce and devise stress coping to deal well with stressors in order to avoid exhaustion due to excessive stress or chronic stress. Also, if you cannot control the stressor, you need to be aware that various effects and symptoms are likely to appear as maladaptation on the physical side. Specifically, in addition to hypertension, bronchial asthma, peptic ulcer, etc. as psychosomatic diseases, reactive mental disorders (depression, anxiety, confusion, etc.), post-traumatic stress disorder (PTSD), etc., may be presented. Furthermore, if such stress symptoms are severe and persist for a long period of time, consider consulting or visiting medical institutions such as psychiatry and psychosomatic medicine. Also, the main methods for alleviating and eliminating stress include exercise, relaxation, and biofeedback. For exercise, aerobic exercises such as walking, light running, dancing, and cycling are recommended based on the individual's age and physical strength. If it is difficult to exercise outdoors, try to incorporate stretching or yoga indoors. Relaxation methods for relieving physical and mental tension include taking a slow bath and resting, listening to your favorite music, adjusting your breathing, relaxing in nature, autogenic training, meditation, etc. Biofeedback is a method of self-control of the body that aims to control the autonomic nervous system by feeding back normally perceptible and difficult to control biological information (blood pressure, brain waves, heartbeats, electromyograms, etc.) to the subject through sensors and other equipment, and is also used to alleviate and eliminate stress.

References

1. Ministry of Health, Labor and Welfare. Overview of the results of the Reiwa 2nd year labor safety and health survey (actual situation survey). Individual survey. https://www.mhlw.go.jp/toukei/list/dl/r02-46-50_kekka-gaiyo02.pdf.
2. Selye H. A syndrome produced by diverse nocuous agents. Nature. 1936;138:32.

101 Stress Management and Anti-aging: Aging and Stress

Sumito Ogawa

101.1 Stress and Adaptation Associated with Aging

As people age, they encounter and respond to many changes, including changes in household structure, living and working environments, health conditions, and the separation or death of close ones. These changes can often become sources of stress. In such cases, stress may manifest as a depressive state, and it becomes increasingly important to establish various support systems, including social support and health support, and to learn how to coexist with stress. Additionally, it is important to consider how to maintain the ability to maintain functions against aging phenomena and biological damage (robustness) and the ability of the organism to recover to its intended state when subjected to external stress (stability). For many species, aging is an important element in maintaining the population at an appropriate size, but unlike starvation, infection, and predators, it does not act as a driving force for evolution. All life functions are designed to maintain the homeostasis of the organism and are originally defensive. However, with the development of civilization, the lifespan has surpassed the limits of nature, and the damaging aspects of the organism have become apparent. Such cases, where the damaging nature becomes more apparent than the defensive nature due to aging exist in many forms, including oxidative stress, endoplasmic reticulum stress, and other stresses, apoptosis, and aging changes in immune function.

101.2 Aging and Stress Adaptation

Oxidative stress is understood as "harmful effects caused by oxidation reactions" within the organism, and it is believed that damage to the biological constituents caused by various stresses, including oxidative stress and glycation stress, and their chronological accumulation are closely related to aging. In addition to oxidative stress and glycation stress, there are also calorie restriction and starvation stress, telomere shortening, and the hypothalamic-pituitary-adrenal axis; HPA axis. The involvement of the neuro-immune-endocrine axis is also considered important. In living organisms, where temporal aging is observed, the rate of aging is influenced by genetic factors and external environmental factors, and individual differences are recognized. In cells and invertebrates, there are physical stressors (temperature changes, ultraviolet rays, radiation, osmotic pressure, pressure, etc.), chemical stressors (heavy metals, pollutants, etc.), and physiological stressors (starvation, infection,

S. Ogawa (✉)
Department of Geriatric Medicine, The University of Tokyo, Tokyo, Japan
e-mail: suogawa@m.u-tokyo.ac.jp

H. Yamada, Y. Naito (eds.), *Anti-Aging Medicine*, https://doi.org/10.1007/978-981-97-8514-8_101

etc.). In humans, stressors are more diverse, including environmental physical stressors (heat, cold, noise, congestion, etc.), chemical stressors (harmful substances, air pollution, drugs, alcohol, tobacco, carbon monoxide, etc.), psychological stressors (anxiety, dissatisfaction, anger, sadness, etc.), social stressors (conflicts in human relationships, work, and family problems, etc.), and biological stressors (bacteria, mold, viruses, pollen, etc.). Various factors can become stressors. In addition, it has been confirmed that genetic mutations in invertebrates can extend lifespan and increase resistance to environmental stress such as ultraviolet rays and reactive oxygen species. In nematodes with mutations in factors involved in the insulin/insulin-like growth factor 1 (IGF-1) signaling pathway, such as daf-2 (insulin/IGF-1-like receptor) and age-1 (PI3 kinase), oxidative stress resistance is acquired. Also, in nematodes, research has shown that activation of heat shock factor-1 (HSF-1) leads to stress resistance and longevity. Furthermore, in nematodes, when the histone methylation (H3K4me3) modification factor ash-2 gene, which is involved in stress responses such as oxidative stress, is knocked down, stress resistance increases and this stress resistance is inherited by the offspring from the parent [1]. Thus, genetic and environmental factors thought to be related to longevity are shown to potentially lead to stress resistance, and it is thought that maintaining robustness and stability contributes to aging control.

101.3 Aging and Hormesis

Various environmental changes surrounding organisms can cause stress, but different biological responses are activated depending on the intensity of the stress. It is known that there are adaptive responses that work favorably for survival when the stress intensity is weak (hormesis). Hormesis refers to the phenomenon in which a stimulus that is harmful when Severe brings beneficial effects to the organism when mild (Fig. 101.1) [2]. Stressors related to hormesis include chemicals, bioactive substances, heat, ionizing radiation, etc., and it has been pointed out that various stimuli may have beneficial effects on the body, such as aging/lifespan control and anti-aging. The extension of lifespan and delay of aging-related phenotypes associated with calorie restriction are also considered beneficial effects related to hormesis. Resveratrol, a type of polyphenol, is known to interact with the factor SIRT1 involved in lifespan extension during calorie restriction, by increasing SIRT1 activity. In groups of mice given a moderate dose of

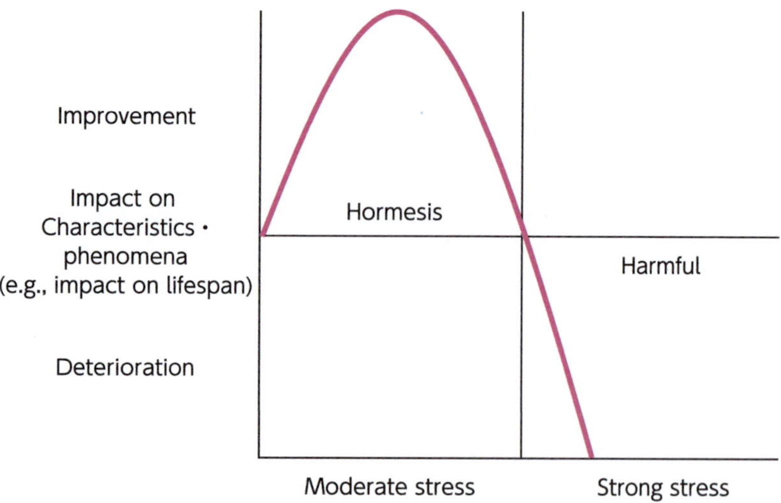

Fig. 101.1 Hormesis [2]. When the stress intensity is moderately weak, it exhibits an anti-aging effect, and when the stress intensity is strong, it promotes aging

resveratrol, it has been reported that they activate AMP-activated protein kinase (AMPK) in a SIRT1-dependent manner and exhibit anti-aging effects. Furthermore, L-deprenyl, a type of MAO inhibitor used as an antidepressant and anti-Parkinson's drug, has been shown to extend lifespan in rodents, etc., while in rats, the group given a low dose of L-deprenyl showed a longer lifespan extension effect and higher induction of antioxidant enzyme (manganese superoxide dismutase, MnSOD) expression in the brain than the group given a high dose. In addition, regarding mitochondrial stress, the relationship between mitochondrial function decline and lifespan extension has been observed in some organisms, such as nematodes, suggesting a connection with hormesis.

References

1. Nono M, Kishimoto S, et al. Intestine-to-germline transmission of epigenetic information Intergenerationally ensures systemic stress resistance in C. elegans. Cell Rep. 2020;30:3207–17.
2. Gems D, Partridge L. Stress-response hormesis and aging: "that which does not kill us makes us stronger". Cell Metab. 2008;7:200–3.

102 Stress Management and Anti-aging: Mental Anti-aging

Takehiro Tamura and Hidehiko Takahashi

102.1 Aging Changes in the Mind

The brain undergoes quantitative and qualitative changes with aging. The former involves a decrease in brain weight due to a reduction in the dendritic protrusions of nerve cells, while the latter involves changes in neurofibrils, senile plaques, and a decrease in vascular density. In addition to these biological changes in the brain, it is believed that physical frailty centered on sarcopenia, which comes with aging, and an increase in psychosocial stress with themes such as "loss," interact with each other, leading to changes in mental function and the emergence of psychiatric symptoms. Changes in mental function include a decline in cognitive processes such as perception, memory, learning, thinking, and judgment (i.e., cognitive decline), mood disorders, apathy, and personality changes, while psychiatric symptoms include depression, anxiety, insomnia, and delusions. Cognitive decline and depression, which can be called pathological aging of the mind, have been proposed to interact with frailty and promote the frailty cycle [1], and mental anti-aging is an essential element for achieving successful aging. In this section, we will first provide an overview of cognitive decline and depression in relation to the frailty cycle. Then, we will discuss stress management as a means of preventing depression and touch on interventions using positive psychology.

102.2 Frailty and Cognitive Decline and Depression

A simplified representation of the relationship between the frailty cycle and cognitive decline and depression is shown (Fig. 102.1). Depression strongly promotes the frailty cycle by causing malnutrition due to loss of appetite and a decrease in activity due to a lack of interest and concern. In addition, cognitive decline interacts with depression and can lead to chronic malnutrition through disruptions in eating habits such as picky eating and refusal to eat. On the other hand, a decrease in activity, which is a component of the frailty cycle, acts as a promoting factor for cognitive decline, and it is evident that these vicious cycles become more entrenched when they persist. It goes without saying that both cognitive decline and depression require multifaceted interventions based on the bio-psycho-social model, but depression is considered to be a reversible condition and psychosocial interventions for its prevention are thought to be more effective (compared to cognitive decline). Depression often occurs as a result of exposure to stressors,

T. Tamura (✉) · H. Takahashi
Department of Psychiatry and Behavioral Sciences, Graduate School of Medical and Dental Sciences
Tokyo Medical and Dental University, Tokyo, Japan
e-mail: hidepsyc@tmd.ac.jp

H. Yamada, Y. Naito (eds.), *Anti-Aging Medicine*, https://doi.org/10.1007/978-981-97-8514-8_102

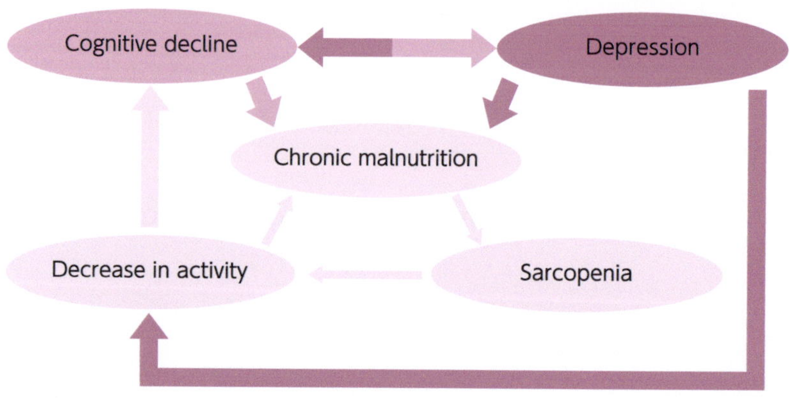

Fig. 102.1 The relationship between the frailty cycle, cognitive decline, and depression. (Adapted from [1] with permission)

and the risk of onset increases under chronic stress conditions. In other words, daily stress management is important in the prevention of depression.

102.3 What Is Stress Management?

Stress management refers to activities aimed at minimizing the negative impact on health brought about by stress responses. Various methodologies are adopted, but a meta-analysis on workplace stress management reports that interventions based on cognitive behavioral therapy and relaxation techniques show particularly high effectiveness [2]. Generally, stress management is a combination of "reducing stress factors" and "strengthening stress resilience," and it is said to be summarized in the reduction in stressors, correction of cognitive evaluation, control of stress responses, development of social support, and development of new coping skills [3]. Coping refers to the cognitive and behavioral efforts of individuals to adapt to stress, and it typically focuses on problems to be solved (problem-focused coping) and negative emotions (emotion-focused coping). On the other hand, social support refers to tangible and intangible support obtained from human relationships surrounding individuals, and it is classified into instrumental (tools/information) and emotional types.

102.4 Stress Management and Positive Psychology

According to the homepage of the Japanese Positive Psychology Association, positive psychology is a discipline that scientifically studies positive human characteristics such as feelings of happiness and gratitude, well-being, strengths, and virtues. In fact, it has been shown by meta-analysis that interventions using positive psychology improve well-being and reduce depressive symptoms [4]. However, in traditional stress science, negative emotions were mainly dealt with, and it was once thought that positive emotions did not bring about appropriate coping [5]. Now, as shown in the revised stress coping theory by Folkman et al., positive emotions occur when a desirable resolution is reached as a result of problem-focused coping, and they are obtained as a result of meaning-based coping. These positive emotions are thought to maintain adaptive coping and lead to the development of more adaptive coping skills. This "meaning-based coping" is to replace unrealistic goals with goals that are realistic and meaningful for the current patient, which is an important element in generating positive emotions during exposure to large stressors. A specific example might be aiming not for the cure of an incurable disease, but for celebrating the next anniversary. Generally, it is considered important to cultivate a habit of consciously looking at positive emotions because ambivalent emotions can exist under stress. However, it is important to emphasize that it is not just blindly focusing on positive emotions, but understanding both positive and negative emotions from a broad perspective.

References

1. Robertson DA, Savva GM, et al. Frailty and cognitive impairment–a review of the evidence and causal mechanisms. Ageing Res Rev. 2013;12:840–51.
2. Richardson KM, Rothstein HR. Effects of occupational stress management intervention programs: a meta-analysis. J Occup Health Psychol. 2008;13:69–93.
3. Fukunaga T. Stress management in the elderly. J Geriatr Psychiatry. 2000;11:1347–52.
4. Sin NL, Lyubomirsky S. Enhancing well-being and alleviating depressive symptoms with positive psychology interventions: a practice-friendly meta-analysis. J Clin Psychol. 2009;65:467–87.
5. Shimai T. Stress management from the perspective of positive psychology. Stress Sci. 2018;32:292–302.

Stress Management and Anti-aging: The Practice of Relaxation

103

Yuzan Shimoguchi

103.1 Stress Removal! Live Long and Healthy! Mental Fitness Method [1, 2]

"The era of health is prevention! The era of self-management!" has been proposed by the Ministry of Health, Labor, and Welfare, advocating "Healthy Japan 21 (Second)." Modern society is an era of anxiety and stress, where stress accumulates unknowingly, people suffer from psychological troubles, interpersonal relationships and work do not go well, many people fall into psychosomatic diseases, and the number of people who are reclusive due to depression, school refusals, and work refusals is rapidly increasing. There were more than 30,000 suicides every year (90 per day) for 14 consecutive years. In this troubled era, there is a scientific and simple self-care method that can eliminate stress and recover vitality, the mental fitness method by autogenic training, and the author strongly feels its necessity. By acquiring this self-care method, everyday life will be full of vitality, it will become a powerful ally in living and a great asset. It is one of the secrets of being active for life and living a long and vibrant life. Autogenic training was publicly introduced by Dr. Schultz, a German psychiatrist, in 1932.It was introduced in Japan in the 1950s by Dr. Gotsaku Naruse, a professor at Kyushu University, under the name "Autogenic Training." In 1961, Dr. Torujiro Ikemi opened the first psychosomatic medicine clinic in Japan and popularized autogenic training. Autogenic training is a method for removing the tension of the whole body by self-suggestion, etc., and to make it an ideal physical and mental state, the "head cold feet hot" state. The author has devised this autogenic training to make it easier to experience, and as a "mental fitness method," he is in charge of stress management and mental health care lectures and training in government offices, companies, medical institutions, and educational institutions more than 200 times a year, contributing to "Healthy Japan 21" (http://www.mental-fitness.jp/). The mental fitness training program devised by the author is implemented in the following four steps. (1) Relax the whole body and remove the power from the body. Utilize the relaxing song "Relaxia" (composed by H. YOSHIKO) and the muscle relaxation method "TORETA" exercise. (2) Practice focusing on one point. (3) Become good at abdominal breathing. (4) With the mental fitness method based on autogenic training theory, both body and mind become smiling, and more positive and vibrant.

Y. Shimoguchi (✉)
Japan Mental Fitness Association, Nara, Japan

Shimoguchi Yuzan Mental Fitness Association, Nara, Japan

H. Yamada, Y. Naito (eds.), *Anti-Aging Medicine*, https://doi.org/10.1007/978-981-97-8514-8_103

103.2 The Formula of Autogenic Training

Formula 0… Tranquility "Feeling calm."
Formula 1… Heaviness "Both hands and feet are relaxed and heavy. Feeling at ease." Formula 2… Warmth "Both hands and feet are warm."
Formula 3… Heart Regulation "The heart is beating quietly."
Formula 4… Breathing Regulation "Breathing is easy."
Formula 5… Abdominal Warmth "The area around the stomach (solar plexus) is warm."
Formula 6… Forehead Coolness "The forehead is cool."

The autogenic training method consists of these formulas 0–6. Most of the effects obtained by the autogenic training method can be experienced through practice up to the second formula. The author has devised formulas 0–2 to be more easily felt and calls it "Mental Fitness Method ①".

103.3 Experiencing Mental Fitness Method ①

It is a rational stress relief method, and the following sensations can be experienced.

- Brain waves become alpha waves, and accumulated fatigue can be removed.
- Irritation and frustration are removed, feelings become calm, and human relationships go smoothly.
- Self-control is well managed, and impulsive actions decrease.
- Concentration increases, and work or learning efficiency improves.
- Physical pain such as insomnia, shoulder stiffness, and irregular physical condition, and mental pain decrease.
- Self-improvement increases, and goal achievement and problem-solving abilities improve.

103.4 When Practicing Mental Fitness Method ①

Location: Start in a quiet and calming place. Posture: Sit deeply in a chair, with both feet forward of the knees. Practice: Practicing at wake-up and bedtime is highly effective.

103.5 The Reality of Mental Fitness Method ①

The method of Mental Fitness Method ① is shown in Fig. 103.1. Which of the following describes your feelings after finishing? (1) No change. (2) A little sleepy. (3) Body feels lighter. (4) Head feels refreshed. Those who are (1) should continue to practice. Those who are (2) will feel refreshed if they practice again.

Fig. 103.1 Mental fitness method ①. (1) Sit comfortably in a chair and close your eyes. Move your toes forward of your knees. Lower both hands to your sides. (2) Lift your toes, clench your hands, lift your shoulders. Close your mouth and take a deep breath through your nose, then hold it. (3) Drop your toes, open your mouth slightly and exhale while lowering your shoulders and opening your hands. (4) Next lift your heels, clench your hands, lift your shoulders. Close your mouth and take a deep breath through your nose, then hold it. (5) Drop your heels, open your mouth slightly and exhale while lowering your shoulders and opening your hands. (6) Shake both hands and feet loosely, and relax the power of both hands and feet. (7) Rotate both shoulders, also do the opposite rotation. (8) Swing both hands in front of your face, then place both hands on your knees. (9) Rotate your head largely, also in the opposite direction. (10) Close your eyes, mouth, and hands tightly (twice). (11) With your eyes closed, open your mouth and hands widely (twice). (12) Perform abdominal breathing, exhaling long breaths and relaxing both hands and feet three times. (13) Focus on the warmth of both hands and feet. "Both hands and feet are warm." (14) Self-suggest. "I am relaxed. I can concentrate. I can do it easily at my own pace. Everything is getting better day by day." (15) Awaken. "I feel good." Think, open your eyes, and stretch twice

References

1. Shimokuchi Y. Relaxation is the path to success. Osaka: Japan Mental Fitness Association; 1995.
2. Shimokuchi Y. Shower of tranquility/desire · goal achievement image training method. (CD).

104 Stress Management and Anti-aging: Happiness and Anti-aging Medicine—Happy People Live Longer!

Kazuo Tsubota

There are two aspects to the relationship between health and happiness. One is the idea that being healthy and living a long life makes you happy, with health being the cause. The other is the idea that being happy leads to health and a long life. The latter idea is encapsulated in the title "Happy People Live Longer." [1] Various epidemiological studies have confirmed that this pathway indeed exists [2, 3]. Anti-aging medicine is not just a theoretical discipline but a new field of medicine that includes "practice based on science." It requires the will to practice and the passion to continue. Happiness, the title of this section, is also a purpose of life, and it can be thought that people aim for longevity because they are happy. If life is boring, it is better to quit early, but if life is happy, let us enjoy it for a long time. Among smokers, there are those who say, "If life without smoking is boring, it's not worth living. I'm going to live a short and fun life." Even in response to calorie restriction, a fundamental part of anti-aging medicine, many people think, "If life without being able to eat delicious food to my heart's content (because it's not happy), there's no point in living." Happiness and anti-aging medicine can be said to have a close relationship in various aspects. In states of depression or excessive stress, mortality rates increase, and the practice of anti-aging medicine becomes difficult. Conversely, if one is more than ordinarily happy, physical functions improve, and the practice of anti-aging medicine becomes even easier. For a long time, happiness was difficult to study scientifically [4]. This was because it was difficult to measure happiness objectively, making a scientific approach was challenging. However, this trend is changing. Recently, there has been a movement to evaluate happiness using scientific methods. Economists have begun to study happiness and economic activity, and neuroscientists have begun to study the connection between brain activity, emotions, and happiness [5]. Although it is still a developing field, in this section, I would like to discuss anti-aging medicine and happiness. How can people become happy? And can we increase the success rate of anti-aging medicine?

104.1 We Are Beginning to Be Able to Objectively Measure Happiness

A method has been developed to understand happiness in real-time by reporting to a research center via mobile devices for each event, which is considered the most reliable way to investigate happiness. This involves converting the sense of

K. Tsubota (✉)
Department of Ophthalmology, Keio University School of Medicine, Tokyo, Japan

Tsubota Laboratory, Inc., Tokyo, Japan
e-mail: tsubota@tsubota-lab.com

H. Yamada, Y. Naito (eds.), *Anti-Aging Medicine*, https://doi.org/10.1007/978-981-97-8514-8_104

Table 104.1 Example of a description by DRM. For each event, the positive and negative degrees are graded from 0 to 6, and the difference is converted into "happiness." For example, if the positive degree is 6 but there is a negative degree of 2, the difference is 4 (this number is an example from the author)

Event	Time (min)	Positive	Negative	Difference
Waking up and stretching in bed	30	6	0	6
Changing clothes	5	4	2	2
Breakfast	15	6	0	6
Dash to the station	5	6	0	6
Commuting train	45	6	2	4

happiness at each moment into a numerical value and reporting it. For example, if you commute by train, you might report that the event was negative, or if you had a meal with your family, you might report that it was perfect. If these data are accumulated over a long period of time with many examples, it will be possible to investigate factors related to the sense of happiness. However, this method is costly. It was a drawback that the burden on the subjects was large because they had to report every time. Therefore, a method called the daily reconstructive method (DRM) was developed. This involves dividing the day into events (about 24 on average), and describing the positive and negative values of each in stages with numerical values from 0 to 6 [6]. These data are then collected at the end of the day. It has been found that there is not much difference in the results compared to the real-time method, so DRM has become more efficient and realistic. Let us give an example of a description by DRM (Table 104.1). In this way, by recalling your own sensations and expressing them in seven stages of numbers, and multiplying the difference subtracted from positive to negative by time, you can quantitatively evaluate the happiness of the day. In addition to this, separate from DRM, interesting things have been found by asking various questions about overall life satisfaction, economic satisfaction, etc. For example, it has been found that DRM tends to be low in the morning and high at night, especially when young, the DRM in the morning is low. Also, among various parameters of daily life, it has been found that the length of sleep time correlates greatly with DRM. Ensuring sleep time is an important lifestyle in anti-aging, and it is an interesting issue whether happiness is involved in the relationship between sleep time and longevity. Also, the pressure from the boss to "have to work quickly" correlates with DRM. On the other hand, it is interesting to find out that things generally considered important, such as "I might lose my job," actually do not have much impact on DRM. Although it seems like a simple method, by expressing happiness, which is originally subjective, in numerical terms, a method is being developed to treat happiness as a science.

104.2 Two Types of Happiness

Research has revealed something interesting. It is that the level of happiness measured by DRM (Day Reconstruction Method) does not necessarily correlate with the overall happiness in life. For instance, if there has been a life event like "divorce," when asked about the overall quality of life, the evaluation tends to be low because people are often bound by the concept of "I have failed." They convince themselves that their life is unhappy. However, when evaluated with DRM, the positive values are quite high. When you break up with someone you dislike or are freed from troublesome issues, there is a high possibility that your daily life becomes happier. This is easy to understand. If there was dissatisfaction in married life, enduring it would lower the DRM. However, if evaluated within the story of "I married the person I love and now have children," you can affirm your current self, and it becomes a successful life overall, raising the overall evaluation. It is understood that there are two types of happiness: one that can be evaluated by DRM in each situation, and the overall image of happiness in life.

104.3 Ways to Increase Happiness

The level of happiness that can be measured by the DRM can be significantly improved by getting a good night's sleep. This information is valuable because many people do not think there is a significant correlation between their happiness and sleep time. In order to feel happiness and properly practice anti-aging medicine, it is necessary to sleep sufficiently first. So, are there any other methods? Recently, the concept of "focus illusion" has been attracting attention. First, there is an interesting experiment. Suppose we ask two questions: (1) Have you been dating a lot recently? (2) Are you generally happy? If we ask (1) before (2), a correlation arises (correlation coefficient 0.66), suggesting a relationship between the number of dates and happiness. If you are asked if you have been dating and you think, "I haven't been dating recently," and then you are asked, "Are you happy?" you might think you are not so happy. However, if we reverse the order of these questions and ask (2) before (1), there is no correlation at all. The same results are obtained even if we ask about marriage or money instead of dating. What does this suggest? One hypothesis is that people lack the ability to judge whether they are generally happy. If the purpose of life is survival, there may have been no need to constantly monitor whether one is happy. It seems that when something suddenly catches our attention, we judge our happiness based on that thing (focus on it). This thing could be marriage, promotion, or money, according to the hypothesis. This is called focus illusion. Dr. Kahneman and his colleagues conducted a meticulous study to investigate whether "money" is really just a focus illusion and what really relates to happiness. Instead of asking, "Are you generally happy?" they asked based on a new measure, "What percentage of your waking time were you in a bad mood?" They also made predictions about how much each factor would influence the focus illusion and compared them with the actual survey. The survey items were as follows: (1) Annual income below $20,000 and above $100,000 (money), (2) Whether women over 40 are married or not (marriage), (3) Whether the level of supervision by the boss is low or high (freedom), and (4) Whether welfare such as health insurance is well-provided (welfare). Interestingly, it was found that all four had much less impact than expected (Table 104.2). From these survey results, it can be concluded that these four factors were focus illusions (as a result, the boss's supervision does affect "moodiness," but less than expected). Especially for women over 40, despite expecting that being married would undoubtedly lead to happiness, when examining the number of hours of moodiness, it turns out that those who are married are moody for longer, completely contrary to expectations. Money does have some impact, especially for those earning less than $20,000 and more than $100,000, but not as much as one might think. Looking back at history, in our country, the average annual income increased fivefold from 1958 to 1987, but it is said that there was no change in

Table 104.2 Unhappy time (%) [4]

		Actual	Expected (focus illusion)
Annual income	Less than 20,000 dollars	32.0	57.7
	More than 100,000 dollars	19.8	25.7
Marriage	Single	21.4	41.1
	Married	23.1	27.9
Marriage	Strict supervision	36.5	64.3
	Not strict	19.1	22.3
Welfare	Without health insurance	26.6	49.7
	With health insurance	22.2	19.2

There is no difference in "unhappy time" whether you are married or have health insurance. The one that most affects your "unhappiness" is your boss. (Quoted from Kahneman D, Krueger AB, et al.: Would you be happier if you were richer? A focusing illusion. Science 2006; 312: 1908–10)

the happiness of the people. However, globally, there are studies that show a strong correlation between money and happiness in countries with an average income of less than $12,000 (about 1.4 million yen per year). Certainly, it is impossible to be happy in a situation where even food is a problem. At the very least, if you do not have enough money to satisfy your hunger, you cannot be happy. However, in our country, where there are hardly any people dying of starvation, the conclusion is, "Money doesn't affect your happiness that much. It's a focus illusion." If it is a focus illusion, you should be able to feel happiness by getting rid of that focus from your mind. For example, if you have a focus illusion that you cannot be happy unless you become rich, you should think of the theory that "there is no correlation between annual income and happiness." Also, recent studies have found that how much money you spend on others, rather than how much money you earn, is related to happiness, and research on the relationship between money and happiness is expected to progress further in the future [7, 8].

104.4 Happy People Live Longer

Also, in the last decade or so, new data have emerged that happy people live longer and are less likely to get sick. Traditionally, the idea was that "you're happy because you're healthy and live a long life" (of course, that path exists too), but now the reverse flow of "happy people live longer" has emerged [1, 9]. Andrew Steptoe, an epidemiologist in London, has beautifully demonstrated this relationship in the London Longevity Study [10]. Also, research by Shoichiro Tsugane of the National Cancer Center and Hiroyasu Iso of Osaka University has shown that people who enjoy life are less likely to get sick [11]. In our research, we have found that happiness and dry eye symptoms are inversely correlated in dry eye [12] (Fig. 104.1). It is unclear whether people with dry eye are unhappy or whether unhappy people are more likely to feel dry eye symptoms, but it is interesting that the relationship with disease has become so clear. Also, related research suggests that being relaxed and dominated by the parasympathetic nervous system is associated with happiness. It is said that deep breathing relaxes tension, and in the same study on dry eye, it was found that just doing abdominal breathing for 3 min makes the parasympathetic nervous system dominant and increases tear production [13] (Fig. 104.2). The fact that tears can come out just by relaxing for 3 min indicates the importance of this kind of research on happiness, relaxation, and parasympathetic dominance for various diseases. Indeed, the positive elements of health through yoga and exercise are thought to be not only physical activ-

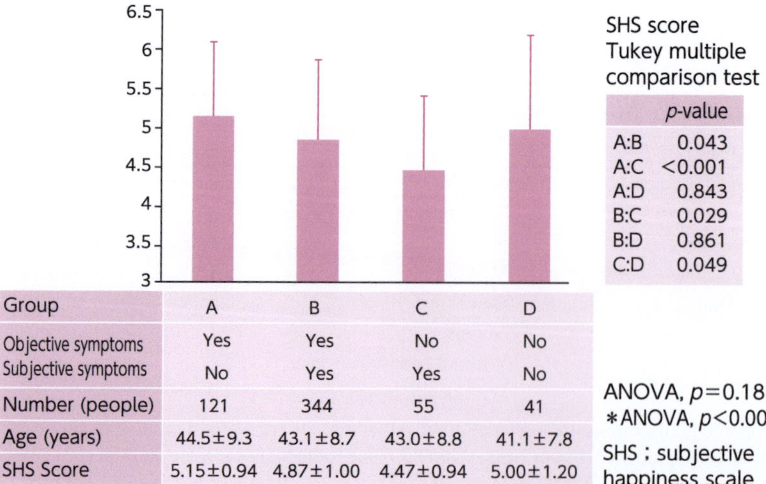

	p-value
A:B	0.043
A:C	<0.001
A:D	0.843
B:C	0.029
B:D	0.861
C:D	0.049

Group	A	B	C	D
Objective symptoms	Yes	Yes	No	No
Subjective symptoms	No	Yes	Yes	No
Number (people)	121	344	55	41
Age (years)	44.5±9.3	43.1±8.7	43.0±8.8	41.1±7.8
SHS Score	5.15±0.94	4.87±1.00	4.47±0.94	5.00±1.20

Fig. 104.1 Comparison of happiness by objective/subjective symptoms of dry eye [12]

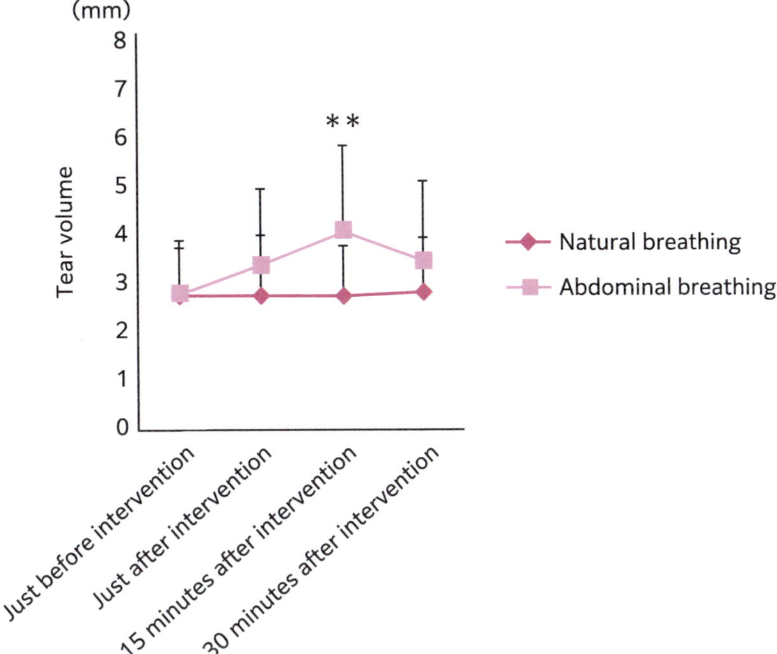

Fig. 104.2 Increase in tear volume by abdominal breathing [13]. Abdominal breathing (relaxation) increases tear secretion

ity but also a significant mental aspect. Furthermore, the authors are conducting research by creating cheerful mice, ordinary mice, and grumpy mice to investigate the relationship between cheerfulness, longevity, and disease. It is interesting that the concentration of brain derived neurotrophic factor (BDNF) in the brain changes and various positive effects occur [14]. The Japan Positive Psychology Medical Association was established in 2012, and research is progressing. It is a field where future development is expected.

104.5 Happiness (Cheerfulness) Can Be Chosen

If you can free yourself from focus illusion, you can realize the essence of life and get closer to a sense of happiness. Regardless of annual income, appearance, or health condition, if you can achieve a sense of happiness just by changing your way of thinking and feeling, it could be said, in extreme terms, that "happiness (cheerfulness) can be chosen" [1]. Of course, simply choosing happiness is psychologically impossible. Understand the science-based theory of happiness, acknowledge the focus illusion, and change your lifestyle that affects happiness. All values are not in the thing itself, but are created by our brains that accept it, which has become clear with the progress of neuroscience. In recent interesting research, it has been proven that when using marketed products (advertised products), the actual blood flow in the brain changes and the same thing feels better [15]. Even if influenced by advertising information, everything is determined by our own thoughts and sensibilities. The purpose of anti-aging medicine is to achieve healthy longevity. Considering this, happiness is not only the goal of anti-aging medicine, but also a powerful force to achieve anti-aging (Fig. 104.3). These ideas, which were once merely theoretical, have been substantiated by the development of research in this field. Recently, there has been growing skepticism towards the notion that positive people are special and that those who accurately perceive reality are normal [16]. The hypothesis is that even normal people have a positive bias and perceive their current situation positively. Conversely, if one perceives the world too accurately, they may glimpse a miserable situation and become prone to depression. The reason we can keep a straight

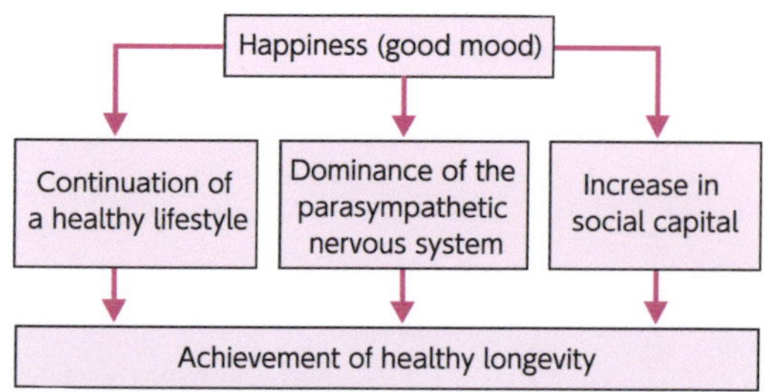

Fig. 104.3 Relationship between happiness (good mood) and healthy longevity

face and stay cheerful even in difficult situations is because of this positive bias. This mechanism is embedded down to the dopamine levels in the corpus callosum, and it has been found that a lack of dopamine reduces positive bias. This science shows that human cognition is inherently biased [17], and it is biased in a positive direction. The Homo sapiens, who were born in East Africa 200,000 years ago, were able to cross vast seas and spread because they had a thought pattern like "There must be something interesting beyond this sea. And I can cross this sea (unfounded confidence)." The elucidation of these essential human thought patterns is expected to have a significant impact on anti-aging medicine as a cheerful science in the future.

References

1. Diener E, Chan MY. Happy people live longer: subjective well-being contributes to health and longevity. Appl Psychol Health Well Being. 2011;3:1–43.
2. Steptoe A, Deaton A, et al. Stone, subjective wellbeing, health, and ageing. Lancet. 2015;385:640–8.
3. Telzer EH, Fuligni AJ, et al. Neural sensitivity to eudaimonic and hedonic rewards differentially predict adolescent depressive symptoms over time. Proc Natl Acad Sci USA. 2014;111:6600–5.
4. Kahneman D, Krueger AB, et al. Would you be happier if you were richer? A focusing illusion. Science. 2006;312:1908–10.
5. Canli T, Sivers H, et al. Amygdala response to happy faces as a function of extraversion. Science. 2002;296:2191.
6. Kahneman D, Krueger AB, et al. A survey method for characterizing daily life experience: the day reconstruction method. Science. 2004;306:1776–80.
7. Dunn EW, Aknin LB, et al. Spending money on others promotes happiness. Science. 2008;319:1687–8.
8. Vohs KD, Mead NL, et al. The psychological consequences of money. Science. 2006;314:1154–6.
9. Frey BS. Psychology. Happy people live longer. Science. 2011;331:542–3.
10. Chida Y, Steptoe A. Positive psychological well-being and mortality: a quantitative review of prospective observational studies. Psychosom Med. 2008;70:741–56.
11. Shirai K, Iso H, et al. Perceived level of life enjoyment and risks of cardiovascular disease incidence and mortality: the Japan public health center-based study. Circulation. 2009;120:956–63.
12. Kawashima M, Uchino M, et al. Associations between subjective happiness and dry eye disease: a new perspective from the Osaka study. PLoS One. 2015;10:e0123299.
13. Sano K, Kawashima M, et al. Abdominal breathing increases tear secretion in healthy women. Ocul Surf. 2015;13:82–7.
14. Sano K, Kawashima M, et al. Enriched environment alleviates stress-induced dry-eye through the BDNF axis. Sci Rep. 2019;9:3422.
15. Plassmann H, O'Doherty J, et al. Marketing actions can modulate neural representations of experienced pleasantness. Proc Natl Acad Sci USA. 2008;105:1050–4.
16. Yamada M, Uddin LQ, et al. Superiority illusion arises from resting-state brain networks modulated by dopamine. Proc Natl Acad Sci USA. 2013;110:4363–7.
17. Taylor SE, Brown JD. Illusion and well-being: a social psychological perspective on mental health. Psychol Bull. 1988;103:193–210.

105 Functional Foods and Anti-aging: Classification and System of Functional Foods

Ryuichi Morishita

105.1 What Are Functional Foods?

To be precise, there is no classification called functional foods. What is commonly referred to as health foods may feel like functional foods, but legally, only "pharmaceuticals and other foods regulated by the Act on Securing Quality, Efficacy and Safety of Products Including Pharmaceuticals and Medical Devices" (Pharmaceuticals and Medical Devices Act) exist. However, there are foods for health use that are allowed to display functionality under certain conditions, and if anything, functional foods could be said to be foods for health use. Prior to 2015, in addition to pharmaceuticals regulated by the Pharmaceuticals and Medical Devices Act, there were nutritionally functional foods defined by the Health Promotion Act and Foods for Specified Health Use (FOSHU), and these two were health functional foods, with so-called health foods in the other category. So-called health foods are no different from ordinary foods from a legal perspective, and words like functionality and effect cannot be written. Therefore, abstract advertisements such as image advertisements that suggest knee rotation and personal impressions were the main focus, and consumers were not provided with the information necessary for purchase. In foreign countries such as the United States and the EU, functional displays are common, and there are few countries where functional displays cannot be made. As a result, it was prohibited in our country at that time, and we were in a situation where we could not tell what was real, and the overwhelming majority of inquiries from consumers were about the ingredients and intake methods of products, which did not match consumer needs. In a survey by the Consumer Committee, it was found that 60% of people wanted a functional display, and in our country, a new system of functional display foods, which allows the display of functionality based on notifications from companies, started in April 2015 (Fig. 105.1). Functional display foods have been accepted for over 6000 notifications by 2023, and it is said that they have formed a market of over 500 billion yen. With the advent of functional display foods, consumers can easily obtain foods that suit them, and self-care and self-medication have become more enriched.

R. Morishita (✉)
Department of Clinical Gene Therapy, School of Medicine, Osaka University, Osaka, Japan
e-mail: morishit@cgt.med.osaka-u.ac.jp

H. Yamada, Y. Naito (eds.), *Anti-Aging Medicine*, https://doi.org/10.1007/978-981-97-8514-8_105

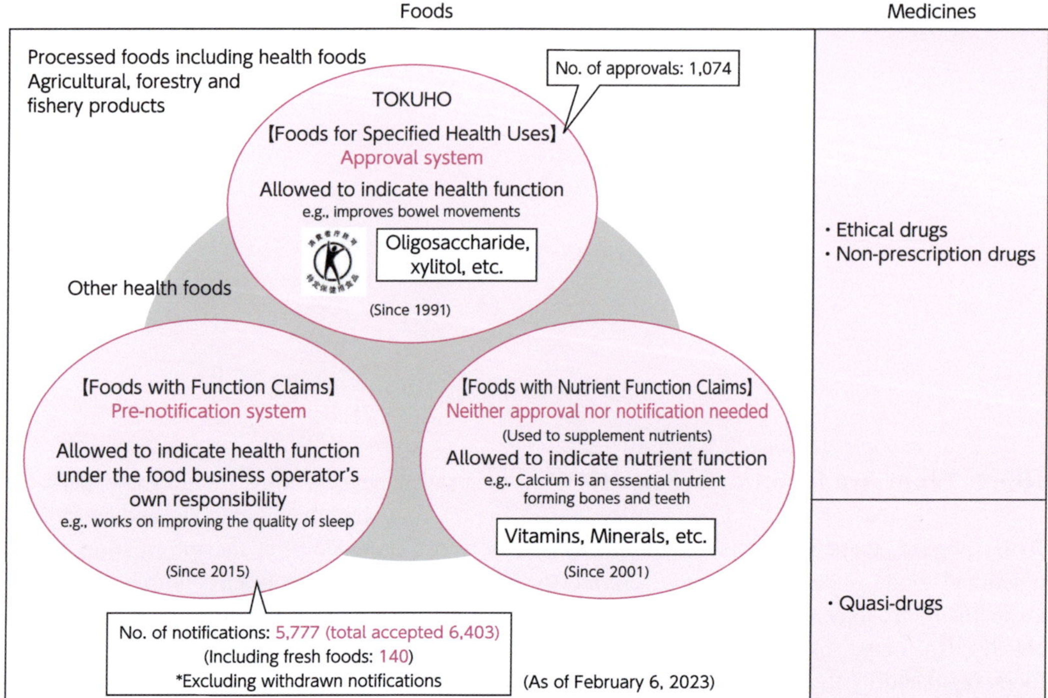

Fig. 105.1 Food functional labeling system (Consumer Affairs Agency)

105.2 FOSHU and Nutritionally Functional Foods

Foods for Specified Health Use (FOSHU) are those for which the government has conducted a review of the effects and safety indicated, and for each food, the Commissioner of the Consumer Affairs Agency gives permission, which takes time and costs. Therefore, since the advent of functional display foods, the number of permissions has decreased significantly (Fig. 105.2). On the other hand, nutritionally functional foods, if they contain a certain standard amount of nutrients whose scientific evidence has been confirmed, can display their functionality without any particular notification, but they can only do so in standardized expressions set by the government, and currently, only 5 types of minerals and 12 types of vitamins are eligible for such display. The standardized display is also quite limited, and displays related to immunity that are permitted in the West are still not allowed. Changes in nutritionally functional foods due to the advancement of scientific research are necessary, but the current system is rigid. Based on these circumstances, the Functional Labeling Food System, which started in April 2015, is a new system to increase the options of products that clearly display functionality and promote a better understanding of health foods for consumers, referring to the functional labeling system in the United States.

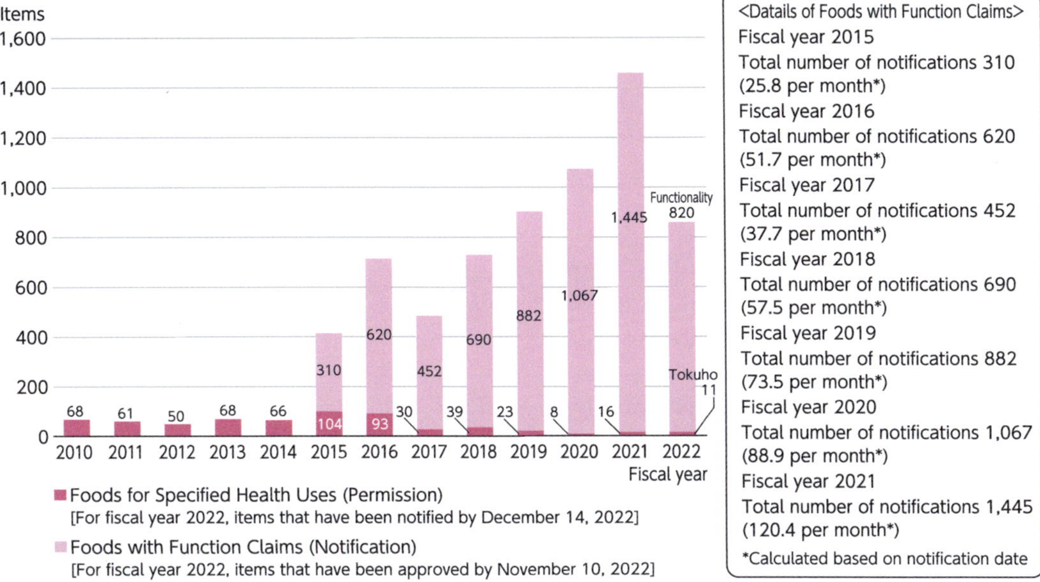

Fig. 105.2 Transition of foods that can display functionality (Consumer Affairs Agency) (Notification)

*The total number of permissions and publications in each fiscal year, including items that have expired or been withdrawn

105.3 Functional Labeling Foods

In the Functional Labeling Foods (for details, refer to "IV. E. 2. What are Functional Labeling Foods"), based on the rules set by the Consumer Affairs Agency [1], businesses can display functionality by notifying the Commissioner of the Consumer Affairs Agency of necessary matters such as scientific evidence related to the safety and functionality of food before selling it. This is a groundbreaking system based on post-regulation, where companies display appropriately based on scientific evidence at their own responsibility, and the Consumer Affairs Agency checks afterward. On June 25, 2013, a speech was made by then Prime Minister Abe, who was the first among successive prime ministers to mention health foods, stating that the functional display of health foods would be lifted. In functional display, it is a condition that there is scientific evidence, and in the guidelines of the Consumer Affairs Agency, human trials equivalent to FOSHU in the final product or systematic review (SR) are conditions. In SR, it is stipulated to evaluate papers of clinical trials targeting healthy people or people in the borderline area, and data using patients who are sick cannot be used. Data from Japanese people are prioritized, but data from overseas can also be used if the extrapolation to Japanese people is evaluated [for details, refer to the Consumer Affairs Agency's website (HP)]. In functional labeling health foods, the adoption of product specification standards is mandatory, in addition to quality control in production such as Good Manufacturing Practice (GMP) and Hazard Analysis and Critical Control Point (HACCP). It has also been decided that it is necessary to organize in a form that is easy for consumers to understand, such as unifying the window for reporting health damage information. Functional labeling is also allowed for agricultural, forestry, and fishery products and their processed products. Various functional labeling foods of agricultural, forestry, and fishery products have already appeared.

105.4 Foods for Special Dietary Uses

Special purpose foods (excluding Tokuhos) are foods that are labeled for special purposes such as infant development, pregnant and nursing women, people with difficulty swallowing, patients, etc., and are necessary for maintaining and restoring health. They require the permission of the Commissioner of the Consumer Affairs Agency for their sale and labeling (Health Promotion Act). Special purpose foods can be broadly divided into foods for patients, powdered milk for pregnant and nursing women, prepared powdered milk for infants, and foods for people with difficulty swallowing (Tokuhos are also considered special purpose foods in a broad sense). Special purpose foods, such as low-protein foods and allergen-free foods, are examined for compliance with the standards and receive the approval of the Commissioner of the Consumer Affairs Agency. Those without specific approval standards are evaluated individually for approval. Special purpose foods can be marked to indicate that they are special purpose foods. Among them are foods for patients, including low-protein foods, allergen-free foods, lactose-free foods, total nutrition foods, and others. Low-protein foods are suitable for people with kidney diseases, and the main ones are rice with low protein content. Also, allergen-free foods are suitable for people with specific food allergies, such as prepared powdered milk and infant milk for patients with milk allergies. Foods for people with difficulty swallowing are jellies for hydration, and from 2018, thickening foods have been added. There are foods for patients that are individually evaluated, such as oral rehydration fluids for dehydration and foods for patients with ulcerative colitis. The development of new foods for patients is expected in the future.

105.5 Conclusion

Functional foods do not have a legally defined definition, but Tokuhos, functional labeling foods, and foods for patients serve as potential foods for anti-aging in the future. The Dietary Supplement Health and Education Act in the United States, which regulates the functional labeling of supplements, has a major purpose of enlightening and educating the public, as the word “education” suggests. It is important to make good use of these functional foods, learn health information on their own, increase interest in health, and extend healthy life expectancy.

Reference

1. Consumer Affairs Agency. Food labeling planning. https://www.caa.go.jp/policies/policy/food_labeling/.

Functional Foods and Anti-aging: What Are Foods with Functional Claims

106

Ryuichi Morishita

106.1 Foods with Functional Claims

Originally, describing functionality such as effects and efficacy in foods was prohibited under the "Act on Securing Quality, Efficacy and Safety of Products Including Pharmaceuticals and Medical Devices" (Pharmaceuticals and Medical Devices Act). The only exception was health foods approved by the Consumer Affairs Agency, which were traditionally limited to Foods for Specified Health Uses (FOSHU) and Foods with Nutrient Function Claims. However, the development costs for FOSHU are said to be several hundred million yen or more, and the review period can take several years, which has been heavily criticized. Therefore, Foods with Functional Claims, which allow companies to express functionality on the condition that evidence has been obtained in clinical trials, and which can be reported to the Consumer Affairs Agency under the responsibility of the company, were initiated in April 2015 to utilize research results conducted at universities and companies. By 2023, over 6000 cases have already been accepted, and the market size is estimated to be approaching 500 billion yen. The outline of Foods with Functional Claims is described in the "Guidelines for Notification of Foods with Functional Claims" posted on the Consumer Affairs Agency's website (Fig. 106.1). Notifications are made online, and the content and clinical evidence are also described. The Consumer Affairs Agency checks the notification documents and, if there are no omissions or formal defects, issues a notification number and discloses the information on the Consumer Affairs Agency's website, etc [1–3].

R. Morishita (✉)
Department of Clinical Gene Therapy, School of Medicine, Osaka University, Osaka, Japan
e-mail: morishit@cgt.med.osaka-u.ac.jp

H. Yamada, Y. Naito (eds.), *Anti-Aging Medicine*, https://doi.org/10.1007/978-981-97-8514-8_106

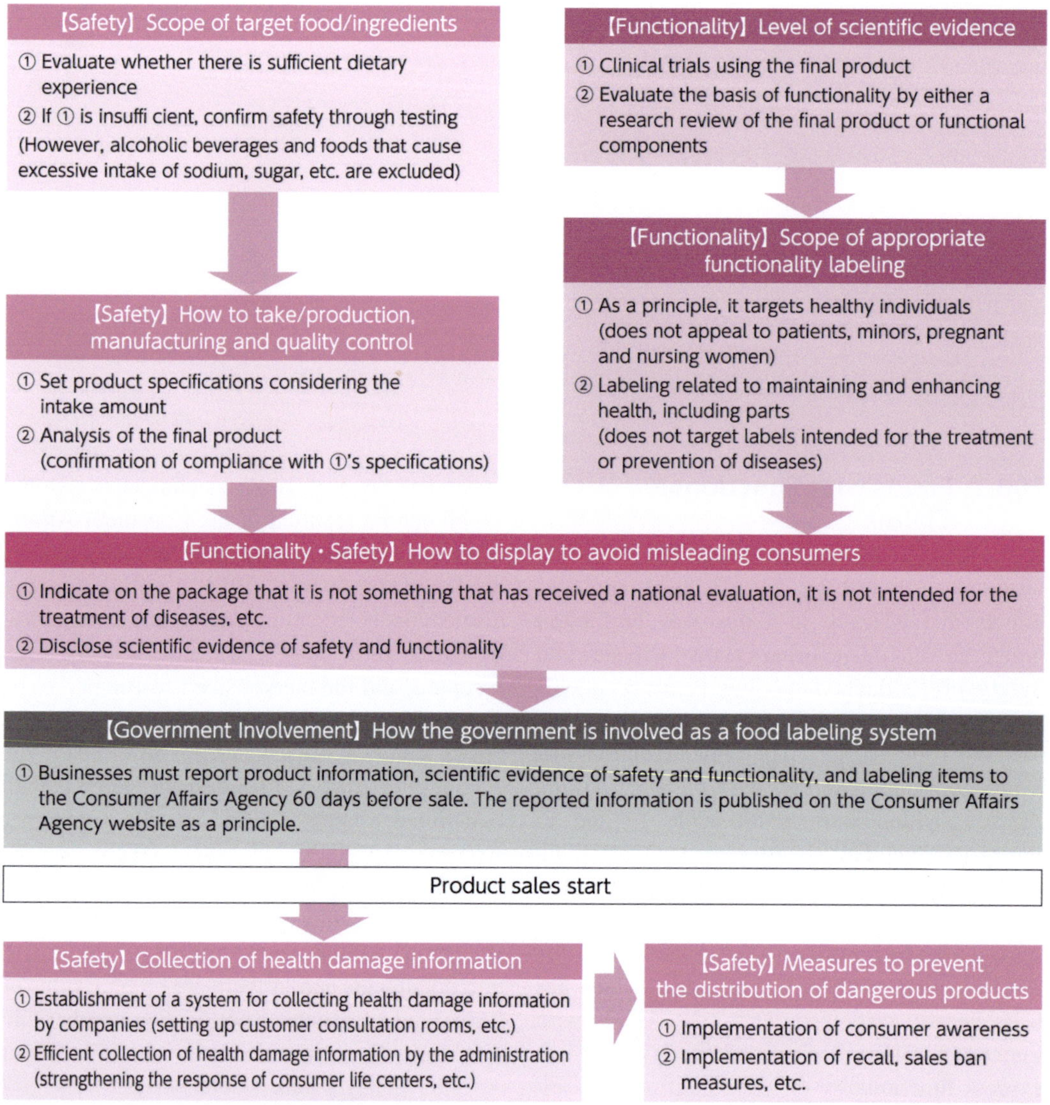

【Safety】Scope of target food/ingredients

① Evaluate whether there is sufficient dietary experience
② If ① is insuffi cient, confirm safety through testing
(However, alcoholic beverages and foods that cause excessive intake of sodium, sugar, etc. are excluded)

【Safety】How to take/production, manufacturing and quality control

① Set product specifications considering the intake amount
② Analysis of the final product (confirmation of compliance with ①'s specifications)

【Functionality】Level of scientific evidence

① Clinical trials using the final product
② Evaluate the basis of functionality by either a research review of the final product or functional components

【Functionality】Scope of appropriate functionality labeling

① As a principle, it targets healthy individuals (does not appeal to patients, minors, pregnant and nursing women)
② Labeling related to maintaining and enhancing health, including parts (does not target labels intended for the treatment or prevention of diseases)

【Functionality・Safety】How to display to avoid misleading consumers

① Indicate on the package that it is not something that has received a national evaluation, it is not intended for the treatment of diseases, etc.
② Disclose scientific evidence of safety and functionality

【Government Involvement】How the government is involved as a food labeling system

① Businesses must report product information, scientific evidence of safety and functionality, and labeling items to the Consumer Affairs Agency 60 days before sale. The reported information is published on the Consumer Affairs Agency website as a principle.

Product sales start

【Safety】Collection of health damage information

① Establishment of a system for collecting health damage information by companies (setting up customer consultation rooms, etc.)
② Efficient collection of health damage information by the administration (strengthening the response of consumer life centers, etc.)

【Safety】Measures to prevent the distribution of dangerous products

① Implementation of consumer awareness
② Implementation of recall, sales ban measures, etc.

Fig. 106.1 Overview of the Foods with Function Claims system

106.2 Conditions for Foods with Functional Claims

Foods with Functional Claims are foods for those who are not suffering from diseases (excluding minors, pregnant women including those planning to become pregnant, and nursing mothers). Also, Foods with Functional Claims need to have clear functional components. Recently, ingredients that were traditionally not subject to functional claims, such as γ-oryzanol and taurine, which are listed in the food-drug classification, have been decided to be newly targeted as Foods with Functional Claims. The target of Foods with Functional Claims includes not only supplements but all foods, including processed foods and fresh foods, and the range is wider than FOSHU. In fact, many functional claims for agricultural and marine products have been reported, and there are already over 140 cases. The safety of Foods with Functional Claims needs to be evaluated by the notifier themselves. First, they evaluate the eating experience based on the consumption record of the final product or similar food, and consider whether the safety has been sufficiently confirmed. If safety cannot be confirmed in such a review, they evaluate the eating experience and safety tests using existing information collected from databases, etc., for the final product or functional component, and consider whether the safety has been sufficiently confirmed. If safety cannot be confirmed even then, they consider whether the safety has been sufficiently confirmed by conducting in vitro tests and in Conduct in vivo tests or clinical trials for verification. Also, to prevent the occurrence and spread of health damage caused by functional foods, it is mandatory to collect information on health damage from consumers and medical practitioners, and to establish a system for reporting to administrative agencies. Within the company, it is necessary to detail who is in charge and how to report, depending on the actual situation. As an explanation of the scientific basis of the functionality to be displayed, either ① conducting clinical trials using the final product or ② a research review on the final product or functional component is required. Clinical trials using the final product comply with the test methods for Foods for Specified Health Uses (FOSHU), but double-blind trials are required, and a significant difference from the placebo group is necessary. An important point is that the participants in the clinical trials are those who are not suffering from diseases, and how to handle the borderline subjects varies depending on the target area. Clinical trial results must be published as peer-reviewed articles in a format compliant with internationally recognized guidelines [Consolidated Standards of Reporting Trials (CONSORT) statement] to ensure appropriate evaluation by all stakeholders. It is also crucial that the selected journal avoids conflicts of interest with the authors. Below are the key features of Foods with Functional Claims:

- Notification-based system (post-market verification) instead of prior approval
- Disclosure of notification documents as a general principle
- Scientific evidence for functionality may include literature reviews in addition to clinical trials
- Evidence for safety can be based on dietary history
- Fresh foods, including agricultural and marine products, are eligible
- Periodic review and revision of the system as necessary
- Regulations on advertising are also stipulated. Notification documents are publicly available and searchable online

Table 106.1 lists the functional ingredients with the highest acceptance rates to date and their associated functionalities, highlighting the broad range of current knowledge. In 2020, a new functionality claim related to immunity was accepted. This claim states: "This product contains plasma lactic acid bacteria (L. lactis strain Plasma). Plasma lactic acid bacteria interact with plasmacytoid dendritic cells (pDCs) and have been reported to contribute to the maintenance of immune function in healthy individuals." In 2021, the Japan Anti-Aging Association, a sister organization of the Japan Society for Anti-Aging Medicine, summarized and published a perspective on the scientific basis for functionality claims related to immunity in Foods with Functional Claims. The Consumer Affairs Agency also participated as an observer, presenting views on "immune markers," "innate and adaptive immunity," "clinical trials," and "sample sizes" for immune-related functionality claims. In the future, we hope to see an increase in functional components with evidence related to immunity under the corona. Note that functional foods do not have a mark set by the Consumer Affairs Agency like FOSHU because they are not approved by the Consumer Affairs Agency. As a side note, the sale of supplements at medical institutions is not prohibited and there is no problem. In the future, the role of doctors knowledgeable about pharmaceuticals is expected to increase. In the USA, which was referred to in designing the system, the Dietary Supplement Health and Education Act stipulates the functional display of supplements, and as the word "education" suggests, the main purpose is to enlighten the public. The deregulation of functional display in Japan also aims to increase public interest in health and

Table 106.1 Functional components and their functionality

(a) Ranking by functional component		
1	GABA (γ-aminobutyric acid)	601
2	Indigestible dextrin	412
3	Lactic acid bacteria and Bifidobacteria	380
4	DHA (including EPA)	259
5	Lutein (including Zeaxanthin)	209
6	Derived from Ginkgo leaves	158
7	Derived from black ginger	154
8	Pueraria flower isoflavones	130
9	Glucosylceramide	129
10	L-theanine	118
11	Astaxanthin	109
12	Catechins	101
13	Inulin	96
14	Bilberry-derived anthocyanins	95
15	Glucosamine (hydrochloride)	87
16	Hyaluronic acid	81
17	Citric acid	79
	Soy isoflavones	
19	Ellagic acid	76
	Monoglucosyl hesperidin	
	HMB	
22	Acetic acid	74
23	Tiliroside derived from rosehip	73
24	Isomaltodextrin	71
25	Derived from raphanus	67
26	Salacinol derived from salacia	65
27	Proteoglycan	64
28	Piperine derived from long pepper	58
29	Type II collagen	57
30	N-acetylglucosamine	55
(b) Ranking by appeal points		
1	Fat (body fat, visceral fat, neutral fat, etc.)	1356
2	Sugar (sugar absorption, post-meal blood sugar levels, etc.)	638
3	Condition of the stomach, intestinal environment, bowel movements	590
4	Blood pressure	466
5	Skin (moisture, elasticity, UV rays, glycation stress, etc.)	455
6	Cognitive function, memory	399
7	Stress	362
8	Sleep	347
9	Fatigue (excluding "eye fatigue")	342
10	Eyes (fatigue, focus adjustment, glare recovery, moisture, etc.)	312
11	Joints	246
12	Exercise, muscles, walking (excluding "post-exercise fatigue")	167
13	Peripheral blood flow, peripheral body temperature, coldness	154
14	Bone health	124
15	Cholesterol	114
16	Uric acid levels	65
17	Swelling	35
18	Immune function	30
19	Oral cavity, gums, teeth	28
20	Flexibility and suppleness of blood vessels	16

extend healthy life expectancy by learning health information through functional display, and it is expected to utilize knowledge to extend healthy life expectancy through food in the medical field.

References

1. Consumer Affairs Agency. Guidelines for notification of foods with function claims. https://www.caa.go.jp/policies/policy/food_labeling/foods_with_function_claims/assets/foods_with_function_claims_210322_0002.pdf.
2. Consumer Affairs Agency. Database of notifications for foods with function claims system. https://www.caa.go.jp/policies/policy/food_labeling/foods_with_function_claims/search/.
3. Consumer Affairs Agency. Food labeling standards Q&A. https://www.caa.go.jp/policies/policy/food_labeling/food_labeling_act/assets/food_labeling_cms101_210317_12.pdf.

Functional Foods and Anti-aging: Types and Functions of Functional Fresh Foods

107

Mari Maeda-Yamamoto

107.1 System of Foods with Function Claims

The systems that allow functional claims on foods include Foods for Specified Health Uses, Foods with Nutrient Function Claims, and Foods with Function Claims. The system of Foods with Function Claims was implemented in April 2015. It is unique in that it applies not only to processed foods and supplements, but also to agricultural, forestry, and fishery products (fresh foods). It is the responsibility of the business operator, not the government, to ensure the safety and functionality of the products. It is a notification system, and the information reported is made public on the Consumer Affairs Agency's website before the product goes on sale [1]. The system allows for claims such as "specific body part expressions" and "evaluations based on subjective indicators." One major difference from Foods for Specified Health Uses is that the method of analyzing functional components (functional ingredients) must be disclosed. The system that allows fresh foods and other agricultural, forestry, and fishery products to be labeled with functional claims is rare worldwide and is currently attracting attention from various Asian countries, with discussions on how to coordinate with each country's system.

M. Maeda-Yamamoto (✉)
Institute of Food Research, National Agriculture and Food Research Organization, Ibaraki, Japan
e-mail: marimy@affrc.go.jp

In order to make a notification, it is necessary to ensure the safety of the food, disclose the method of analyzing the functional ingredients, clarify the scientific basis through clinical trials or research reviews, consider the mechanism of action of the functional ingredients, provide experience of eating the food or results of safety tests, and ensure that the recommended daily intake is a normally consumable amount. However, it is often difficult for agricultural operators and others to establish the scientific basis and mechanism of action. Therefore, the National Agriculture and Food Research Organization has published 17 types of research reviews and examples of mechanisms of action of functional ingredients that can be freely used for notifications for fresh foods and other products [2]. Currently, about half of the notifications for fresh foods are utilizing this research review. The targets of the functionality of the Foods with Function Claims that are on sale are body fat, stomach condition, blood pressure, blood sugar, cholesterol, vascular flexibility, eyes, discomfort in eyes and nose, skin, muscles, urination, swelling, bones, knees, peripheral body temperature, sleep, physical fatigue, mental stress, cognitive function, uric acid, liver, oral care, immune function, and promotion of mineral absorption. In addition, as of July 7, 2022, 150 items of fresh foods have been notified and accepted, including vegetables, fruits, soybeans, mushrooms, meat, fish, rice, barley, and nuts, and are on sale (Table 107.1).

H. Yamada, Y. Naito (eds.), *Anti-Aging Medicine*, https://doi.org/10.1007/978-981-97-8514-8_107

Table 107.1 Functional display food (fresh food)

Functional component	Function	Recommended daily intake	Fresh food
β-Cryptoxanthin	Helps maintain bone health	3 mg	Mandarin oranges, persimmons
γ-aminobutyric acid (GABA)	Lowers high blood pressure	12.3–20 mg	Tomatoes, kale, grapes, enoki mushrooms, bell peppers, soybean sprouts, bananas, mandarin oranges, sponge gourd, mulberries
GABA	Alleviates temporary mental stress caused by work or study	28 mg	Melon
Lutein	Increases retinal (macular) pigment that is believed to protect the eyes from light stimulation	10 mg	Spinach, pumpkin, kale, eggs
Lycopene	Lowers LDL cholesterol in the blood	22.8 mg	Tomato
Isoflavones (aglycone)	Contributes to bone health	23–35 mg	Soybean sprouts
Isoflavones	Maintains skin moisture in middle-aged and older women	26 mg	Soybean sprouts
Choline ester	Improves high blood pressure	2.3 mg	Eggplant
Sulforaphane glucosinolate	Lowers slightly elevated blood liver function enzymes (ALT)	24 mg	Broccoli sprouts
Sulforaphane glucosinolate	Increases moisture content in skin that is concerned with dryness	20 mg	Broccoli sprouts
Rosmarinic acid	Reduces discomfort in the eyes caused by pollen or house dust	50 mg	Perilla leaves, large leaves
Luteolin	Suppresses post-meal blood sugar rise	5 mg	Chili pepper
Proanthocyanidin	Reduces visceral fat	110 mg	Apple
Proanthocyanidin	Helps protect skin from UV stimulation	55 mg	Apple
Anthocyanin	Helps alleviate eye fatigue by maintaining the moisture in the eyes, which tends to temporarily decrease when using smartphones and computers, and by making it easier for the eyes to focus.	43.5–57.6 mg	Bilberry
Docosahexaenoic acid (DHA), eicosapentaenoic acid (EPA)	Lower high blood triglyceride levels	DHA230–620 mg・EPA20–600 mg	Kanpachi, sardine, sea bream, roe, whale, yellowtail
Eicosapentaenoic acid (EPA) and docosahexaenoic acid (DHA)	Maintains memory, a part of cognitive function that declines with age.	DHA900mg・EPA300mg, DHA450 mg	Yellowtail, whale

Table 107.1 (continued)

Functional component	Function	Recommended daily intake	Fresh food
Imidazole dipeptide	Reduce temporary fatigue in daily life	400 mg	Chicken breast, whale red meat, pork
Imidazole dipeptide	Maintains memory (the ability to remember and recall words), a part of cognitive function that declines with age.	1000 mg	Pork, chicken
1	Supports memory for relatively recent events experienced by the individual, a part of cognitive function that tends to decline with age	Anserine 750 mg, carnosine 250 mg	Chickensa
β-glucan	Moderates the rise in blood glucose levels after eating	3.1 g	Barley (Hordeum vulgare)
Inulin	Increase the number of bifidobacteria and lactobacilli, which are good bacteria, to improve the condition of the stomach.	5 g	Garlic
Citric acid	Reduce temporary fatigue after daily office work or light exercise	1000 mg	Lemon, tomato
Oleic acid	Supports normal cholesterol (LDL cholesterol and total cholesterol)	21.65~32.4 g	Pecan nuts, pork
Xanthophyll	Helps protect the skin from ultraviolet light irritation. Helps reduce body fat and improves an elevated BMI	9 mg	Bell pepper

107.2 Utilization of Fresh Foods with Function Claims

In this way, by making functional claims on fresh foods, for example, the sales of vegetables have increased by 20 It increased by approximately 50%. This is due to the opening of new sales channels and increased sales volume by displaying functionality, and the selling price is being higher than regular products. However, fresh foods and primary processed foods such as vegetables and fruits vary in functional component content depending on the variety, production area (field), cultivation period, and cultivation method and may decrease during the sales period at retail stores. Therefore, it is required to minimize the variation of functional components and to ship fresh foods that exceed the lower limit [3]. Also, if the principle is to directly ingest functional components from agricultural products, the amount of food that can be ingested in a day is fixed, so in order to ingest an amount of various types of functional components that show effects, each agricultural product must contain more functional components than usual. Therefore, it is necessary to develop varieties of agricultural products that contain a lot of functional components. In addition, since processing and cooking are always necessary for actual eating, it is also necessary to consider the decomposition and denaturation of functional components in the processing and cooking process. Fresh foods are mixed component-containing foods with various components. Currently, nutritional components and functional components are clearly separated and displayed. However, in many places overseas, nutritional components and functional components are treated as nutritional components without distinction. In Japan as well, both components should be treated similarly as components that can maintain and enhance health, and it is important to redesign the functional display system so that consumers can easily understand it by organizing the relationship with the specific health food system and the nutritional function food system. Furthermore, in the future, it is necessary to conduct research on the combined effects of food combinations, clarify the importance of meals, and have consumers understand that the health functionality of agricultural, forestry, and fishery products and meals, including fresh foods, is demonstrated by long-term continuous eating.

References

1. Consumer Affairs Agency, Cabinet Office. Search for information on foods with function claims. https://www.caa.go.jp/policies/policy/food_labeling/foods_with_function_claims/search/
2. National Research and Development Agency for Agriculture and Food. Research review of agricultural, forestry and fishery products (systematic review). https://www.naro.affrc.go.jp/org/nfri/yakudachi/sys-review/index.html
3. Ministry of Agriculture, Forestry and Fisheries, Agricultural, Forestry and Fisheries Technology Council. Technical response to functional display of agricultural, forestry and fishery products. https://www.affrc.maff.go.jp/kinousei/gijyututekitaio.htm.

108 Functionality of Supplements and Anti-aging: Functionality of Supplements—Sensory Organs

Tomoko Tsuji

Supplements are not used for the purpose of treating diseases, but research on their use for maintaining and restoring biological functions that decline due to aging and lifestyle habits is widely conducted. Here, we will explain the functional components and their functionality and the scientific basis, focusing on the field of ophthalmology, where the effectiveness of supplements has been reported in large numbers, and taste, which is prone to disorders due to a deficiency of trace nutrients, among sensory organs.

108.1 Lutein and Zeaxanthin

Lutein and zeaxanthin are types of carotenoids found in dark green vegetables such as kale and spinach and in marigold flowers, and they coexist in food as structural isomers of each other. Lutein and zeaxanthin cannot be synthesized in the human body and can only be replenished by intake through food or supplements, but once they enter the body, they are selectively incorporated into the macula of the eyeball. The mechanism of this incorporation has been substantiated by binding to the xanthophyll binding proteins glutathione S-transferase P (GSTP1) and StAR-related lipid transfer domain containing (StARD3) in the human retina [1]. Lutein and zeaxanthin exist as pigments in the macular region of the human eye after being incorporated and are thought to protect the eyes from photooxidation by capturing free radicals emitted from ultraviolet and blue light, and many trials have been conducted to prove their effectiveness in humans. A comprehensive literature search using PubMed, Cochrane, and medical journals has reported that daily intake of 10 mg of lutein and 2 mg of zeaxanthin for 1-year results in increased macular pigment density, improved color contrast sensitivity, and enhanced recovery from light stress compared to a placebo group [2]. These effects are thought to be due to the increased macular pigment density from lutein and zeaxanthin intake acting as a filter to protect cone and rod cells from short-wavelength light, promoting recovery from light stress and leading to improved color contrast sensitivity. On the other hand, N-retinylidene-N-retinylethanolamine (A2E), a specific constituent of lipofuscin, an aging pigment accumulated in the retina, emits spontaneous fluorescence under blue light and promotes the production of free radicals, inducing apoptosis in retinal pigment epithelial cells. Lutein and zeaxanthin present in the macula and lens absorb short-wavelength (400–500 nm), especially 446 nm wavelength blue light, suppressing the generation of free radicals and protecting the retina [3]. Thus, in addition to their direct anti-

T. Tsuji (✉)
Materials Development Department, Yoshinoya Holdings Co., Ltd., Tokyo, Japan

H. Yamada, Y. Naito (eds.), *Anti-Aging Medicine*, https://doi.org/10.1007/978-981-97-8514-8_108

oxidant effects through free radical scavenging, lutein and zeaxanthin absorb blue light, which has a high cell-damaging ability, and protect eye tissues from oxidative stress.

108.2 Astaxanthin

Astaxanthin is a red pigment found in the shells of shrimp and crabs, and in the flesh of salmon, and is a type of carotenoid. Currently, astaxanthin derived from artificially cultivated hematococcus (algae) is widely distributed as a supplement. Astaxanthin is known to exhibit effects such as inhibiting lipid peroxidation of red blood cell membranes, improving red blood cell deformability and oxygen transport capacity, and increasing retinal capillary blood flow due to its powerful antioxidant action. It is thought that astaxanthin may increase ciliary body blood flow, as the retina and ciliary body are under the same vascular control, and in animal experiments, orally ingested astaxanthin is distributed to the ciliary body. Furthermore, it is also known to inhibit the signaling system of tumor necrosis factor-α (TNF-α) by inhibiting the activation of nuclear factor κB (NF-κB) [4]. Many clinical studies have been conducted on individuals engaged in visual display terminal (VDT) work, hoping for the effect of astaxanthin on improving the focus adjustment function of the ciliary muscle through such mechanisms. A systematic review submitted as the basis for multiple products in the Consumer Affairs Agency's functional food database states that continuous intake of 6–9 mg/day of astaxanthin reduces eye fatigue from daily VDT work such as computers, and reduces the burden on shoulders and waist from tired eyes.

108.3 n-3 Highly Unsaturated Fatty Acids

n-3 highly unsaturated fatty acids refer to fatty acids that have multiple double bonds within the molecule, starting from the third carbon position from the methyl terminal, and are also called ω3 fatty acids. Particularly, eicosapentaenoic acid (EPA) and docosahexaenoic acid (DHA), which are abundant in fish oil, have been studied for their diverse functionalities. These ω3 fatty acids are contained in fish oil, but neither exists alone; they coexist. The sources of these ω3 fatty acids for supplement ingredients include marine organisms such as blue fish and krill, as well as cultivated algae, and are also known as pharmaceutical ingredients that lower blood triglycerides. Dry eye is defined as a "chronic disease" of tear fluid and corneal epithelium due to various factors. In recent years, as VDT work permeates people's daily lives, the number of people who are aware of dry eye has dramatically increased, leading to a decrease in labor productivity. An observational study of American women in 2005 reported results suggesting that fish consumption and ω3 intake may reduce the risk of developing dry eye. In 2019, a meta-analysis of 17 placebo-controlled randomized controlled trials (RCTs, total number of subjects 3363) targeting dry eye patients showed that dry eye symptoms, tear film breakup time (breakup time; BUT, it has been reported in "Cornea" that Schirmer's test and the degree of corneal fluorescein staining significantly improved with ω3 intake [5]. ω3 intake not only increases the amount of ω3 fatty acids in the red blood cell membrane, but also in the meibomian glands and the fat in the tear fluid, which has been shown to stabilize the tear fluid layer by inhibiting evaporation from the tear glands and suppress corneal inflammation through anti-inflammatory effects. ω3 supplements are known to improve lifestyle-related diseases and cognitive function and are often added to ophthalmic supplements such as lutein, zeaxanthin, and astaxanthin mentioned earlier.

108.4 Zinc

Taste cells contained in taste buds are said to be replaced every 30 days, and zinc deficiency has long been known to disrupt this cell regeneration and cause taste disorders. The diagnostic criterion for taste disorders is a serum zinc concentration of 75 μg/dL, and it is thought that nearly half of the people who develop taste disorders have a base of zinc deficiency, and zinc supplementation is the only treatment for taste disorders with evidence. Zinc is abundant in animal foods, but it is thought to be mutually involved with frailty due to malnutrition seen in elderly people with low dietary intake [6]. Therefore, prevention of deficiency by supplements is effective, and it is considered effective to take products with complex prescriptions like multivitamins & minerals on a daily basis. Especially in single people, the awareness of taste disorders may be delayed, and if the deficiency continues for a long time, recovery becomes difficult. According to the Japanese nutritional intake standards, the recommended amount of zinc intake for adults is about 10 mg/day, on the other hand, if excessive intake of 45 mg/day or more continues, side effects such as nausea, headache, and anemia occur, and caution is necessary.

References

1. Li B, Vachali P, et al. Human ocular carotenoid-binding proteins. Photochem Photobiol Sci. 2010;9:1418–25.
2. Hammond BR, Fletcher LM, et al. A double-blind, placebo-controlled study on the effects of lutein and zeaxanthin on photostress recovery, glare disability, and chromatic contrast. Invest Ophthalmol Vis Sci. 2014;55:8583–9.
3. Snodderly DM, Auran JD, et al. The macular pigment. II. Spatial distribution in primate retinas. Invest Ophthalmol Vis Sci. 1984;25:674–85.
4. Suzuki Y, Ohgami K, et al. Suppressive effects of astaxanthin against rat endotoxin-induced uveitis by inhibiting the NF-κB signaling pathway. Exp Eye Res. 2006;82:275–81.
5. Giannaccare G, Pellegrini M, et al. Efficacy of Omega-3 fatty acid supplementation for treatment of dry eye disease: a meta-analysis of randomized clinical trials. Cornea. 2019;38:565–73.
6. Nin T. Diagnosis and treatment of taste disorders. J Otolaryngol Jpn. 2019;122:738–43.

109 Functionality and Anti-aging of Supplements: Functionality of Supplements—Oral and Dental

Kenji Mishima and Takayoshi Sakai

109.1 Functional Foods in the Oral and Dental Field

The oral cavity is the entrance to the digestive tract, respiratory system, and sensory organs, and it connects to the esophagus and airway through the pharynx and larynx. In addition to having the function of eating and swallowing food, it also plays a role in respiration, speech, taste, and other life-sustaining and communication senses. As aging progresses, the function of the oral cavity declines due to age-related diseases, and various measures such as exercise and nutrition are attempted, and many people are practicing them. Among them, the popularity of general foods and health foods (functional foods, specified health foods, nutritional functional foods) as so-called health foods in daily life is thriving. The functional display of food introduced in April 2015 presupposes scientific evidence from human research, but ultimately the choice is left to the consumer. Therefore, in this section, we will introduce some foods related to the maintenance of oral diseases and oral function from the reports related to the oral cavity that have been analyzed so far.

K. Mishima (✉)
Division of Pathology, Department of Oral Diagnostic Sciences, Showa University School of Dentistry, Tokyo, Japan
e-mail: sakai.takayoshi.dent@osaka-u.ac.jp

T. Sakai
Department of Rehabilitation for Orofacial Disorders, Osaka University Graduate School of Dentistry, Osaka, Japan

109.2 Caries

Caries (tooth decay) refers to a typical dental disease in which the surface of the hard tissue of the teeth is destroyed by bacterial acid production, progressing from enamel and cementum to dentin, forming substantial defects. The prevention of caries is based on inhibiting bacterial growth, and one of the measures is the intake of xylitol by gum or tablets [1]. Xylitol is a sweet carbohydrate called sugar alcohol, the same as sorbitol and maltitol. It is found in many fruits and vegetables in nature, for example, strawberries contain about 300 mg of xylitol per 100 g of dry weight. Also, about 15 g of xylitol is produced in the human liver per day. The xylitol we consume is industrially produced using xylan hemicellulose extracted from trees such as birch and oak as raw materials. The only sweeteners proven to prevent caries in many long-term clinical studies are xylitol and sorbitol. Many research results have been reported, and when xylitol is used, it shows a caries prevention effect with a probability of 30–80%. The reasons for preventing caries include promoting saliva secretion and remineralization, like other sugar alcohols. Also, it does

H. Yamada, Y. Naito (eds.), *Anti-Aging Medicine*, https://doi.org/10.1007/978-981-97-8514-8_109

not produce acid, promotes the neutralization of acid in dental plaque, and inhibits the metabolism of mutans bacteria.

109.3 Periodontal Disease

The two major causes of tooth loss are caries and periodontal disease, both of which are bacterial infections. As periodontal disease, a lifestyle-related disease, progresses, the periodontal pocket deepens, and the alveolar bone that supports the teeth is absorbed, leading to tooth loss. The most effective prevention and symptom improvement is oral cleaning with a toothbrush and dental floss, but as an auxiliary measure, the effect of coenzyme Q10 (CoQ10) is being studied. CoQ10 is known as a coenzyme involved in the production of adenosine triphosphate (ATP), and its effects have been reported in various clinical studies due to its antioxidant action in pharmaceuticals, health foods, etc [2]. In human clinical studies of periodontal disease with CoQ10, administration of reduced CoQ10 (150 mg/day) or placebo for 8 weeks significantly reduced plaque adhesion and bleeding during probing, and the antioxidant power in saliva tended to increase, and halitosis tended to decrease by about half, suggesting that intake of reduced CoQ10 may improve the oral environment.

109.4 Dry Mouth (Oral Dryness)

Dry mouth is a symptom characterized by a dry mouth sensation, discomfort in the mouth, and taste disorders. The causes of dry mouth are diverse and include side effects of drugs, lifestyle-related diseases such as diabetes, decreased muscle strength around the mouth, and stress, among other complex factors, and the involvement of oxidative stress in its pathogenesis has been reported. Most of these are due to aging-related diseases that are the target of anti-aging medicine, such as lifestyle-related diseases, menopausal disorders, and stress. Since many cases develop from daily lifestyle habits, it is important to propose a lifestyle that includes exercise guidance and dietary guidance, and an approach that understands anti-aging medicine is necessary. Here, we introduce isoflavones, resveratrol, and quercetin, which have been attracting attention in recent years.

- Isoflavones

  Isoflavones are a type of flavonoid that are abundant in soybean germ and are absorbed as aglycone forms due to the action of intestinal bacteria when ingested. They are similar in chemical structure to female hormones (estrogens) and have antioxidant effects, suggesting potential for inhibiting the absorption of alveolar bone due to osteoporosis, promoting the remineralization of teeth with calcium, and preventing bad breath. In postmenopausal women, oral-related complaints such as tongue pain and dry mouth are known, and the intake of isoflavones is considered as one of the means to prevent and alleviate menopausal symptoms. Therefore, when aglycone (25 mg/day) was taken for 2 months in patients with dry mouth, significant improvements were observed in the subjective items of "feeling thirsty" and "needing water during meals" and in the amount of saliva secretion. Furthermore, oxidative stress markers in saliva showed a decreasing trend, and genistein, a soy metabolite in saliva, showed a significant increase. From these results, it was suggested that the intake of isoflavones may be an effective means to improve the symptoms of salivary secretion disorders [3].

- Resveratrol

  Resveratrol is a natural polyphenol produced by plants in response to environmental stress. In recent years, it has been shown to have pharmacological effects on a wide range of diseases, including neurological disorders, liver diseases, cardiovascular diseases, and autoimmune diseases. Nonobese diabetic (NOD) mice, in which the function of the lacrimal and salivary glands is lost, are studied as animal models of Sjögren's syndrome, but in this study, it was shown that administration of resveratrol increased saliva secretion in NOD mice [4]. Resveratrol enhanced Sirt1 activity, but did not affect inflammatory cell infiltration. Furthermore, the expression of the anti-inflammatory cytokine, interleukin 10 (IL-10), in the salivary glands was enhanced in the resveratrol administration group. The new therapeutic effect of resveratrol on salivary secretion dysfunction in Sjögren's syndrome was confirmed.

- Quercetin

  In pursuit of an effective treatment for dry mouth, we analyzed the effects of quercetin on saliva secretion and its mechanism of action. We created a model mouse with saliva secretion disorder due to radiation exposure and discovered that the secretion disorder was suppressed by quercetin intake. Furthermore, saliva secretion increased in normal mice administered quercetin. To elucidate the mechanism of these effects on saliva secretion, we conducted analyses using mouse submandibular gland tissue, human salivary gland epithelial cell line (HSY), and mouse aortic endothelial cells (MAEC). As a result, quercetin was found to increase the expression of aquaporin 5 (aquaporin 5; AQP5) expression and enhances calcium uptake, and it has been shown to suppress oxidative stress and inflammatory responses induced by radiation exposure [5]. Aspiration pneumonia in the elderly is a leading cause of death, and it has been reported that the cause is due to bacteria in the oral cavity and pharynx. Aspiration pneumonia is a disease that occurs when food and saliva flow into the lungs due to eating and swallowing disorders, and the body's immune defense mechanism does not work properly due to a decrease in overall immunity. It is said that a variety of resident bacteria live in the oral cavity. It is clear that dry mouth occurs due to a decrease in saliva secretion, and the environment in the oral cavity deteriorates, causing resident bacteria to proliferate. By working on oral care and sufficient saliva secretion as a bacterial exclusion mechanism, it can prevent aspiration pneumonia. Indeed, the practice of anti-aging medicine from the oral cavity leads to the prevention of systemic diseases. We look forward to future research results.

References

1. Wu YF, Salamanca E, et al. Xylitol-containing chewing gum reduces cariogenic and periodontopathic bacteria in dental plaque-microbiome investigation. Front Nutr. 2022;9:882636.
2. Kanno N, Fujii T, et al. Improvement of oral environment in periodontal disease patients by supplement containing ubiquinol (reduced coenzyme Q10). Jpn J Conserv Dent. 2013;56:385–9.
3. Ryo K, Takahashi A, et al. Therapeutic effects of isoflavones on impaired salivary secretion. J Clin Biochem Nutr. 2014;55:168–73.
4. Inoue H, Kishimoto A, et al. Resveratrol improves salivary dysfunction in a non-obese diabetic (NOD) mouse model of Sjögren's syndrome. J Clin Biochem Nutr. 2016;59:107–12.
5. Takahashi A, Inoue H, et al. Evaluation of the effects of quercetin on damaged salivary secretion. PLoS One. 2015;10:e0116008.

Functionality of Supplements and Anti-aging: Functionality of Supplements—Endocrine Metabolism

110

Sho-ichi Yamagishi

110.1 Supplements Related to Endocrine Metabolism

Among the supplements related to endocrine metabolism, there are those that can be expected to have anti-obesity effects, corrective effects on postprandial hyperglycemia and hypertriglyceridemia, and cholesterol-lowering effects, and the areas they target are broad. It has been clearly shown that a balanced diet that restricts the excessive intake of refined carbohydrates, salt, and saturated fats, and abundantly includes vegetables and dietary fiber, leads to anti-obesity and anti-diabetic measures. However, in busy daily life, it is not always easy to consume sufficient amounts of vegetables and fruits and to adequately intake phytochemicals, vitamins, and minerals, which are functional chemical components contained in them. In this section, we will focus on supplements related to phytochemicals, among the many supplements, that have undergone double-blind trials, and explain their functionality.

110.2 Sulforaphane

Glucosinolates are contained in cruciferous plants such as broccoli, kale, Brussels sprouts, and cauliflower. Glucosinolates are hydrolyzed into isothiocyanates by myrosinase contained in these plants when chewed in the mouth and by the action of intestinal bacteria present in the body. Sulforaphane is a type of isothiocyanate that activates phase 2 enzymes and has detoxifying effects, and it has been reported to have antioxidant, anti-glycation, and anti-inflammatory effects [1]. In a double-blind trial that randomly divided 130 volunteers into a group that added 400 g of standard broccoli to their usual diet and a group that consumed broccoli (HG broccoli) containing a high dose of glucoraphanin, one of the glucosinolates, it was reported that LDL cholesterol levels significantly decreased with the intake of HG broccoli. It has also been revealed that the intake of broccoli sprouts containing a high dose of glucoraphanin reduces LDL cholesterol and advanced glycation end-product (AGEs) levels, and decreases body weight and waist circumference. Furthermore, in a double-blind trial targeting patients with type 2 diabetes, it has been shown that consuming powdered broccoli sprouts improves insulin resistance [2].

S.-i. Yamagishi (✉)
Department of Medicine, Showa University School of Medicine, Tokyo, Japan
e-mail: shoichi@med.showa-u.ac.jp

H. Yamada, Y. Naito (eds.), *Anti-Aging Medicine*, https://doi.org/10.1007/978-981-97-8514-8_110

110.3 Quercetin

Quercetin is a yellow flavonol found in many types of vegetables and fruits such as onions and buckwheat, and is known to have antioxidant and anti-glycation effects [1]. A randomized double-blind placebo-controlled crossover trial showed that when 22 healthy men took 500 mg of quercetin per day for 4 weeks, their blood uric acid levels significantly decreased. Furthermore, a double-blind trial targeting healthy humans diagnosed with dyslipidemia revealed that taking quercetin improved dyslipidemia, and when overweight to obese hypertensive patients took quercetin, their blood pressure significantly decreased and their oxidized LDL levels decreased [3]. Methylglyoxal, a highly reactive dicarbonyl compound that causes AGEs' formation, is also known to decrease with the intake of quercetin, and a supplement called d-Gs (deegies) has been released along with other ingredients whose effects have been confirmed in randomized double-blind placebo-controlled trials against AGEs. On the other hand, quercetin can affect the activity of drug-metabolizing enzymes CYP2C9 and CYP3A4 and may therefore increase or decrease the blood levels and effects of drugs metabolized by these enzymes. Special caution is required when using quinolone antibiotics or the anticancer drug paclitaxel.

110.4 Catechin

Catechins are a type of flavanol polyphenol mainly found in tea, with four main types of catechins present in green tea: epicatechin, epigallocatechin, epicatechin gallate, and epigallocatechin gallate. These catechins have various biological effects, including antioxidant, anti-glycation, and anti-inflammatory properties. In addition to green tea, it is known that epicatechin is also abundant in blackberries, cocoa, and chocolate. When 20 volunteers were administered 1 mg of epicatechin per 1 kg of body weight, it was found that fat oxidation was enhanced and postprandial hyperglycemia and hypertriglyceridemia were suppressed. Also, when patients with coronary artery disease were given green tea containing 400 mg of catechins at breakfast, postprandial antioxidant capacity increased, and the rise in triglycerides and uric acid levels was suppressed. Similarly, when postmenopausal women were made to drink green tea containing 615 mg of catechins continuously for 4 weeks, postprandial antioxidant capacity improved, and the rise in blood glucose levels was suppressed. Furthermore, a double-blind placebo-controlled crossover trial revealed that daily intake of 100 mg of epicatechin supplements improved insulin resistance after 4 weeks, and drinking green tea containing about 600 mg of catechins significantly reduced waist circumference and visceral fat area in obese and diabetic patients. Meta-analysis results have also shown that intake of green tea catechins lowers blood pressure and total cholesterol and LDL cholesterol levels [4]. However, in Europe, there are scattered reports of liver damage due to the intake of large amounts of green tea catechins.

110.5 Resveratrol

Resveratrol is a polyphenol found in red wine, grape skins, and Japanese knotweed, among others. It activates the longevity gene sirtuin and has been reported to extend the lifespan in many species, including fruit flies and mice. A double-blind placebo-controlled crossover trial revealed that when 11 healthy obese men were administered 150 mg of resveratrol for 30 days, insulin resistance improved, and markers of systolic blood pressure and liver dysfunction decreased. Also, when resveratrol was given to patients with type 2 diabetes, it was found that administering ~250 mg increases insulin sensitivity and decreases HbA1c and systolic blood pressure. Furthermore, in a double-blind trial, it has been shown that the intake of 300 mg of resveratrol improves insulin resistance and liver damage in patients with non-alcoholic fatty liver disease. On the other hand, according to a meta-analysis of 11 randomized trials, the intake of resveratrol improves blood glucose levels, HbA1c, and HOMA-IR index in patients with type 2 diabetes, but such effects are not observed in non-diabetic subjects [5]. In fact, it has been reported that there is no change in metabolic markers even if 75 mg of resveratrol is ingested by non-obese women without glucose tolerance abnormalities, and there is no effect on insulin resistance or glucose metabolism markers even if a high dose of resveratrol is taken by obese men with mild insulin resistance. The effect of resveratrol may be strongly manifested in patients with a certain degree of underlying metabolic abnormalities.

References

1. Yamagishi S, Matsui T, et al. Phytochemicals against advanced glycation end products (AGEs) and the receptor system. Curr Pharm Des. 2017;23:1135–41.
2. Bahadoran Z, Tohidi M, et al. Effect of broccoli sprouts on insulin resistance in type 2 diabetic patients: a randomized double-blind clinical trial. Int J Food Sci Nutr. 2012;63:767–71.
3. Brüll V, Burak C, et al. Effects of a quercetin-rich onion skin extract on 24 h ambulatory blood pressure and endothelial function in overweight-to-obese patients with (pre-)hypertension: a randomised double-blinded placebo-controlled cross-over trial. Br J Nutr. 2015;114:1263–77.
4. Khalesi S, Sun J, et al. Green tea catechins and blood pressure: a systematic review and meta-analysis of randomised controlled trials. Eur J Nutr. 2014;53:1299–311.
5. Liu K, Zhou R, et al. Effect of resveratrol on glucose control and insulin sensitivity: a meta-analysis of 11 randomized controlled trials. Am J Clin Nutr. 2014;99:1510–9.

111 Functionality of Supplements and Anti-aging: Functionality of Supplements—Digestive System

Kenichi Ikejima

The digestive system is a group of organs responsible for the process from nutrient intake, digestion, and absorption to metabolism and excretion, and its functional preservation occupies a very important position from the perspective of anti-aging. The digestive tract is in contact with the outside world through the oral cavity and anus, and a symbiotic relationship is established with a microbial community, including intestinal bacteria, formed in its lumen. In addition, the liver and biliary system and the pancreas are groups of organs derived from the primitive gut, mainly responsible for endocrine and metabolic functions, in addition to exocrine functions as aids to digestion and absorption. This paper will provide an overview of the benefits and issues of supplements in the digestive system, focusing mainly on the digestive tract and liver.

111.1 Digestive Tract and Supplements

111.1.1 Probiotics and Functional Digestive Diseases

The representative supplements for the digestive system are probiotics and prebiotics. Various probiotic formulations are frequently used as therapeutic drugs for digestive tract conditions, especially functional digestive diseases such as diarrhea and constipation, even as prescription drugs. Dysbiosis, an imbalance of the intestinal bacterial flora, not only directly causes digestive symptoms, but also has been highlighted for its association with organ correlations such as intestinal-liver correlation, and even systemic diseases. Therefore, the benefits of probiotics are expected to extend beyond simply improving digestive symptoms to enhancing immune function, allergy countermeasures, metabolic improvement, cancer prevention, and a wide range of areas. There are a wide variety of probiotic products on the market as supplements or health foods, many of which highlight the characteristics of each species and strain. While it is difficult to say that objective scientific verification has been sufficiently conducted for the individual specificity of these products, it is necessary to make judgments carefully, but all of them can be considered to positively affect the maintenance of a good balance of intestinal bacterial flora, or eubiosis. Therefore, from the perspective of anti-aging, it would be reasonable to recommend the use of probiotic formulations, especially for those with some digestive symptoms.

K. Ikejima (✉)
Department of Gastroenterology, Juntendo University Graduate School of Medicine, Tokyo, Japan
e-mail: ikejima@juntendo.ac.jp

H. Yamada, Y. Naito (eds.), *Anti-Aging Medicine*, https://doi.org/10.1007/978-981-97-8514-8_111

111.1.2 *Helicobacter pylori* Infection and Probiotics

Our country still has a high incidence of stomach cancer, most of which are cancers arising from atrophic gastritis associated with *Helicobacter pylori* (HP) infection. Eradication therapy for HP infection involves the use of amoxicillin, clarithromycin, and proton pump inhibitors (PPI) or potassium-competitive acid blockers. The standard treatment for insurance application is the triple therapy of primary eradication using Proton Pump Inhibitors (PPI) and Potassium-Competitive Acid Blockers (PCAB), and secondary eradication regimen using amoxicillin, metronidazole, and PPI/PCAB. However, the bacteriostatic activity of probiotics against *Helicobacter pylori* (HP) is attracting attention. Not only for the prevention of antibiotic-induced enteritis and diarrhea in primary and secondary eradication, but also for the enhancement of eradication effects, the usefulness of probiotics alone is considered, although complete eradication cannot be expected. However, caution is required as it can also cause false negatives in HP presence diagnosis such as urea breath test.

111.2 Liver and Supplements

111.2.1 Viral Chronic Liver Disease and Hepatoprotective Drugs

In the treatment of viral hepatitis, various adjunctive therapies such as hepatoprotective drugs like ursodeoxycholic acid (UDCA) and nutritional management have been performed in addition to antiviral therapy. Hepatitis C has achieved an astonishing efficacy rate of over 95% in terms of virus elimination since direct-acting antivirals (DAAs) have been used in clinical practice, but recovery from advanced liver fibrosis is not always easy in cases of cirrhosis progression, and it is often difficult to avoid the risk of liver cancer. In Asian countries other than Japan, the prevalence of hepatitis B is high, and although mother-to-child infection measures are effective in our country, the problem of imported and re-emerging infections is feared in the future. Traditionally, various hepatoprotective drugs have been used for chronic hepatitis, but their efficacy is insufficient and their use has narrowed now that antiviral therapy has been established. Therefore, the usefulness of supplements for hepatoprotection is considered limited.

111.2.2 Fatty Liver, Steatohepatitis, and Supplements

The causes of fatty liver that have been known for a long time are nutritional imbalance (overnutrition, malnutrition, and dysnutrition) and excessive alcohol intake. Alcohol-related fatty liver progresses from steatohepatitis to cirrhosis, and the pathogenesis is deeply involved with nutritional status and dysbiosis of the intestinal microbiota. Excessive alcohol drinkers are prone to vitamin deficiency, and especially vitamin B1 deficiency can lead to severe mental and neurological symptoms such as Wernicke's encephalopathy, so vitamin B1 supplementation is necessary. In alcohol-related hepatitis, correction of dysbiosis is effective for improving the disease state, and the usefulness of probiotics such as Lactobacillus has been proven in animal experiments. Alcohol weakens the barrier mechanism of the intestinal mucosa, so it is believed that the intestinal permeability of toxins derived from intestinal bacteria, represented by endotoxin (pathogen-associated molecular patterns; PAMPs), increases, causing damage to the liver through the portal vein (Fig. 111.1) [1]. Zinc supplementation is effective in preserving the intestinal barrier mechanism, and its inhibitory effect on alcohol-related hepatitis is expected [2]. On the other hand, what has been attracting attention in recent years is non-alcoholic fatty liver disease (NAFLD), which is closely related to metabolic syndrome, especially non-alcoholic steatohepatitis (NASH), which shows progressive liver pathology. Vitamin E shows some effectiveness against NAFLD/NASH, but since prescription drugs are not covered by insurance, the use of supplements is worth considering. Vitamin E is thought to improve fatty hepatitis due to its

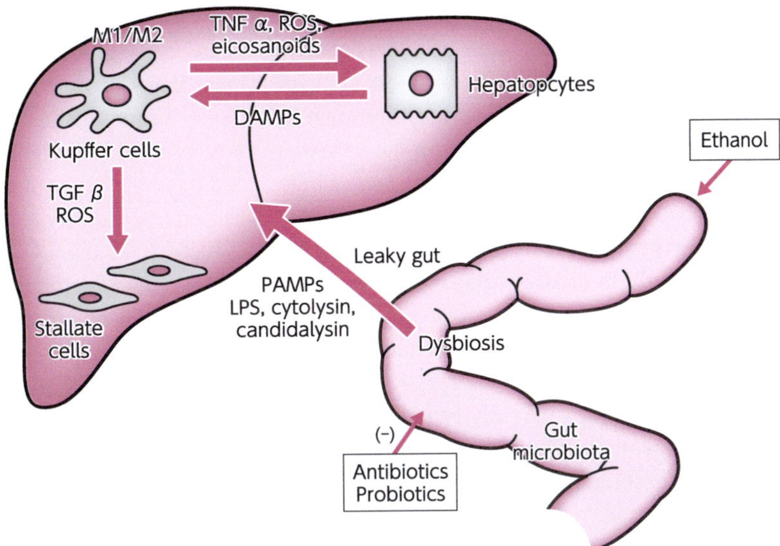

Fig. 111.1 Involvement of intestinal bacteria in the pathogenesis of alcohol-related liver disease

antioxidant effects, but not all supplements with antioxidant effects necessarily have a fatty liver improvement effect. Although various trials exist, the reality is that there are none with a high level of evidence. Some amino acids, such as *N*-acetyl cysteine, *S*-adenosylmethionine, and betaine, contribute to the reduction of oxidative stress and endoplasmic reticulum (ER) stress through their metabolic systems. In fact, these substances have been reported to show a fatty liver improvement effect in animal experiments, but the evidence for their usefulness in humans has not been established. On the other hand, NASH, like alcohol-related liver disease, is attracting attention for the role it plays in the formation of intestinal liver-related pathology, and the effectiveness of probiotics and zinc is also expected.

111.2.3 Utilization of Supplements as Adjunctive Therapy for Progressive Liver Disease

In cases of progression to cirrhosis regardless of cause, abnormal amino acid balance is one of the factors exacerbating the pathology, so branched-chain amino acids (BCAA) are used in the treatment of hypoalbuminemia and hepatic encephalopathy. Early intervention has been suggested to be beneficial, but insurance coverage is limited to cases where hypoalbuminemia is significant (less than 3.5 g/dL), so the use of amino acid supplements primarily composed of BCAA may be useful depending on the situation. In cases of chronic liver disease progression, abnormalities in energy metabolism in the liver are likely to occur, and the organ correlation between the liver and muscles, especially the importance of sarcopenia, is also being noted. Carnitine plays an important role in the uptake of fatty acids in mitochondria and is involved in the rate-limiting stage of β-oxidation, suggesting that carnitine deficiency could be a factor in worsening liver conditions. Carnitine supplementation is thought to be effective in improving liver metabolism and preventing sarcopenia and is also used in the prevention and treatment of painful muscle cramps (charley horses) associated with cirrhosis [3]. In progressive liver conditions, imbalances in minerals are also likely to occur. In particular, zinc deficiency not only affects metabolic functions such as the urea cycle in the liver, but has also been found to be related to the barrier mechanism in the intestines, making zinc supplementation important in cirrhosis. Zinc preparations can be used as prescription drugs, but there are also issues with polypharmacy, so the use of supplements as part of nutritional management is also considered. On the other hand, iron is not always deficient, and there are often problems due to excess, requiring caution. In progressive liver diseases, unless there are complications such as gastrointestinal bleeding, it is often the case that conditions indicating iron overload, such as high serum ferritin levels, are shown, and diffuse iron deposition (hemosiderosis) occurs in the liver tissue. Especially in alcohol-related liver diseases, secondary hemochromatosis may be presented, and anemia is often not due to iron deficiency but due to vitamin deficiencies such as folic acid or hemolysis. Therefore, casually using iron supplements for anemia can be harmful in some cases, and it is appropriate to consider vitamin supplementation instead.

111.2.4 Supplements and Drug-Induced Liver Injury

The benefits of supplements are diverse, but on the other hand, careless use of supplements can also cause drug-induced liver damage, so caution is necessary. Risk assessment is necessary for all supplements and health foods, but it is particularly problematic that there are many products imported from overseas with unknown safety profiles. It is often difficult to judge whether the ingredients listed on the product are correct, and there are many cases where drugs are intentionally mixed in addition to the effects of impurities. Also, even products believed to have effective ingredients can sometimes cause health damage. For example, turmeric is believed to have liver protective effects, but the effects vary by product, and there are cases of drug-induced liver damage caused by turmeric. The diagnosis of drug-induced liver damage is surprisingly difficult in actual clinical settings, and one of the reasons is the problem of polypharmacy. It is not uncommon to have difficulty identifying the suspected drug when drug-induced liver damage is suspected, such as when multiple supplements are used in addition to overlapping prescription drugs. Therefore, we should caution against blindly accepting the benefits and using supplements carelessly, and guidance on proper use and safety monitoring is desirable.

References

1. Ikejima K, Kon K, et al. Nonalcoholic fatty liver disease and alcohol-related liver disease: from clinical aspects to pathophysiological insights. Clin Mol Hepatol. 2020;26:728–35.
2. Singal AK, Shah VH. Current trials and novel therapeutic targets for alcoholic hepatitis. J Hepatol. 2019;70:305–13.
3. Hiramatsu A, Aikata H, et al. Levocarnitine use is associated with improvement in sarcopenia in patients with liver cirrhosis. Hepatol Commun. 2019;3:348–55.

112 Functionality of Supplements and Anti-aging: Functionality of Supplements—Immunity and Allergies

Tomohisa Takagi and Yuji Naito

The market for health foods, including supplements, has grown to a massive scale exceeding one trillion yen. Focusing on the functionality of immunity and allergies, there are no notifications related to anti-allergy functions in the database of functional food labeling information publicly available from the Consumer Affairs Agency based on the functional food labeling system, but plasma lactic acid bacteria and L-92 lactic acid bacteria, which have the mechanism of "activation of dendritic cells," have been notified as functional foods related to immune function. In addition, in September 2021, the Japan Anti-Aging Association presented a "view on the scientific basis for the functionality of immune-related functions," and in addition to the already accepted activation of dendritic cells, the usefulness of phagocyte activity, natural killer (NK) cell activity, T cell (CD4 T cell) proliferation/activation, secretory type IgA antibody concentration, etc., as immune indicators was shown, and further expansion is expected in the future. In this section, we would like to introduce representative supplements and functional ingredients related to immunity and anti-allergies.

T. Takagi (✉)
Molecular Gastroenterology and Hepatology, Graduate School of Medical Science, Kyoto Prefectural University of Medicine, Kyoto, Japan

Department for Medical Innovation and Translational Medical Science, Graduate School of Medical Science, Kyoto Prefectural University of Medicine, Kyoto, Japan
e-mail: takatomo@koto.kpu-m.ac.jp

Y. Naito
Department of Human Immunology and Nutrition Science, Graduate School of Medical Science, Kyoto Prefectural University of Medicine, Kyoto, Japan
e-mail: ynaito@koto.kpu-m.ac.jp

112.1 Lactic Acid Bacteria

In the aforementioned database of functional food labeling information from the Consumer Affairs Agency, more than 50 notifications have been made for plasma lactic acid bacteria, and their functionality is reported to "work on pDC (plasmacytoid dendritic cells) and help maintain the immune function of healthy people." As for their action on the immune mechanism, macrophages, and dendritic cells, which play a role in antigen presentation to the acquired immune system, are mentioned as the most important cells in the natural immune system, which is the front line of infection defense mechanisms. Dendritic cells are classified into two types: myeloid dendritic cells (mDC) and plasmacytoid dendritic cells (pDC). Although pDCs are a very minor subset, accounting for less than 1% of human peripheral blood mononuclear cells, they play a role as the command center for virus infection defense, highly express TLR7 and TLR9, which recognize viral nucleic acids, inside cells, recognize viral infections, release large amounts of

H. Yamada, Y. Naito (eds.), *Anti-Aging Medicine*, https://doi.org/10.1007/978-981-97-8514-8_112

interferon-α (IFN-α) and IFN-β, type I interferons, and exert direct antiviral effects. Furthermore, pDCs activate the acquired immune system, which recognizes specific viruses and eliminates them from the body, such as CD8+ T cells, CD4+ T cells, and B cells. *Lactococcus lactis* subsp. *lactis* JCM 5805 strain has been selected as a lactic acid bacterium with this pDC activation effect, and its maintenance effect on pDC in humans has been clarified [1], and it is expected to contribute to reducing the risk of viral infection. Also, it has been shown that L-92 lactic acid bacteria (Lactobacillus acidophilus L-92), improves the Th1/Th2 cytokine balance, induces apoptosis of activated T cells through pDC, and induces regulatory T cells. Through human clinical trials, it has been shown to be effective in improving symptoms of allergic diseases such as hay fever, perennial allergic rhinitis, and atopic dermatitis. This bacterium also received a new "immune" functional display in November 2022, and it is expected to alleviate allergy symptoms and reduce the risk of viral infection. *Lactobacillus brevis* KB290 (commonly known as Lovele bacteria), which was isolated from the traditional Kyoto pickle "suguki," is a lactic acid bacterium with excellent digestive juice resistance, including bile acids. The intestinal mucosal immune mechanism is tightly controlled by various immune cells, and CD11c+ macrophages and CD103− dendritic cells are important for inflammation homeostasis of the intestinal mucosa. CD11c+ macrophages contribute to inflammation suppression through the anti-inflammatory cytokine interleukin-10 (IL-10), and CD103− dendritic cells produce inflammatory cytokines such as tumor necrosis factor-α (TNF-α), IL-6, and IL-12 in response to microbial stimuli, which are involved in the differentiation of lymphocyte Th1 and Th17. We found that the number of CD103− dendritic cells decreases and the ratio of CD11c+ macrophage cells/CD103− dendritic cells significantly increases with co-stimulation of *L. brevis* KB290 and retinoic acid. Using the mouse 2,4,6-trinitrobenzenesulfonic acid (TNBS) induced colitis model, which is commonly used as a human inflammatory bowel disease model, we found that the ratio of CD11c+ macrophage cells/CD103− dendritic cells significantly decreases with the onset of colitis, but by co-administering *L. brevis* KB290 and vitamin A, the decrease in the ratio of CD11c+ macrophage cells/CD103− dendritic cells is suppressed, and the weight loss and formation of colonic ulcer lesions due to the progression of TNBS-induced colitis are suppressed both macroscopically and histologically [2]. Furthermore, through human trials, it has been confirmed that Lovele bacteria have a bowel-regulating effect that alleviates constipation and diarrhea symptoms [3] (Fig. 112.1), and it has also been shown to have immune-enhancing effects such as increased INF-α production and natural killer activity, and a reduction in influenza incidence and symptom relief has been confirmed. From these findings, Lovele bacteria are expected to be used not only in food but also in the medical field as probiotics.

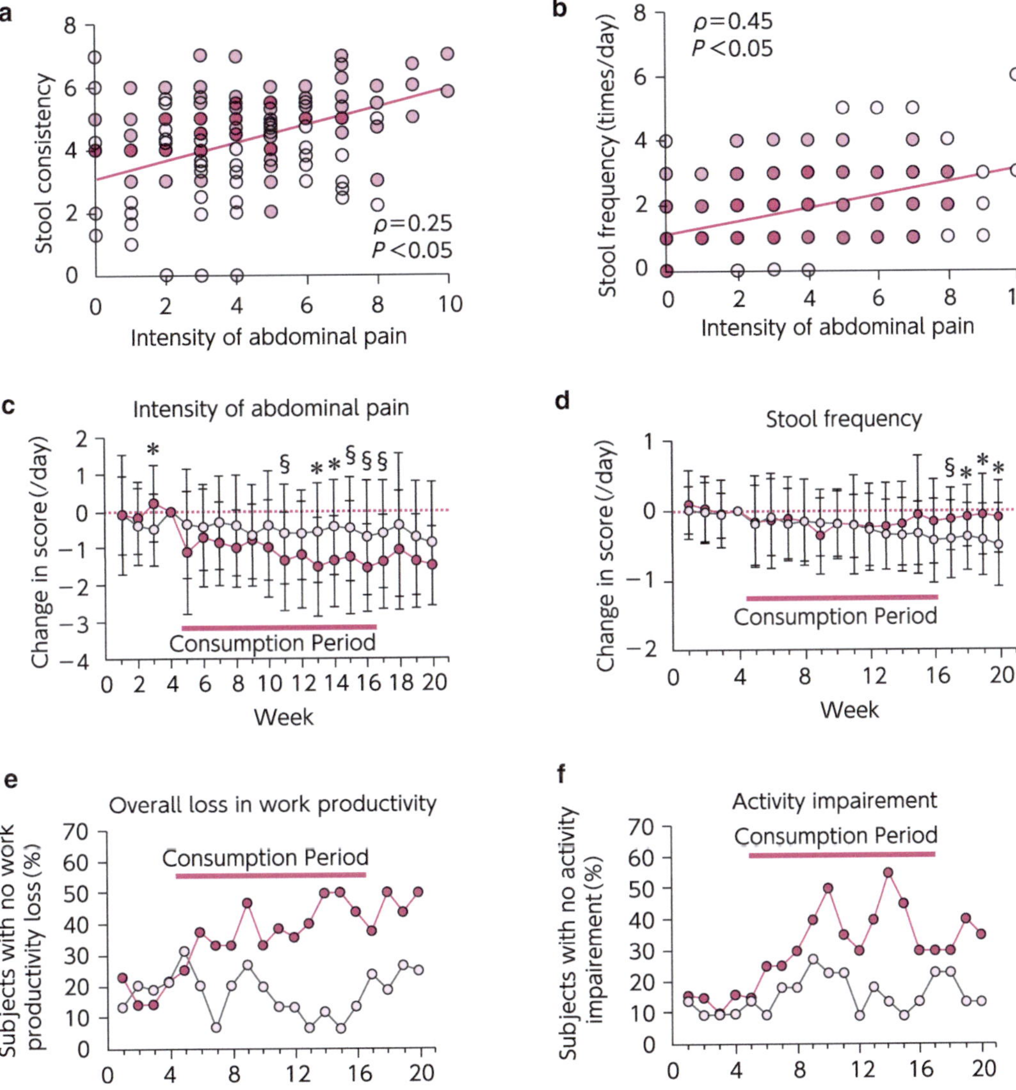

Fig. 112.1 Effectiveness of *L. brevis* KB290 and β-carotene combination therapy for diarrhea-predominant irritable bowel syndrome-like symptoms (Adapted and reprinted from [3]). In the subjects (44 healthy men and women aged 20–49 years who suffer from diarrhea with abdominal pain and discomfort, so-called diarrhea-predominant irritable bowel syndrome-like symptoms), the intensity of abdominal pain increased as the stool became watery and the frequency of bowel movements increased (**a**, **b**). The intake of *L. brevis* KB290 and β-carotene resulted in a decrease in the intensity of abdominal pain and the frequency of bowel movements (**c**, **d**) and further improved the decrease in labor productivity due to abdominal symptoms (**e**) and the hindrance to daily life (**f**)

112.2 Methylated Catechins

Methylated catechins are epigallocatechin gallate (EGCG) and epicatechin. Tea catechins, in which part of the gallate of epicatechin gallate (ECG) is methylated, such as *O*-methylated epigallocatechin-3-*O*-[3-*O*-methyl] gallate (*O*-methylated EGCG), have been shown to suppress the activation of mast cells and the release of histamine in studies using mice. This *O*-methylated EGCG is abundant in green tea varieties such as "Benifuuki" and "Benihomare," and the alleviating effect of "Benifuuki" green tea on cedar pollen allergy has been verified in a double-blind randomized placebo-controlled trial [4]. As a placebo, "Yabukita" green tea, which does not contain *O*-methylated EGCG, was used, and when subjects with cedar pollen allergy were studied, nasal and eye symptoms such as runny nose, itchy eyes, and tears during the pollen dispersal peak were suppressed with "Benifuuki" green tea, which contains a lot of *O*-methylated EGCG, and the increase in peripheral eosinophil count associated with pollen dispersal was also suppressed. From these results, it was suggested that "Benifuuki" green tea, which contains a lot of *O*-methylated EGCG, has potential as a complementary and alternative medicine for seasonal allergic rhinitis. In patients with food allergies, a phenomenon in which harmful symptoms are induced through an antigen-specific immunological mechanism caused by food, it has been reported that regulatory T cells (Treg cells), one of the immune cells, are reduced. It is believed that enhancing the function of Treg cells quantitatively and functionally is a key to the prevention and treatment of food allergies. Especially, food allergies are common in children, and high safety is required for their prevention and treatment. It is well known that retinoic acid, a metabolite of vitamin A, promotes the differentiation of Treg cells. On the other hand, it was not clear whether carotenoids, which are precursors of vitamin A and pigments contained in vegetables and fruits, promote the differentiation of Treg cells, but the authors evaluated the number of Treg cells in the mouse colon mucosa with the intake of various carotenoids (carotenes: lycopene, β-carotene, xanthophylls: astaxanthin, lutein) and revealed that the intake of lycopene induces Treg cells in the colon mucosa and improves food allergy symptoms in ovalbumin (OVA) allergy model mice [5]. As a mechanism, it was suggested that lycopene may be involved in the suppression of mast cell proliferation through the suppression of IL-9 production by controlling excessive immune responses of Th2 and Th9 types with the increase of regulatory T cells in the colon mucosa by lycopene. Although it is necessary to verify the effect of lycopene on food allergy suppression in humans in the future, it has been reported that healthy children consume more vegetables and fruits after weaning compared to children with food allergies, and it is expected that this will lead to the prevention and improvement of food allergies.

References

1. Sugimura T, Jounai K, et al. Immunomodulatory effect of Lactococcus lactis JCM5805 on human plasmacytoid dendritic cells. Clin Immunol. 2013;149:509–18.
2. Fuke N, Takagi T, et al. Lactobacillus brevis KB290 with vitamin A ameliorates murine intestinal inflammation associated with the increase of CD11c+ macrophage/CD103− dendritic cell ratio. Inflamm Bowel Dis. 2018;24:317–31.
3. Fuke N, Aizawa K, et al. Effect of combined consumption of Lactobacillus brevis KB290 and β-carotene on minor diarrhea-predominant irritable bowel syndrome-like symptoms in healthy subjects: a randomised, double-blind, placebo-controlled, parallel-group trial. Int J Food Sci Nutr. 2017;68:973–86.
4. Masuda S, Maeda-Yamamoto M, et al. 'Benifuuki' green tea containing o-methylated catechin reduces symptoms of Japanese cedar pollinosis: a randomized, double-blind, placebo-controlled trial. Allergol Int. 2014;63:211–7.
5. Ushiroda C, Takagi T, et al. Lycopene intake induces colonic regulatory T cells in mice and suppresses food allergy symptoms. Pediatr Allergy Immunol. 2022;33:e13691.

113 Functionality of Supplements and Anti-aging: Functionality of Supplements—Skin

Sawako Hibino

It is suggested that the intake of vitamins, minerals, or essential fatty acids through diet can improve the condition of the skin, but it is difficult to supplement everything from diet alone. Since April 2015, the food industry has initiated a system of functional food labeling, and many functional foods with the theme of "skin moisturizing, maintaining skin barrier function" have been accepted, and the use of food ingredients and supplements that claim to reduce the risk of skin diseases and alleviate skin aging has increased. It is important to thoroughly evaluate the various ingredient groups used in supplements, including such functional foods.

113.1 Defense of the Skin Aging Process [1]

The best way to defend against the skin aging process, such as loss of elasticity, thinning, dryness, fine wrinkles, and atrophy of the dermis, is "prevention", and the three principles of prevention are sun protection, stress reduction, and a balanced diet containing antioxidants, and comprehensive measures including calorie restriction and exercise. In particular, excessive sugar in daily meals accelerates the aging process through the production of advanced glycation end products that inhibit the proper repair of collagen fibers, so carbohydrate control in the diet is important. On the other hand, in terms of lipids, monounsaturated fatty acids and ω3 polyunsaturated fatty acids are said to be inversely correlated with severe photoaging, and antioxidants such as vitamin C, A, E, carotenoids, flavonoids, and plant-based antioxidants such as resveratrol, curcumin, and green tea polyphenols effectively slow down this process. Minerals such as zinc, selenium, and copper are coenzymes of metallothionein and glutathione, reduce intracellular oxidative stress, and have been reported to protect the skin as a result.

113.2 Gut–Brain–Skin Axis (Brain–Gut–Skin Theory) [2]

In 1930, John Stokes and Donald Pillsbury of Duke University Medical School proposed a unified theory of the gut–brain–skin axis. They linked emotional states such as anxiety and depression to changes in the gut microbiota, theorizing that these changes promote local and systemic inflammation. As a result, they hypothesized that this could also affect the skin. The health of the brain and skin are closely related, and it has been epidemiologically shown that mental stress can exacerbate skin diseases such as atopic dermatitis. The impact of foods and nutrients

S. Hibino (✉)
Department of Future Medical Science, Osaka University Graduate School of Medicine, Osaka, Japan

H. Yamada, Y. Naito (eds.), *Anti-Aging Medicine*, https://doi.org/10.1007/978-981-97-8514-8_113

(including supplements) that affect the gut environment on the skin is suggested to be significant. In particular, specific strains of probiotics such as lactic acid bacteria have been shown to limit UV damage and reduce skin hypersensitivity.

113.3 Nutrients Expected to Be Effective for the Skin [3]

Products containing α-tocopherol (vitamin E), L-ascorbic acid (vitamin C), retinol (vitamin A), and niacinamide (vitamin B3) are antioxidants and are effective in treating photoaging. In particular, recent research has focused on nicotinamide mononucleotide (NMN), a precursor of the coenzyme nicotinamide adenine dinucleotide (NAD) produced in the body from vitamin B3, which activates the sirtuin gene. This NMN, in the skin area, by inhibiting the cAMP/Wnt signal and suppressing the melanin production of aged melanocytes, has been reported to prevent age-related pigmentation and prevent skin aging. Resveratrol, a natural polyphenol found in grape skins (not in the flesh), which is known to be naturally produced by plants to protect against bacterial and fungal infections, is also known to activate the sirtuin gene like NMN. Resveratrol has been shown to inhibit various human cancer cells, including melanoma in the skin, and prevent their metastasis to other tissues, and it has anti-inflammatory, antioxidant, and anti-collagenase activity, and is known to have a skin-protecting effect.

113.4 Particularly Good Supplements for the Skin [4]

Monounsaturated fatty acids (MUFA) have been shown in previous studies to reduce insulin resistance and inflammatory processes, thereby protecting the skin from external aging. And, polyunsaturated fatty acids such as α-linolenic acid (ALA), eicosapentaenoic acid (EPA), and docosahexaenoic acid (DHA) have been shown to have a significant impact on skin health. In polyunsaturated fatty acids (PUFA), it has been reported that they have an effect on photoaging, and ω3 fatty acids have been shown to reduce the production of inflammatory compounds, insulin, and insulin-related compounds, which can also cause a decline in skin function. These essential fatty acids have been shown to be beneficial for specific skin conditions such as eczema, psoriasis, acne, and skin cancer. ω3 fatty acids are thought to delay the visible aging caused by aging. Lycopene, a type of carotenoid, is a powerful antioxidant and has been confirmed to have strong photoprotective properties. Curcumin, the main active ingredient in turmeric, has been shown to improve wound healing, psoriasis, and skin cancer. Flavonoids are powerful antioxidants, and green tea contains epigallocatechin-3-gallate (EGCG), which not only has antioxidant and photoprotective effects, but also has many anticancer effects. Clinical studies in humans are lacking, and the ideal dosage as a supplement is not known. Multivitamins include vitamins A, B group, C, D, E, biotin, and minerals such as chromium, iron, selenium, and zinc, which have been proven to play an important role in maintaining skin health. On the other hand, iron tends to be deficient in young women, but especially, excessive intake of iron during menopause has been found to promote aging. There are many reports on the effects of collagen, which has been shown to have a positive effect on wound healing and reduce signs of aging, such as elasticity, dermal collagen density, and improved skin moisture retention. Cannabidiol (CBD), a non-psychoactive compound of cannabis, has been attracting great attention in recent years as a potential treatment for various pathologies in the skin and beauty field. CBD is already used for various medical and mental disorders such as depression, anxiety neurosis, post-traumatic stress disorder, Alzheimer's disease and other cognitive disorders, pain, and not only for prevention of SARS-CoV-2 infection [5] but also for skin diseases and a wide range of fields, it is a component that will be watched in the future. Supplements can be expected to have various effects by efficiently ingesting nutrients that cannot be obtained from the diet, but they are not without side effects. Persistent high-dose intake must be avoided. To maintain youthful skin, first of all, improving the living environment, especially, actively taking a balanced diet is the basic.

References

1. Das L, Bhaumik E, et al. Role of nutraceuticals in human health. J Food Sci Technol. 2012;49:173–83.
2. Stokes JH, Pillsbury D. The effect on the skin of emotional and nervous states: theoretical and practical consideration of a gastrointestinal mechanism. Arch Dermatol Syphilol. 1930;22:962–93.
3. Krutmann J, Humbert P, editors. Nutrition for healthy skin: strategies for clinical and cosmetic practice. Heidelberg: Springer; 2011.
4. Cao C, Xiao Z, et al. Diet and skin aging—from the perspective of food nutrition. Nutrients. 2020;12:870.
5. Nguyen LC, Yang D, et al. Cannabidiol inhibits SARS-CoV-2 replication and promotes the host innate immune response. bioRxiv. 2021; 2021.03.10.432967

Functionality of Supplements and Anti-aging: Functionality of Supplements—Male Medical

114

Akira Tsujimura

114.1 Supplements Expected to Improve Erectile Function

Phosphodiesterase (PDE) 5 inhibitors maintain erectile ability by increasing the concentration of cyclic guanosine monophosphate (cGMP) in penile corpus cavernosum smooth muscle cells. Nitric oxide (NO), synthesized from L-arginine, is involved in the increase of cGMP. Therefore, arginine, the substrate, has been attracting attention as a supplement to improve erectile dysfunction (ED). Also, citrulline, which is converted from arginine by NO synthase, is easily reconverted to arginine, and improvement of ED by citrulline intake has been reported (Fig. 114.1). In addition, resveratrol, a type of polyphenol, induces an increase in NO synthase activity by enhancing the activity of one of the sirtuin/sirtuin family, SIRT1, and contributes to the improvement of erectile ability, as reported in animal experiments and clinical studies in humans (Fig. 114.1). The author has reported the improvement of erectile ability by a combination of citrulline and resveratrol in a placebo-controlled double-blind crossover trial [1] (Fig. 114.2). In addition, maca, a species of plant in the Brassicaceae family that grows in Peru, has been

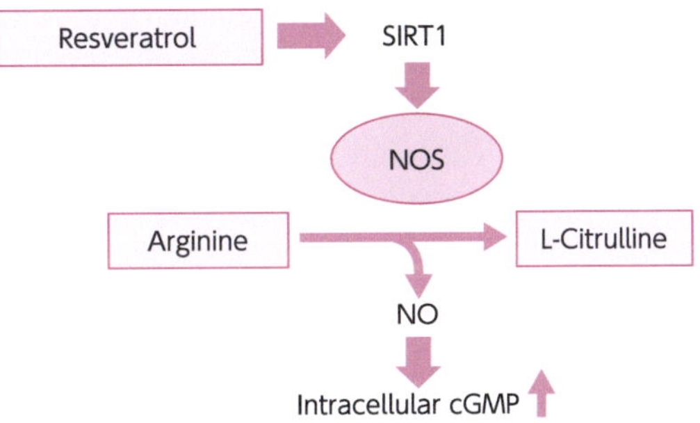

Fig. 114.1 Involvement of resveratrol, arginine, L-citrulline in erectile ability

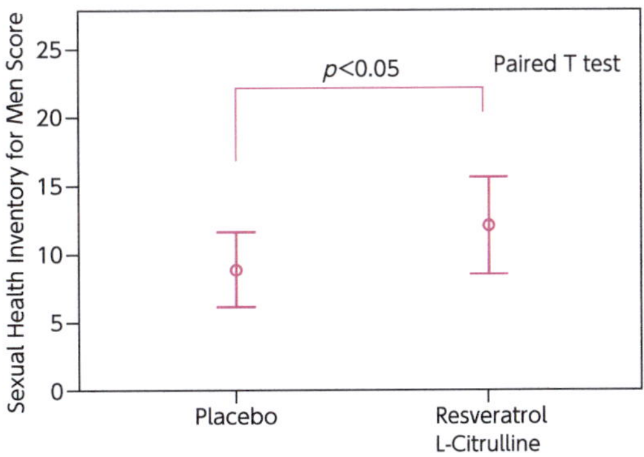

Fig. 114.2 Effect of resveratrol, citrulline compound. Twenty ED patients who are taking PDE5 inhibitors on demand but are not sufficiently effective were given placebo and actual medicine in a crossover for 1 month

A. Tsujimura (✉)
Department of Urology, Juntendo University Urayasu Hospital, Chiba, Japan
e-mail: atsujimu@juntendo.ac.jp

H. Yamada, Y. Naito (eds.), *Anti-Aging Medicine*, https://doi.org/10.1007/978-981-97-8514-8_114

reported to be useful for ED in a few studies, although its mechanism is not yet clear, and it is a familiar supplement in Japan.

114.2 Supplements Expected to Improve Reproductive Function

More than 80% of male infertility is due to spermatogenic dysfunction, and more than half of these cases are idiopathic. There is no specific treatment for this idiopathic spermatogenic dysfunction, and vitamins and supplements with antioxidant properties have been mainly used. In fact, coenzyme Q10, L-carnitine, and zinc are frequently used. There are reports that sperm concentration and motility significantly improved with the intake of coenzyme Q10 (26 weeks). L-Carnitine is used for patients with poor semen (sperm concentration 10,000,000–20,000,000/mL), sperm motility 10. An increase in total motile sperm count has been shown with administration (for 2 months) to about 30% of the population, compared to placebo. Zinc is an essential mineral indispensable for life support, and it has been speculated to be involved in reproductive function since the zinc concentration in semen is 100 times that of serum. In fact, significant improvements in semen volume and sperm progressive motility have been shown with zinc administration (for 3 months) [2]. Vitamins B12, E, and C in supplements also have antioxidant effects. Vitamin E is often used in combination with endocrine therapy (clomiphene administration), and significant improvements in sperm concentration and sperm motility have been shown with combination therapy (for 6 months) compared to placebo. Vitamin C is a representative vitamin with antioxidant effects. Significant improvements in sperm concentration, sperm motility, and normal morphology rate have been shown with administration (for 2 months) to oligospermia patients. The author has experienced significant improvement in total motile sperm count by administering a combination of vitamins B12, E, and C, coenzyme Q10, L-carnitine, astaxanthin, and zinc (for 3 months) to patients with poor semen quality [3]. However, it should be emphasized that no prospective comparative studies have been conducted in the majority of cases for any of these supplements, and they are still being used empirically.

114.3 Supplements for Improving Low Testosterone Syndrome

Zinc is a supplement that can be expected to increase blood testosterone levels, which is of interest in aging male hypogonadism syndrome. That is, zinc is expected to have not only reproductive function improvement but also endocrinological improvement effects. In a zinc-deficient animal model, Leydig cell function decreases without affecting the hypothalamic-pituitary-testis system, and conversely, blood testosterone levels increase without involvement in gonadotropin secretion by zinc administration. In humans as well, significant increases in blood testosterone levels have been shown with zinc administration (for 4 weeks). The next one is Tongkat Ali. This is a plant of the Simaroubaceae family that grows naturally in the tropical rainforests of Malaysia, the Indochina Peninsula, Sumatra, and Borneo, and can be purchased in Japan. It increases blood testosterone levels by promoting gonadotropin secretion through direct action on the hypothalamus-pituitary and by inhibiting aromatase activity. In elderly men with symptoms of aging male hypogonadism syndrome, blood testosterone levels increased and symptoms improved with the administration of 200 mg/day of Tongkat Ali (for 1 month) [4]. Testofen™, which is available in Japan, has also been reported to have similar effects. Testofen™ is derived from fenugreek, a type of spice and an annual legume plant. In a placebo-controlled double-blind trial, significant increases in blood testosterone levels were shown with the administration of Testofen™ (for 12 weeks) to healthy men [5].

References

1. Shirai M, Hiramatsu I, et al. Oral L-citrulline and transresveratrol supplementation improves erectile function in men with phosphodiesterase 5 inhibitors: a randomized, double-blind, placebo-controlled cross-over pilot study. Sex Med. 2018;6:291–6.
2. Hadwan MH, Almashhedy LA, et al. Oral zinc supplementation restores superoxide radical scavengers to normal levels in spermatozoa of Iraqi asthenospermic patients. Int J Vitam Nutr Res. 2015;85:165–73.
3. Terai K, Horie S, et al. Combination therapy with antioxidants improves total motile sperm counts: a preliminary study. Reprod Med Biol. 2019;19:89–94.
4. Tambi MIBM, Imran MK, et al. Standardised water-soluble extract of Eurycoma longifolia, Tongkat ali, as testosterone booster for managing men with late-onset hypogonadism? Andrologia. 2012;44(Suppl 1):226–30.
5. Rao A, Steels E, et al. Testofen, a specialised Trigonella foenum-graecum seed extract reduces age-related symptoms of androgen decrease, increases testosterone levels and improves sexual function in healthy aging males in a double-blind randomised clinical study. Aging Male. 2016;19:134–42.

115 Functionality of Supplements and Anti-aging: Functionality of Supplements—Women's Health Care

Masakazu Terauchi

With the transient decline in support for menopausal hormone therapy, once considered the ultimate anti-aging intervention for middle-aged and older women, in the early 2000s, the importance of complementary and alternative medicine in this field, especially supplements, has increased. There are countless supplements that have proven effective in the field of women's healthcare, but in this section, we will focus on soy isoflavones and grape seed proanthocyanidins.

115.1 Soy Isoflavones

Isoflavones are a type of flavonoid polyphenol that is abundant in leguminous plants. In addition to their antioxidant properties, they bind to estrogen receptors (ER) and act as agonists or antagonists. They have a higher affinity for ERβ than for ERα. The ability of isoflavones to bind to ERs is explained by the similarity between the isoflavone skeleton and estrogen (Fig. 115.1). Representative isoflavones include genistein, daidzein, and glycitein. These are easily absorbed into the body when their sugar chains are removed (= aglycone formation) by the action of intestinal bacteria from their respective precursors, genistin, daidzin, and glycitein. Daidzein is further activated by the action of intestinal bacteria to form equol (Fig. 115.1). Soy and soy products are the most well-known sources of isoflavones, but red Clover is sometimes used. There have been various studies on the effects of soy isoflavones on health indicators in middle-aged and older women, and meta-analyses of these studies have shown effects such as (1) improvement of menopausal symptoms, (2) increase in lumbar spine bone density, and (3) reduction in the incidence of breast cancer, which are thought to be due to ER agonist action. One issue with soy isoflavones is the "Basic Approach to Safety Evaluation of Foods for Specified Health Use Containing Soy Isoflavones" published by the Food Safety Commission of the Cabinet Office in 2006. This document stipulates a safe additional intake of about 30 mg/day from sources other than food, based on an Italian study that diagnosed 4% of women who consumed 150 mg/day of soy isoflavones for 5 years with endometrial hyperplasia. However, all of the above studies target cases that consumed more than this amount. Taking this into account, a randomized double-blind placebo-controlled trial was conducted on 90 women aged 40 to under 60, and it was shown that even an intake of 25 mg/day significantly improved menopausal symptoms [1]. There have also been many studies on S-equol, and it has been shown to have effects such as (1) improvement of meno-

M. Terauchi (✉)
Department of Women's Health, Tokyo Medical and Dental University, Tokyo, Japan
e-mail: teragyne@tmd.ac.jp

H. Yamada, Y. Naito (eds.), *Anti-Aging Medicine*, https://doi.org/10.1007/978-981-97-8514-8_115

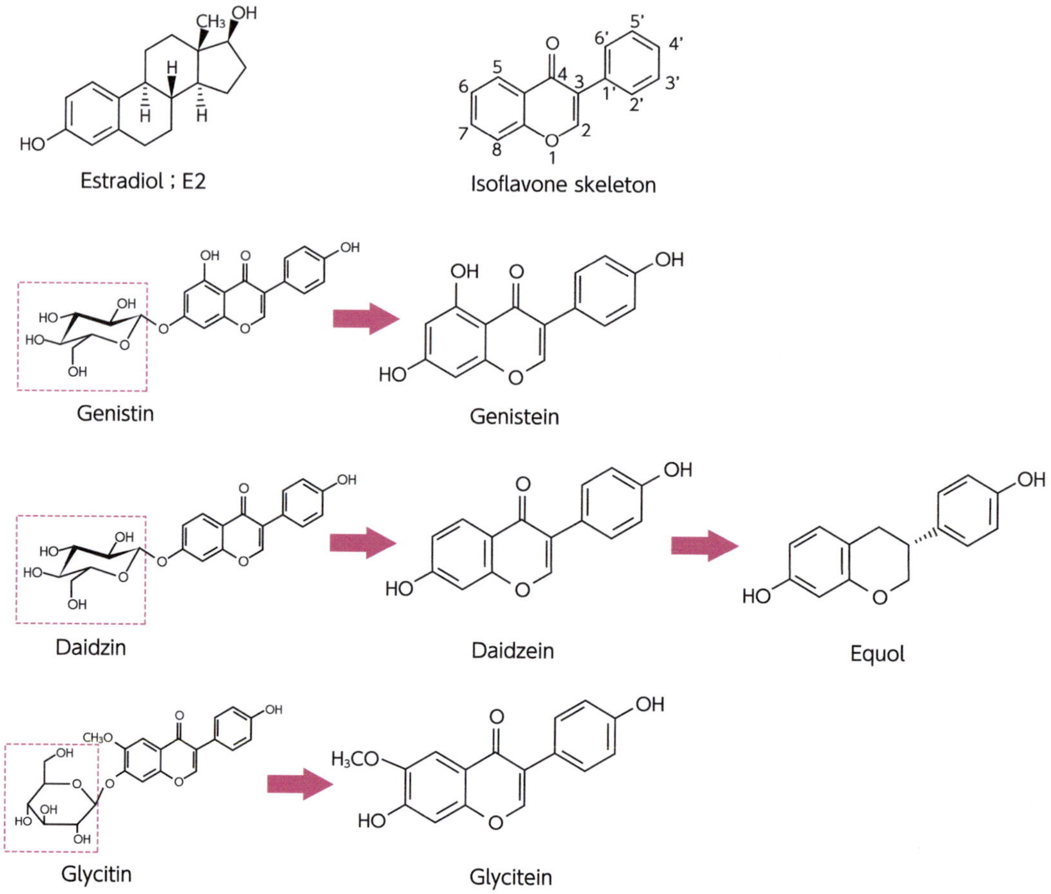

Fig. 115.1 Isoflavone skeleton and representative isoflavones

pausal symptoms, (2) increase in bone density, and (3) reduction of cardiovascular risk factors (LDL cholesterol and CRP) [2].

115.2 Grape Seed Proanthocyanidin

Polyphenols, which have multiple phenolic hydroxyl groups within the same molecule, have excellent antioxidant effects due to the strong radical scavenging activity associated with the electron-donating nature of the phenolic hydroxyl groups. Polyphenols are components of plant pigments and bitterness, and there are over 5000 types, but among the flavonoids that make up the majority, flavanols composed of catechins contained in green tea and proanthocyanidins contained in grapes are attracting attention. Proanthocyanidins are oligomers of flavan-3-ols (mainly [+]-catechin and [−]-epicatechin) and are found in plant fruits, bark, leaves, and seeds. A randomized double-blind placebo-controlled trial was conducted on 96 women aged 40 to under 60 using grape seed proanthocyanidin extract (GSPE), and it was revealed that GSPE has effects such as (1) improvement of physical and mental symptoms of menopause, (2) increase in muscle mass, and (3) reduction in blood pressure (Fig. 115.2) [3]. Furthermore, a randomized, double-blind, placebo-controlled trial was conducted on 30 men and women aged 40 to under 65 who were pointed out to have normal high blood pressure, consisting of a control group, a low-dose GSPE group (200 mg/day), and a high-dose group (400 mg/day). In 21 non-smokers, systolic and diastolic blood pressure significantly decreased and the incremental elasticity coefficient (E_{inc}) and pulse wave velocity (PWV) significantly improved 12 weeks after the intake of

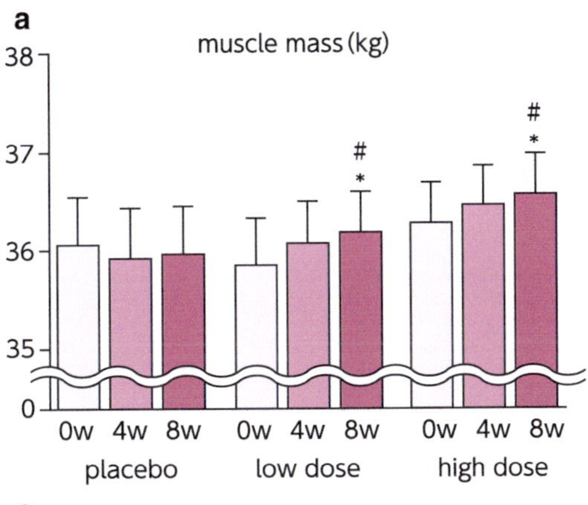

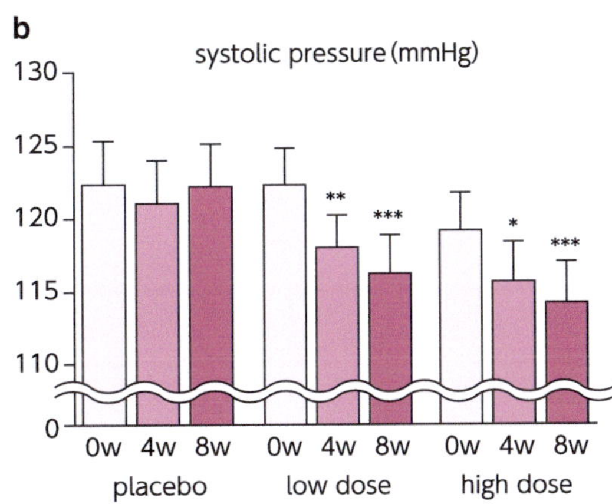

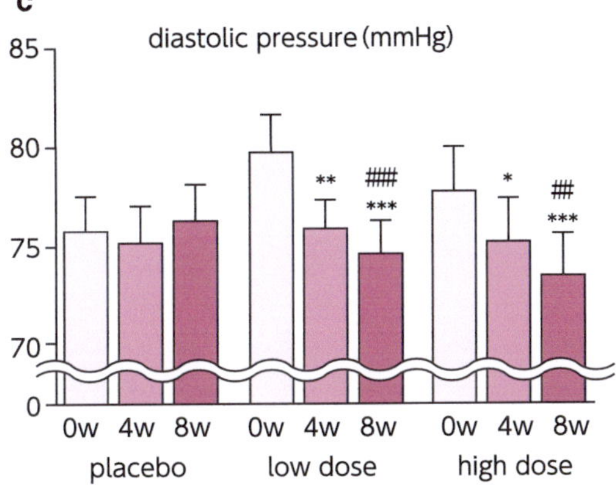

Fig. 115.2 Effects of grape seed proanthocyanidin extract [3]. (**a**) Effect on muscle mass, (**b**) effect on systolic blood pressure, (**c**) effect on diastolic blood pressure. $*P < 0.05$, $**P < 0.01$, $***P < 0.001$, vs. pre-administration, paired *t* test. $\#P < 0.05$, $\#\#P < 0.01$, $\#\#\#P < 0.001$, change from pre-administration vs. placebo, unpaired t test

high-dose GSPE compared to the control group, suggesting that high-dose GSPE intake is useful for improving arteriosclerosis in individuals with normal high blood pressure [4]. For more information on the various physiological activities of GSPE and its application to middle-aged women, please refer to a recently published review [5].

References

1. Hirose A, Terauchi M, et al. Low-dose isoflavone aglycone alleviates psychological symptoms of menopause in Japanese women: a randomized, double-blind, placebo-controlled study. Arch Gynecol Obstet. 2016;293:609–15.
2. Jackson RL, Greiwe JS, et al. Emerging evidence of the health benefits of S-equol, an estrogen receptor β agonist. Nutr Rev. 2011;69:432–48.
3. Terauchi M, Horiguchi N, et al. Effects of grape seed proanthocyanidin extract on menopausal symptoms, body composition, and cardiovascular parameters in middle-aged women: a randomized, double-blind, placebo-controlled pilot study. Menopause. 2014;21:990–6.
4. Odai T, Terauchi M, et al. Effects of grape seed proanthocyanidin extract on vascular endothelial function in participants with prehypertension: a randomized, double-blind, placebo-controlled study. Nutrients. 2019;11:2844.
5. Izumi T, Terauchi M. The diverse efficacy of food-derived proanthocyanidins for middle-aged and elderly women. Nutrients. 2020;12:3833.

Functionality of Supplements and Anti-aging: Functionality of Supplements—Fertility

116

Kuniaki Ota

116.1 Overview

In recent years, due to societal factors such as the advancement of women in society, the trend of women marrying later, and the aging of the desired child-rearing age, age-related infertility has been increasing [1]. In fact, the Centers for Disease Control and Prevention (CDC) in the United States reported that the average age of first childbirth was 21.4 years in 1970, but it was 25.1 years in 2008 [2]. Furthermore, the same agency points out that the number of women giving birth to their first child after the age of 35 has increased from 1 in 100 in 1970 to 1 in 12 in 2006, indicating the aging of women at the time of childbirth [3, 4]. On the other hand, the number of infertility patients due to the aging of women's desire to have children is rapidly increasing in our country, and the number of babies born in 2018 due to assisted reproductive technology (ART) treatments such as in vitro fertilization and embryo transfer is 56,979, with 1 in 16 being an in vitro fertilization baby [5]. Infertility can indeed be called a national disease related to aging, and the use of supplements as part of integrated medicine is frequently performed in addition to ART. However, among the pregnancy activity (hereinafter, fertility) supplements, there are those on the market that lack evidence, such as "improving the quality of eggs," "improving sperm motility," and "increasing implantation power." One of the reasons for this is the reality that supplements whose safety and effectiveness have not been proven are circulating without public regulation by the Food and Drug Administration (FDA) and others [6]. On the other hand, evidence of nutritional interventions centered on supplements for fertility is accumulating due to recent literature and statistical methods and cohort studies. In this section, we will introduce and overview the latest evidence, focusing on folic acid and vitamin D, which are particularly becoming established as evidence for contributing to fertility.

K. Ota (✉)
Department of Obstetrics and Gynecology, Kawasaki Medical School, Okayama, Japan

116.2 Supplements for Fertility (Women's Edition)

116.2.1 Folic Acid

It can be easily guessed that the supplement with the highest recognition among women seeking to conceive is folic acid. However, the pre-pregnancy intake rate of folic acid supplements in Japan is as low as 8%, which is very low compared to other countries, and its evidence has not penetrated [7]. Recent research results have shown that folic acid supplements are also effective in the field of reproduc-

H. Yamada, Y. Naito (eds.), *Anti-Aging Medicine*, https://doi.org/10.1007/978-981-97-8514-8_116

tive medicine. Recently, a prospective cohort study [Environment and Reproductive Health (EARTH) Study] conducted in Boston, USA, investigated the dietary intake [(food frequency questionnaire (FFQ)] of 353 patients before the start of in vitro fertilization, and the ART performance and pregnancy outcomes were prospectively tracked. No correlation was found between the amount of folic acid intake from diet and both ART performance and pregnancy outcomes. However, in an analysis limited to folic acid supplements, it was reported that the group consuming 800 μg/day or more of folic acid supplements had significantly higher fertilization rates in ART performance and significantly higher implantation rates, clinical pregnancy rates, and birth rates compared to the group consuming 400 μg/day or less of folic acid supplements [8]. Furthermore, from the same EARTH study, it was found that the adjusted birth rate, corrected for race and BMI, has a positive correlation up to 1200 μg/day of folic acid supplements, but the number of preantral follicles reflecting ovarian function shows a positive correlation up to 800 μg/day of folic acid supplements, and at 1200 μg/day, the number of preantral follicles decreases [8, 9] (Fig. 116.1). In another prospective cohort study (The Nurses' Health Study-II), 11,072 people (number of pregnancies: 15,950) were tracked from 1992 to 2009, and when the amount of folic acid supplement intake was 730 μg/day or more (median: 1000 μg), compared to the group not taking it, the miscarriage risk was found to be reduced by 20% [10]. On the other hand, regardless of the amount of folic acid intake from food, it was not possible to reduce the risk of miscarriage [10]. Thus, these two prospective cohort studies had results proving how effective folic acid supplements are as fertility supplements. From the ongoing 100,000-person birth cohort in Norway: Norwegian Mother and Child Cohort Study (MoBa), it was reported that taking folic acid supplements from before pregnancy (early pregnancy) reduced the onset of severe language development delay at age 3 by 50% and the onset of autism at age 6 by 40% [11, 12]. Also, in a birth cohort in Israel, it was reported that folic acid supplements prevented the onset of autism by nearly 40% in the long term when tracked up to age 15 [13]. Furthermore, in an interesting report, when the blood folic acid concentration in early pregnancy and the telomere length of lymphocytes collected from newborns were measured, there was a positive correlation between blood folic acid concentration and telomere length, suggesting that the mother's folic acid intake status may be involved in the child's lifespan [14] (Fig. 116.2). Thus, the intake of folic acid supplements is improving reproductive medical outcomes and involvement in the child's future, as evidenced by the latest evidence.

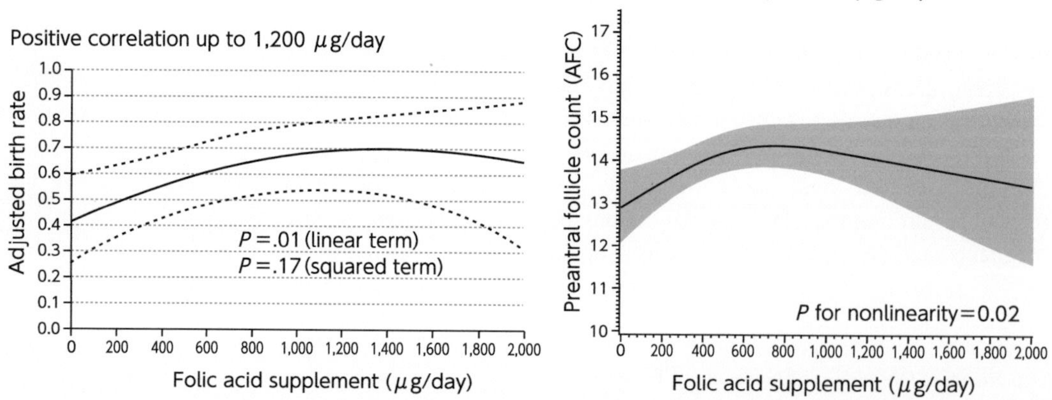

Fig. 116.1 Influence of folic acid supplement intake on birth rate and ovarian function [8, 9]

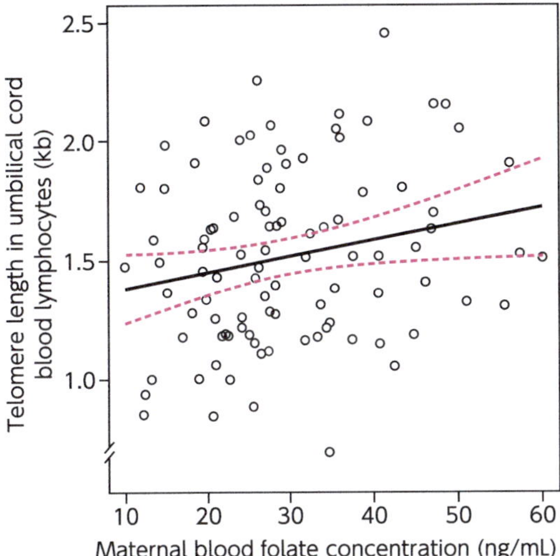

Fig. 116.2 Maternal folate concentration and newborn telomere length [14]

116.2.2 Vitamin D

It has been reported for a long time that animals with vitamin D deficiency or vitamin D receptor deficiency become infertile [15–20]. Furthermore, it has been reported that rats raised on a diet without vitamin D are infertile [21]. Therefore, it was expected that clinical data following the animal model would appear in humans, but it took time for it to appear. In humans, in 2010, Ozkan et al. reported that in patients undergoing in vitro fertilization, the concentration of vitamin D in the blood and follicular fluid was significantly higher in the group that became pregnant compared to the group that did not [22]. Since then, many reports on vitamin D and infertility have been recognized. In particular, in recent years, the increase in ultraviolet rays due to the destruction of the ozone layer has become a global environmental problem, and as a result, there is a tendency to avoid sunlight exposure necessary for the synthesis of vitamin D, making vitamin D deficiency a global issue. An interesting report is that in Denmark, a Nordic country with short daylight hours throughout the year, the government legally obligated margarine manufacturers to add vitamin D (50 IU per 100 g of margarine) to all products from 1962 to May 1985. A retrospective cohort study based on this policy revealed that the birth rate was higher in the era when vitamin D was legally enhanced (Exposed) compared to the era after the legal enhancement ended (Wash-out, Non-exposed) in patients with infertility [23] (Fig. 116.3). This can be said to be the result of a valuable epidemiological study that supports the relationship between vitamin D and fertility. More recently, a prospective cohort study showed that by taking vitamin D before pregnancy and maintaining a serum 25(OH)D concentration of 30 ng/mL or more, the pregnancy rate and birth rate were significantly higher than those less than 30 ng/mL, suggesting that maintaining a blood vitamin D concentration of 30 ng/mL or more may have benefits for pregnancy and childbirth [24]. On the other hand, in three groups divided by blood vitamin D concentration (vitamin D deficiency group: <20 ng/mL, vitamin D insufficiency group: 20~30 ng/mL, vitamin D sufficiency group: >30 ng/mL), the correlation with clinical pregnancy rate, birth rate, and miscarriage rate due to in vitro fertilization was analyzed, indicating that no effect was observed in any of the vitamin D blood concentration groups [25]. Furthermore, a recent prospective cohort study (2569 people) reported that there was no correlation between blood vitamin D concentration and clinical outcomes such as implantation rate, clinical pregnancy rate, and miscarriage rate due to fresh embryo transfer [26]. The detailed mechanism between vitamin D and fertility is not understood in the aforementioned animal models, and it remains unknown in humans as well. The effect of vitamin D intake on patients undergoing in vitro fertilization is unclear, but there is no high-evidence study such as a placebo-controlled randomized comparative trial, so future research results are expected. However, considering from the cohort studies so far, it is clear that "it is better to take vitamin D than not to take it," and it is thought that actively measuring blood vitamin D and taking vitamin D supplements at a confirmed safe dose when the blood vitamin D concentration is 30 ng/mL or less may contribute to improving fertility. On the other hand, since the expression of the vitamin D receptor decreases with age [27] (Fig. 116.4), it might be advisable to consider age-tailored vitamin D intake when using vitamin D for anti-aging purposes.

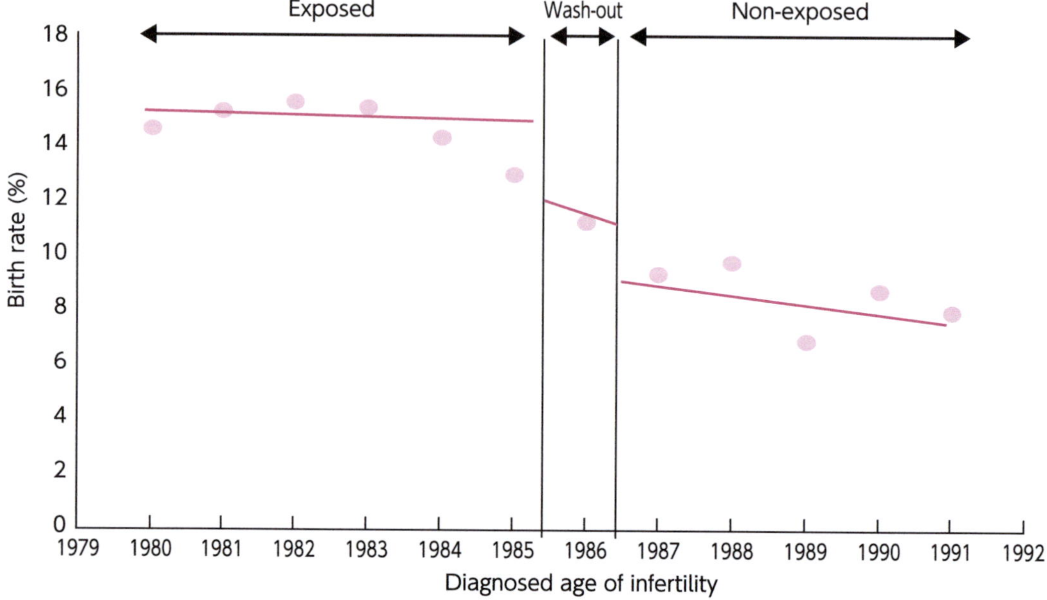

Fig. 116.3 Impact on birth rate due to Denmark's vitamin D fortification policy diagnosed age of infertility [23]

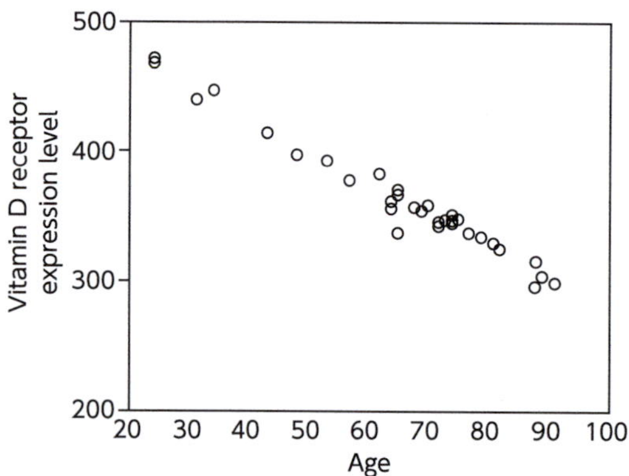

Fig. 116.4 Correlation between age and vitamin D receptor expression level [27]

116.3 Supplements for Fertility (For Men)

Various fertility supplements for sperm have been proposed so far. Folic acid and zinc are always included in the most widely available supplements. Sperm are extremely sensitive to oxidative stress, which can cause chromatin damage, sperm membrane peroxidation, decreased motility, and increased apoptosis [27]. Folic acid provides carbon for important DNA synthesis and methylation in sperm formation, and it itself has antioxidant effects, eliminating free radicals [28]. Zinc, on the other hand, is essential for sperm formation as a component of metal enzymes involved in steroid receptors and DNA transcription [29], and the high concentration of zinc in semen (30 times higher than in blood [30]) is avoiding exposure to oxidative stress in sperm due to its antioxidant effects [31]. Irani et al. conducted a meta-analysis after extracting six papers through a systematic review and reported that the combined use of folic acid and zinc supplements improves semen quality, especially sperm concentration [32]. Furthermore, Azizollahi et al. reported in a placebo-controlled randomized comparative trial that the combined supplement of folic acid and zinc reduced sperm DNA fragmentation and increased the amount of protamine, which is essential for mature sperm formation [33]. Recently, a large-scale placebo-controlled randomized comparative trial published in the JAMA journal analyzed parameters related to semen quality, pregnancy rate, birth rate, and perinatal prognosis in 2370 couples undergoing fertility treatment, including in vitro fertilization, when 1185 men took a supplement containing 5 mg of folic acid and 30 mg of zinc for 6 months (the control group of 118 people took a placebo for 6 months) [34]. It was not possible to improve the parameters related to semen quality with folic acid and zinc, nor to confirm an increase in pregnancy and birth rates. On the other hand, an increased risk of preterm birth was confirmed in the group taking the folic acid and zinc supplement. Also, like folic acid, vitamin E, which has antioxidant stress effects, is often included in male fertility supplements, but in a placebo-controlled randomized comparative trial using vitamin E, when 400 mg/day of vitamin E was administered for 3 months to men undergoing in vitro fertilization, no improvement was observed in semen findings, but the birth rate was reported to be 2.75 times higher (odds ratio: 2.75, 95% confidence interval: 1.05–7.19, $p = 0.04$) [35]. Further accumulation of data is expected regarding the effects of vitamin E.

116.4 Conclusion

In recent years, there have been scattered reports suggesting that the Mediterranean diet is extremely beneficial in terms of nutrition and disease prevention. The Mediterranean diet refers to the eating habits of Mediterranean coastal countries such as Spain, Italy, Greece, Turkey, and Morocco, characterized by a high intake of vegetables, fruits, herbs, and seafood, moderate consumption of red meat and processed foods, and the use of olive oil, resulting in a high intake ratio of unsaturated fatty acids. Furthermore, the Mediterranean diet was registered as an intangible cultural heritage by UNESCO in 2010. In fact, in the PREvención con DIeta MEDiterránea (PREDIMED) study, a clinical trial involving parallel group comparisons, multiple facilities, simple blind, and randomization conducted by 16 research groups in 7 regions of Spain, it was reported that the Mediterranean diet reduced the risk of cardiovascular diseases (myocardial infarction, stroke, cardiovascular death) by 30% and the risk of stroke by 49% compared to a low-fat diet [36]. In the field of reproductive medicine, there have been reports that the Mediterranean diet has improved the clinical pregnancy rate and birth rate of in vitro fertilization patients [37]. Recently, an interesting report from the EARTH study examined ART performance and pregnancy outcomes [38]. Among the 357 in vitro fertilization patients who were traceable among the 608 people registered in the EARTH study, those who consumed a high level of pro-fertility diet enriched with monoglutamic acid folic acid, vitamin B12, and vitamin D had improved implantation rates, clinical pregnancy rates, and birth rates, but there was no change in these rates even if the intake level of the Mediterranean diet was increased. This suggests that a diet rich in B vitamins including folic acid and vitamin D may improve fertility more than the Mediterranean diet, which has been considered good during pregnancy. Many medical professionals and patients believed that it was important for couples trying to conceive to practice diets and habits (such as quitting smoking and exercising) that make it easier to get pregnant, but the expression was very vague and far from evidence. However, with the accumulation of recent evidence, this ambiguity is about to change to clarity. Nearly 50 years have passed since the advent of in vitro fertilization, and there is no doubt that the clinical application of fertility supplements will be included in the next development.

References

1. Balasch J. Ageing and infertility: an overview. Gynecol Endocrinol. 2010;26:855–60.
2. Martin JA, Hamilton BE, et al. Births: final data for 2008. Natl Vital Stat Rep. 2010;59:3–71.
3. Cil AP, Turkgeldi L, et al. Oocyte cryopreservation as a preventive measure for age-related fertility loss. Semin Reprod Med. 2015;33:429–35.
4. Matthews TJ, Hamilton BE. Delayed childbearing: more women are having their first child later in life. NCHS Data Brief. 2009;21:1–8.
5. Ishihara O, Jwa SC, et al. Assisted reproductive technology in Japan: a summary report for 2018 by the Ethics Committee of the Japan Society of Obstetrics and Gynecology. Reprod Med Biol. 2020;20:3–12.
6. Incze M, Katz MH. Regulating the dietary supplement industry: the taming of the slew. JAMA Intern Med. 2018;178:1723.
7. Ishikawa T, Obara T, et al. Update on the prevalence and determinants of folic acid use in Japan evaluated

with 91,538 pregnant women: the Japan Environment and Children's Study. J Matern Fetal Neonatal Med. 2020;33:427–36.

8. Gaskins AJ, Afeiche MC, et al. Dietary folate and reproductive success among women undergoing assisted reproduction. Obstet Gynecol. 2014;124:801–9.
9. Kadir M, Hood RB, et al. Folate intake and ovarian reserve among women attending a fertility center. Fertil Steril. 2022;117:171–80.
10. Gaskins AJ, Rich-Edwards JW, et al. Maternal prepregnancy folate intake and risk of spontaneous abortion and stillbirth. Obstet Gynecol. 2014;124:23–31.
11. Roth C, Magnus P, et al. Folic acid supplements in pregnancy and severe language delay in children. JAMA. 2011;306:1566–73.
12. Surén P, Roth C, et al. Association between maternal use of folic acid supplements and risk of autism spectrum disorders in children. JAMA. 2013;309:570–7.
13. Levine SZ, Kodesh A, et al. Association of maternal use of folic acid and multivitamin supplements in the periods before and during pregnancy with the risk of autism spectrum disorder in offspring. JAMA Psychiatry. 2018;75:176–84.
14. Entringer S, Epel ES, et al. Maternal folate concentration in early pregnancy and newborn telomere length. Ann Nutr Metab. 2015;66:202–8.
15. Halloran BP, DeLuca HF. Effect of vitamin D deficiency on fertility and reproductive capacity in the female rat. J Nutr. 1980;110:1573–80.
16. Hickie JP, Lavigne DM, et al. Reduced fecundity of vitamin D deficient rats. Comp Biochem Physiol A Comp Physiol. 1983;74:923–5.
17. Kwiecinksi GG, Petrie GI, et al. 1,25-Dihydroxyvitamin D3 restores fertility of vitamin D-deficient female rats. Am J Phys. 1989;256:E483–7.
18. Yoshizawa T, Handa Y, et al. Mice lacking the vitamin D receptor exhibit impaired bone formation, uterine hypoplasia and growth retardation after weaning. Nat Genet. 1997;16:391–6.
19. Kinuta K, Tanaka H, et al. Vitamin D is an important factor in estrogen biosynthesis of both female and male gonads. Endocrinology. 2000;141:1317–24.
20. Panda DK, Miao D, et al. Targeted ablation of the 25-hydroxyvitamin D 1α-hydroxylase enzyme: evidence for skeletal, reproductive, and immune dysfunction. Proc Natl Acad Sci USA. 2001;98:7498–503.
21. Johnson LE, DeLuca HF. Reproductive defects are corrected in vitamin D-deficient female rats fed a high calcium, phosphorus and lactose diet. J Nutr. 2002;132:2270–3.
22. Ozkan S, Jindal S, et al. Replete vitamin D stores predict reproductive success following in vitro fertilization. Fertil Steril. 2010;94:1314–9.
23. Jensen A, Nielsen ML, et al. Chances of live birth after exposure to vitamin D-fortified margarine in women with fertility problems: results from a Danish population-based cohort study. Fertil Steril. 2020;113:383–91.
24. Mumford SL, Garbose RA, et al. Association of preconception serum 25-hydroxyvitamin D concentrations with livebirth and pregnancy loss: a prospective cohort study. Lancet Diabetes Endocrinol. 2018;6:725–32.
25. Cozzolino M, Busnelli A, et al. How vitamin D level influences in vitro fertilization outcomes: results of a systematic review and meta-analysis. Fertil Steril. 2020;114:1014–25.
26. Cai S, Li J, et al. Impact of vitamin D on human embryo implantation—a prospective cohort study in women undergoing fresh embryo transfer. Fertil Steril. 2021;115:655–64.
27. Aitken RJ, Smith TB, et al. Oxidative stress and male reproductive health. Asian J Androl. 2014;16:31–8.
28. Forges T, Monnier-Barbarino P, et al. Impact of folate and homocysteine metabolism on human reproductive health. Hum Reprod Update. 2007;13:225–38.
29. Favier AE. The role of zinc in reproduction. Hormonal mechanisms. Biol Trace Elem Res. 1992;32:363–82.
30. Xu B, Chia SE, et al. Trace elements in blood and seminal plasma and their relationship to sperm quality. Reprod Toxicol. 1993;7:613–8.
31. Zago MP, Oteiza PI. The antioxidant properties of zinc: interactions with iron and antioxidants. Free Radic Biol Med. 2001;31:266–74.
32. Irani M, Amirian M, et al. The effect of folate and folate plus zinc supplementation on endocrine parameters and sperm characteristics in sub-fertile men: a systematic review and meta-analysis. Urol J. 2017;14:4069–78.
33. Azizollahi G, Azizollahi S, et al. Effects of supplement therapy on sperm parameters, protamine content and acrosomal integrity of varicocelectomized subjects. J Assist Reprod Genet. 2013;30:593–9.
34. Schisterman EF, Sjaarda LA, et al. Effect of folic acid and zinc supplementation in men on semen quality and live birth among couples undergoing infertility treatment: a randomized clinical trial. JAMA. 2020;323:35–48.
35. Matorras R, Pérez-Sanz J, et al. Effect of vitamin E administered to men in infertile couples on sperm and assisted reproduction outcomes: a double-blind randomized study. F S Rep. 2020;1:219–26.
36. Estruch R, Ros E, et al. Primary prevention of cardiovascular disease with a Mediterranean diet. N Engl J Med. 2013;368:1279–90.
37. Karayiannis D, Kontogianni MD, et al. Adherence to the Mediterranean diet and IVF success rate among non-obese women attempting fertility. Hum Reprod. 2018;33:494–502.
38. Gaskins AJ, Nassan FL, et al. Dietary patterns and outcomes of assisted reproduction. Am J Obstet Gynecol. 2019;220(567):e1–18.

Functionality of Supplements and Anti-aging: Functionality of Supplements—Neurology

117

Koji Abe

117.1 Anti-aging of the Brain and Lifestyle Disease Prevention

Prevention of lifestyle diseases is a fundamental principle for achieving anti-aging of the brain. In our dementia clinic, 66.6% of 968 Alzheimer's disease (AD) patients had some form of lifestyle disease, with hypertension (HT) at 54.1%, dyslipidemia (DL) at 28.4%, and diabetes mellitus (DM) at 18.2%. It was found that just under 70% of AD patients have a lifestyle disease [1]. Furthermore, 21.6% of patients had two diseases, and 6.3% had three diseases, indicating that major internal diseases, lifestyle diseases, are highly prevalent in AD patients [2]. Among these 968 AD patients, 570 who could also have their waist circumference measured, 36.7% met the criteria for metabolic syndrome (men ≥85 cm, women ≥90 cm), and of these, 75.6% had some form of lifestyle disease. Among these, 59.8% had HT, 37.3% had DL, and 26.8% had DM, indicating that over 75% of AD patients with metabolic waist circumference have a lifestyle disease. Furthermore, 30.1% of patients had two diseases, and 9.1% had three diseases, indicating that the prevalence of lifestyle diseases is higher in AD patients with metabolic waist circumference than in all AD patients. In other words, 14.0% of all AD patients have metabolic syndrome [1].

K. Abe (✉)
National Center of Neurology and Psychiatry (NCNP), Tokyo, Japan

Arts Ginza Clinic, Tokyo, Japan

117.2 Anti-aging of the Brain and Dementia Prevention

Recent research suggests that AD is a disease that progresses over 35 years, with the first 25 years being a silent progression in the brain, leading to death about 10 years after symptoms appear. Therefore, there is a great opportunity for anti-aging preventive intervention during this long course, and the importance of preventive strategies has been pointed out. The usefulness of lifestyle diseases such as hypertension, hyperlipidemia, diabetes, lifestyle, supplements, various dementia prevention programs, and dementia prevention cafes has been pointed out. In our research, when lipophilic statins such as atorvastatin and pitavastatin were administered to Alzheimer's disease model mice Tg2567 from 5 months of age, the subsequent deposition of senile plaques in the mouse brain was suppressed at 10, 15, and 20 months of age [2]. In our research on diabetes, 3.8% of 182 outpatient diabetes patients, mainly in their 50s to 70s, had dementia, and mild cognitive impairment (mild cognitive Impairment; MCI) was found in 7.7% of the cases [3]. Of these 182 diabetic outpatients,

H. Yamada, Y. Naito (eds.), *Anti-Aging Medicine*, https://doi.org/10.1007/978-981-97-8514-8_117

27% had HT, 6% had DL, and 17% had both. Furthermore, according to recent research in Hisayama, Japan, insulin resistance is a risk factor for increased senile plaques, which is attracting attention. In the brains of diabetic mice, it is becoming clear that neurovascular unit (NVU) dysfunction, such as the breakdown of occludin, is occurring, and this is attracting attention as a common molecular mechanism in various neurological diseases.

117.3 Brain Anti-aging and the Functionality of Supplements

Supplements, as shown in Fig. 117.1, are classified as functional foods among health foods, and although they are a notification system by companies, companies are required to present scientific evidence, so some evidence is necessary. The Japan Brain Supplement Society and the Japan Dementia Prevention Society are certifying and creating evidence for brain health enhancement effects and the like by such supplements. With the super-aging of Japanese society, the brain pathology of AD patients is also changing, and the presence of vascular risk factors (VRF), which were previously rare, is becoming commonplace. Such multimodality interventions, including VRF, are attracting attention as they have been shown to be effective in reducing the onset of AD (FINGER study) [4].

Therefore, the characteristics required for supplements to promote brain anti-aging, as shown in Fig. 117.2, are extremely important, such as improving chronic cerebral hypoperfusion, reducing oxidative stress, and reducing inflammation in the brain. Examples of such supplements include fish oil such as EPA and DHA, polyphenols such as rosmarinic acid, antioxidant substances such as vitamins and Oxicut, Cordyceps with strong anti-inflammatory effects, plasmalogens contained in sea squirts and scallops, imidazole dipeptides contained in chicken breast, pork, and whale meat, and the anti-aging effects of the brain are expected.

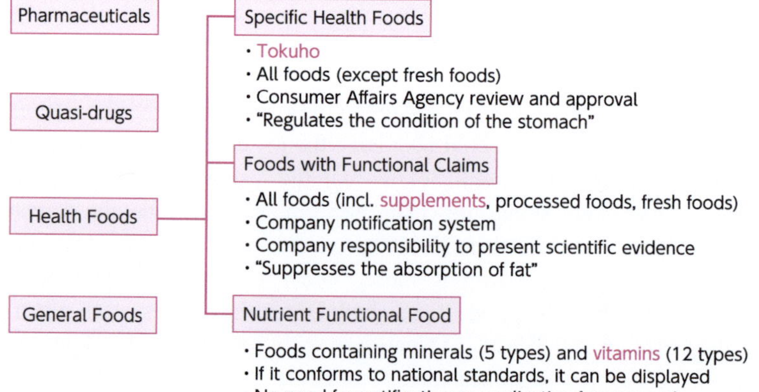

Fig. 117.1 Classification of pharmaceuticals and foods

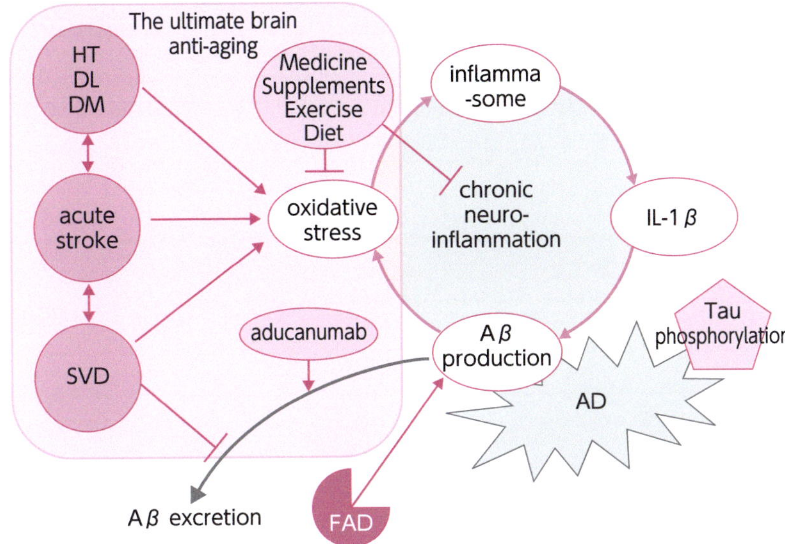

Fig. 117.2 Oxidative stress, neuroinflammation, and AD pathophysiology functional display food nutrition functional food

References

1. Kurata T, Miyazaki K, et al. Atorvastatin and pitavastatin reduce senile plaques and inflammatory responses in a mouse model of Alzheimer's disease. Neurol Res. 2012;34:601–10.
2. Hishikawa N, Fukui Y, et al. Cognitive and affective functions in Alzheimer's disease patients with metabolic syndrome. Eur J Neurol. 2016;23:339–45.
3. Hishikawa N, Yamashita T, et al. Cognitive and affective functions in diabetic patients associated with diabetes-related factors, white matter abnormality and aging. Eur J Neurol. 2015;22:313–21.
4. Kivipelto M, Solomon A, et al. The Finnish Geriatric Intervention Study to Prevent Cognitive Impairment and Disability (FINGER): study design and progress. Alzheimers Dement. 2013;9:657–65.

118 Functionality of Supplements and Anti-aging: Functionality of Supplements—Musculoskeletal and Sports

Wataru Aoi

Maintaining musculoskeletal function is important in preventing diseases and care, and extending healthy lifespan. In modern life, where transportation and communication technologies have developed, factors that reduce physical activity and physical strength are accumulating, and anti-aging measures for the musculoskeletal system are desired. Supplements have been considered to complement nutritional intake from meals to enhance the function of skeletal muscles and bones. Traditionally, in the field of sports, functional materials and supplements have been verified to enhance musculoskeletal function, but this evidence can also be applied to anti-aging (Fig. 118.1).

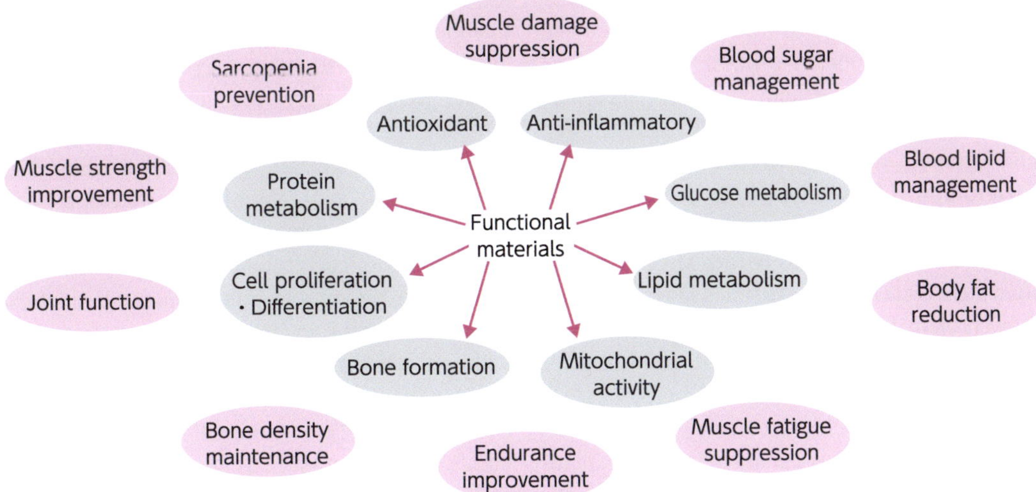

Fig. 118.1 Musculoskeletal function targeted by functional materials

W. Aoi (✉)
Laboratory of Nutrition Science, Graduate School of Life and Environmental Sciences, Kyoto Prefectural University, Kyoto, Japan
e-mail: waoi@kpu.ac.jp

H. Yamada, Y. Naito (eds.), *Anti-Aging Medicine*, https://doi.org/10.1007/978-981-97-8514-8_118

118.1 Sports and Physical Abilities and Supplements

Engaging in sports requires one to be healthy, as infections, musculoskeletal disorders, and circulatory disorders can hinder good performance. Therefore, athletes and sports enthusiasts need to pay attention to their diet to maintain and improve their health. Also, during sports, it is necessary to secure energy substrates for muscle contraction and to pay attention to the intake of water and minerals to prevent body temperature rise. In other words, with the increase in physical activity, it is necessary to secure the intake of energy and nutrients according to physiological changes. Various supplements have been developed to supplement nutrients that are likely to be deficient in sports activities. The usefulness of consuming foods with added ingredients that are not easily obtained from meals, or in the form of powders, tablets, capsules, etc., is widely recognized. Their portability, preservability, and ease of intake are advantages in dealing with diverse nutritional supplementation in sports activities. For conditioning and performance improvement, it is necessary to consider the timing of nutrient intake [1]. Body protein synthesis increases immediately after exercise and just before sleep when growth hormone secretion is prominent. Therefore, it is beneficial to supply amino acids to skeletal muscles by consuming protein immediately after exercise and before sleep, but it is often difficult to eat at these times in reality. Also, many protein foods contain animal fats, so it is not rational to consume them before sleep. With high-protein, low-fat materials like protein powder, it is easier to achieve convenient and timed supplementation. Also, a characteristic of supplements is that their digestion and absorption are fast, causing a temporary increase in blood concentration. Fast-absorbing protein supply can quickly reach the target organ. Furthermore, it is known that protein synthesis increases by exposing muscle tissue to high concentrations of amino acids, and it is possible to find value in utilization conscious of blood concentration. On the other hand, when taking supplements, the blood concentration peaks temporarily and then quickly decreases. In contrast, intake from food is slow to digest and absorb, so it can maintain blood amino acid concentration for a long time. Therefore, one should use supplements knowing their advantages and disadvantages.

118.2 Skeletal Muscle Anti-aging and Supplements

To prevent muscle weakening associated with aging, it is necessary to ensure an intake of dietary protein that matches body protein synthesis. In the "Dietary Reference Intakes for Japanese (2020 Edition)" [2], the recommended protein intake for the elderly is 60 g/day for men and 50 g/day for women, which exceeds the intake per body weight in adulthood. This takes into account the fact that muscle protein synthesis is induced in response to protein intake, considering measures against frailty and sarcopenia. Components known to stimulate muscle protein synthesis are also known [3]. Branched-chain amino acids (valine, leucine, isoleucine) are metabolized in muscles not only as energy substrates but also have the effect of promoting protein synthesis and inhibiting degradation. The effectiveness of β-hydroxy-β-methylbutyrate, a metabolite of leucine, and glutamine, which is abundant in the muscle amino acid pool, in inhibiting skeletal muscle atrophy has been shown. In addition, oxidative stress is involved in the protein degradation of skeletal muscles due to aging and inactivity [4], so it might be a good idea to utilize antioxidants such as vitamins, carotenoids, and polyphenols. The decrease in the ability to metabolize sugars and fats in skeletal muscles promotes the accumulation of body fat. Therefore, from the perspective of preventing metabolic syndrome, food components that enhance the metabolism of skeletal muscles may be useful. Oxidative stress caused by aging and inactivity is involved in the decrease in insulin sensitivity and mitochondrial metabolic capacity. Therefore, the intake of antioxidants may be beneficial in maintaining skeletal muscle metabolic capacity. Also, minerals such as zinc, manganese, iron, copper, and selenium function as components of antioxidant enzymes. There are also minerals such as chromium and vanadium that may improve the insulin sensitivity of skeletal muscles.

118.3 Bone Anti-aging and Protein/Amino Acids

The decrease in bone mass with aging can be attributed to a decrease in sex hormone secretion, oxidative stress, and a decrease in physical activity leading to a decrease in bone formation and an increase in bone resorption. It is important to maintain the intake of calcium, a component of bone, and vitamin D, which enhances bone formation. If it is difficult to intake from a normal diet, the intake of supplements should be considered. Also, it is important to maintain the amount of bone collagen, which serves as a base for calcium binding. Therefore, maintaining dietary protein and enhancing protein synthesis through exercise also contribute to maintaining bone mass. In postmenopausal women, a rapid decrease in estrogen secretion is accompanied by a decrease in bone mass, increasing the risk of fractures. Isoflavones, which are highly contained in soybeans and have an estrogen-like structure, are known to be useful for maintaining bone mass [5].

118.4 Future Prospects

Extending healthy life expectancy involves not only preventing lifestyle-related diseases, but also preventing the need for care to avoid falling into a state of needing assistance. In particular, preventing the decline of musculoskeletal function is key to achieving these two preventions. On the other hand, it should be kept in mind that there is little evidence about the benefits of further intake by supplements when intake is satisfied by diet. In the future, we hope that research on aging and skeletal muscles will further advance, and strategies for diet and supplements to efficiently maintain muscle function will be established.

References

1. Kerksick CM, Arent S, et al. International society of sports nutrition position stand: nutrient timing. J Int Soc Sports Nutr. 2017;14:33.
2. "Dietary Reference Intakes for Japanese" Review Committee. Dietary reference intakes for Japanese (2020 edition). Ministry of Health, Labour and Welfare; 2019.
3. Cruz-Jentoft AJ, Hughes BD, et al. Nutritional strategies for maintaining muscle mass and strength from middle age to later life: a narrative review. Maturitas. 2020;132:57–64.
4. Aoi W. Oxidative stress-induced signal transduction in skeletal muscle. Basic biology and current understanding of skeletal muscle. New York: Nova Science Publishers; 2013. p. 123–46.
5. Xiao Y, Zhang S, et al. Comprehensive evaluation of the role of soy and isoflavone supplementation in humans and animals over the past two decades. Phytother Res. 2018;32:384–94.

119 Functionality of Supplements and Anti-aging: Functionality of Supplements—Cardiovascular

Ken Shinmura

In 2018, the functionality of ω3 polyunsaturated fatty acids as supplements was discussed in the recommendations from the American Heart Association (AHA) [1]. However, in the 2019 guidelines for primary prevention of cardiovascular disease by the American Heart Association/American College of Cardiology, only precautions about nutrition and diet were mentioned, and no specific supplements were recommended. Currently, in addition to ω3 fatty acids, resveratrol, coenzyme Q10, and others are frequently used as supplements in the cardiovascular field, and a lot of evidence is being accumulated. In Japan, since the introduction of the "Foods with Function Claims" system in 2015, there has been an increasing trend in the number of functional ingredients applied for in the cardiovascular field. In this section, we would like to provide an overview of supplements in the cardiovascular field.

119.1 ω3 Polyunsaturated Fatty Acids

ω3 polyunsaturated fatty acids include eicosapentaenoic acid (EPA) and docosahexaenoic acid (DHA), which are derived from fish, and α-linolenic acid, which is derived from plants. After intake, about 10% of α-linolenic acid is converted to EPA and about 1% is converted to DHA. ω3 fatty acids reduce the synthesis of serum triglycerides and very low-density lipoprotein (VLDL), and have beneficial effects on blood pressure, insulin signaling, and inflammatory responses. As a component of cell membrane phospholipids, they also have anti-arrhythmic effects and prevent sudden death and atrial fibrillation [1]. Meta-analyses of various clinical studies have confirmed their beneficial effects on cardiovascular outcomes [1]. However, the 2018 AHA recommendation stated that while fish oil-derived ω3 fatty acid supplements are reasonable for primary prevention of cardiovascular disease, their usefulness is not sufficient for patients with heart disease who are receiving appropriate treatment [1]. A meta-analysis from the Cochrane Library examining the effects of increasing intake of ω3 fatty acids, regardless of whether they are derived from fish or plants, concluded that there is either no or only a slight preventive effect on mortality and cardiovascular disease from supplementing with EPA/DHA [2]. On the other hand, although the evidence is limited, it was concluded that there might be a slight possibility of preventing cardiovascular disease and reducing arrhythmia risk with increased intake of α-linolenic acid [2].

K. Shinmura (✉)
Department of General Internal Medicine, Hyogo Medical University, Nishinomiya, Hyogo, Japan
e-mail: ke-shimmura@hyo-med.ac.jp

H. Yamada, Y. Naito (eds.), *Anti-Aging Medicine*, https://doi.org/10.1007/978-981-97-8514-8_119

119.2 Resveratrol

The life-extending effects of resveratrol in mice were reported in "Nature" in 2006, and since then, resveratrol has become one of the most highly regarded supplements. It is a polyphenol found in the skin of red grapes, peanuts, blueberries, and blackberries, and while it exists in both cis and trans forms, the trans form is more active. So far, the reported side effects are minimal, with mild nausea and digestive disorders being the most common [3]. The effects of resveratrol vary depending on the dosage, with low to medium doses of 20–50 mg/day often used for maintaining and enhancing health and preventing aging, and high doses of 250–5000 mg/day used for treating diabetes, cognitive impairment, cancer, etc. [3]. In vitro, direct activation of sirtuins has been reported, but many reports suggest that the sirtuin-activating effect is indirect, and the target molecules of resveratrol include AMP-activated protein kinase (AMPK), nuclear factor κB (NF-κB), nuclear factor erythroid 2-related factor 2 (Nrf2), and endothelial nitric oxide synthase (eNOS) [2]. Many polyphenols other than resveratrol are used as supplements in the cardiovascular field. Polyphenols such as quercetin, hydroxytyrosol, epigallocatechin-3-gallate, and curcumin are included in CR mimicry compounds [4]. They are reported to act on mitochondria, reducing oxidative stress and promoting autophagy (mitophagy) [4].

119.3 Coenzyme Q10 (CoQ10)

Coenzyme Q10 (CoQ10) acts as a coenzyme that enhances the efficiency of ATP synthesis by participating in electron transfer in the electron transport chain in mitochondria. CoQ10 also acts as an antioxidant in conjunction with vitamin E. In Japan, it has been used as a treatment for congestive heart failure since 1979, but since 2001, it has been available as a health food. The usefulness of CoQ10 in the cardiovascular field has been expected because it decreases in the body with aging and heart failure, and its concentration decreases in statin users due to reduced synthesis [5]. However, the results of clinical studies so far have been inconsistent [5]. The dosage varies greatly from 20 to 600 mg, making it difficult to evaluate CoQ10, but it has been reported to be safe even at high doses [5].

119.4 Functional Ingredients Reported

Below are some of the functional ingredients that have been reported.

- Oat-Derived β-Glucan: LDL Cholesterol-Lowering Effect

  This is a dietary fiber found in oats. High molecular weight (2200 kg/mol) and medium molecular weight (850 kg/mol) oat-derived β-glucans have a high viscosity and bind with bile acids in the intestine, inhibiting the reabsorption of bile acids and promoting the synthesis of bile acids from cholesterol in the liver [6]. In addition to this action, it may also suppress cholesterol synthesis by increasing the proportion of deoxycholic acid in total bile acids [6].
- Hydroxytyrosol Derived from Olive Fruit: Inhibits Oxidation of LDL Cholesterol and Improves Vasodilation Function of Damaged Vessels

  In addition to its antioxidant and anti-inflammatory effects, it exhibits endothelial protective effects and inhibits the proliferation and migration of vascular smooth muscle cells [6]. It is a phenolic compound found in olive oil and no side effects have been reported [6].
- Quercetin: Improves Blood Flow through Vasodilation and Lowers Blood Pressure in Hypertensive Individuals

  It shows vasodilatory effects that are both nitric oxide-dependent and independent, and antioxidant effects have also been observed [6]. No side effects have been reported within the usage range of 200–1095 mg/day [6].
- γ-Aminobutyric Acid (GABA) Derived from Red Yeast Rice: Lowers Blood Pressure in Hypertensive Individuals

It is believed to work through peripheral vasodilation and inhibition of the sympathetic nervous system [6]. It has been used as a traditional Chinese medicine in China [6].

- L-Arginine: Restores Vascular Elasticity that Decreases with Age

  It is a substrate for nitric oxide synthesis and is expected to improve endothelial function, have anti-inflammatory effects, and inhibit the progression of atherosclerosis [6]. It is a non-essential amino acid, and no serious side effects were observed in a clinical study of 7 g/day for 3 months [6].

- 6-Gingerol, 6-Shogaol: Maintains Surface Temperature of Peripheral Skin

  These are polyphenols derived from ginger, and are thought to improve peripheral cold sensation by increasing body heat production through activation of the temperature-sensitive receptor TRPV1 (transient receptor potential vanilloid 1). 6-Shogaol is also thought to exert a warming effect through vasodilation mediated by promoting the secretion of calcitonin gene-related peptide.

As the saying goes, "People age from their blood vessels," maintaining and promoting cardiovascular and vascular health is a very important goal in anti-aging medicine. Therefore, supplements in the cardiovascular field should be actively researched in the future, and it is considered a high-priority area to establish a system to provide them to the public with appropriate information.

References

1. Siscovick DS, Barringer TA, et al. Omega-3 polyunsaturated fatty acid (fish oil) supplementation and the prevention of clinical cardiovascular disease: a science advisory from the American Heart Association. Circulation. 2017;135:e867–84.
2. Abdelhamid AS, Brown TJ, et al. Omega-3 fatty acids for the primary and secondary prevention of cardiovascular disease. Cochrane Database Syst Rev. 2018;11:CD003177.
3. Pannu N, Bhatnagar A. Resveratrol: from enhanced biosynthesis and bioavailability to multitargeting chronic diseases. Biomed Pharmacother. 2019;109:2237–51.
4. Davinelli S, Stefani DD, et al. Polyphenols as caloric restriction mimetics regulating mitochondrial biogenesis and mitophagy. Trends Endocrinol Metab. 2020;31:536–50.
5. Rabanal-Ruiz Y, Llanos-González E, et al. The use of coenzyme Q10 in cardiovascular diseases. Antioxidants (Basel). 2021;10:755.
6. Keinori N, Ryuichi M. Understanding the evidence on supplements and cardiovascular diseases. Heart View. 2017;21:630–5.

120 Functionality and Anti-aging of Supplements: Safety of Supplements

Koya Nakashima

Supplements, mainly in the form of tablets or capsules, are made by adding additives to concentrated nutrient ingredients. Therefore, safety is required in the ingredients containing additives and the manufacturing process. Also, when taking supplements, it is necessary to avoid excessive intake and consider the presence or absence of underlying diseases, allergies, and interactions.

120.1 Safety of Ingredients

The ingredients of supplements are obligated by the Food Sanitation Act to not include components that correspond to pharmaceuticals or genetically modified foods that have not undergone safety inspections. The following are examples of components that correspond to pharmaceuticals.

1. Those containing the essence of ingredients (raw materials) used as pharmaceuticals (antipyretic analgesics, hormones, antibiotics, digestive enzymes, etc.)
2. Those containing components equivalent to designated poisonous drugs
3. Narcotics, psychotropic drugs, those with stimulant-like effects
4. Those containing components equivalent to designated pharmaceuticals, prescription pharmaceuticals

The most important consideration in health foods is the difference from pharmaceuticals, and the Ministry of Health, Labour and Welfare publishes information on unapproved and unlicensed pharmaceuticals. Ingredients that have actually been cracked down on include those for strengthening and invigorating purposes (sildenafil), obesity suppression purposes (*N*-nitrosophenfluramine, sibutramine, mazindol, thyroid powder, senna leaves), blood sugar lowering purposes (glibenclamide), anti-inflammatory purposes (dexamethasone, indomethacin), etc [1]. Genetically modified foods have standards set based on Article 13 of the Food Sanitation Act, and those that have not undergone safety inspections are prohibited from manufacturing, importing, and selling. If products that do not conform to the standards are distributed, they will be subject to administrative measures such as disposal, recall, and return. Genetically modified foods that have undergone safety inspections are potatoes, soybeans, sugar beets, corn, rapeseed, cotton, alfalfa, papaya, and mustard greens (as of June 2022). Additives, like the ingredients of capsules, are necessary in the manufacturing pro-

K. Nakashima (✉)
Nakashima Koya Clinic, Tokyo, Japan
e-mail: nakashimako@k9.dion.ne.jp

H. Yamada, Y. Naito (eds.), *Anti-Aging Medicine*, https://doi.org/10.1007/978-981-97-8514-8_120

cess of supplements, but sweeteners, colorings, flavorings, and preservatives should be minimized. Standards for the use of additives and ingredient specifications are stipulated by Article 13 of the Food Sanitation Act. The standard for the use of additives is set at a level significantly below the acceptable daily intake (ADI), i.e., the amount that has no effect on health even if consumed every day for a lifetime. Ingredient specifications include the purity of additives, the upper limit of by-products and harmful heavy metals produced during manufacturing, etc. The safety of 24 types of genetically modified additives has been confirmed (as of June 2022).

120.2 Safety of Manufacturing Process

The Ministry of Health, Labour and Welfare established guidelines for Good Manufacturing Practice (GMP) in February 2005 to ensure the quality of supplements. Businesses are encouraged to voluntarily engage in safety management, standardizing products and ensuring safety from both operational management (GMP software) and structural facilities (GMP hardware). Specifically, the guidelines aim to (1) prevent human errors in the manufacturing process, (2) prevent contamination and quality degradation of the product itself due to factors other than human error, and (3) build a management system aimed at ensuring consistent quality throughout the entire manufacturing process. Self-inspection guidelines have been created for these purposes [2]. As of June 2020, GMP has been made mandatory for foods containing special ingredients such as alkaloids and hormones, which can have harmful effects on the human body (designated ingredient-containing foods). As of now, *Coleus forskohlii*, rhubarb, *Pueraria mirifica*, and black cohosh have been designated. From June 2021, all food businesses are required to implement hygiene management in accordance with Hazard Analysis and Critical Control Point (HACCP) to prevent the inclusion of harmful microorganisms and foreign substances.

120.3 Health Damage Caused by Supplements

Factors causing health damage from supplements include product-related factors such as ingredients, quality, misrepresentation, and excessive advertising, and user-related factors such as inappropriate use, user constitution, and interactions with drugs and other health foods. Many adverse events caused by supplements are gastrointestinal symptoms and skin symptoms, and attention should also be paid to liver damage. Past cases where the Ministry of Health, Labour and Welfare issued warnings include deaths due to dimethylamylamine (hypertensive heart attack), ibogaine (arrhythmia), ABC Dophilus® Powder (mucormycosis), and OxyElite Pro™ (liver damage), and caution is needed when purchasing via the internet or personal import [3]. Supplements are not originally manufactured for patients with diseases, so people with underlying diseases need to check the raw materials. Foods with hormone-like effects may increase the risk of cancer, foods with anticoagulant effects may increase the tendency to bleed, and foods with hypoglycemic effects may increase the risk of hypoglycemia. Combinations to be careful of include turmeric (gallstones), iron supplements (chronic hepatitis), aloe (pregnant women), Korean ginseng (pregnant women), and psyllium (bowel irregularities). The Natural Medicines Database can be referred to for the effectiveness and safety of raw materials [4]. Information on supplement damage and material information is posted on the web by the National Institute of Health and Nutrition [5].

120.4 Safe Intake of Supplements

When using supplements, it is necessary to ensure that they are not consumed excessively. According to the food labeling standards of the Food Sanitation Act, the ingredients that can be labeled as nutritional functional foods are 13 vitamins, 6 minerals, and n-3 fatty acids. The Ministry of Health, Labour and Welfare has set the estimated average requirement, recommended amount, guideline amount, and tolerable upper intake level for these ingredients [6]. The tolerable upper intake level is the maximum habitual intake that is considered to have no risk of health damage, and it is believed that exceeding this amount increases the risk of potential health problems. For other foods and nutrients, no upper limit for safe intake has been set, so special care is needed, especially for children, the elderly, pregnant women, and nursing mothers. To choose a safe supplement, it is important to select manufacturers and sellers who disclose information about the ingredients and origin of the raw materials, allergenic substances, additives, the risk of excessive intake, and GMP certification. Most of the health damage caused by supplements is due to personal imports and internet sales. Medical professionals who use supplements are required to not be misled by the seller's advertising and to use them while verifying their effectiveness and safety in the literature.

References

1. Ministry of Health, Labour and Welfare. Information on unapproved and unauthorized medicines. https://www.mhlw.go.jp/stf/kinkyu/diet/musyounin.html.
2. Director of the Food Safety Department, Pharmaceutical and Food Bureau, Ministry of Health, Labour and Welfare. "Basic concept for the proper manufacturing of tablet, capsule, etc. foods" and "Self-inspection guidelines for the safety of raw materials for tablet, capsule, etc. foods". https://www.mhlw.go.jp/topics/bukyoku/iyaku/syoku-anzen/hokenkinou/dl/kankeihourei_01.pdf.
3. Ministry of Health, Labour and Welfare. Information on the safety of so-called "health foods" (notifications, etc.). https://www.mhlw.go.jp/stf/seisakunitsuite/bunya/kenkou_iryou/shokuhin/hokenkinou/houreituuti.html.
4. Japan Medical Association/Japan Dental Association/Japan Pharmaceutical Association General supervision. All about health foods and supplements [ingredients]. 7th ed. Dobunshoin, Tokyo: Natural Medicine Database Japan version; 2022.
5. National Institute of Health and Nutrition, National Research and Development Corporation. Information on the safety and effectiveness of "health foods". https://hfnet.nibiohn.go.jp.
6. "Dietary Reference Intakes for Japanese" Review Committee. Dietary reference intakes for Japanese (2020 edition). Ministry of Health, Labour and Welfare; 2019. https://www.mhlw.go.jp/content/10904750/000586553.pdf.

Functionality and Antiaging of Supplements: Interactions Between Supplements and Medicines

121

Kiyomitsu Ikeoka

121.1 Current State of Concurrent Use of Medicines and Supplements

The demand for supplements and health foods is increasing as people's interest in health grows, and they are becoming more accessible due to advertising and sales in various media. According to statistics, it is estimated that 30–40% of the entire population of Japan has experience using them. What should be noted is that, according to a document from the Cabinet Office Food Safety Commission (July 27, 2015), 36.8% of outpatients with some kind of disease are taking prescription drugs and "health foods" together, but 70% of them have not consulted a doctor. This is due to a lack of awareness that supplements can interact with medicines, and a lack of questioning from doctors. Although there are few supplements with evidence at the same level as medicines, many people consider them as a placebo, but some scientific reaction can occur once they enter the body. It should be assumed that all supplements have the potential to interact with medicines. There are also reports that about half of those with experience using supplements have used ones that carry some risk of drug interaction [1].

K. Ikeoka (✉)
Ikeoka Clinic, Ikeoka Medical Corporation, Osaka, Japan
e-mail: holy@ikeoka.net

121.2 What Is a Drug Interaction?

A drug interaction is defined in a 2001 notification from the Ministry of Health, Labour and Welfare's Pharmaceutical Affairs Bureau as "a phenomenon that occurs between a drug's pharmacokinetic profile and/or the effects/side effects of the drug and concomitant drugs, dietary factors or lifestyle habits (such as smoking and drinking)." Simply put, it is a phenomenon where supplements, etc., enhance or weaken the effects of medicines. Among the interactions, 53% are related to the pharmacokinetic profile, and 35% are pharmacodynamic ones such as binding and sensitivity that affect effects and side effects, according to a report [2].

121.3 Interactions Mediated by Cytochrome P450

The pharmacokinetic profile includes absorption, distribution, metabolism, and excretion, and interactions can occur at each stage. Among them, those caused by the metabolic process are the most common, and several enzymes are involved, but the main one is the interaction mediated by Cytochrome P450 (CYP). CYP is mainly present in the liver but is also expressed in the epithelial cells of the small intestine and is involved in the metabolism of drugs after oral administration. There are many subtypes of

H. Yamada, Y. Naito (eds.), *Anti-Aging Medicine*, https://doi.org/10.1007/978-981-97-8514-8_121

Table 121.1 Representative drugs that interact frequently with grapefruit juice

Type of drug	Generic name
Calcium antagonists	Nifedipine, Nicardipine, Verapamil, etc.
α, β Blockers	Carvedilol
Antiarrhythmic drugs	Amiodarone, Quinidine, Disopyramide, Procainamide
Lipid disorder treatment drugs	Atorvastatin, Fluvastatin, Rosuvastatin, Simvastatin
Antibiotics	Erythromycin, Rifampicin
Antifungal drugs	Ketoconazole, Itraconazole
Antiallergy drugs	Fexofenadine
Antiplatelet drugs	Cilostazol, Clopidogrel
Antiepileptic drugs	Carbamazepine
Antidepressants	Fluvoxamine
Sedatives	Diazepam, Midazolam, Triazolam
Erectile dysfunction drugs	Sildenafil, Tadalafil
Hormones	Estradiol, Methylprednisolone
Immunosuppressants	Cyclosporine, Tacrolimus, Methotrexate

Adapted with permission from [3]

CYP. More than 90% are metabolized by five types of enzymes: CYP1A2, CYP2C9, CYP2C19, CYP2D6, CYP3A4, with CYP3A4 involved in the metabolism of many drugs and controlling drug blood concentrations by suppressing excessive gastrointestinal absorption. Grapefruit juice increases the absorption in the digestive tract by inhibiting this mechanism, leading to an increase in drug blood concentration. The drug interactions of grapefruit juice are diverse, as shown in Table 121.1 and summarized in.

121.4 Pharmacodynamic Mechanism

The mechanism that affects the effects and side effects of drugs involves how the supplements used in combination may change the binding and sensitivity to the site of action of the drug. In this case, it is broadly divided into interactions at the same site of action (such as receptors) and those due to different sites of action. However, in many cases, the mechanism of action and active ingredients of supplements are often unknown, and the mechanism of interaction is often unclear. Examples of this type of interaction include the blood clotting inhibition effect of ginkgo leaf extract enhancing the action of warfarin, and the blood clotting promotion effect of vitamin K in green juice, chlorella, etc. weakening the action of warfarin. There are many individual cases of interactions between supplements and drugs, so please refer to other books and reviews. As a representative, I would like to mention the Natural Medicine Database Japan Version [3]. Please refer to it. Here, the main drug interactions with vitamins and minerals, which are often used as supplements, are summarized in Table 121.2.

Table 121.2 Interactions between vitamins, minerals, and drugs

Vitamins, minerals	Drugs	Interactions
Vitamin A	Retinoids Warfarin	Risk of enhanced side effects Enhanced anticoagulant effect
Vitamin B_6	Phenytoin Angiotensin receptor antagonists, Calcium channel blockers and other antihypertensive drugs Amiodarone	Metabolism is promoted, reducing effectiveness Excessive drop in blood pressure Worsening of photosensitivity
Vitamin D	Atorvastatin Thiazide diuretics Calcium channel blockers Digoxin	Reduced absorption, reducing effectiveness Occurrence of hypercalcemia Reduced effectiveness due to enhanced action, inducing arrhythmia
Vitamin K	Warfarin	Reduced effectiveness
Calcium	Thiazide diuretics Quinolone antibiotics Bisphosphonate drugs Levothyroxine	Occurrence of hypercalcemia Binding in the digestive tract reduces absorption Reduced absorption, reducing effectiveness Reduced absorption, reducing effectiveness
Magnesium	Levodopa-carbidopa combination drug Bisphosphonate drugs Gabapentin	Reduced effectiveness Reduced absorption, reducing effectiveness Reduced absorption, reducing effectiveness
Zinc	Quinolone antibiotics Cephalexin	Reduced absorption, reducing effectiveness Reduced absorption, reducing effectiveness
Iron	Quinolone antibiotics Bisphosphonate drugs Levothyroxine	Reduced absorption, reducing effectiveness Reduced absorption, reducing effectiveness Reduced absorption, reducing effectiveness

Adapted with permission from [3]

121.5 Nutrient Deficiency Caused by Drugs

Although it is the opposite of the definition of drug interaction, nutrient deficiency can also occur due to drug use [4]. Vitamin B_{12} deficiency occurs with metformin, various antibiotics, and proton pump inhibitors. Aspirin inhibits the absorption of vitamins C and E. There are events such as zinc deficiency due to estrogen, corticosteroids, and potassium-sparing diuretics. These are also important issues that should not be overlooked. As mentioned above, there are some interactions between supplements and drugs that cannot be ignored. However, they are often not well known in general. In the future, the number of elderly people using drugs will increase, and at the same time, the demand for supplements in society in general is expected to increase. Knowing drug interactions is essential for healthcare professionals. First of all, it is necessary to always ask patients about the use of supplements in interviews, and to always have supplements listed in the medication notebook in cooperation with pharmacists.

References

1. Kanamori M, Tomizawa T, et al. A survey on the interaction risk between health foods/supplements and drugs in hospitalized patients. J Jpn Soc Geriatr Pharmacol. 2019;2:9–18.
2. Chiba H. Drug interactions mediated by cytochrome P450. Pharmacia. 1995;31:992–6.
3. Japan Medical Association, Japan Dental Association, among others General Supervision. All about health foods & supplements [ingredients]. 7th ed. Dobunshoin, Tokyo: Natural Medicine Database Japan Version; 2022.
4. Morimoto Y. Translation supervision: handbook of drug-food interactions. 2nd ed. Tokyo: Maruzen Publishing; 2011.

Preferences and Antiaging: Alcohol Consumption and Antiaging

122

Kenichi Ikejima

Alcoholic beverages have a long history of being used not only as a mere luxury but also as a medicine, and various benefits have been expected. On the other hand, it has long been well known that excessive drinking can cause a variety of organ damage, including addiction, and can harm health. This section will discuss the pros and cons of drinking, incorporating the perspective of antiaging.

122.1 Alcohol Consumption and Life Prognosis: Is the J-Curve Phenomenon Real?

The so-called "J-curve phenomenon," where the total mortality rate of light drinkers is lower than that of nondrinkers, is often presented as epidemiological data supporting the benefits of alcohol. This phenomenon is recognized not only in the West but also in Japan [1], and it is the basis for setting the appropriate amount of alcohol for Japanese people at 20 g/day in terms of pure ethanol in Healthy Japan 21. The J-curve phenomenon is most clearly observed in mortality rates due to cardiovascular events, and the main cause is changes in serum lipid profiles, especially increases in HDL cholesterol levels, associated with drinking. However, drinking not only increases serum triglyceride levels but also raises blood pressure in a dose-dependent manner, so the J-curve phenomenon is unclear for cerebrovascular disorders, and the mortality rate of cardiovascular diseases increases with moderate or higher amounts of alcohol. In recent years, it has been shown that the J-curve phenomenon in the cardiovascular system is quite limited and is at a very small level [2] (Fig. 122.1). There is no doubt that the mortality rate due to liver disease increases with excessive drinking. In studies on non-alcoholic fatty liver disease (NAFLD), there are reports that light drinkers (men 30 g/day, women 20 g/day or less) have milder fatty hepatitis than nondrinkers, but in drinkers with metabolic factors, even mild to moderate drinking tends to worsen liver damage, and this tendency is more pronounced with excessive drinking. Mortality due to malignant neoplasms also increases with drinking, but there is no clear J-curve phenomenon. Excessive drinking is listed as a risk factor not only for the liver but also for pancreatic cancer, esophageal and head and neck cancer, and breast cancer. Also, both alcohol and metabolic factors increase the risk of liver cancer, and a synergistic effect has been shown. One thing to note when discussing the J-curve phenomenon is that there are no data to suggest that non-drinkers improve their life prognosis by drinking a small amount. To begin with, it is

K. Ikejima (✉)
Department of Gastroenterology, Juntendo University Graduate School of Medicine, Tokyo, Japan
e-mail: ikejima@juntendo.ac.jp

H. Yamada, Y. Naito (eds.), *Anti-Aging Medicine*, https://doi.org/10.1007/978-981-97-8514-8_122

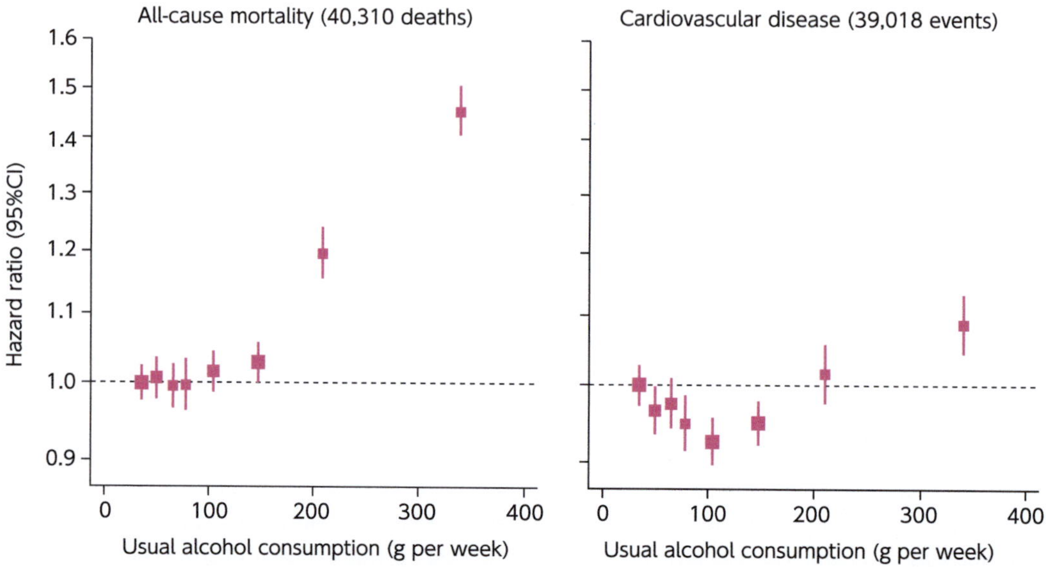

Fig. 122.1 The reality of the J-curve phenomenon [2]

necessary to consider whether nondrinkers are alcohol intolerant (nondrinkers) or whether they used to drink and have quit, and there are various confounding factors between nondrinkers and light drinkers. For example, whether or not one is alcohol intolerant is related to aldehyde dehydrogenase (aldehyde). The genetic polymorphism of aldehyde dehydrogenase (ALDH) determines the metabolism of ethanol, but since ALDH is involved in the detoxification of various harmful substances in the body besides ethanol metabolism, it is simplistic to conclude that moderate drinking is good. In January 2023, the WHO issued a statement warning that "there is no safe level of alcohol consumption for health" (https://www.who.int/europe/news/item/04-01-2023-no-level-of-alcohol-consumption-is-safe-for-our-health).

122.2 The Impact of Excessive Drinking on the Digestive Organs

It is known that excessive alcohol consumption forms dependence and causes various organ damage systemically (Fig. 122.2), and the representative is the lesion of the digestive system. Alcohol-related liver diseases progress from the initial image of alcohol-related fatty liver to repeated alcohol-related hepatitis and cirrhosis, and some of them are complicated by hepatocellular carcinoma (Fig. 122.3). The mechanism of the onset of alcohol-related liver disease is thought to involve a redox shift and increased oxidative stress due to the production of reactive oxygen species (ROS) associated with ethanol metabolism, as well as microcirculation disorders and activation of the innate immune system. In recent years, it has been noted that abnormalities in the intestinal microbiota (dysbiosis) associated with excessive drinking are deeply involved in the pathogenesis of alcohol-related hepatitis. In addition, excessive drinking is a major cause of acute and chronic pancreatitis and is also considered a risk factor for pancreatic cancer. Digestive tract disorders are also common in excessive drinkers, and in addition to upper digestive tract mucosal damage and ulcers, and esophageal and gastric varices associated with alcohol-related cirrhosis, attention should also be paid to the high incidence of esophageal cancer.

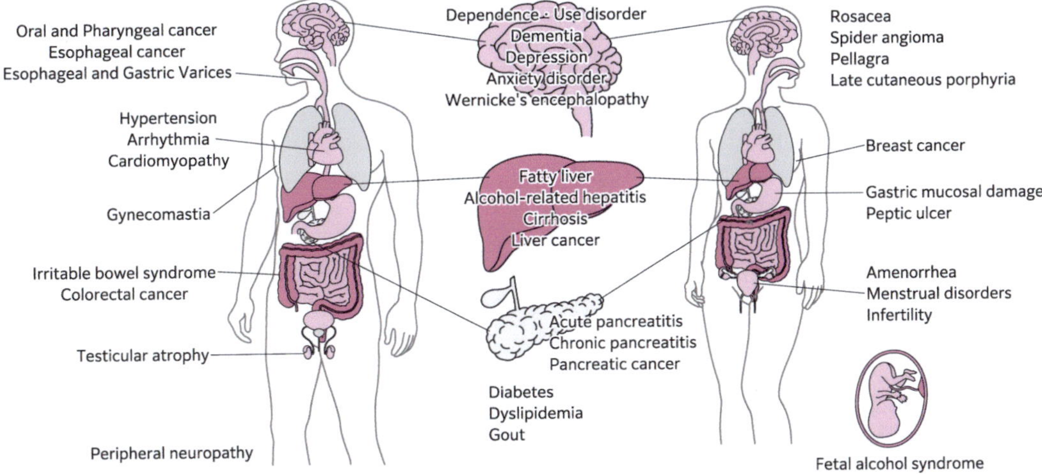

Fig. 122.2 Various organ damage caused by excessive drinking

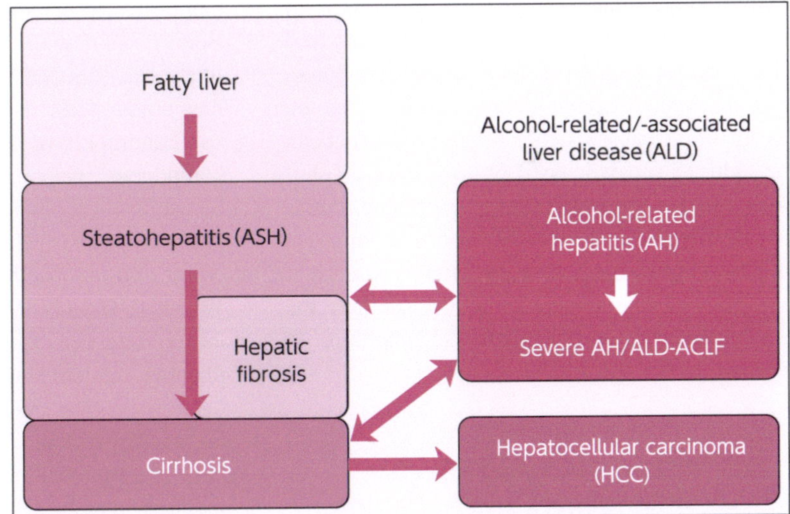

Fig. 122.3 Types and progression of alcohol-related liver diseases

122.3 Sex Differences in Alcohol-Related Liver Disease and the Effects of Alcohol on the Reproductive System

There is a clear sex difference in the onset and progression of alcohol-related liver disease, with women known to develop alcohol-related hepatitis and progress to cirrhosis with less and shorter-term drinking than men. One of the causes is believed to be the difference in physique between men and women, but even when this is corrected, it has been shown that women are more likely to develop organ damage, and the influence of sex hormones has been pointed out. In fact, in animal models, there is a clear sex difference in liver damage caused by chronic alcohol administration, and it has been reported that the progression of liver damage is suppressed by ovariectomy in females and worsened by estrogen supplementation. On the other hand, long-term excessive drinking can cause a decrease in gonadal function in both men and women. In men, long-term excessive drinking can lead to signs of feminization such as testicular atrophy and gynecomastia, along with the progression of organ damage, and a decrease in libido. The causes are thought to include direct damage to the testes by ethanol, and an increase in blood estrogen levels due to steroid hormone metabolic disorders associated with the progression of liver damage. In women with alcohol use disorders, menstrual irregularities and amenorrhea can occur. This is due to alcohol-induced pituitary dysfunction, and sometimes it is related to eating disorders, which requires attention.

122.4 Excessive Alcohol Intake and Skin

From an antiaging perspective, the effects of alcohol on the skin are also important. In patients with alcohol-related cirrhosis, it is common to have skin capillary dilation such as flushing (red face), spider angiomas, and palmar erythema, and abnormal estrogen metabolism is suggested as one of the causes. Other skin diseases that can be induced and worsened by excessive alcohol intake include psoriasis and pellagra due to niacin deficiency. It is rare for a normal Japanese diet to cause niacin deficiency, but it can occur in the case of excessive alcohol intake and the accompanying malnutrition, so caution is needed.

122.5 Pharmacological Effects of Congeners and Antiaging

Alcoholic beverages are not just ethanol solutions, but form their unique aroma and taste by mixing various substances during the brewing process. Also, impurities and additives that mix in during the manufacturing process can sometimes cause health damage. These nonethanol components contained in alcoholic beverages are collectively called congeners. It has been revealed that congeners have various pharmacological effects, which is interesting from an antiaging perspective. However, it should be fully recognized that the benefits of these congeners do not outweigh the risks of excessive drinking.

- Wine: Resveratrol, found in red wine, is a representative of polyphenols with antioxidant effects, originating from the French Paradox, and is also widely used as a supplement. The pharmacological effects of resveratrol are not only due to its antioxidant activity, but also through the activation of sirtuin (SIRT1) by deacetylation. In addition, it has been reported to have effects such as the activation of AMP-activated protein kinase (AMPK) and nuclear factor (erythroid-derived 2)-like 2 (Nrf2), and the inhibition of phosphodiesterase (PDE) 1, 3, 4. These molecular mechanisms contribute to cardiovascular and renal protective effects such as inhibition of arteriosclerosis and myocardial injury, as well as metabolic improvement effects and antitumor effects, as demonstrated in various animal experiments.
- Beer: It is known that beer contains estrogen-like substances derived from hops, and prenylflavonoids such as 8-prenylnaringenin and xanthohumol have been identified as the main components. Hop extracts have been shown to be useful for women's menopausal symptoms and osteoporosis [3], and there are also reports on the anti-tumor effects of these components.
- Whiskey, Brandy: Whiskey and brandy contain a variety of components, including polyphenols, that are leached from the oak barrels used for aging. Whiskey's congeners have high tyrosinase inhibitory activity, and it has been reported that they suppress the amount of melanin synthesis in mouse B16 melanoma cells in vitro [4].
- Gin: It has long been known that the components of juniper berries, used for flavoring gin, have various pharmacological effects such as antibacterial and diuretic effects. In fact, gin is said to have been made as a medicinal liquor based on the diuretic effect of juniper berries. On the other hand, it has been clarified that the intake of juniper berry oil suppresses hepatic ischemia-reperfusion injury, and that this mechanism involves a decrease in prostaglandin production associated with the inhibition of hepatic macrophage (Kupffer cell) activation [5].

122.6 Reduction in Alcohol Consumption and Harm Reduction

Traditionally, abstinence has been considered the only option for alcohol dependence and alcohol-related organ damage. In recent years, the option of reducing alcohol consumption has been presented for mild dependence and organ damage, and the concept of harm reduction has been introduced. Harm reduction originally means taking measures to reduce health damage even if it is not possible to completely eliminate dependence in the field of drug addiction. For example, promoting the use of disposable syringes to prevent the spread of infectious diseases due to shared injections of addictive drugs. Therefore, although the meaning is slightly different, it is used with the intention of minimizing health damage by reducing alcohol consumption. From the perspective of antiaging, excessive drinking should be avoided, and reducing alcohol consumption is recommended for healthy longevity.

122.7 Conclusion

Alcohol is said to be the leader of all medicines, and its health benefits have been explored from various perspectives, but at present, it is hard to say that there is sufficient scientific evidence that alcoholic beverages actively contribute to health. In our country, the Basic Law on Alcohol Health Damage Measures has been enacted and implemented in 2014, and it is required to explore how to interact with alcohol that contributes to health promotion. There are guidelines and guidance on the management of alcohol-related liver diseases in Europe and the United States, but in our country, the "Clinical Guide for Alcoholic Liver Disease (Alcohol-Related Liver Disease)" was published by the Japan Liver Society in the fall of 2022. Experts in antiaging medicine must also refrain from commercially promoting the benefits of alcohol and supplements, and strive to disseminate correct knowledge based on solid scientific evidence verification.

References

1. Tsugane S, Fahey MT, et al. Alcohol consumption and all-cause and cancer mortality among middle-aged Japanese men: seven-year follow-up of the JPHC Study Cohort I. Am J Epidemiol. 1999;150:1201–7.
2. Wood AM, Kaptoge S, et al. Risk thresholds for alcohol consumption: combined analysis of individual-participant data for 599,912 current drinkers in 83 prospective studies. Lancet. 2018;391:1513–23.
3. Erkkola R, Vervarcke S, et al. A randomized, double-blind, placebo-controlled, cross-over pilot study on the use of a standardized hop extract to alleviate menopausal discomforts. Phytomedicine. 2010;17:389–96.
4. Ohguchi K, Koike M, et al. Inhibitory effects of whisky congeners on melanogenesis in mouse B16 melanoma cells. Biosci Biotechnol Biochem. 2008;72:1107–10.
5. Jones SM, Zhong Z, et al. Dietary juniper berry oil minimizes hepatic reperfusion injury in the rat. Hepatology. 1998;28:1042–50.

Preferences and Antiaging: Coffee and Antiaging

123

Yuji Naito

Many metaanalyses are proving positive health outcomes for habitual coffee intake, and the perception of coffee is shifting from a beverage with a stimulating effect to an antiaging beverage when consumed regularly within the usual intake range. The antiaging effects include preventive effects against type 2 diabetes, kidney stones, Parkinson's disease, gout, liver fibrosis, nonalcoholic fatty liver disease, cirrhosis, liver cancer, and chronic liver disease. These effects remain even after adjusting for a number of potential confounding factors, according to comprehensive reviews of meta-analyses, and the EPIC (European Prospective Investigation into Cancer and Nutrition) results also suggest a correlation between coffee intake and a reduction in overall mortality rate [1]. The molecular mechanisms causing the antiaging effects of coffee intake have not yet been fully elucidated. While epidemiological studies cannot prove causality, the observed reduction in type 2 diabetes risk due to coffee intake in various regions is noteworthy. This article explains the antiaging effects of coffee.

Y. Naito (✉)
Department of Human Immunology and Nutrition Science, Kyoto Prefectural University of Medicine, Kyoto, Japan
e-mail: ynaito@koto.kpu-m.ac.jp

123.1 Coffee Intake and Genetic Background

In Japanese people, the top two sources of polyphenol intake from beverages are coffee and green tea. The intake of coffee and green tea varies greatly among individuals, and this individual difference in intake may affect health. The possibility that genetic background explains better health outcomes cannot be ignored, and drinking several cups of coffee a day may simply be a marker of a favorable genetic background. Genome-wide association studies have confirmed the influence of several genetic polymorphisms on caffeine or coffee intake. Genetically faster caffeine metabolism may be associated with higher caffeine or coffee intake. However, there is no consistent association between genetic polymorphisms for faster caffeine metabolism (and higher caffeine/coffee consumption) and health effects.

123.2 Coffee Intake and Epidemiological Studies

Many epidemiological studies often find significant associations in different cohorts with the intake of two or more cups of coffee [2] (Fig. 123.1). Among people with the same genotype of CYP1A2 (Cytochrome P450 Isozyme 1A2), which is involved in the liver metabolism

H. Yamada, Y. Naito (eds.), *Anti-Aging Medicine*, https://doi.org/10.1007/978-981-97-8514-8_123

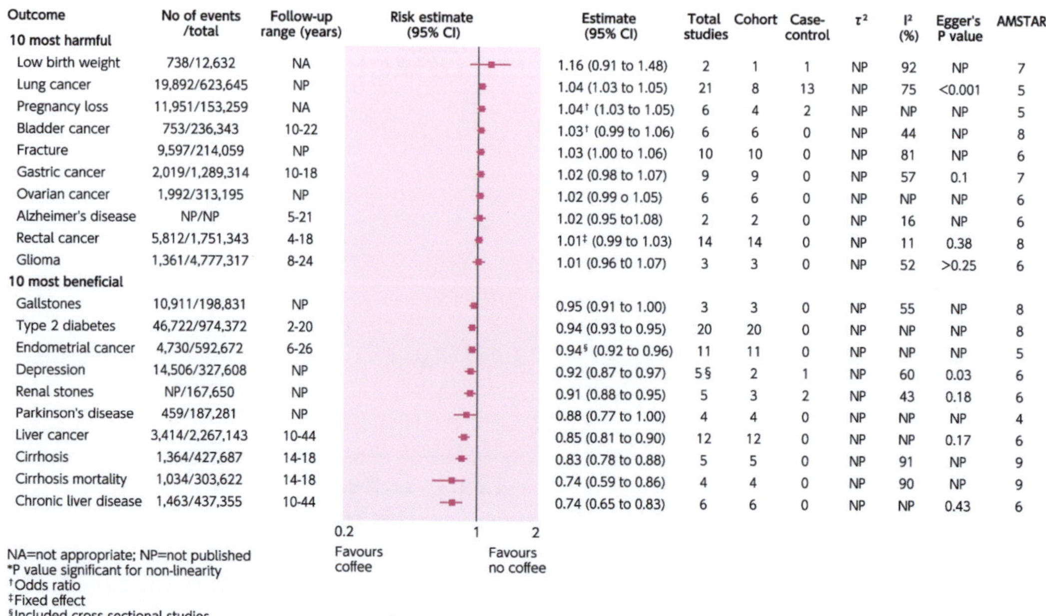

Outcome	No of events /total	Follow-up range (years)	Estimate (95% CI)	Total studies	Cohort	Case-control	τ²	I² (%)	Egger's P value	AMSTAR
10 most harmful										
Low birth weight	738/12,632	NA	1.16 (0.91 to 1.48)	2	1	1	NP	92	NP	7
Lung cancer	19,892/623,645	NP	1.04 (1.03 to 1.05)	21	8	13	NP	75	<0.001	5
Pregnancy loss	11,951/153,259	NA	1.04† (1.03 to 1.05)	6	4	2	NP	NP	NP	5
Bladder cancer	753/236,343	10-22	1.03† (0.99 to 1.06)	6	6	0	NP	44	NP	8
Fracture	9,597/214,059	NP	1.03 (1.00 to 1.06)	10	10	0	NP	81	NP	6
Gastric cancer	2,019/1,289,314	10-18	1.02 (0.98 to 1.07)	9	9	0	NP	57	0.1	7
Ovarian cancer	1,992/313,195	NP	1.02 (0.99 o 1.05)	6	6	0	NP	NP	NP	6
Alzheimer's disease	NP/NP	5-21	1.02 (0.95 to1.08)	2	2	0	NP	16	NP	6
Rectal cancer	5,812/1,751,343	4-18	1.01‡ (0.99 to 1.03)	14	14	0	NP	11	0.38	8
Glioma	1,361/4,777,317	8-24	1.01 (0.96 t0 1.07)	3	3	0	NP	52	>0.25	6
10 most beneficial										
Gallstones	10,911/198,831	NP	0.95 (0.91 to 1.00)	3	3	0	NP	55	NP	8
Type 2 diabetes	46,722/974,372	2-20	0.94 (0.93 to 0.95)	20	20	0	NP	NP	NP	8
Endometrial cancer	4,730/592,672	6-26	0.94§ (0.92 to 0.96)	11	11	0	NP	NP	NP	5
Depression	14,506/327,608	NP	0.92 (0.87 to 0.97)	5 §	2	1	NP	60	0.03	6
Renal stones	NP/167,650	NP	0.91 (0.88 to 0.95)	5	3	2	NP	43	0.18	6
Parkinson's disease	459/187,281	NP	0.88 (0.77 to 1.00)	4	4	0	NP	NP	NP	4
Liver cancer	3,414/2,267,143	10-44	0.85 (0.81 to 0.90)	12	12	0	NP	NP	0.17	6
Cirrhosis	1,364/427,687	14-18	0.83 (0.78 to 0.88)	5	5	0	NP	91	NP	9
Cirrhosis mortality	1,034/303,622	14-18	0.74 (0.59 to 0.86)	4	4	0	NP	90	NP	9
Chronic liver disease	1,463/437,355	10-44	0.74 (0.65 to 0.83)	6	6	0	NP	NP	0.43	6

NA=not appropriate; NP=not published
*P value significant for non-linearity
†Odds ratio
‡Fixed effect
§Included cross sectional studies

Fig. 123.1 The association between consuming one or more cups of coffee per day and multiple health outcomes [2]

of caffeine, those who drink more coffee show better health outcomes in relation to diseases such as Parkinson's disease and breast cancer. An epidemiological study involving approximately 500,000 participants from the UK Biobank reported a correlation between the number of cups of coffee consumed per day and a decrease in mortality from all causes, regardless of the genetic caffeine metabolism score, i.e., the circadian level of caffeine in circulation [3]. It is also suggested that polyphenols and other bioactive substances contained in decaffeinated coffee may demonstrate these effects on health. It has been concluded that the main role of caffeine action is not included in the possible mechanisms of health effects through coffee. The health-promoting effects of vitamins and minerals in coffee brewing seem to be negligible, and the health effects of coffee are associated with other active ingredients, including chlorogenic acid, trigonelline, *N*-methylpyridinium, diterpenes kahweol and cafestol, polysaccharides, peptides, and melanoidins. Some of these ingredients have been reported to have radical scavenging or antiinflammatory activity, mainly in in vitro tests, but the association with antiaging effects is insufficient. Rather, the beneficial effects of coffee are thought to possibly induce adaptive responses to cellular aging in the body. Furthermore, the indigestible components of coffee may regulate the composition and function of the microbiota, as is known for other plant-based foods, and are attracting attention.

123.3 Radical Scavenging by Coffee Components

Some phenolic components of coffee brew increase in concentration during roasting and show radical scavenging activity. The radical scavenging activity of ingested coffee components is still one of the mechanisms explaining the functionality of coffee, but there is evidence from animal and human studies indicating that the concentration of coffee components reaching the plasma is too low for efficient radical scavenging. After coffee intake, the peak plasma concentration of phenolic metabolites ranges from 0.01 to 6 μmol/L, and the caffeine concentration can reach 30 μmol/L [4]. Such concentrations are far below the levels of endogenous antioxidant systems such as uric acid, ascorbic acid, α-tocopherol, glutamine, and glutathione. Furthermore, the main components of coffee, caffeoylquinic acid, trigonelline, and caffeine, are weak antioxidants compared to vitamin C or E, or glutathione, and cannot effectively reduce/regenerate oxidized forms of antioxidant vitamins or glutathione. Therefore, radical removal by coffee components in vivo is limited and is thought to contribute little to the health effects of coffee.

123.4 Weak Antiinflammatory Effects of Coffee

Cross-sectional studies on blood immune or inflammatory markers have reported a weak anti-inflammatory effect associated with habitual coffee intake, but the consistency is not established. Randomized comparative trials of several weeks of coffee intake have also been conducted, with slight decreases in some immune/inflammatory mediator concentrations observed, but the results are not consistent. No inhibitory effect of habitual coffee intake on the risk of rheumatoid arthritis, an inflammatory disease, has been observed [5]. In the future, it should be evaluated the effects of coffee on the intestinal environment, the impact on chronic inflammation by more sensitive inflammatory marker measurements, and the effects of aging.

123.5 Coffee's Phenolic Components Activate the Nrf2 Pathway

In addition to caffeine, the main components of coffee are phenolic compounds. These include chlorogenic acid (caffeoylquinic acid), trigonelline, and its roasting product, the roasting-induced decomposition product of *N*-methylpyridinium. For those who habitually consume coffee, coffee is a major dietary source of plant chemicals such as phenolic acids and polyphenols. When cells or animals are exposed to coffee extracts, an increase in the expression of a group of cell-protective genes involved in antioxidant defense, drug detoxification mechanisms, or repair responses has been observed [6, 7]. The cell-protective response observed in mice after coffee intake is suppressed after inactivation of the Nrf2 (nuclear factor erythroid 2-related factor 2) gene, suggesting that many components of coffee may activate the Nrf2 pathway, with melanoidins, diterpenes cafestol, kahweol, and others contributing to these effects. Numerous trials have shown that coffee or some of its isolated components are potent activators of the expression of antioxidants or cell-protective enzymes. It has been reported that spontaneous DNA strand breaks were 16–35% less with the intake of dark roast Arabica coffee blend, which has been shown to activate Nrf2.

123.6 Antiaging Effects of Coffee Through the Intestine

The health-promoting effects of coffee may occur in the gut, and the prebiotic effects of coffee intake, such as the increase in bifidobacteria in humans and mice, and changes in the ratio of Firmicutes to Bacteroidetes in rats, have been observed in both animals and humans. When coffee is administered to mice, the concentrations of short-chain fatty acids such as acetic acid, propionic acid, and butyric acid increase, and the candidates for prebiotic components of coffee are soluble arabinogalactan and galactomannan, melanoidin, and polyphenols [8, 9]. The impact on the gut microbiota is also evaluated in studies assessing the health effects of Japanese food. Seura et al. [10] calculated a Japanese food score from nine foods (rice, miso soup, pickles, green and yellow vegetables, seaweed, fish, green tea, meat, and coffee) targeting young people, and reported a correlation with the gut microbiota. Bifidobacteria, which have a high occupancy rate in Japanese people, have been reported to have a positive correlation with the intake of green tea.

In our Kyoto Tango Longevity Cohort Study, in a survey of the BDHQ (brief-type self-administered diet history questionnaire) and the relationship with the gut microbiota in 318 people over 65 years old, the occupancy rate of bifidobacteria was not related to green tea intake, but was correlated with coffee intake. Jaquet et al. [11] conducted an intervention trial of three cups of coffee a day for 3 weeks in 16 healthy people, and revealed that Bifidobacterium significantly increased. The growth-stimulating effect of chlorogenic acid on bifidobacteria has also been confirmed in vitro. In a study on the relationship between coffee intake and gut microbiota in 147 healthy people, the occupancy rates of *Bacteroides*, *Prevotella*, and *Porphyromonas* were higher in coffee drinkers, and methoxyphenols, alkylphenols, and caffeine were correlated with the occupancy rate of Bacteroides. In order to confirm the long-term effects of coffee on health, it is necessary to conduct cohort studies observing inflammation markers along with changes in the microbiota.

123.7 Conclusion

Coffee has shown beneficial antiaging effects. In developed countries, coffee provides more phenolic plant chemicals to people than vegetables or fruits. Polyphenols in coffee, such as polysaccharides and melanoidins, and other indigestible components, may be causing changes in the composition and metabolic function of the gut microbiota.

References

1. Gunter MJ, Murphy N, et al. Coffee drinking and mortality in 10 European countries: a multinational cohort study. Ann Intern Med. 2017;167:236–47.
2. Poole R, Kennedy OJ, et al. Coffee consumption and health: umbrella review of meta-analyses of multiple health outcomes. BMJ. 2017;359:j5024.
3. Loftfield E, Cornelis MC, et al. Association of coffee drinking with mortality by genetic variation in caffeine metabolism: findings from the UK Biobank. JAMA Intern Med. 2018;178:1086–97.
4. Lang R, Dieminger N, et al. Bioappearance and pharmacokinetics of bioactives upon coffee consumption. Anal Bioanal Chem. 2013;405:8487–503.
5. Lamichhane D, Collins C, et al. Coffee and tea consumption in relation to risk of rheumatoid arthritis in the women's health initiative observational cohort. J Clin Rheumatol. 2019;25:127–32.
6. Boettler U, Sommerfeld K, et al. Coffee constituents as modulators of Nrf2 nuclear translocation and ARE (EpRE)-dependent gene expression. J Nutr Biochem. 2011;22:426–40.
7. Kalthoff S, Ehmer U, et al. Coffee induces expression of glucuronosyltransferases by the aryl hydrocarbon receptor and Nrf2 in liver and stomach. Gastroenterology. 2010;139:1699–710, 1710.e1–2.
8. Gniechwitz D, Brueckel B, et al. Coffee dietary fiber contents and structural characteristics as influenced by coffee type and technological and brewing procedures. J Agric Food Chem. 2007;55:11027–34.
9. Pérez-Burillo S, Rajakaruna S, et al. Bioactivity of food melanoidins is mediated by gut microbiota. Food Chem. 2020;316:126309.
10. Seura T, Fukuwatari T. Japanese diet score is associated with gut microbiota composition in young Japanese adults. J Nutr Sci Vitaminol (Tokyo). 2019;65:414–20.
11. Jaquet M, Rochat I, et al. Impact of coffee consumption on the gut microbiota: a human volunteer study. Int J Food Microbiol. 2009;130:117–21.

124 Preferences and Antiaging: Green Tea and Antiaging

Mari Maeda-Yamamoto

124.1 Characteristics of Green Tea

Green tea contains many functional ingredients, including catechins, flavonols, caffeine, theanine, minerals, dietary fiber, vitamin A, and vitamin E. The content of these ingredients varies depending on the type of tea, variety, and cultivation method. The generally recognized functions of tea include antioxidant activity, lowering blood cholesterol, lowering blood pressure, regulating bone metabolism, regulating brain and nerve function, antiallergic effects, anticancer effects, antibacterial effects, preventing tooth decay, and deodorizing effects.

124.2 The Functionality of Green Tea Related to Antiaging

Antiaging involves improving lifestyle habits to aim for extending healthy life expectancy. From such a perspective, the functions of tea apply to metabolic diseases (diabetes, lipid disorders, and obesity), circulatory diseases (hypertension, heart disease), osteoporosis, dementia, periodontitis, and stress. Here are some research examples. While reports have been made on the reduction of body fat and the activation of energy metabolism by tea, there is a metaanalysis of six randomized controlled trials (RCTs) (921 people) investigating whether epigallocatechin gallate (EGCG), a major catechin in tea, reduces the risk of metabolic syndrome related to visceral fat. According to this, if you continue to drink a beverage containing 540–588 mg of green tea catechins for 12 weeks, visceral fat significantly decreases (an average reduction of 17.7 cm^2 in total area) compared to drinking a control beverage without catechins [1]. In a metaanalysis of eight RCTs examining the effect on energy expenditure, EGCG had the property of increasing the metabolic rate even at low doses (about 300 mg per day). Continuous consumption of green tea is considered effective in improving metabolic syndrome, and 540 mg of green tea catechins can be ingested with 5–6 cups of strong tea. A metaanalysis of 13 RCTs (total of 925 people, aged 18–73) was conducted on the relationship between green tea consumption and blood pressure. Drinking beverages or capsules containing 200–1207 mg of green tea polyphenols for 3. By continuously drinking for ~12 weeks, both systolic and diastolic blood pressure significantly decreased compared to the control group that did not contain green tea polyphenols [2]. Particularly, the effect was high when the amount of polyphenols was less than 582.8 mg per day, and the duration was about 12 weeks or more. In another meta-analysis (total of 942 cases) [3], the effect of lowering blood pressure was high in

M. Maeda-Yamamoto (✉)
Institute of Food Research, National Agriculture and Food Research Organization, Ibaraki, Japan
e-mail: marimy@affrc.go.jp

H. Yamada, Y. Naito (eds.), *Anti-Aging Medicine*, https://doi.org/10.1007/978-981-97-8514-8_124

Table 124.1 Foods with functional indications using tea's ingredients as functional components

Functionality	Item	Functional component
Function to reduce body fat and LDL cholesterol in people with higher BMI	Green tea (powder, bottled drinks)	Gallate-type catechins 150–340 mg
Function to reduce visceral fat in overweight people	Green tea (powder, bottled drinks)	Tea catechins 400–540 mg
Function to moderate the rise in blood sugar levels after meals	Sun Rouge green tea (powder, tea bags)	EGCG 140.2 mg
Function to moderate the rise in blood sugar levels after meals	Green tea (powder)	EGCG 140.2 mg
Function to maintain a good oral environment (suppresses the formation of dental plaque)	Green tea extract (processed food)	EGCG 18 mg
Function to facilitate fat burning	Green tea extract (supplement)	EGCG 300 mg
Function to support a healthy sleep at night, and to reduce feelings of fatigue and sleepiness upon waking	Green tea extract (tea bags, supplement)	L-Theanine 200 mg
Function to enhance the accuracy of attention (the ability to sustain attention and continue a single action) and judgment (the accuracy and speed of judgment, the ability to appropriately process according to changing situations), which are cognitive functions that decline with age, and to maintain language fluency (the ability to quickly think of and utter many words), which is a part of cognitive function	Green tea extract (bottled drinks, supplement)	Theanine 50.3 mg + tea catechins 171 mg, L-Theanine 200 mg
Function to alleviate discomfort in the eyes and nose caused by pollen, dust, house dust, etc.	Benifuuki green tea (powder, tea bags, latte, bottled drinks)	Methylated catechins 34 mg

people with systolic blood pressure below 130 mmHg, and at the same time, the effect of improving total cholesterol and LDL cholesterol in the blood was also observed. Furthermore, in a metaanalysis of 20 RCTs (total of 1536 cases), diastolic blood pressure and LDL cholesterol significantly decreased by drinking green tea, but no significant effect was observed in HDL cholesterol or systolic blood pressure [4]. The concentration of green tea catechins is recognized to have many effects at around 200 mg as EGCG, and it is expected that drinking 5–6 cups of green tea every day can improve blood pressure and lipid metabolism. In a metaanalysis of 29 prospective cohort studies in the United States, Japan, China, and Europe, drinking more than one cup of green tea a day reduced the risk of cognitive impairment by 6%, and drinking more than 2.3 cups of coffee a day did not contribute to reducing the onset. Also, from a 5.7-year-follow-up study of 13,645 Japanese people who participated in the Osaki Cohort Study, the proportion of people who developed dementia was 8.7%, and those who drink more than five cups of green tea every day had a 27% lower risk of onset than those who drink less than one cup [5].

124.3 Foods with Functional Indications Using Green Tea Components

The tea's functional foods using the tea's ingredients are shown in Table 124.1. There are those that use green tea itself and those that use green tea extracts. As a functional component that has the function of reducing body fat, there are tea catechins (daily intake guideline: 400~540 mg), gallate-type catechins (150~340 mg), EGCG (300 mg), and as a component that suppresses postprandial blood glucose rise, EGCG (140.2 mg). As a component that suppresses the formation of dental plaque, EGCG (18 mg), as a component that maintains cognitive function, L-theanine (200 mg), theanine (50.3 mg) plus tea catechins (171 mg), as a component that supports a healthy sleep at night and reduces fatigue and sleepiness upon waking, L-theanine (200 mg), as a component that reduces discomfort in the eyes and nose due to pollen and house dust, there is methylated catechin (34 mg).

References

1. Hibi M, Takase H, et al. Efficacy of tea catechin-rich beverages to reduce abdominal adiposity and metabolic syndrome risks in obese and overweight subjects: a pooled analysis of 6 human trials. Nutr Res. 2018;55:1–10.
2. Peng X, Zhou R, et al. Effect of green tea consumption on blood pressure: meta-analysis of 13 randomized controlled trials. Sci Rep. 2014;4:6251.
3. Khalesi S, Sun J, et al. Green tea catechins and blood pressures: a systematic review and meta-analysis of randomized controlled trials. Eur J Nutr. 2014;53:1299–311.
4. Onakpoya I, Spencer E, et al. The effect of green tea on blood pressure and lipid profile: a systematic review and meta-analysis of randomized clinical trials. Nutr Metab Cardiovasc Dis. 2013;24:823–36.
5. Tomata Y, Sugiyama K, et al. Green tea consumption and the risk of incident dementia in elderly Japanese: the Ohsaki cohort 2006 study. Am J Geriatr Psychiatry. 2016;24:881–9.

125 Preferences and Antiaging: Smoking Cessation and Antiaging

Miwako Watanabe

In the practice of anti-aging medicine, which aims to extend healthy life expectancy and improve the quality of life, "smoking cessation" is the most important first step. In July 2018, the "Revised Health Promotion Act" was enacted, which stipulates indoor smoking bans in public facilities and public transportation. Ahead of this, the Japan Anti-Aging Medical Association declared a "smoking cessation declaration" in 2017. Smoking is considered the single largest risk factor for cancer worldwide, and according to a report by the International Agency for Research on Cancer (IARC) of the WHO, it is definitively associated with 19 primary cancers, not just lung cancer. Furthermore, it is an established risk factor for many diseases, including ischemic heart disease, cerebrovascular disease, and chronic obstructive pulmonary disease. However, awareness of the impact of smoking on health is still insufficient, and further measures are needed to achieve the ultimate goal of Japan's smoking control, which is "reduction of disease and death." In this section, we will summarize the knowledge needed to correctly convey the harm of smoking and correct the cognitive distortions that can occur due to withdrawal symptoms, with the aim of guiding smoking cessation in antiaging medical practice.

M. Watanabe (✉)
Tokyo Midtown Clinic, Tokyo, Japan
e-mail: m-watanabe@medical-farm.jp

125.1 "Not Just Life Expectancy, But Reduction in Healthy Life Expectancy" Due to Smoking

According to Health Japan 21 by the Ministry of Health, Labour and Welfare, the mortality rate of smokers is higher than that of nonsmokers, and it is estimated that there are 120,000–130,000 smoking-related deaths in Japan and over five million worldwide each year. The cost of measures against diseases and deaths caused by tobacco was estimated to be 1.2 trillion yen per year, equivalent to 5% of the national medical expenses in 1993, and the total loss to society was at least 4 trillion yen. In the case of Japanese people, it has been reported that starting smoking before the age of 20 shortens the lifespan of men by 8 years and women by 10 years [1], and in the United States, smokers have a lifespan shortened by more than 10 years [2], but if they quit smoking between the ages of 35 and 40, they can regain their remaining life expectancy before quitting [3]. For smokers, the cliché is "I'd rather live as I please and die suddenly than endure and live long," but according to NIPPONDATA90, smokers who continue to smoke after the age of 60 have a healthy life expectancy that is 4 years shorter than nonsmokers. The impact on health problems that force suffering in old age, such as "bedridden," needs to be more widely known.

H. Yamada, Y. Naito (eds.), *Anti-Aging Medicine*, https://doi.org/10.1007/978-981-97-8514-8_125

125.2 Correcting Misunderstandings About Health Damage

The relationship between smoking and the onset of lung cancer is relatively well known, but awareness of its impact on other diseases is not high. In a 1999 survey by the Ministry of Health, Labour and Welfare on smoking and health issues, 84.5% of respondents recognized that "smoking makes it easier to get lung cancer," but the rates were lower for "heart disease" at 40.5% and "stroke" at 35.1%. It is said that tobacco smoke contains more than 70 types of carcinogens recognized by IARC and over 200 harmful substances. Even the Japan Tobacco Industry alone discloses that more than 190 types of additives are added, some of which amplify the addictiveness of nicotine and induce continued use. Generally, cigarettes labeled as "low nicotine" and "low tar" give the impression of being less harmful, but in reality, they may not only reduce the impact on health, but may even increase health damage. The amount of nicotine and tar displayed on the package may be underestimated compared to the actual amount due to adjustments in the number of air holes in the filter and the use of easy-to-breathe rolling paper during machine measurement. Also, since the amount of nicotine needed to alleviate withdrawal symptoms is about 1 mg, even with low nicotine and low tar cigarettes, smokers may engage in compensatory smoking, such as inhaling deeply, smoking to the root, or blocking air holes, which may increase the relative amount of secondhand smoke. Nicotine has an extremely high degree of dependence, and several studies using past DSM and ICD diagnostic criteria have reported that the prevalence of tobacco and nicotine is significantly higher than alcohol, stimulants, marijuana, and cocaine. In a follow-up study comparing lung cancer mortality rates by brand, the risk of lung cancer death was higher for smokers who smoked ultrahigh tar brands, but the risk for smokers who smoked low tar brands was slightly higher compared to medium and high tar brands [4].

125.3 About Heated Tobacco

According to the Japan Tobacco Association's announcement for the fiscal year 2020, the sales volume of rolled cigarettes decreased by 16.3% from the previous fiscal year to 988 billion, the first time in the last 30 years that it fell below 1000 billion (the peak was 3483 billion in 1996). In contrast, heated tobacco has rapidly become popular, and its share of the total has expanded to 31.7%. The intake of nicotine is almost the same or slightly less than traditional rolled cigarettes, and it cannot be said that the disease risk is lower than rolled cigarettes, nor can it be said that there is no harm from secondhand smoke [5]. Rather, there is also a risk of exposure to flavors and cyanide compounds that are not found in rolled cigarettes.

125.4 The Reality of Smoking Cessation Guidance

125.4.1 Psychology of Smokers

Habitual smoking is a nicotine addiction (mental illness), which has two aspects: physical dependence and psychological dependence. In physical dependence, symptoms such as irritability, restlessness, and mood depression due to withdrawal symptoms are prominent. The brain of a smoker is in a state of functional decline on brain waves when not smoking, and when nicotine enters through smoking, it temporarily becomes equivalent to a nonsmoker, causing cognitive distortions such as "stress is relieved" and "thinking becomes clear." In psychological dependence, the brain is mistaken for the utility of tobacco to alleviate withdrawal symptoms, and it can be said that it is in a state of misunderstanding that it has the effect of relieving work and private stress. Understanding the psychology of smokers, such as denial of reality, such as "there are long-lived smokers and non-smokers who get lung cancer," and rationalization for smoking, such as "the harm of stress is greater than the harm of tobacco," it is necessary to patiently guide them to the correct understanding while empathizing with the conflict of addiction.

125.4.2 Benefits of Quitting Smoking

1. Reduction of personal disease risk
2. Reduction of risk to cohabiting family members due to secondhand smoke
3. Reduction of stress due to liberation from nicotine dependence
4. Disappearance of unpleasant respiratory symptoms such as coughing and phlegm
5. Improvement of taste and smell (e.g., scent of natural flowers)
6. Improvement of bad breath and body odor
7. Improvement of financial balance (a burden of about 10 million yen for 50 years of smoking one pack a day at 500 yen)
8. Effective use of time (no need to search for smoking areas)

125.4.3 Points to Promote Behavior Change

In interviews to promote quitting smoking, there is likely to be psychological resistance to pointing out only problems. Care should be taken with the "error pointing reflex" that reflexively corrects the patient's incorrect understanding or determination for the patient. While showing empathy with "asking back" to ambivalent statements such as "I want to smoke even though I know it's bad for my body" and "I know but I can't change," it is desirable to guide them to increase statements that lead to behavior change themselves, such as "quitting smoking by not buying cigarettes" and "aiming to stop antihypertensive drugs by quitting smoking". After strengthening the motivation to quit smoking, set a deadline and make a specific plan. It is also important to properly evaluate and motivate when things go as planned.

125.4.4 Current Status of Smoking Cessation Drugs and Apps

The smoking cessation clinic has been applicable for insurance medical treatment since the establishment of the nicotine dependence management fee in 2006. The details of the treatment drugs and smoking cessation support methods are left to the standard procedure manual, but smoking cessation aids are essential drugs for the smoking cessation clinic, and should be introduced promptly when the patient has clearly expressed the intention to quit smoking. In recent years, the scope of utilization has expanded, with remote medical treatment in insurance medical treatment and the use of treatment apps being permitted.

References

1. Sakata R, McGale P, et al. Impact of smoking on mortality and life expectancy in Japanese smokers: a prospective cohort study. BMJ. 2012;345:e7093.
2. Jha P, Ramasundarahettige C, et al. 21st-Century hazards of smoking and benefits of cessation in the United States. N Engl J Med. 2013;368:341–50.
3. Doll R, Peto R, et al. Mortality in relation to smoking: 50 years' observations on male British doctors. BMJ. 2004;328:1519.
4. Harris JE, Thun MJ, et al. Cigarette tar yields in relation to mortality from lung cancer in the cancer prevention study II prospective cohort, 1982–8. BMJ. 2004;328:72.
5. Tabuchi T, Gallus S, et al. Heat-not-burn tobacco product use in Japan: its prevalence, predictors and perceived symptoms from exposure to secondhand heat-not-burn tobacco. Tob Control. 2018;27:e25–33.

126 Traditional Chinese Medicine, Alternative Medicine, and Antiaging: Traditional Chinese Medicine and Antiaging

Kenji Watanabe

126.1 Traditional Chinese Medicine and Antiaging

The "Huangdi Neijing Lingshu," written 2000 years ago, states that "a good doctor treats pre-disease, not post-disease." In our country, the "Nourishing Life Teachings" (Kaibara Ekiken) argue that to stay healthy in old age, it is important to "imagine yourself when you are sick and be cautious." Aging is inevitable, but the quality of health that accompanies it can change depending on how we live now, a fact known since ancient times. "Treating pre-disease" means predicting future diseases such as vascular aging, brain aging, and cancer, and considering what can be done now. In other words, anti-aging in traditional Chinese medicine is nothing but conducting Chinese medicine treatment and nourishing life (improving lifestyle) to stop the progression of potential health problems in the future.

K. Watanabe (✉)
Otsuka Kampo Clinic, Tokyo, Japan

Yokohama University of Pharmacy, Yokohama, Kanagawa, Japan

126.2 Traditional Chinese Medicine's Perspective on Aging

In traditional Chinese medicine, aging is described as a state of "kidney deficiency" when the "kidney qi" weakens [1]. In this case, "kidney" does not refer to the organ, but to a hypothetical place that houses the life energy that a person is born with. When this weakens, it becomes "kidney deficiency," which refers to aging. Traditionally, the state of kidney deficiency indicates various symptoms associated with aging, such as decreased vitality, lower limb muscle weakness, vision loss, hair loss, urinary abnormalities, impotence, and tinnitus. There are various patterns of aging, but broadly speaking, there are patterns where the gastrointestinal system remains strong while the vascular system ages, and patterns where the gastrointestinal system weakens. Metabolic syndrome is representative of the former, and in such cases, the Chinese medicine Hachimijiogan is often used. Hachimijiogan is covered by insurance for diabetes, hypertension, prostate enlargement, back pain, impotence, cataracts, tinnitus, and more. These disease names were applied later, and in terms of Chinese medicine, Hachimijiogan is a medicine for various symptoms caused by aging. On the other hand, for those who age from the gastrointestinal system and whose digestive absorption ability weakens, Shin'ibuto is used. When gastrointestinal function decreases, they

H. Yamada, Y. Naito (eds.), *Anti-Aging Medicine*, https://doi.org/10.1007/978-981-97-8514-8_126

feel cold and are prone to diarrhea from overeating. Typical diarrhea occurs when they feel the urge to defecate as they get cold in the early morning. Even if it's not in the early morning, they usually have diarrhea 2–3 times in the morning. Shin'ibuto is often used for elderly people of this type. Both Hachimijiogan and Shin'ibuto contain aconite (torikabuto), which has analgesic effects and warms the body, making it a useful herbal medicine for the elderly with decreased metabolism.

126.3 Practical Use of Traditional Chinese Medicine in the Elderly

126.3.1 Back Pain and Numbness

For lower back pain and numbness, Hachimijiogan, Goshajinkigan, and Sokeikakketsuto are used. Goshajinkigan is a combination of Hachimijiogan with two herbal medicines, Niu Xi and Che Qian Zi. Hachimijiogan and Goshajinkigan are used for various complaints associated with aging, but it is a prerequisite that the stomach and intestines are strong. This is because the iridoid glycosides contained in "Ji Huang" can impair gastrointestinal function and potentially cause loss of appetite and indigestion. If the stomach and intestines are not strong, Sokeikakketsuto is chosen. In addition, Shakuyakukanzoto may be used. Also known as "Kyojotou," it is effective for lower back pain when muscle tension is strong. Shakuyakukanzoto is often used as a medicine for cramps. It is a fast-acting medicine, but if taken three times a day, the licorice can become excessive, lowering serum potassium levels and potentially causing pseudoaldosteronism, so it is advised to take it only when the pain is severe. The onset of pseudoaldosteronism due to licorice depends on the composition of the intestinal flora, so there is a large individual difference, and regularly checking serum potassium levels leads to proper use.

126.3.2 Osteoarthritis of the Knee

Eppikajutsuto and Bohiogito are often used. If you have physical strength and feel heat in your knee due to local inflammation, use Eppikajutsuto. If water accumulates and needs to be drained regularly, Bohiogito is good. Eppikajutsuto contains 6 g of Ephedra, which can cause palpitations and loss of appetite, so in that case, use Bohiogito. If the stomach and intestines are strong, these two prescriptions may be combined. In that case, since licorice is included in both, it is better to reduce the dose a little, such as twice a day.

126.3.3 Urinary Disorders

If you have frequent urination when you are cold, use Hachimijiogan if your stomach and intestines are strong, and Shinbuto if they are weak. It increases metabolism and has the effect of removing body chill. Especially, Hachimijiogan is often used for nocturia and urinary disorders due to prostate enlargement in men. For urinary disorders with a large psychological element, such as wanting to go to the toilet when going out, Seishinrenshoin is used. In addition, for urinary tract infections, Inrin-to and Rindan-shakogan are used.

126.3.4 Tinnitus

Age-related tinnitus is often difficult to cure, but in Chinese medicine, if the stomach and intestines are strong, you can try Hachimijiogan. Symptoms such as urinary disorders often improve first, but it is necessary to continue for a little longer. If you have high blood pressure, Chototusan is also often used. Chototusan has the effect of improving headaches in the morning due to arteriosclerosis. Also, it is sometimes good to combine the two.

126.3.5 Insomnia

If you have moderate physical strength and tend to be irritable or hypersensitive, Yokukansan is good. If your physical strength is slightly reduced, Suanzaorentang is used, and if you have a loss of appetite, Guipi Tang is used. These are not sleep inducers or sedatives, and they do not induce daytime sleepiness even if taken three times a day. If you are taking other Chinese medicines, you can take them only before bedtime.

126.3.6 Constipation

The number of people complaining of constipation increases with age. As we age, the cells in our intestines lack water, tend to dry out, and many cases become constipated. In such cases, Run Chang Tang and Mashinin-gan are often used. It is often used when constipated, and small, dry stools come out little by little.

126.4 The Future of Antiaging Chinese Medicine Research

Traditional medicine has long had treatments for aging. However, there is still much unknown about modern interpretations that are beginning to scientifically explain the phenomena of aging, and there is room for research. In past research, studies have been conducted on the prevention of complications of diabetes [2], prevention of brain aging, prevention of arteriosclerosis, etc., but they have not been systematized as antiaging Chinese medicine. Many Chinese medicines have been shown to have strong antioxidant abilities [3] and suppressive effects on advanced glycation end products (AGEs). There are also Chinese medicines that contribute to the improvement of mitochondrial function [4]. For chronic kidney disease, a risk factor for vascular aging, intervention trials with Chinese medicine have been conducted, and the improvement effect on serum creatinine has been recognized [5]. By integrating these findings and further developing them, we expect that antiaging Chinese medicine will be systematized.

References

1. Watanabe K. Understand with matrix! The secret of how to use Chinese medicine. Tokyo: Nanjiangdo; 2013.
2. Watanabe K, Shimada A, et al. Long-term effects of goshajinkigan in prevention of diabetic complications: a randomized open-labeled clinical trial. Evid Based Complement Alternat Med. 2014;2014:128726.
3. Nishimura K, Osawa T, et al. Evaluation of oxygen radical absorbance capacity in Kampo medicine. Evid Based Complement Alternat Med. 2011;2011:812163.
4. Sato N, Seiwa C, et al. Administration of chinpi, a component of the herbal medicine ninjin-youei-to, reverses age-induced demyelination. Evid Based Complement Alternat Med. 2011;2011:617438.
5. Zhang HW, Lin ZX, et al. Astragalus (a traditional Chinese medicine) for treating chronic kidney disease. Cochrane Database Syst Rev. 2014;10:CD008369.

127 Traditional Chinese Medicine, Alternative Medicine, and Antiaging: Medical Herbs and Antiaging

Tadashi Mitsuo

The history of herbal (herb) treatment is old, with hundreds of herbs described on clay tablets left by the Sumerians about 5000 years ago. This section discusses 11 types of medical herbs that are mainly used in modern medicine. Each name is written in the order of common name, Japanese name, English name, and scientific name.

127.1 Ginkgo, Ginkgo Leaf, Ginkgo Tree, *Ginkgo biloba*

It is attracting attention as a treatment for peripheral circulatory insufficiency mainly in the nervous system, and memory disorders, dizziness, tinnitus, headaches, and anxiety disorders are indications. There are reports that administering 120 mg of ginkgo leaf extract per day to Alzheimer's patients improved symptoms [1]. It is said that about 1.7% of users have mild side effects such as nausea and gastrointestinal disorders.

T. Mitsuo (✉)
Mitsuo Clinic, Tokyo, Japan

127.2 Echinacea, Purple Cornflower, Echinacea

From around 1980, its immune activation function began to attract attention, and its application to various diseases such as candidiasis, AIDS, chronic fatigue syndrome, and cancer has been attempted. The active ingredients are mainly contained in the rhizomes, and it is effective to take 150–300 mg of dried powder extract (3.5%). Echinacea is said to have the effect of preventing and treating colds, and also effective against viral upper respiratory infections [2]. There are no reports of significant side effects.

127.3 Milk Thistle, *Silybum marianum*

Since the Greek era, it was known that the seeds of Milk thistle have a therapeutic effect on jaundice. Recent studies have clarified that silymarin, a component contained in the seeds, has medicinal effects. It has been reported to have active oxygen removal and lipid peroxidation prevention effects, and also inhibits the activity of 5-lipooxygenase, and is mainly used for the treatment and prevention of liver diseases [3]. Therefore, it is effective for the treatment of alcoholic liver damage and viral hepatitis, and is also effective for drug-induced liver damage. The usual dose is 200–400 mg per day. There are rare

H. Yamada, Y. Naito (eds.), *Anti-Aging Medicine*, https://doi.org/10.1007/978-981-97-8514-8_127

cases of loose stools, but no notable side effects have been reported.

127.4 St. John's Wort (SJW), *Hypericum perforatum*

It has an antidepressant effect, and it is known that the extract inhibits the reuptake of norepinephrine, serotonin, and dopamine [4]. It has also been reported to suppress the secretion of adrenocorticotropic hormone (ACTH) by binding to GABA receptors. In clinical studies, selective serotonin reuptake inhibitors (SSRIs) have been shown to inhibit the reuptake of serotonin. In a double-blind comparison with selective serotonin reuptake inhibitors (SSRIs), St. John's Wort (SJW) has been recognized to have an effect equivalent to SSRIs. SJW has the ability to induce the drug-metabolizing enzyme P450 in the liver, which may weaken the effect of concomitant drugs. Therefore, caution is required when used in combination with other drugs.

127.5 Saw Palmetto, *Serenoa repens*

It is known as a supplement effective for benign prostatic hyperplasia. Fatty acids and their esters extracted from the fruit of saw palmetto have been reported to have anti-inflammatory effects, growth factor inhibitory effects, and anti-androgen effects. In particular, these extracts have been reported to inhibit 5-α-reductase and block the conversion from testosterone to dihydrotestosterone (DHT). In Europe, it is often the first to be prescribed for benign prostatic hyperplasia (BPH). No clear side effects are known.

127.6 Bilberry, *Vaccinium myrtillus*

It is considered effective for ophthalmic diseases such as night blindness, glaucoma, and retinal degenerative diseases. Its history is old, and it has been used as a folk medicine for the treatment of scurvy and urinary tract infections. Its efficacy is said to be derived from the anthocyanosides it contains, which are thought to suppress bleeding from capillaries by maintaining collagen fibers. From these effects, it is also said to be effective for the prevention of diabetic complications and arthritis. A dose of 60–100 mg of anthocyanidin is recommended. No significant side effects have been reported.

127.7 Reishi Mushroom, *Ganoderma lucidum*

It refers to a perennial mushroom of the Ganoderma family. It is described as a medicine for longevity in the "Shennong Ben Cao Jing", the oldest Chinese pharmacopoeia, which is said to have been compiled in 2000 BC. Its effects are known to include pain relief, anti-inflammatory, anti-coagulant effects, and the enhancement of immune function such as the activation of natural killer (NK) cells, and it is sometimes used in cancer treatment.

127.8 Ginseng, *Panax ginseng*

It has been used as a folk medicine for the purpose of nourishing and strengthening the body since ancient times. Its main component, Ginsenoids, has immune-enhancing and hormone-balancing effects, and epidemiological surveys suggest that people who habitually take ginseng have a lower incidence of cancer. On the other hand, because it has a metabolic activation effect, it can raise blood pressure, so caution is needed when used in patients with hypertension.

127.9 Turmeric, *Curcuma longa*

Turmeric is a perennial herb of the ginger family, known as a representative spice contained in curry powder, and its medicinal ingredient contained in the rhizome is called curcumin. There are different types of turmeric, such as autumn, spring, and purple turmeric, and the amount of medicinal ingredients varies. The medicinal effects of curcumin are known to include anti-inflammatory effects and liver function enhancement.

127.10 Ashwagandha, *Withania somnifera* Dunal

A medicinal herb used in Indian Ayurvedic medicine since ancient times. Also known as Indian ginseng, it has a nourishing and invigorating effect. In recent years, there has been an increase in cases of consumption for stress relief, immune enhancement, and fatigue recovery.

127.11 Cannabis, *Cannabis sativa*

Cannabis contains hundreds of medicinal ingredients, and the "Shennong Ben Cao Jing" states that consuming large amounts can lead to the risk of neurological damage, but consuming small amounts has the effect of longevity [5]. In December 2020, the United Nations Narcotics Commission removed cannabis from the list of most dangerous drugs. As a result, the market for medical cannabis is expanding, mainly in Western countries. In Japan, the use of the addictive ingredient tetrahydrocannabinol (THC) is illegal, but the use of cannabidiol (CBD) is legal. Its medicinal effects include sedative effects on the nervous system, such as insomnia and pain management, anti-anxiety, and anti-convulsant effects, and it is considered valuable for use in cases where normal analgesics are ineffective, such as cancer pain.

References

1. LeBars PL, Katz MM, et al. A placebo-controlled, double-blind, randomized trial of an extract of Ginkgo biloba for dementia. North American EGb Study Group. JAMA. 1997;278:1327–32.
2. Ross SM. Echinacea purpurea: a proprietary extract of Echinacea purpurea is shown to be safe and effective in the prevention of the common cold. Holist Nurs Pract. 2016;30:54–7.
3. Dehmlow C, Murawski N, et al. Scavenging of reactive oxygen species and inhibition of arachidonic acid metabolism by silibinin in human cells. Life Sci. 1996;58:1591–600.
4. Müller WE, Singer A, et al. Hyperforin represents the neurotransmitter reuptake inhibiting constituent of hypericum extract. Pharmacopsychiatry. 1998;31(Suppl 1):16–21.
5. Pizzorno JE, Murray MT. Textbook of natural medicine. 5th ed. Elsevier Health Sciences; 2021. E-book. Kindle No. 42204-6.

128 Traditional Chinese Medicine, Alternative Medicine, and Antiaging: Acupuncture and Moxibustion and Antiaging

Masato Egawa

128.1 What Is Acupuncture and Moxibustion: Therapeutic Tools and Clinical Applications

Acupuncture and moxibustion is a traditional Japanese medical practice that uses needles or moxibustion to physically and thermally stimulate therapeutic points called meridians, and aims to alleviate symptoms, treat and prevent diseases, and promote health through the activation of the body's natural healing response. The typical acupuncture needle is made of stainless steel or silver, with a diameter of 0.14–0.20 mm, and is inserted from about 1 mm to several centimeters depending on the condition. There are also non-insertion acupuncture methods that apply low-frequency electricity, attach small metal pieces (granular acupuncture), or press or rub the skin with a spatula-like tool. Moxibustion involves burning a lump of mugwort (moxa) collected from the glandular hairs of mugwort leaves on the skin. In recent years, warm moxibustion, which maintains a distance between the burning part and the skin and provides a warm sensation, has become common (Fig. 128.1). In the acupuncture and moxibustion medicine, there are 361 defined meridians throughout the body, which connect the six organs and six viscera through 14 lines called meridians, and each point has its own indications. However, the specificity of the meridians is not clear, and in actual treatment, the treatment points are often determined based on reaction points such as tenderness and muscle tension, and the condition. Acupuncture and moxibustion have traditionally played a significant role in the health management of the elderly, and in recent years, evidence is accumulating as a complementary and alternative medicine. Also, the effects of cosmetic acupuncture and moxibustion (health and beauty acupuncture and moxibustion) in the beauty field are being recognized for their value in "visual anti-aging." This section discusses acupuncture and moxibustion treatments for symptoms and diseases characteristic of the elderly and frailty.

M. Egawa (✉)
Department of Acupuncture & Moxibustion, Niigata University of Health and Welfare, Niigata, Japan
e-mail: masato-egawa@nuhw.ac.jp

H. Yamada, Y. Naito (eds.), *Anti-Aging Medicine*, https://doi.org/10.1007/978-981-97-8514-8_128

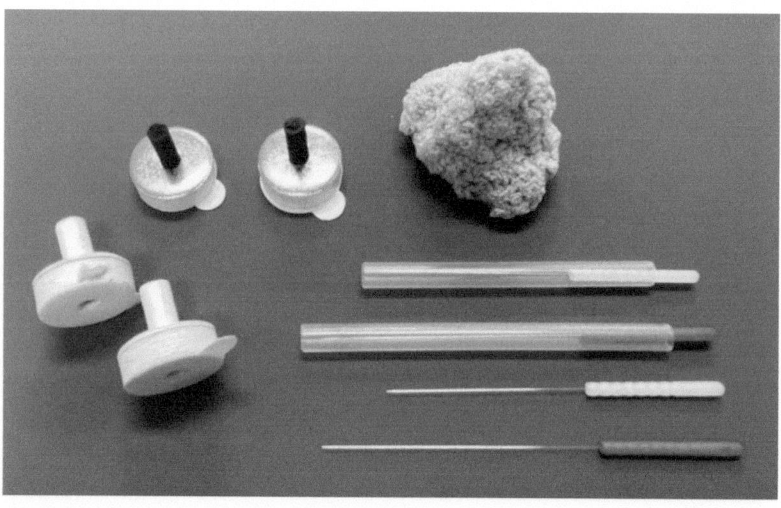

Fig. 128.1 Pedestal moxibustion, mugwort, therapeutic needle

128.2 The Effectiveness of Acupuncture and Moxibustion Treatment for Symptoms and Diseases in the Elderly

128.2.1 Back Pain

Back pain is one of the most widely clinically applied symptoms for acupuncture and moxibustion treatment, and it is common to use tender points as treatment points. In studies comparing the treatment of chronic back pain in the elderly using tender points as treatment points with a sham acupuncture group, the effectiveness of acupuncture treatment has been recognized in terms of pain evaluated by the visual analogue scale (VAS) and specific quality of life evaluated by the Roland Morris Disability Questionnaire [1].

128.2.2 Spinal Stenosis

Acupuncture treatment on peripheral nerves and intervertebral joints at the same dermatome level as the pain and numbness, as well as the lumbar multifidus muscle, can improve blood flow in the affected nerve root and alleviate pain and difficulty walking [2]. Compared to drug therapy and exercise therapy, acupuncture has a higher effect on pain relief and improvement of daily living activities (activities of daily living; ADL), suggesting its effectiveness as a conservative treatment for this common disease in the elderly.

128.2.3 Chronic Obstructive Pulmonary Disease (COPD)

Suzuki et al. [3] compared the effects of acupuncture aimed at reducing exertional dyspnea in chronic obstructive pulmonary disease (COPD) with a sham acupuncture group. The treatment points were bilateral Zhongfu, Taibai, Futu, Guanyuan, Zhongwan, Zusanli, Taixi, Wan'gu, Feishu, Pishu, and Shenshu. As a result, the effectiveness of acupuncture was recognized in improving dyspnea evaluated by the modified Borg scale, walking distance and minimum arterial oxygen saturation during the 6-min walk test, one-second volume, and QOL (St. George's Respiratory Questionnaire; SGRQ). It is believed that the recovery of respiratory auxiliary muscle function such as the sternocleidomastoid muscle by acupuncture, airway dilation mediated by the autonomic nervous system, and physical recovery due to improved digestive absorption function lead to improved respiratory function and exercise endurance.

128.2.4 Parkinson's Disease

Fukuda et al. [4] compared the treatment effects of a standard acupuncture group (once a week × 3 months = 12 times in total) and a low-frequency acupuncture group (once a month × 3 months = 3 times in total) in patients with Parkinson's disease, with the treatment points being Quchi, Hegu, Zusanli, Taichong, Ganshu, and Shenshu. Although there was no significant difference in treatment effects between the two groups, significant improvements were observed in the standard acupuncture group in terms of changes in mental symptoms and daily living activities evaluated by the Unified Parkinson's Disease Rating Scale (UPDRS), and balance function evaluated by the functional reach test (FRT). High evidence has not been obtained in clinical studies of acupuncture and moxibustion treatment for Parkinson's disease. However, clinical reports have repeatedly recognized improvements in walking, posture, and mental state, and reductions in necessary medication amounts due to acupuncture and moxibustion treatment, and it is expected to be positioned as a treatment method that can be used in conjunction with drug treatment centered on L-DOPA in the future.

128.3 Acupuncture and Moxibustion Treatment for Frailty [5]

Compared to elderly people living in the community, elderly people who visit acupuncture clinics have a higher rate of frailty, with physical and oral function decline and depressive mood being the factors. In addition, muscle strength (dominant hand grip strength), walking ability [3m-TUGT (time up and go test)], and balance ability (functional reach test; FRT), a decrease in oral function (tongue pressure) was also observed. It is within the expected range that elderly people who come to the hospital with the aim of alleviating some symptoms tend to be frail, but it is considered that acupuncture and moxibustion treatment is positioned as a prevention of frailty. In addition, the impact of a single acupuncture and moxibustion treatment on frailty factors in the elderly: muscle strength, walking ability, balance ability, and oral function, the proportion of those who showed improved function after treatment was higher in the frail group compared to the same age group. Differences in response were also observed depending on the underlying disease. The effect of acupuncture and moxibustion treatment as a prevention of frailty is suggested, and its medical positioning and application are expected in the future.

References

1. Katsumi Y, Itoi M, et al. Ah's acupuncture treatment for chronic low back pain in the elderly. Rehabil Med. 2004;41:824–9.
2. Kasuya D. Acupuncture and moxibustion treatment for low back and leg pain—lumbar spinal stenosis (syndrome). Pain Clin. 2011;32:519–28.
3. Suzuki M, Muro S, et al. A randomized, placebo-controlled trial of acupuncture in patients with chronic obstructive pulmonary disease (COPD): the COPD-acupuncture trial (CAT). Arch Intern Med. 2012;172:878–86.
4. Fukuda S, Egawa M, et al. A study on the clinical effect of acupuncture treatment for Parkinson's disease—a randomized controlled trial (RCT). Meiji Int Med Univ J. 2012;6:21–45.
5. Egawa M, Fukuda S. On the frailty situation of the elderly who visit the acupuncture and moxibustion clinic. Abstracts of the 21st Japan Anti-Aging Medical Association; 2021. p. 61.

129 Antiaging Checkup: What Is an Antiaging Checkup

Yoshikazu Yonei

129.1 Antiaging Checkup

The body is composed of various tissues, organs, and organs, but they do not all age uniformly. There are individual differences and differences depending on the part. The risk factors that promote aging also vary from person to person. From the late 30s, pathological degenerative changes due to aging occur in some parts of the body, leading to diseases and adversely affecting other healthy parts. A summary of the survey results of centenarians, who are achievers of healthy longevity, shows that centenarians have uniformly balanced aging throughout the body and have few aging risk factors. Therefore, diagnosing the weaknesses of aging early with an antiaging checkup, balancing the whole, and correcting aging risk factors will be the royal road to healthy longevity [1, 2] (Fig. 129.1). Before entering the antiaging therapy, evaluate aging, that is, the degree of aging and the risk factors that promote aging. The antiaging checkup is something that shows the future direction of human dock and health checkups [3]. Regular human docks and health checkups focus on the prevention, early detection, and treatment of cancer and lifestyle-related diseases, but the antiaging checkup adds the detection, prevention, and treatment of pathological aging. Many lifestyle-related diseases are positioned as aging risk factors.

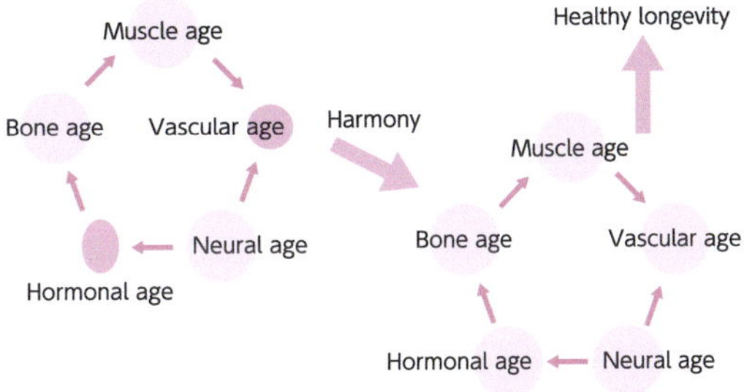

Fig. 129.1 The journey to a healthy long life [1]. Identify the weaknesses of aging early and balance the body. The goal is to eliminate the difference between average lifespan and healthy lifespan

Y. Yonei (✉)
Anti-Aging Medical Research Center/Glycative Stress Research Center, Graduate School of Life and Medical Sciences, Kyoto, Japan
e-mail: yyonei@mail.doshisha.ac.jp

H. Yamada, Y. Naito (eds.), *Anti-Aging Medicine*, https://doi.org/10.1007/978-981-97-8514-8_129

129.2 Test Items: Degree of Aging and Aging Risk Factors

While the number of medical institutions implementing antiaging checkups is increasing, it is hard to say that the public fully understands their content. One of the reasons for this is that medical institutions have been operating on disparate standards over the past few years. There is a need to unify the standards and more clearly communicate to the public what can be learned from antiaging checkups. Aging-related test items can be divided into two categories: aging degree and aging risk factors. In antiaging checkups, the degree of aging is evaluated in five items: muscle age, blood vessel age, nerve age, hormone age, and bone age. Also, five items related to aging risk factors, namely, immune stress, oxidative stress, glycation stress, psychosomatic stress, and lifestyle habits, are evaluated. When explaining the results of the aging degree test, it is easier for the examinee to understand if the functional age is shown to correspond to what average age, rather than explaining that the numbers have "increased or decreased" or "increased or decreased." By scoring risk factors and showing which ones account for a large proportion, it is easier to prioritize treatment and guidance. If antiaging could be expressed in one word, it would be "prevention of functional age aging, rejuvenation." By correcting the two items of the weakest point of aging (the item with the highest functional age) and the largest risk factor among all ten items, 80% of the whole can be achieved. This is an application of "Pareto's law." There is currently no absolute evaluation method for functional age. If it is a relative evaluation, a simple method is to "create an aging standard curve for Japanese people regarding a specific parameter and determine what age level it corresponds to." The target value of functional age is about 80% of the actual age, and the test value is plus 0.5–1.0 SD (standard deviation) of the age average value would be appropriate. For currently available representative measurement methods, biomarkers, aging standard curves, and optimal values, please refer to each specialized item. For software to support the operation of antiaging checkups, a support system (AAD Life Works®) has been developed through hearings and discussions with members of this society and the Japan Society of Ningen Dock [4–7] (Fig. 129.2). It is extremely useful for accumulating large-scale data.

ADD Life Works

Date of implementation 2022 / 06 / 14 ID 25

Basic items

ID	25	Height	161	cm
Name 1	Yamada Hanako	Weight	55	kg
Name 2	Yamada Hanako	Body fat percentage	25	%
Date of birth	1972 / 11 / 11	BMI	21.2	
Gender	Female	Blood pressure	100/69	mmHg
Date of implementation	2022 / 06 / 14			

Aging judgment graph

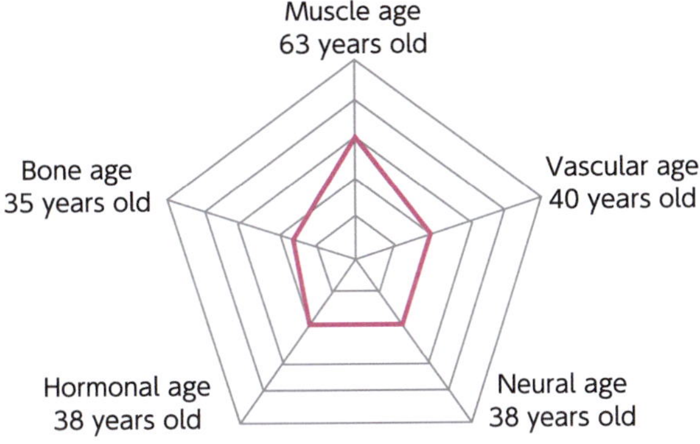

Risk factor judgment graph

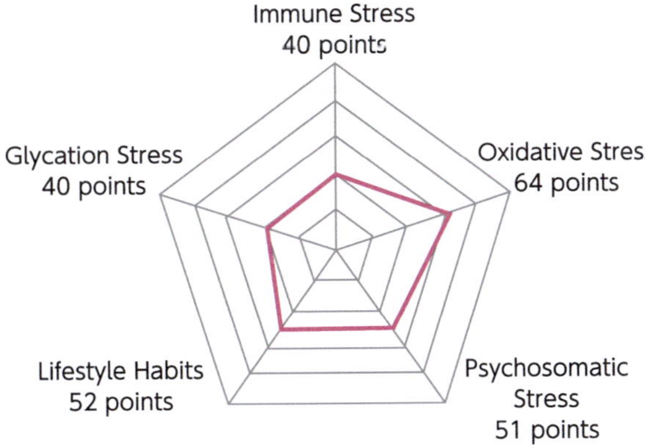

Fig. 129.2 Display of antiaging dock (antiaging dock; AAD) results. Case: 49-year-old female. AAD support software "AAD Life Works®" used

129.3 Guidance Based on Results

For example, if the bone density of a 55-year-old man is close to the average bone density of a 60-year-old, the bone age is evaluated as 60 years old. The functional age can be calculated by comparing it with the database. Emphasizing primary prevention as preventive medicine, we provide guidance on improving lifestyle habits based on the results of anti-aging checkups (food education, physical education, and intellectual education). In addition, medical treatment (drug treatment, hormone replacement therapy, and regenerative medicine) is performed as needed. Motivation and behavior change are important in guidance, and the discretion of antiaging specialists and instructors (guidance skills, knowledge, and enthusiasm) is questioned.

129.4 The Significance of Accumulating Data for Hypothesis Verification

I propose the hypothesis that "By identifying and correcting the most aged functional age (weakness of aging) and the most significant risk factors early on, and if the whole body ages evenly and in balance, the difference between average lifespan and healthy lifespan can be reduced, and the healthy lifespan can be extended." The evaluation of functional age and aging risk factors by antiaging doctors, and the accumulation of large-scale data through the use of support software, should play a crucial role in demonstrating this hypothesis. The use of this system is designed for multifacility participation in clinical research aimed at hypothesis verification.

129.5 Conclusion

Antiaging is about preventing aging and rejuvenating functional age. It is important to diagnose functional age and risk factors early and intervene through an antiaging doctor to achieve a healthy long life. Also, large-scale data is necessary for demonstrating the pursuit of a healthy long life. I sincerely hope that more medical institutions will practice anti-aging medicine and that many citizens will receive examinations.

References

1. Yonei Y. Introduction to anti-aging medicine. 3rd ed. Tokyo: Keio University Press; 2022.
2. Toshikazu Y, editor. Anti-aging dock. Tokyo: Diagnosis and Treatment Society; 2007.
3. Yonei Y, Mizuno Y. The human dock of tomorrow: annual health check-up for anti-aging. Ningen Dock. 2005;19:5–8.
4. Yonei Y, Takabe W. Aging assessment by anti-aging medical checkup. Compreh Med Checkup. 2015;42:459–64.
5. Yonei Y, Takabe W, et al. What does the anti-aging medical checkup show?: data presentation. Compreh Med Checkup. 2017;44:600–5.
6. Yonei Y. Practice and problems—anti-aging dock. Prog Med. 2017;261:698–704.
7. Yonei Y. Diagnosis and treatment of aging degree. Health Sci. 2022;64:585–90.

Antiaging Checkup: Vascular Age 130

Yukihito Higashi

Thomas Sydenham, a seventeenth-century chemist, left behind the words "A man is as old as his arteries." Endothelial dysfunction is the beginning of arteriosclerosis. Furthermore, when smooth muscle dysfunction occurs, it becomes a state of organic arteriosclerosis, leading to the maintenance and progression of arteriosclerosis, and eventually leading to cardiovascular complications. Aging is the greatest risk factor for arteriosclerosis and is an independent risk factor for endothelial dysfunction and smooth muscle dysfunction [1]. Vascular age includes both physiological aging (calendar age) and aging under the presence of coronary risk factors (Fig. 130.1). Evaluating vascular age can be an important item in evaluating anti-aging through various interventions. Currently, not only invasive vascular function tests but also many noninvasive vascular function tests are being performed to evaluate vascular age.

Y. Higashi (✉)
Department of Regenerative Medicine, Research Institute for Radiation Biology and Medicine, Hiroshima University, Hiroshima, Japan
e-mail: yhigashi@hiroshima-u.ac.jp

H. Yamada, Y. Naito (eds.), *Anti-Aging Medicine*, https://doi.org/10.1007/978-981-97-8514-8_130

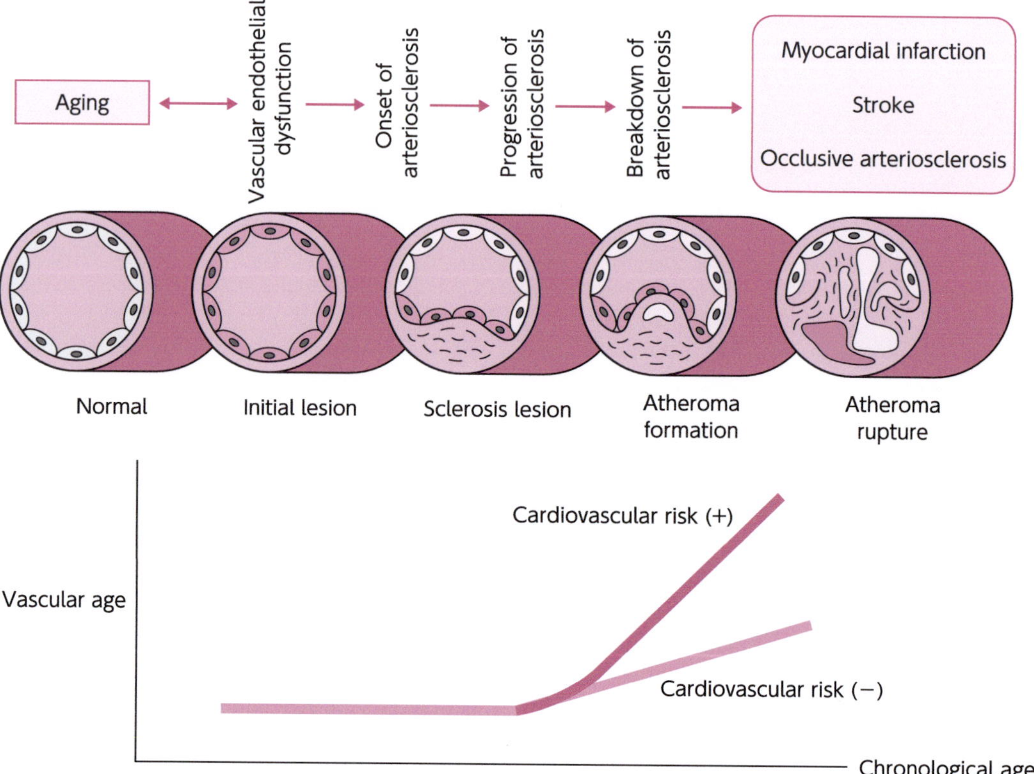

Fig. 130.1 The process from aging to vascular endothelial dysfunction, onset of cardiovascular complications, and vascular age

130.1 Vessels as Targets for Antiaging

The vascular system is composed of the circulatory system and the lymphatic system. Vessels not only serve as mere conduits for oxygen, nutrients, and waste products, but it has become clear that they perform many functions necessary for life support. Normally, vessels consist of a three-layer structure, with the lumen of the vessel covered by a single layer of endothelial cells that form the inner layer of both arteries and veins, a media composed of smooth muscle cells surrounding the endothelial cells, and an outer layer located outside. The vascular structure is made robust by connective tissue and extracellular matrix. The three-layer structure of the vessels not only supports the structure of the vessels themselves but also has various functions, including contraction and dilation of the vessels, and plays an important role in the onset, maintenance, and progression of arteriosclerosis. Therefore, it is important to evaluate the function and structure of the vessels. Endothelial function and smooth muscle function are particularly important. Classically, the venous system is considered as a capacity vessel that stores blood, but the veins themselves also have various functions. Ideally, it is important to evaluate the function of all vessels that make up the vascular system.

130.2 Using Endothelial Function to Evaluate Vascular Age

It has become clear that many physiologically active substances, including vasodilators and vasoconstrictors, are produced and secreted from the vascular endothelium [2]. In particular, nitric oxide (NO) plays a very important role in arterio-

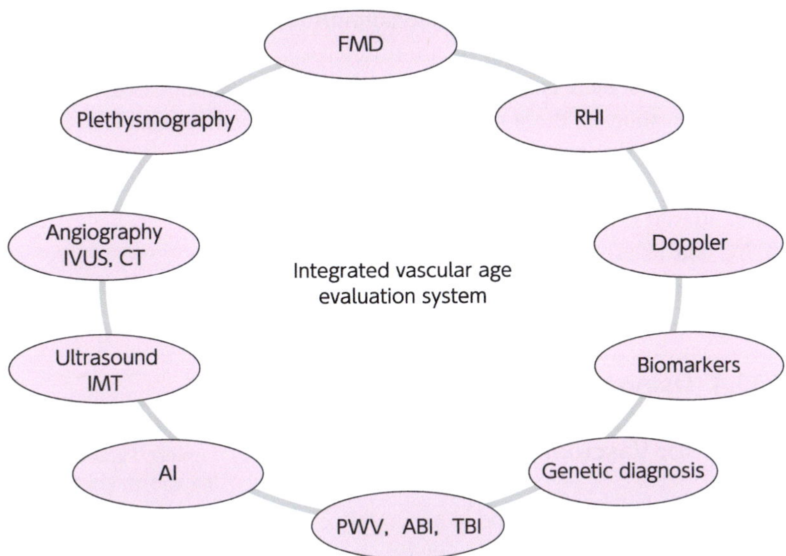

Fig. 130.2 Integrated vascular function test system

sclerosis. Normal vascular endothelium has vasodilation and vasoconstriction, proliferation and antiproliferation of vascular smooth muscle cells, coagulation and anticoagulation effects, inflammation and antiinflammatory effects, oxidation and antioxidation effects, and these balances work to regulate and maintain vascular tone and structure. When the vascular endothelium is damaged, these balances are disrupted, leading to breakdown of vascular tone and structure. Arteriosclerosis develops with endothelial dysfunction as the first stage and progresses. If it progresses further, it can cause cardiovascular complications [3]. Metaanalysis has confirmed that endothelial function is an independent predictor of cardiovascular complications [4]. Endothelial function can be considered as a therapeutic target for arteriosclerosis. Endothelial dysfunction is not irreversible and can be improved with appropriate interventions such as drug therapy, supplementation therapy, and lifestyle improvements [5]. Currently, methods for evaluating endothelial function include strain gauge plethysmography, flow-mediated vasodilation (FMD), and reactive hyperemia peripheral arterial tonometry (RH-PAT) (Fig. 130.2).

130.3 Using Vascular Smooth Muscle Function for Vascular Age Evaluation

Healthy vascular smooth muscle cells work to maintain vascular homeostasis by synthesizing extracellular matrix and extracellular matrix degrading enzymes, but when they lean towards arteriosclerosis, they produce growth factors and migratory factors, proliferate and migrate themselves, leading to structural changes in the vessels (pathological remodeling). Vascular smooth muscle cells undergo phenotypic transformation due to external stimuli such as reactive oxygen species and inflammation, and become directly involved in the formation of arteriosclerotic foci. Arterial stiffness, represented by indicators such as pulse wave velocity (PWV), cardioankle vascular index, and augmentation index (AI), is determined by the amount of vascular smooth muscle, the amount of elastic fibers (elastin), the amount of calcium deposition, and the presence or absence of atheroma. In a broad sense, methods for evaluating vascular smooth muscle function include ankle-brachial pressure index (ABI), stiffness β, intima-media thickness (IMT), and vascular reactivity tests with nitrate administration. In the advanced stages of arteriosclerosis, which causes organic changes such as thickening and calcification of the arterial wall and atheroma, there are noninvasive imaging diagnoses such as cardiac echo, carotid echo, limb vascular echo, arterial CT, coronary CT, CT/MR angiography, and invasive tests such as coronary angiography (Fig. 130.2).

130.4 Using Venous Function and Lymphatic Function for Vascular Age Evaluation

At present, there is no method to accurately evaluate venous function and lymphatic function. Therefore, the relationship between vascular age and venous function and lymphatic function is not clear. There is no room for debate about the importance of the clinical significance of evaluating vascular age. Many methods of measuring vascular age are used in clinical settings. Some have been standardized, but unfortunately, many have not been standardized. It is desirable that the evaluation of vascular age is made from various indicators. The establishment of an integrated vascular age evaluation system is also awaited as an evaluation of aging (Fig. 130.2).

References

1. Maruhashi T, Soga J, et al. Relationship between flow-mediated vasodilatation and cardiovascular risk factors in a large community-based study. Heart. 2013;99:1837–42.
2. Lüscher TF. Imbalance of endothelium-derived relaxing and contracting factors: a new concept in hypertension? Am J Hypertens. 1990;3:317–30.
3. Higashi Y, Noma K, et al. Endothelial function and oxidative stress in cardiovascular diseases. Circ J. 2009;73:411–8.
4. Lerman A, Zeiher AM. Endothelial function: cardiac events. Circulation. 2005;111:363–8.
5. Higashi Y, Yoshizumi M. Exercise and endothelial function: role of endothelium-derived nitric oxide and oxidative stress in healthy subjects and hypertensive patients. Pharmacol Ther. 2004;102:87–96.

131 Antiaging Checkup: Brain Age

Koji Abe

131.1 Brain Age and Appearance

Vascular age, which is used as an indicator of arteriosclerosis, has been utilized in regular health checkups, but the indicators for neural age or brain age are not well established. According to a study comparing monkeys that were fed a restricted diet for 20 years with those that were allowed to eat freely, the overfed monkeys showed signs of aging, such as a wrinkled face and significant hair loss, after 20 years (Fig. 131.1) [1]. On the other hand, the restricted diet monkeys had sharp eyes, a robust face, and thick fur, clearly showing that aging was suppressed and antiaging effects were evident. In a study of ten pairs of identical twins in Denmark, composite photos were made from ten people who looked younger and ten people who looked older, and despite being genetically identical, clear differences in appearance were reproduced [2]. In a study of about 600 middle-aged and elderly men and women in the Netherlands (average age 60), it was reported that the higher the blood sugar level, the older the perceived age, regardless of whether they had diabetes [3]. Furthermore, a study in North Carolina, USA, reported that visceral age and perceived age are proportional, and cognitive function is also proportional [4]. These studies did not sufficiently examine brain age, but if the brain ages along with the blood vessels, it is speculated that brain age will also correlate with perceived age.

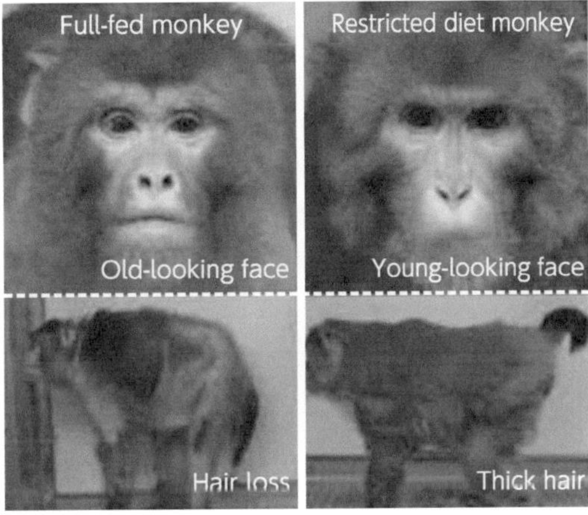

Fig. 131.1 Same 20-year-old captive monkeys (full-fed and restricted diet). (Reprinted with permission from [7])

131.2 Functional and Structural Tests to Evaluate Brain Age

Brain age can be indirectly measured functionally by cognitive function tests such as the mini-mental scale examination (MMSE) and Hasegawa's dementia scale (HDS), and emotional function tests such as Abe's BPSD score (ABS) [5, 6]. That is, anything exceeding the nor-

K. Abe (✉)
National Center of Neurology and Psychiatry (NCNP), Tokyo, Japan

Arts Ginza Clinic, Tokyo, Japan

H. Yamada, Y. Naito (eds.), *Anti-Aging Medicine*, https://doi.org/10.1007/978-981-97-8514-8_131

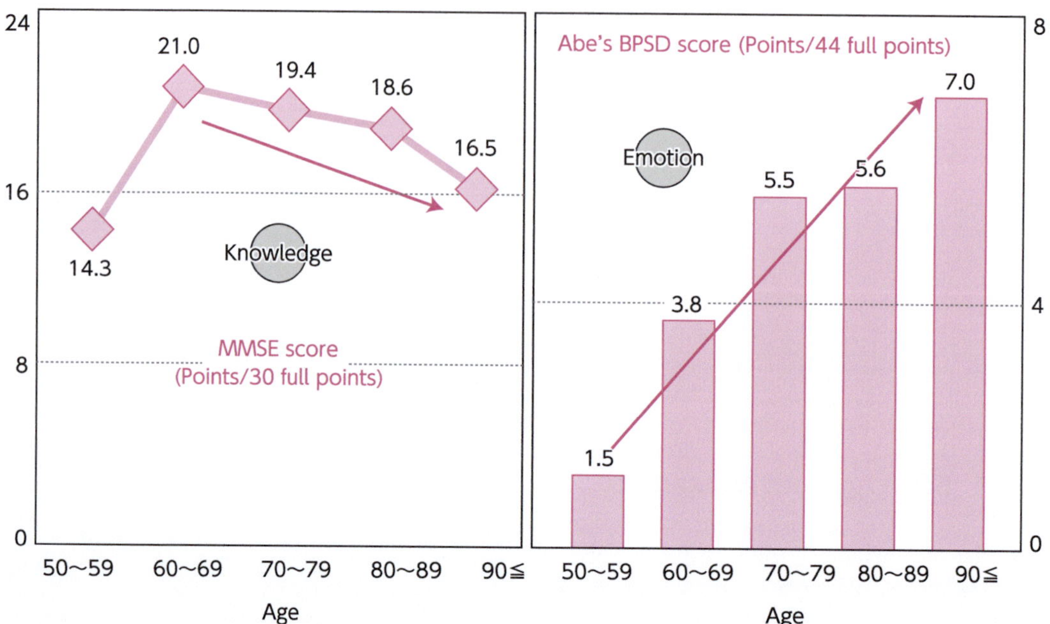

Fig. 131.2 Age-related changes in MMSE and Abe's BPSD score (ABS). (AD patients + MCI patients = 838 people)

mal range is judged as a decline in function based on changes in clinical scores. However, there is no system to express this in terms of brain age. As shown in Fig. 131.2, which shows the age-specific scores of 838 Alzheimer's disease patients + mild cognitive impairment (MCI) patients in our dementia outpatient clinic, the intellectual MMSE score decreases with age, and at the same time, emotional function as shown by Abe's BPSD score also worsens. On the other hand, to evaluate brain age structurally, it can be indirectly indicated by the occupancy ratio of brain volume to the cranial cavity, and in the dementia outpatient clinic, the volume of the hippocampus can also be evaluated by brain MRI using the voxel-based specific regional analysis system for Alzheimer's. It is commonly used in daily practice with Voxel-based Specific Regional Analysis System for Alzheimer's Disease (VSRAD). This is an MRI image processing and statistical analysis software used to read the degree of atrophy of the parahippocampal gyrus seen in early Alzheimer's disease, expressing the strength of atrophy in four levels of numbers. For example, if the degree of atrophy is 0–1, there is almost no brain atrophy, if it is 1–2, there is some atrophy, and if it is 2–3, there is considerable atrophy. However, in individual cases, the proportion of brain volume and this structural atrophy of the hippocampus do not always coincide with clinical findings as brain functions, so careful response is required.

References

1. Mattison JA, Colman RJ, et al. Caloric restriction improves health and survival of rhesus monkeys. Nat Commun. 2017;8:14063.
2. Christensen K, Thinggaard M, et al. Perceived age as clinically useful biomarker of ageing: cohort study. BMJ. 2009;339:b5262.
3. Noordam R, Gunn DA, et al. High serum glucose levels are associated with a higher perceived age. Age (Dordr). 2013;35:189–95.
4. Belsky DW, Caspi A, et al. Quantification of biological aging in young adults. Proc Natl Acad Sci USA. 2015;112:E4104–10.
5. Abe K, Yamashita T, et al. A new simple score (ABS) for assessing behavioral and psychological symptoms of dementia. J Neurol Sci. 2015;350:14–7.
6. For inquiries about the application. https://prtimes.jp/main/html/rd/p/000000065.000013467.html.
7. Colman RJ, Anderson RM, et al. Caloric restriction delays disease onset and mortality in rhesus monkeys. Science. 2009;325:201–4.

132 Antiaging Checkup: Bone Age and Osteoporosis

Hiroaki Ohta

132.1 Bone Age and Antiaging Checkup

"Bone age" is a term used to express the degree of bone maturity in terms of age. In other words, it is used to determine whether the growth and development of bones is age-appropriate, delayed, or precocious. Therefore, the bone age derived from the bone age evaluation criteria by hand X-ray is a good indicator that finely and objectively expresses the degree of maturity, but it is not an indicator of bones used in anti-aging checkups aiming to delay or prevent aging phenomena that progress with age after adulthood. On the other hand, in the body, aging cells are routinely removed and replaced with new cells, a metabolic process known as antiaging. While the intestinal epithelial cells, which are replaced in a few days, take the shortest time, it takes the longest time, several months, to replace the new bone tissue to make it a sturdy bone that can withstand load. This is the most prominent antiaging in the body, especially called bone remodeling. However, no indicator has yet been found to capture the entire picture of this bone remodeling, which is the therapeutic area [1]. After reaching the young adult mean (YAM), bone density decreases with age and menopause in women, but the degree of this phenomenon allows estimation of the degree of bone aging and fragility. Bone aging can be quantitatively detected through bone mineral density (BMD) measurements, making it a suitable indicator for use in anti-aging medical checkups.

132.2 Measurement of Osteoporosis

Osteoporosis is a condition primarily associated with aging, defined by a reduction in bone mass, deterioration of bone microarchitecture, and compromised bone matrix, leading to increased susceptibility to fractures [2]. Among the three factors contributing to osteoporosis, only bone mass—represented by bone mineral density—can be clinically measured. The deterioration of bone microarchitecture and bone matrix remains beyond clinical evaluation. Early-stage osteoporosis without fractures is identified solely by a decrease in bone mineral density as measured by BMD tests, without any accompanying clinical symptoms. This characteristic parallels other lifestyle-related diseases such as hypertension, diabetes mellitus, and dyslipidemia, wherein diagnosis is impossible without specific testing.

Even if you are careful about bone aging by measuring bone density, even if you exercise moderately, even if you have no symptoms, you

H. Ohta (✉)
Department of Obstetrics and Gynecology, Kawasaki Medical Center, Okayama, Japan
e-mail: ohta-h@feel.ocn.ne.jp

H. Yamada, Y. Naito (eds.), *Anti-Aging Medicine*, https://doi.org/10.1007/978-981-97-8514-8_132

cannot deny the possibility of osteoporosis. First, you need to measure bone density in osteoporosis screening and check the current state of your bones. The standard method for measuring bone density is the DXA (dual-energy X-ray absorptiometry) method for the lumbar spine or proximal femur total, and this should be used as much as possible. Since vertebral deformities appear with age, the accuracy is lacking, so the proximal femur total (not the neck), which is the international standard as a measurement site, is desirable after the 70s. Forearm DXA method and second metacarpal MD (microdensitometry) method can also diagnose, but there is a possibility of delay in early diagnosis because they are peripheral bones. Also, the QUS (quantitative ultrasound) method, which measures the calcaneus with an ultrasound index, can screen, but cannot diagnose. The fact that the metacarpal and calcaneus are not bones that fracture due to osteoporosis speaks to their diagnostic value. In the case of Japanese women over 50, the lifetime fracture rate is highest at 37% for the vertebrae [3], followed by 22% for the proximal femur [4]. This shows the significance of measurements in the lumbar spine and proximal femur, which have the highest fracture probability due to osteoporosis.

132.3 Evaluation of Measurement Results

In explaining the results of bone density measurements, the most frequently asked question is, "What age does my bone density correspond to?" Since the average Japanese woman has a YAM (Young Adult Mean) of 70% or less at the age of 75 (half of them develop osteoporosis) (Figs. 132.1 and 132.2), almost everyone falls into this category, so the situation is not serious. However, it becomes significantly serious for those under the age of 50. Young people who develop osteoporosis can easily understand that their bone density is low, but as they age, they tend to think that their values are appropriate for their age because everyone in their age group has osteoporosis, which is a big mistake. Bone density does not decrease indefinitely, it stops decreasing at a certain point and does not decrease any further. This is because bones need to maintain a certain level of bone density regardless of age in order to function as the body's support structure. There are two standardized ways to express bone density. The WHO standard sets the cut-off value as the young adult mean minus 2.5SD [2]. On the other hand, in our country, the cut-off value is set as less than 70% of the young

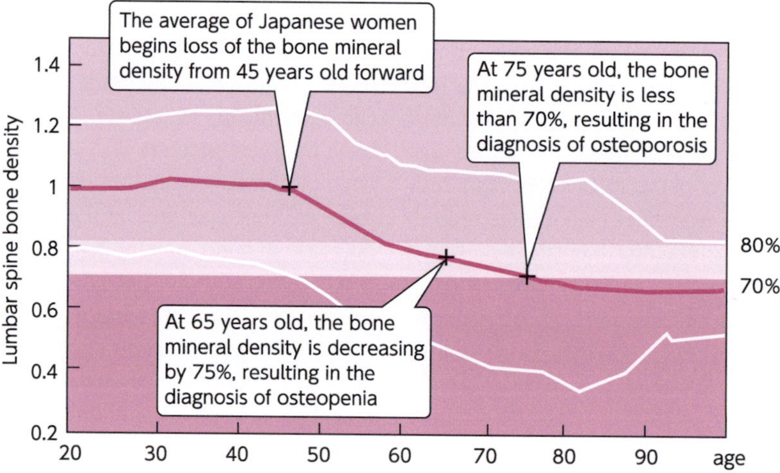

Fig. 132.1 Lumbar spine bone density measurement results table. The average Japanese woman starts to lose bone density from the age of 45. At 65, bone mass decreases by 75%. At 75, they develop osteoporosis with less than 70% bone density. Lumbar spine bone density. [Created by the author from the Diagnostic Criteria for Primary Osteoporosis (1996 Revised Edition) by the Japan Bone Metabolism Society Osteoporosis Diagnostic Criteria Study Committee]

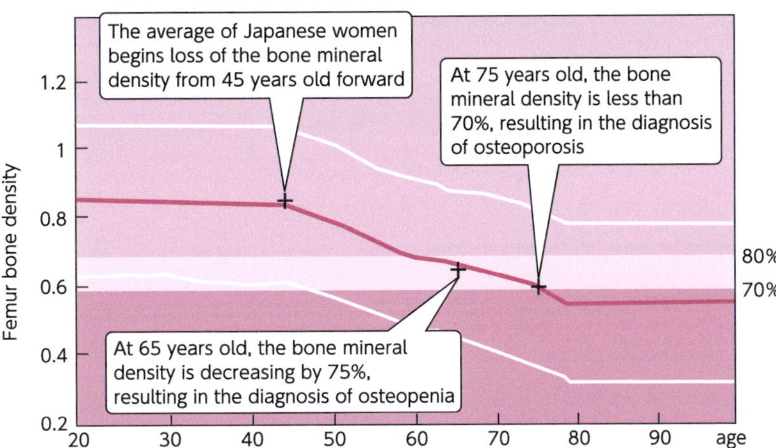

Fig. 132.2 Femur bone density measurement results table. Guide to Bone Salt Quantitative Testing Equipment 2000 Edition. Osteoporosis Japan: Journal of the Japanese Osteoporosis Society 2000; 8: 309–325. (Created by the author)

adult mean (−2.3~−2.6SD of young adults). This is because the frequency of fractures increases when the value falls below this level, so anything below this is defined as osteoporosis. The reason why our country sought a cut-off value in a different way from the WHO is that at that time, many different bone density measuring devices had already been introduced in our country, and if SD was adopted, there would be a problem that more people would be diagnosed with osteoporosis than actually. However, if it is %YAM, there is no such disadvantage, it is possible to set a common standard for all models, and moreover, it is easier to explain %YAM than SD, so we decided to set %YAM. Furthermore, regarding the *T*-score (value) and *Z*-score (value), the former set the YAM as 20–44 years old in the lumbar spine, but in the proximal femur, our research showed that the rate of bone density decrease after the 20s is larger than that in the lumbar spine [4, 5], and we decided to set the YAM standard as 20~29 years old for the purpose of aligning with international standards. The *T*-value is a comparison number with the maximum bone mass and the young adult mean that is almost maintaining it, and the *Z*-value explains how it compares with people of the same generation (Table 132.1). The former is essential for the diagnosis of osteoporosis, but the *Z*-value shows the positioning within the same generation.

Table 132.1 Terms for describing bone density measurement results

%YAM (% young adult mean)	Describes what percentage of the young adult average. The variation in the measurement site can be ignored. The cut-off value is 70%
YAM-2.5SD	The idea of using the value subtracted 2.5SD from the young adult average as the cut-off value (WHO standard). The number of patients below the cut-off value varies greatly depending on the measurement site
T-score	(Subject's measured value − YAM value) ÷ (Young adult's SD value) The value obtained by dividing the bone density by the standard deviation (SD value) of young adults compared to the young adult average
Z-score	(Subject's measured value − same age average value) ÷ (Same age SD value) The value obtained by dividing the bone density by the standard deviation (SD value) of the same age compared to the average value of the same age as the subject

References

1. Ota H, Ota I. Please tell me about the future prospects of bone metabolism markers using specific examples. Bone metabolism marker handbook. Tokyo: Medical Review Company; 2022.
2. NIH Consensus Development Panel on Osteoporosis Prevention, Diagnosis, and Therapy

(Klibanski A, Adams-Campbell L, et al). Osteoporosis prevention, diagnosis, and therapy. JAMA. 2001;285:785–95.

3. Fujiwara S, Kasagi F, et al. Fracture prediction from bone mineral density in Japanese men and women. J Bone Miner Res. 2003;18:1547–53.
4. Hagino H, Yamamoto K, et al. Changing incidence of hip, distal radius, and proximal humerus fractures in Tottori Prefecture, Japan. Bone. 1999;24:265–70.
5. Orito S, Kuroda T, et al. Age-related distribution of bone and skeletal parameters in 1,322 Japanese young women. J Bone Miner Metab. 2009;27:698–704.

133 Antiaging Checkup: Hormonal Age

Shigeo Horie

133.1 Hormonal Age

The endocrine system, which is responsible for the production and secretion of hormones, is involved in the activities of all parts of the body, performing control and coordination. The main glands of the endocrine system include the hypothalamus, pituitary gland, thyroid gland, parathyroid gland, islet cells, adrenal gland, testes, and ovaries. As we age, the blood concentration of many hormones decreases, but there are also hormones that maintain the same concentration or even increase [1]. Sensitivity of hormone receptors also decreases with age, so it is considered that overall endocrine function decreases. In clinical practice, it is difficult to evaluate the sensitivity of receptors, so the evaluation of hormonal age is basically done by measuring the blood concentration of hormones. Hormones known to decrease secretion with age include estrogen, testosterone, dehydroepiandrosterone (DHEA), growth hormone, and melatonin. This section will explain these representative hormones.

S. Horie (✉)
Department of Urology, Advanced Informatics of Genetic Diseases, Digital Therapeutics, Juntendo University Graduate School of Medicine, Tokyo, Japan
e-mail: shorie@juntendo.ac.jp; secretary-uroprof@juntendo.ac.jp

133.2 Sex Hormones

133.2.1 Estrogen

Estrogen levels peak in the 20s, fluctuating periodically in relation to the menstrual cycle, and then drop sharply when menopause occurs around the age of 45–50. When menopause occurs, in addition to cosmetic problems such as skin and vaginal atrophy, dyslipidemia, bone loss, hypertension, and obesity may appear. Also, as ovarian function decreases, the state of decreased hormone secretion continues, leading to an increase in follicle-stimulating hormone (FSH) and luteinizing hormone (LH) from the pituitary gland due to positive feedback. Estradiol (E2) concentration measurement is mainly used to understand estrogen levels, but the significance of measuring E2 levels is limited, as E2 levels in postmenopausal women are often significantly reduced, except for understanding menopause and menopausal states, and the relationship with aging and disease. In men, testosterone is converted to E2 by aromatase. It is said that E2 levels remain unchanged or decrease slowly with age, but the clinical significance of measurement is limited.

H. Yamada, Y. Naito (eds.), *Anti-Aging Medicine*, https://doi.org/10.1007/978-981-97-8514-8_133

133.2.2 Testosterone

Testosterone plays many important physiological roles, affecting muscles, bones, central nervous system, prostate, bone marrow, skin, and sexual function. According to a previous survey by the Japanese Urological Association [2], the serum free testosterone concentration decreases by 1.6 pg/mL (9.2%) every 10 years after the 20s, and since the publication of the "Guide to the Treatment of Late-Onset Hypogonadism (LOH Syndrome)" in 2007 [3], late-onset hypogonadism (late onset in the diagnosis of hypogonadism and Late-Onset Hypogonadism (LOH), it has been recommended to measure free testosterone levels in the diagnosis of LOH). On the other hand, in the West, total testosterone levels have been used in the diagnosis of LOH syndrome. In recent years, there has been much discussion about diagnostic guidelines in our country, and in the 2022 revision of the "Guide to LOH Syndrome (Aging Male/Hypogonadism) Treatment," [4] a total testosterone level of 2.5 ng/mL has been established as the standard. In LOH syndrome, patients often experience symptoms such as irritability, insomnia, anxiety, depression, fatigue, tiredness, lethargy, erectile dysfunction, and decreased libido. The diagnostic criteria vary from country to country, and a consistent view has not been reached. Serum testosterone levels should be used as a reference, and appropriate treatment should be performed based on individual patient symptoms and signs. It is known that testosterone levels fluctuate throughout the day, and it is recommended to take blood samples between 7 and 11 a.m., regardless of meals.

133.3 DHEA

DHEA is a hormone mainly secreted from the adrenal glands, which is metabolized to testosterone and further converted to estrogen by aromatase. Along with cortisol, another adrenal cortical hormone, it is also known as an antistress hormone. Cortisol secretion increases with stress, which generates reactive oxygen species and is associated with the oxidation of deoxyribonucleic acid (DNA). DHEA plays a role in preventing this oxidation, which is why it is known to the world as an anti-aging hormone. In the blood, DHEA mostly exists as a sulfate (DHEA-sulfate; DHEA-S). Serum DHEA-S concentration increases significantly after puberty, peaks in the 20s, and then decreases linearly with age (Table 133.1). Regression formulas for men: $Y = -31.922X + 3346$, and women: $Y = -19.031X + 1974$ [Y: serum DHEA-S concentration (ng/mL), X: age (years)] have also been shown.

Table 133.1 Age-related changes in serum DHEA-S concentration [5]

Age group	Mean		95% CI	*n*
Men (*n* = 2006)				
40–49	1894.70	±	88.5	331
50–59	1576.80	±	55.5	659
60–69	1322.30	±	49.7	713
70–79	995.1	±	59.9	303
Women (*n* = 1307)				
40–49	1128.00	±	75.5	234
50–59	928.2	±	44.9	419
60–69	739.1	±	36.6	473
70–79	605.2	±	51.5	181

CI confidence interval

133.4 Growth Hormone

Growth hormone (GH) is regulated by somatostatin and GH-releasing hormone in the hypothalamus, and through the production of insulin-like growth factor 1 (IGF-1), it exerts growth-promoting effects, insulin-like effects, and promotes cell proliferation and differentiation in various organs. Both GH secretion ability and blood IGF-1 concentration decrease slowly with age. Measurement of GH is useful in the diagnosis of growth hormone deficiency short stature and acromegaly, but its clinical significance as an evaluation of aging is limited.

133.5 Melatonin

Melatonin is a hormone secreted from a part of the brain called the pineal gland, and it is known to decrease from morning to daytime and increase at night. It is also one of the typical hormones whose secretion decreases with age. It is well known as a sleep hormone because it plays a significant role in sleep rhythms. Sleep disorders associated with aging are related to circadian rhythm disorders due to decreased melatonin, and are significantly involved in serious problems in geriatric medicine such as urinary disorders and falls. The specificity of the test is low, and it is rarely measured alone. When measuring, it is necessary to confirm the history of supplement and sleep inducer intake.

References

1. Takayanagi R. Endocrine and metabolic diseases. In: Geriatric medicine text. 3rd ed. Tokyo: Medical View Society; 2008. p. 472–4.
2. Iwamoto A, Yanase T, et al. Setting the reference values for total testosterone and free testosterone in Japanese adult males. J Jpn Soc Endocrinol. 2004;95:751–60.
3. The Japan Urological Association/Japan Men's Health Medical Association "LOH Syndrome Treatment Guidelines" Study Working Committee, editor. Guide to the treatment of aging male gonadal dysfunction syndrome (LOH syndrome). Tokyo: Jihou; 2007.
4. The Japan Urological Association/Japan Men's Health Medical Association LOH Syndrome (Aging Male Gonadal Dysfunction) Treatment Guide Creation Committee, editor. Guide to the treatment of LOH syndrome (aging male gonadal dysfunction). Tokyo: Medical Book Publishing; 2022.
5. Nomoto K, Arita S, et al. Development of a model of functional endocrine age in Japanese people-serum dehydroepiandrosterone-sulfate (DHEA-s) concentration as an index of aging-anti-aging. Medicine. 2011;8:69–74.

134 Antiaging Dock: Muscle Age

Hiroaki Ohta

134.1 Basic Knowledge of Muscles

There are over 600 muscles of various sizes in the human body, playing a crucial role in maintaining life activities. Based on their structure and function, they are classified into three types: skeletal muscles, which make up about 40% of the muscles and can be increased through exercise; cardiac muscles, which move the heart; and smooth muscles, which move the digestive tract and blood vessels, promoting digestion and blood flow. Muscles not only move the body, such as walking, running, and sitting, through the contraction and relaxation of skeletal muscles, but also protect organs and bones from external shocks, generate heat to increase basal metabolism and prevent lifestyle-related diseases, maintain body temperature, and promote the circulation of blood and lymph fluid, among other functions. The calf contains a large amount of muscle and has the function of pushing back the blood that has circulated to the leg, which is the farthest from the heart. For this reason, the calf is also referred to as the second heart. Muscles are also related to immunity, and immune cells, including lymphocytes, are activated by the amino acid glutamine, which is stored in large amounts in muscles. Therefore, it is said that if muscle mass decreases, immune function may decline. Consequently, a decrease in muscle mass may lead to physical discomfort. Skeletal muscles produce several types of hormones, collectively referred to as myokines. The first myokine was discovered in 2003, and more than 50 types of myokines have been identified to date, all of which are secreted from skeletal muscles through exercise and exert various systemic effects [1]. Their typical functions include (1) formation and regeneration of muscles and bones, (2) antiinflammatory effects, (3) involvement in carbohydrate and lipid metabolism, and (4) protection of cardiac muscle cells and vascular endothelial cells.

H. Ohta (✉)
Department of Obstetrics and Gynecology, Kawasaki Medical Center, Okayama, Japan
e-mail: ohta-h@feel.ocn.ne.jp

134.2 Muscle Evaluation Indicators: Muscle Mass and Strength

Initially, only the decrease in muscle mass was considered as an evaluation indicator for muscles, but with the establishment of measuring skeletal muscle mass from lean body mass using the bone density measurement method, dual-energy X-ray absorptiometry (DXA), it became possible to observe the temporal changes in skeletal muscle mass. As a result, it was found that the age-related changes in muscle mass reduction and strength decline do not necessarily progress in parallel, and the age-related changes in muscle

H. Yamada, Y. Naito (eds.), *Anti-Aging Medicine*, https://doi.org/10.1007/978-981-97-8514-8_134

strength are more pronounced than those in muscle mass [2]. Therefore, the traditional muscle function evaluation standard was only the decrease in skeletal muscle mass, but with the accumulation of clinical research, it is now considered that in addition to the age-related decrease in skeletal muscle mass, there is also a decrease in skeletal muscle strength. The European Working Group on Sarcopenia in Older The concept of the European Working Group on Sarcopenia in Older People (EWGSOP) considers the reduction of skeletal muscle mass as an essential factor, along with either a decrease in muscle strength (measured by grip strength) or a decrease in physical function (measured by walking speed) [3]. Methods for measuring muscle mass and fat mass include computed tomography (CT), magnetic resonance imaging (MRI), DXA method, and bioelectrical impedance analysis (BIA method). Among these, the measurement of muscle mass by the BIA method using a body composition analyzer has been improving in accuracy in recent years, and because it is simple and noninvasive, it can be measured anywhere [4]. Therefore, data obtained from body composition analyzers, including muscle mass, are contributing in various clinical settings such as antiaging clinics.

134.3 Age-Related Changes in Muscle Mass by Body Part

There are racial differences in the changes in muscle mass with aging, so it is necessary to clarify this from studies targeting Japanese people. If we can clarify the changes in muscle mass, it will not only enable efforts to prevent muscle mass reduction, but also be useful for diagnosing sarcopenia and levels related to decreased life function. In Japan, Tanimoto and others have used the multifrequency body composition analyzer MC-190 (Tanita Corporation) to survey the muscle mass of the upper limbs, lower limbs, trunk, and whole body in a standing position for over 4000 subjects, including 1702 men and 2301 women [5]. As a result, the coefficient of determination (R2) was larger for the quadratic regression equation than for the linear regression equation in all parts for both men and women, so they showed the relationship between muscle mass and age using the quadratic regression equation. The negative slope of the regression equation was larger for men than for women in all parts. In other words, the decrease in muscle mass with aging is greater in men than in women. Furthermore, it was shown that the negative slope of muscle mass reduction increases with age for both men and women. When looking at each part, Fig. 134.1 shows that the decrease in upper limb muscle mass is relatively gentle, and in particular, women remained flat until around 60 years old. Figure 134.2 shows that the muscle mass of the lower limbs decreased with a large slope from the 20s for both men and women. Figure 134.3 shows that the trunk muscle mass increased gently until around 45 years old in men and around 50 years old in women, and then started to decrease. Figure 134.4 shows that the total body muscle mass increased slightly until around 40 years old in men, then decreased, and in women, it remained flat until around 50 years old, and then decreased. As shown above, it has been found that the age-related changes in muscle

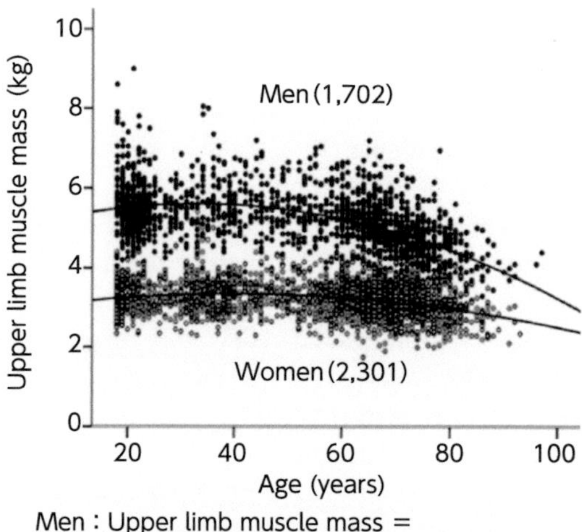

Fig. 134.1 Changes in upper limb muscle mass with age. (Reprinted with permission from [5])

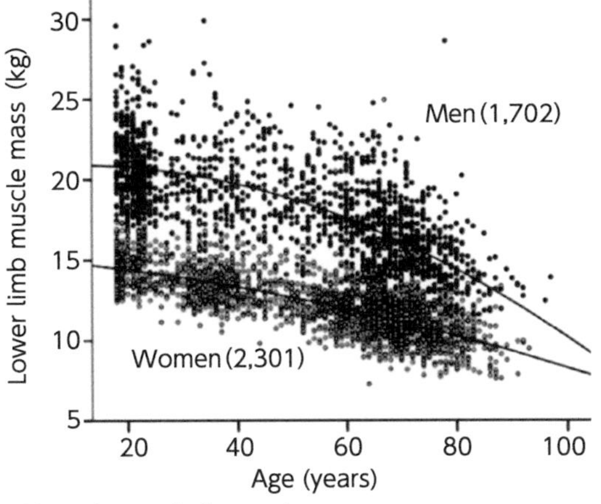

Men : Lower limb muscle mass = $0.025(\text{age}) - 0.0013(\text{age}^2) + 20.79$

$R^2 = 0.491$ ($p < 0.001$)

Women : Lower limb muscle mass = $-0.027(\text{age}) - 0.0004(\text{age}^2) + 15.08$

$R^2 = 0.553$ ($p < 0.001$)

Fig. 134.2 Changes in lower limb muscle mass with age. (Reprinted with permission from [5])

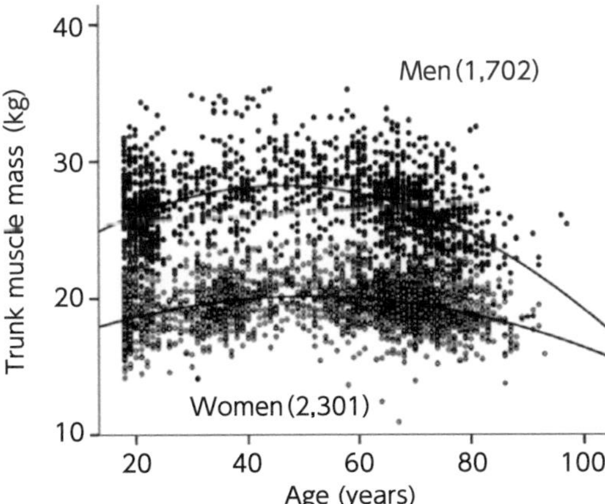

Men : Trunk muscle mass = $0.288(\text{age}) - 0.0031(\text{age}^2) + 21.54$

$R^2 = 0.153$ ($p < 0.001$)

Women : Trunk muscle mass = $0.162(\text{age}) - 0.0016(\text{age}^2) + 16.03$

$R^2 = 0.553$ ($p < 0.001$)

Fig. 134.3 Changes in trunk muscle mass with age. (Reprinted with permission from [5])

mass vary by gender and by part. In particular, the muscle mass of the lower limbs decreases from the 20s for both men and women, and it is

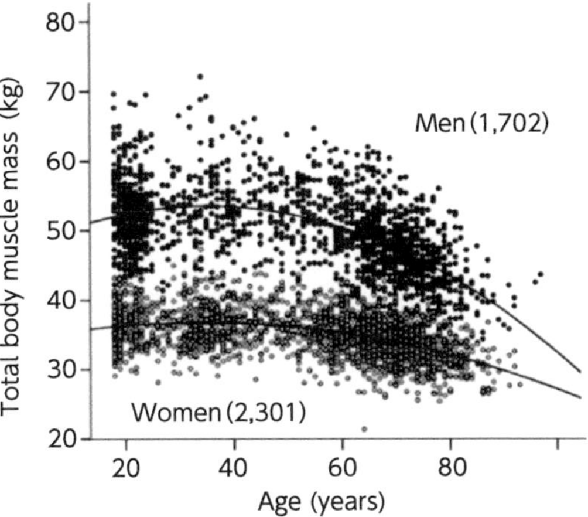

Men : Total body muscle mass = $0.352(\text{age}) - 0.0050(\text{age}^2) + 47.28$

$R^2 = 0.303$ ($p < 0.001$)

Women : Total body muscle mass = $0.154(\text{age}) - 0.0022(\text{age}^2) + 34.07$

$R^2 = 0.229$ ($p < 0.001$)

Fig. 134.4 Changes in total body muscle mass with age. (Reprinted with permission from [5])

necessary to deal with this in anti-aging clinics. The Locomo Challenge! Promotion Council recommends the following as training for this lower limb muscle strength: (1) Balance ability: standing on one leg with eyes open, (2) Lower limb strength: squats, (3) Calf muscle strength: heel (calf) raises, and (4) Lower limb flexibility, balance ability, muscle strength: front lunges.

References

1. Lee JH, Jun HS. Role of myokines in regulating skeletal muscle mass and function. Front Physiol. 2019;10:42.
2. Morley JE, Baumgartner RN, et al. Sarcopenia. J Lab Clin Med. 2001;137:231–43.
3. Cruz-Jentoft AJ, Baeyens JP, et al. Sarcopenia: European consensus on definition and diagnosis: report of the European Working Group on Sarcopenia in Older People. Age Ageing. 2010;39:412–23.
4. Janssen I, Heymsfield SB, et al. Estimation of skeletal muscle mass by bioelectrical impedance analysis. J Appl Physiol. 2000;89:465–71.
5. Tanimoto Y, Watanabe M, et al. Characteristics of aging muscle mass in Japanese. J Jpn Geriatr Soc. 2010;47:52–7.

135 Antiaging Checkup: Perceived (Skin) Age

Hiroyuki Ohjimi

Biomarkers of aging are used to quantify changes in the aging process and determine biological age. Among them, perceived age is an easily accessible biomarker of aging, and it has been shown to correlate with biological age and DNAm (DNA methylation) age. In twin studies, perceived age was significantly associated with survival rate. It also correlated with important functional and molecular aging phenotypes such as muscle strength, cognitive function, and the length of leukocyte telomeres, a molecular biomarker of aging [1]. About 40% of perceived age is said to vary due to nongenetic factors. Sun exposure, smoking, and low BMI promote facial aging, while high social status, low depression scores, and being married are reported to delay it. However, the strength of these associations varies by gender. Many research studies suggest that perceived age is a better biomarker of aging than chronological age. Women who look younger for their age have larger lips, avoid sun exposure, and have genetic factors that prevent the onset of gray hair and skin wrinkles. Signs of facial aging are expressed as "spots, wrinkles, and sagging." Spots refer to changes in skin tone on the surface, indicating acquired melanin pigmentation disorders, including solar lentigo (senile pigmentation spots), melasma, freckles, bilateral late-onset Ota's nevus (dermal melanocytosis), and postinflammatory hyperpigmentation. Wrinkles indicate changes in skin texture (pores) on the surface, and sagging indicates drooping due to gravity. In this section, we will discuss the terms related to "wrinkles and sagging," which are signs of aging, and then describe the classification methods and degree classification for wrinkles and sagging.

135.1 Terms for Wrinkles and Sagging

We will explain the terms related to "wrinkles and sagging," which represent signs of facial aging [2] (see Fig. 135.1). In the upper face, horizontal forehead lines, glabellar frown lines, horizontal procerus wrinkles, sunken eyes, and dermatochalasis occur. In Japanese (East Asian) people, there is a tendency for blepharoptosis to occur with aging, narrowing the palpebral fissure. In the mid-face, a nasojugal groove (tear trough depression) occurs on the inner lower eyelid, and a palpebromalar groove forms in an arc along the lower edge of the orbit. The protrusion above this is called a baggy eyelid. On the outer lower eyelid, crow's feet appear. A mid-cheek junction (Glogau's line) may appear on the front of the cheekbone. With aging, nasolabial folds (smile lines) become deeper and clearer, and perioral wrinkles (mouth wrinkles) appear around

H. Ohjimi (✉)
Department of Plastic, Reconstructive and Aesthetic Surgery, Kitasato University, Kanagawa, Japan
e-mail: plast1980@ohjimi-med.com

H. Yamada, Y. Naito (eds.), *Anti-Aging Medicine*, https://doi.org/10.1007/978-981-97-8514-8_135

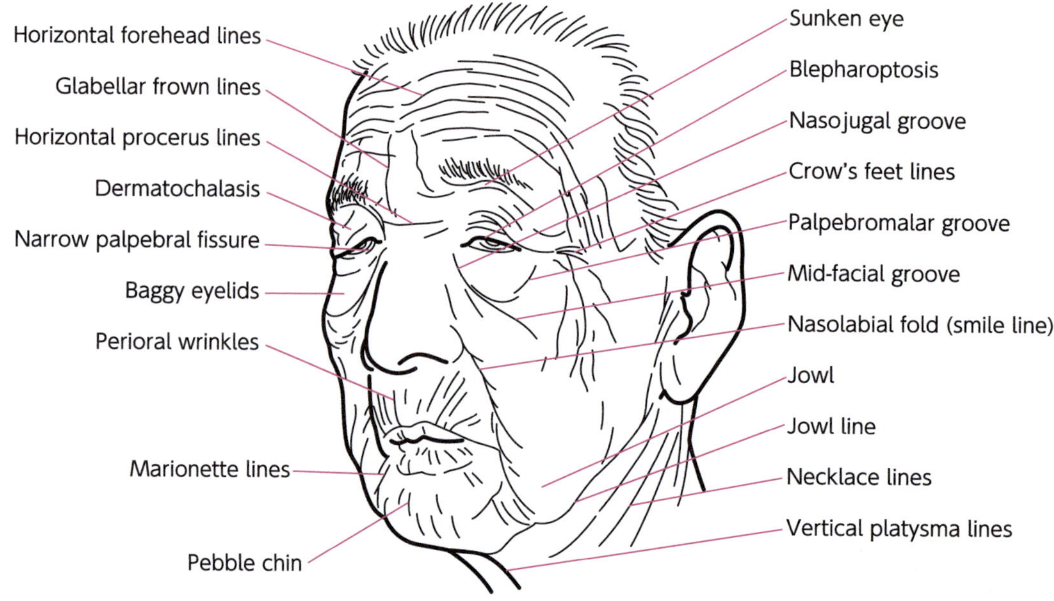

Fig. 135.1 Terms for facial wrinkles and sagging. (Modified with permission from [2])

the lips lines occur. In the lower face and neck, melomental folds, also known as marionette lines, appear from the corners of the mouth to the chin, and on the outside, a bulge called a jowl forms, creating a jowl line on the contour of the lower jaw. In the chin area, a pebble chin or orange peel chin may occur. In the neck, characteristic features are necklace lines and vertical platysma lines, also known as turkey gobble neck [2] (Fig. 135.1).

135.2 Types and Classification of Wrinkles and Sagging

Facial wrinkles and sagging are classified by depth, shape, cause, and whether they are dynamic or static. The general classification by depth is "disorder of skin texture, fine lines, wrinkles, deep wrinkles, sagging." The classification by shape includes Kligman's "crinkle wrinkle, glyphic wrinkle, linear wrinkle." The classification by cause includes Pierard's "atrophy wrinkles, elastic wrinkles, expression wrinkles, gravity wrinkles". Atrophy wrinkles are biomarkers of aging, used to quantify changes in the aging process and determine biological age. Among them, perceived age is an easily accessible biomarker of aging, and it has been shown to correlate with biological age and DNAm (DNA methylation) age. In twin studies, perceived age was significantly associated with survival rate. It also correlated with important functional and molecular aging phenotypes such as muscle strength, cognitive function, and the length of leukocyte telomeres, a molecular biomarker of aging. About 40% of perceived age is thought to vary due to nongenetic factors. Sun exposure, smoking, and low BMI promote facial aging, while high social status, low depression scores, and being married delay it. However, the strength of these associations varies by gender. Many research studies suggest that perceived age is a better biomarker of aging than chronological age. Women who look younger for their age have larger lips, avoid sun exposure, and have genetic factors that prevent the onset of gray hair and skin wrinkles. The signs of facial aging are expressed as "spots, wrinkles, sagging." Spots refer to changes in skin tone, indicating acquired melanin pigmentation disorders, including solar lentigo (senile lentigo), melasma, freckles, bilateral delayed Ota's nevus (dermal melanocytosis), and postinflammatory hyperpigmentation. Wrinkles indicate changes in skin texture, and sagging indicates drooping due to gravity. This section discusses the terms related to "wrinkles and sagging," which are signs of aging, and then discusses the classification and degree of wrinkles and sagging. Terms for Wrinkles and Sagging. This section explains the terms related to "wrinkles and sagging," which are signs of facial aging (Fig. 135.1). In the upper face, there are horizontal forehead lines and glabellar frown lines, horizontal procerus wrinkles at the root of the nose, sunken upper eyelids, and dermatochalasis (loose skin on the upper eyelids) occur. In Japanese (East Asians), there is a tendency for blepharoptosis (drooping of the eyelids) with aging, and the palpebral fissure narrows. In the mid-face, a nasojugal groove (tear trough depression) occurs on the inner side of the lower eyelid, and a palpebromalar groove is formed in an arc along the lower edge of the orbit. The protrusion above it is called a baggy eyelid. On the outer side of the lower eyelid, crow's feet appear. On the anterior surface of the cheekbone, a mid-cheek junction (Glogau line) may appear. With aging, nasolabial folds (smile lines) become deeper and clearer, and perioral wrinkles (mouth frown) occur. In the lower face and neck, melomental folds (marionette lines) appear from the corners of the mouth to the jaw, and a bulge called a jowl occurs on the outside, creating a jowl line on the contour of the lower jaw. In the chin area, a pebble chin or orange peel chin may occur. In the neck, necklace lines and vertical platysma lines (turkey gobble neck) are characteristic (Fig. 135.1). Types and Classification of Wrinkles and Sagging Facial wrinkles and sagging are classified by depth, shape, cause, and whether they are dynamic or static. The general classification by depth is "disorder of skin texture, fine wrinkles, wrinkles, deep wrinkles, sagging." The classification by shape includes Kligman's "crinkle wrinkle, glyphic wrinkle, linear wrinkle." The classification by cause includes Pierard's "atrophy wrinkles, elastic wrinkles, expression wrinkles, gravity wrinkles". Atrophy wrinkles are due to skin atrophy, elastic wrinkles are due to solar elastosis, expression wrinkles are due to contraction of the facial muscles, and gravity wrinkles are due to skin laxity caused by gravity. Glogau distinguishes between wrinkles caused by facial muscle movement and those that exist even at rest. The Fitzpatrick classification, Alexiades-Armenakas classification, and Callaghan classification classify wrinkles and sagging by degree. Bazin's classification, the Cosmetic Science Society of Japan's Cosmetic Function Evaluation Guidelines, and Lemperle's facial wrinkle classification numerically classify each wrinkle and sagging by degree, using it to evaluate aging, treatment, and preventive effects [3].

135.3 Relationship Between Age and Spots, Wrinkles, and Sagging

Bazin and others conducted a study on 400 Asian women aged 18 to 80, quantified facial aging using a photo scale, and examined its relationship with age. Seventeen signs of facial aging related to wrinkles, sagging, and pigment uniformity (spots) were evaluated. Wrinkles were evaluated in terms of horizontal forehead lines, vertical lines between the eyebrows, crow's feet, vertical lines on the upper lip, marionette lines, under the eyes (nasolabial and eyelid-cheek grooves), between the inner corners of the eyes (nasal root), nasolabial folds (laugh lines), necklace lines, and neck skin atrophy. Sagging was evaluated in terms of eye bags, lower face sagging (jowls), sagging on the outer upper eyelid (upper eyelid skin laxity), neck texture, and neck sagging (submandibular neck angle). For spots, the density of cheek spots and the contrast of spots near the cheekbones were evaluated. Each of the aging signs evaluated changed with age, but the rate of progression varied by sign. Signs appeared earliest on the upper part of the face, especially around the eyes. Signs of aging on the lower part of the face appeared after the age of 45, but then progressed rapidly. Also, it was shown that spots and neck aging begin gradually from an early age of 30. Nasolabial and eyelid-cheek grooves, which are wrinkles on the lower eyelid, appear from the early 20s. Also, East Asians have characteristics such as more pronounced aging changes on the upper part of the face compared to Westerners. This is mainly due to two factors: upper eyelid skin laxity and regressive eyelid ptosis, the former due to excess skin on the upper eyelid and the latter due to laxity of the levator muscle of the upper eyelid. As both progress, the upper eyelid covers the cornea when the eye is open, limiting the field of vision. As a result, the frontalis muscle is unconsciously tensed, leading to the appearance of characteristic upper facial aging features in East Asians, such as horizontal forehead lines, eyebrow elevation, and upper eyelid hollowing.

References

1. Christensen K, Thinggaard M, et al. Perceived age as clinically useful biomarker of ageing: cohort study. BMJ. 2009;339:b5262.
2. Japan Anti-Aging Medical Association Specialist and Instructor Certification Committee. The basics and clinical practice of anti-aging medicine. 3rd ed. Tokyo: Medical View Co., Ltd; 2015. p. 146–53.
3. Yamada H. Types and treatment methods of wrinkles. Cosmetic dermatology guidebook. 2nd ed. Tokyo: Chugai Medical Co., Ltd.; 2019. p. 62–75.
4. Bazin R, Flament R. Introduction to the atlas. Asian ed, Skin aging atlas, vol. 2. Paris: Publications Puce et Plume; 2013. p. 17–28.
5. Daijiya H, et al. Skillful aging around the eyes. Tokyo: All Japan Hospital Publishing Association; 2021. p. 5–12.

Antiaging Checkup: Evaluation of Oral Age

136

Kenji Matsushita and Takayoshi Sakai

136.1 Current Status and Challenges of Oral Age Evaluation Method

The word "loss of teeth" appears in Confucius's "Analects." It is a term that signifies the end of life. Since the era before the Common Era, losing teeth was synonymous with impending death. The positive correlation between the number of teeth retained and lifespan, and the increase in the incidence of periodontal disease with age, led to the development of an oral age index that comprehensively evaluates teeth and periodontal tissues. This index is used to assess an individual's oral health status and to evaluate dental health services at each workplace. The index is calculated by applying the number of healthy teeth of the survey subject and the number of segments with a Community Periodontal Index (CPI) code of 0 to a formula. This allows for the understanding of regional disparities in oral health status and the characteristics of dental patients in each region. It also raises awareness about oral health among patients, leading to an increase in dental visits. Liang and others have used the current number of teeth, CPI, biting force, and repetitive saliva swallowing test to evaluate oral age.

The Resting Salivary Secretion Test (RSST), quiet saliva, stimulated saliva, and the correlation between the number of tongue coating fungi (Candida count) and age are each analyzed, and based on these data, a formula for calculating oral age is derived, and the degree of oral aging of the patient is evaluated [1]. This method is an attempt to evaluate the overall health of the oral cavity, including oral function. In recent years, the Japan Gerodontology Society has proposed that "oral frailty" and "oral functional decline" exist in the process of transitioning from health to oral functional impairment, and that seven tests (oral uncleanliness, oral dryness, decreased biting force, decreased tongue-lip movement function, low tongue pressure, decreased mastication function, and decreased swallowing function) are conducted, and if three or more of these are applicable, it is diagnosed as "oral functional decline." And, a method has been proposed to easily calculate the oral functional age (mouth age) from the test items (biting pressure, remaining tooth number, oral diadochokinesis, tongue pressure, and mastication efficiency) that had a significant correlation between the test results of oral functional decline in the elderly and age [2]. This method is

K. Matsushita
Department of Oral Disease Research, National Center for Geriatrics and Gerontology, Obu-shi, Aichi, Japan

Department of Dental Hygiene, Ogaki Women's College, Ogaki, Gifu, Japan

T. Sakai (✉)
Department of Rehabilitation for Orofacial Disorders, Osaka University Graduate School of Dentistry, Osaka, Japan
e-mail: sakai.takayoshi.dent@osaka-u.ac.jp

H. Yamada, Y. Naito (eds.), *Anti-Aging Medicine*, https://doi.org/10.1007/978-981-97-8514-8_136

considered effective as a way to increase interest in the oral health of the elderly and to encourage them to seek medical attention. On the other hand, many issues to be resolved are also visible. In subjects before middle age, the correlation between each test item and age is low, so it is not possible to derive a valid oral age for this generation of subjects. Also, because there are items where the speed of functional decline differs by gender, it is necessary to derive a valid formula after examining the correlation for each gender. Furthermore, there are many issues with each oral function test method, so it is difficult to calculate an accurate oral function age with the current oral function measurement method.

136.2 The Possibility of Periodontal Age Evaluation from the Perspective of Geroscience

In recent years, it has become clear that chronic inflammation and infection due to periodontitis are independent risk factors for age-related diseases and mortality. On the other hand, the progression of aging varies among individuals, so actual age does not serve as an indicator of each person's physical condition. Recent advances in geroscience have shown that various biomarkers of biological aging are associated with declines in physical function, disease incidence, and mortality due to various age-related diseases. By examining the qualitative and quantitative changes in such biomarkers, it may be possible to assess oral age, such as oral function and periodontal tissues. Aging markers are "biological parameters that predict physical function in the absence of disease, disease onset, and mortality more sensitively than actual age." Candidate biomarkers include telomere length, proteomics and transcriptomics predictive factors, metabolomic age scores, epigenetic clocks, and composite biomarker estimation factors [3]. Among these, the ones clinically verified to be associated with periodontal disease are the composite biomarker estimation factors and telomere length. Meisel et al. proposed a biological aging score to predict individual risk of tooth loss [4]. They showed that by modeling biological age based on indicators of periodontal disease, metabolic syndrome, cardiovascular disorders, and systemic inflammation, they could predict tooth loss more accurately than actual age. In addition, they predicted the risk of periodontitis using a composite age prediction factor based on LDL cholesterol levels, fibrinogen, glycated hemoglobin, blood pressure, waist circumference, and white blood cell count. These factors can cause alveolar bone resorption by inducing mild chronic inflammation throughout the body, which can lead to periodontitis and tooth loss. Telomeres, the terminal noncoding nucleotide sequences of chromosomes, serve as indicators of aging and cell death as they shorten with each cell division. The rate of telomere shortening in peripheral T cells is higher in patients with moderate to severe periodontitis compared to those with healthy or mild periodontitis. The increase in aging T cells with short telomeres may reduce resistance to periodontal bacterial infection, potentially exacerbating periodontitis. It is thought that the accumulation of cellular aging leads to the progression of individual aging. Cellular aging is a complex intracellular mechanism characterized by permanent cell cycle arrest. The increase in aging cells has been observed in the lesion areas of age-related diseases such as cancer, osteoporosis, atherosclerotic diseases, and osteoarthritis, and they promote inflammation through the expression of the senescence-associated secretory phenotype (SASP). Recently, the relationship between periodontal disease and aging cells is becoming clear through studies using in vitro and mice experiments [5, 6]. In the alveolar bone of aged mice, an increase in the messenger RNA (mRNA) of p16Ink4a, a cellular aging marker, was observed. In addition, an increase in the expression of various inflammation-related factors was also observed in the same sample. When lipopolysaccharide (LPS) was added to the aging osteoblast culture system, the expression of interleukin-1α (IL-1α), IL-1β, and IL-6 increased. Cellular aging may enhance inflammation caused by periodontal bacteria. By measuring changes in aging markers related to oral diseases and oral function, including periodontal disease, it may be possible to estimate an accurate biological oral age. However, to do this, it is necessary to clarify the causal relationship between markers and diseases/functions from large-scale epidemiological studies. If age can be estimated based on molecular data, it is possible to detect early deviations from the trajectory of physiological aging and correct them promptly. Also, since periodontal disease is a complication of other chronic inflammatory diseases, determining the periodontal age can also be used for the prevention of these diseases. Evaluating periodontal and oral age from the perspective of geroscience can be a trigger to prevent age-related diseases at the predisease stage and can become a new tool for medical and dental collaboration.

References

1. Yoshikawa T. Anti-aging medicine for the whole body from the mouth. In: Anti-aging medicine—its theory and practice. Tokyo: Diagnosis and Treatment Company; 2006. p. 236–41.
2. Sato Y, Shida T, et al. Proposal of oral function age (mouth age) using the test results of oral function decline. Showa Sch J. 2022;82:104–11.
3. Jylhävä J, Pedersen NL, et al. Biological age predictors. eBioMedicine. 2017;21:29–36.
4. Meisel P, Pink C, et al. Construction of a biological age score to predict tooth loss over 10 years. J Dent Res. 2019;98:1096–102.
5. Furukawa M, Matsuda K, et al. Analysis of senescence in gingival tissues and gingival fibroblast cultures. Clin Exp Dent Res. 2022;8:939–49.
6. Aquino-Martinez R, Khosla S, et al. Periodontal disease and senescent cells: new players for an old oral health problem? Int J Mol Sci. 2020;21:7441.

Antiaging Checkup: Gut Age

137

Yuji Naito

137.1 Factors Affecting the Gut Microbiota

Omics analysis combining metagenomic information on the gut microbiota and various background information is providing important information on what factors influence the gut microbiota and how it is formed. The results of a study conducted in the Netherlands are significant [1]. They performed shotgun metagenomic sequence analysis on the gut microbiota of 8208 Dutch people from a three-generation cohort including 2756 families, collecting information on classification, functional pathways, antibiotic resistance, pathogenic factors, disease status, geographical and social environment, drug use, blood collection, urinalysis findings, etc., and analyzed the relationship between each item and the bacterial flora. As a result, when clustering was performed using principal coordinate analysis (PCoA), it was observed that *Prevotella copri* (*P. copri*) had a significant influence, and it was also shown that there was a positive correlation with general health and a lower risk of irritable bowel syndrome when *P. copri* was abundant. The impact on the composition of the gut microbiota was explained by 48.6% due to cohabitation, while genetic factors accounted for only 6.6%, and the contribution of stool characteristics, past medical history, used drugs, physical characteristics, etc. was high. Summarizing the factors related to the composition of the gut microbiota (Table 137.1), it is clear that food and nutrition including alcohol, environmental factors in childhood and currently, and socio-economic factors are evident [1]. In particular, as pointed out in previous studies, it is clear that carbohydrates such as sugars, fats, and proteins affect the gut microbiota, and in addition, birth conditions, breastfeeding, and childhood living environment are important environmental factors in the formation of resident bacterial flora. Such basic information provides important insights to consider background factors when performing gut microbiota analysis in relation to disease association, drug susceptibility, health status evaluation, etc. It shows that it is difficult to interpret without excluding the contribution of confounding factors in the analysis where each information is not available, starting from the condition of the stool at the time of collection, the situation in childhood, and the history of medication.

Y. Naito (✉)
Department of Human Immunology and Nutrition Science, Kyoto Prefectural University of Medicine, Kyoto, Japan
e-mail: ynaito@koto.kpu-m.ac.jp

H. Yamada, Y. Naito (eds.), *Anti-Aging Medicine*, https://doi.org/10.1007/978-981-97-8514-8_137

Table 137.1 Environmental factors including diet determine gut microbiota [1]

Diet	Early childhood environmental factors	Adult environmental factors	Socioeconomic factors
Glycemic load Total calorie intake Alcohol Carbohydrates Fat Animal protein Plant protein Probiotics Rural diet score	Maternal smoking during pregnancy Premature birth Mode of delivery Breastfeeding Urban environment in early childhood Parental smoking in early childhood Pet ownership in early childhood	Greenery Vegetation Index Pet ownership NO_2 concentration PM2.5 concentration Urbanity Passive smoking Smoking Past smoking	Income After retirement Full-time housewife or househusband Employed High income

137.2 Characteristics of the Gut Microbiota in the Elderly

Research is underway to clarify the characteristics of the bacterial flora associated with longevity, the characteristics of bacterial gene functions, and the intestinal environment by comparing metagenomic analysis data of the gut microbiota of young people and long-lived people. The gut microbiota is mainly composed of four phyla: Actinobacteria, Bacteroidetes, Firmicutes, and Proteobacteria, which account for more than 90%, and the results comparing the average occupancy rate by age of Japanese people are, for adults in their 20s. In the 60s, the Firmicutes phylum occupies more than 70% of the population, and the ratio of the four phyla shows no significant difference in adults [2, 3]. Previous reports have revealed that the characteristics of the gut microbiota in the elderly, compared to young adults, are characterized by a decrease in the occupancy rate of the Actinobacteria and Firmicutes phyla, an increase in the Bacteroidetes and Proteobacteria phyla, a decrease in diversity, a decrease in carbohydrate-degrading bacteria, and an increase in protein-degrading bacteria [2, 4]. Furthermore, when a principal component analysis of the gut microbiota of 453 healthy Japanese people aged 0–104 was conducted, they were classified into six clusters with age [5]. Cluster 6 is a group of super-elderly people with a median age of 87.0 years, but within the super-elderly group, there are adult groups of clusters younger than the calendar age and elderly groups of age-appropriate clusters, and their comparison was analyzed. As a result, the characteristics of the adult group considered to be healthy were shown to have a high occupancy rate of *Bifidobacterium pseudocatenulatum* and *Blautia wexlerae*, and a low occupancy rate of *Lactobacillus salivarius*, *Streptococcus salivarius*, and *Escherichia coli* [5]. This result means that the colonic environment of the adult group has an extremely low oxygen concentration, and facultative anaerobic bacteria such as *Streptococcus salivarius*, oral bacteria, and *Escherichia coli* (Proteobacteria phylum) are few. We believe this is important information for measuring gut age.

137.3 Gut Age

From the latest academic background on biological age and gut microbiota, we proposed “gut age” (gAge^{TM}: gut clock of aging) as an indicator to estimate biological age from the gut environment [6] (Fig. 137.1). gAge^{TM} can be used as a biomarker in clinical intervention trials targeting humans. Furthermore, by defining gAge^{TM}, it is possible to propose a new strategy for extending healthy life expectancy. Specifically, we are con-

Factors to maintain/restore gAge™

- Increase in gut bacterial diversity
- Bacterial metabolites
  - Polyamines
  - Short-chain fatty acids
  - Bile acids
  - Amino acids
- Healthy diet
  - Mediterranean diet
  - Japanese diet
  - Anti-inflammatory diet
  - Calorie restriction
- Physical activity level

Factors that advance gAge™

- Decrease in gut bacterial diversity
- Decrease in butyrate-producing bacteria
- Decrease in bifidobacteria
- Unhealthy diet
  - High-fat diet
  - High-monosaccharide diet
  - High-salt diet
- Antibiotics, gastric acid secretion inhibitors
- Environmental pollutants
- Food additives, artificial sweeteners
- Urbanization, industrialization

Fig. 137.1 gAge™ (gut clock of aging) [6]

sidering gAge™-clinic, which is a clinical gut age, gAge™-data, which is a data type gut age, and gAge™, which integrates them. Each has items that promote gAge™ and items that inhibit it. The data type gut age is looking to the future where it can be determined by big data such as metagenomic analysis information on gut microbiota and quantitative information on gut metabolites by mass spectrometry. Decreases in α diversity and β diversity of the gut microbiota, increases in the Proteobacteria phylum, increases in the Bacteroides/Prevotella ratio, decreases in butyric acid-producing bacteria, and decreases in bifidobacteria are reported as brake factors, and increases in uniqueness, increases in the Firmicutes/Bacteroides ratio, increases in short-chain fatty acids in feces, and increases in polyamines are reported, and each item is verified and criteria are set to assign −1 (brake), 0, +1 (accelerator) points to each item. Ultimately, it is calculated as follows.

$$\text{Data type gut age}\left(\text{gAge}^{\text{TM}}-\text{data}\right)=\text{Calendar age}+\Sigma\left(\text{each factor point}\right).$$

Although the clinical intestinal age is conceptual at present, Fig. 137.1 suggests possible accelerator and brake items. We plan to verify each one and set criteria for assigning −1 (brake), 0, +1 (accelerator) points to each item. For example, the item "constipation" has been confirmed in both Western and Japanese literature to increase mortality rates, so we assign "+1" to bowel movements less than three times a week and "−1" to bowel movements four or more times a week. Regarding diet, the Mediterranean diet index and antiinflammatory diet index have already been proposed, and some have been proven to improve gut microbiota, so we will assign −1, 0, +1 points based on these indices. Ultimately, the clinical intestinal age (gAge™-clinic) is calculated as chronological age +Σ (each factor point). We plan to evaluate in various fields, revise, and bring it closer to the final form. Considering more clinical use, Fig. 137.2 suggests that an intestinal age check sheet could be useful. Although many items have not been fully

Factors to maintain/restore gAge™ (-1 for each item)	Factors that advance gAge™ (+1 for each item)
( ) I like tofu and deep-fried tofu ( ) I limit my salt intake ( ) I eat whole grains like brown rice and barley at least once in three meals ( ) I often have a bowel movement after breakfast ( ) I am often told I look young ( ) I like fermented foods ( ) Prefers miso soup over other soups ( ) Comes from a rural or regional area ( ) Has three or more siblings ( ) Exercises more than three times a week ( ) Goes to bed before midnight	( ) Skips breakfast four or more days a week ( ) Eats out four or more times a week ( ) Adds sugar to coffee ( ) Drinks alcohol four or more times a week ( ) Thinks they don't eat enough vegetables ( ) Likes meat such as beef, pork, and lamb ( ) Suffers from constipation ( ) Often can't defecate without straining ( ) Often has hard, pellet-like stools ( ) Has been told their flatulence or feces smell bad ( ) Doesn't get enough exercise, whether at work or on days off ( ) Smokes cigarettes ( ) Feels stressed ( ) Doesn't get enough sleep ( ) Struggles with skin problems or acne ( ) Takes medication to suppress stomach acid ( ) Often takes antibiotics ( ) Frequently uses convenience stores ( ) Uses a car for work and shopping

Fig. 137.2 Gut age check sheet [6]

verified, we hope to have many people use it, discuss it, and bring it closer to the final form.

137.4 Conclusion

We believe that the realization of antiaging medicine using biological age is an important issue for our society. The measurement of epigenetic age by quantifying DNA methylation, which is the gold standard, has been delayed due to cost and ethical issues. In such a situation, we believe that intestinal age could be an option as a biological age.

References

1. Gacesa R, Kurilshikov A, et al. Environmental factors shaping the gut microbiome in a Dutch population. Nature. 2022;604:732–9.
2. Odamaki T, Kato K, et al. Age-related changes in gut microbiota composition from newborn to centenarian: a cross-sectional study. BMC Microbiol. 2016;16:90.
3. Takagi T, Naito Y, et al. Differences in gut microbiota associated with age, sex, and stool consistency in healthy Japanese subjects. J Gastroenterol. 2019;54:53–63.
4. Bischoff SC. Microbiota and aging. Curr Opin Clin Nutr Metab Care. 2016;19:26–30.
5. Yoshimoto S, Mitsuyama E, et al. Enriched metabolites that potentially promote age-associated diseases in subjects with an elderly-type gut microbiota. Gut Microbes. 2021;13:1–11.
6. gAge. https://www.gut-clock-gage.com.

138 Antiaging Checkup: Age Evaluation of the Lungs

Koichi Tomoda

138.1 What Is Lung Age?

Various ages such as muscle age, blood vessel age, nerve age, hormone age, etc., are measured in medical checkups, especially in antiaging checkups. These ages, which are quantified as ages compared to actual age to evaluate the decline in function due to aging, are easier to understand than specialized measurement results and are also linked to motivation for health. They are used as evaluation methods for various physical functions from the perspective of preventive medicine. Lung age was proposed by the Japanese Respiratory Society for the purpose of recognizing the importance of lung health and the prevention and treatment of respiratory diseases.

138.2 Respiratory Function Test and Aging

The respiratory function test includes the lung volume fraction measured at rest and the forced vital capacity (FVC) measured by forced exhalation. In the lung volume fraction, the residual volume (RV) increases with age. The total lung capacity (TLC) does not change much with age, and the vital capacity (VC), which is the difference between TLC and RV, decreases with age, and decreases more in men who have a larger increase in RV [1]. The one-second volume (forced expiratory volume in one; FEV1) measured when measuring FVC decreases with age. Smoking accelerates this decline, and when the one-second rate (FEV1%) which is FEV1/FVC falls below 70%, chronic obstructive pulmonary disease (COPD) develops. If you quit smoking, the accelerated decline in FEV1 recovers to the same level as healthy people, and if the decline in FEV1 is mild in the early stages of smoking, it can recover to the level of healthy people by quitting smoking [2]. The awareness of COPD, which is mainly caused by smoking, is less than 30%, and the current situation is that the awareness of the impact of smoking on respiratory diseases and the importance of quitting smoking is not sufficient. Considering these circumstances, lung age was proposed by the Japanese Respiratory Society in 2008 to promote prevention and awareness of respiratory diseases with many undiagnosed and untreated cases, such as COPD, and to help early detection and treatment. Lung age is also printed on the result sheet of spirometry used in respiratory function tests.

138.3 Actual Lung Age [3] (Fig. 138.1)

The regression formula of FEV1 for each gender issued by the Japanese Respiratory Society: Men: FEV1 (L) = 0.036 × height (cm) −

K. Tomoda (✉)
Department of General Internal Medicine I, Kawasaki Medical School, Okayama, Japan

H. Yamada, Y. Naito (eds.), *Anti-Aging Medicine*, https://doi.org/10.1007/978-981-97-8514-8_138

0.028 × age − 1.178; Women: FEV1 (L) = 0.022 × height (cm) − 0.022 × age − 0.005. Based on gender, height, and FEV1, lung age is calculated by the inverse calculation method of the standard regression formula. Men: Lung Age = 0.036 × height (cm) − 1.178 − FEV1 (L)/0.028; Women: Lung Age = 0.022 × height (cm) − 0.005 − FEV1 (L)/0.022. This regression formula is intended for ages 18–95, so any results outside this range will be less than 18 or over 95. If the FEV1% is less than 70% and COPD is suspected, the predicted one-second volume ratio (FEV1%) may exceed 100%, resulting in a lung age younger than the actual age. To prevent misunderstanding, it is corrected to the actual age and a comment equivalent to mild COPD is displayed. Not only lung age, but also measured FEV1% and FEV1% are used to classify into five groups, and evaluation comments are added for each group (Fig. 138.2, Table 138.1). Lung age is used to evaluate obstructive ventilation disorder due to decreased FEV1, and restrictive ventilation disorder due to decreased VC is not evaluated. If there is no decrease in FEV1% and FEV1% is decreased (Fig. 138.2, Table 138.1 Group C), there may be a restrictive disorder, and the evaluation comment states that there may be other lung diseases.

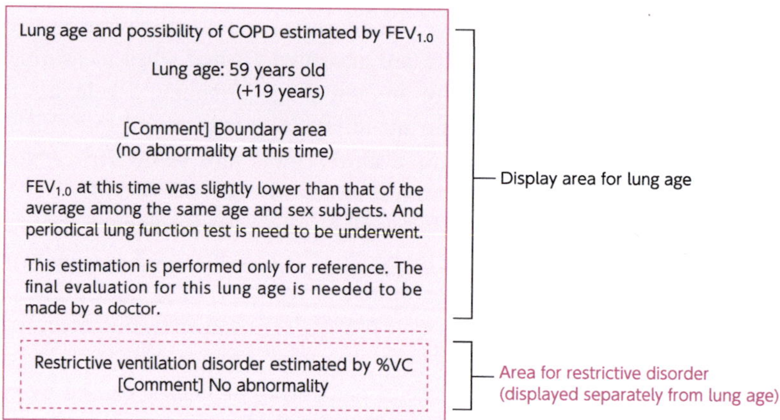

Fig. 138.1 Lung age result. (Reprinted with permission from [3])

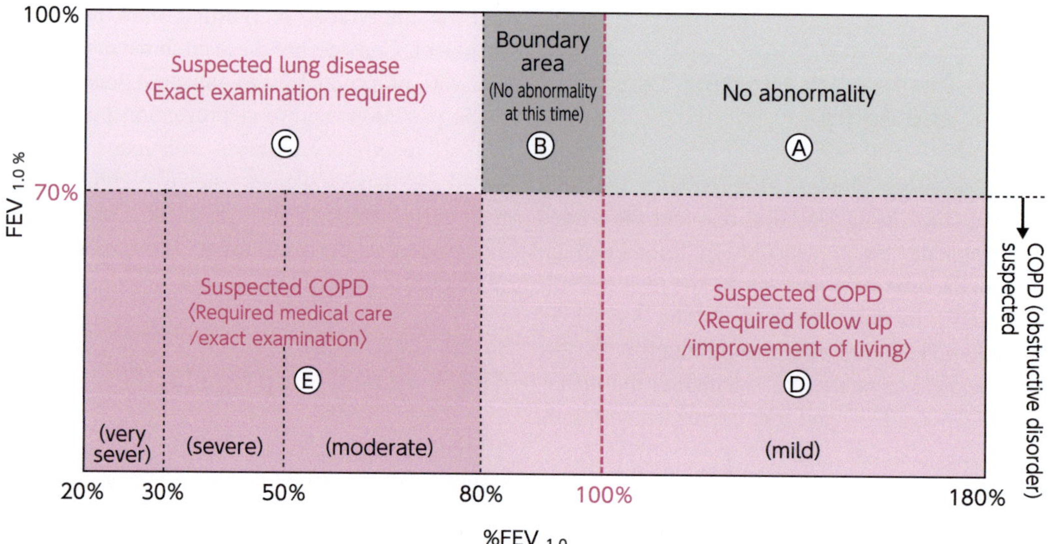

Fig. 138.2 Grouping by FEV1 ratio and %FEV1. (Modified with permission from [3])

Table 138.1 Evaluation comments for each group

Ⓐ No abnormality	The likelihood of lung disease is low. Your values are better than the average for individuals of the same gender and age group, so please continue to undergo regular respiratory function tests to maintain your health	The forced expiratory volume in one second (FEV1) is over 70%, and the percentage of FEV1 is over 100%
Ⓑ Boundary area (no abnormality at this time)	Your values are somewhat below average compared to individuals of the same gender and age group, indicating a necessity to continue regular respiratory function tests in the future	The forced expiratory volume in one second (FEV1) is above 70%, and the percentage of FEV1 is 80% or higher but less than 100%
© Suspected lung disease <Exact examination required>	The possibility of COPD is low; however, your values are worse than the average for individuals of the same gender and age group, raising suspicions of other lung diseases. A re-examination by a specialist is necessary	The forced expiratory volume in one second (FEV1) is over 70%, and the percentage of FEV1 is under 80%
Ⓓ Suspected COPD <Required medical care/exact examination>	Suspected mild COPD. Even if you currently have no noticeable symptoms, leaving it untreated can lead to severe progression. A re-examination by a specialist is necessary	The forced expiratory volume in one second (FEV1) is below 70%, and the percentage of FEV1 is 80% or higher
Ⓔ Suspected COPD <Required follow up/improvement of living>	Suspected moderate or more severe COPD. Re-examination by a specialist is essential. By initiating appropriate treatment early, symptoms can be improved and the progression of the disease can be controlled	The forced expiratory volume in one second (FEV1) is below 70%, and the percentage of FEV1 is below 80%

Reprinted with permission from [3]

When measuring VC, there may be a comment about restrictive ventilation disorder in the comment section below the lung age result display (Fig. 138.1).

138.4 Future of Lung Age [4]

Due to the characteristics of the regression formula estimating FEV1, about half of healthy individuals have a "lung age" higher than their actual age, and there is variation in the one-second measurement value even among healthy individuals of the same age, and the acceptable age range has not been considered, which has been a problem. New reference values for Japanese spirometry were announced in 2014, and it is necessary to introduce a "new lung age" to replace the conventional "lung age" in the lung age calculation formula. However, it is difficult to correct spirometers using the conventional lung age, and it has not been implemented at this time. The Japan Respiratory Society, which proposed lung age, has commented that "lung age" should be used as a reference for disease awareness and smoking cessation guidance, without sticking to how much it deviates from the actual age. If you have respiratory symptoms such as sputum, cough, and shortness of breath, and these respiratory symptoms have reduced your physical activity compared to your peers, it is important to visit a specialist institution without thinking it's due to age, and undergo various tests including respiratory function tests.

References

1. Chan ED, Welsh CH. Geriatric respiratory medicine. Chest. 1998;114:1704–33.
2. Fletcher C, Peto R. The natural history of chronic airflow obstruction. Br Med J. 1997;1:1645–8.
3. Japanese Respiratory Society Lung Age Promotion Office. About the use of lung age—"Lung Age Concept" specification. 2008. http://www.hainenrei.net/pdf/hainenreishiyousyo 11b.pdf.
4. Japanese Respiratory Society Pulmonary Physiology Committee: Statement on lung age. 2019. https://www.jrs.or.jp/activities/guidelines/statement/20191217170833.html.

Antiaging Checkup: Age Evaluation of the Eyes

139

Nobuyoshi Kitaichi

Frailty is defined as "a state where various functions of the body decline with aging, making it easier to fall into health disorders," in other words, it is positioned as an intermediate stage between health and a state requiring care. The Ministry of Health, Labor and Welfare has set the goal of "extending healthy life expectancy and reducing health disparities" in the "Basic Policy for Promoting Comprehensive Promotion of National Health" in "Health Japan 21 (2nd)." By preventing frailty and avoiding the need for care, the aim is to extend healthy life expectancy. The Japan Ophthalmology Awareness Conference has started "Eye Frailty" awareness activities.

139.1 Eye Frailty

Eye frailty refers to "a state where the eyes have deteriorated with age and various external stresses have caused a decline in eye function, or a state with a high risk of this." With aging, the eyeball structurally and functionally deteriorates in various ways. If any stress is added to this state, visual function impairment occurs. Initially, there may be difficulty seeing or discomfort, but if left untreated, the decline progresses and the decrease in vision is constantly perceived. Further progression can lead to severe impairment, making recovery difficult. If detected early, appropriate prevention and treatment are possible, and it is expected that progression can be slowed and symptoms can be alleviated. The Eye Frailty Checklist allows you to self-check for the possibility of early symptoms of age-related diseases such as presbyopia, dry eye, cataracts, glaucoma, diabetic retinopathy, age-related macular degeneration, and macular pucker by checking 10 items (Table 139.1).

Table 139.1 Eye frailty self-checklist

If you recognize two or more of the following 10 items, it is recommended to see an ophthalmologist
1. Eyes get tired easily
2. Sometimes it becomes difficult to see in the evening
3. Spending long hours reading newspapers or books has decreased
4. Sometimes spill food on the table while eating
5. Often feel that even with glasses on, vision is not clear
6. Easily feel dazzled
7. Sometimes need to blink to see clearly
8. Sometimes straight lines appear wavy
9. Have felt that steps or stairs are dangerous
10. Have missed seeing traffic signals or road signs

N. Kitaichi (✉)
Department of Ophthalmology, Health Sciences University of Hokkaido Hospital, Ishikari, Hokkaido, Japan
e-mail: nobukita@hoku-iryo-u.ac.jp

H. Yamada, Y. Naito (eds.), *Anti-Aging Medicine*, https://doi.org/10.1007/978-981-97-8514-8_139

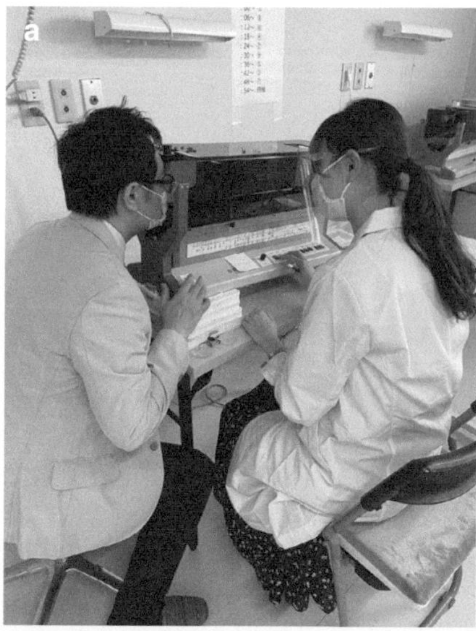

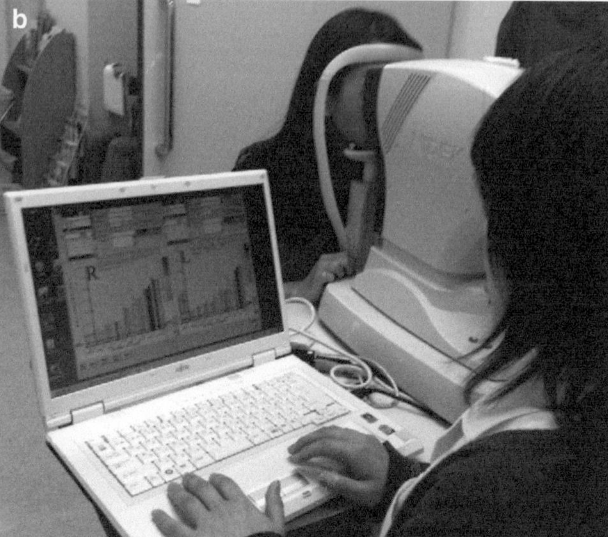

Fig. 139.1 Measurement of adjustment power. There are several methods to measure the objective adjustment power for evaluating conditions such as presbyopia. (**a**) Fixed refraction near point meter. (**b**) Adjustment function measuring device

139.2 Presbyopia

Inside the eye is a lens tissue called the crystalline lens, which focuses on distant and near objects. However, as we age, the elasticity of the crystalline lens decreases and it becomes impossible to change the thickness of the lens, making it impossible to focus on close objects, a condition known as presbyopia. The ability to focus from far to near is called accommodation, and it decreases with age. To measure accommodation, measuring instruments such as a fixed refraction near point meter, near point response measuring device, and accommodation function measuring device are used (Fig. 139.1).

139.3 Dry Eye

Dry eye is a major cause of chronic eye discomfort and, while it does not lead to blindness, it is a disease that impairs quality of life over the long term. Symptoms include dry eyes, gritty eyes, difficulty opening eyes, and tired eyes. It increases with age, and mental stress, the use of antianxiety drugs, antipsychotics, and antihistamines can also reduce tear volume. Environmental factors include the "three cons" of computers, contact lenses, and air conditioning, which are representative exacerbating factors.

139.4 Cataracts

Cataracts are caused by the clouding of the lens due to aging and other factors. There is no medication for treatment, and surgical treatment is performed. The clouded lens is emulsified and aspirated under local anesthesia, and an artificial intraocular lens is inserted. Refractive errors such as farsightedness and nearsightedness that existed before the surgery can also be corrected.

139.5 Glaucoma

Glaucoma is an age-related disease in which the optic nerve is damaged and the visual field defect gradually expands. The prevalence among Japanese people over 40 years old is about 5%. Symptoms are often not noticeable until the late stages. The three tests necessary for diagnosis are intraocular pressure measurement, fundus examination, and visual field test. It is not a disease that can be completely cured, and the purpose of treatment is to slow the progression and maintain visual function throughout life, so it is important to continue treatment. Controlling intraocular pressure with eye drops, etc. is effective.

139.6 Diabetic Retinopathy

Retinopathy is one of the most serious complications of diabetes, and it progresses without being noticed, and when symptoms are noticed, it is often already at risk of blindness. Retinopathy progresses in three stages: simple retinopathy, preproliferative retinopathy, and proliferative retinopathy. Treatments such as laser photocoagulation, intravitreal administration of vascular endothelial growth factor (VEGF) antibodies, and vitreous surgery are performed.

139.7 Age-Related Macular Degeneration

With aging, blood and edema occur in the macula, the center of the retina, causing a decrease in visual acuity. As the ability to process waste products in the retina decreases with age, new blood vessels are induced from the choroid. Because the visual cells are damaged by bleeding and exudation, treatments such as intravitreal administration of anti-VEGF antibodies, photodynamic therapy, and laser photocoagulation are performed.

139.8 Epiretinal Membrane

An epiretinal membrane is a transparent membrane that forms over the macula, the central part of the retina. It is common in middle-aged and older people, and many people notice it when things appear distorted (metamorphopsia). It is a slow-progressing benign disease, but if it progresses, vitreous surgery is necessary.

139.9 Conclusion

Among the physical changes that come with aging, presbyopia, which makes close-up work difficult, is considered the first symptom that makes one aware of functional decline. There are cases where other diseases are overlooked due to incorrect self-judgment without acknowledging one's own aging. We encourage you to visit a medical institution for an appropriate age assessment of your eyes and body, and to have regular check-ups to ensure there is no disease, in order to maintain your QOL (quality of life) and QOV (quality of vision).

Antiaging Checkup: Evaluation of Glycation Stress

140

Sho-ichi Yamagishi

Glycation in the body generally refers to the reaction where reducing sugars in the blood, such as glucose (blood sugar) and fructose (fruit sugar), cause a non-enzymatic chemical reaction with proteins, resulting in glycated proteins. These proteins then go through glycation intermediates, mainly carbonyl compounds, to eventually become advanced glycation end products (AGEs). Recent studies have shown that AGEs are known to exacerbate factors such as diabetic complications, arteriosclerosis, osteoporosis, and dementia, including Alzheimer's disease [1]. Glycation stress is a concept that comprehensively captures the stress on the body caused by reducing sugars and aldehyde loads and the subsequent reactions.

140.1 Evaluation of Glycation Stress

Various substances involved in the production of AGEs can be used as markers for the evaluation of glycation stress. Among them, HbA1c (a type of glycated hemoglobin where glucose is bound to the protein of red blood cells, hemoglobin) and glycoalbumin (where glucose is bound to albumin, a type of protein) are frequently used because they can easily evaluate the glycation stress of diabetic patients in daily clinical practice. On the other hand, AGEs are a representative marker for comprehensively evaluating glycation stress. In particular, the measurement of blood pentosidine concentration is covered by insurance for the diagnosis of early kidney disease in nondiabetic patients. Specifically, it applies to those suspected of having decreased kidney function in nondiabetics [urinary protein (+) or more, or urinary protein (−) with urinary occult blood (±) or more], with blood creatinine values <1.5 mg/dL, and normal urea nitrogen values. Note that for the early diagnosis of diabetic nephropathy, HbA1c, which has high sensitivity, is already covered by insurance, so AGEs are not covered.

140.2 Measurement of AGEs Accumulation in the Skin

AGEs are generated and accumulated by uncontrolled, nonenzymatic reactions with proteins, not only from reducing sugars but also from lipids and aldehyde groups derived from alcohol. This leads to cross-linking degeneration and functional degradation of proteins, activation of intracellular signals, and tissue damage. AGEs have been shown to play a significant role in the pathogenesis of atheromatous arteriosclerosis.

S.-i. Yamagishi (✉)
Department of Medicine, Showa University School of Medicine, Tokyo, Japan
e-mail: shoichi@med.showa-u.ac.jp

H. Yamada, Y. Naito (eds.), *Anti-Aging Medicine*, https://doi.org/10.1007/978-981-97-8514-8_140

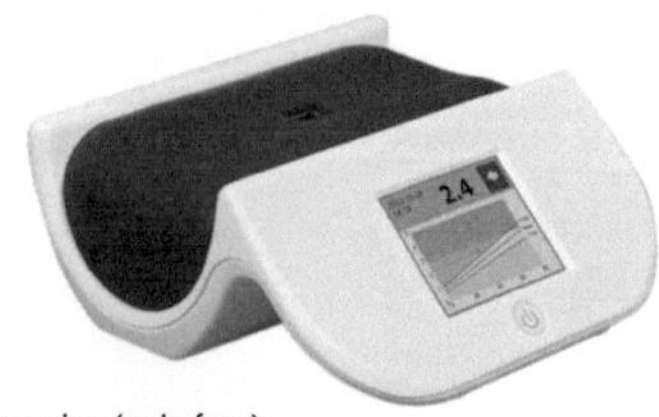

- Non-invasive (pain-free)
- Short time (within 3 minutes)
- Accurate with high reproducibility
- Detects AGE accumulation through autofluorescence of skin, tissue, and vascular wall

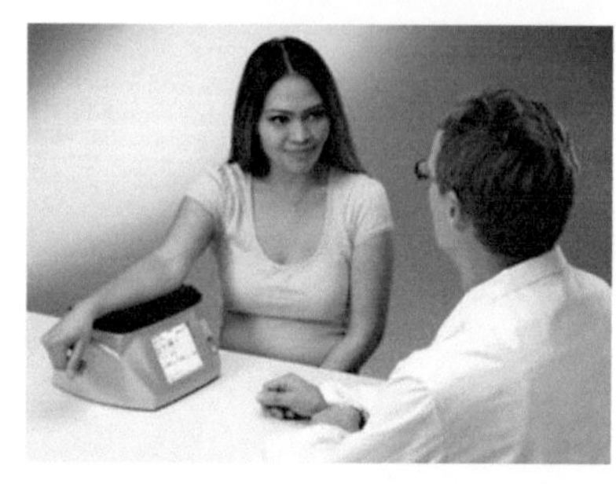

Fig. 140.1 AGE Reader mu

For example, LDL cholesterol, which has been oxidized and glycated by AGEs, is recognized by scavenger receptors and promotes the foaming of macrophages. Foamed macrophages secrete various chemokines and growth factors, promoting the migration and proliferation of smooth muscle cells and further infiltration of monocytes into the endothelium. Recently, a Dutch-made device, the AGE Reader, has been able to noninvasively measure the accumulation levels of self-fluorescent AGEs in skin tissue. The AGE Reader mu (imported and sold by Serista Corporation) (Fig. 140.1) has been approved as a general medical device. The amount of AGEs accumulated in the skin, which is noninvasively measured by placing the subject's forearm on the device, is quantified as skin autofluorescence (SAF). SAF is related to functional and morphological changes of atherosclerosis, such as brachial-ankle pulse wave velocity (baPWV) and carotid intima-media thickness (CIMT) [2], making it suitable as a test item in antiaging checkups. In recent years, it has been reported that SAF can also be used as an indicator of cognitive decline [3].

140.3 Relationship Between SAF and Cognitive Function

In a study examining the relationship between SAF and mild cognitive impairment (MCI), evaluated by the "brain health check," in 226 antiaging checkup participants, the group evaluated as having MCI (18 people, 7.9%) had significantly higher SAF values, in addition to kidney function, brain natriuretic peptide (BNP), baPWV, and brain atrophy (lateral ventricle subangle area), compared to the group evaluated as not having MCI (MCI group: 2.56 ± 0.55 AU, normal group 2.10 ± 0.41 AU, <0.001). The proportion of participants evaluated as having MCI was significantly different between the SAF tertile groups (Fig. 140.2). Even after adjusting for various confounding factors in logistic regression analysis, a high SAF value (≧2.27 AU) was significantly associated with the presence of MCI. From the perspective of dementia prevention, it is considered beneficial to measure SAF values in antiaging checkups. Various clinical studies targeting antiaging checkup participants have also provided beneficial information [4].

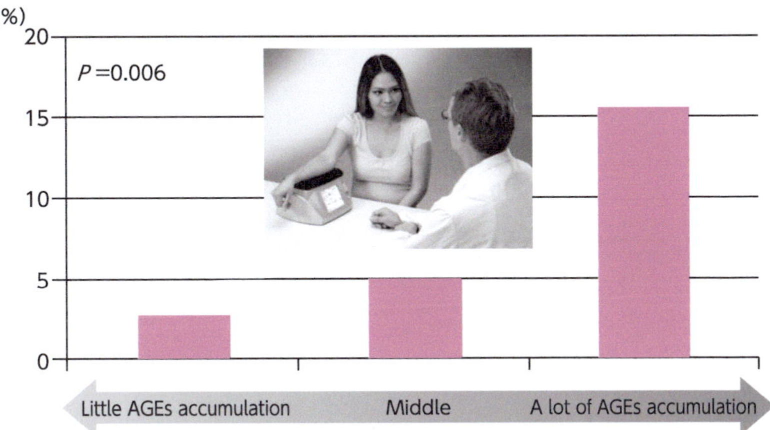

Fig. 140.2 Frequency of MCI in SAF (accumulation of AGEs in the skin) tertiles. Reprinted with permission from [3]. Significant difference. The higher the SAF (the more AGE accumulation), the higher the frequency of MCI (the association remains even after adjusting for confounding factors)

References

1. Kuzan A. Toxicity of advanced glycation end products (review). Biomed Rep. 2021;14:46.
2. Choi LS, Ahmed K, et al. Skin accumulation of advanced glycation end products and cardiovascular risk in Korean patients with type 2 diabetes mellitus. Heliyon. 2022;8:e09571.
3. Igase M, Ohara M, et al. Skin autofluorescence examination as a diagnostic tool for mild cognitive impairment in healthy people. J Alzheimers Dis. 2017;55:1481–7.
4. Igase M, Okada Y, et al. Casein hydrolysate containing milk-derived peptides reduces facial pigmentation partly by decreasing advanced glycation end products in the skin: a randomized double-blind placebo-controlled trial. Rejuvenation Res. 2021;24:97–103.

141 Antiaging Dock: Evaluation of Psychosomatic Stress

Takehiro Tamura and Hidehiko Takahashi

The examination items of the anti-aging dock are broadly divided into "degree of aging" and "aging risk factors." Biological aging is a unidirectional and inevitable phenomenon, but its speed is not uniform in all humans and is considered to be elastic [1]. In other words, the degree of aging is regulated by genes, exposure to stressors, and their interactions, and is particularly thought to depend heavily on exposure to stressors. It has become clear that the overactivity of the hypothalamic–pituitary–adrenal (HPA) axis, a major neuroendocrine response mediating stress responses, contributes to the deterioration of nerve cells and peripheral nerves with aging, and exposure to stressors may accelerate aging [2]. In other words, chronic high-stress conditions are one of the aging risk factors. Stressors in humans can be replaced with life/environmental factors that evoke stress responses, and include not only physical stressors such as the natural environment, but also psychosocial stressors that occur in interpersonal relationships. This section provides an overview of the psychosomatic stress response caused by exposure to stressors from the perspective of HPA axis activity, and discusses psychosocial stressors and psychological stress theory, stress-level evaluation scales, and biomarkers of psychosomatic stress.

T. Tamura · H. Takahashi (✉)
Department of Psychiatry and Behavioral Sciences, Graduate School of Medical and Dental Sciences
Tokyo Medical and Dental University, Tokyo, Japan
e-mail: hidepsyc@tmd.ac.jp

141.1 Stressors and Psychosomatic Stress Response

For survival and well-being, it is necessary to continuously adapt to external stressors and internal environmental changes, and this adaptation is achieved by the coordinated activation of behaviors, autonomic nerves, and endocrine responses that maintain homeostasis [2]. Specifically, the activation of the HPA axis, a major neuroendocrine response to stress, occurs, leading to an increase in glucocorticoid secretion from the adrenal cortex through the release of corticotropin-releasing hormone (CRH), vasopressin (VP), and adrenocorticotropic hormone (ACTH). CRH and glucocorticoids are collectively referred to as stress hormones, and their basal secretion is essential for neuronal plasticity and normal brain function. Also, the transient increase in stress hormones associated with exposure to stressors is heavily involved in stress adaptation, and hence necessary behaviors and autonomic nerve responses, and is also considered essential. However, excessive exposure to a state of persistently elevated stress hormones can be harmful, leading to predisposing factors for mental disorders, metabolic disorders, and

H. Yamada, Y. Naito (eds.), *Anti-Aging Medicine*, https://doi.org/10.1007/978-981-97-8514-8_141

cardiovascular diseases, so it is considered important to stop the stress response promptly.

141.2 Psychosocial Stressors and Psychological Stress Theory

The highly complex lives of modern people are rife with various psychosocial stressors, which are becoming increasingly chronic. Chronic psychological stress promotes changes in the autonomic nervous system, neuroendocrine system, metabolic system, and immune system, creating a dramatically different negative biochemical environment compared to organisms not exposed to stress [1]. Already, large-scale database studies have shown that exposure to chronic adversity, such as job-related tension, stress and isolation, adversity or trauma in early childhood, leads to physiological exhaustion of the stress response regulation system, promoting the onset of various functional abnormalities and diseases, as well as mortality. Also, when considering psychosocial stressors, a psychological perspective is useful. In psychology, stressors are defined as "demands that burden or exceed an individual's resources," and this includes not only the intensity of the psychological burden that a life event carries, but also the individual's cognitive appraisal of it [3]. Furthermore, cognitive appraisal is divided into primary and secondary appraisal, the former relates to how an individual perceives their environment, and the latter relates to what kind of coping (response) is possible, and whether good results can be expected from its execution. In other words, the intensity of a stressor is ultimately relative, and it is important how much an individual evaluates a life event as stressful.

141.3 Stress-Level Assessment Scale for Psychosocial Stressors

There seems to be a consensus that there is no established method for assessing the psychosomatic stress response to psychosocial stressors. There are many scales, some targeting depression that arises secondarily from exposure to chronic psychosocial stressors (as a surrogate indicator), others targeting only psychological symptoms as a stress response, but it is thought desirable to have an assessment scale that targets both psychological and physical symptoms [4]. As a comprehensive assessment scale, there are the Public Health Research Foundation Stress Checklist Short Form (PHRF–SCL[SF]) and the Inventory to Measure Psychological Stress (IMPS) [4]. Also, the Anti-Aging Medical Association's common questionnaire for anti-aging QOL, which can be downloaded from their website, has items related to psychological and physical symptoms. Being aware of the temporal relationship with life events, and repeatedly implementing the QOL common questionnaire over time, self-monitoring the stress responses occurring in oneself is thought to be very useful from the perspective of stress management.

141.4 Biomarkers of Psychosomatic Stress

Cortisol, a type of glucocorticoid, is understood as a stress marker because its production and secretion increase due to the activation of the HPA axis [5]. Moreover, cortisol can be detected in many specimens such as blood, urine, saliva, hair, and fingernails. Generally, blood, saliva, and urine are used as specimens in acute stress responses, while hair and fingernails are used in chronic stress, capturing exposure to stressors experienced 1–6 months prior. However, as cortisol plays many roles in the body beyond responding to psychosocial stressors, it is important to note that it may also be chronically elevated in response to other tasks related to maintaining homeostasis. Also, it is known that dehydroepiandrosterone sulfate (DHEA-s) has an antagonistic effect on cortisol. A high cortisol/DHEA-s ratio indicates an increased risk to mental and physical health due to stress responses, and it has been reported to correlate with an increased risk of depression in children and metabolic syndrome in adult males [5].

References

1. Epel ES, Lithgow GJ. Stress biology and aging mechanisms: toward understanding the deep connection between adaptation to stress and longevity. J Gerontol A Biol Sci Med Sci. 2014;69(Suppl 1):S10–6.
2. Aguilera G. HPA axis responsiveness to stress: implications for healthy aging. Exp Gerontol. 2011;46:90–5.
3. Otsuka Y, Shimazu A. Psychological stress theory by Richard S. Lazarus. Job Stress Res. 2019;26:389–93.
4. Yamamoto K. Development of a questionnaire to evaluate the degree of psychosocial stress and decrease in stress resistance. J Physiol Anthropol. 2005;10:67–77.
5. DeCaro JA, Helfrecht C. Applying minimally invasive biomarkers of chronic stress across complex ecological contexts. Am J Hum Biol. 2022;34:e23814.

Antiaging Dock: Evaluation of Oxidative Stress

142

Hiroshi Ichikawa

142.1 To Properly Understand the Body's Antioxidant Capacity

Oxidative stress refers to a situation where the balance of oxidation and reduction in cells is disrupted due to factors such as the overproduction of reactive oxygen species (ROS) due to endogenous or exogenous factors, or the breakdown of antioxidant defense mechanisms. However, antioxidant capacity does not simply refer to the ability to directly eliminate ROS and free radicals. It should be evaluated from various angles, including the control of ROS production from neutrophils and macrophages in the body, the suppression of Fenton reactions by chelating transition metals, the suppression of ROS production by maintaining the quality of mitochondria, the maintenance of cell function by stabilizing cell membranes, the impact on the arachidonic acid cascade, the effect on enzyme activity in the NADPH oxidase system and the hypoxanthine–xanthine oxidase system, the indirect ROS production suppression effect via other signal transduction, and the effect of enhancing the expression of antioxidant enzyme genes. However, what is intuitively easy to understand is the evaluation of the direct scavenging activity of ROS and free radicals, so the current situation is that the body's antioxidant capacity also depends solely on the analysis of antioxidant components, as shown below.

H. Ichikawa (✉)
Major of Medical Life Systems, Graduate School of Life and Medical Sciences, Doshisha University, Kyoto, Japan
e-mail: hichikaw@mail.doshisha.ac.jp

142.2 Methods to Evaluate Oxidative Stress

The methods to evaluate the extent to which the body has been exposed to oxidative stress are generally chosen from three methods: ① measurement of reactive oxygen itself, ② measurement of endogenous products (oxidatively modified substances) produced by reactive oxygen, ③ measurement of antioxidants (see Table 142.1).

142.2.1 Measurement of Reactive Oxygen Species (ROS) Itself

ROS produced in the body easily reacts with biomolecules, affecting biological functions. However, their lifespan is very short, and therefore, most of the ROS generated in the biological system cannot be directly observed. Therefore, methods have been attempted to detect ROS by reacting them with specific compounds using their high reactivity, converting ROS into a more

H. Yamada, Y. Naito (eds.), *Anti-Aging Medicine*, https://doi.org/10.1007/978-981-97-8514-8_142

Table 142.1 Oxidative stress markers

	Measurement items	Oxidative stress markers
Measurement of reactive oxygen species	Reactive oxygen species	Superoxide, hydroxyl radical, hydrogen peroxide, singlet oxygen
Measurement of biological products (oxidatively modified substances) produced by reactive oxygen species	Lipids Proteins DNA Others	Malondialdehyde, 4-HNE, 8-isoPGF20 Carbonylated proteins, Dityrosine 8-OHdG d-ROMs test
Measurement of antioxidants	Antioxidant enzymes Antioxidants	SOD, GSHPx, Catalase Vitamin C, Vitamin E, Glutathione, Thioredoxin, Ubiquinone, Bilirubin, Uric acid, Albumin

stable compound. One method is to trap ROS/free radicals and convert them into more stable radical compounds, a technique called "spin trapping" [1]. Another method uses compounds (chemical probes) that are converted into molecules with characteristics such as fluorescence or chemiluminescence by specific chemical reactions with specific ROS, a method called "chemical probe method."

142.2.2 Measurement of Biological Products (Oxidatively Modified Substances) Caused by Reactive Oxygen

When oxidative stress is excessive, biological components such as proteins, nucleic acids, and lipids are oxidatively modified by ROS. Many amino acid residues in proteins are carbonylated by ROS [2]. Furthermore, oxidative DNA damage is observed in many diseases, and oxidatively damaged bases can cause mutations and cell death. Therefore, 8-oxoguanosine (8-oxodG), 8-hydroxydeoxyguanosine (8-OHdG), etc., are used as excellent biomarkers for evaluating oxidative stress exposure [3]. It should be noted that the concentration of oxidatively modified substances, an indicator of oxidative stress, in tissues is primarily thought to be determined by the balance between ROS production and ROS elimination. In other words, oxidatively modified substances are produced in the body only when exposed to oxidative stress that exceeds the body's ROS elimination activity. In other words, it should be recognized that measuring only oxidatively modified substances cannot accurately evaluate the degree of potential oxidative stress exposure.

142.2.3 Measurement of Antioxidants

As a defense mechanism against oxidative stress, there is an antioxidant network in the body, equipped with antioxidant enzymes such as superoxide dismutase, glutathione peroxidase, and catalase, which remove ROS, forming an antioxidant network as preventive antioxidants. In addition, general antioxidants themselves directly react with and eliminate ROS, so they are called radical scavenging antioxidants. There are water-soluble ones such as vitamin C and fat-soluble ones such as vitamins A, E, and coenzyme Q10, each forming a clever network to protect the body from oxidative damage. As such, it is a very useful method to evaluate the body's antioxidant capacity by measuring the activity of these preventive antioxidant enzymes and the concentration of radical scavenging antioxidants. However, because there are countless antioxidants in the body, it is impossible to measure all antioxidant enzymes and antioxidants. From the above perspective, as a method to evaluate the antioxidant capacity of the body, we propose the following MULTIS (MULTIple free-radical). We use the scavenging method to quantify the elimination activity against the six types of ROS that can be produced in the body.

142.3 What Is the MULTIS Method?

The first ROS produced in the body is the superoxide radical. However, as mentioned earlier, the half-life of ROS is extremely short and its reactivity is very high, so it immediately transforms into various ROS and changes its form. At the same time, it is involved in the production of various types of lipid radicals in the body. In other words, ROS in the body can instantly change into various forms regardless of place and time, and unless antioxidants are present at the site of ROS production, there is a possibility that ROS cannot be efficiently eliminated. Therefore, when we evaluate the antioxidant capacity of the body, we do not measure individual antioxidant enzymes or antioxidants, but adopt a method to quantify the elimination activity for each of the six types of ROS that can be produced in the body. Specifically, we generate six different types of ROS/free radicals (hydroxyl radicals, superoxide radicals, alkyl oxy radicals, alkyl peroxyl radicals, methyl radicals, singlet oxygen) by light irradiation, and measure the ROS/free radical elimination activity of various foods and biological samples of patients with oxidative stress-related diseases by the ESR spin trap method [4]. Already, the ROS elimination activity of the body by the MULTIS method has been reported in diseases such as diabetes, chronic obstructive pulmonary disease, and autism [5], but it has been found that the pattern of elimination activity varies depending on the disease, even though they are all oxidative stress-related diseases. Figure 142.1 shows the measurement of blood ROS elimination activity in senescence-accelerated mice (SAMP8), which are established as early learning/memory impairment models. It has been found that blood ROS elimination activity is significantly reduced compared to SAMR1, which undergoes natural aging, even before the onset of dementia, and that if food factors that suppress cognitive decline are introduced, these ROS elimination activities are almost completely improved. These results suggest that dementia is also one of the oxidative stress-related diseases, and that by evaluating the decrease in the body's antioxidant capacity through interventions such as food using the MULTIS method, it may be possible to prevent these diseases.

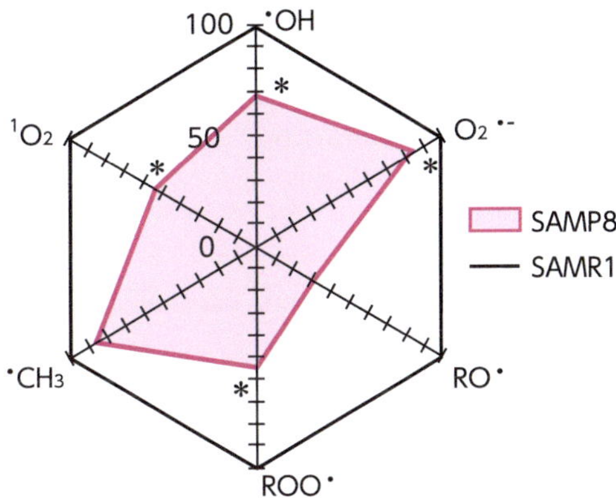

※Values are mean±S.E. (n=6) * p<0.05 vs. R1

Fig. 142.1 Reactive oxygen species and free radical scavenging activity in the SAMP8 group. Aging-accelerated mice show a decrease in antioxidant capacity. The SAMP8, an aging-accelerated mouse, is used as a model for Alzheimer's disease. It has been reported that sarcopenia develops in this SAMP8. The figure shows the change in the weight of the gastrocnemius muscle with age, decreasing after peaking at 7 months. In our laboratory, we measured the plasma antioxidant capacity of SAMP8 using the Multis method. The figure shows the plasma antioxidant capacity at 30 weeks of age in SAMP8. As shown in the figure, SAMP8 shows a significant decrease in six types of reactive oxygen species scavenging activity compared to the control, SAMR1

References

1. Buettner GR. Spin trapping: ESR parameters of spin adducts. Free Radic Biol Med. 1987;3:259–303.
2. Fedorova M, Bollineni RC, et al. Protein carbonylation as a major hallmark of oxidative damage: update of analytical strategies. Mass Spectrom Rev. 2014;33:79–97.
3. Barregard L, Møller P, et al. Human and methodological sources of variability in the measurement of urinary 8-oxo-7,8-dihydro-2′-deoxyguanosine. Antioxid Redox Signal. 2013;18:2377–91.
4. Oowada S, Endo N, et al. Multiple free-radical scavenging capacity in serum. J Clin Biochem Nutr. 2012;51:117–21.
5. Hirayama A, Wakusawa K, et al. Simultaneous evaluation of antioxidative serum profiles facilitates the diagnostic screening of autism spectrum disorder in under-6-year-old children. Sci Rep. 2020;10:20602.

143 Antiaging Checkup: Evaluation of Immunity

Mitsuo Maruyama

143.1 About the Evaluation of Immune Function

The immune system, which protects us from infectious diseases caused by pathogens and transformed cells like cancer cells that occur within our bodies, is an important function for maintaining health. The immune system operates throughout the body, and immune cells are basically present in all organs, repeatedly carrying out immune responses to correctly recognize self and nonself and eliminate nonself. Immune function is known to be affected by multiple endogenous and exogenous factors, such as gender, age, exposure to harmful chemicals, stress, exercise, and nutrition, in addition to immune abnormalities where the incidence and severity of various infectious diseases increase due to changes in the number or proportion of specific immune cells, which are diseases of the immune system. It is difficult to evaluate immune function by understanding the immune status of healthy people clinically because the immune parameters to be evaluated vary considerably among healthy individuals, depending on which factors most affect immune function. The correlation between the factors mentioned above and immune function is organized below based on previous findings.

M. Maruyama (✉)
Geroscience Research Center, National Center for Geriatrics and Gerontology, Obu-shi, Aichi, Japan
e-mail: michan@ncgg.go.jp

143.1.1 Aging and Gender Differences

In terms of endogenous factors, changes with aging are summarized separately (II-G-1) (Chap. 33). As aging changes in the immune system, but the proportion of immature lymphocytes decreases, resulting in a decrease in new immune responses, and the effectiveness of vaccines, the attenuation of immune memory, and as a result, susceptibility to infectious diseases increase. Also, receptors for sex steroid hormones such as estrogen, progesterone, and androgens are expressed in many cells of the innate and adaptive immune systems. As a result, immune function is affected by changes in sex hormone levels due to maturity, pregnancy, aging, or immunosenescence. This effect may support the fact that immune function in infections and cancer is superior in women compared to men, except for the elderly whose sex hormones decrease. In fact, a report examining the immune function of healthy Japanese people from their 20s to 90s by activating T cell proliferation and cytokine production with anti-CD3 antibodies found that the decline in immune indicators in women was slower than in men, and this may be related to the gender difference in average lifespan [1].

H. Yamada, Y. Naito (eds.), *Anti-Aging Medicine*, https://doi.org/10.1007/978-981-97-8514-8_143

143.1.2 Stress

Stress is also an important parameter factor for evaluating immune function. It is associated with the hypothalamic–pituitary–adrenal (HPA) axis and the sympathetic nervous system, which control stress hormones such as catecholamines and glucocorticoids. Acute stress mainly involves the production of inflammatory cytokines and neutrophils and natural killer (natural kinase) pathway, which induces the redistribution of natural killer (NK) cells, triggers a fight-or-flight response, whereas chronic stress weakens immune responses such as T cell proliferation and NK cell cytotoxicity, promoting immunosenescence, which is relevant to the evaluation of immune function.

143.1.3 Exposure to Harmful Chemicals

Phthalate esters known as plasticizers and dibutyltin affect the function of NK cells, T cells, and macrophages. Also, nitrosamines contained in tobacco smoke have been shown to inhibit the cytotoxic function of lung macrophages, potentially affecting the immune system. The impact of these exogenous and endogenous factors on the immune system is not dramatically acute but usually gradual. Therefore, to obtain reliable data to evaluate changes in immune function due to factors such as aging, sex differences, stress, and exposure to chemicals introduced here, a broad analysis of immune parameters is necessary. The following section summarizes knowledge about these parameters, particularly in the elderly.

143.2 Immune Function and Its Evaluation in the Elderly

The immune system plays a crucial role in maintaining daily health, but like the muscle strength of limbs and many other biological functions of the body such as eyes and ears, it undergoes aging changes and its function declines. So, what does the commonly heard term "immune power" evaluate in terms of immune function? From the perspective of self and non-self-recognition functions, it includes defense against infections by foreign pathogens, which are non-self, or the elimination of cancer cells that originally are self but mutate and proliferate. It has been reported that the immune power of the elderly is about 20–30% compared to young people in their 20s, and it is clear from many studies using aging mice as well as human reports that the decline in function due to aging is significant among the factors mentioned above. One of the first recognizable functional changes in the aging of the immune system is a decrease in antibody production in response to primary immune responses from vaccination and secondary immune responses including boost effects, which includes functional impairments seen in germinal centers centered on B cells and a decrease in somatic hypermutation [2]. Also, the decrease in hematopoietic stem cells in the bone marrow not only shows a decrease in naive T and B cells with aging but also further reduces the pool of naive lymphocytes as it is involved in the proliferation and differentiation of lymphocytes. This decrease in the number of naive T and B cells reduces the diversity of antigen receptors, which are central to the humoral immune response of the acquired immune system, but there are also many reports [3] on the functional decline of cellular immune responses. The function of memory T cells, which are at the final stage of differentiation, remains stable throughout aging, so naive CD4 positive or CD8 positive T cells and aging-associated pSTAT signals and MAPK (mitogen-activated protein kinase)/ERK (extracellular signal-regulated The function of the kinase signaling pathway tends to decline due to the effects of repeated encounters with antigens, leading to the manifestation of cellular immune aging in the elderly [4]. This trend is typical of immune aging associated with aging, where aged lymphocytes are often seen, and chronic antigen stimulation leads to a bias in antigen specificity and an increase in the clonality of lymphocytes. On the other hand, cytokines produced by immune cells are essential for intercellular activation and inhibitory signal transmission that control immune responses, but in particular, cytokines that cause inflammation are attracting attention in recent years as SASP (senescence-associated secretory phenotype) factors, which are molecules that cause aging, not only in the immune system. So far, we have first organized various factors that can be indicators of immune function evaluation, and among those factors, we have dug a little deeper into the changes based on aging at the cellular and molecular levels. As a result, to evaluate the changes in immune function with aging, it is most appropriate to comprehensively analyze the immune parameters (such as phospho-epitope analysis by cytokine stimulation and gene expression analysis) that are used as indicators, and to use a database that can judge the immunological health status. In particular, it is important to have no obvious age-related diseases clinically to correctly evaluate the immune function of each elderly person, and it is necessary to combine data from multiple assays. Another factor that complicates this evaluation is the individual differences, which can be called "diversity of aging." Even in experimental data using model animals where both genetic background and environment are controlled, the difference in biological function with aging, not limited to immune function, is significant. Currently, many of the available databases that summarize immunological values related to healthy aging are cross-sectional analyses, and therefore, it is important how to exclude people at risk of disease or those who have developed it in the evaluation.

143.3 Relationship Between Daily Life and Immunity

When trying to predict the age of immune function using immunological and clinical data for the evaluation of immunity, we are reminded that the correlation between the indicators of daily life activities and immune measurements is surprisingly high. This, in addition to the analysis using the previous model animals, is again convinced of the large impact of the environment on the immune system, as has been clearly shown in studies of the immune characteristics of twins with high genetic homology [5]. Then, it is also possible to maintain our own immune function by a little ingenuity in our daily life, Fig. 143.1 shows the "5 points" to keep in mind, and it is important to maintain immunity and resistance every day to prevent the onset of infectious diseases and the proliferation of tumor cells.

Fig. 143.1 Advice to maintain immunity

References

1. Hirokawa K, Utsuyama M, et al. Slower immune system aging in women versus men in the Japanese population. Immun Ageing. 2013;10:19.
2. Yang X, Stedra J, et al. Relative contribution of T and B cells to hypermutation and selection of the antibody repertoire in germinal centers of aged mice. J Exp Med. 1996;183:959–70.
3. Lang A, Nikolich-Zugich J. Functional CD8 T cell memory responding to persistent latent infection is maintained for life. J Immunol. 2011;187:3759–68.
4. Goronzy JJ, Weyand CM. Understanding immunosenescence to improve responses to vaccines. Nat Immunol. 2013;14:428–36.
5. Brodin P, Jojic V, et al. Variation in the human immune system is largely driven by non-heritable influences. Cell. 2015;160:37–47.

144 Antiaging Checkup: Evaluation of Physical Activity

Motohiko Miyachi

144.1 What Is Physical Activity?

The amount of physical activity is related to the onset of lifestyle diseases, premature death, and also to physical strength and life functions [1, 2]. Therefore, it is important to understand the state of physical activity when promoting health and anti-aging. Physical activity refers to "the activity of skeletal muscles that increases energy consumption," in simpler terms, it refers to all actions that move the body. Based on the concept of METs (metabolic equivalents for tasks), which is an index of how many times the energy consumption is when the state of being still is set to 1, actions of 1.5 METs or more correspond to physical activity. Actions less than 1.5 METs are not included in physical activity and are called rest, most of which are sedentary behaviors such as sleep and sitting down. Physical activity can be classified into two categories according to its purpose. One is exercise, which refers to activities performed with a purpose during leisure time. Specifically, it includes enjoying sports, working out at the gym, etc. The other is daily activities, which refer to activities necessary for daily life. Specifically, it includes housework, physical labor, movement, etc. Also, physical activity is classified by the difference in intensity (hardness). 1.5–2.9 METs are classified as low intensity, 3.0–5.9 METs as medium intensity, and 6 METs or more as high intensity. Low-intensity activities include stretching and standing, medium-intensity activities include table tennis, weight training, walking, vacuuming, etc., and high-intensity activities include running, carrying heavy loads, etc.

M. Miyachi (✉)
Faculty of Sport Sciences, Waseda University, Saitama, Japan

144.2 Energy Consumption and METs

Physical activity is accompanied by an increase in energy consumption. The more physical activity, the more energy consumption. The amount of physical activity is calculated as the product of the intensity of physical activity and the implementation time. As mentioned earlier, the amount of physical activity is measured in "MET hours," a unit derived from the unit of intensity of physical activity, "METs," and the unit of time, "hours." If you walk at an intensity of 3.5 METs for 2 h, it is 3.5 METs × 2 h = 7.0 MET hours, and if you continue running at 7 METs for 30 min, it is 7.0 METs × 0.5 h = 3.5 MET hours. The amount of physical activity expressed in MET hours can be converted into energy consumption in "kcal" by multiplying the performer's weight and a coefficient of 1.05. For example, if you run at 7 METs for 30 min, it can be calculated as 3.5 MET hours × 65 kg × 1.05 = 239 kcal.

H. Yamada, Y. Naito (eds.), *Anti-Aging Medicine*, https://doi.org/10.1007/978-981-97-8514-8_144

144.3 Standard Method

Measuring energy consumption is the most accurate way to measure the amount of physical activity. The double-labeled water method is used to measure energy consumption under free-living conditions in humans. After drinking heavy water composed of the stable isotopes ^{3}H and ^{18}O, ^{3}H and ^{18}O, the amount of excretion in urine is measured continuously for about 2 weeks, and it can be quantified that the faster the excretion rate, the higher the total energy consumption. From there, by subtracting the resting energy consumption, the physical activity energy consumption can be quantified. Alternatively, the total energy consumption can be divided by the basal metabolic rate to calculate the physical activity level (PAL). A PAL of less than 1.5 is considered inactive, 1.5–1.74 is standard, and 1.75 or more is considered active. This is a very accurate and excellent method, but the cost of heavy water and isotope analysis is about 200,000 yen per person, making it difficult to implement for the general public.

144.4 Questionnaire Method

When evaluating the activity level of a relatively large group, a questionnaire method is used. The questionnaire method makes it relatively easy to understand the situation of habitual physical activity. While it is simple, there are not many people who accurately perceive and remember physical activity, and it cannot be said that it is highly quantitative or valid. Many questionnaires have been developed so far, but the simplified version of the physical activity questionnaire used in the cohort study of the National Cancer Research Center (Japan Public Health Center; JPHC) is a physical activity questionnaire consisting of three items asking about the time to move the body on a usual day. Specifically, it asks about ① muscle labor or intense sports (none, less than 1 h, 1 h or more), ② sitting time (3 h or less, 3–8 h, 8 h or more), and ③ walking or standing time (less than 1 h, 1–3 h, 3 h or more). The total energy consumption estimated by this very simple questionnaire, compared to the total energy consumption measured by the doubly labeled water method, has a relatively good estimation accuracy with an average error of +49 kcal/day and a correlation coefficient of 0.73 [3].

144.5 Activity Meter and Wearable

Activity meters can measure physical activity just by wearing the device, except for special activities such as bathing, so they do not impose a burden that interferes with the subject's activities. In recent years, activity meters incorporating a three-dimensional accelerometer using a piezoelectric element have become common, and they can objectively estimate not only the number of steps but also the energy consumption and activity intensity (METs) from the size and frequency of shocks (acceleration) in the front-back, left-right, and up-down directions that occur when humans are active. The total energy consumption estimated by the activity meter tends to be lower than that measured by the doubly labeled water method, but the correlation coefficient between the two is 0.80–0.88 [4], and the accuracy has been improving in recent years. Activity meters are evolving into wearables and smartwatches, and not only can they evaluate the intensity, time, and amount of physical activity, but they can also evaluate sedentary behavior and its interruption, distinction of means of transportation (train, car, bicycle, walking, and running), and sleep status by utilizing not only acceleration but also pulse rate and GPS information. In the future, the evaluation of physical activity using activity meters is expected to become indispensable in our lives, not only for research and surveys, but also as a tool for health promotion.

References

1. Murakami H, Tripette J, et al. "Add 10 min for your health": the new Japanese recommendation for physical activity based on dose-response analysis. J Am Coll Cardiol. 2015;65:1153–4.
2. Miyachi M, Tripette J, et al. "+10 min of physical activity per day": Japan is looking for efficient but

feasible recommendations for its population. J Nutr Sci Vitaminol (Tokyo). 2015;61(Suppl):S7–9.
3. Sasai H, Nakata Y, et al. Simultaneous validation of seven physical activity questionnaires used in Japanese cohorts for estimating energy expenditure: a doubly labeled water study. J Epidemiol. 2018;28:437–42.
4. Murakami H, Kawakami R, et al. Accuracy of wearable devices for estimating total energy expenditure: comparison with metabolic chamber and doubly labeled water method. JAMA Intern Med. 2016;176:702–3.

145 Antiaging Dock: Evaluation of Well-Being

Hidemichi Suga

The term "well-being" is often heard recently, but it has already been used in the definition of health in the WHO Charter of 1946. "Health is a state of complete physical, mental and social well-being and not merely the absence of disease or infirmity." This is a profound concept that goes beyond medical care and health, stating that health is not the opposite of illness or weakness, but is physical, mental, and social well-being. Psychologically, Diener has defined subjective well-being and psychological well-being. Subjective well-being is the basis of human pleasure-seeking from sensory pleasures, joys, and fun, derived from the fulfillment of pleasurable desires. On the other hand, psychological well-being is a sense of happiness derived from the fulfillment of joys and fun other than sensory fulfillment, such as self-realization and purpose in life. In the Greek era, Aristotle defined the former as hedonia and the latter as eudaimonia. Positive psychology demonstrates the effects of improving mental health through such well-being, and includes it in positive psychosocial factors (PPSFs). Positive psychology was proposed by Seligman [1], and its effectiveness has been demonstrated using experimental psychology for specific methods. It is based on the well-being theory of PERMA, as shown in Fig. 145.1. PERMA stands for *P* positive emotion, *E* engagement (active involvement, flow), *R* relationships (relationships, human connections), *M* meaning (meaning of life), and *A* achievement (sense of goal achievement).

Figure 145.1 is based on the well-being theory called PERMA. PERMA stands for *P* positive emotion, *E* engagement (flow, active involvement), *R* relationships (connections with others), *M* meaning (purpose in life), and *A* achievement (sense of accomplishment), which are the five elements. Positive emotion is the fundamental element of well-being, representing subjective pleasure in the current state. While negative emotions serve as a warning of danger, leading to risk-avoidant behavior, positive emotions signal safety and lead to expanded behavior, making them more evolved emotions than negative ones. Furthermore, these are not fleeting pleasures, but emotions that persist from beginning to end. Specifically, they are expressed as feelings of joy, pleasure, interest, satisfaction, love, ecstasy, comfort, warmth, etc. In surveys measuring positive emotions, subjective evaluations such as "How happy do you think you are?" are used. Engagement is the subjective pleasure derived from reflecting on a state of immersion in something. Csikszentmihalyi has mentioned the well-being effects of this active involvement in his flow theory. That is, it is a process that leads to

H. Suga (✉)
Junior College, Ryukoku University, Kyoto, Japan

H. Yamada, Y. Naito (eds.), *Anti-Aging Medicine*, https://doi.org/10.1007/978-981-97-8514-8_145

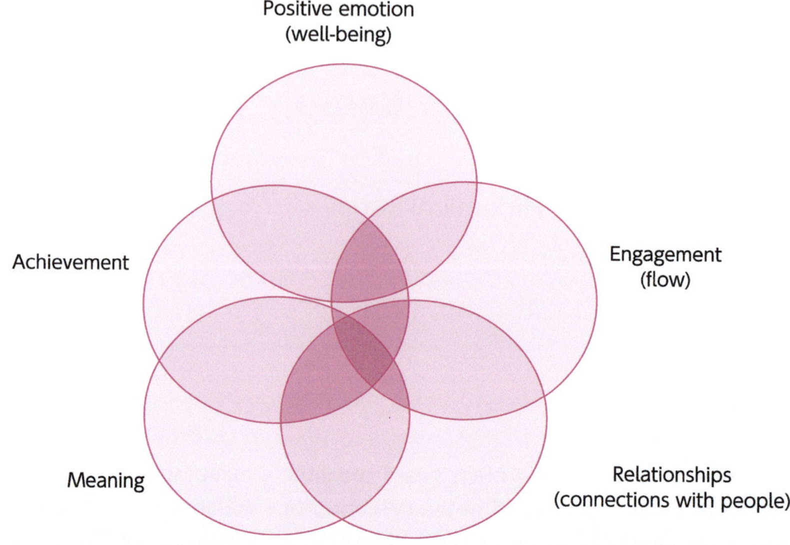

Fig. 145.1 Seligman's PERMA model. (Reprinted with permission from [1])

mood improvement through a sense of accomplishment, a sense of completion, and a sense of liberation by working hard on specific things. In surveys measuring engagement, subjective evaluations such as "Do you like to get involved and immerse yourself in something?" are used. Relationships are the awareness of living in connection with others, not in isolation. One's well-being is recognized through connections with others. Altruistic acts of kindness are also a form of well-being that arises from interactions with others. In surveys measuring relationships, evaluations such as "Do you think there is someone who truly cares about you?" are used. Meaning is the significance and purpose of one's life, which is connected to one's values. Here, objective judgments such as self-evaluation from justice, ethics, logic, and social evaluation are also considered. In surveys measuring the purpose of life, evaluations such as "Do you think what you are doing is beneficial and valuable?" are used. Achievement is the consciousness of having a goal and winning it. Here, motivation is increased by temporary achievements and achievements are accumulated. This leads to a life of achievement. In surveys measuring the sense of achievement, evaluations such as "Do you have a goal to achieve a sense of accomplishment?" are used. These elements are said to lead to the Flourish (prosperity) of a person's life when they increase. In various experimental studies on healthy individuals, the positive psychological social factors of each element have been measured, and the effectiveness of each element in improving subjective happiness has been demonstrated. In addition, Seligman has clinically applied positive psychology therapy [positive The effects of Positive Psychotherapy (PPT)] have been demonstrated. When comparing a group of major depressive patients who underwent PPT with a group who underwent pharmacotherapy with antidepressants plus psychotherapy, the PPT group showed significant improvement. In particular, tools related to positive emotions, active involvement, and finding meaning in life were found to be effective in improving depression. In addition to Seligman, Peterson also mentions the sustainability of the effects of gratitude work in positive psychology. Gratitude work involves imagining someone to whom you have never directly expressed gratitude, writing words of gratitude, visiting that person, and expressing your gratitude. After implementing this work, happiness significantly increased. Based on the backing of these positive psychology theories, the author implements the following methods in clinical practice [2]. Each method is related to PERMA, and its effectiveness is suggested.

1. Enjoyable conversation and smiles (exciting conversation): PERMA's positive emotions, relationships (connections with people).
2. Setting short-term goals and self-affirmation through their achievement: PERMA's active involvement (flow), sense of goal achievement.
3. Affirming oneself by recognizing strengths (recording in a strength notebook): PERMA's positive emotions, finding meaning in life, sense of goal achievement.
4. Gratitude work (Thank you diary): Gratitude work effect.

References

1. Seligman M. (Supervised translation by Kaori Uno): The challenge of positive psychology: from "happiness" to "sustained happiness". Tokyo: Discover Twenty One; 2014.
2. Suga H. Utilization of positive psychiatry: how mental health care will change in 10 years! Tokyo: Seiwa Shoten; 2019.

146 Antiaging Dock: Evaluation of Sarcopenia

Yuki Ito and Kojiro Ishii

146.1 History of the Diagnostic Criteria for Sarcopenia

In 1989, Rosenberg advocated the importance of the decrease in skeletal muscle mass with aging, and proposed the term "sarcopenia," a compound word from the Greek "sarx (muscle)" and "penia (loss)" [1]. Subsequently, in 2010, the European Working Group on Sarcopenia in Older People (EWGSOP) presented the world's first consensus on sarcopenia [2]. This includes not only the decrease in muscle mass, which had been traditionally treated, but also the evaluation of muscle strength and physical function. Since then, various research groups such as the Asian Working Group for Sarcopenia (AWGS) have created diagnostic criteria for sarcopenia, and these criteria commonly include the evaluation of muscle mass, muscle strength, and physical function.

Y. Ito
Department of Lifelong Sports and Health Sciences, College of Life and Health Sciences, Chubu University, Kasugai, Aichi, Japan

Graduate School of Health and Sports Science, Doshisha University, Kyoto, Japan

K. Ishii (✉)
Faculty of Health and Sports Science, Doshisha University, Kyoto, Japan
e-mail: kishii@mail.doshisha.ac.jp

146.2 Diagnostic Criteria by EWGSOP2

Currently, various diagnostic criteria exist worldwide, but the criteria of EWGSOP are widely used. EWGSOP announced a new consensus, "EWGSOP2," in 2018 [3]. EWGSOP2 defines sarcopenia as a "progressive and systemic skeletal muscle disease associated with an increased likelihood of adverse outcomes including falls, fractures, physical disability and mortality." The diagnostic algorithm consists of four stages: ① screening using self-administered questionnaires, ② diagnosis of the possibility of sarcopenia by evaluating muscle strength, ③ definitive diagnosis of sarcopenia by evaluating muscle mass or quality, and ④ diagnosis of the severity of sarcopenia by evaluating physical function (Fig. 146.1). In other words, EWGSOP2 makes a definitive diagnosis of sarcopenia with a decrease in muscle strength and muscle mass or quality, and diagnoses the severity of sarcopenia with a decrease in physical function. A notable point of this criterion is that it emphasizes the decrease in muscle strength as a mandatory item. In previous criteria, the evaluation of muscle mass was a mandatory item, but various studies have shown that a decrease in muscle strength is strongly associated with the occurrence of harmful outcomes [4]. Therefore, in EWGSOP2, if a decrease in muscle strength is observed, it is evaluated as "sarcopenia probable," and it is recommended to

H. Yamada, Y. Naito (eds.), *Anti-Aging Medicine*, https://doi.org/10.1007/978-981-97-8514-8_146

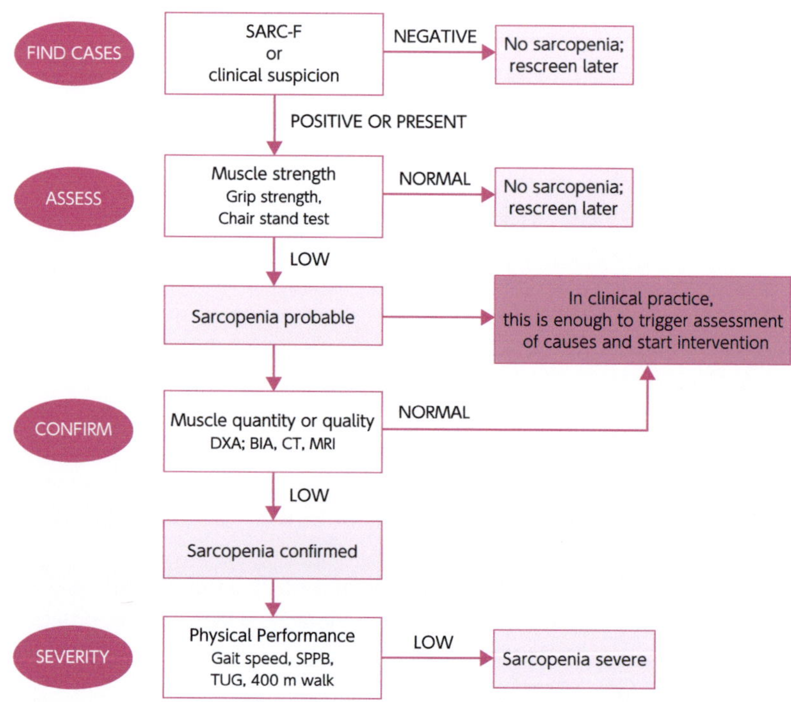

Fig. 146.1 EWGSOP2's sarcopenia diagnostic algorithm. (Cited with permission from [3])

start evaluation and intervention in daily clinical practice. Also, in EWGSOP2, the application range of sarcopenia has been expanded. In the criteria before the revision of EWGSOP, the target of sarcopenia was the elderly over 65 years old. In EWGSOP2, it is considered that sarcopenia can occur not only in the elderly, but also in younger people, and it is divided into "primary sarcopenia" due to aging and "secondary sarcopenia" due to factors such as disease, decreased activity, and nutritional disorders, regardless of age.

146.3 Diagnostic Criteria by AWGS2019

The diagnostic criteria of AWGS are recommended for use in diagnosing sarcopenia in Japanese people, as they are targeted at Asians. AWGS released AWGS2019 as a revised version of the sarcopenia diagnostic criteria announced in 2014 [5]. In AWGS2019, sarcopenia is defined as "a decrease in skeletal muscle mass due to aging, in addition to decreased muscle strength and/or physical function." In other words, unlike EWGSOP2, it includes a decrease in physical function in the definitive diagnosis of sarcopenia. Also, if all of the following apply: decreased muscle mass, decreased muscle strength, and decreased physical function, it is diagnosed as severe sarcopenia. A major feature of AWGS2019 is that the diagnostic algorithm is largely divided into "community and primary care settings" and "hospitals and research facilities" (Fig. 146.2). This is intended to enable the diagnosis of sarcopenia even in community and primary care settings where it is difficult to measure skeletal muscle mass due to the need for expensive dedicated equipment, and to identify and intervene in at-risk individuals at an early stage. Therefore, in community and primary care settings, screening is performed using SARC-F, calf circumference, or a combination of these in SARC-CalF. If a low value is recognized, an evaluation of muscle strength (grip strength) or physical function (5 times chair stand-up test) is performed, and if either is decreased, a diagnosis of possible sarcopenia is made. On the other hand, in hospitals and research facilities where skeletal muscle mass can be measured, the measurement of skeletal muscle mass by dual-energy X-ray absorptiometry (DXA) or bioelectrical impedance analysis (BIA) is performed in addition to the evaluation of mus-

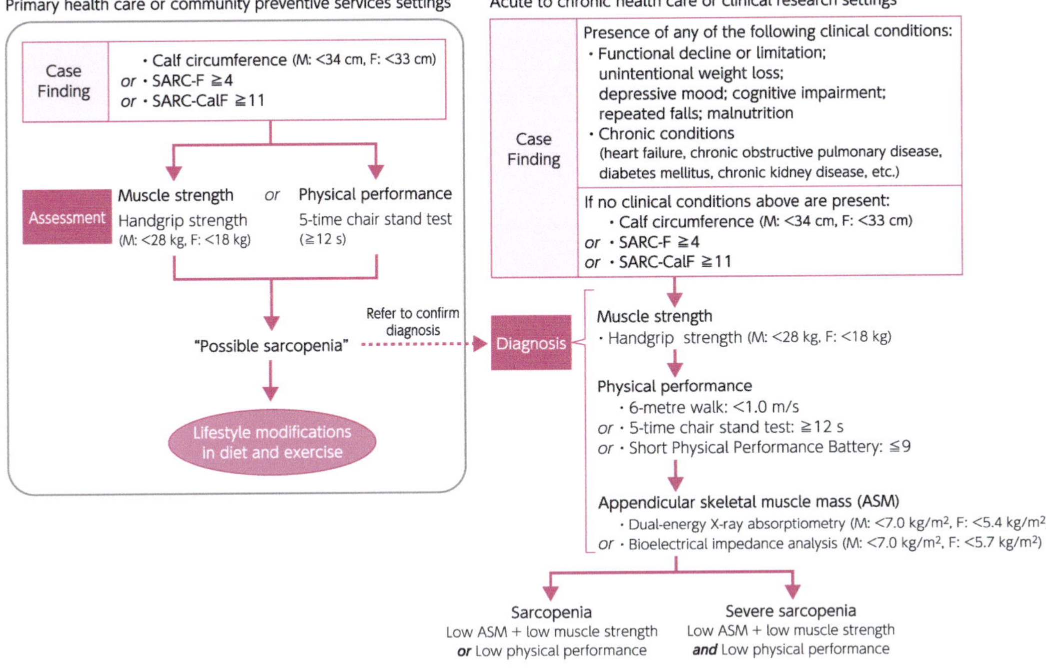

Fig. 146.2 AWGS2019 sarcopenia diagnostic algorithm. (Cited with permission from [4])

cle strength and physical function. Here, muscle strength is evaluated using grip strength as well, and for the evaluation of physical function, in addition to the 5 times chair stand-up test, the choice of 6 m walking or short physical performance battery (SPPB) is also available.

References

1. Rosenberg IH. Summary comments: Epidemiological and methodological problems in determining nutritional status of older persons. Am J Clin Nutr. 1989;50:1231–3.
2. Cruz-Jentoft AJ, Baeyens JP, et al. Sarcopenia: European consensus on definition and diagnosis: report of the European working group on sarcopenia in older people. Age Ageing. 2010;39:412–23.
3. Cruz-Jentoft AJ, Bahat G, et al. Sarcopenia: revised European consensus on definition and diagnosis. Age Ageing. 2019;48:16–31.
4. Leong DP, Teo KK, et al. Prognostic value of grip strength: findings from the prospective urban rural epidemiology (PURE) study. Lancet. 2015;386:266–73.
5. Chen LK, Woo J, et al. Asian Working Group for Sarcopenia: 2019 consensus update on sarcopenia diagnosis and treatment. J Am Med Dir Assoc. 2020;21:300–7.

147 Antiaging Dock: Evaluation of Locomotive Syndrome

Kenji Takahashi and Naoki Okubo

In our country, the rapid progression of an aging society with fewer children is a significant issue, and extending the "healthy lifespan," where one can live independently without the need for care, is a crucial challenge. In 2007, the Japanese Orthopaedic Association named the condition of reduced mobility due to musculoskeletal disorders as Locomotive Syndrome (hereafter, Locomo). Initially, a simple check called Loco-check was used to detect reduced mobility, but in 2013, the Locomo degree test was developed. The Locomo degree test consists of three parts: the stand-up test, the 2-step test, and the Locomo 25 questionnaire (Fig. 147.1). In 2015, a three-tier classification of non-Locomo, Locomo degree 1, and Locomo degree 2 was announced as clinical judgment values (Locomo degrees) for Locomo. Furthermore, in 2020, the most severe Locomo degree 3, which could hinder participation in social activities, was announced, and it was classified into four stages (Table 147.1). The Locomo degree is determined by the results of the three tests, and the most severe Locomo degree is applied.

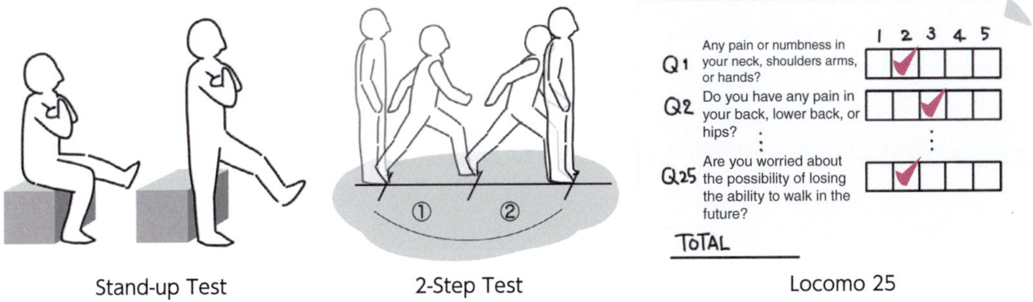

Fig. 147.1 Locomo level test

K. Takahashi (✉) · N. Okubo
Department of Orthopaedics, Kyoto Prefectural University of Medicine, Kyoto, Japan
e-mail: kenji-am@koto.kpu-m.ac.jp

H. Yamada, Y. Naito (eds.), *Anti-Aging Medicine*, https://doi.org/10.1007/978-981-97-8514-8_147

Table 147.1 Locomo level test and clinical judgment values of locomo

	Stand-up test	2-Step test	Locomo 25
Locomo level 1	Unable to stand up from a 40 cm platform with either leg	2-step value is 1.1 or more but less than 1.3	7 points or more but less than 16 points
Locomo level 2	Unable to stand up from a 20 cm platform with both legs	2-step value is 0.9 or more but less than 1.1	16 points or more but less than 24 points
Locomo level 3	Unable to stand up from a 30 cm platform with both legs	2-step value is less than 0.9	24 points or more

147.1 Practical Evaluation of Locomo Degree

147.1.1 Stand-Up Test

Evaluate whether it is possible to stand up from a platform of four different heights (10–40 cm) with both legs or one leg. The basic posture is to cross your arms in front of your chest and stand up without momentum. First, stand up with both legs from a 40 cm platform. If you cannot stand up, it is Locomo degree 3. If you can stand up, perform the 40 cm test with one leg. If you can stand up with one leg on both sides, perform a one-leg stand in the order of 30 cm, 20 cm, 10 cm. On the other hand, if you cannot stand up with one leg at 40 cm, perform a two-leg stand at 30 cm, and if possible, evaluate a two-leg stand at 20 cm and 10 cm in order.

147.1.2 2-Step Test

It is calculated as the maximum two-step length divided by height. The measurement is taken by measuring the maximum two-step length that can be performed without losing balance from a stationary standing position.

147.1.3 Locomo 25 Questionnaire (Locomo 25)

It consists of 25 self-administered questions, each scored on a 5-point scale (0 points for no disability, 4 points for the most severe disability), totaling 0–100 points. The questions cover a wide range of issues, including pain, difficulty in indoor activities/daily life activities/participation in social activities, and anxiety.

147.2 Epidemiology of Locomo

2017 In 2019, the "Locomo 10,000 Nationwide Survey" was conducted targeting regional residents over 20 years old who can walk on their own. In all of the stand-up test, 2-step test, and Locomo 25, the scores of those in their 20s were the best, and significant deterioration was observed from those in their 30s or 40s, and this change progressed with age [1]. In the ROAD study, a large-scale resident cohort study conducted in Japan, the prevalence of Locomo (the proportion of Locomo level 1 or higher) is reported to be 69.8% (men: 68.4%, women: 70.5%), and the number of patients with Locomo level 1 in Japanese people over 40 years old is estimated to be 45.9 million [2]. Although the prevalence of Locomo increases with age, in contrast to the fact that the prevalence of frailty and sarcopenia increases sharply after the age of 70 [3], even under 40 years old, the prevalence of Locomo exceeds 20% for both men and women, which can be said to be a characteristic of Locomo level 1 [2] (Fig. 147.2). From this point, Locomo level 1 is considered to play an important role in early prevention. On the other hand, at the stage of falling into Locomo level 1, it is said that the health QOL evaluated by the patient-based evaluation is not lower than that of non-Locomo [4], and many patients are still unaware of the decrease in mobility, so how quickly Locomo can be detected was a challenge.

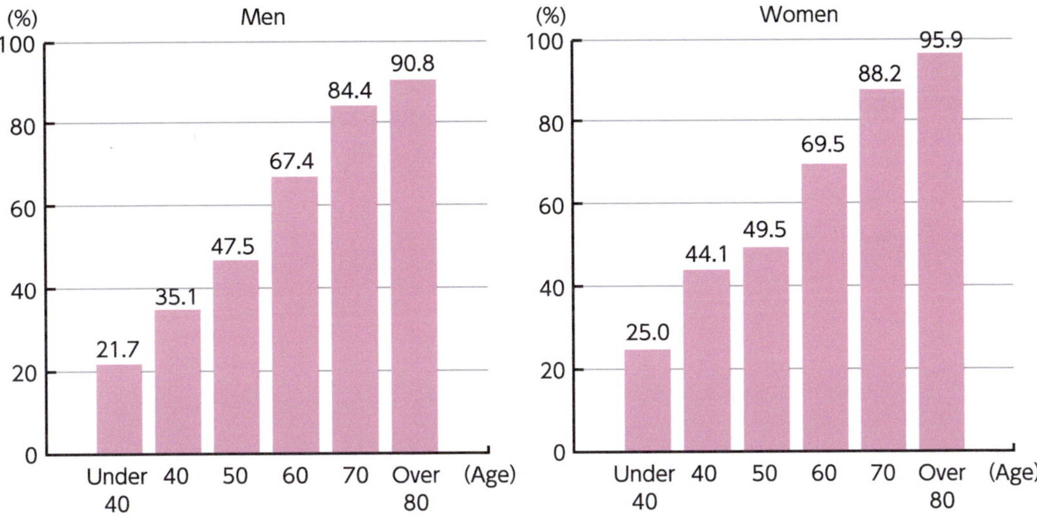

Fig. 147.2 Age-specific changes in the prevalence of Locomo. (Cited with permission from [2])

147.3 Locomo Age

In 2022, the "Locomo Age" judgment tool was released from the Locomo Challenge! Promotion Council. As mentioned earlier, the results of the Locomo level test are calculated based on results that significantly correlate with aging. The biggest feature of the Locomo age measurement is that it can be measured anywhere at any time without visiting a hospital if you have a smartphone. Locomo age measurement is possible either by answering questions about Locomo 25 and standing up and walking, or by answering all three types of Locomo level tests. Although the awareness of Locomo continues to struggle in the 40% range, it is expected that if the awareness improves to the generation with a high penetration rate of smartphones through the visualization of mobility using an age that is generally familiar, it will lead to the prevention of Locomo, and further, the prevention of frailty and sarcopenia.

References

1. Yamada K, Ito YM, et al. Reference values for the locomotive syndrome risk test quantifying mobility of 8681 adults aged 20–89 years: a cross-sectional nationwide study in Japan. J Orthop Sci. 2020;25:1084–92.
2. Yoshimura N, Muraki S, et al. Epidemiology of the locomotive syndrome: the research on osteoarthritis/osteoporosis against disability study 2005–2015. Mod Rheumatol. 2017;27:1–7.
3. Yoshimura N, Muraki S, et al. Is osteoporosis a predictor for future sarcopenia or vice versa? Four-year observations between the second and third ROAD study surveys. Osteoporos Int. 2017;28:189–99.
4. Shigematsu H, Tanaka M, et al. Affirmative answers on loco-check as a predictor of health-related quality of life and locomotive syndrome progression in the elderly: a cross-sectional study. Mod Rheumatol. 2020;30:580–5.

148 Antiaging Dock: Evaluation of Frailty

Sumito Ogawa

Frailty and its concept are summarized in a statement released by the Japan Geriatrics Society in 2014 as "a state that is prone to outcomes such as functional impairment, need for care, and death due to the decline in physiological reserve capacity in old age, which enhances vulnerability to stress." Frailty has an intermediate, reversible nature that is located between health and the need for care, and is distinguished from states that already have physical functional impairment or comorbidities. In our country, it is considered important to evaluate frailty and implement appropriate interventions based on it in order to further extend healthy life expectancy and promote prevention of care. In fact, frail elderly people are more likely to have health problems such as daily life functional impairment, falls, hospitalization, and a higher mortality rate. The evaluation of frailty in Japan is also important for estimating the life and functional prognosis of the elderly and practicing comprehensive medical care.

148.1 Frailty and Its Evaluation Methods

While there are no unified criteria for evaluating frailty, the main methods are the cardiovascular health study criteria (CHS criteria) based on the phenotype model by Fried et al. and the frailty index based on the accumulated deficit model by Rockwood et al. The CHS criteria are positioned as a typical diagnostic method for physical frailty, and in our country, the revised Japanese version of the CHS (revised J-CHS) criteria has been proposed (Table 148.1) [1, 2]. Also, as simple evaluation methods, the simple FRAIL questionnaire, Edmonton frail scale (EFS), Tilburg frailty indicator (TFI), basic checklist, simple frailty index, etc., are known, and their validity has been shown. Thus, various scales and evaluation methods have been proposed for frailty indicators, and it is often the case that multiple items are evaluated together for components such as cognitive function, mobility, muscle strength, balance ability, endurance, nutritional status, and physical activity. In the frailty index by Rockwood et al. one of the representative frailty evaluation methods, the evaluation is based on the concept of comprehensive geriatric assessment (CGA), which accumulates diseases, physical functional impairments, and life functional impairments with aging [3, 4]. There, based on the 7-stage clinical frailty scale, they are classified into ① very fit, ② well, ③ well with treated comorbid dis-

S. Ogawa (✉)
Department of Geriatric Medicine, The University of Tokyo, Tokyo, Japan
e-mail: suogawa@m.u-tokyo.ac.jp

H. Yamada, Y. Naito (eds.), *Anti-Aging Medicine*, https://doi.org/10.1007/978-981-97-8514-8_148

Table 148.1 Revised J-CHS criteria

Item	Evaluation criteria
Weight loss	More than 2 kg of (unintentional) weight loss in 6 months (basic checklist #11)
Decreased muscle strength	Grip strength: men <28 kg, women <18 kg
Fatigue	(Past 2 weeks) feeling tired for no reason (basic checklist #25)
Walking speed	Normal walking speed <1.0 m/s
Physical activity	① Are you doing light exercise or gymnastics? ② Are you doing regular exercise or sports? Answered "not even once a week" for both of the above

(Quoted from Satake S, Arai H: The revised Japanese version of the Cardiovascular Health Study criteria (revised J-CHS criteria). Geriatr Gerontol Int 2020; 20:992–3.)

ease, ④ apparently vulnerable, ⑤ mildly frail, ⑥ moderately frail, and ⑦ severely frail categories, and are used to estimate and evaluate the functional and life prognosis of the elderly. Mildly frail is a state that requires slight assistance in instrumental activities of daily living (IADL), moderately frail is a state that requires assistance in both activities of daily living (ADL) and IADL, and severely frail is a state that requires full assistance in daily activities or is close to the final stage of life. Clinical frailty When the category is high according to the scale, it was shown that it could lead to lower rates of avoidance of facility admission and survival, and it could be useful for evaluating the function of the elderly and predicting their prognosis [4]. In addition, the concept of frailty is thought to include not only the physical aspects of the elderly but also the mental and psychological aspects (cognitive decline, depression, etc.) and social aspects (living alone, economic power, etc.), but the index by Fried et al. mainly defines the phenotype of physical function [1]. There, five symptoms are extracted: ① weight loss, ② subjective decrease in vitality, ③ decrease in grip strength, ④ decrease in walking speed, and ⑤ decrease in activity level, and if three or more items apply, it is considered frail, and if one or two items apply, it is considered prefrail. According to the large-scale resident cohort study ROAD (research on osteoarthritis against disability) in our country, the frequency of frailty was 5.6% (men 3.8%, women 6.6%), and the incidence of frailty increased by 1.2% per year (men 0.8%, women 1.3%) [5].

148.2 Frailty Assessment Using a Questionnaire for the Elderly

In our country, the evaluation of frailty and appropriate intervention are becoming increasingly important in promoting preventive care and further extending healthy life expectancy, and the questionnaire for the elderly, so-called "frailty health checkup," which has started nationwide from 2020, is one of them [6] (Table 148.2). In the questionnaire for the elderly, the purpose is to comprehensively understand the health status, taking into account the characteristics of the elderly such as frailty, and it consists of 10 types: ① health status, ② mental health status, ③ eating habits, ④ oral function, ⑤ weight change, ⑥ exercise/fall, ⑦ cognitive function, ⑧ smoking, ⑨ social participation, and ⑩ social support, considering the knowledge so far and the burden on the elderly respondents, it consists of 15 questions. The evaluation using the questionnaire for the elderly is expected to be implemented and utilized in various situations, such as at health checkups, at local salons and other places to visit, and at medical institutions of family doctors.

Table 148.2 Questionnaire for the elderly

Question	Answer
1. How is your current health condition?	① Good ② fairly good ③ average ④ not so good ⑤ bad
2. Are you satisfied with your daily life?	① Satisfied ② somewhat satisfied ③ somewhat dissatisfied ④ dissatisfied
3. Do you eat three meals a day properly?	① Yes ② no
4. Compared to half a year ago, has it become difficult to eat hard things? *Such as squid, pickled radish, etc.	① Yes ② no
5. Do you choke on tea or soup, etc.?	① Yes ② no
6. Have you lost 2–3 kg or more in weight over the past 6 months?	① Yes ② no
7. Do you think your walking speed has become slower than before?	① Yes ② no
8. Have you fallen in the past year?	① Yes ② no
9. Do you do exercises like walking more than once a week?	① Yes ② no
10. Have people around you told you that you forget things, like asking the same question over and over?	① Yes ② no
11. Do you ever not know what month or day it is?	① Yes ② no
12. Do you smoke?	① Yes, I do smoke ② no, I don't smoke ③ I quit
13. Do you go out more than once a week?	① Yes ② no
14. Do you usually socialize with family or friends?	① Yes ② no
15. Do you have someone close to consult when you are not feeling well?	① Yes ② no

Ministry of Health, Labour and Welfare: 34th Study Group on Health Checkups and Health Guidance by Insurers, Document 4 About the health business for the elderly. March 28, 2019. (Quoted from)

References

1. Fried LP, Tangen CM, et al. Frailty in older adults: evidence for a phenotype. J Gerontol A Biol Sci Med Sci. 2001;56:M146–56.
2. Sateke S, Arai H. The revised Japanese version of the cardiovascular health study criteria (revised J-CHS criteria). Geriatr Gerontol Int. 2020;20:992–3.
3. Rockwood K, Song X, et al. A global clinical measure of fitness and frailty in elderly people. CMAJ. 2005;173:489–95.
4. Rockwood K, Stadnyk K, et al. A brief clinical instrument to classify frailty in elderly people. Lancet. 1999;353:205–6.
5. Yoshimura N, Muraki S, et al. Do sarcopenia and/or osteoporosis increase the risk of frailty? A 4-year observation of the second and third ROAD study surveys. Osteoporos Int. 2018;29:2181–90.
6. Ministry of Health, Labour and Welfare: 34th Study Group on Health Checkups and Health Guidance by Insurers, Document 4 About the health business for the elderly. 2019. https://www.mhlw.go.jp/content/12401000/000494444.pdf.

149 Antiaging Dock: New Appearance Evaluation by AI

Koji Abe

149.1 AI's Medical Applications

The application of artificial intelligence (AI) in medicine is primarily advancing in the field of diagnosis. Its utility is determined by the combination of the hardware, such as the memory capacity and processing speed of the computer itself, and the software, such as AI programs, target medical fields, symptom diagnosis, image diagnosis, etc. Starting with AI for medical interviews that understands the patient's chief complaint and medical history as divided keywords and lists potential diagnoses, there are rapidly spreading technologies such as AI for diagnostic support through the analysis of examination data already applied in electrocardiograms and blood tests, automatic reading and image diagnosis AI for chest X-ray photos and brain MRI images, skin diagnosis AI in the field of dermatology, and face diagnosis AI (Fig. 149.1) developed from face recognition technology. The benefits of using such AI include speed, accuracy, and independence from the diagnostician's ability. Although not yet perfect, its usefulness is quite high, and it is bringing about a revolutionary change in medical diagnosis, similar to the transition from individual combat with swords to group combat with guns on the battlefield. AI technology is expected to further expand into treatment, suggesting standard internal medicine treatments based on guidelines, proposing surgical procedure choices, and postoperative management. Naturally, the application range of AI technology is also expanding in anti-aging medicine.

K. Abe (✉)
National Center of Neurology and Psychiatry (NCNP), Tokyo, Japan

Arts Ginza Clinic, Tokyo, Japan

H. Yamada, Y. Naito (eds.), *Anti-Aging Medicine*, https://doi.org/10.1007/978-981-97-8514-8_149

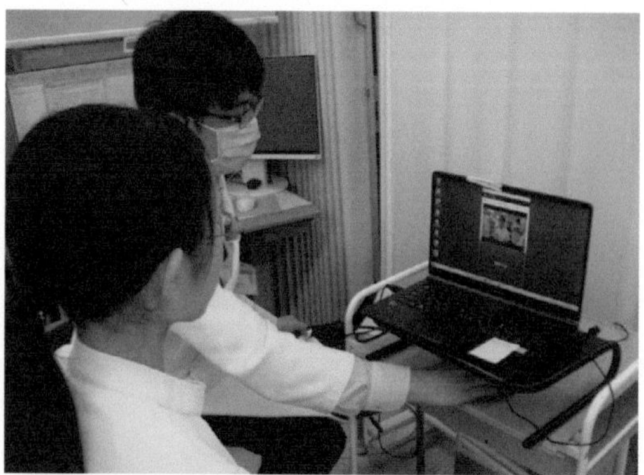
Facial diagnosis AI implementation scene

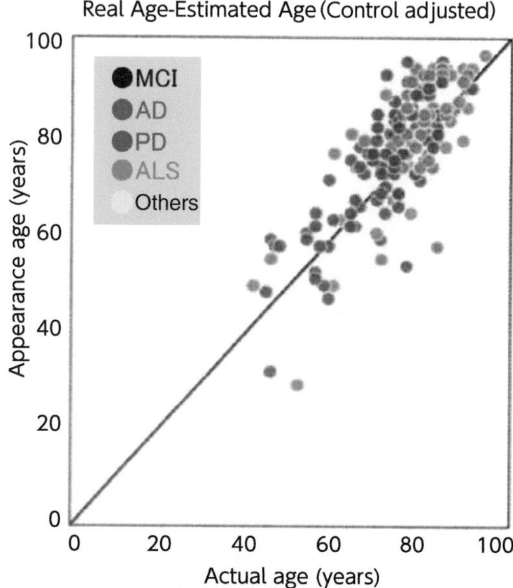

Fig. 149.1 Age and neurological diseases as seen by facial diagnosis AI. Scene of facial diagnosis AI implementation (Right figure: Quoted from with permission [1])

149.2 AI Technology and Antiaging

With the super-aging of Japanese society, the number of dementia patients is rapidly increasing, and the importance of screening tests before onset of dementia or in its early stages is increasing more and more. We have been developing such screening tests using touch panels [2] and eye trackers [3]. Recently, we have started to study aging of the brain and its pathological result, dementia, incorporating the innovative features of AI technology such as speed, accuracy, safety, economy, and independence from the diagnostician's ability [4]. That is, we are advancing verification research with the hypothesis that aging of the brain appears on the face and that screening for dementia is possible with face diagnosis AI. Using the Face API (application programming interface) developed by Microsoft in the United States, we measured the apparent age and emotional expression in about 2 s for various neurological diseases, including dementia. As a result, it was found that patients with Alzheimer's disease and Parkinson's disease looked older than their actual age, and a statistically significant difference was confirmed for Parkinson's disease [4]. Similar studies are also underway in Japan [5]. Furthermore, using this method, we conducted cosmetic beauty therapy for residents of dementia care facilities, and after 3 months, it was clear that the intellectual function, the mini-mental state. The Mini-Mental State Examination (MMSE) score improved, and it was found that the apparent age also rejuvenated in subjects without depressive symptoms, and it is attracting attention as the first anti-aging study using AI technology [1] (Fig. 149.2).

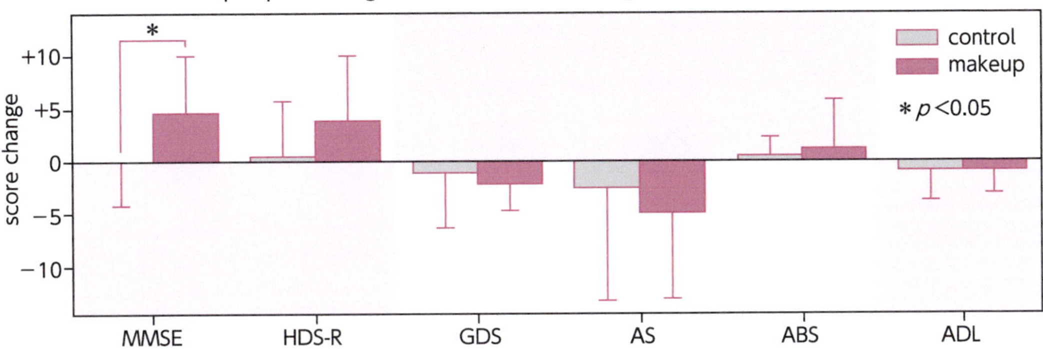

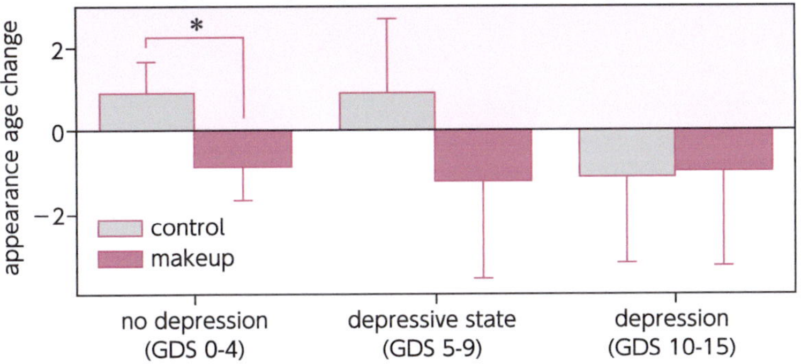

Fig. 149.2 Results of rejuvenation and cognitive function improvement after 3 months of cosmetic beauty therapy. Result 1. Makeup improved cognitive function MMSE (long-term effect). Result 2. Apparent age rejuvenated in the makeup group without depression (Quoted with permission from [1])

References

1. Tadokoro K, Yamashita T, et al. Immediate beneficial effect of makeup therapy on behavioral and psychological symptoms of dementia and facial appearance analyzed by artificial intelligence software. J Alzheimers Dis. 2021;83:57–63.
2. Fukui Y, Yamashita T, et al. Computerized touch-panel screening tests for detecting mild cognitive impairment and Alzheimer's disease. Intern Med. 2015;54:895–902.
3. Tadokoro K, Yamashita T, et al. Early detection of cognitive decline in mild cognitive impairment and Alzheimer's disease with a novel eye tracking test. J Neurol Sci. 2021;427:117529.
4. Tadokoro K, Yamashita T, et al. Detecting facial characteristics of Parkinson's disease by novel artificial intelligence (AI) softwares. Brain Suppl. 2021;3:1–7.
5. Umeda-Kameyama Y, Kameyama M, et al. Screening of Alzheimer's disease by facial complexion using artificial intelligence. Aging (Albany NY). 2021;13:1765–72.

150 Antiaging Interventions: Weight Loss and Metabolic Improvement Surgery and Antiaging

Akira Sasaki

150.1 Surgical Treatment for Severe Obesity

The goal of obesity treatment is not to eliminate obesity, but to reduce obesity and improve obesity-related health disorders. In the treatment of obesity, nutritional, exercise, cognitive-behavioral, and pharmacological therapies have been implemented after understanding the patient's complex pathology, but in severe obesity with a BMI (body mass index) of 35 kg/m^2 or more, it is often extremely difficult to reduce and maintain weight through internal medicine treatment. Surgical treatment for severe obesity has been reported both overseas and in our country to be able to maintain effective weight loss in the long term compared to internal medicine treatment, and the improvement of obesity-related health disorders is also good [1]. Obesity surgery has been generally referred to as weight loss and metabolic improvement surgery, as metabolic improvements and various changes in gastrointestinal hormones are confirmed from the early postoperative period before weight loss occurs [1].

150.2 Types of Weight Loss and Metabolic Improvement Surgery

There are adjustable gastric banding and sleeve gastrectomy as surgical methods to limit the intake of food. In addition to surgeries that limit food intake, there are gastric bypass surgery and sleeve gastrectomy and duodenal jejunal bypass surgery as surgical methods that suppress digestion and absorption by bypassing the small intestine (Fig. 150.1). The only procedure covered by insurance in our country is laparoscopic sleeve gastrectomy (LSG), which is a method of leaving the stomach on the lesser curvature side in a banana shape (about 100 mL) by removing the greater curvature side. This procedure is recommended for obese T2D (type 2 diabetes) patients with a short history of disease and well-preserved insulin secretion [1, 2]. In 2021, LSG accounted for 93.1% of the 890 weight loss and metabolic improvement surgeries performed in our country.

A. Sasaki (✉)
Department of Surgery, School of Medicine, Iwate Medical University, Iwate, Japan
e-mail: sakira@iwate-med.ac.jp

H. Yamada, Y. Naito (eds.), *Anti-Aging Medicine*, https://doi.org/10.1007/978-981-97-8514-8_150

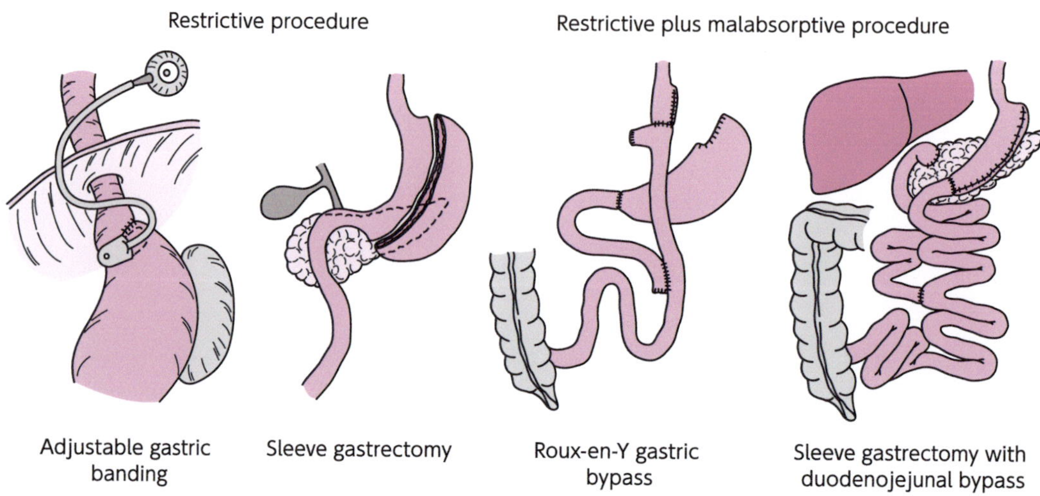

Fig. 150.1 Main weight loss/metabolic improvement surgery dietary intake restriction surgery. (Modified with permission from [3])

150.3 Surgical Indication Criteria

The indication criteria for LSG in our country are severe obesity with a BMI of 35 or more, which does not achieve sufficient effects even with internal medicine treatment for 6 months or more, and is complicated by one or more of T2D, hypertension, dyslipidemia, or sleep apnea syndrome. In the 2020 medical fee revision, it is obesity with a BMI of 32.5–34.9 and hemoglobin A1c (HbA1c) 8.4, which does not achieve sufficient effects even with internal medicine treatment for 6 months or more. Additional calculations were applied to more than a percentage of T2D patients. The calculation conditions are, ① hypertension (limited to those with systolic blood pressure of 160 mmHg or more, which is difficult to manage even with antihypertensive drug treatment for 6 months or more), ② dyslipidemia (limited to those with LDL cholesterol of 140 mg/dL or more or non-HDL cholesterol of 170 mg/dL or more, which is difficult to manage even with statin drug treatment for 6 months or more), and ③ obstructive sleep apnea syndrome [limited to severe cases with an apnea hypopnea index (AHI) of 30 or more] were limited to patients with one or more of these complications. The fact that the applicable obesity-related disorders are limited and the application conditions are strict, and there are many problems such as the lack of other surgical options, is one of the reasons why weight loss and metabolic improvement surgery is not widespread in our country. Against this background, a consensus statement on weight loss and metabolic improvement surgery for obese T2D patients in Japan was published by the Joint Committee of the three academic societies (Japan Society for the Treatment of Obesity, Japan Diabetes Society, and Japan Obesity Society) [1]. In the surgical indication criteria, ① in T2D with a BMI of 35 or more at the time of consultation, if the BMI remains 35 or more even after 6 months of treatment by a diabetes specialist or obesity specialist, weight loss and metabolic improvement surgery was recommended as a treatment option regardless of blood glucose control (recommendation grade: recommendation). ② In T2D with a BMI of 32 or more at the time of consultation, if a weight loss of 5% or more cannot be achieved within 6 months of treatment by a diabetes specialist or obesity specialist, or if blood glucose control is HbA1c8.0% or more, surgery should be considered as a treatment option (recommendation grade: consideration). Based on the consensus statement created this time, if appropriate informed consent is given to obese T2D patients, there is a possibility that the number of patients who are candidates for weight loss and metabolic improvement surgery will increase.

150.4 Effects on Obesity-Related Health Disorders

150.4.1 Effect on T2D

As a result in our country, there is the J-SMART (Japanese Survey of Morbid and Treatment-Resistant Obesity) study by 10 facilities certified by the Japan Obesity Treatment Society. Among 831 severely obese people (average BMI 43.7), the total weight loss rate (TWL) was 29.9% and the complete remission rate of T2D (HbA1c less than 6.0% and no diabetes medication) was 75.6% after 2 years of LSG3) [4] (Fig. 150.2). In the group with a preoperative BMI of 32.0–34.9, the preoperative HbA1c and visceral/subcutaneous fat area ratio were higher compared to the group with a higher BMI, and the T2D resolution rate after 2 years of surgery was 52.4%, and the total weight loss rate was 22%, although inferior, it has been reported that the amount of decrease in HbA1c value was superior, and the decrease in the number of T2D treatment drugs and the rate of insulin use was equal or more [1]. As a prediction method for T2D remission after weight loss and metabolic improvement surgery, the ABCD score, which combines age, BMI, fasting C-peptide, and duration of diabetes, is recommended [5] (Table 150.1). The ABCD score is classified from 0 to 10 points, with higher scores indicating a higher rate of T2D remission. In our country, a correlation has been found with the ABCD score in the results of weight loss/metabolic improvement surgery, and the T2D remission rate of LSG is reported to be 47% for an ABCD score of 5 points or less, and 91% for 6 points or more [1].

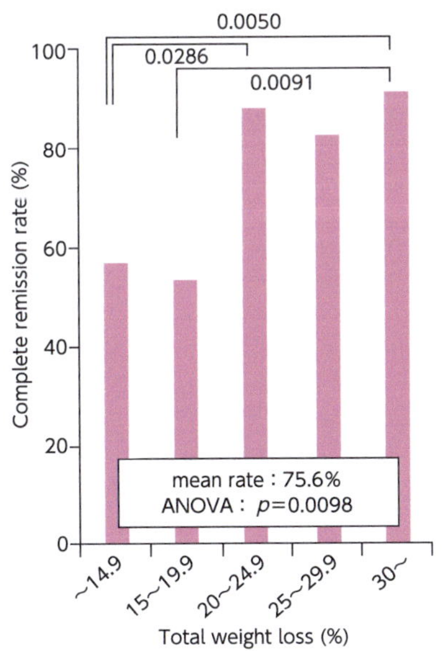

Fig. 150.2 T2D remission rate at 2 years after LSG [4]

150.4.2 Effect on Non-Alcoholic Steatohepatitis (NASH)

In the "NAFLD/NASH Clinical Guidelines 2020", in response to the background question "Is weight loss surgery useful for severe obesity NAFLD/NASH?", it is answered that "Weight loss surgery is useful for improving hepatic steatosis and inflammation in patients with severe obesity NASH", and the recommendation level of surgical treatment is said to be undeterminable. In our study, the NAFLD (nonalcoholic fatty liver disease) activity score showed significant improvement after LSG, and in multivariate analysis, the prognostic factor for hepatic fibrosis improvement was HbA1c ($p = 0.013$, odds ratio 0.440, 95% confidence interval 0.229–0.842, cut-off value 8.1%) [6]. Also, at the time of LSG, histopathologically NASH (nonalcoholic steatohepatitis) was found in 35 people (56%), but in the follow-up liver biopsy, it decreased to

Table 150.1 ABCD score [5]

Score	0 Point	1 Point	2 Points	3 Points
Age (years old)	≧40	<40		
BMI (kg/m²)	<27	27–34.9	35–41.9	>42
CPR: C-peptide in blood (ng/mL)	<2.0	2–2.9	3–4.9	≧5
Duration: duration of diabetes (years)	>8	4–8	1–3.9	<1

43% at 6 months after LSG and 26% at 1 year [7], and in the future, the expansion of indications for LSG for NASH is expected.

References

1. Joint committee on the criteria for weight loss and metabolic improvement surgery for obese type 2 diabetic patients in Japan edited. In: Consensus statement on weight loss and metabolic improvement surgery for obese type 2 diabetic patients in Japan. Tokyo: Compass Publishing; 2021. p. 29–52.
2. Sasaki A, Yokote K, et al. Metabolic surgery in treatment of obese Japanese patients with type 2 diabetes: a joint consensus statement from the Japanese Society for Treatment of obesity, the Japan diabetes society, and the Japan Society for the Study of obesity. Diabetol Int. 2021;13:1–30.
3. Obesity Surgery Treatment Guidelines Committee of the Japan Society for the Treatment of Obesity: Guidelines for safe and excellent surgical treatment for severe obesity in Japan (2013 edition), p. 9.
4. Saiki A, Yamaguchi T, et al. Background characteristics and postoperative outcomes of insufficient weight loss after laparoscopic sleeve gastrectomy in Japanese patients. Ann Gastroenterol Surg. 2019;26:638–47.
5. Lee WJ, Chong K, et al. Metabolic surgery for diabetes treatment: sleeve gastrectomy or gastric bypass? World J Surg. 2017;41:216–23.
6. Nikai H, Ishida K, et al. Effects of laparoscopic sleeve gastrectomy on non-alcoholic steatohepatitis and liver fibrosis in Japanese patients with severe obesity. Obes Surg. 2020;30:2579–87.
7. Sasaki A, Umemura A, et al. The concept of indeterminable NASH inducted by preoperative diet and metabolic surgery: analyses of histopathological and clinical features. Biomedicines. 2022;10:453.

Antiaging Interventions: Aesthetic Medicine

151

Hiroyuki Ohjimi

Aging-related changes in appearance and body shape are treated in plastic surgery, dermatology, cosmetic surgery, and aesthetic dermatology. In this context, aesthetic medicine (cosmetic surgery, aesthetic dermatology), which is a self-pay medical service, includes two types of techniques: surgical procedures and non-surgical procedures. The former includes facelift surgery, eyelid surgery, facial bone contouring, fat injection, autologous hair transplantation, rhinoplasty, breast augmentation, breast lift/reduction, feminization breast surgery, abdominal wall formation, and liposuction. The latter includes injection treatments using botulinum toxin and hyaluronic acid preparations, facial rejuvenation treatments using chemical peels and lasers, and other treatments such as hair removal, nonsurgical fat removal slimming, sclerotherapy, and regenerative therapy. According to the annual national survey of aesthetic medicine conducted by the Japan Society of Aesthetic Plastic Surgery (JSAPS), approximately two million aesthetic medical procedures were performed at aesthetic medical facilities nationwide in 2019. Of these, surgical treatments accounted for 13.1%, and nonsurgical treatments accounted for 86.9%. Aesthetic medicine is increasing in nonsurgical treatments in many countries, but in our country, the proportion is particularly high. Also, in our country, the face is the main treatment area, which is also a characteristic [1]. In aesthetic medicine, unlike general insurance medical treatment, there is a situation where many unapproved drugs, medical materials, and medical devices are used. These are used for patients by individual import by doctors, and some are used without knowing their effectiveness and safety. Many adverse events, such as granulomas after filler injection, have occurred, and have been a concern for professional societies and administrative agencies. In this context, the Ministry of Health, Labor and Welfare's scientific research project on aesthetic medicine was launched in the first year of Reiwa (2019). The purpose of this is to enhance the safety and improve the quality of aesthetic medicine by the five major societies related to aesthetic medicine jointly conducting a survey of complications related to aesthetic medicine and creating treatment guidelines. The aesthetic medicine treatment guidelines examined the effectiveness and safety of nonsurgical treatments, which are frequently performed in aesthetic medicine. The first year of Reiwa (2019) edition has already been published [2]. In this section, we will explain antiaging aesthetic medicine for the face and trunk according to the first year of Reiwa (2019) edition and the third year of Reiwa (2021) edition of the aesthetic medicine treatment guidelines.

H. Ohjimi (✉)
Department of Plastic, Reconstructive and Aesthetic Surgery, Kitasato University, Sagamihara, Kanagawa, Japan
e-mail: plast1980@ohjimi-med.com

H. Yamada, Y. Naito (eds.), *Anti-Aging Medicine*, https://doi.org/10.1007/978-981-97-8514-8_151

151.1 Antiaging Aesthetic Medicine for the Face

Aging changes in the face are expressed in general terms as spots, wrinkles, and sagging. Spots are changes in skin tone, and in medical terms, they refer to acquired melanin pigmentation. This includes solar lentigo (senile pigmentation), melasma, freckles, bilateral delayed Ota's nevus (dermal melanocytosis), and postinflammatory hyperpigmentation. Laser and light therapy are effective against the most common solar lentigo. The problem is differentiating it from malignant tumors. In doubtful cases, it is necessary to consider thorough examination by dermoscopy or biopsy before laser treatment. In the case of melasma, tranexamic acid treatment is performed with attention to shading. Laser and light therapy [IPL (intense pulsed light) irradiation] for melasma is recommended in the guidelines for cosmetic medical treatment as a combined therapy with conservative treatment [3]. Wrinkles are changes in the texture of the skin, and sagging is the drooping of tissues due to gravity. Devices used for the purpose of improving facial wrinkles and sagging include fractional laser therapy, radiofrequency (RF) treatment, and high-intensity focused ultrasound (HIFU) therapy. Although these do not match the effects of surgical procedures or injectable treatments, they are performed as one of the options for patients who wish to improve wrinkles and sagging by non-surgical methods, as scar formation is rare and no foreign bodies are left in the body [2]. Absorbable filler (filling agent) injection therapy for wrinkles and sagging is currently dominated by hyaluronic acid preparations. It aims to improve facial wrinkles, replenish bone and fat volume loss, and correct shape. Absorbable fillers are physiologically absorbed, so the treatment is reversible. Serious complications include blindness and skin necrosis due to circulatory disorders, and it is important for a doctor who understands facial anatomy and has the correct technique to perform it [3]. There are various types of injection materials for fillers, and non-absorbable fillers have also been used. In recent years, there are high polymer processed products represented by polyacrylamide (PAAG). All of these have the risk of causing late complications such as granuloma formation. In addition, because removal is difficult and long-term safety has not been established, the guidelines for cosmetic medical treatment strongly recommend not performing it [2]. Botulinum toxin treatment for facial wrinkles and sagging is also performed frequently. The same preparation injection can be expected to improve facial expression wrinkles. There have been reports of local reactions at the injection site and eyelid ptosis, but serious complications have not been observed and it can be said to be a relatively safe treatment [2]. Platelet-rich plasma (PRP) therapy for wrinkles and sagging is also performed frequently. PRP monotherapy has been reported to be effective for wrinkles and is weakly recommended. However, human basic fibroblast growth factor (basic fibroblast growth Regarding the addition of platelet-rich plasma (PRP) therapy with basic fibroblast growth factor (bFGF), there have been many reports of complications such as induration and swelling at the injection site, and it is recommended not to perform it because its safety cannot be guaranteed [3]. In recent years, thread lift treatments have been frequently performed for facial wrinkles and sagging. Thread lift treatment is a method of lifting the soft tissues of the face using specially processed threads. Thread lifts can temporarily improve facial wrinkles and sagging, but the effect is not significant, and it is recommended in the cosmetic medical treatment guidelines to perform it after fully explaining that the duration of the effect is short-term [3].

151.2 Antiaging Cosmetic Medicine for the Trunk

Breast augmentation surgery (breast augmentation) is commonly performed with breast implants or fat injections overseas. However, in our country, injection treatments using non-absorbable or absorbable fillers are often performed. In particular, breast augmentation surgery using non-absorbable fillers has repeatedly caused similar health damage in the past, regardless of the change in materials such as paraffin, vaseline, silicone, polyacrylamide hydrogel, and aquafilling. Considering these circumstances, it is not recommended to perform breast augmentation surgery with non-absorbable filler injections [2]. For hyaluronic acid preparations, which are absorbable fillers, the conclusion was reached in the 2021 treatment guidelines that it is recommended not to perform them. The reason is that hyaluronic acid injections have been pointed out as potentially causing capsular contracture and hindering breast cancer screening, and they are not approved for use in the United States Food and Drug Administration (FDA) or Europe [3]. Regarding breast augmentation by fat injection, it is recommended to perform it because effective results can be expected while avoiding complications by steadily selecting appropriate patients, preoperative evaluation, surgical techniques by experienced surgeons, and postoperative follow-up [2].

References

1. Japan Society of Aesthetic Plastic Surgery Survey Committee: 3rd National Cosmetic Medical Practice Survey Final Report (Public Version). 2020. https://www.jsaps.com/jsaps_explore_3.html.
2. Yamada H, Hashimoto I, et al. Cosmetic medical treatment guidelines fiscal 2019 Ministry of Health, Labor and Welfare special research project. J Jpn Soc Aesthet Plast Surg. 2020;42:91–139.
3. Ohjimi H, Yamada H, et al. Survey of complications in cosmetic medicine and creation of treatment guidelines and issues for building a system for ensuring medical safety. Report on the Guidelines for Cosmetic Medical Treatment. Ministry of Health, Labor and Welfare Scientific Research Results Database. 2021. https://mhlw-grants.niph.go.jp/project/158585.

Antiaging Interventions: Chelation Therapy

152

Tadashi Mitsuo

152.1 Overview

Chelation therapy is a general term for treatment methods that use chelating agents to form complexes with metals and excrete them from the body. The term "chelation" is said to originate from the Greek word "chele," meaning "crab's claw." The human body maintains its physiological functions through the action of various metal atoms, but the presence of metals in excess of the necessary amount or harmful metals is not desirable as it produces excessive free radicals and inhibits cellular metabolic functions. Chelation therapy is attracting attention in antiaging medicine because it is expected to have therapeutic and preventive effects on arteriosclerosis and to normalize biological functions by removing harmful metals. Its origin dates back to the metal ligand theory proposed by Alfred Werner in 1893. In the 1940s, Na_2EDTA (ethylenediaminetetraacetic acid), one of the chelating agents, was used as a treatment for lead poisoning and hypercalcemia. Furthermore, in the 1950s, it was reported that the symptoms of ischemic heart disease improved with the administration of Na_2EDTA, and it was effective in the treatment of angina pectoris [1]. Since then, EDTA chelation therapy has been used as a noninvasive treatment for atherosclerotic cardiovascular disease in a wide range of countries, including the United States, Europe, Brazil, and Southeast Asia. However, there was no solid evidence that chelation therapy was beneficial for atherosclerotic diseases, and the evaluation of this treatment had not been settled for many years. Therefore, the National Center for Complementary and Alternative Medicine (NCCAM), a related agency of the U.S. government, started a 5-year clinical trial, the TACT (trial to assess chelation therapy) study, in the spring of 2003, with an investment of 30 million US dollars. This study is a double-blind trial to evaluate the therapeutic effect of EDTA chelation therapy on patients with a history of myocardial infarction. The results were published in the JAMA journal in 2013, and it was reported that it has a preventive effect on myocardial infarction [2]. After more than half a century, the effect of chelation therapy has finally been properly evaluated. In addition, this clinical trial highlighted the significant effect of preventing recurrent myocardial infarction in diabetic patients, and since 2017, a clinical study focusing on the effect of chelation therapy in diabetic patients, TACT2, has been conducted [3].

T. Mitsuo (✉)
Mitsuo Clinic, Tokyo, Japan

H. Yamada, Y. Naito (eds.), *Anti-Aging Medicine*, https://doi.org/10.1007/978-981-97-8514-8_152

152.2 The Practice of Chelation Therapy

EDTA chelation therapy involves repeatedly infusing Na_2EDTA along with vitamins and minerals. The standard for administration is the American College for Advancement in This is provided by the American College for Advancement in Medicine (ACAM, www.acam.org). The major mechanism of action is believed to be the strong antioxidant power and calcium excretion effect of EDTA. It is thought that one of the mechanisms of arterial effect treatment is to suppress oxidative stress in the body caused by active oxygen by administering EDTA intravenously. Another mechanism of atherosclerosis treatment is to promote the excretion of free calcium in the blood, which is pointed out to affect the body's calcium metabolism.

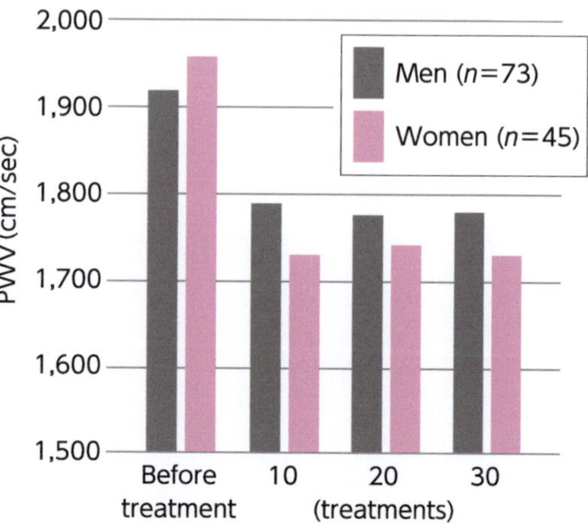

Fig. 152.1 Changes in PWV values due to chelation therapy. As a result of repeatedly performing chelation therapy, it was confirmed how the PWV values change. The PWV values tended to decrease after the completion of 10 treatments

152.3 Changes in PWV Values Due to EDTA Chelation Therapy

One of the methods to objectively measure arteriosclerotic changes is the pulse wave velocity (PWV) measurement method. This shows the changes in PWV values of 118 people (73 men, 45 women) who received chelation therapy more than 30 times at our hospital (Fig. 152.1). Along with chelation therapy, there is a tendency for PWV values, which indicate the degree of arteriosclerosis, to decrease. PWV values are easily affected by blood pressure, but no significant difference in blood pressure before and after treatment has been confirmed in the control group. These results suggest that chelation therapy is an effective treatment for arteriosclerosis.

152.4 Contraindications and Side Effects of EDTA Chelation Therapy

Since EDTA is known to have nephrotoxicity, caution is needed when administering it to patients with renal failure. Recent reports have confirmed that in patients with a high accumulation of lead in the body and renal failure, it is possible to prevent the deterioration of renal function by chelation therapy [4]. This suggests the possibility that EDTA prevented kidney damage caused by lead, and we should consider the indication for chelation therapy for such patients in the future. Hypocalcemia, hypoglycemia, and fatigue are reported side effects after chelation therapy, but their frequency is rare.

152.5 Harmful Metals and Chelation Therapy

The issue of harmful metal contamination can be cited as one of the causes of undefined complaints. Health damage caused by harmful metals such as lead is an important theme that should not be forgotten clinically [5]. Chelation therapy, which started as a treatment for lead poisoning patients, is indispensable in the diagnosis and treatment of harmful metal contamination. Harmful metals that modern people should be aware of include arsenic, lead, mercury, and cadmium. Hair tests and spectrophotometric tests are used as screening tests for diagnosis, but the most reliable test is to collect urine after dripping a chelating agent and examine the metal concentration excreted in the urine. If a high concentration of metal excretion is confirmed, repeated chelation therapy is required to promote the excretion of harmful metals from the body. The improvement of clinical symptoms such as neurological symptoms along with the excretion of harmful metals can be said to prove that chelation therapy is beneficial.

References

1. Clarke CN, Clarke NE, et al. Treatment of angina pectoris with disodium ethylene diamine tetraacetic acid. Am J Med Sci. 1956;232:654–66.
2. Lamas GA, Goertz C, et al. Effect of disodium EDTA chelation regimen on cardiovascular events in patients with previous myocardial infarction: the TACT randomized trial. JAMA. 2013;309:1241–50.
3. Lamas GA, Navas-Acien A, et al. Heavy metals, cardiovascular disease, and the unexpected benefits of chelation therapy. J Am Coll Cardiol. 2016;67:2411–8.
4. Lin JL, Lin-Tan DT, et al. Environmental lead exposure and progression of chronic renal diseases in patients without diabetes. N Engl J Med. 2003;348:277–86.
5. Lanphear BP, Rauch S, et al. Low-level lead exposure and mortality in US adults: a population-based cohort study. Lancet Public Health. 2018;3:e177–84.

Antiaging Interventions: Global Clinical Trials on Antiaging Treatments

153

Goro Katsuumi and Tohru Minamino

Until now, the field of lifespan extension and aging research, which has seen remarkable developments in basic research, is gradually being applied clinically. In recent years, numerous clinical trials aimed at antiaging treatments have been conducted, mainly in the United States. This section introduces the main clinical trials on antiaging treatments and discusses prospects and issues.

153.1 Metabolic-Related Drugs

Many of the drugs used as antiaging drugs are metabolic-related drugs and dietary interventions, mainly those that can forcibly create a state of calorie restriction or starvation or something similar. The first to mention would be metformin. Metformin, an old drug derived in the 1960s, still holds a prominent position as a diabetes treatment. From subanalyses of trials such as the United Kingdom Prospective Diabetes Study (UKPDS) and metaanalyses, it is expected that metformin has the potential as an antiaging drug, as metformin users were found to have healthier longevity. In fact, small-scale trials have already shown the effect of improving physical and metabolic functions in elderly people over 65 years old who are not diabetic by taking metformin, and the Targeting Aging with Metformin (TAME) trial, involving 3000 people, is currently underway [1]. The mTOR (mammalian target of rapamycin) inhibitor rapamycin is also a promising life-extending and antiaging treatment, and abundant evidence of antiaging effects has been accumulated in animal experiments. It has already been marketed as a treatment for lymphangioleiomyomatosis in our country and others. The advantage is that clinical application through drug repositioning can be easily performed in conjunction with metformin. Currently, a large-scale phase II trial called the Participatory Evaluation (of) Aging (With) Rapamycin (for) Longevity Study (PEARL) (NCT04488601) is underway. Clinical trials are also being conducted on nicotinamide mononucleotide (NMN) and nicotinamide riboside (NR), which are NAD (nicotinamide adenine dinucleotide) metabolites. Results of small-scale clinical trials of NMN and NR have been reported, showing improvements in muscle function and insulin sensitivity in middle-aged and older people [2]. Larger trials are not yet underway, but are expected to be conducted soon. Resveratrol, a sirtuin activator, has long been expected as an anti-aging drug, and lifespan extension effects and metabolic improvement effects have been obtained in animal experiments using mice and marmosets. However, a clinical trial in humans reported in 2008 did not

G. Katsuumi (✉) · T. Minamino
Department of Cardiovascular Biology and Medicine, Juntendo University School of Medicine, Tokyo, Japan
e-mail: g.katsuumi.qv@juntendo.ac.jp

H. Yamada, Y. Naito (eds.), *Anti-Aging Medicine*, https://doi.org/10.1007/978-981-97-8514-8_153

show a lifespan extension effect, and subsequent research has waned. The intake necessary to show anti-aging effects is difficult from normal dietary intake, and it is necessary to supplement it as a supplement, but on the other hand, high doses can cause side effects such as kidney damage and CYP (cytochrome P450). The enzyme can be affected, and if there are other drugs being taken, it can cause drug interactions. However, the potential benefits of resveratrol may be recognized if targeted at specific attributes such as obesity, and there is still the possibility of promising findings being generated in the future [3]. In the past, randomized controlled trials using vitamin D and omega-3 unsaturated fatty acids have been conducted, but neither was able to significantly suppress cardiovascular events, cancer onset, or death. Recently, it has been reported that acarbose, an alpha-glucosidase inhibitor, and canagliflozin, a type of SGLT2 (sodium-glucose cotransporter 2) inhibitor, have been found to extend lifespan in mouse experiments, and these diabetes drugs are expected to undergo larger clinical trials as anti-aging drugs in the future.

153.2 Dietary Intervention

Caloric restriction, "time-restricted eating," ketogenic diet, and Mediterranean diet are mentioned as dietary interventions that are expected to have lifespan extension and age-related disease prevention effects. All of them are thought to work through the enhancement of hunger signals, similar to metabolic-related drugs, and many findings on lifespan extension and age-related disease prevention effects have been accumulated in experiments using mice. However, while clinical trials are being conducted worldwide, there are few large-scale ones evaluating lifespan and disease onset. A clinical trial of time-restricted eating was conducted on about 100 healthy individuals, but no clear antiaging effect was observed [4]. The only one that showed promising effects was the Mediterranean diet. The Mediterranean diet, which has been practiced in southern Europe since the 1960s, refers to a dietary habit of limiting meat, eggs, and dairy products, and focusing on vegetables, beans, grains, olive oil, and fish. The PREDIMED (Prevención con Dieta Mediterránea) trial, a large-scale clinical trial practicing this diet, was conducted mainly by Harvard University, and it was shown to significantly reduce the incidence of major cardiovascular events (myocardial infarction, stroke, cardiovascular death) by about 30% in middle-aged and older people with high cardiovascular risk [5]. Although it is not strictly a trial targeting aging, and the reliability of the paper itself has been shaken due to problems with statistical methods, including a retraction, it may be premature to consider this as an anti-aging effect, but the significant results of a large-scale clinical trial on dietary intervention would have had a major impact.

153.3 Anticell Aging Treatment: Telomere Extension, Senescent Cell Removal Drugs

Several clinical applications targeting the cellular aging mechanism have been attempted, and I will introduce the main ones. First, there are therapies targeting telomeres, particularly the introduction or activation of telomerase, which is expected to be a promising treatment. TA-65 (cycloastragenol) is a natural compound extracted from the root of a plant called Kibanaougi, and since the 2000s, it has been reported to have telomerase activating effects in various types of cells. Not only in animal experiments but also in clinical trials in humans, the telomere lengthening effect in leukocytes has been confirmed in healthy middle-aged and older individuals. Although small-scale, improvements in lipids and inflammation markers have been observed in patients with metabolic syndrome [6]. Currently, no large-scale clinical trials have been conducted, but there is a possibility that they will be planned in the future. In addition, a clinical trial (NCT04133454) is underway to introduce hTERT (human telomerase reverse transcriptase) using an adenovirus vector. Antiaging treatments through the control of SASP (senescence-associated secretory phenotype), including inflammatory cytokines, are also promising. Many reports at the animal experiment level have shown effectiveness, and it is considered one of the promising treatment strategies. Antibodies and compounds that suppress cytokines have already been marketed for autoimmune diseases such as collagen diseases and cytokine storms caused by COVID-19, and clinical application is relatively easy. In a large-scale clinical trial of myocardial infarction patients using canakinumab, a neutralizing antibody for IL-1β (interleukin-1β), the CANTOS (Canakinumab Anti-inflammatory Thrombosis Outcomes Study) trial showed that the drug is effective in preventing the recurrence of cardiovascular diseases, which has increased the promise of antiaging treatments targeting SASP [7]. However, many drugs have the aspect of immunosuppressants, and in fact, the CANTOS trial showed an increase in the incidence of infectious diseases. The price of the drug itself is often high, and there are many problems to be solved before it can be widely used for prevention. Senolytics, drugs that remove senescent cells, are recently gaining attention as next-generation anticell aging therapies. Currently, more than 20 candidate drugs have been derived, and more than 20 clinical trials are underway. Among them, the combination therapy of dasatinib and quercetin is probably the closest to clinical application. This combination therapy was derived by the research group of Kirkland at the Mayo Clinic, and clinical trials are currently underway in the United States for patients with chronic kidney disease (diabetic nephropathy) and pulmonary fibrosis, and promising initial treatment effects have been shown. Also, the SToMP-AD (Senolytic Therapy to Modulate the Progression of Alzheimer's Trials targeting Alzheimer's disease, also known as "Disease", have also begun. In addition, clinical trials are being conducted on fisetin, with the former planning trials for osteoarthritis and COVID-19. Clinical trials were conducted on UBX0101 for osteoarthritis, but unfortunately, it was unable to demonstrate significant short-term therapeutic effects [8]. In addition to compounds, new senescent cell removal therapies such as senescent cell removal vaccines, CAR-T therapy, and antibody drugs are being developed, but clinical trials have not yet been conducted, and further progress is awaited.

153.4 Conclusion

Antiaging treatments range from the application of existing drugs to the removal of senescent cells, and the results of future clinical trials are very exciting. However, anti-aging treatments target a more diverse population than disease treatments and essentially involve healthy subjects, so the number of subjects needed to achieve a certain detection power in clinical trials for "hard endpoints" such as lifespan and disease onset inevitably increases. At present, intervention trials are limited to a few thousand cases at most, with short observation periods and outcomes limited to improvements in physical function and surrogate markers of age-related diseases. It is somewhat difficult to verify more important outcomes at the current scale. To realize this, it is desirable to build a system that supports anti-aging research, as in recent years in the United States, where aging is considered a "treatable disease" and industry, academia, and government work together, requiring more financial and human resources.

References

1. Mohammed I, Hollenberg MD, et al. A critical review of the evidence that metformin is a putative anti-aging drug that enhances Healthspan and extends lifespan. Front Endocrinol (Lausanne). 2021;12:718942.
2. Irie J, Inagaki E, et al. Effect of oral administration of nicotinamide mononucleotide on clinical parameters and nicotinamide metabolite levels in healthy Japanese men. Endocr J. 2020;67:153–60.
3. Zhou DD, Luo M, et al. Effects and mechanisms of resveratrol on aging and age-related diseases. Oxidative Med Cell Longev. 2021;2021:9932218.
4. Lowe DA, Wu N, et al. Effects of time-restricted eating on weight loss and other metabolic parameters in women and men with overweight and obesity: the

TREAT randomized clinical trial. JAMA Intern Med. 2020;180:1491–9.
5. Estruch R, Ros E, et al. Primary prevention of cardiovascular disease with a Mediterranean diet supplemented with extra-virgin olive oil or nuts. N Engl J Med. 2018;378:e34.
6. Fernandez ML, Thomas MS, et al. TA-65, a telomerase activator improves cardiovascular markers in patients with metabolic syndrome. Curr Pharm Des. 2018;24:1905–11.
7. Ridker PM, Everett BM, et al. Antiinflammatory therapy with Canakinumab for atherosclerotic disease. N Engl J Med. 2017;377:1119–31.
8. Chaib S, Tchkonia T, et al. Cellular senescence and Senolytics: the path to the clinic. Nat Med. 2022;28:1556–68.

154 Antiaging Interventions: The Forefront of Cancer Chemoprevention

Michihiro Mutoh and Gen Fujii

154.1 Cancer Prevention Strategy

In 2015, President Obama's State of the Union address highlighted the Precision Medicine Initiative, which seemed to renew the concept of cancer prevention as a form of preemptive medicine (predicting the onset of disease and taking preventive measures in advance to prevent or delay the onset of disease). However, this concept has actually been around for a long time. Cancer prevention strategies are basically divided into two categories: the "population strategy," which assumes that a large number of people carry a small but potential risk, and the "high-risk strategy," which takes into account the existence of a small number of individuals with high risk. The former is the foundation of traditional cancer prevention policies such as primary and secondary prevention, while the latter is a strategy close to "diagnosis and treatment" in hospitals and can be considered an aggressive measure in line with personalized medicine.

M. Mutoh (✉) · G. Fujii
Department of Molecular-Targeting Cancer Prevention, Kyoto Prefectural University of Medicine, Kyoto, Japan
e-mail: mimutoh@koto.kpu-m.ac.jp

154.2 What Is Cancer Chemoprevention?

Cancer chemoprevention is defined by Dr. Sporn in 1976 as "the inhibition of cancer growth using compounds," and is essentially one of the primary methods of cancer prevention. On the other hand, cancer chemoprevention uses drugs with anti-inflammatory and cell proliferation inhibitory effects, and cell differentiation inducing effects, and should be implemented as a high-risk strategy targeting people with high cancer risk. This is a different approach from the general primary prevention. In other words, it is a method that is difficult to understand from the standpoint of preventive medicine, which aims to minimize risk based on the population strategy. However, with the advent of genomic medicine and the manifestation of cancer susceptibility risk, "individualized responses based on diversity" are now required, and the "practical application of cancer chemoprevention drugs" is an important issue that needs to be implemented promptly. This is because the drugs that are clinically applied (those approved by the Food and Drug Administration; FDA) are limited to a very small number, such as tamoxifen/raloxifene for breast cancer, and in Japan, there have been no insurance-covered drugs so far, which is a barrier to the realization of cancer chemoprevention. The fact that the commercialization of cancer chemoprevention is not currently feasible makes it

H. Yamada, Y. Naito (eds.), *Anti-Aging Medicine*, https://doi.org/10.1007/978-981-97-8514-8_154

difficult for pharmaceutical companies to enter the market, and the lack of development funds and researchers is a major problem.

154.3 Colorectal Cancer Chemoprevention Using Aspirin

We aim to break through this situation and bring the "first cancer prevention drug to the world," starting from basic research and epidemiological research, and currently conducting activities up to the exit strategy (dissemination and implementation research) in parallel. This is because we believe that if even one cancer prevention drug listed in insurance could be released to the world, it will pave the way for commercialization and lead to the expansion of new business models and academic fields. The first drug we chose as our target is aspirin. Aspirin is a drug with anti-inflammatory, antipyretic, analgesic, and platelet aggregation inhibitory effects, but its evidence has been accumulating since its colorectal cancer inhibitory effect was epidemiologically shown about 30 years ago, and since Dr. Rothwell and others continued to publish papers from 2010 [1], it has been considered that the benefits outweigh the side effects. They also report that aspirin administration has a cancer onset and death inhibitory effect not only in colorectal cancer but also in various solid cancers. This drug, which has been used for over 120 years, may not be attractive to cutting-edge researchers and young doctors, but because it has a long history of use, the side effects are well known, and the drug price is also cheap at 5.6 yen/tablet, its potential as a cancer prevention drug is very high. Therefore, we first targeted those who are considered to be moderate risk holders (lifetime colorectal cancer incidence rate 10–20%), those with colorectal polyps, and had them take low-dose aspirin enteric-coated tablets (100 mg/day), and examined the presence or absence of new colorectal polyp occurrence in the second year colonoscopy in a double-blind randomized comparative trial (Japan Colon Aspirin Polyp Prevention Study; J-CAPP Study) [2]. The trial participants were 311 people aged 40–70 who had one or more colorectal polyps (intramucosal cancer/adenoma) and were able to remove all of them endoscopically. The odds ratio corrected for age, sex, and the number of colorectal polyps before entry was 0.60 (95% CI: 0.36–0.98), and we were able to show for the first time in Asians that aspirin significantly inhibits the occurrence of new colorectal polyps. This figure is lower than the polyp occurrence risk ratio of 0.83 reported from meta-analysis of similar trials in the West [3], and the inhibitory effect of aspirin intake on the occurrence of colorectal cancer was expected in Asians as well. Next, we conducted a preventive clinical trial for familial adenomatous polyposis (FAP), a hereditary disease with a high cancer risk (lifetime colorectal cancer incidence rate 40–100%) (J-FAPP Study IV) [4] was advanced. The number of FAP patients in our country is currently estimated to be about 7300, many of whom develop colorectal cancer at a high probability in their youth due to germline variants of the cancer suppressor gene APC. The standard cancer prevention method is prophylactic total colectomy, but in recent years, the number of patients who do not want surgery has been increasing due to the significant decrease in the patient's quality of life and the possibility of inducing desmoid tumors by surgery. After removing as many polyps as possible with colonoscopy (intensive downstaging polypectomy; IDP) [5], low-dose aspirin was administered. The trial randomly assigned 104 FAP patients with completed removal of polyps larger than 5.0 mm and preserved colon to a group receiving low-dose aspirin enteric-coated tablets and a placebo tablet group, and performed colonoscopy 8 months later. As a result, the adjusted odds ratio for patients with newly found colorectal polyps larger than 5.0 mm was 0.37 (95% CI: 0.16–0.86), showing that low-dose aspirin enteric-coated tablets strongly suppress the growth of colorectal polyps. In a secondary analysis, it was found that low-dose aspirin enteric-coated tablets showed a stronger inhibitory effect on left-sided colorectal polyps. This was an expected result as sporadic colorectal cancer often occurs in the left colon, suggesting that aspirin's preventive effect could also be expected for the occurrence of sporadic colorectal cancer.

154.4 Future Prospects

The J-FAPP Study V, a new multicenter single intervention clinical trial aimed at building the final evidence leading to insurance coverage for aspirin therapy in FAP patients, is currently underway as Advanced Medical B. Here, low-dose aspirin enteric-coated tablets will be administered to 200 FAP patients who do not wish to have a colectomy after IDP for 2 years, and the cumulative incidence of polyps larger than 5.0 mm over 2 years and adverse events will be comprehensively evaluated. It is not difficult to imagine that after the "Cancer Genome Medicine Era," the "Cancer Prevention Medicine Era" will come, where the occurrence of cancer is "scientifically" and "individually" suppressed "actively and positively" using cancer chemopreventive agents. We must not stop the flow of "realizing preemptive medicine through prevention", a concept that includes aging.

References

1. Rothwell PM, Wilson M, et al. Long-term effect of aspirin on colorectal cancer incidence and mortality: 20-year follow-up of five randomized trials. Lancet. 2010;376:1741–50.
2. Ishikawa H, Mutoh M, et al. The preventive effects of low-dose enteric-coated aspirin tablets on the development of colorectal tumors in Asian patients: a randomised trial. Gut. 2014;63:1755–9.
3. Cole BF, Logan RF, et al. Aspirin for the chemoprevention of colorectal adenomas: meta-analysis of the randomized trials. J Natl Cancer Inst. 2009;101:256–66.
4. Ishikawa H, Mutoh M, et al. Chemoprevention with low-dose aspirin, mesalazine, or both in patients with familial adenomatous polyposis without previous colectomy (J-FAPP study IV): a multicentre, double-blind, randomized, two-by-two factorial design trial. Lancet Gastroenterol Hepatol. 2021;6:474–81.
5. Ishikawa H, Yamada M, et al. Intensive endoscopic resection for downstaging of polyp burden in patients with familial adenomatous polyposis (J-FAPP study III): a multicenter prospective interventional study. Endoscopy. 2023;55:344–52.

Part V

Anti-aging and Environment

Housing and Antiaging Medicine 155

Hiroyoshi Inoue

Research on aging is showing remarkable progress, and it has become clear that the speed of aging is somewhat controlled by genetic pathways and biochemical processes. As indicated by "The Hallmarks of Aging" by López-Otín et al. there are nine common factors related to the aging of mammals [1]. Factors accelerating these include genetics, lifestyle, social systems, and in recent years, the impact from the environment, including its countermeasures, has been significantly increasing. In this section, we will discuss the impact of air quality in the environment, particularly PM2.5, on aging. Furthermore, the average time spent at home in our country is 17 h and 23 min, and since we spend more than 70% of the day at home (according to the 2021 Ministry of Health, Labour and Welfare White Paper), we will also mention measures focusing on housing. PM2.5 is a particulate matter with a diameter of 2.5 μm or less, and due to its small size, its impact on organisms is a concern. Regarding aging, PM2.5 affected the length of telomeres and the activity of telomerase, which are among the aging factors mentioned above [2]. In this study, primary human bronchial epithelial cells derived from healthy subjects or patients with chronic obstructive pulmonary disease (COPD) were differentiated and repeatedly exposed to PM2.5 concentrated from the atmosphere. As a result, although PM2.5 did not cause DNA strand breaks or chromosomal damage, as shown in Fig. 155.1, telomere length was found to shorten in response to exposure concentration and number of exposures. In addition to telomeres, research on DNA methylation, one of the epigenetic changes, is also progressing. A study was conducted on changes in mRNA expression in the hearts of zebrafish exposed to extracted organic matter (EOM) derived from PM2.5 [3]. As a result, exposure to EOM reduced SAM/SAH, a potential methylation indicator, and also suppressed the expression of DNA methyltransferase. In other words, EOM induced a wide range of DNA methylation changes. Research on the relationship between PM2.5 and allergies is progressing both experimentally and epidemiologically. For example, an in vitro study was conducted to evaluate the association between PM2.5 exposure and the onset of asthma [4]. The PM2.5 used in this study was collected in Nanjing, China, and made into a soluble substance. Human lung epithelial cells (BEAS-2B) were treated with soluble PM2.5, and cell survival rate and other factors were detected using Annexin V-FITC staining (apoptosis detection). As a result, exposure to soluble PM2.5 in vitro reduced survival rate and increased apoptosis of airway epithelial cells. Soluble PM2.5 reduced airway barrier function, accompanied by an increase in NF-κB (nuclear factor-κB) and MAPK (mitogen-activated protein

H. Inoue (✉)
Department of Chemistry, Keio University School of Medicine, Fujisawa, Kanagawa, Japan
e-mail: hiroin@keio.jp

H. Yamada, Y. Naito (eds.), *Anti-Aging Medicine*, https://doi.org/10.1007/978-981-97-8514-8_155

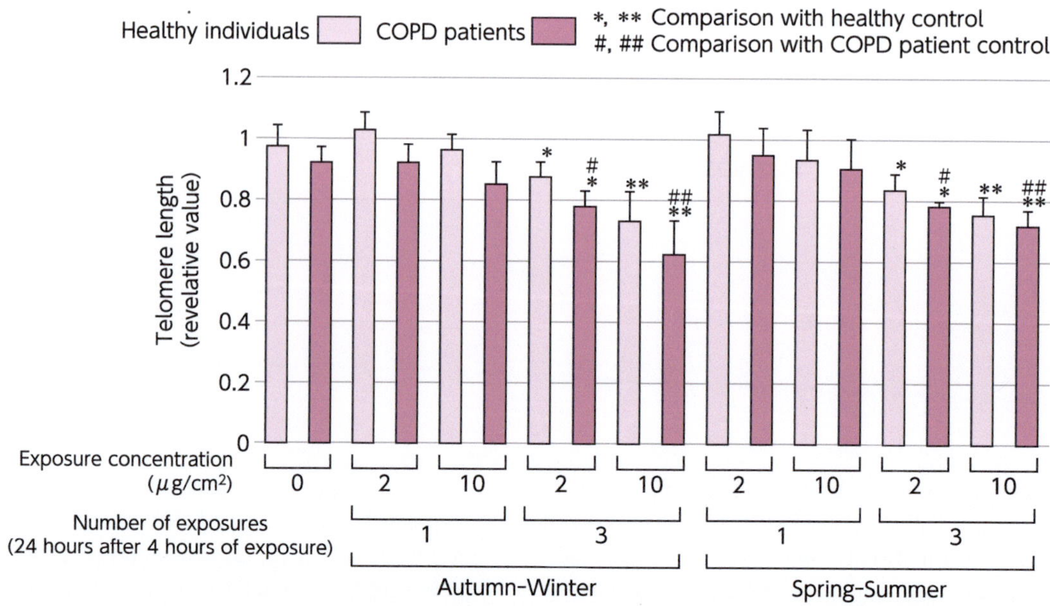

Fig. 155.1 Effect of PM2.5 on the telomeres of bronchial epithelial cells (cultured) derived from healthy individuals and COPD patients. (Modified with permission from [2])

By activating the kinase signaling pathway, it induces oxidative stress and enhances the expression of inflammation-inducing factors. It was shown that PM2.5 exposure increases the impact of allergen sensitization after respiratory exposure to allergens, leading to the onset of asthma. In response to these, attempts are now being made to suppress aging promotion and the onset of allergies and hatred by improving the living environment. In our country today, due to the improvement of building materials and insulation materials, and the advancement of construction methods and technologies, houses with high airtightness and good ventilation efficiency are widespread. In such a situation, houses equipped with air conditioning equipment using a HEPA filter (high efficiency particulate air filter) showed a high PM2.5 removal effect [5]. In this house, as shown in Fig. 155.2, during the intermediate period (when the air conditioner is not in use), PM2.5 was reduced by 96% in the living room and 97% in the master bedroom. During the air conditioning period (when the air conditioner is in use), PM2.5 was reduced by 99% in the living room and 98% in the master bedroom. Also, when mold was measured in this house equipped with a HEPA filter, as shown in Fig. 155.3, it was 2452 particles/m^3 in the outside air, but achieved an air environment equivalent to a clean room (200 particles/m^3 or less) such as a food factory indoors at 116 particles/m^3. By the way, the same results were obtained for pollen. In current homes, not only air quality, but also stable temperature by whole-house air conditioning, and constant humidity achieved by collecting water with a desiccant (zeolite) and heating it are realized.

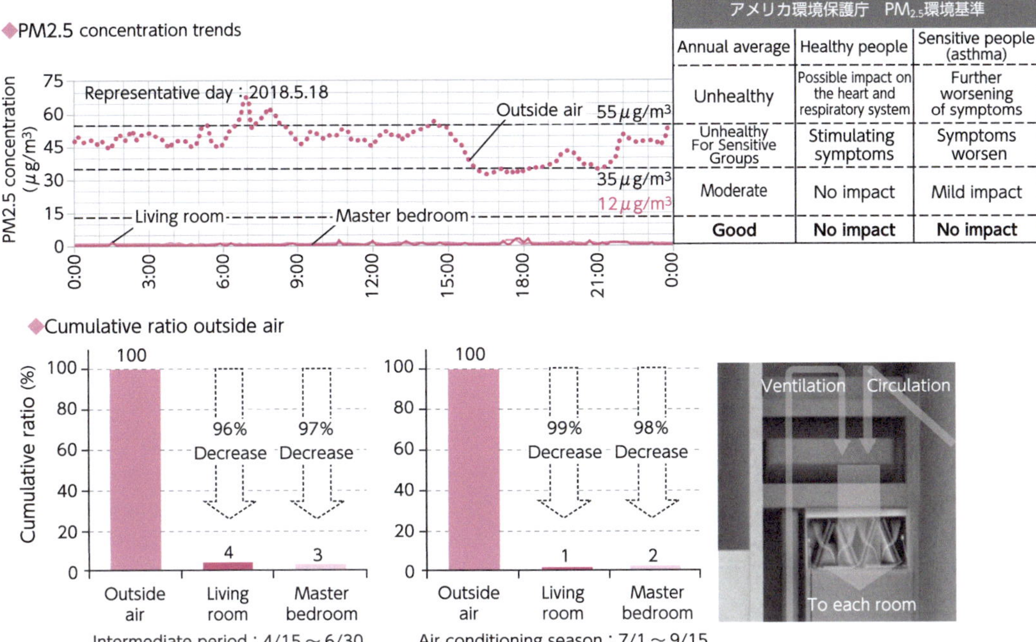

Fig. 155.2 PM2.5 removal by HEPA filter in actual house

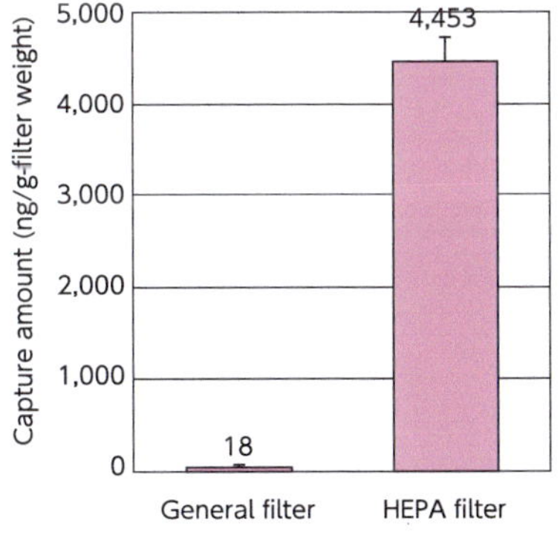

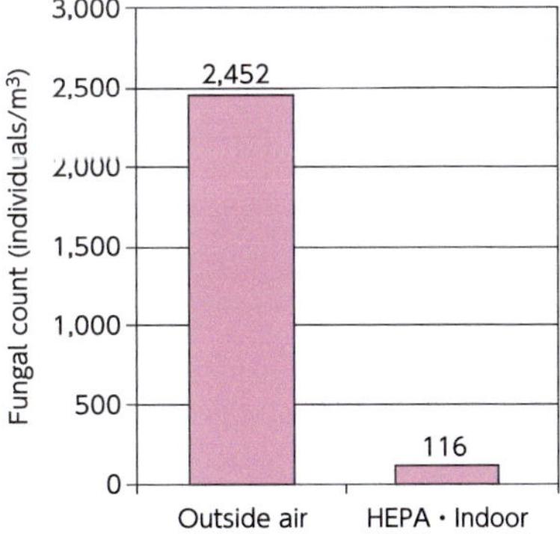

Fig. 155.3 Capture of cedar antigen by HEPA filter (in vitro test) (left) and capture of mold spores by HEPA filter (actual measurement test in a test house) (right)

References

1. López-Otín C, Blasco MA, et al. The hallmarks of aging. Cell. 2013;153:1194–217.
2. Leclercq B, Platel A, et al. Genetic and epigenetic alterations in normal and sensitive COPD-diseased human bronchial epithelial cells repeatedly exposed to air pollution-derived PM2.5. Environ Pollut. 2017;230:163–77.
3. Jiang Y, Li J, et al. PM2.5-induced extensive DNA methylation changes in the heart of zebrafish embryos and the protective effect of folic acid. Environ Pollut. 2019;255(Pt 3):113331.
4. Zhao C, Wang Y, et al. Respiratory exposure to PM2.5 soluble extract disrupts mucosal barrier function and promotes the development of experimental asthma. Sci Total Environ. 2020;730:139145.
5. Daijin I, et al. The effect of PM2.5 removal equipment in the living environment on health status. Ther Res. 2021;42:791–9.

Architecture, Indoor Environment, and Antiaging Medicine

156

Atsushi Iwamae

156.1 Sick House Syndrome and Residential Airtightness

Buildings can be broadly divided into commercial buildings, represented by offices, and residences. The former are often active during weekdays, while the latter are often passive (relaxing) during nights and holidays. Recent research has revealed the impact on the health of residences where people spend a lot of time in a passive state. One major example is the Sick House Syndrome, which became a social issue in the 1990s. From the 1980s, there were reports of people falling ill inside houses and buildings, and this phenomenon began to be publicized. In the 1990s, it was covered in general newspapers and quickly became a social issue. In 2000, a large-scale epidemiological survey conducted by the government showed that chemical substances in indoor air were causing health problems. More precisely, it was inferred that these were the cause because a large number of volatile substances were detected at much higher concentrations than outside. In 2003, when the Building Standards Law was revised, measures against Sick House Syndrome were introduced, and the installation of mechanical ventilation systems and the restriction of the use of building materials containing formaldehyde were explicitly stated. Since then, the number of people suffering from Sick House Syndrome has decreased, suggesting that these measures have had some effect. In fact, the airtightness of houses had started in the 1970s. Windows that used to be made of wood frames with glass were replaced with aluminum sashes, and earthen walls and wooden floors and ceilings were replaced with plywood and gypsum board, greatly reducing gaps (unintentional airtightness), but this was not widely recognized. There is a value called "ventilation rate (times/hour)" that shows how much the air inside a building is replaced in an hour. In traditional earthen-wall houses in our country, it is said to be more than 10 times/hour, but in concrete apartments, it can be less than 0.3 times/hour, and the same amount of indoor generation as before can easily exceed the allowable amount. Since 2003, intentional airtightness has been implemented, and on top of that, the installation of planned ventilation, a device that forcibly exchanges indoor and outdoor air (0.5 times/hour), has been made mandatory. It can be said that unintentional airtightness is the main cause of Sick House Syndrome, and the solution is intentional airtightness. Currently, there are about 60 million existing houses in Japan, but only about 10% of them have been intentionally made airtight. Approximately 80% are unintentionally airtight (the remaining 10% are as leaky as houses in the Edo period).

A. Iwamae (✉)
Faculty of Architecture, Kindai University, Osaka, Japan
e-mail: ai@arch.kindai.ac.jp

H. Yamada, Y. Naito (eds.), *Anti-Aging Medicine*, https://doi.org/10.1007/978-981-97-8514-8_156

156.2 Health Problems Caused by Low Temperatures and Their Countermeasures

The authors have been focusing on the possibility that factors other than air quality, such as temperature, humidity, noise, and illuminance, may affect human health, and have started to clarify the correlation between the actual conditions in residential interiors and the health of the residents. This was initiated in the UK in 2000 with the "Housing Health and Safety Rating The influence of the "Housing Health Watch System" [1] implemented in the UK in 2000, and a large-scale survey of over 1000 cases of health changes in residents due to environmental improvements in New Zealand in the same year [2] is significant. The former quantified various factors and their effects from the annual 500,000 medical data in the UK and the living conditions known from the ZIP code of the address, and constructed a system to score the "healthiness of the house." The latter showed that the effect of health improvement was twice the cost of insulation, which was carried out as an environmental improvement, by examining medical expenses before and after the improvement. Our survey results also reveal that low temperatures in homes in daily life are deteriorating human health in 2010 (Fig. 156.1). In a questionnaire survey of 20,000 people nationwide who had experienced moving to a newly built detached house [3], the improvement rate of daily symptoms (the proportion of people who had symptoms before moving and did not have symptoms after moving) was divided into three groups according to the insulation performance of the building after moving. The higher the insulation performance, the higher the improvement rate. These include blood pressure surges caused by touching a cold floor with bare feet, drying near the trachea due to inhaling cold air during sleep, and excessive pressure on the skin due to heavy clothing for cold protection, along with unavoidable human exposure to low temperatures during bathing and toilet use. In November 2018, the WHO issued the "Housing and Health Guideline," strongly recommending that indoor temperatures be kept above 18 °C for health reasons. In response to this, our country revised the Building Environmental Health Management

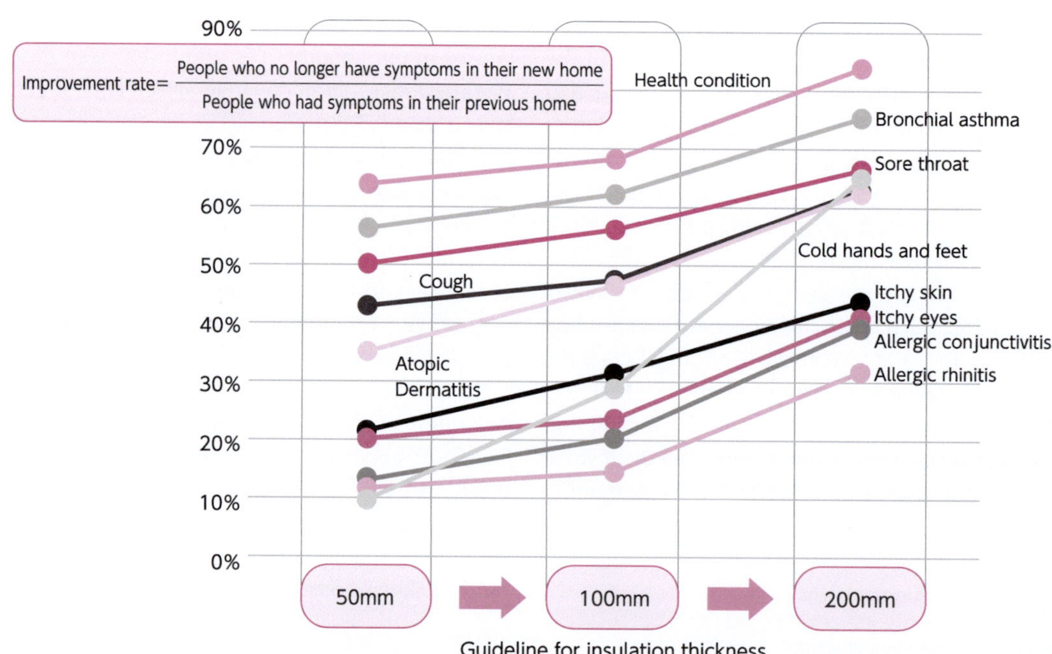

Fig. 156.1 Insulation performance and health improvement

Standards in April 2022, changing the indoor temperature from 17 to 18 °C. It can be said that it is a very clear fact that low temperatures impair human health.

156.3 Health Factors Other Than Low Temperature

Apart from low temperature, humidity is said to be an important factor. It's an interesting phenomenon for the human body, which does not have a mechanism to directly perceive humidity. In experiments where subjects were asked to report the degree of humidity in multiple laboratories with constant humidity, no correlation was seen between the reported values and the actual values. In a case where the transition of facial skin moisture content was measured [4], it was shown that the moisture content decreases during sleep in the summer, while it increases during sleep in the winter. This is thought to be due to the fact that the dryness and moisture of the skin during the day is compensated for at night, but if it decreases during sleep in the winter, it is thought that the cause is the over-drying of the room, and the skin deteriorates. Moisturizing during sleep is important, but in rooms with many gaps, it is almost impossible to humidify the space, so it is prioritized to protect the skin directly with moisturizing cream, etc. Noise is also an important factor in health impairment. The European Directive [5] explicitly states the health hazards of noise, particularly requesting each country to reduce road traffic noise. Traditionally, in Japan, the discomfort of summer nights was dealt with by opening windows and letting the wind pass through, but with the recent increase in nighttime temperatures, it has become extremely difficult to maintain health in this way. Closing the windows and turning on the air conditioner not only eliminates high temperatures, but also contributes to the suppression of external noise and the improvement of sleep satisfaction.

References

1. GOV.UK: Collection: Housing health and safety rating system (HHSRS) guidance. https://www.gov.uk/government/collections/housing-health-and-safety-rating-system-hhsrs-guidance.
2. Howden-Chapman P, and Housing and Heating Research Team. Housing, heating and health study: 2005/2006 report one. Aims and methods. 2006.
3. Iwamae A. Large-scale questionnaire survey on the health improvement effects of housing insulation. In: Architectural Institute of Japan Environmental Engineering Committee Thermal Environment Management Committee: 43rd thermal symposium "coldness and health/comfort in living environment", 2013, p. 87–90.
4. Isobe M, Iwamae A. Study on the effect of temperature and humidity fluctuations on skin moisture. Architect Inst Jpn Acad Lect Abstr Collect Environ Eng. 2010;2:609–10.
5. European Commission. Environmental Noise Directive. https://environment.ec.europa.eu/topics/noise/environmental-noise-directive.

157 Global Environment, COP, and Antiaging Medicine

Hiroyuki Nakamura

157.1 Global Environment

As a result of changes in the global environment caused by human activities, mankind is facing a serious situation due to many environmental problems surrounding the earth. These problems include climate change, represented by global warming, which is causing the earth's temperature to rise, the loss of biodiversity due to changes in living environments, and environmental pollution of the oceans, soil, and atmosphere. Environmental issues related to climate change include sea-level rise, forest fires, desertification, deforestation, destruction of the ozone layer, and other extremely serious crises on earth, primarily caused by global warming, which have been addressed first and foremost at the COP (Conference of the Parties). The Kyoto Protocol, which became the international rule for global warming measures until 2020, was adopted at COP3 held in Kyoto in 1997, in the history of COP.

H. Nakamura (✉)
Department of Environmental and Preventive Medicine, Kanazawa University School of Medicine, Kahoku, Ishikawa, Japan
e-mail: hiro-n@po.incl.ne.jp

157.2 COP

COP refers to the Conference of the Parties, and the most recent COP held in the fall of 2021 was the 26th meeting of the United Nations Framework Convention on Climate Change (UNFCCC), hence it is called COP26. Positioned as the highest decision-making body of the treaty, it is a large-scale international conference where all treaty signatories (197 countries and regions as of November 2021) participate to discuss international rules for global warming measures. The health effects of temperature are left to the previous section, and this section focuses on environmental issues related to ecosystems or biodiversity, which is one of the imminent crises for mankind, as a global environmental issue affecting human health.

157.3 Biodiversity and Health

In recent years, biodiversity on earth has been rapidly lost, and this in itself is considered an environmental problem. Biodiversity was defined by Edward Osborne Wilson [E. O. According to Dr. Wilson (1992)], "Biodiversity refers to the diversity of life considered at all hierarchical levels, from the genetic diversity within a species, through the diversity of species, genera, and families, to higher taxonomic groups. It also includes the diversity of ecosystems, which consist of

H. Yamada, Y. Naito (eds.), *Anti-Aging Medicine*, https://doi.org/10.1007/978-981-97-8514-8_157

biological communities in specific habitats and the physical conditions that affect the life of organisms" [1]. It is said that 60% of biodiversity has been lost in the past 40 years. Considering that the history of the Earth is 4.6 billion years, this is a dramatic event that occurred in just a moment of 40 years. Some of this can be attributed to climate change, but much of it is due to human activities such as overfishing, pesticide use, increased demand for biomass to replace aviation fuel and chemical materials, food issues in agricultural land conversion, and development of forests and unused land. The maintenance and recovery of biodiversity has been delayed compared to the issue of climate change because it is difficult to quantify. In response to these issues, the first part of COP15 was held in Kunming, China in October, and the 15th Conference of the Parties (COP15) to the Convention on Biological Diversity (CBD), the tenth Meeting of the Parties to the Cartagena Protocol, and the fourth Meeting of the Parties to the Nagoya Protocol were held in October 2021. The representative example of the impact of biodiversity loss on health is the increase in infectious diseases. The loss of species diversity mainly causes a significant impact on ecosystems and leads to an increase in infectious diseases. According to the National Science Foundation's "Biodiversity Loss: Detrimental to Your Health" (2010) [2], it was known that the decrease in species can cause the occurrence of infectious diseases such as West Nile virus and Hantavirus, but it was found that this is a universal pattern. Due to environmental destruction such as deforestation, species that played a buffering function to prevent the expansion of pathogens are extinct, while hosts and vectors that carry pathogens survive, resulting in an increase in infectious diseases and promoting health damage. It is unclear why organisms that proliferate pathogens are more likely to survive, but the best way to prevent this is to conserve natural habitats. Also, according to Pieter T. J. Johnson [3], there is evidence to suggest that when biodiversity decreases in an ecosystem, pathogens may become dominant, affecting the transmission of pathogens and the diseases of hosts. Pathogens tend to infect multiple host species, but there may be differences in the ability to maintain and transmit infections among each host species.

157.4 One Health

The above biodiversity is also connected to the concept of One Health. One Health has been gaining global attention in the wake of the novel coronavirus infection, but the idea of connecting the health of ecosystems, animals, and humans has been passed down along with the history of medicine (Fig. 157.1). This concept recognizes that the health of humans, livestock, and animals

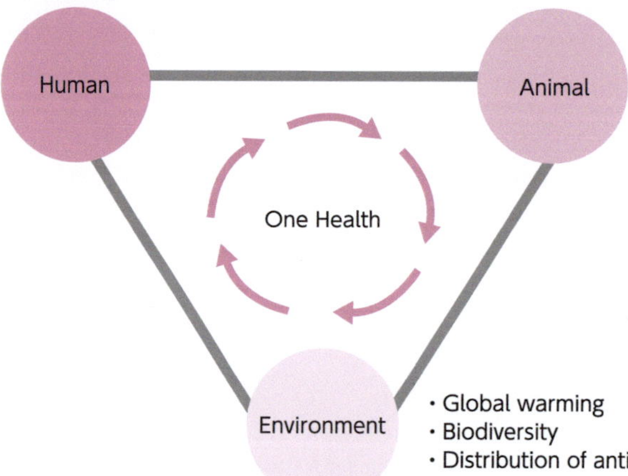

Fig. 157.1 Conceptual diagram of One Health

is interconnected, and that infectious diseases pose a threat to food, economy, and biodiversity. When formulating measures against infectious diseases, it is fundamental to consider the complete integration of biodiversity conservation and human needs (including the health of livestock). In other words, it has become a common understanding that unless the "three healths"—the health of the natural environment (ecosystem) that causes outbreaks, the health of livestock and wild animals that carry and transport pathogens, and human health—are all kept sound, it is impossible to prevent the increasingly frequent animal-borne infectious diseases. On the other hand, One Health is thought to apply not only to infectious diseases but also to health in general. That is, the immune and antioxidant stress responses that can counteract diseases are common to humans, animals, and plants, and it is not hard to imagine that the role One Health plays in anti-aging is enormous, considering that human nutrition depends solely on ecosystems such as animals and plants.

References

1. Wilson EO. The diversity of life. Cambridge: Harvard University Press; 1992.
2. https://www.nsf.gov/news/news_summ.jsp?cntn_id=118114&org=NSF.
3. Johnson PTJ, Preston DL, et al. Biodiversity decreases disease through predictable changes in host community competence. Nature. 2013;494:230–3.

158 Medical, Food, and Agricultural Cooperation and Antiaging Medicine

Sakae Shibusawa

158.1 Precision Medicine and Precision Agriculture

From the 1990s to the 2000s, precision medicine and precision agriculture emerged and spread globally almost simultaneously. This expanded the scope of medical and agricultural cooperation, and increased the momentum for the development of food ingredients based on medical data. Both were triggered by the introduction of cutting-edge technologies in information science and bioscience, and the sharing of specialized knowledge became possible. Precision medicine is a medical management concept that focuses on the individual variability of human pathological conditions, analyzing down to the cellular and genetic level, and providing appropriate treatment prescriptions for each individual [1]. In 2015, President Obama of the United States announced the "Precision Medicine Initiative," which attracted attention. Especially in the field of preventive medicine, when creating prescriptions for lifestyle-related diseases, it is necessary to propose improvements in dietary habits that are tailored to the individual's actual situation, and food ingredients with functions that respond to these requests are needed. Precision agriculture is a farm management concept that aims to propose prescriptions that achieve both productivity improvement and environmental conservation by tracking, recording, analyzing, and visualizing the variability of farms down to the individual crop level [2, 3]. In recent years, precision agriculture is also referred to as smart agriculture. For example, a group at Tsukuba University has developed a system that can stably produce crop varieties that contain a large amount of the functional substance gamma-aminobutyric acid (GABA) through genome editing and plant factories, and it has become possible to reorganize and adjust production processes using advanced technologies in response to strict environmental constraints and market needs. The industrial ethics of good medical practice (GMP) and good agricultural practice (GAP) are also being encouraged, and a situation has arisen where the concept of medicine and food being of the same origin, which has been said from the past, can be implemented backed by the latest scientific achievements and ethics.

S. Shibusawa (✉)
Institute of WISE, Tokyo University of Agriculture and Technology, Tokyo, Japan
e-mail: sshibu@cc.tuat.ac.jp

158.2 Agro-medical Foods (AMF)

Agro-medical foods (AMF) refer to agricultural, forestry and fishery products (including primary processed products such as tea and fruit juice) that contain functional ingredients with health-promoting effects recognized by scientific evidence, and are defined by the Agro-Medical

H. Yamada, Y. Naito (eds.), *Anti-Aging Medicine*, https://doi.org/10.1007/978-981-97-8514-8_158

(By disease)

	Medical Verification				Standardization of Analysis		Verification of Agriculture		
Disease	Medical verification of functionality				Analytical Evaluation Techniques		Precision Agriculture & Supply of Agricultural Products		
	Human (Epidemiological Survey)	Human (Intervention Trial)	Animal	Cell	Agricultural products・foods	Biological samples	Breeding	Cultivation	Processing・Cooking
Metabolic Syndrome	△Quercetin ○Isoflavone ○Proanthocyanidin ○β-Cryptoxanthin	△Quercetin △Proanthocyanidin	○Quercetin ○Isoflavone	○Quercetin	○Onion ○Soybean ○Apple	○Onion ○Apple	○Onion ○Soybean ○Apple ○Mandarin	○Onion ○Soybean ○Apple ○Mandarin	○Onion ○Apple ○Mandarin Juice
Osteoporosis	○Proanthocyanidin ○β-Cryptoxanthin	○Isoflavone	○Isoflavone ○β-Cryptoxanthin	○Isoflavone	○Soybean ○Apple	○Apple	○Soybean ○Apple ○Mandarin	○Soybean ○Apple ○Mandarin	○Apple ○Mandarin Juice
Non-alcoholic Fatty Liver		△Quercetin △Isoflavone △β-Cryptoxanthin	△β-Cryptoxanthin	△β-Cryptoxanthin	○Onion ○Soybean	○Onion	○Onion ○Soybean ○Mandarin	○Onion ○Soybean ○Mandarin	○Onion ○Mandarin Juice

○In development △Not started

(By disease, created on November 28, 2012)

Fig. 158.1 Example of research and development of agro-medical foods through medical-agricultural collaboration

Foods Research Institute (AMF Research Institute) as one of the goals of medical, food, and agricultural cooperation [3]. Unlike industrial products, agricultural products have natural characteristics such as variations in shape and quality. Among these variations, agricultural products produced by producers who have functional ingredients that are effective for health exceeding the regional average level, and whose production process is risk-managed by global GAP, etc., and incorporate a production system with less environmental impact, were targeted for AMF. Production roadmaps for AMF were created for each crop, disease, and functional substance, and research and development to confirm existing results and fill in the gaps were promoted. Figure 158.1, is an example of a disease-specific roadmap for the production of AMF (Agro-Medical Foods). In the figure, the names of diseases are listed in the leftmost column, followed by a column listing medical verification items such as human epidemiological surveys, human intervention trials, animal experiments, and cell tissue experiments, indicating the status of the investigation of the intended functional substances. The next column is for the verification of analytical techniques, showing the status of the verification of the analysis methods for the intended agricultural products and biological samples. The rightmost column is for agricultural verification, showing the status of the verification of variety characteristics, cultivation methods, and processing cooking methods. When these columns are filled with verification facts, AMF for the targeted disease is born. To make it easier for researchers and technicians related to medical, agricultural, and food fields to participate in joint research from their respective specialties, we also created production roadmaps for each crop and functional substance. Furthermore, in 2012, the Ministry of Agriculture, Forestry and Fisheries' "Development Project for Agricultural, Forestry and Fisheries Products and Foods with Functionality" was carried out, and many functional agricultural products were commercialized.

158.3 Recommended Agricultural Products for Agro-medical Foods

The functional agricultural products and foods currently on the market are of three types: Foods for Specified Health Uses (FOSHU) (individual examination type, 2009), Foods with Function Claims (pre-notification system, 2015), and Nutrient Function Foods (standard type, 2001), but there are very few fresh foods. Although there are registrations of fresh foods in Foods with Function Claims (108 cases as of July 14, 2021) [3], this is still small compared to the total number of notifications of 5000. It is quite difficult to manage the functionality of fresh foods according to a certain standard, as their form and quality vary throughout the year and by region.

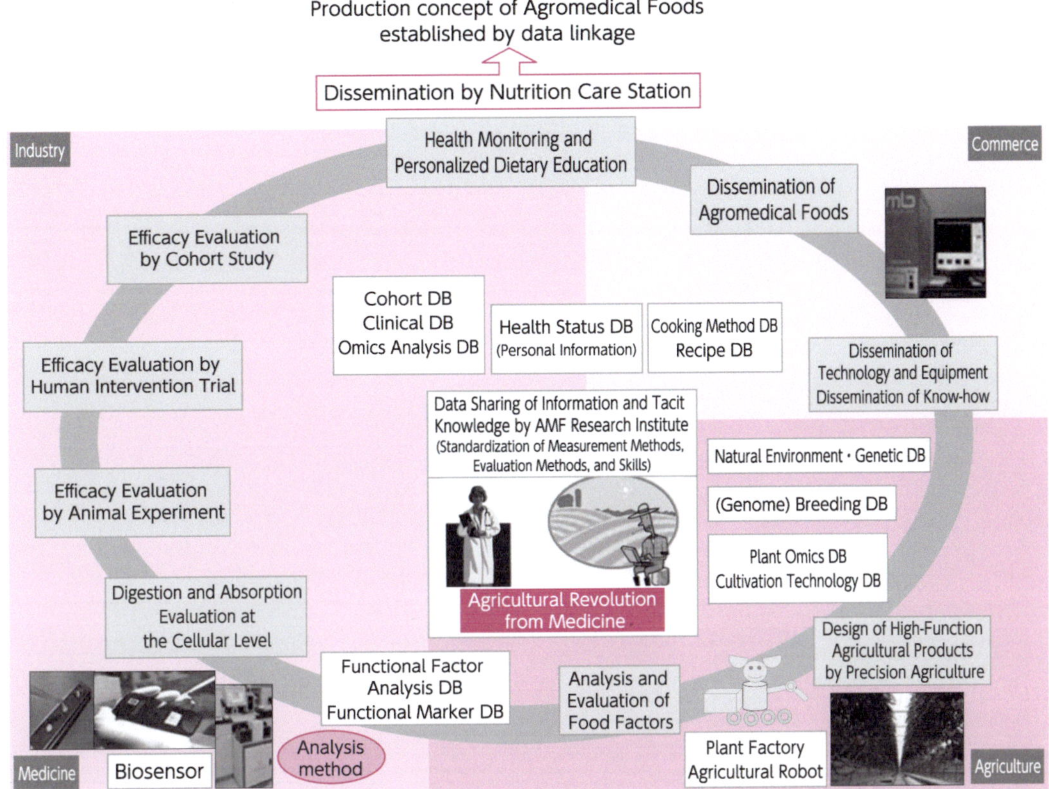

Fig. 158.2 Production concept of agro-medical foods made possible by data linkage

Guidelines are needed to promote the production of functional agricultural products and foods that match the characteristics of agricultural, forestry and fishery products. Therefore, the AMF Research Institute has established AMF recommended products [3]. The aim is to contribute to the food system that connects the value chain from production to consumption of those AMFs that are particularly expected to have health-promoting effects. The certification criteria for AMF recommended products are as follows: ① The producer and variety can be identified. ② It contains a lot of functional components that have shown scientific evidence of health effects, or it contains a little. ③ It is expected to have stable production volume and quality. ④ Efforts are being made to ensure safety. In order to stably produce and supply AMF, data linkage that enables wide cooperation of medical, agricultural, commercial and industrial stakeholders is necessary [2] (see Fig. 158.2). At AOI-PARC (Agri Open Innovation Practical and Applied Research Center) in Shizuoka Prefecture, where the AMF Research Institute is located, cultivation experiments of functional plants are underway, and products such as Soft Kale GABA and Quick-Acting Chitin Fertilizer are being developed [3].

References

1. NHK Special. The cancer treatment revolution has begun—the impact of precision medicine. 2016.
2. Shibusawa E. Production concept of agromedical foods and new developments in community-based precision agriculture. Mutual Aid Compr Res. 2011;62:48–65.
3. Yoshikawa T, Shibusawa E. Dialogue on medical-agricultural cooperation and agromedical foods. Med Food. 2021;13(5):4–12.

Part VI

Anti-aging and Society

159 Medical Systems (Digital Medicine and Telemedicine) and Antiaging Medicine

Tetsuya Tomita

A human-centered society that achieves both economic development and the resolution of social issues through a system that highly integrates cyberspace (virtual space) and physical space (real space) has been proposed by the Cabinet Office as the future society (Society5.0) that our country should aim for [1]. With the Internet of Things (IoT), all people and things are connected, various knowledge and information are shared, and necessary information is provided when needed by artificial intelligence (AI), realizing a society that overcomes current issues and difficulties. Furthermore, the necessity and importance of digitalization in medicine has been recognized in general society and is accelerating due to the confusion in the medical field caused by the COVID-19 infection.

159.1 What Is Digital Medicine?

Generally, "digital healthcare" is understood as enhancing the effects of disease diagnosis, treatment, or health maintenance by utilizing the latest digital technologies such as AI, IoT, wearable devices, and virtual reality (VR). This includes a wide range of technologies (products) from health promotion applications for the general public, including healthy people, to medical devices that have shown scientific evidence of therapeutic intervention effects for treating patients. It is a direction from treatment to prevention, or from uniform medical care to personalized medical care. From the perspective of prevention and diagnosis, AI medical image analysis, genomic medicine, liquid biopsy, wearable devices, diagnostic support (AI doctor), etc., and from the therapeutic perspective, brain-machine interface, cybernetics, neurostimulation therapy, digital therapy, etc. are mentioned [2]. Also, the implementation of these requires the preparation of an environment such as hardware, software, big data analysis, and services.

- AI Medical Image Analysis

  Since the automatic diagnosis system for early diabetic retinopathy was approved as a medical device by the Food and Drug Administration (FDA) in the United States in 2018, various image analysis support systems have been developed and clinically applied, and in our country, endoscopic images, head MRA images, and chest CT images have been approved as computer diagnostic support devices.
- Genomic Medicine

  By detecting multiple genetic changes (gene mutations, deletions, insertions, gene fusions, copy number abnormalities, etc.) at the same time, it is becoming possible to pro-

T. Tomita (✉)
Graduate School of Health Sciences, Morinomiya University of Medical Sciences, Osaka, Japan
e-mail: tetsuya_tomita@morinomiya-u.ac.jp

H. Yamada, Y. Naito (eds.), *Anti-Aging Medicine*, https://doi.org/10.1007/978-981-97-8514-8_159

vide the best cancer treatment for each patient's condition. In our country, cancer genome profiling tests have been covered by insurance since 2019. Furthermore, efforts are being made towards personalized medicine and preventive medicine in areas such as diabetes, dementia, and hypertension.

- Liquid Biopsy

  Early diagnosis, screening, cancer staging, prognosis prediction, treatment stratification, and applications for diseases other than cancer, such as dementia, are expected through diagnostic and testing methods using biologically derived biomarkers collected noninvasively or minimally invasively from blood and body fluids.
- Wearable Devices

  Technological development is advancing to utilize physiological data collected and analyzed continuously during daily life as digital biomarkers. For the prevention of disease recurrence, a system of PHR (personal health record) is expected, where individuals collect health and medical information about themselves over their lifetime, with their consent, and reference, share, and utilize this information through a network. The analysis by AI is fed back, leading to the expectation of early diagnosis and treatment intervention, and even predisease.
- Diagnostic Support (AI Doctor)

  This is diagnostic support that accumulates vast amounts of clinical data and performs AI-based diagnoses to support actual doctors. It is clinically applied in China, where there is a shortage of doctors relative to the huge population, and where the use of personal information is easy. The efficiency of medical care through online and remote medical care is a benefit.
- Digital Therapy

  Therapeutic interventions based on evidence of effectiveness are called digital therapy. It differs from health promotion applications in that it is approved as a subject of the Pharmaceutical and Medical Device Act. Clinical applications are advancing in chronic diseases and mental disorders that require long-term treatment and continuous management by the patient themselves, specifically in diabetes, attention deficit hyperactivity disorder in children, nicotine dependence, etc.

159.2 Online Medical Care/Telemedicine

Telemedicine refers to the practice of using information and communication technologies such as the internet to provide medical care from a distance between a doctor and a patient. It is also referred to as remote medical care or online medical care [3]. In our country, it has been used as a means to alleviate the shortage of doctors in remote islands and isolated areas since 1997, and in 2003, it became possible for some patients with chronic diseases in the stable phase to receive telemedicine. Until 2018, it was called telemedicine, but the term was changed to online medical care, and the guidelines for online medical care were clarified for the first time. In our country, medical practices are divided into online medical care, online medical consultation recommendations, and remote health medical consultations, all of which are collectively referred to as telemedicine. However, the government has been promoting online medical care to solve problems such as the super-aging society and labor shortage in our country, but its use has been sluggish. Since 2020, due to the spread of COVID-19, the demand for it has rapidly increased due to the surge in patients who want to visit medical institutions in urban areas but cannot. In January 2022, the guidelines were revised [4]. Until now, online medical care from the first visit was basically not allowed, but if the primary care doctor who has the medical information in principle can judge it, it became possible. There are no restrictions on diseases, and the proportion of online medical care in all medical care at medical institutions, which had been less than 10% until now, has been lifted. Prescriptions are still required to be less than 90 days in principle, and it is required to refer to the treatment guidelines set by the related academic societies, such as the "drugs

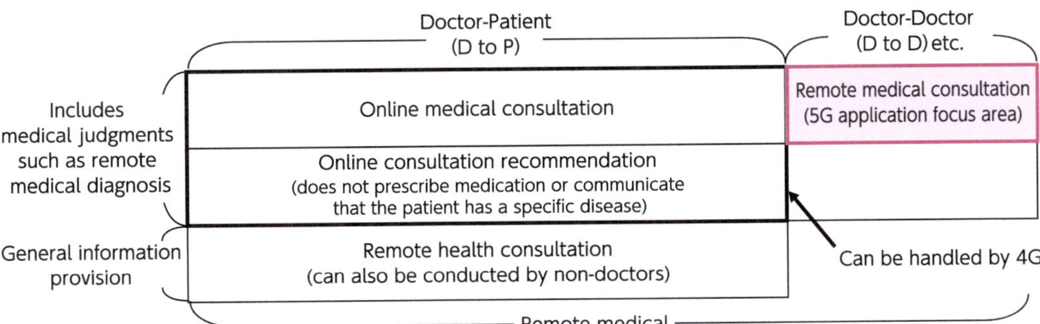

Fig. 159.1 Application areas of 5g in telemedicine. (Reused with permission from [4])

that require sufficient consideration for administration at the first online medical consultation" created by the Japan Medical Association Federation. The requirement to take online medical care training, which was not mandatory in the temporary measures during the Corona disaster, has been changed to mandatory. At present, it is a big challenge whether all departments can accurately grasp the patient's condition through the monitor, and the establishment of infrastructure for equipment and communication is necessary for medical institutions, and there are many challenges. On the other hand, telemedicine between doctors is expected to develop greatly in the future. If it becomes possible to connect primary care doctors and specialists and create an environment where information can be shared and direct communication can be made between doctors, it will be possible to provide specialized medical care under the primary care doctor even in remote islands and isolated areas, with the patient's primary care doctor present, sharing test results and image information between doctors (Fig. 159.1). Furthermore, if it is not the first consultation, certain medical acts can be performed by instructing the nurse to perform medical assistance acts in online medical care with the nurse present. As the practical application of 5G progresses in the future, the application of virtual reality technology will become possible, and not only diagnosis and internal treatment, but also remote operation using surgical support robots will be possible if communication delay is eliminated, and high-quality surgical operations can be expected to be possible nationwide even in areas where there are not enough specialists.

References

1. Cabinet Office. Society_5.0. https://www8.cao.go.jp/cstp/society5_0/index.html.
2. Japan Science and Technology Agency (JST) Center for Research and Development Strategy (CRDS). Overview report of research and development. Life science and clinical medicine (2021). Tokyo: Nikkei Printing; 2021. p. 230–76.
3. Free Encyclopedia "Wikipedia" Telemedicine.
4. Ministry of Health, Labour and Welfare. Guidelines for the proper implementation of online medical consultation. 2018. https://www.mhlw.go.jp/content/000889114.pdf.

Antiaging Medicine from Infancy 160

Keisuke Sugimoto

It is estimated that half of the children born in Japan in 2007 will live longer than 107 years, and from reports that the healthy lifespan will extend by about 3–4 years in 10 years, it can be said that half of the newborns currently born will live in a world where they live over 110 years. It is very important to prevent the onset of lifestyle-related diseases and other diseases in order to achieve the extension of the healthy lifespan of children living in the 100-year era, and it is desirable to be in a healthy state both physically and mentally.

160.1 DOHaD

DOHaD (developmental origins of health and disease) is a theory that "the susceptibility to future health and specific diseases is strongly influenced and determined by the fetal and early postnatal environment," and is particularly involved in the onset of lifestyle-related diseases and the fetal environment, especially the low-nutrition environment [1]. In addition, factors such as maternal aging, smoking, and drug use can change the placental environment and result in excessive glucocorticoid exposure to the fetus. Furthermore, it changes the expression amount of DNA methylase in the fetal brain, causing changes in epigenetic modifications, and even after the exposure to environmental factors during the fetal period is removed, it is involved in the growth of the child and changes in the internal environment, and becomes a disease factor (first hit). Further exposure to negative environmental factors after birth (infections, drugs, allergens, rapid weight gain due to high-fat and high-protein diets, obesity, etc.) (second hit) is involved in the onset and progression of various diseases, including lifestyle-related diseases. Currently, diseases with a clear association with low birth weight include hypertension, coronary artery disease, type 2 diabetes, cerebral infarction, dyslipidemia, and diseases with a high association include chronic obstructive pulmonary disease, chronic kidney disease, metabolic syndrome, depression, schizophrenia, precocious puberty, breast cancer, prostate cancer, etc.

160.2 Delivery Style and Chronic Diseases

Although cesarean section is important from the perspective of saving the lives of mothers and children, the current cesarean section rate in advanced countries, including the United States, exceeds 30%, and it is also increasing in our country, exceeding 20%. Children born by cesarean section have a higher risk of developing obesity, metabolic syndrome including type 2

K. Sugimoto (✉)
Department of Pediatrics, Faculty of Medicine, Kindai University, Osaka, Japan
e-mail: ksugimo@med.kindai.ac.jp

H. Yamada, Y. Naito (eds.), *Anti-Aging Medicine*, https://doi.org/10.1007/978-981-97-8514-8_160

diabetes, ulcerative colitis, food allergies, immune diseases such as bronchial asthma, and mental developmental disorders including attention deficit hyperactivity disorder (ADHD) and autism spectrum disorder (ASD). Neutrophils and monocytes, which should be secreted into the blood during normal delivery, Natural killer cells and other various mediators are cancelled in cesarean section delivery [2], and the loss of opportunity to acquire Bifidobacterium, Bacteroides, and facultative anaerobic bacteria such as Lactobacillus from the mother, which results in the failure to construct a normal intestinal bacterial flora, is considered to be involved in postnatal immune dysfunction [3].

160.3 Child Abuse and Aging

In recent years, child abuse has been increasing annually and has become a serious social issue. Adverse childhood experiences (ACE) pose significant health risks during the growth process, as well as in various developmental areas. It is known that children with ACE have higher rates of ADHD, depression, suicide attempts, alcohol dependence, and drug abuse. Furthermore, patients who have experienced sexual, physical, or psychological abuse in childhood not only suffer from psychological trauma, but also have shorter telomere lengths compared to healthy individuals and patients without a history of abuse in childhood [4]. Moreover, in children who have been abused/neglected, the weight of the thymus significantly decreases, and histological examination of the thymus reveals marked atrophy of the cortex, a decrease in the number of lymphocytes, and nuclear condensation, suggesting that the thymus may begin to atrophy from early childhood in abused children, possibly related to immune dysfunction [5]. In the future, deterioration of mental health and developmental disorders may be major factors accelerating aging, and support for caregivers with risk factors leading to a decline in healthy parenting functions can contribute to anti-aging, not only for abused children.

160.4 Obesity Measures to Be Implemented from an Earlier Stage

Lifestyle-related diseases are the biggest obstacles to healthy longevity and also have a significant impact on national medical expenses. The medical expenses of metabolic syndrome are over 100,000 yen per year more than those without it, and in the United States, the medical expenses of obesity in 6–17 year olds have tripled over about 20 years. In our country, the prevalence of childhood obesity, which was 2–3% in the 1960s, has increased significantly to about five times due to the westernization of diet and the decrease in physical activity due to lifestyle changes. In our country's birth cohort, 84 out of 143 obese children pointed out at the age of 5 (59%) became obese with an obesity rate of 20% in the first year of junior high school, indicating that childhood obesity often transitions to obesity in adolescence and adulthood. Furthermore, the phenomenon of adiposity rebound, where the BMI observed in infancy transitions from decreasing to increasing, It has been made clear that the earlier the onset of adiposity rebound (AR), the higher the risk of obesity and lifestyle-related diseases later on. In other words, children who show signs of AR at an early age have higher resistance to insulin and leptin. Furthermore, if AR is observed in children under the age of 4, it has been revealed that the risk of cardiovascular metabolic events, such as high levels of neutral fat, apolipoprotein B, arteriosclerosis index, and blood pressure at the age of 12 in boys, increases [6]. Obesity is a central item of metabolic syndrome and is also an important factor in arteriosclerosis. In children with obesity, thickening of the carotid artery wall, which causes high blood pressure, has already begun, and vascular lesions that form the basis of myocardial infarction and cerebrovascular disorders are gradually progressing from a younger age, necessitating early intervention.

160.5 Gut Microbiota and Lifestyle Diseases

Many studies have shown that an imbalance in the gut microbiota (dysbiosis) is associated with the onset and progression of chronic inflammatory diseases such as allergic diseases, inflammatory bowel diseases, autoimmune diseases, and non-infectious diseases (cancer, cardiovascular diseases, diabetes, and chronic respiratory diseases). In the intestines of obese individuals, the decrease in bacteria from the Bacteroidetes phylum and the increase in bacteria from the Firmicutes phylum lead to a higher calorie recovery rate from dietary components, resulting in weight gain and accumulation of body fat [7]. A large-scale cohort study in the UK reported that children who were administered antibiotics within the first 2 years of life had an increased risk of obesity [8]. In East Asia, including Japan, the rate of penicillin-resistant pneumococcus is high, inevitably increasing the use of antibiotics. Therefore, it is not possible to simply compare the use of antibiotics with Western countries, but appropriate use is desired not only from the perspective of increasing resistant bacteria but also from the perspective of dysbiosis. Recent studies using mouse models have shown that the antitumor effect of immune checkpoint inhibitors [anti-PD-1 (programmed cell death-1) antibody, anti-CTLA-4 (cytotoxic T-lymphocyte associated antigen-4) antibody] is higher when gut microbiota such as *Bifidobacterium* and Bacteroidales are established. In Japan, the number of diseases for which these drugs are indicated and the number of patients using them are likely to increase in the future. As we become a longevity society, it is expected that the children of today will inevitably have more opportunities to use these drugs in the future, so it is necessary to create an environment that does not cause dysbiosis from an early age as much as possible.

160.6 Immune Aging Due to Drug Use

In recent years, various immunosuppressive drugs have been used from early childhood, and the increase in treatment options has contributed to the induction of disease remission and disease control. On the other hand, there is concern that the use of immunosuppressive drugs during periods when the immune system is not yet mature may cause T cell dysfunction and induce immunosenescence. In vitro, renal tubular cells exposed to cyclosporine, a calcineurin inhibitor, show clear signs of cellular aging [9]. Also, it has been reported that mycophenolic acid (MPA), an immunosuppressant widely used in the treatment of patients with rheumatoid arthritis, dose-dependently reduces the proliferation, clonal formation ability, and angiogenic function of endothelial colony-forming cells, and increases the expression of aging-related β-galactosidase, p21 gene expression, and p53 phosphorylation, indicating cellular aging. These results suggest that these drugs have a potent anti-angiogenic and aging-promoting effect on vascular precursor cells [10]. Currently, in the pediatric field as well, multiple immunosuppressive drugs are often used from an early stage in patients with nephrotic syndrome and autoimmune diseases. The use of these drugs improves the child's quality of life from the perspective of early steroid reduction and steroid-free, but in the future, a treatment plan that takes into account the healthy lifespan after adulthood is necessary.

References

1. Bianco-Miotto T, Craig JM, et al. Epigenetics and DOHaD: from basics to birth and beyond. J Dev Orig Health Dis. 2017;8:513–9.

2. Thilaganathan B, Meher-Homji N, et al. Labor: an immunologically beneficial process for the neonate. Am J Obstet Gynecol. 1994;171:1271–2.
3. Kristensen K, Henriksen L. Cesarean section and disease associated with immune function. J Allergy Clin Immunol. 2016;137:587–90.
4. Aas M, Elvsåshagen T, et al. Telomere length is associated with childhood trauma in patients with severe mental disorders. Transl Psychiatry. 2019;9:97.
5. Fukunaga T, Mizoi Y, et al. Thymus of abused/neglected children. Forensic Sci Int. 1992;53:69–79.
6. Koyama S, Ichikawa G, et al. Adiposity rebound and the development of metabolic syndrome. Pediatrics. 2014;133:e114–9.
7. Ley RE, Turnbaugh PJ, et al. Microbial ecology: human gut microbes associated with obesity. Nature. 2006;444:1022–3.
8. Scott FI, Horton DB, et al. Administration of antibiotics to children before age 2 years increases risk for childhood obesity. Gastroenterology. 2016;151:120–9.
9. Koppelstaetter C, Kern G, et al. Effect of cyclosporine, tacrolimus and sirolimus on cellular senescence in renal epithelial cells. Toxicol In Vitro. 2018;48:86–92.
10. Go E, Tarnawsky SP, et al. Mycophenolic acid induces senescence of vascular precursor cells. PLoS One. 2018;13:e0193749.

161 Antiaging from a Demographic Perspective

Ayano Takeuchi

Demography, or population statistics, is a field of study that focuses on changes and fluctuations in population. In a narrower sense, it refers to the field that quantitatively analyzes the state and changes of the population at a certain point in time [1]. Amid the outcry of a declining birthrate and aging population, Japan's population experienced its first post-war decrease in 2005, and has been declining since its peak of 128.08 million in 2008 [2]. Many people may have seen predictions about Japan's future population, such as "Japan's population will drop below 50 million by the twenty-second century (2101)" [3]. These estimates are calculated by determining the changes that occur with aging in the population by age, for each related factor (death, birth, population movement). Even if unprecedented disasters, economic crises, or wars occur, it can be said that there have been few instances where future population estimates, which are estimated with a range of scenarios from optimistic to pessimistic, have been off. For example, in a long-term series of articles titled "Alarm from 2020" published in the Nihon Keizai Shimbun on New Year's Day 1997, future predictions were made for various fields such as economics, finance, education, and environmental issues, but there were hardly any that were significantly off based on future population predictions [4]. There are several methods used for future population predictions, but here we will briefly introduce the "cohort component method". In the cohort component method, the future population is determined by considering deaths that occur with aging and international population movements (outflows) for the existing population. For the population that increases due to births, etc., the number of births and population movements (inflows) according to the reproductive age population are incorporated as the population for the next year. To calculate this, it is necessary to make assumptions for the future based on ① the standard population classified by gender and age, ② birth rate, ③ survival rate, and ④ international population movement rate [5]. Specifically, for example, for ② birth rate, estimates are made for the distribution of first marriage age, the unmarried rate at age 50, the number of completed births (the average number of births at the wife's age of 50 in a first-marriage couple), and the effect of divorce/death/remarriage on the number of completed births, and in some cases, multiple optimistic/pessimistic scenarios are set and multiple estimates are made based on statistical mathematical models [5]. Of course, when estimating the future population, the trend of the average lifespan, which has been somewhat leveling off in recent years, is taken into account (Fig. 161.1). Consideration has been given to the extension of life expectancy, with certain assumptions in place, but there seems to

A. Takeuchi (✉)
Epidemiology and Biostatistics, Faculty of Science and Engineering, Chuo University, Tokyo, Japan
e-mail: atakeuchi883@g.chuo-u.ac.jp

H. Yamada, Y. Naito (eds.), *Anti-Aging Medicine*, https://doi.org/10.1007/978-981-97-8514-8_161

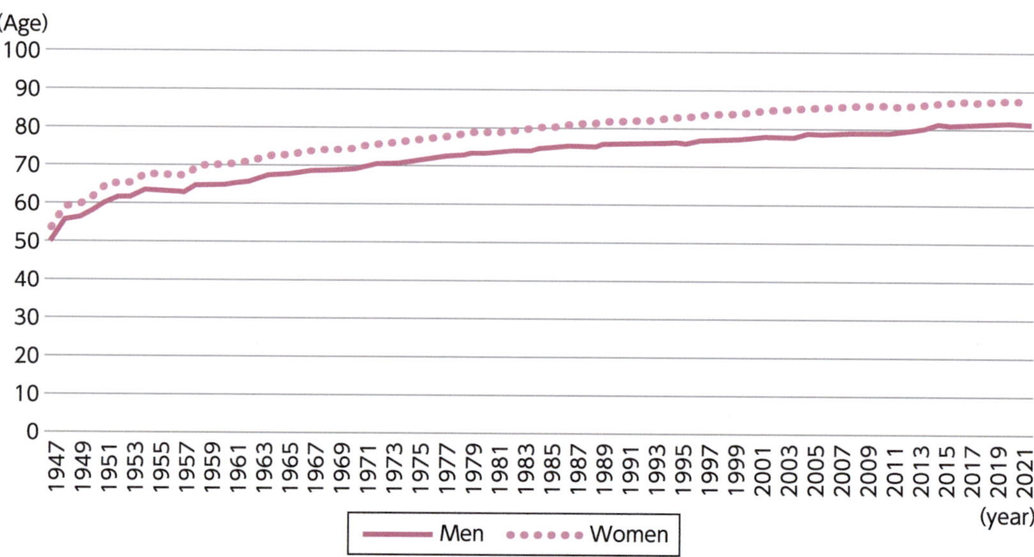

Fig. 161.1 Transition of average lifespan from 1947 to 2021

be no consideration of the factor of "anti-aging". As mentioned above ②, in order to increase the birth rate, a Child and Child-rearing Headquarters has been established and extraordinary measures against declining birth rates are being implemented, and there is a lot of reaction to the news about this in the public, but considering that the second baby boom generation (baby boom junior generation) is already around 50 years old, it is unlikely that this will have a direct impact that would affect demographic assumptions and necessitate further revision of optimistic scenarios. Therefore, from a demographic perspective, if we want to maintain the future population, the options left are to increase the ③ survival rate and the international population ④ migration rate. For ④, a shift in immigration policy based on data is necessary, and there are strong concerns about the impact on society, but if the effect of stopping the decrease in the existing population by the penetration and spread of anti-aging medicine increases, it will be possible to contribute to ③ without significantly changing the social structure from a policy perspective, so the expectation for anti-aging medicine is high from the perspective of maintaining the future population. Also, it goes without saying that careful discussion and accumulation of evidence are necessary for various risks associated with late childbirth, but the age at which pregnancy and childbirth are possible may be raised by anti-aging medicine, and anti-aging medicine may contribute to the ② birth rate to a significant extent. In addition, while demography considers the "population" purely in terms of numbers by considering births and deaths, it is highly likely that anti-aging medicine will also affect the extension of so-called healthy life expectancy, not just the extension of survival time. Specifically, in considering the sustainable maintenance and development of society, how to reduce the burden of the population onus period [populationonus]. A period when the proportion of the working population decreases compared to the population of children and the elderly, and the economic burden increases. The opposite of the "population bonus period" when the productive age population (15 to under 65) is more than twice the rest of the population] that Japan has already entered is a problem in various situations. By extending the time that one can engage in social activities in good health, it is possible to put a brake on the actual decrease in the working population, and anti-aging medicine may have the potential to reduce the various economic burdens that will increase during the population onus period, such as the increase in social security costs. It should be noted that while the productive age population is defined purely by age, the working population refers to people over the age of 15 who have the intention and ability to work, and there is no upper age limit. Although accurate surveys are difficult, it is thought that the greatest contribution that anti-aging medicine can make from a demographic perspective is the maintenance and increase of this working population. This section has discussed the expectations and possibilities of antiaging medicine from a demographic perspective, but "living healthily in mind and body" is a common wish for all people. The classification of primary prevention (prevention of disease occurrence), secondary prevention (early detection, early treatment, and prevention of severity), and tertiary prevention (restoration of function, social return, and prevention of recurrence) has been established for diseases, and there has been a long-standing emphasis on prevention. Anti-aging medicine can contribute to all of these stages, and it is hoped that it also has the potential to prevent causes and factors related to diseases, which could be called zero prevention.

References

1. Okazaki Y. The advancement of population statistics. J Jpn Statist Soc. 1982;12:77–84.
2. Statistics Bureau of the Ministry of Internal Affairs and Communications. 2010 census results. https://www.stat.go.jp/data/kokusei/2010/. Accessed 1 Feb 2023.
3. National Land Council Policy Committee Long-term Outlook Committee. Long-term outlook for national land. Interim Summary; 2011.
4. Nikkei Inc. Alarm from 2020—Japan disappears. Tokyo: Nikkei BP Marketing; 1997.
5. National Institute of Population and Social Security Research. Report on Japan's future estimated population (2017 estimate). Population Problem Research Material No. 336. 2017.

162 Women's Comprehensive Outpatient Clinic and Antiaging

Sumiyo Akazawa

The Women's Comprehensive Outpatient Clinic is a division of obstetrics and gynecology, divided into four departments: ① Obstetrics and perinatal care, ② Reproductive medicine, ③ Obstetrics and gynecology tumors, and ④ Women's healthcare. In terms of internal medicine, the concept of gender-specific medicine was established in the United States' review of women's healthcare, and the Women's Comprehensive Outpatient Clinic became a place to practice women's medicine and gender-specific medicine. Gender-specific medicine considers diseases that men and women are prone to, taking into account ① Sex (biological differences), ② Gender (psychological and sociological differences, cultural differences in the case of doctors), and reflects these considerations in medical care and policy, as learned from Dr. Keiko Amano. Women were often associated with "bikini medicine" due to the focus on breasts, uterus, and ovaries. Men are easily associated with the prostate and testicles, but it's not just the reproductive system. Recently, it has been found that there are gender differences in the pathology and treatment of diseases common to both men and women, and that fluctuations in sex hormone secretion at each life stage are important for the onset and progression of diseases. This concept of gender-specific medicine (considering biological and sociological differences and the impact of life stages) was introduced a quarter of a century ago, and recently, with the advent of femtech (products that can solve various issues at women's life stages such as "menstruation," "fertility and pregnancy," "pregnancy and postpartum," "pre-menopause and menopause"), gendered innovation is attracting attention again, with the male version being called maletech. Now, let's consider diseases that women are prone to at each life stage [1–5].

S. Akazawa (✉)
Department of General Internal Medicine, Kanazawa Medical University, Kanazawa, Japan
e-mail: sumiyok@kanazawa-med.ac.jp

162.1 Women's Life Cycle and Health (Fig. 162.1)

As diseases prone to occur with fluctuations in female hormones change, the Women's Comprehensive Outpatient Clinic checks where in life the patient is and selects test items so that appropriate medical care can be provided. It is important to overlay things other than diseases, such as career at work or child-rearing, on the chart of life stages and hormones. The Anti-Aging Society values positive thinking, so we suggest thinking positively about how to achieve slightly ambitious goals. Take menopause as an example. The period around 50 years old is considered the perimenopausal period, and hot flashes are a famous symptom of menopausal dis-

H. Yamada, Y. Naito (eds.), *Anti-Aging Medicine*, https://doi.org/10.1007/978-981-97-8514-8_162

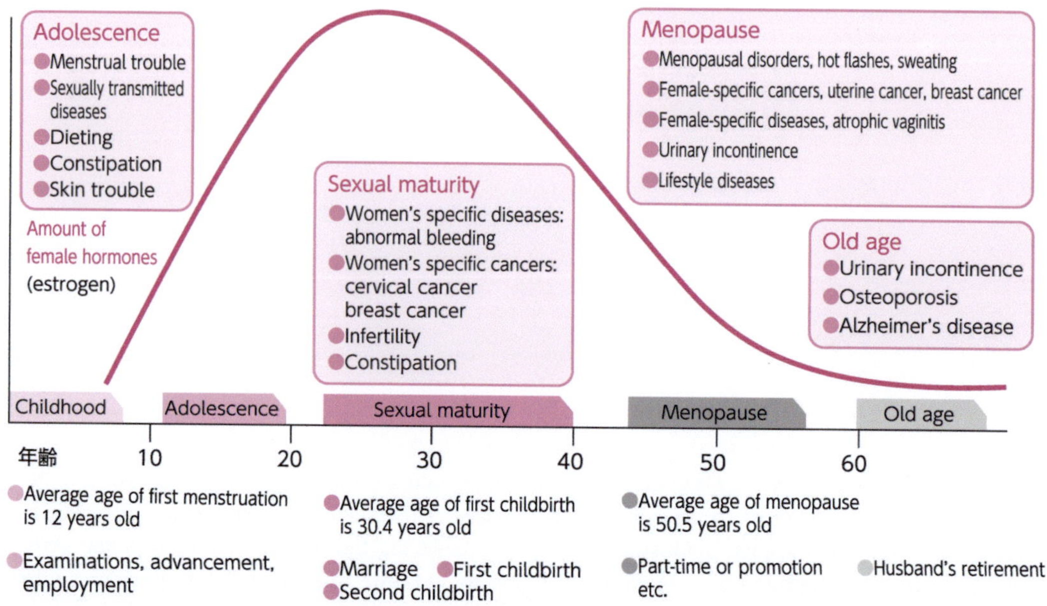

Fig. 162.1 Women's life cycle and health. (Reprinted from [6])

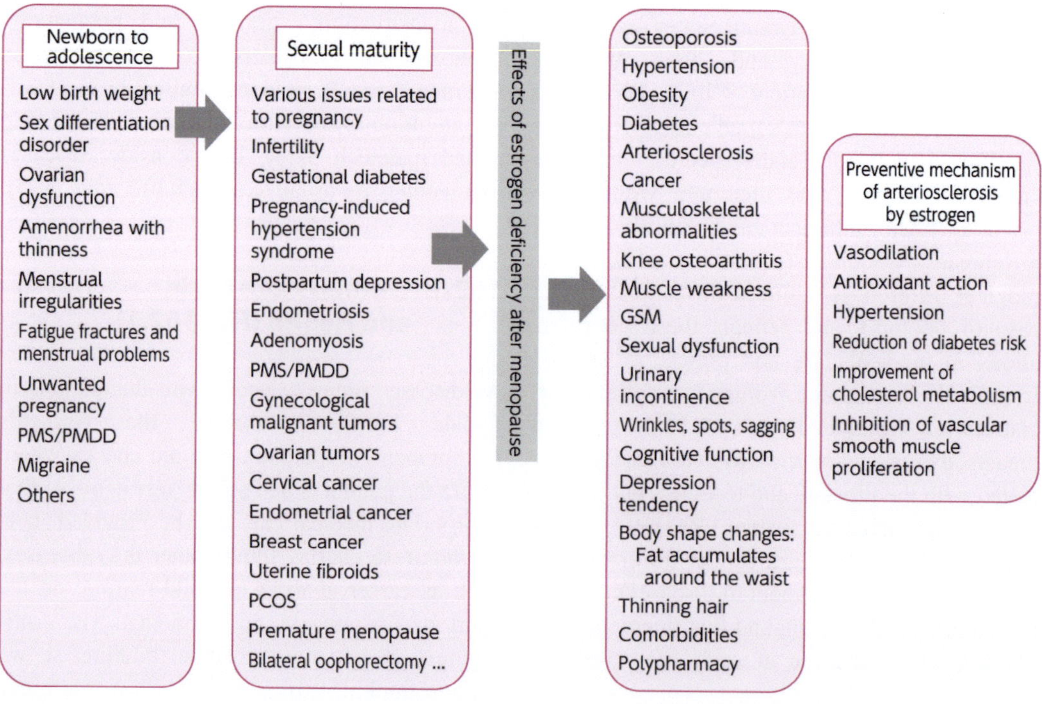

Fig. 162.2 The impact of estrogen deficiency after menopause throughout a woman's life. Comprehensively understand women's health and solve problems for each generation. *PMS* premenstrual syndrome; *PMDD* premenstrual dysphoric disorder; *PCOS* polycystic ovarian syndrome; *GSM* genitourinary syndrome of menopause

orders, but similar symptoms can occur in hyperthyroidism, so we exclude hyperthyroidism and other diseases before finally diagnosing menopausal disorders. Figure 162.2 shows the impact of estrogen deficiency after menopause throughout a woman's life, and Table 162.1 shows symp-

Table 162.1 Differential diagnosis of confusing symptoms that appear early in menopausal disorders and other diseases

Differential diagnosis is necessary for menopausal disorders—symptoms that appear early		
Autonomic nervous system disorder symptoms	Psychiatric and neurological symptoms	Psychiatric and neurological symptoms
• Flushes • Hot flushes • Sweating and palpitations • Panic attacks • Chest tightness • Difficulty breathing • Blood pressure fluctuations • Increase in body temperature, feeling of heat	• Easily irritated and excited • Inability to concentrate • Worrying and feeling restless • Anxiety and depression • Insomnia (difficulty falling asleep, waking up in the middle of the night) • Feeling of a lump in the throat • Forgetfulness • Sensitive to noise, unable to distinguish tastes or smells	• Shoulder stiffness • Neck stiffness • Numbness in hands and feet • Dizziness (feeling light-headed) • Tinnitus • Headache, feeling of heaviness in the head • Easily fatigued • Joint pain

Table 162.2 Simplified Menopausal Index (SMI). Reprinted with permission from [7]

① Face feels hot	10	6	3	0
② Easily sweat	10	6	3	0
③ Lower back and limbs easily get cold	14	9	5	0
④ Shortness of breath, palpitations	12	8	4	0
⑤ Difficulty falling asleep, light sleep	14	9	5	0
⑥ Easily angered, irritable	12	8	4	0
⑦ Often worry or feel depressed	7	5	3	0
⑧ Often have headaches, dizziness, nausea	7	5	3	0
⑨ Easily tired	7	4	2	0
⑩ Have shoulder stiffness, back pain, pain in hands and feet	7	5	3	0
Total points				

Depending on the severity of the symptoms (even if only one symptom is severe, it is considered severe), score yourself and evaluate based on the total score of the self-scoring method of the menopausal index check

0–25 points: No abnormalities, 26–50 points: Pay attention to diet and exercise, 51–65 points: Visit menopause and menopause outpatient clinic, 66–80 points: Long-term planned treatment, 81–100 points: Detailed examination by each department, long-term planned response

toms that appear early in menopause and are easily mistaken for other diseases. Also, Table 162.2 The simple menopausal index to confirm the trigger for hospital visits is shown in Fig. 162.3, which illustrates the general treatment for menopausal disorders.

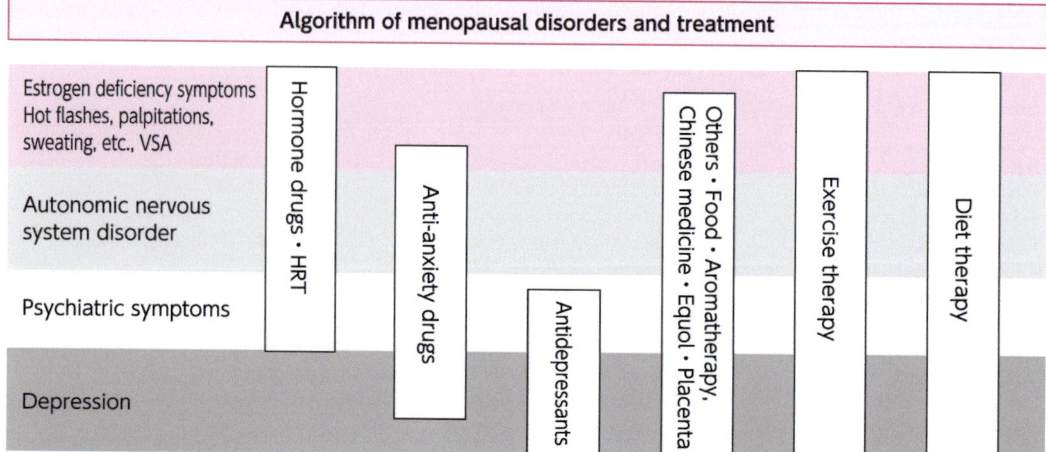

Symptoms of depression

- Decreased motivation
- Anxiety, Irritability
- Inability to make decisions
- Sleep disorders
- Palpitations, chest pressure
- Headache, dizziness
- Decreased libido
- Decreased appetite, Weight loss
- Diurnal variation

Menopausal disorders
The biggest differential diagnosis is "depression"

Symptoms of menopausal disorders

- Depression
- Lack of motivation
- Anxiety, Irritability
- Disorganized thinking
- Sleep disorders
- Palpitations, chest pressure
- Headache, feeling of light-headedness
- Decreased libido

Treatment of menopausal disorders

When various symptoms due to estrogen deficiency appear, confirm that there is no breast cancer or uterine cancer and start treatment:
Refer to the guidelines Q&A

Hormone replacement therapy (HRT/ This topic's natural progesterone produced domestically)
Psychotherapy (antidepressants, anti-anxiety drugs, sleeping pills)
Alternative medicine
Supplements: Equol, Reduced coenzyme Q10, Pycnogenol, etc.
Chinese medicine: Danggui Shaoyao Powder, Kamishoyosan, Keishibukuryogan, Rokumigan, Tohaku-joki-to, Kakkonto, Saikokeishikankyoto, Goreisan, etc.
Massage, Acupuncture, Meditation, etc.
Counseling (improvement of social environmental factors), Quality of living space, Room temperature, etc.
Evaluation of health risks that are likely to appear after menopause and review of lifestyle habits
Cancer, Lifestyle-related diseases (diabetes, hypertension, dyslipidemia, etc.), Obesity, Osteoporosis, Quality of sleep, etc.
Blood test data, Thyroid hormone, Autoantibodies, RA, IN, BODY, Visceral fat
Carotid artery echo, Endothelial function, and other imaging diagnoses for the exclusion of organic diseases

Fig. 162.3 Overview of treatment for menopausal disorders. *VSA* vasospastic angina; *HRT* hormone replacement therapy; *RA* rheumatoid arthritis, IN, BODY

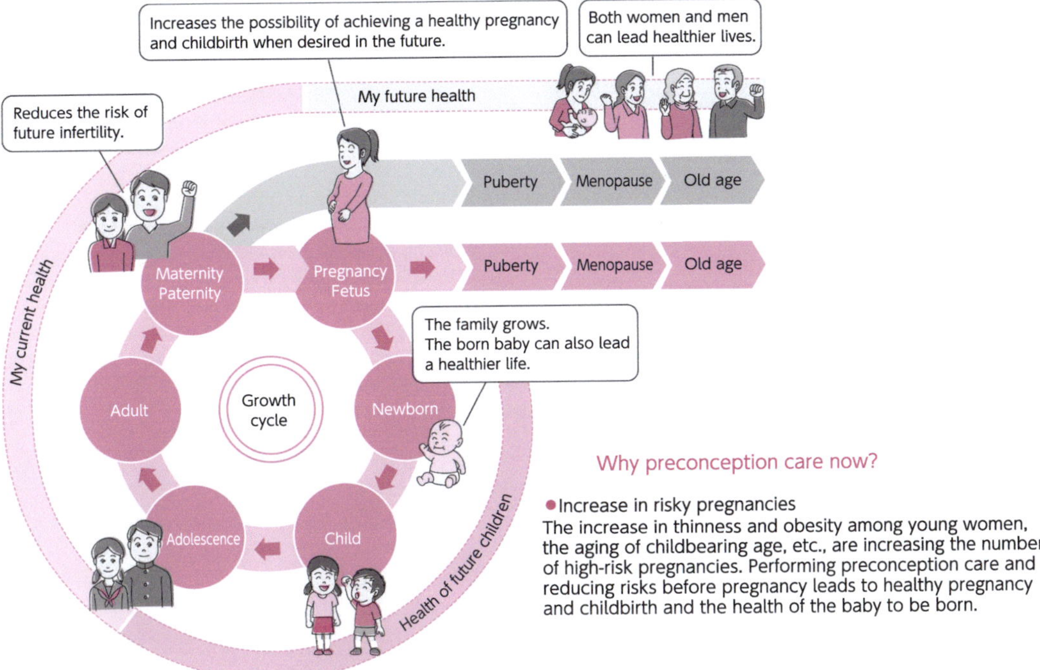

Fig. 162.4 Preconception care. (Reprinted with permission from [8])

162.2 Improvement of Lifestyle Habits

One of the goals of the Women's General Outpatient Clinic is to improve lifestyle habits. Exercise, sleep, nutrition, thinking, and preferences are important for health, and in Japan, this content is introduced as the concept of preconception care by Dr. Atsuko Murashima and Dr. Naoko Arata of the National Center for Child Health and Development (Fig. 162.4). Conception refers to fertilization, or the bestowing of new life in the womb. In preconception care, it is important for women and couples to face their own lives and health while considering future pregnancies, and it is important to re-recognize unconscious biases and to be aware of them. This concept represents pre-pregnancy care, but it also indicates that lifestyle habits affect three generations. There are checklists for women and men (https://www.ncchd.go.jp/hospital/about/section/preconception/pcc_check-list.html), so please take a look.

162.3 The Relationship Between the Women's General Outpatient Clinic and Antiaging

The relationship between the Women's General Outpatient Clinic and anti-aging is very deep, and the relationship between aging, the decrease in the number of blood vessels, and "appearance" has been proven. When the skin's blood vessels age, the phenotype leads to wrinkles and sagging. In menopause, when female hormones decrease, the skin sags about 30% more than when young. Cosmetic surgical treatment is important, but the choice of diet, exercise, nutrition, thinking, and preferences (tobacco, alcohol) for antiaging is directly linked to the concept that "aging is a disease." It is essential for antiaging. Through the Women's General Outpatient Clinic, I would be happy if you could practice improving your lifestyle habits, including your living environment, from the perspective of gender differences, so that the individual called a human can live healthily, rather than the word "Women's General Outpatient Clinic."

References

1. Japan Society of Obstetrics and Gynecology, Japan Association of Obstetricians and Gynecologists. Obstetrics and gynecology clinical guidelines outpatient gynecology 2020. Tokyo: Japan Society of Obstetrics and Gynecology Secretariat; 2020.
2. Japan Society for Women's Health. Women's medicine guidebook menopause medical edition 2019. Tokyo: Kanehara Publishing; 2019.
3. Japan Society of Obstetrics and Gynecology, Japan Society for Women's Health. Hormone replacement therapy guidelines 2017. Tokyo: Japan Society of Obstetrics and Gynecology; 2017.
4. Aso T, Uchiyama S. Supplements in Women's healthcare. The role of soy Isoflavone metabolite Equol. J Japan Soc Women's Health. 2012;20:313–32.
5. Uchiyama S. About the new soy-derived ingredient "Equol" that is expected to prevent and improve menopausal symptoms. Menopause Aging Healthcare. 2008;7:26–31.
6. Beautiful Aging/Kanazawa City official website Iine Kanazawa. https://www4.city.kanazawa.lg.jp.
7. Koyama T, et al. Kampo treatment in menopausal women: evaluation by a simplified menopausal index. Progress Obstetric Kampo Res. 1992;9:30–4.
8. National Center for Child Health and Development Arata Team. Precon Note. https://www.ncchd.go.jp/hospital/about/section/preconception/preconnote.

163 Antiaging from a Business Perspective

Tsutomu Tatemiya

163.1 Investment and Return on Investment from a Business Evaluation Perspective

Success in business refers to a state where one can continuously generate cash flow (flow of cash) that exceeds the expectations of each money lender, against the cash collected through borrowings from banks, capital injections from business owners, and capital injections through shares from investors. Cash flow does not refer to the revenue from selling some kind of medical-related goods or medical devices, or the revenue from providing medical services such as treatment and care, but rather the amount of profit that remains after deducting the cost of establishing the business from that sales or revenue, and which can actually be collected in cash. The anti-aging business contains significant business opportunities, but here we will organize each opportunity and risk from the perspective of investors and accounting.

T. Tatemiya (✉)
Graduate School of Social and Cultural School, Nihon University, Saitama, Japan
e-mail: tatemiya.tsutomu@nihon-u.ac.jp

163.2 Business Opportunities to Increase Sales

Sales are the total of unit price x number of purchases, so being able to sell a large number of items at a high unit price leads to the highest effect. The same effect can be achieved by increasing the number of purchases, the number of purchase times, and the number of customers without incurring a lot of costs. In anti-aging medicine, there are three stages: the "phase of maintaining youth and health," the "diagnosis and treatment phase," and the "prognosis and care phase." High unit prices include the sale of the latest medical devices and high-priced new drugs in the "diagnosis and treatment phase," and the provision of the latest medical technologies to the wealthy. However, these provisions include the risk of large research and development costs, long research hours, and low success rates, and from a business perspective, they are high-risk, high-return unless they are based on an insurance system. On the other hand, in the first stage, the "phase of maintaining youth and health," all citizens in their 30s and beyond are potential customers, and it is certain that the number will increase every year. Even if the unit price is low, it is expected that the stability of cash flow can be secured by the large number of customers and the small regular purchases of the subscription model. This is embodied in a customer database-type business that combines dig-

H. Yamada, Y. Naito (eds.), *Anti-Aging Medicine*, https://doi.org/10.1007/978-981-97-8514-8_163

ital health technology such as wearable devices and medical apps with individual advice. In the field of anti-aging, this includes nicotinamide mononucleotide (NMN) and antioxidant supplements, which have medical evidence. The challenges of these businesses are establishing a trust axis in terms of effectiveness, publicity, and adult education, and it will be even more necessary in the future to show an easy-to-understand and trustworthy anti-aging judgment axis that even the general public can understand, centered on specialists and instructors, and incorporate it into publicity and adult education [1].

163.3 Business Opportunities to Lower Costs and Increase Profits

The most advanced aspect in the anti-aging business currently is a system that reduces costs and increases profits. The largest cost in the medical business is research and development, followed by personnel costs and communication costs, including travel. Utilizing machinery and IT to separate time and space of limited human assets, and organizing them so that they can function even if they are not in the same place at the same time, leads to improved labor productivity, and hence a decrease in costs and an increase in profits. In the rapidly evolving field of medical DX (digital transformation), online medical consultations, medication guidance, image diagnosis AI, interview AI, remote surgery support robots, etc. have emerged mainly in the "diagnosis and treatment phase," reducing communication costs of spatial movement and personnel costs for diagnosis. Also, by separating repetitive tasks that can be patterned to some extent by machines from doctors and medical personnel and processing them up to the point of decision, business productivity is also being enhanced [2].

163.4 Consideration of Business Portfolio for Sustainable Growth Cash Flow

When considering the antiaging business, people tend to focus on the latest technologies, DX, AI, and online medicine, but from the perspective of accounting success, it is important to maximize early returns on investment and continue. The latest technologies, DX, AI, etc. require large initial investments, but they are mainly effective for reducing costs, not for maximizing profitability and hence cash flow at an early stage. Therefore, a combination with a business that can recover cash flow at a low cost and increase sales is necessary. Such a business is the "increase in the number of customers for products that have already been developed and commercialized" in the "phase to maintain youth and health." In this field, supplements and beauty-related products that do not rely on medical evidence are creating a large market, but the knowledge that is common sense among medical professionals and academic societies is not yet widely accepted in reality. In other words, there is a large knowledge gap. By automating education and evidence-based guidance to fill this gap in combination with the web and wearable devices, it is necessary to consider creating a business that generates profits without initial investment from existing cost-free knowledge and experience, and generates cash flow. And it may be necessary to consider a business portfolio that reinvests that cash flow into the development of new drugs and medical devices. Specifically, antiaging supplements, subscription-based individual medical advice, and antiaging education for adults in cooperation with the government and administration would be realistic (Figs. 163.1 and 163.2) [3].

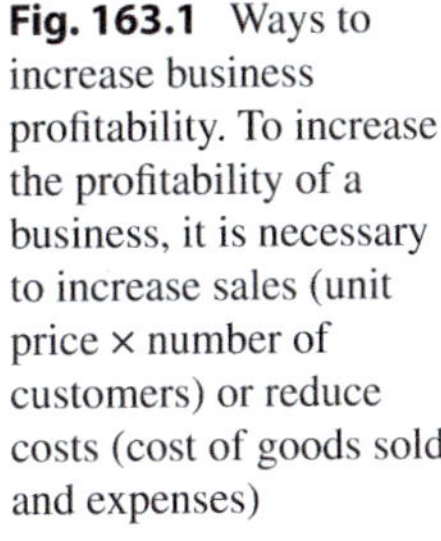

Fig. 163.1 Ways to increase business profitability. To increase the profitability of a business, it is necessary to increase sales (unit price × number of customers) or reduce costs (cost of goods sold and expenses)

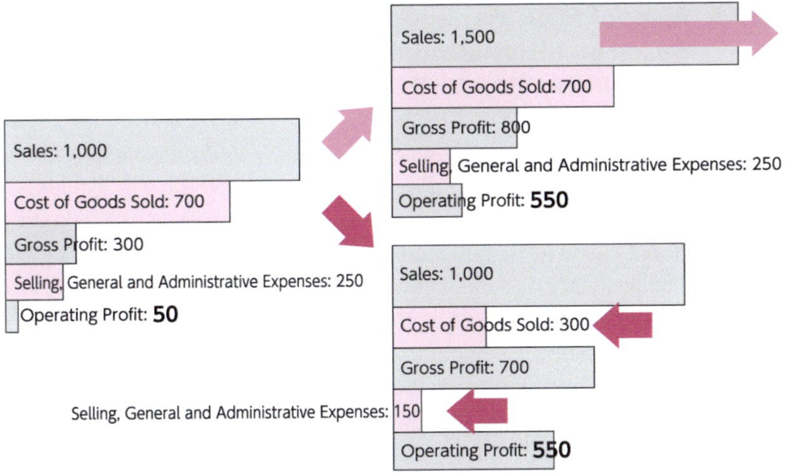

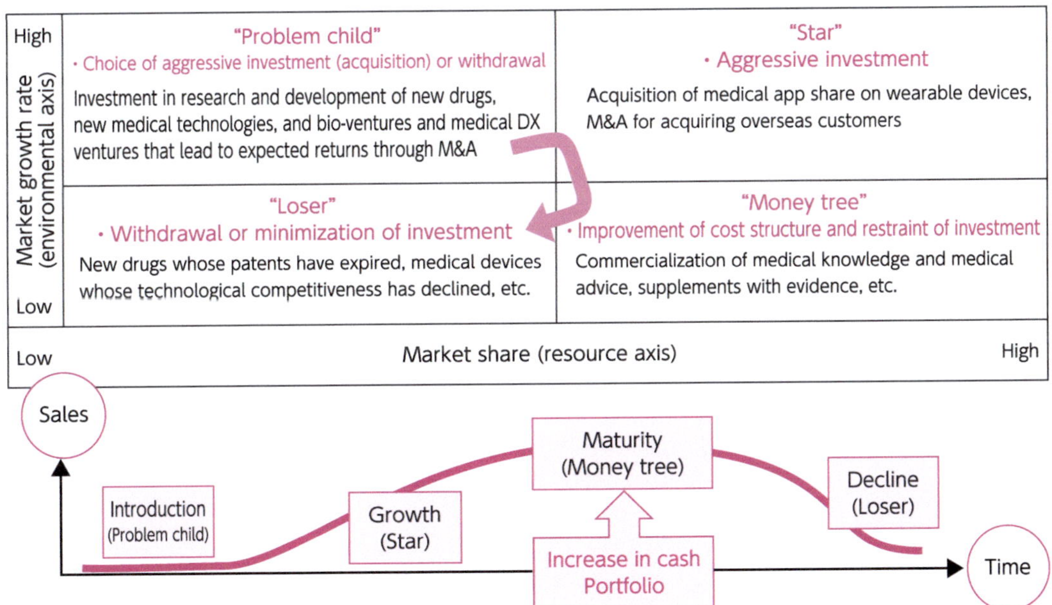

Fig. 163.2 Consideration of the business portfolio of the medical business

163.5 Risks of M&A Management Based on International Accounting Standards

From an accounting perspective, when considering the anti-aging business, there are risks such as the impairment risk inherent in goodwill, an asset item of global medical manufacturers, and the risk of a management crisis due to the short-term occurrence of huge debts caused by this. Many of Japan's global medical manufacturers have repeatedly engaged in mergers and acquisitions (M&A), buying time with money. This was aimed at reducing research investment risks, realizing implementation speed, and acquiring global customer routes in the development of new drugs in the "diagnosis, treatment phase" and the development of cutting-edge revolutionary medical devices. As a result, a huge amount of goodwill has accumulated as an asset on the balance sheet. Goodwill is the difference when a company is purchased at a higher price than its book value. According to accounting rules, if the performance of the acquired company deteriorates, the value of goodwill decreases rapidly, and there is a risk of causing a huge corporate bankruptcy that suddenly surfaces and involves the world. Japanese global medical manufacturers have adopted the International Financial Reporting Standards (IFRS), an international accounting standard, but the IFRS does not have a mechanism to amortize goodwill regularly and expense it like Japanese accounting standards, and to soft-land the risk. It adopts a hard landing method that processes the loss all at once when a huge loss occurs. It is convenient for management that emphasizes speed because the cost does not increase even if M&A is repeated, but there is always a big risk. From a business perspective, it would be important to always consider such impacts and think about future strategies.

References

1. Ministry of Economy, Trade and Industry. About healthcare industry policy in the Ministry of Economy, Trade and Industry; 2019. https://www.meti.go.jp/policy/mono_info_service/healthcare/01metihealthcarepolicy.pdf. Accessed 1 Aug 2022.
2. Research and Development Promotion Division, Medical Affairs Bureau, Ministry of Health, Labour and Welfare. About the roadmap for data health reform and future considerations. 2021. https://www.mhlw.go.jp/content/12601000/000815999.pdf. Accessed 7 Feb 2023.
3. Cabinet Office. 2022 Edition of the White Paper on Aging Society (Whole Edition). 2022. https://www8.cao.go.jp/kourei/whitepaper/w-2022/zenbun/04pdf_index.html. Accessed 7 Feb 2023.

Correction to: Anti-Aging Medicine

Hidekazu Yamada and Yuji Naito

Correction to:
H. Yamada, Y. Naito (eds.), *Anti-Aging Medicine*,
https://doi.org/10.1007/978-981-97-8514-8

This book was recently published. The editor's name Yuji Naito was incorrectly captured and should be updated as follows in the components listed below:

Current Output:

Naito Yuji,
Department of Human Immunology and Nutrition Science, Kyoto Prefectural University of Medicine, Kamigyoku, Kyoto, Japan

Required Output:

Yuji Naito,
Department of Human Immunology and Nutrition Science, Kyoto Prefectural University of Medicine, Kamigyoku, Kyoto, Japan

- Cover Page (Editor Name)
- In FM
- Title/Copyright Page (Editor Name)

The updated version of this book can be found at https://doi.org/10.1007/978-981-97-8514-8

H. Yamada, Y. Naito (eds.), *Anti-Aging Medicine*, https://doi.org/10.1007/978-981-97-8514-8_164

Correction to: Brain and Antiaging Medicine: Antiaging of Cognitive Function: Exercise

Masahiro Okamoto and Hideaki Soya

Correction to:
Chapter 38: H. Yamada, Y. Naito (eds.), *Anti-Aging Medicine*, https://doi.org/10.1007/978-981-97-8514-8_38

Current Output:

Hideaki Soya,
Laboratory of Exercise Biochemistry and Neuroscience, Faculty of Health & Sport Sciences, University of Tsukuba, Ibaraki, Japan. e-mail: soya.hideaki.gt@u.tsukuba.ac.jp

Required Output:

Masahiro Okamoto (should be placed first as per editor's comments)
Laboratory of Exercise Biochemistry and Neuroendocrinology, Sport Neuroscience Division, Advanced Research Initiative for Human High Performance (ARIHHP), Institute of Health and Sport Sciences, University of Tsukuba, Ibaraki, Japan. E-mail: okamoto.masahiro.ga@u.tsukuba.ac.jp

Hideaki Soya (Corresponding author)

Laboratory of Exercise Biochemistry and Neuroendocrinology, Sport Neuroscience Division, Advanced Research Initiative for Human High Performance (ARIHHP), Institute of Health and Sport Sciences, University of Tsukuba, Ibaraki, Japan. E-mail: soya.hideaki.gt@u.tsukuba.ac.jp

- In FM (TOC)
- Chapter opening page and footnote

The updated version of this chapter can be found at https://doi.org/10.1007/978-981-97-8514-8_38

H. Yamada, Y. Naito (eds.), *Anti-Aging Medicine*, https://doi.org/10.1007/978-981-97-8514-8_165

MIX
Papier aus verantwortungsvollen Quellen
Paper from responsible sources
FSC® C105338

If you have any concerns about our products,
you can contact us on
ProductSafety@springernature.com

In case Publisher is established outside the EU,
the EU authorized representative is:
Springer Nature Customer Service Center GmbH
Europaplatz 3, 69115 Heidelberg, Germany

Printed by Libri Plureos GmbH
in Hamburg, Germany